Encyclopedia of Molecular Cell Biology and Molecular Medicine

Edited by Robert A. Meyers

Volume 7
Innate Immunity to Mass Spectrometry, High Speed DNA Fragment Sizing

Encyclopedia of Molecular Cell Biology and Molecular Medicine

Encyclopedia of Molecular Cell Biology and Molecular Medicine

Edited by Robert A. Meyers

Second Edition

Volume 7

Innate Immunity to Mass Spectrometry, High Speed DNA Fragment Sizing

WILEY-VCH Verlag GmbH & Co. KGaA

Editor:

Dr. Robert A. Meyers
President, Ramtech Limited
7 Harbor Point Drive, 7A
Mill Valley, CA 94941
USA

Library of Congress Card No.: applied for

British Library Cataloguing-in-Publication Data: A catalogue record for this book is available from the British Library.

Bibliographic information published by Die Deutsche Bibliothek
Die Deutsche Bibliothek lists this publication in the Deutsche Nationalbibliografie; detailed bibliographic data is available in the internet at http://dnb.ddb.de.

Printed in the Federal Republic of Germany.
Printed on acid-free paper.

Composition: Laserwords Private Ltd, Chennai, India
Printing: Druckhaus Darmstadt GmbH, Darmstadt
Bookbinding: Litges & Dopf Buchbinderei GmbH, Heppenheim
ISBN-13: 978-3-527-30549-0
ISBN-10: 3-527-30549-1

Preface

The *Encyclopedia of Molecular Cell Biology and Molecular Medicine*, which is the successor and second edition of the *Encyclopedia of Molecular Biology and Molecular Medicine* (VCH Publishers, Weinheim), covers the molecular and cellular basis of life at a university and professional researcher level. The first edition, published in 1996–97, was very successful and is being used in libraries around the world. This second edition will almost double the first edition in length and will comprise the most detailed treatment of both molecular cell biology and molecular medicine available today. The Board Members and I believe that there is a serious need for this publication, even in view of the vast amount of information available on the World Wide Web and in text books and monographs. We feel that there is no substitute for our tightly organized and integrated approach to selection of articles and authors and implementation of peer review standards for providing an authoritative single-source reference for undergraduate and graduate students, faculty, librarians, and researchers in industry and government.

Our purpose is to provide a comprehensive foundation for the expanding number of molecular biologists, cell biologists, pharmacologists, biophysicists, biotechnologists, biochemists, and physicians, as well as for those entering molecular cell biology and molecular medicine from majors or careers in physics, chemistry, mathematics, computer science, and engineering. For example, there is an unprecedented demand for physicists, chemists, and computer scientists who will work with biologists to define the genome, proteome, and interactome through experimental and computational biology.

The Board Members and I first divided the entire study of molecular cell biology and molecular medicine into primary topical categories and further defined each of these into subtopics. The following is a summary of the topics and subtopics:

- *Nucleic Acids:* amplification, disease genetics overview, DNA structure, evolution, general genetics, nucleic acid processes, oligonucleotides, RNA structure, RNA replication and transcription.
- *Structure Determination Technologies Applicable to Biomolecules:* chromatography, labeling, large structures, mapping, mass spectrometry, microscopy, magnetic resonance, sequencing, spectroscopy, X-ray diffraction.
- *Biochemistry:* carbohydrates, chirality, energetics, enzymes, biochemical genetics, inorganics, lipids, mechanisms, metabolism, neurology, vitamins.

Encyclopedia of Molecular Cell Biology and Molecular Medicine, 2nd Edition. Volume 7
Edited by Robert A. Meyers.

ISBN: 3-527-30549-1

- *Proteins, Peptides, and Amino Acids:* analysis, enzymes, folding, mechanisms, modeling, peptides, structural genomics (proteomics), structure, types.
- *Biomolecular Interactions:* cell properties, charge transfer, immunology, recognition, senses.
- *Cell Biology:* developmental cell biology, diseases, dynamics, fertilization, immunology, organelles and structures, senses, structural biology, techniques.
- *Molecular Cell Biology of Specific Organisms:* algae, amoeba, birds, fish, insects, mammals, microbes, nematodes, parasites, plants, viruses, yeasts.
- *Molecular Cell Biology of Specific Organs or Systems:* excretory, lymphatic, muscular, nervous, reproductive, skin.
- *Molecular Cell Biology of Specific Diseases:* cancer, circulatory, endocrinal, environmental stress, immune, infectious, neurological, radiational.
- *Pharmacology:* chemistry, disease therapy, gene therapy, general molecular medicine, synthesis, toxicology.
- *Biotechnology:* applications, diagnostics, gene-altered animals, bacteria and fungi, laboratory techniques, legal, materials, process engineering, nanotechnology, production of classes or specific molecules, sensors, vaccine production.

We then selected some 400 article titles and author or author teams to cover the above topics. Each article is designed as a self-contained treatment which begins with a keyword section including definitions, to assist the scientist or student who is unfamiliar with the specific subject area. The Encyclopedia includes more than 3000 key words, each defined within the context of the particular scientific field covered by the article. In addition to these definitions, the glossary of basic terms found at the back of each volume, defines the most commonly used terms in molecular cell biology. These definitions, along with the reference materials (the genetic code, the common amino acids, and the structures of the deoxyribonucleotides) printed at the back of each volume, should allow most readers to understand articles in the Encyclopedia without referring to a dictionary, textbook, or other reference work. There is, of course, a detailed subject index in Volume 16 as well as a cumulative table of contents and list of authors, as well as a list of scientists who assisted in the development of this Encyclopedia.

Each article begins with a concise definition of the subject and its importance, followed by the body of the article and extensive references for further reading. The references are divided into secondary references (books and review articles) and primary research papers. Each subject is presented on a first-principle basis, including detailed figures, tables and drawings. Because of the self-contained nature of each article, some articles on related topics overlap. Extensive cross-referencing is provided to help the reader expand his or her range of inquiry.

The articles contained in the Encyclopedia include core articles, which summarize broad areas, directing the reader to satellite articles that present additional detail and depth for each subject. The core article Brain Development is a typical example. This 45-page article spans neural induction, early patterning, differentiation, and wiring at a molecular through to cellular and tissue level. It is directly supported, and cross-referenced, by a number of molecular neurobiology satellite articles, for example, Behavior Genes, and further supported by other core presentations, for example,

Developmental Cell Biology; Genetics, Molecular Basis of, and their satellite articles. Another example is the core article on Genetic Variation and Molecular Evolution by Werner Arber. It is supported by a number of satellite articles supporting the evolutionary relatedness of genetic information, for example, Genetic Analysis of Populations.

Approximately 250 article titles from the first edition are retained, but rewritten, half by new authors and half by returning authors. Approximately 80 articles on cell biology and 70 molecular biology articles have been added covering areas that have become prominent since preparation of the first edition. Thus, we have compiled a totally updated single source treatment of the molecular and cellular basis of life.

Finally, I wish to thank the following Wiley-VCH staff for their outstanding support of this project: Andreas Sendtko, who provided project and personnel supervision from the earliest phases, and Prisca-Maryla Henheik and Renate Dötzer, who served as the managing editors.

November 2003

Robert A. Meyers
Editor-in-Chief

Editor-in-Chief

Robert A. Meyers

Dr. Meyers earned his Ph.D. in organic chemistry from the University of California Los Angeles, was a post-doctoral fellow at California Institute of Technology and manager of chemical processes for TRW Inc. He has published in *Science*, written or edited 12 scientific books and his research has been reviewed in the *New York Times* and the *Wall Street Journal*. He is one of the most prolific science editors in the world having originated, organized and served as Editor-in-Chief of three editions of the *Encyclopedia of Physical Science and Technology*, the *Encyclopedia of Analytical Chemistry* and two editions of the present *Encyclopedia of Molecular Cell Biology and Molecular Medicine*.

Encyclopedia of Molecular Cell Biology and Molecular Medicine, 2nd Edition. Volume 7
Edited by Robert A. Meyers.

ISBN: 3-527-30549-1

Editorial Board

Werner Arber
Biozentrum, University of Basel, Switzerland
Nobel Prize in Physiology/Medicine for the discovery of restriction enzymes and their application to problems of molecular genetics

David Baltimore
California Institute of Technology, Pasadena, USA
Nobel Prize in Physiology/Medicine for the discoveries concerning the interaction between tumor viruses and the genetic material of the cell

Günter Blobel
The Rockefeller University, New York, USA
Nobel Prize in Physiology/Medicine for the discovery that proteins have intrinsic signals that govern their transport and localization in the cell

Martin Evans
Cardiff University, United Kingdom
Lasker Award for the development of a powerful technology for manipulating the mouse genome, which allows the creation of animal models of human disease

Contents

Encyclopedia of Molecular Cell Biology and Molecular Medicine, 2nd Edition. Volume 7
Edited by Robert A. Meyers.

ISBN: 3-527-30549-1

List of Contributors

Shizuo Akira
Osaka University,
Osaka,
Japan

and

ERATO,
Japan Science and Technology Agency,
Osaka, Japan

Rengul Cetin-Atalay
Bilkent University,
Bilkent, Ankara,
Turkey

Chung-Hsuan Chen
Oak Ridge National Laboratory,
Oak Ridge, TN,
USA

Frank H. Collins
University of Notre Dame,
Notre Dame, IN,
USA

David M. de Kretser
Monash Institute of Reproduction and
Development, Monash University,
Clayton, Victoria, Australia

and

The Australian Research Council Centre of
Excellence in Biotechnology and
Development
Canberra
Australia

Robert A. Holt
Canada's Michael Smith
Genome Science Centre,
Vancouver, BC,
Canada

Robert Huber
Max-Planck-Institut für Biochemie,
Martinsried,
Germany

Dieter F. Hülser
University of Stuttgart,
Stuttgart,
Germany

John P. Kane
University of California,
San Francisco, CA,
USA

Naoka Komori
The University of Oklahoma Health
Sciences Center,
Oklahoma City, OK,
USA

Encyclopedia of Molecular Cell Biology and Molecular Medicine, 2nd Edition. Volume 7
Edited by Robert A. Meyers.

ISBN: 3-527-30549-1

Sadamu Kurono
The University of Oklahoma Health Sciences Center,
Oklahoma City, OK,
USA

Ulf Landegren
Department of Genetics and Pathology,
Uppsala University,
Sweden

Brian D. Lehmann
The Brody School of Medicine at East Carolina University,
Anatomy & Cell Biology,
Greenville, NC,
USA

William Lesser
Department of Applied Economics and Management,
Cornell University,
Ithaca, NY,
USA

Gertz I. Likhtenshtein
Ben-Gurion University of the Negev,
Beer-Sheva,
Israel

Kate A.L. Loveland
Monash Institute of Reproduction and Development,
Monash University,
Clayton, Victoria,
Australia

and

The Australian Research Council Centre of Excellence in Biotechnology and Development
Canberra
Australia

Hiroyuki Matsumoto
The University of Oklahoma Health Sciences Center,
Oklahoma City, OK,
USA

Masaomi Matsumoto
Oklahoma Department of Agriculture,
Oklahoma City, OK,
USA

James A. McCubrey
The Brody School of Medicine at East Carolina University,
Microbiology & Immunology,
Greenville, NC,
USA

and

Leo Jenkins Cancer Center,
Greenville, NC,
USA

Lindsay McDermott
University of Glasgow,
Glasgow,
UK

Albrecht Messerschmidt
Max-Planck-Institut für Biochemie,
Martinsried,
Germany

Dominic S. Ng
University of Toronto,
Toronto, Ontario,
Canada

Mats Nilsson
Department of Genetics and Pathology,
Uppsala University,
Sweden

Mehmet Ozturk
Bilkent University,
Bilkent, Ankara,
Turkey

Clive R. Pullinger
University of California,
San Francisco, CA,
USA

Colin Ratledge
Department of Biological Sciences,
University of Hull, Hull,
UK

Alan T. Remaley
National Institutes of Health,
Bethesda, MD
USA

Clemens von Sonntag
Leibniz-Institut für
Oberflächenmodifizierung (IOM),
Leipzig,
Germany

Johan Stenberg
Department of Genetics and
Pathology,
Uppsala University,
Sweden

John M. Stewart
Mount Allison University,
Sackville, NB,
Canada

Judith Storch
Rutgers University,
New Brunswick, NJ,
USA

Osamu Takeuchi
Osaka University,
Osaka,
Japan

and

ERATO,
Japan Science and Technology Agency,
Osaka,
Japan

Nancy Smyth Templeton
Department of Molecular and
Cellular Biology,
Baylor College of Medicine,
Houston, TX,
USA

David M. Terrian
The Brody School of Medicine at East
Carolina University,
Anatomy & Cell Biology,
Greenville, NC,
USA

and

Leo Jenkins Cancer Center,
Greenville, NC,
USA

Chittam U. Thakore
The Brody School of Medicine at East
Carolina University,
Anatomy & Cell Biology,
Greenville, NC,
USA

G.Russell Warnick
Pacific Biometrics Research Foundation
Issaquah, WA,
USA

John Lewis Williams
Roslin Institute,
Roslin,
Midlothian, Scotland,
UK

Color Plates

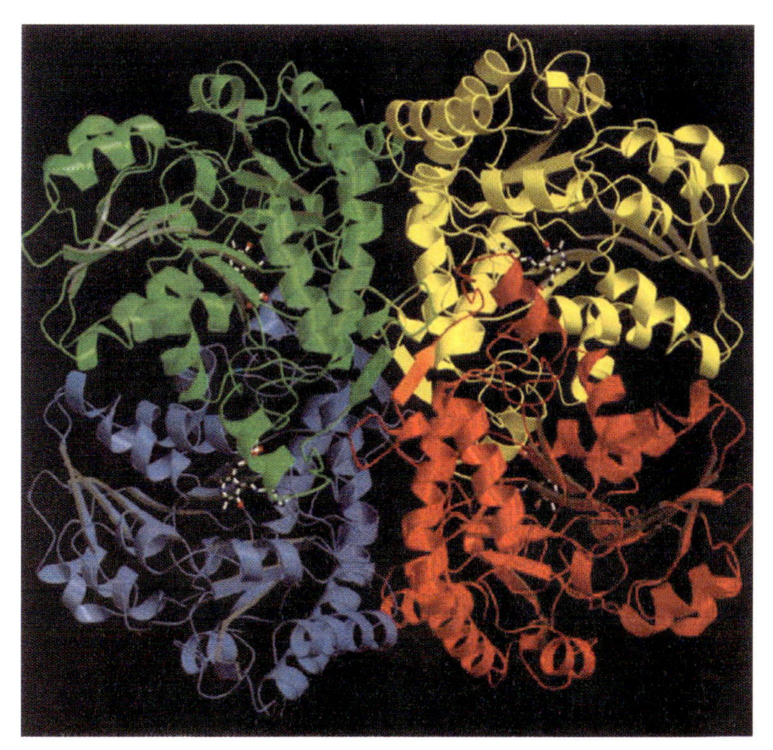

Fig. 39 (p. 452) Ribbon plot of the CBL tetramer viewed along the *x*-axis. The monomers are colored differently. The blue- and green-colored monomers, which are related by a crystallographic axis (horizontal, in the plane of the paper), build up one catalytic active dimer, and the yellow and red ones the other. The location of the PLP-binding site is shown in a ball-and-stick presentation; MOLSCRIPT and RASTER3D. (Reproduced by permission of Academic Press, Ltd., from Clausen, T. et al. (1996) *J. Mol. Biol.* **23**, 202–224.)

Encyclopedia of Molecular Cell Biology and Molecular Medicine, 2nd Edition. Volume 7
Edited by Robert A. Meyers.

ISBN: 3-527-30549-1

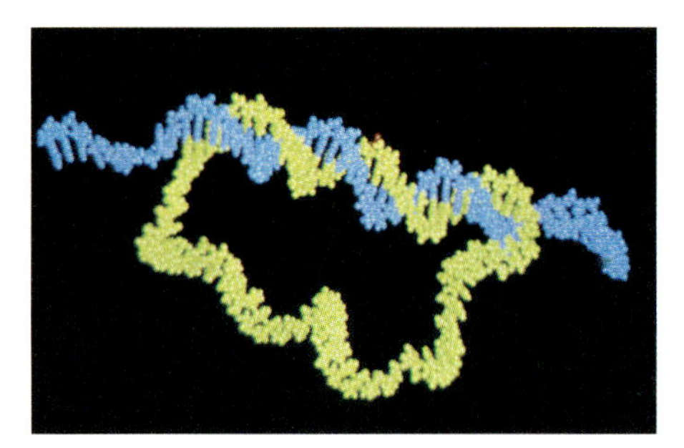

Fig. 5 (p. 188) A padlock probe, designed to be ligated into a circle upon interaction with the proper target sequence. The two ends of the linear probe (yellow) hybridize in juxtaposition on the target sequence (blue). The 5′-phosphate group on one end of the probe about to be joined by ligation to the 3′-hydroxyl at the opposite end is shown in red. Circularized molecules are wound around the target strand and can detach only by strand breakage or by sliding off a nearby end on the target molecule.

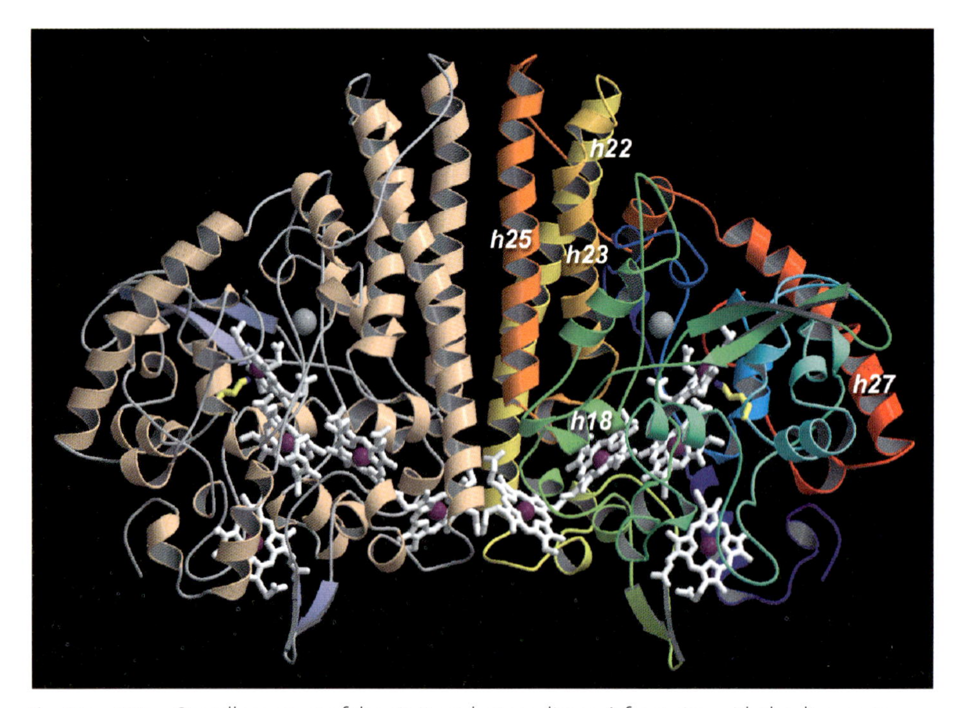

Fig. 49 (p. 461) Overall structure of the nitrite reductase dimer. A front view with the dimer axis oriented vertically, five hemes in each monomer (white), the Ca^{2+} ions (grey), and Lys133 that coordinates the active-site iron (yellow). The dimer interface is dominated by three long α-helices per monomer. All hemes in the dimer are covalently attached to the protein. (Reproduced by permission of Macmillan Magazines, Ltd., from Einsle, O. et al. (1999) *Nature* (*London*) **400**, 476–480.)

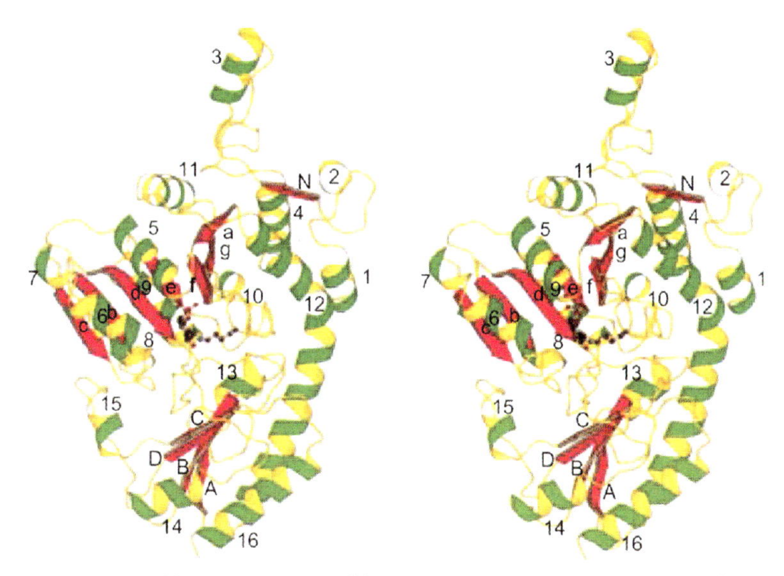

Fig. 40 (p. 453) Stereo ribbon presentation of the CBL monomer, emphasizing secondary structure elements. α-Helices are drawn as green spirals, ß-strands as magenta arrows. PLP and PLP-binding Lys210 are shown in a ball-and-stick representation; MOLSCRIPT and RASTER3D. (Reproduced by permission of Academic Press, Ltd., from Clausen, T. et al. (1996) *J. Mol. Biol.* **23**, 202–224.)

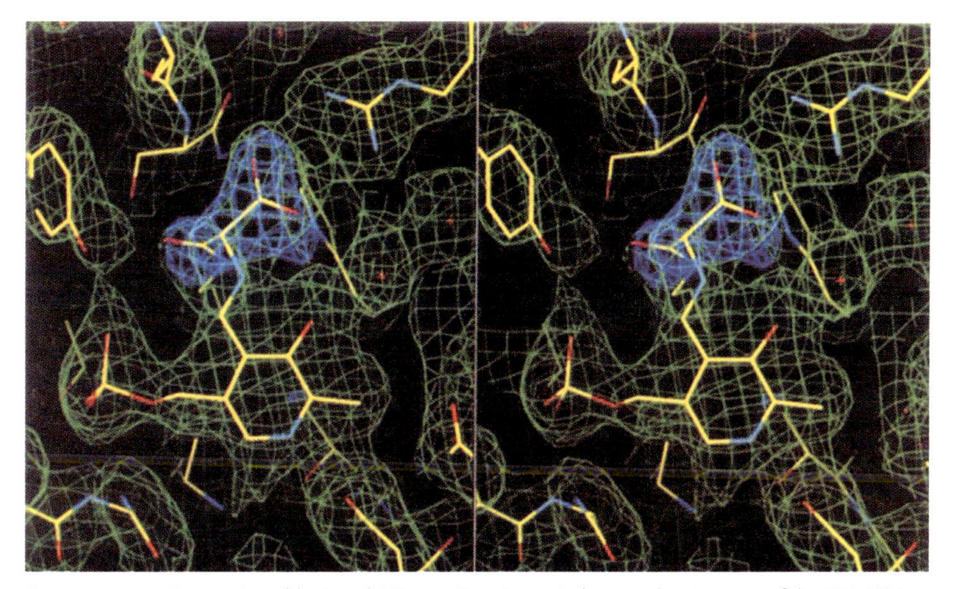

Fig. 42 (p. 454) $F_{obs} - F_{calc}$ (blue) and $2F_{obs} - F_{calc}$ (green) electron density map of the CBL/TFA complex around the active site contoured at 3.5σ and 1.0σ respectively, at 2.3-Å resolution. (Reproduced by permission of Academic Press, Ltd., from Clausen, T. et al. (1996) *J. Mol. Biol.* **23**, 202–224.)

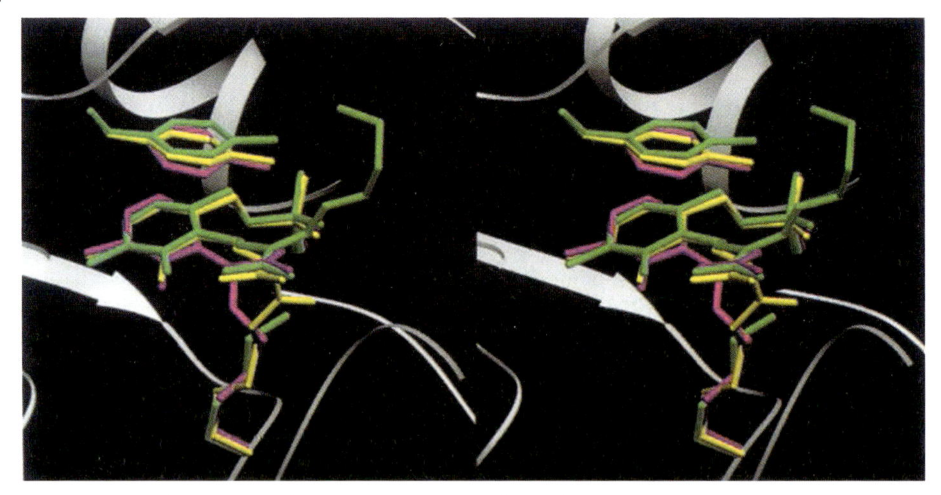

Fig. 43 (p. 455) Stereo view of a superposition of Tyr111 and the PLP derivative of the unliganded enzyme (magenta), the CBL/AVG adduct (green), and the TFA-inactivated enzyme (yellow); SETOR. (Reproduced by permission of the American Chemical Society, Clausen, T. et al. (1997) *Biochemistry* **36**, 12633–12643.)

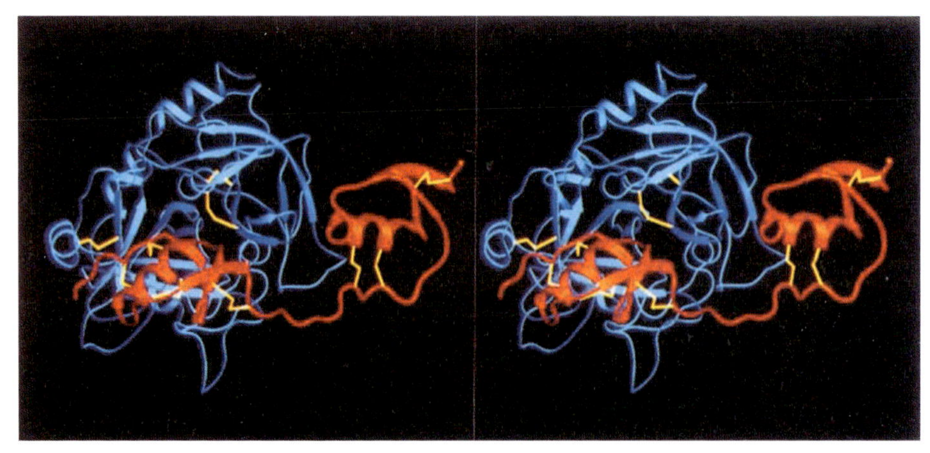

Fig. 44 (p. 456) Stereo view of the complex formed between thrombin (blue) and rhodniin (red) in the thrombin standard orientation, that is, with the active-site cleft facing the viewer and a bound inhibitor chain from left to right. Yellow connections indicate disulfide bridges. Rhodniin interacts through its N-terminal domain in a canonical manner with the active site and through its C-terminal domain with the fibrinogen recognition exosite of thrombin; SETOR. (Reproduced by permission of Oxford University Press, from van de Locht, A. et al. (1995) *EMBO J.* **14**, 5149–5157).

Encyclopedia of Molecular Cell Biology and Molecular Medicine

Second Edition

Innate Immunity

Osamu Takeuchi[1,2] *and Shizuo Akira*[1,2]
[1]*Osaka University, Osaka, Japan*
[2]*ERATO, Japan Science and Technology Agency, Osaka, Japan*

Encyclopedia of Molecular Cell Biology and Molecular Medicine, 2nd Edition. Volume 7
Edited by Robert A. Meyers.

ISBN: 3-527-30549-1

Keywords

Innate Immunity
An evolutionally conserved immune system utilizing germline-encoded pattern-recognition receptors.

PAMPs
Pathogen-specific molecular patterns recognized by host pattern-recognition receptors.

TLRs
Pattern-recognition receptors that detect pathogen-derived components and are composed of extracellular leucine-rich repeats and a cytoplasmic TIR domain.

TIR Domain
A domain shared by the cytoplasmic portions of TLRs, IL-1R family members, and adaptor proteins that mediate signal transduction.

NOD-LRR Family
Pattern-recognition receptors composed of C-terminal leucine-rich repeats, a central nucleotide-binding oligomerization domain, and N-terminal protein–protein interaction motifs, such as caspase recruitment domains, pyrin domains, or a TIR domain.

The innate immune system utilizes a limited number of germline-encoded pattern-recognition receptors to sense invading pathogens, and is evolutionally conserved from *Drosophila* to vertebrates. The mammalian innate immune system is subdivided into three different mechanisms. The first mechanism is the lectin-complement pathway, which recognizes invading microorganisms using plasma proteins such as lectin. The second mechanism is the Toll-like receptor (TLR) pathway, which detects pathogen-specific molecular patterns (PAMPs) shared by broad classes of microorganisms via TLRs on the membranes of innate immune cells. Each TLR activates a distinct signaling pathway by recruiting different TIR-domain containing adaptor molecules. The pathways lead to activation of the transcription factors NF-κB, AP-1, and interferon regulatory factor 3 (IRF3), followed by rapid induction of proinflammatory cytokines and type I interferons (IFNs). The third mechanism is mediated by cytoplasmic pattern-recognition receptors that detect microorganisms in the cytosol, and also leads to the production of cytokines and IFNs. These innate responses also instruct acquired immunity by expressing costimulatory molecules and presenting antigens for recognition by lymphocytes.

1 Innate Immunity

In vertebrates, the immune system consists of both innate and acquired immunity, which are defined on the basis of their mechanisms for detecting the presence of infectious organisms. The innate immune system is characterized by the use of a limited number of germline-encoded receptors that recognize diverse pathogens invading the host. The innate system, therefore, targets a set of molecular structures that are absent from host cells, but are unique to microorganisms and shared by various pathogens. By recognizing these "pathogen-specific" patterns, the innate system is able to prevent autoimmune responses. For instance, lipopolysaccharide (LPS), bacterial lipoprotein, peptidoglycan (PGN), lipoteichoic acid (LTA) and double-stranded RNA (dsRNA) are synthesized by bacteria or viruses, but not by host cells. Furthermore, carbohydrates such as mannose and *N*-acetylglucosamine on the surface of pathogens also represent targets for recognition.

The innate immune recognition is mediated by three different mechanisms. The first mechanism is the lectin-complement pathway, which is mediated by lectins and complement proteins secreted into the extracellular space. Binding of lectin to a microorganism triggers a chain of reactions on the surface of that microorganism leading to its destruction or opsonization. The second mechanism is the Toll-like receptor (TLR) pathway, which is mediated by transmembrane receptors on innate immune cells such as macrophages and dendritic cells (DCs). Detection of microorganisms by TLRs activates the cells to produce proinflammatory cytokines and chemokines to evoke inflammation and to recruit immune cells to the site of the infection. In addition, TLR pathways can induce maturation of DCs to instruct and adequately activate acquired immunity. The third mechanism is mediated by cytoplasmic pattern-recognition receptors that detect viruses and bacteria that have successfully invaded cells.

In contrast, the acquired immune system generates a highly diverse repertoire of antigen receptors, T- and B-cell receptors, via DNA rearrangement then followed by clonal selection that eliminates self-reacting cells. After encountering a pathogen, the lymphocytes bearing appropriately high affinity antigen receptors for that pathogen expand and eliminate the pathogens in the late stage of infection and contribute to the establishment of immunological memory. However, recent studies have revealed that adequate instruction by innate immune cells is a prerequisite for the activation of lymphocytes.

The innate immune system is evolutionally conserved from *Caenorhabditis elegans* and *Drosophila melanogaster* to vertebrates. In fact, it should be noted that *Toll* was originally identified during research into *Drosophila* developmental biology, and its role in *Drosophila* immune responses has been extensively studied. Moreover, the knowledge gained from that research has had a tremendous influence on advancing research into mammalian innate immunity.

2
Drosophila Immune Response

Drosophila does not possess the adaptive immune system, and mounts its host defense by discriminating between classes of pathogens and inducing immune responses. An important part of *Drosophila* immune response is the synthesis of antimicrobial peptides against microbial infection. *Drosophila* has seven antimicrobial peptides, and among those, Drosomycins and Metchnikowin have activity against fungi, Defensin acts against gram-positive bacteria and Attacins, Cecropins, Drosocin, and Diptericins have spectra against gram-negative bacteria. The induction of appropriate antimicrobial peptides is regulated by two distinct signaling pathways, *Toll* and *Imd*. The *Toll* pathway is activated in response to infections by fungi and gram-positive bacteria. Toll is a receptor that contains extracellular leucine-rich repeats and a cytoplasmic Toll/IL-1R homology domain. It was originally identified as an essential component of the pathway that determines the dorsal-ventral axis in early *Drosophila* embryogenesis, but *Toll*-mutant flies were also found to be susceptible to fungal infection. Invading gram-positive bacteria are recognized by PGN-recognition proteins (PGRP) followed by activation of a Toll ligand, Spaetzle. Binding of Spaetzle to Toll activates the *Drosophila*homolog of MyD88 and a serine/threonine kinase, Pelle, followed by the phosphorylation and degradation of Cactus, an IκB like inhibitory protein of Rel-type transcription factors. The degradation of Cactus releases the transcription factors Dorsal and Dorsal-related immunity factor (DIF) and initiates their nuclear translocation to induce the expression of Drosomycin genes. The pathway closely resembles the mammalian TLR- and interleukin-1 receptor (IL-1R)-mediated signaling pathway. In *Drosophila*, the Toll family consists of nine members, but the other Toll family members may not be involved in antimicrobial immune response.

The *Imd* pathway is involved in host defense against gram-negative bacteria by inducing Attacins, Cecropins, Drosocin, and Diptericins. The IMD protein contains a death domain (DD) that has high homology to that of mammalian TNF-receptor interacting protein (RIP). *Drosophila* homologs of FADD, TGF-β-activated kinase 1 (TAK1), IκB kinase (IKK) β are involved in the *Imd* pathway. Activated *Drosophila* IKKβ phosphorylates another NF-κB like protein, Relish, which results in its cleavage and subsequent nuclear translocation. In turn, Relish induces the expression of genes encoding antimicrobial peptides against gram-negative bacteria. Gram-negative-binding proteins (GNBPs) and a member of the PGRP family containing a transmembrane domain have been implicated in the *Imd* pathway, but the precise mechanisms for *Imd* pathway activation by these molecules remain to be clarified.

Taken together, *Drosophila* distinguishes among different types of invading microorganisms by activation of distinct signaling pathways, resulting in the production of appropriate antimicrobial peptides for the particular type of invading pathogen.

3
The Lectin-complement Pathway

In mammalian innate immunity, the lectin-complement pathway is an important mechanism for recognizing a wide range of pathogens outside of host cells, and is mediated by a number of proteins that are secreted into the serum. The lectin pathway involves carbohydrate recognition by mannose-binding lectin (MBL) and ficolins (Fig. 1). MBL and ficolins specifically recognize *N*-acetyl-D-glucosamine (GlcNAc), mannose, and fucose, and do not interact with D-galactose or sialic acid, which are abundant on mammalian cells. Binding of MBL and ficolins to microbial carbohydrates activates members of the MBL-associated serine protease (MASP) family on the surface of bacterial cells. MASPs, which form complexes with MBL, are enzymes that cleave complement components to activate the complement pathway (Fig. 1). The cleavage leads to direct destruction of the invading bacteria or to their opsonization, which facilitates phagocytosis by macrophages, and in human, MBL-deficiency was shown to increase susceptibility to a variety of infectious diseases. Among MASPs, MASP-2 is responsible for activation of C4 and C2, which leads to C4bC2a enzyme complex formation. C4bC2a then acts as a C3 convertase and cleaves C3, generating products that can bind and destroy bacteria. In contrast, MASP-1 is able to cleave C3 directly (Fig. 1). Together, the lectin-complement pathway plays an important role in detecting microorganisms in the bloodstream and in extracellular space.

4
Toll-like Receptors (TLRs)

In 1997, Janeway and Medzhitov identified a mammalian homolog of *Drosophila* Toll and showed that it was involved in the innate immune response. Eleven mammalian TLRs have been reported to

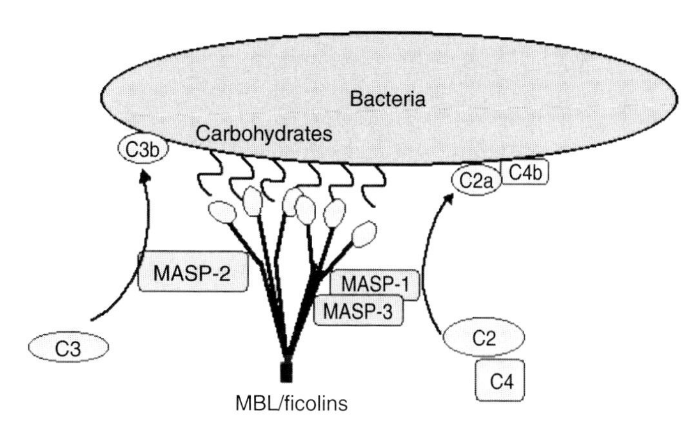

Fig. 1 The lectin-complement pathway. MBL or ficolin binds bacterial carbohydrates on the surface of a bacterial cell. This results in the activation of MASP family members that associate with MBL or ficolin. MASP-2 is responsible for catalyzing cleavage of C4 and C2 to generate C4b and C2a. A C4bC2a complex catalyzes the cleavage of C3, resulting in the generation of C3b that binds to and destroys the bacteria. In contrast, MASP-1 and -3 cleaves C3 directly.

date, and 10 of them have been shown to recognize specific PAMPs. TLRs (Toll-like receptors) are expressed not only on macrophages, DCs, and B cells, but also on fibroblasts, vascular endothelial cells, and intestinal epithelial cells. The TLR family proteins consist of extracellular leucine-rich repeat (LRR) motifs, a transmembrane region, and a cytoplasmic tail containing a Toll/IL-1 receptor homology (TIR) domain. The LRR motifs are responsible for ligand recognition and the TIR domain is essential for triggering intracellular signaling pathways. The activation of TLR signaling pathways leads to the expression of proinflammatory cytokine and chemokine genes. This induces migration of immune cells from peripheral blood to the site of infection. DCs play a pivotal role in T cell activation and differentiation into T helper 1 cells. DCs phagocytose pathogens, process them, and present the antigenic peptides on MHC molecules. TLR signaling is critical for the maturation of DC by regulating phagosome maturation, costimulatory molecule expression and cytokine production.

Homeostasis of acquired immunity is controlled by regulatory T cells, which can suppress self-reacting T cells. However, regulatory T cells do not interfere with T cell activation in the course of pathogenic infection. Recent studies revealed that TLR signaling is involved in suppression of the inhibitory function of regulatory T cells. This is controlled by IL-6 secreted by TLR stimulation.

In addition, ligands for TLR3, TLR4, TLR7, and TLR9, but not for TLR2 and TLR5, are known to upregulate type I interferons (IFNs) and IFN-inducible genes. These TLRs recognize viral as well as bacterial components, indicating that TLR signaling is also involved in antiviral immune responses. Roles of each TLR in the recognition of bacterial components are described in the following section (Fig. 2).

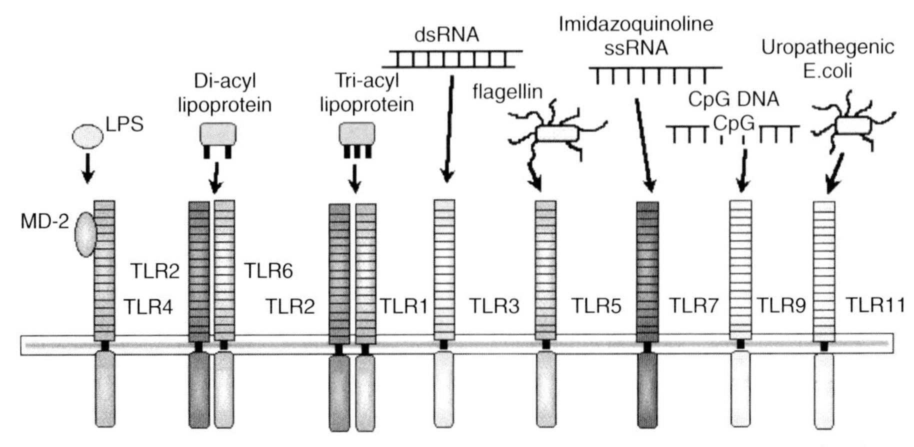

Fig. 2 TLRs recognize various PAMPs. TLR2 recognizes bacterial lipoproteins in concert with TLR1 or TLR6. TLR1 and TLR6 discriminate subtle difference in the lipid moiety of tri-acyl and di-acyl lipoproteins, respectively. TLR4 forms a complex with MD-2 and detects LPS. TLR3 is involved in extracellular recognition of viral dsRNA and TLR5 recognize bacterial flagellin. TLR7 and TLR9 are responsible for viral ssRNA and unmethylated CpG-DNA, respectively. TLR11 detects uropathogenic bacteria in mice.

4.1 TLR2

TLR2 recognizes various PAMPs from bacteria, fungi, and parasites including bacterial lipoprotein, PGN, *Saccharomyces cerevisiae* zymosan, and glycosylphosphatidylinositol (GPI)-anchors from *Trypanosoma cruzi*. In addition, TLR2 is involved in the recognition of LPS from *Porphyromonas gingivalis* and *Leptospira interrogans*, which have different structures to LPS from enterobacteria. Macrophages derived from TLR2$^{-/-}$ mice did not produce proinflammatory cytokines in response to TLR2 ligands, such as PGN and bacterial lipoproteins, and TLR2 signaling did not induce type I IFNs. Corresponding to the failure to detect these components through TLR2, TLR2$^{-/-}$ mice were susceptible to infections by various gram-positive bacteria and fungi, including *Staphylococcus aureus*, Group B *Streptococcus, Borrelia burgdorferi, Chlamydia trachomatis, Candida albicans, Streptococcus pneumoniae*, and so on. On the other hand, recent reports have shown that TLR2 signaling suppresses immunity against *Candida* infection. The population of regulatory T cells and IL-10 secretion were partially reduced in TLR2$^{-/-}$ mice, which may explain the improved resistance to *Candida* infection. Although the precise mechanism of this suppression is to be clarified, this is an interesting phenomenon that TLR signaling can work as both stimulatory and inhibitory manner.

4.2 TLR1 and TLR6

TLR1 and TLR6 are highly homologous with each other and are also homologous to TLR2. Heterodimerization of TLR2 with TLR1 or TLR6 determines the ligand specificity. Although expression of either TLR1 or TLR6 alone failed to confer any response, coexpression of TLR1 or TLR6 with TLR2 resulted in cell activation in response to triacyl- or diacyl-lipopeptide, respectively. The cellular response to mycoplasmal diacyl-lipopeptide was abolished in TLR6$^{-/-}$ mice. In contrast, TLR1$^{-/-}$ mice showed an impaired response to triacyl-lipopeptide, while the response to diacyl-lipopeptide was normal. Overexpression studies revealed that TLR1 and TLR6 each interacted with TLR2 in cells. These observations suggest that TLR1 and TLR6 recognize the difference in the lipid portion of lipoproteins through heterodimerization with TLR2.

4.3 TLR3

Infection with RNA viruses leads to dsRNA generation in the cytoplasm of infected cells. Such dsRNA is a virus-specific molecular pattern that can be recognized by the host to stimulate immune responses. Cells expressing TLR3 respond to dsRNA or its synthetic mimic, polyinosinic-polycytidylic acid (poly I:C), and activate an intracellular signaling pathway leading to upregulation of type I IFNs and proinflammatory cytokines. Myeloid DCs and lung fibroblasts derived from TLR3$^{-/-}$ mice were hyporesponsive to poly I:C stimulation, and TLR3$^{-/-}$ mice were reported to be susceptible to cytomegalovirus (CMV) infection. However, inoculation of poly I:C into TLR3$^{-/-}$ mice resulted in normal production of serum IFNα, indicating that other molecule(s) are also involved in the dsRNA-induced immune responses.

4.4 TLR4

Two strains of mice, C3H/HeJ and C57BL/10ScCr, were known to be hyporesponsive to LPS, a component of outer cell membrane of gram-negative bacteria. Positional cloning of the gene responsible for LPS recognition in these mice identified TLR4. In C3H/HeJ mice, a missense point mutation resulted in replacement of a proline in the cytoplasmic domain of the TLR4 protein with histidine, and created a dominant-negative form of the protein. In C57BL/10ScCr mice, the *Tlr4* locus was entirely deleted. TLR4$^{-/-}$ mice generated by homologous recombination of embryonic stem (ES) cells were highly resistant to LPS-induced shock, and macrophages and B cells derived from these mice were unresponsive to LPS in terms of proinflammatory cytokine production and the proliferative response. The activation of NF-κB and MAP kinase in response to LPS was also abrogated. Moreover, C3H/HeJ mice were highly susceptible to infection by *Salmonella* and *Escherichia coli*, indicating that recognition of bacteria by TLR4 is critical to host defense

The entire LPS recognition machinery is more complicated. When LPS is present in the bloodstream, it is immediately captured by LPS-binding protein (LBP), and the LPS–LBP complex is transferred to CD14, a GPI-anchored protein, located on the surface of macrophages. An extracellular protein, MD-2, directly binds TLR4 and enhances LPS-induced NF-κB activation in the TLR4-expressing cells. LBP$^{-/-}$ and CD14$^{-/-}$ mice are hyporesponsive to LPS stimulation. The response of MD-2$^{-/-}$ mice to LPS was also severely impaired. Interestingly, surface expression of TLR4 was abolished in MD-2$^{-/-}$ macrophages. LPS stimulation initiates the dimerization of TLR4 to ignite the downstream signaling pathway. An LRR containing transmembrane protein, RP105, is also involved in LPS recognition in B cells, since targeted disruption of RP105 resulted in an impaired proliferative response of B cells to LPS.

In addition to LPS, TLR4 recognizes several other PAMPs such as respiratory syncytial virus F protein and mouse mammary tumor virus envelope protein. In addition, it has also been reported to recognize endogenous products. Under inflammatory conditions, components of the extracellular matrix are degraded or processed. The products, such as fibronectin fragments and hyaluronic acid degradation products can stimulate cells through TLR4. Furthermore, several studies have demonstrated that both human and chlamydial HSP60 are putative ligands for TLR2 and TLR4. However, it is possible that these ligand preparations were contaminated with LPS or other PAMPs, since most studies utilize recombinant proteins synthesized in *E. coli*.

4.5 TLR5

TLR5 recognizes bacterial flagellin, a unique PAMP composed of protein. Flagellin is a protein component of the flagellum, a structure used by some bacteria to move through liquid media. Chinese hamster ovary (CHO) cells expressing human TLR5 can activate NF-κB in response to flagellin. Moreover, *Salmonella typhimurium* lacking flagellin failed to activate TLR5, indicating that TLR5 is critical for flagellin recognition. A stop codon within the open-reading frame (ORF) of human TLR5 in many individuals renders them incapable of responding adequately to flagellated bacteria. The

MOLF/Ei mouse strain is shown to be susceptible to *S. typhimurium* infection, and one of the MOLF/Ei susceptibility loci was mapped to chromosome 1, which contains the mouse *Tlr5* gene. Furthermore, TLR5 expression was reduced in MOLF/Ei mice.

4.6
TLR7 and TLR8

TLR7 is closely related to TLR8 and shows significant homology with TLR9. TLR7 and TLR8 are both located on the X chromosome. Recent studies have revealed that the ligands of these three TLRs are nucleotides and their derivatives. Initially, imidazoquinoline derivatives, including imiquimod and resiquimod (R-848), were identified as TLR7 ligands in mice. These derivatives are small synthetic compounds that are approved for use as antiviral drugs due to their potency for inducing type I IFNs. TLR7$^{-/-}$ mice failed to respond to these compounds by producing proinflammatory cytokines or type I IFNs. In humans, R-848 strongly activates plasmacytoid DCs (pDCs) to produce IFNα. TLR7 is also involved in the recognition of other synthetic anticancer drugs, such as loxoribine (7-allyl-8-oxoguanosine), bropirimine (2-amin-5-bromo-6-phenyl-4(3)-pyrimidinone) and certain guanosine analogs. Recently, TLR7 was found to be essential for DCs to recognize single-stranded RNAs (ssRNAs) (Fig. 2).

Expression of human TLR8 as well as human TLR7 in HEK293 cells conferred the responsiveness to ssRNAs to activate an NF-κB reporter gene. In contrast, the expression of mouse TLR7, but not TLR8, could restore ssRNA response in culture cells. Moreover, antigen-presenting cells from TLR8$^{-/-}$ mice produced proinflammatory cytokines normally in response to ssRNAs, suggesting that human and mouse TLR8 recognize different PAMPs.

4.7
TLR9

Bacterial DNA is characterized by an abundance of CpG motifs and the lack of methylation. This PAMP is recognized by the host immune cells such as DCs, macrophages, and B cells, and strongly induce T helper 1 development. Furthermore, synthetic oligodeoxynucleotides containing an unmethylated CpG motif (CpG-DNA) show strong immunostimulatory activity.

TLR9 was shown to be essential for CpG-DNA detection, since the immune response to CpG-DNA stimulation was abolished in TLR9$^{-/-}$ mice. TLR9 is localized in the endoplasmic reticulum (ER) where CpG-DNA is transported after its incorporation into a tubular lysosomal compartment. DNA viruses, such as Herpes simplex virus (HSV) types I and II, also contain genomic DNA with unmethylated CpG motifs, and pDCs from TLR9$^{-/-}$ mice failed to produce IFNα upon HSV I or II infection. Moreover, mice with a mutation in the *Tlr9* locus were susceptible to CMV infection, indicating that TLR9 is critically involved in the immune response against DNA viruses.

4.8
TLR11

Mouse TLR11 is expressed in macrophages and liver, kidney, and bladder epithelial cells. Cells expressing TLR11 specifically respond to uropathogenic bacteria, but not to known other TLR ligands. TLR11$^{-/-}$ mice were highly susceptible to kidney infections by uropathogenic bacteria, and macrophages derived from TLR11$^{-/-}$

mice failed to produce proinflammatory cytokines in response to these bacteria. However, human TLR11 is likely to be nonfunctional, since the stop codons were observed in its ORF.

5
The TLR Signaling Pathway

By detecting PAMPs, TLRs trigger downstream Signaling pathways that lead to nuclear translocation of NF-κB and the activation of MAP kinases (Fig. 3). TIR domain-containing adaptors are recruited to the receptors first, and then interact with IL-1R-associated kinases (IRAKs). Activated IRAK1 recruits TNF receptor-associated factor 6 (TRAF6), which activates two ubiquitination proteins, Ubc13 and Uev1A. Ubc13 and Uev1A form a complex, and this UBC complex catalyzes the formation of a lysine 63-linked polyubiquitin chain. Ubiqutinated TRAF6 activates a complex of TGFβ activating kinase 1 (TAK1) and associated proteins, TAB1, TAB2, and TAB3, to directly phosphorylate the IκB kinase (IKK) complex. The activated IKK complex, composed of IKKα,

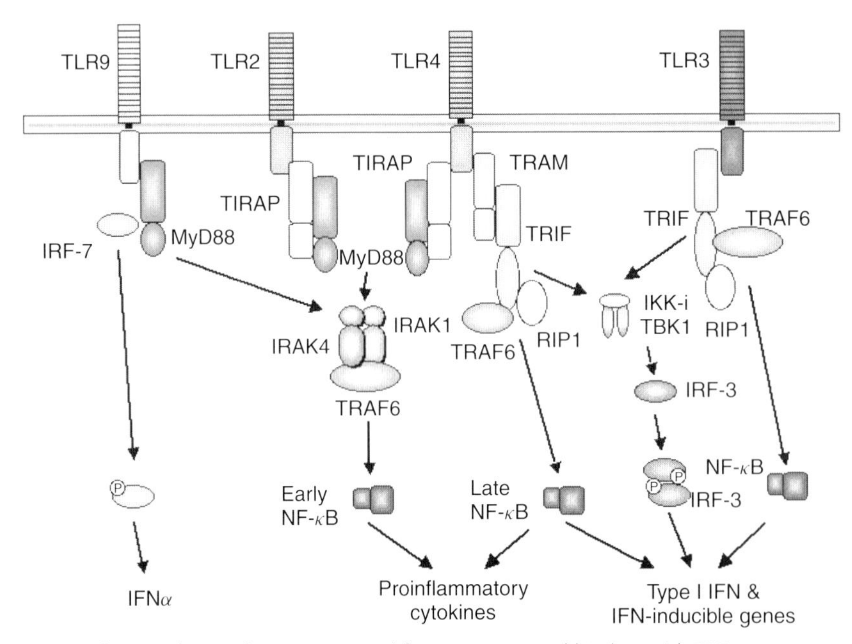

Fig. 3 The signaling pathways emanated from TLRs. Ligand binding with TLRs except TLR3 recruits MyD88 to the cytoplasmic TIR domain. MyD88, in turn, associates with IRAK4 and IRAK1. IRAK1 then activates TRAF6 leading to the nuclear translocation of NF-κB. Activated NF-κB upregulates transcription of proinflammatory cytokine genes. Another TIR-domain adaptor molecule, TIRF, is recruited to TLR3 and TLR4. TRIF activates late NF-κB via RIP1 and TBK1/IKK-i to induce phosphorylation of IRF-3. Subsequently, IRF3 homodimerizes, translocates to the nucleus and upregulates transcription of type I IFNs and IFN-inducible genes. TIRAP is specifically involved in a TLR2- and TLR4-mediated signaling pathway via MyD88, but is dispensable for the activation of TRIF-dependent pathway. TRAM bridges TLR4 and TRIF to activate TRIF-dependent pathway. TLR9 signaling is dependent on MyD88. MyD88 directly associates with IRF7 in pDC to induce phosphorylation. IRF7 subsequently activate IFNα gene expression.

IKKβ, and NF-κB essential modulator (NEMO), phosphorylates IκB, leading to its proteasome-mediated degradation, and the released NF-κB translocates from the cytosol to the nucleus to mediate the expression of proinflammatory cytokines. Next, we focus on the signaling molecules that act adjacent to the receptors, and act on the TIR domain-containing adaptors in particular.

5.1 TIR Domain-containing Adaptor Molecules

All IL-1R and TLR family members possess a TIR domain, a motif of ~160 amino acids, in their cytoplasmic region, which is critical for emanating downstream signaling pathways. Following ligand binding, TLRs dimerize and recruit cytoplasmic adaptors that also contain a TIR domain. Five TIR domain-containing adaptor molecules have been identified in mammals to date, namely, myeloid differentiation factor 88 (MyD88), TIR domain-containing adaptor inducing IFNβ (TRIF)/TIR domain-containing adaptor molecule-1 (TICAM-1), TIR domain-containing adaptor protein (TIRAP)/MyD88 adaptor like (MAL), TRIF-related adaptor molecule (TRAM)/TICAM-2 and sterile α and HEAT-Armadillo motifs (SARM). Individual TLRs can activate distinct signaling pathways through recruiting distinct adaptor molecules.

5.1.1 MyD88

MyD88 is an adaptor molecule consisting of an N-terminal DD and a C-terminal TIR domain. All IL-1R/TLR family members, except for TLR3, share at least one common signaling pathway via MyD88. After ligand stimulation, TLR/IL-1R interacts with MyD88 via the TIR domains, and MyD88 then recruits and associates with IRAK proteins by homophilic interaction via the DD. Analysis of MyD88$^{-/-}$ mice revealed that MyD88 is a gateway for signaling through all the TLR/IL-1R family members, except for TLR3. Cells derived from MyD88$^{-/-}$ mice failed to produce proinflammatory cytokines in response to various TLR ligands, such as bacterial lipoprotein, PGN, LPS, flagellin, ssRNA and CpG-DNA, in addition to IL-1β and IL-18. MyD88$^{-/-}$ cells showed defective activation of NF-κB and MAP kinases in response to TLR ligands, except for the TLR4 ligand (LPS) and TLR3 ligand (poly I:C). When MyD88$^{-/-}$ cells were stimulated with LPS, delayed activation of NF-κB and MAP kinases was observed. The defect in TLR signaling renders MyD88$^{-/-}$ mice highly susceptible to infections with various pathogens, including bacteria, fungi, viruses, and parasites, such as *S. aureus, Mycobacterium tuberculosis, Mycobacterium avium, Listeria monocytogenes*, Group B *Streptococcus, B. burgdorferi, C. albicans* and *T. cruzi*. Taken together, these observations indicate that MyD88 is essential for host defense against various pathogens by activating innate immune responses.

5.1.2 TRIF/TICAM-1

TRIF contains a TIR domain, a TRAF6 binding motif, and a C-terminal receptor interacting protein (RIP) homotypic interaction domain. Overexpression of TRIF activated an NF-κB-dependent reporter gene, and also activated the IFNβ reporter much more strongly than other TIR domain-containing adaptors. TRIF can associate with TRAF6 through its TRAF6 binding motif, and also interact with RIP1 via the RIP homotypic interaction motif. TRIF can also associate with TBK1, a kinase known to phosphorylate IRF3 to induce IFNβ gene induction. Poly(I:C)-mediated NF-κB activation was abolished in RIP$^{-/-}$ cells, indicating that RIP is essential for TLR3-mediated NF-κB activation. TRIF$^{-/-}$ mice were defective in TLR3- and TLR4-mediated proinflammatory cytokine production and the IFN response. Although LPS-mediated initial activation of NF-κB occurred in TRIF$^{-/-}$ cells, it was attenuated more rapidly than that in wild-type cells. LPS-induced signaling pathways were not activated in the absence of both MyD88 and TRIF, indicating that LPS signaling is entirely dependent on these two adaptors. Overexpression of TRIF in cells induced apoptosis via caspase-8 activation, suggesting a role for TRIF in bacteria-induced cell death. Interestingly, neither TLR7- and TLR9-mediated cytokine production nor the IFN response was impaired in TRIF$^{-/-}$ cells, indicating that the function of TRIF is limited to TLR3 and TLR4 signaling.

5.1.3 TIRAP/Mal

TIRAP/Mal was cloned as another TIR domain-containing adaptor molecule from a database search. Although TIRAP/Mal was initially implicated in an MyD88-independent signaling pathway based on *in vitro* studies, generation of TIRAP/Mal$^{-/-}$ mice revealed that TIRAP/Mal was required for an MyD88-dependent pathway emanating from TLR2 and TLR4 but not for MyD88-independent pathway.

5.1.4 TRAM/TICAM-2

TRAM/TICAM-2 is a 235-amino acids protein with a C-terminal TIR domain. Like TRIF, expression of TRAM/TICAM-2 activates IRF-3 and NF-κB-dependent gene induction. In TRAM/TICAM-2$^{-/-}$ cells, LPS-induced cytokine and IFN-inducible gene expression was severely impaired. Furthermore, TRAM/TICAM-2$^{-/-}$ cells failed to sustain LPS-induced NF-κB activation compared to wild-type cells. In contrast, there were no defects in the poly I:C and CpG-DNA responses. TRAM/TICAM-2 can physically interact with both TLR4 and TRIF/TICAM-1, whereas no direct association between TLR4 and TRIF/TICAM-1 was detected. When taken together, TRAM/TICAM-2 functions to bridge TLR4 and TRIF/TICAM-1 to activate the downstream signaling pathway.

5.1.5 Sterile α and HEAT-Armadillo Motifs (SARM)

SARM is a 690-amino acid protein containing a TIR domain, two sterile α-motif (SAM) domains and an Armadillo repeat motif (ARM). SARM is evolutionally conserved from *C. elegans* and *D. melanogaster*. *C. elegans* only contains two genes that encode TIR-domain proteins. One is *tol-1*, a Toll homolog, and the other is *tir-1*, a SARM homolog. *Caenorhabditis elegans* SARM is required for resistance to microbial pathogens. However, overexpression of human SARM in mammalian cells was not sufficient to activate NF-κB and IRF3. The establishment of SARM$^{-/-}$ mice has not yet been reported.

5.2 The IRAK Family

Four IRAK family members including IRAK1, IRAK2, IRAK4, and IRAK-M have been identified in mammals, and are composed of an N-terminal DD and a subsequent serine/threonine kinase domain. While IRAK1 and IRAK4 have bona fide kinase activity, IRAK2 and IRAK-M are both inactive due to mutation of a critical residue in their kinase domains. After ligand stimulation with IL-1R/TLR, IRAK4 is recruited to TLR/IL-1R via the DD to interact with MyD88 or TIRAP/Mal. Subsequently, IRAK4 can associate and directly phosphorylate IRAK1, which triggers the transient recruitment of TRAF6. Studies using IRAK4$^{-/-}$ mice revealed that IRAK4 is essential for activation of the IL-1R/TLR-induced signaling pathway. In contrast, although embryonic fibroblasts (MEFs) derived from IRAK1$^{-/-}$ mice exhibited an attenuated response to stimulation with IL-1β, LPS or other PAMPs, significant cytokine production was observed. This may be due to compensation by IRAK2. It is interesting to explore whether the kinase activity of IRAKs is critical for signaling, since kinase-deficient IRAK1 was still able to restore IL-1β-mediated NF-κB activation in IRAK1-deficient HEK293 cells. The role of kinase activity of IRAK4 is still controversial. Some reports showed that expression of kinase-deficient IRAK4 strongly impaired IL-1-mediated NF-κB activation, while a report showed that a kinase-inactive mutant of IRAK4 partially restored IL-1β-mediated IL-6 production as well as NF-κB activation in IRAK4$^{-/-}$ cells. Future studies will clarify the roles of the kinase activities of IRAK family members. IRAK2 also interacts with MyD88 and TIRAP/Mal, and may mediate the TIRAP/Mal signaling pathway emanating from TLR2 and TLR4. A future IRAK2 knockout study will be necessary to reveal its exact physiological functions. IRAK-M expression is restricted to monocytes/macrophages and is upregulated upon stimulation with TLR ligands. IRAK-M$^{-/-}$ mice showed enhanced production of proinflammatory cytokines in response to TLR ligands. IRAK-M perturbs the association of the IRAK4-IRAK1 complex resulting in the prevention of IRAK1-IRAK4 complex, suggesting that IRAK-M is a negative regulator of TLR signaling pathways.

These observations suggest that IRAKs play distinct roles in IL-1R/TLR signaling. IRAK4 acts upstream for IRAK1, IRAK2 can signal with TIRAP/Mal, and IRAK-M is a negative regulator of the signaling pathway.

5.3
TNF Receptor-associated Factor 6 (TRAF6)

TRAF6 belongs to a TRAF family, which is characterized by an N-terminal RING finger domain and a conserved C-terminal TRAF-domain. The RING finger domain is also found in a large family of E3 ubiquitin ligases, and TRAF6 was recently shown to function as an ubiquitin ligase, along with an E2 ligase complex consisting of Ubc13 and Uev1A/Mms2, to catalyze the formation of a polyubiquitin chain through lysine-63 (K63) of ubiquitin. TRAF6$^{-/-}$ mice exhibited severe osteopetrosis due to a defect in osteoclast differentiation. B cells and fibroblasts from TRAF6$^{-/-}$ mice showed defective responses to IL-1β, TLR ligands, and CD40 ligation. However, LPS-induced IFN-inducible gene expression was not impaired in TRAF6$^{-/-}$ cells, suggesting that TRAF6 is dispensable for the MyD88-independent pathway.

5.4
TAK1 and TAB1, TAB2 and TAB3

TAK1 was originally identified as a TGFβ-activated kinase and subsequently reported to be involved in the IL-1β signaling pathway. Ubiquitin-dependent activation of TAK1 results in phosphorylation of IKKβ and MKK6 leading to NF-κB and MAP kinase activation. Knockdown of TAK1 abolished the IKK activation induced by TNFα and IL-1β stimulation.

TAB1 and TAB2 were identified as adaptor proteins that associate with TAK1. TAB3, a TAB2 homolog, is also involved in IL-1β-mediated NF-κB and MAP kinase activation. TAB2 and TAB3 bind preferentially to lysine-63-linked polyubiquitin chains through a zinc finger domain. However, IL-1β-mediated activation of NF-κB and MAP kinases was not impaired in MEFs from TAB2$^{-/-}$ mice. TAB2 and TAB3 may therefore compensate for each other's functions *in vivo*.

6
Intracellular Recognition of PAMPs

TLRs have a transmembrane domain and are localized on the plasma membrane, ER and lysosome membrane, suggesting that TLRs are not suitable for recognizing pathogens invaded into the cytosol of cells. Various pathogens such as intracellular bacteria and viruses are supposed to be detected by PRRs in the cytosol. In mammals, the NOD-LRR family is thought to play a role in detecting cytosolic pathogens. The members of this family share a tripartite domain structure consisting of a C-terminal LRR motifs, a central nucleotide-binding oligomerization domain (NOD), and N-terminal protein–protein interaction motifs, such as caspase recruitment domains (CARDs), pyrin domains, or a TIR domain. NOD1 and NOD2, containing N-terminal CARD domains have been extensively studied for their ligands. NOD1 detects γ-D-glutamyl-meso diaminopimelic acid (iE-DAP) found in gram-negative bacterial peptidoglycan. NOD1$^{-/-}$ macrophages fail to produce cytokines in response to iE-DAP. In contrast, NOD2 is a receptor for muramyl-dipeptide (MDP) derived from bacterial PGN (Fig. 4). A missense point mutation in the human *Nod2* gene is correlated with susceptibility to Crohn's disease, an inflammatory bowel disease. A recent study revealed that NOD2 functioned to suppress TLR2-mediated activation of NF-κB in response to PGN. NOD2$^{-/-}$ mice showed enhanced IL-12 production in response to PGN stimulation, which may explain the mechanism by which NOD2 controls the susceptibility to the disease. In addition, a mutation in the NOD domain of NOD2 is the cause of Blau disease, an autosomal dominant disorder characterized by granulomatous arthritis, uveitis, and skin rash. Ligand binding of NOD1 and NOD2 causes their oligomerization and results in the activation of NF-κB through recruitment of a serine/threonine kinase, RIP2/RICK. NODs associate with RIP2/RICK through their CARD domains by a homophilic interaction. Although the number of the NOD-LRR family members is growing, the functions of members other than NOD1 and NOD2 have yet to be clarified.

Viral dsRNA is synthesized in infected cells for the viral replication. Viral dsRNA is a virus-specific molecular structure to be recognized by hosts. Most cells infected with virus produce type I IFNs in a TLR3 independent manner, implying that the mechanism recognizing dsRNA in cytoplasm plays an important role in IFN response. Cytoplasmic proteins, protein kinase R (PKR), and retinoic acid inducible protein-I (RIG-I) are implicated in viral dsRNA recognition. PKR belongs to a protein family containing dsRNA-binding domains, and is induced in response to stimulation with IFNs. Activation of PKR results in growth arrest of the cells via phosphorylating eIF2α. Several reports have shown that virus-induced type I IFN production is modestly impaired in PKR$^{-/-}$ mice. In addition, PKR$^{-/-}$ mice show defective induction of apoptosis

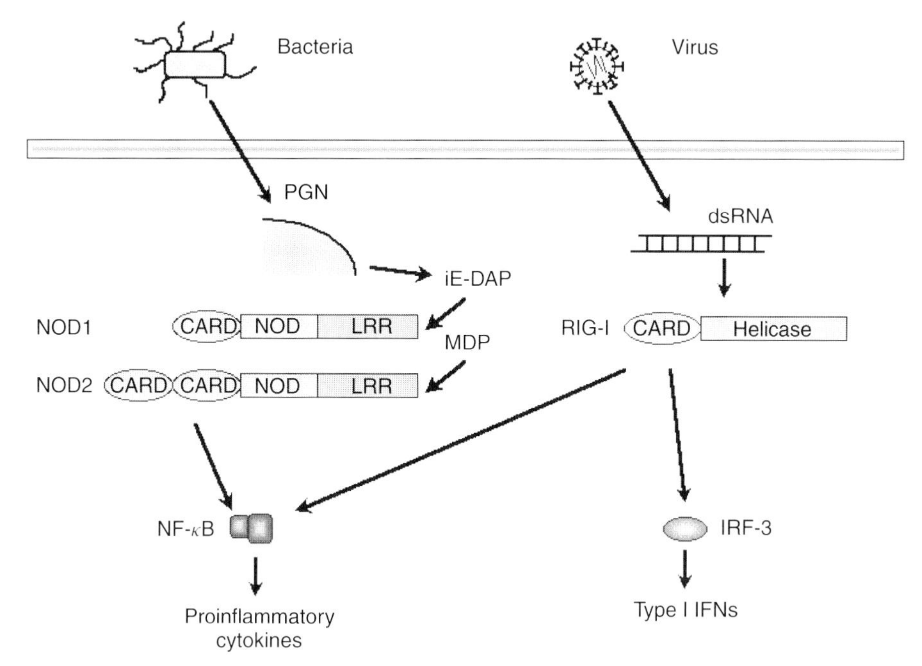

Fig. 4 Cytoplasmic recognition of microorganism. Intracellular bacteria are recognized by NOD-LRR family members in the cytoplasm. Components of PGN, iE-DAP, and MDP, are recognized by NOD1 and NOD2, respectively. NOD1 and NOD2 are composed of C-terminal LRR, central NOD domain, and N-terminal CARD domain. They detect bacterial components via LRR motifs and signal through CARD domain to activate NF-κB. When RNA viruses invade into a cell, dsRNA is generated for viral replication. RIG-I detects viral dsRNA via its helicase domain. CARD domain of RIG-I is required for activating IRF3 and NF-κB.

in response to poly I:C and bacterial burden. The other candidate of dsRNA recognition molecules, RIG-I, is also an IFN-inducible protein. RIG-I contains an N-terminal CARD domain and a DExD/H box RNA helicase domain. RIG-I interacts with poly I:C, and overexpression of RIG-I in cells conferred Newcastle Disease virus (NDV)- and dsRNA-mediated IFN response (Fig. 4). Knockdown of RIG-I by RNA interference abolished the IFN response upon NDV infection, suggesting that RIG-I is responsible for detecting RNA viruses in the cytosol of cells. It will be interesting to explore the relationships between cytoplasmic and extracellular virus recognition in host defense.

7 Pathways Activating IFN Response

Type I IFNs mainly consist of IFNα and IFNβ. The IFNα family is encoded by a cluster of multiple genes, whereas there is only a single *IFNβ* gene. Type I IFNs are strongly induced upon viral infection and stimulation with TLR ligands. Among the TLR ligands, stimulation with the TLR3, TLR4, TLR7, and TLR9 ligands, but not the TLR2 ligand, is known to upregulate type I IFN production.

Recent studies have revealed the involvement of two IKK-related kinases, inducible IκB kinase (IKK-I; also known as IKKε) and TRAF-family member-associated

NF-κB activator (TANK) TANK-binding kinase 1 (TBK1 also known as T2K), in IFN production upon TLR stimulation and viral infection. IKK-i and TBK1 can directly phosphorylate IRF3 and IRF7 *in vitro* and activate the IFNβ promoter. $TBK1^{-/-}$, but not $IKK\text{-}i^{-/-}$, MEFs showed severely impaired induction of IFNβ and IFN-inducible genes in response to LPS, intracellular introduction of poly I:C and RNA virus infection. Activation of IRF3, but not NF-κB, in response to LPS and poly(I:C) was also diminished in $TBK1^{-/-}$ cells. IKK-i/TBK1 double-deficient MEFs failed to express any detectable levels of IFNβ and IFN-inducible genes in response to poly I:C, indicating that both IKK-i and TBK1 contribute to the IFN pathway. Furthermore, these observations suggest that the signaling pathways triggered by TLR stimulation as well as cytoplasmic viral recognition converge at the level of TBK1/IKK-i (Fig. 3).

The transcription factors, IRF3 and IRF7 are essential for the expression of type I IFNs in response to viral infection and TLR stimulation. Phosphorylated IRF3 forms homodimers that translocate into the nucleus and bind to the IFN-stimulated regulatory element (ISRE) present in the promoters of a set of IFN-inducible genes to induce their expression. Initially, secreted type I IFNs activate their receptors in both autocrine and paracrine manners. This activates IFN-stimulated gene factor 3 (ISGF3), which consists of signal transducer and activator of transcription (STAT) 1, STAT2, and p40/IRF9 to amplify the response. IRF7 is also implicated in the production of IFNα. Cells deficient in IRF3 and IRF7 expression are defective in type I IFN production in response to viral infection. Moreover, $IRF3^{-/-}$ mice showed defective induction of IFNβ in response to LPS, indicating a critical role for IRF3 in the TLR4-induced IFN response.

Plasmacytoid DCs (pDCs) are known to be recruited to inflamed lymph nodes and produce large amounts of type I IFNs. pDCs express TLR7 and TLR9, and stimulation by their ligands induces IFNα production. Since this IFNα induction is independent of TBK1, the existence of a novel mechanism is hypothesized. It was revealed that MyD88 and TRAF6 can directly interact with IRF7, and induce its nuclear translocation and IFNα upregulation. (Fig. 3). Since IRF7 constitutively expresses only in pDCs, the interaction can explain the mechanism on how TLR7 and TLR9 signaling produce a large amount of IFNα in pDCs.

In summary, TLR3- and TLR4-induced IFN production is mediated through a TRIF-TBK1/IKK-i-IRF3 signaling pathway. In contrast, TLR7- and TLR9-induced IFNα production in pDC is mediated by direct association of MyD88 and IRF7, suggesting that type I IFN production is regulated by stimulus- and tissue-specific mechanisms.

8 Perspectives

As described above, the mammalian innate immune system comprised of three different mechanisms in the recognition of microorganisms. The lectin-complement pathway can detect microorganisms at the extracellular space to destroy them directly and/or cause their opsonization to facilitate phagocyte recognition. TLRs, located on the cell membrane, are responsible for activating innate immune cells by detecting microorganisms on the cell surface or in the lysosome. Finally, there are cytoplasmic PRRs recognizing viruses infected into cells and some intracellular bacteria. Secreted type I IFNs have a strong antiviral activity by expressing IFN-inducible genes.

Recent extensive studies have revealed ligands of most TLRs and their signaling pathways. Although the cytoplasmic domains of TLRs resemble each other, the adaptor molecule(s) required for the signaling pathways differ depending on the TLR. However, most studies are implemented using PAMPs purified from pathogens to simplify the experimental system. Therefore, for comprehensive understanding of host–pathogen interactions, the next logical step would be to study the host immune responses against whole pathogens in addition to purified PAMPs and to analyze how the responses are modulated by the TLR system.

In contrast, studies on the cytoplasmic recognition of pathogens are just beginning. Although a gene family encoding putative PRR, NOD-LRR proteins, has been identified, the functions of the members are yet to be identified. The response to intracellular bacteria is more complex than previously expected. For example, in *L. monocytogenes* infection, the type I IFN response exacerbates the course of infection. IFNα/β $R^{-/-}$ mice as well as $IRF3^{-/-}$ mice are more resistant to *Listeria* infection compared to wild-type mice. *Listeria*-induced IFN response is activated independent of the TLR pathway, implying that the bacteria may take advantage of host cytoplasmic recognition system for improving their environment in hosts. Virus recognition by the host occurs both inside and outside of cells. Since viruses are complex and diverse, the host must develop an elaborate mechanism(s) to sense the invasion of various viruses.

Together, the innate immune system is not evoking a nonspecific response and random inflammation, but controlling a coordinated and adaptive response depending on the pathogen. Future studies will clarify the intricate pathogen–host relationship, allowing for more targeted therapeutic treatment of infection via modulating the innate immune system.

See also Antigen Presenting Cells (APCs); Autoantibodies and Autoimmunity; Cellular Interactions; Dendritic Cells; Immune Defence, Cell Mediated; Immunologic Memory; Immunology; Pathogens: Innate Immune Reponses.

Bibliography

Books and Reviews

Akira, S., Takeda, K. (2004) Toll-like receptor signalling, *Nat. Rev. Immunol.* **4**, 499–511.

Barton, G.M., Medzhitov, R. (2002) Toll-like receptors and their ligands, *Curr. Top. Microbiol. Immunol.* **270**, 81–92.

Beutler, B., Rietschel, E.T. (2003) Innate immune sensing and its roots: the story of endotoxin, *Nat. Rev. Immunol.* **3**, 169–176.

Hoffmann, J.A. (2003) The immune response of Drosophila, *Nature* **426**, 33–38.

Inohara, N., Nunez, G. (2003) NODs: intracellular proteins involved in inflammation and apoptosis, *Nat. Rev. Immunol.* **3**, 371–382.

Janeway, C.A., Travers, P., Walport, M., Shlomchik, M. (2001) *Immunobiology: The Immune System in Health & Disease*, Garland Publishing, New York.

Turner, M.W. (2003) The role of mannose-binding lectin in health and disease, *Mol. Immunol.* **40**, 423–429.

Primary Literature

Alexopoulou, L., Holt, A.C., Medzhitov, R., Flavell, R.A. (2001) Recognition of double-stranded RNA and activation of NF-kappaB by Toll-like receptor 3, *Nature* **413**, 732–738.

Alexopoulou, L., Thomas, V., Schnare, M., Lobet, Y., Anguita, J., Schoen, R.T., Medzhitov, R., Fikrig, E., Flavell, R.A. (2002) Hyporesponsiveness to vaccination with Borrelia burgdorferi OspA in humans and in TLR1- and TLR2-deficient mice, *Nat. Med.* **8**, 878–884.

Auerbuch, V., Brockstedt, D.G., Meyer-Morse, N., O'Riordan, M., Portnoy, D.A. (2004) Mice lacking the type I interferon receptor are resistant to Listeria monocytogenes, *J. Exp. Med.* **200**, 527–533.

Blander, J.M., Medzhitov, R. (2004) Regulation of phagosome maturation by signals from toll-like receptors, *Science* **304**, 1014–1018.

Carrero, J.A., Calderon, B., Unanue, E.R. (2004) Type I interferon sensitizes lymphocytes to apoptosis and reduces resistance to listeria infection, *J. Exp. Med.* **200**, 535–540.

Chamaillard, M., Hashimoto, M., Horie, Y., Masumoto, J., Qiu, S., Saab, L., Ogura, Y., Kawasaki, A., Fukase, K., Kusumoto, S., Valvano, M.A., Foster, S.J., Mak, T.W., Nunez, G., Inohara, N. (2003) An essential role for NOD1 in host recognition of bacterial peptidoglycan containing diaminopimelic acid, *Nat. Immunol.* **4**, 702–707.

Cheung, P.C., Nebreda, A.R., Cohen, P. (2004) TAB3, a new binding partner of the protein kinase TAK1, *Biochem. J.* **378**, 27–34.

Couillault, C., Pujol, N., Reboul, J., Sabatier, L., Guichou, J.F., Kohara, Y., Ewbank, J.J. (2004) TLR-independent control of innate immunity in Caenorhabditis elegans by the TIR domain adaptor protein TIR-1, an ortholog of human SARM, *Nat. Immunol.* **5**, 488–494.

Diebold, S.S., Kaisho, T., Hemmi, H., Akira, S., Reis e Sousa, C. (2004) Innate antiviral responses by means of TLR7-mediated recognition of single-stranded RNA, *Science* **303**, 1529–1531.

Diebold, S.S., Montoya, M., Unger, H., Alexopoulou, L., Roy, P., Haswell, L.E., Al-Shamkhani, A., Flavell, R., Borrow, P., Reis e Sousa, C. (2003) Viral infection switches non-plasmacytoid dendritic cells into high interferon producers, *Nature* **424**, 324–328.

Fitzgerald, K.A., McWhirter, S.M., Faia, K.L., Rowe, D.C., Latz, E., Golenbock, D.T., Coyle, A.J., Liao, S.M., Maniatis, T. (2003) IKKepsilon and TBK1 are essential components of the IRF3 signaling pathway, *Nat. Immunol.* **4**, 491–496.

Fitzgerald, K.A., Palsson-McDermott, E.M., Bowie, A.G., Jefferies, C.A., Mansell, A.S., Brady, G., Brint, E., Dunne, A., Gray, P., Harte, M.T., McMurray, D., Smith, D.E., Sims, J.E., Bird, T.A., O'Neill, L.A. (2001) Mal (MyD88-adapter-like) is required for Toll-like receptor-4 signal transduction, *Nature* **413**, 78–83.

Georgel, P., Naitza, S., Kappler, C., Ferrandon, D., Zachary, D., Swimmer, C., Kopczynski, C., Duyk, G., Reichhart, J.M., Hoffmann, J.A. (2001) Drosophila immune deficiency (IMD) is a death domain protein that activates antibacterial defense and can promote apoptosis, *Dev. Cell.* **1**, 503–514.

Girardin, S.E., Boneca, I.G., Viala, J., Chamaillard, M., Labigne, A., Thomas, G., Philpott, D.J., Sansonetti, P.J. (2003) Nod2 is a general sensor of peptidoglycan through muramyl dipeptide (MDP) detection, *J. Biol. Chem.* **278**, 8869–8872.

Girardin, S.E., Boneca, I.G., Carneiro, L.A., Antignac, A., Jehanno, M., Viala, J., Tedin, K., Taha, M.K., Labigne, A., Zahringer, U., Coyle, A.J., DiStefano, P.S., Bertin, J., Sansonetti, P.J., Philpott, D.J. (2003) Nod1 detects a unique muropeptide from gram-negative bacterial peptidoglycan, *Science* **300**, 1584–1587.

Hawn, T.R., Verbon, A., Lettinga, K.D., Zhao, L.P., Li, S.S., Laws, R.J., Skerrett, S.J., Beutler, B., Schroeder, L., Nachman, A., Ozinsky, A., Smith, K.D., Aderem, A. (2003) A common dominant TLR5 stop codon polymorphism abolishes flagellin signaling and is associated with susceptibility to legionnaires' disease, *J. Exp. Med.* **198**, 1563–1572.

Hayashi, F., Smith, K.D., Ozinsky, A., Hawn, T.R., Yi, E.C., Goodlett, D.R., Eng, J.K., Akira, S., Underhill, D.M., Aderem, A. (2001) The innate immune response to bacterial flagellin is mediated by Toll-like receptor 5, *Nature* **410**, 1099–1103.

Heil, F., Hemmi, H., Hochrein, H., Ampenberger, F., Kirschning, C., Akira, S., Lipford, G., Wagner, H., Bauer, S. (2004) Species-specific recognition of single-stranded RNA via toll-like receptor 7 and 8, *Science* **303**, 1526–1529.

Hemmi, H., Kaisho, T., Takeuchi, O., Sato, S., Sanjo, H., Hoshino, K., Horiuchi, T., Tomizawa, H., Takeda, K., Akira, S. (2002) Small anti-viral compounds activate immune cells via the TLR7 MyD88-dependent signaling pathway, *Nat. Immunol.* **3**, 196–200.

Hemmi, H., Takeuchi, O., Sato, S., Yamamoto, M., Kaisho, T., Sanjo, H., Kawai, T., Hoshino, K., Takeda, K., Akira, S. (2004) The roles of two IkappaB kinase-related kinases in lipopolysaccharide and double stranded RNA signaling and viral infection, *J. Exp. Med.* **199**, 1641–1650.

Hemmi, H., Takeuchi, O., Kawai, T., Kaisho, T., Sato, S., Sanjo, H., Matsumoto, M., Hoshino, K., Wagner, H., Takeda, K., Akira, S. (2000) A Toll-like receptor recognizes bacterial DNA, *Nature* **408**, 740–745.

Hochrein, H., Schlatter, B., O'Keeffe, M., Wagner, C., Schmitz, F., Schiemann, M., Bauer, S., Suter, M., Wagner, H. (2004) Herpes simplex virus type-1 induces IFN-alpha production via Toll-like receptor 9-dependent and -independent pathways, *Proc. Natl. Acad. Sci. U. S. A.* **101**, 11416–11421.

Hoebe, K., Du, X., Georgel, P., Janssen, E., Tabeta, K., Kim, S.O., Goode, J., Lin, P., Mann, N., Mudd, S., Crozat, K., Sovath, S., Han, J., Beutler, B. (2003) Identification of Lps2 as a key transducer of MyD88-independent TIR signalling, *Nature* **424**, 743–748.

Horng, T., Barton, G.M., Medzhitov, R. (2001) TIRAP: an adapter molecule in the Toll signaling pathway, *Nat. Immunol.* **2**, 835–841.

Horng, T., Barton, G.M., Flavell, R.A., Medzhitov, R. (2002) The adaptor molecule TIRAP provides signalling specificity for Toll-like receptors, *Nature* **420**, 329–333.

Hoshino, K., Takeuchi, O., Kawai, T., Sanjo, H., Ogawa, T., Takeda, Y., Takeda, K., Akira, S. (1999) Cutting edge: Toll-like receptor 4 (TLR4)-deficient mice are hyporesponsive to lipopolysaccharide: evidence for TLR4 as the Lps gene product, *J. Immunol.* **162**, 3749–3752.

Hsu, L.C., Park, J.M., Zhang, K., Luo, J.L., Maeda, S., Kaufman, R.J., Eckmann, L., Guiney, D.G., Karin, M. (2004) The protein kinase PKR is required for macrophage apoptosis after activation of Toll-like receptor 4, *Nature* **428**, 341–345.

Hugot, J.P., Chamaillard, M., Zouali, H., Lesage, S., Cezard, J.P., Belaiche, J., Almer, S., Tysk, C., O'Morain, C.A., Gassull, M., Binder, V., Finkel, Y., Cortot, A., Modigliani, R., Laurent-Puig, P., Gower-Rousseau, C., Macry, J., Colombel, J.F., Sahbatou, M., Thomas, G. (2001) Association of NOD2 leucine-rich repeat variants with susceptibility to Crohn's disease, *Nature* **411**, 599–603.

Ishitani, T., Takaesu, G., Ninomiya-Tsuji, J., Shibuya, H., Gaynor, R.B., Matsumoto, K. (2003) Role of the TAB2-related protein TAB3 in IL-1 and TNF signaling, *EMBO J.* **22**, 6277–6288.

Kanayama, A., Seth, R.B., Sun, L., Ea, C.K., Hong, M., Shaito, A., Chiu, Y.H., Deng, L., Chen, Z.J. (2004) TAB2 and TAB3 activate the NF-kappaB pathway through binding to polyubiquitin chains, *Mol. Cell.* **15**, 535–548.

Kawai, T., Adachi, O., Ogawa, T., Takeda, K., Akira, S. (1999) Unresponsiveness of MyD88-deficient mice to endotoxin, *Immunity* **11**, 115–122.

Kawai, T., Takeuchi, O., Fujita, T., Inoue, J., Muhlradt, P.F., Sato, S., Hoshino, K., Akira, S. (2001) Lipopolysaccharide stimulates the MyD88-independent pathway and results in activation of IFN-regulatory factor 3 and the expression of a subset of lipopolysaccharide-inducible genes, *J. Immunol.* **167**, 5887–5894.

Kobayashi, K., Hernandez, L.D., Galan, J.E., Janeway, C.A. Jr., Medzhitov, R., Flavell, R.A. (2002) IRAK-M is a negative regulator of Toll-like receptor signaling, *Cell* **110**, 191–202.

Krug, A., Luker, G.D., Barchet, W., Leib, D.A., Akira, S., Colonna, M. (2004) Herpes simplex virus type 1 activates murine natural interferon-producing cells through toll-like receptor 9, *Blood* **103**, 1433–1437.

Kurt-Jones, E.A., Popova, L., Kwinn, L., Haynes, L.M., Jones, L.P., Tripp, R.A., Walsh, E.E., Freeman, M.W., Golenbock, D.T., Anderson, L.J., Finberg, R.W. (2000) Pattern recognition receptors TLR4 and CD14 mediate response to respiratory syncytial virus, *Nat. Immunol.* **1**, 398–401.

Lemaitre, B., Nicolas, E., Michaut, L., Reichhart, J.M., Hoffmann, J.A. (1996) The dorsoventral regulatory gene cassette spatzle/Toll/cactus controls the potent antifungal response in Drosophila adults, *Cell* **86**, 973–983.

Liberati, N.T., Fitzgerald, K.A., Kim, D.H., Feinbaum, R., Golenbock, D.T., Ausubel, F.M. (2004) Requirement for a conserved Toll/interleukin-1 resistance domain protein in the Caenorhabditis elegans immune response, *Proc. Natl. Acad. Sci. U. S. A.* **101**, 6593–6598.

Lomaga, M.A., Yeh, W.C., Sarosi, I., Duncan, G.S., Furlonger, C., Ho, A., Morony, S., Capparelli, C., Van, G., Kaufman, S., van der Heiden, A., Itie, A., Wakeham, A., Khoo, W., Sasaki, T., Cao, Z., Penninger, J.M., Paige, C.J., Lacey, D.L., Dunstan, C.R., Boyle, W.J., Goeddel, D.V., Mak, T.W. (1999) TRAF6 deficiency results in osteopetrosis and defective interleukin-1, CD40, and LPS signaling, *Genes Dev.* **13**, 1015–1024.

Lund, J., Sato, A., Akira, S., Medzhitov, R., Iwasaki, A. (2003) Toll-like receptor 9-mediated recognition of Herpes simplex virus-2 by plasmacytoid dendritic cells, *J. Exp. Med.* **198**, 513–520.

Lye, E., Mirtsos, C., Suzuki, N., Suzuki, S., Yeh, W.C. (2004) The role of IRAK-4 kinase activity in IRAK-4-mediated signaling, *J. Biol. Chem.* **279**, 40653–40658.

Nagai, Y., Akashi, S., Nagafuku, M., Ogata, M., Iwakura, Y., Akira, S., Kitamura, T., Kosugi, A., Kimoto, M., Miyake, K. (2002) Essential role of MD-2 in LPS responsiveness and TLR4 distribution, *Nat. Immunol.* **3**, 667–672.

Naito, A., Azuma, S., Tanaka, S., Miyazaki, T., Takaki, S., Takatsu, K., Nakao, K., Nakamura, K., Katsuki, M., Yamamoto, T., Inoue, J. (1999) Severe osteopetrosis, defective interleukin-1 signalling and lymph node organogenesis in TRAF6-deficient mice, *Genes Cells* **4**, 353–362.

Netea, M.G., Sutmuller, R., Hermann, C., Van der Graaf, C.A., Van der Meer, J.W., van Krieken, J.H., Hartung, T., Adema, G., Kullberg, B.J. (2004) Toll-like receptor 2 suppresses immunity against Candida albicans through induction of IL-10 and regulatory T cells, *J. Immunol.* **172**, 3712–3718.

Ninomiya-Tsuji, J., Kishimoto, K., Hiyama, A., Inoue, J., Cao, Z., Matsumoto, K. (1999) The kinase TAK1 can activate the NIK-I kappaB as well as the MAP kinase cascade in the IL-1 signalling pathway, *Nature* **398**, 252–256.

O'Connell, R.M., Saha, S.K., Vaidya, S.A., Bruhn, K.W., Miranda, G.A., Zarnegar, B., Perry, A.K., Nguyen, B.O., Lane, T.F., Taniguchi, T., Miller, J.F., Cheng, G. (2004) Type I Interferon Production Enhances Susceptibility to Listeria monocytogenes Infection, *J. Exp. Med.* **200**, 437–445.

Ogura, Y., Bonen, D.K., Inohara, N., Nicolae, D.L., Chen, F.F., Ramos, R., Britton, H., Moran, T., Karaliuskas, R., Duerr, R.H., Achkar, J.P., Brant, S.R., Bayless, T.M., Kirschner, B.S., Hanauer, S.B., Nunez, G., Cho, J.H. (2001) A frameshift mutation in NOD2 associated with susceptibility to Crohn's disease, *Nature* **411**, 603–606.

Ozinsky, A., Underhill, D.M., Fontenot, J.D., Hajjar, A.M., Smith, K.D., Wilson, C.B., Schroeder, L., Aderem, A. (2000) The repertoire for pattern recognition of pathogens by the innate immune system is defined by cooperation between toll-like receptors, *Proc. Natl. Acad. Sci. U. S. A.* **97**, 13766–13771.

Perry, A.K., Chow, E.K., Goodnough, J.B., Yeh, W.C., Cheng, G. (2004) Differential Requirement for TANK-binding Kinase-1 in Type I Interferon Responses to Toll-like Receptor Activation and Viral Infection, *J. Exp. Med.* **199**, 1651–1658.

Poltorak, A., He, X., Smirnova, I., Liu, M.Y., Van Huffel, C., Du, X., Birdwell, D., Alejos, E., Silva, M., Galanos, C., Freudenberg, M., Ricciardi-Castagnoli, P., Layton, B., Beutler, B. (1998) Defective LPS signaling in C3H/HeJ and C57BL/10ScCr mice: mutations in Tlr4 gene, *Science* **282**, 2085–2088.

Qin, J., Jiang, Z., Qian, Y., Casanova, J.L., Li, X. (2004) IRAK4 kinase activity is redundant for interleukin-1 (IL-1) receptor-associated kinase phosphorylation and IL-1 responsiveness, *J. Biol. Chem.* **279**, 26748–26753.

Sakaguchi, S., Negishi, H., Asagiri, M., Nakajima, C., Mizutani, T., Takaoka, A., Honda, K., Taniguchi, T. (2003) Essential role of IRF-3 in lipopolysaccharide-induced interferon-beta gene expression and endotoxin shock, *Biochem. Biophys. Res. Commun.* **306**, 860–866.

Sato, M., Suemori, H., Hata, N., Asagiri, M., Ogasawara, K., Nakao, K., Nakaya, T., Katsuki, M., Noguchi, S., Tanaka, N., Taniguchi, T. (2000) Distinct and essential roles of transcription factors IRF-3 and IRF-7 in response to viruses for IFN-alpha/beta gene induction, *Immunity* **13**, 539–548.

Sharma, S., tenOever, B.R., Grandvaux, N., Zhou, G.P., Lin, R., Hiscott, J. (2003) Triggering the interferon antiviral response through an IKK-related pathway, *Science* **300**, 1148–1151.

Shimazu, R., Akashi, S., Ogata, H., Nagai, Y., Fukudome, K., Miyake, K., Kimoto, M. (1999) MD-2, a molecule that confers lipopolysaccharide responsiveness on Toll-like receptor 4, *J. Exp. Med.* **189**, 1777–1782.

Suzuki, N., Suzuki, S., Duncan, G.S., Millar, D.G., Wada, T., Mirtsos, C., Takada, H., Wakeham, A., Itie, A., Li, S., Penninger, J.M., Wesche, H., Ohashi, P.S., Mak, T.W., Yeh, W.C. (2002) Severe impairment of interleukin-1 and Toll-like receptor signalling in mice lacking IRAK-4, *Nature* **416**, 750–756.

Takaesu, G., Kishida, S., Hiyama, A., Yamaguchi, K., Shibuya, H., Irie, K., Ninomiya-Tsuji, J., Matsumoto, K. (2000) TAB2, a novel adaptor protein, mediates activation of TAK1 MAPKKK by linking TAK1 to TRAF6 in the IL-1 signal transduction pathway, *Mol. Cell.* **5**, 649–658.

Takeuchi, O., Hoshino, K., Akira, S. (2000) Cutting edge: TLR2-deficient and MyD88-deficient mice are highly susceptible to Staphylococcus aureus infection, *J. Immunol.* **165**, 5392–5396.

Takeuchi, O., Hoshino, K., Kawai, T., Sanjo, H., Takada, H., Ogawa, T., Takeda, K., Akira, S. (1999) Differential roles of TLR2 and TLR4 in recognition of gram-negative and gram-positive bacterial cell wall components, *Immunity* **11**, 443–451.

Takeuchi, O., Kawai, T., Muhlradt, P.F., Morr, M., Radolf, J.D., Zychlinsky, A., Takeda, K., Akira, S. (2001) Discrimination of bacterial lipoproteins by Toll-like receptor 6, *Int. Immunol.* **13**, 933–940.

Takeuchi, O., Sato, S., Horiuchi, T., Hoshino, K., Takeda, K., Dong, Z., Modlin, R.L., Akira, S. (2002) Cutting edge: role of Toll-like receptor 1 in mediating immune response to microbial lipoproteins, *J. Immunol.* **169**, 10–14.

Tanabe, T., Chamaillard, M., Ogura, Y., Zhu, L., Qiu, S., Masumoto, J., Ghosh, P., Moran, A., Predergast, M.M., Tromp, G., Williams, C.J., Inohara, N., Nunez, G. (2004) Regulatory regions and critical residues of NOD2 involved in muramyl dipeptide recognition, *EMBO J.* **23**, 1587–1597.

Underhill, D.M., Ozinsky, A., Hajjar, A.M., Stevens, A., Wilson, C.B., Bassetti, M., Aderem, A. (1999) The Toll-like receptor 2 is recruited to macrophage phagosomes and discriminates between pathogens, *Nature* **401**, 811–815.

Watanabe, T., Kitani, A., Murray, P.J., Strober, W. (2004) NOD2 is a negative regulator of Toll-like receptor 2-mediated T helper type 1 responses, *Nat. Immunol.* **5**, 800–808.

Yamamoto, M., Sato, S., Hemmi, H., Uematsu, S., Hoshino, K., Kaisho, T., Takeuchi, O., Takeda, K., Akira, S. (2003) TRAM is specifically involved in the Toll-like receptor 4-mediated MyD88-independent signaling pathway, *Nat. Immunol.* **4**, 1144–1150.

Yamamoto, M., Sato, S., Hemmi, H., Hoshino, K., Kaisho, T., Sanjo, H., Takeuchi, O., Sugiyama, M., Okabe, M., Takeda, K., Akira, S. (2003) Role of adaptor TRIF in the MyD88-independent toll-like receptor signaling pathway, *Science* **301**, 640–643.

Yamamoto, M., Sato, S., Hemmi, H., Sanjo, H., Uematsu, S., Kaisho, T., Hoshino, K., Takeuchi, O., Kobayashi, M., Fujita, T., Takeda, K., Akira, S. (2002) Essential role for TIRAP in activation of the signalling cascade shared by TLR2 and TLR4, *Nature* **420**, 324–329.

Yang, Y.L., Reis, L.F., Pavlovic, J., Aguzzi, A., Schafer, R., Kumar, A., Williams, B.R., Aguet, M., Weissmann, C. (1995) Deficient signaling in mice devoid of double-stranded RNA-dependent protein kinase, *EMBO J.* **14**, 6095–6106.

Yoneyama, M., Kikuchi, M., Natsukawa, T., Shinobu, N., Imaizumi, T., Miyagishi, M., Taira, K., Akira, S., Fujita, T. (2004) The RNA helicase RIG-I has an essential function in double-stranded RNA-induced innate antiviral responses, *Nat. Immunol.* **5**, 730–737.

Zhang, D., Zhang, G., Hayden, M.S., Greenblatt, M.B., Bussey, C., Flavell, R.A., Ghosh, S. (2004) A toll-like receptor that prevents infection by uropathogenic bacteria, *Science* **303**, 1522–1526.

Insulin Resistance, Diabetes and its Complications

Dominic S. Ng
University of Toronto, Toronto, Ontario, Canada

Encyclopedia of Molecular Cell Biology and Molecular Medicine, 2nd Edition. Volume 7
Edited by Robert A. Meyers.

ISBN: 3-527-30549-1

Keywords

Apoptosis
A cell-death sequence that is largely genetically programmed.

Atherogenesis
The complex array of pathologic processes that result in the development of the atherosclerotic plaques, lesions responsible for the most common type of occlusion arterial diseases.

β-cells
The endocrine cell type in the pancreas that specializes in the secretion of insulin in response to ambient glucose concentrations and other signals.

Dyslipidemia
Altered blood levels of various lipoprotein fractions when compared with normal subjects.

Insulin Resistance
A reduced cellular response of any tissue to the binding of insulin to its cell surface insulin receptors. At the whole organism level, insulin resistance impairs insulin-induced glucose uptake by the tissues, a major predisposing factor to elevated blood glucose.

Insulin Signaling
The binding of insulin to its receptors results in a sequence of biochemical alterations, primarily phosphorylation, of a series of downstream intracellular signaling molecules, which eventually leads to the target cellular events, for example, translocation of the glucose transporter, GLUT4, induction of target gene transcription, etc.

Knockout Mice
Genetically engineered mice with a gene or a selected segment of DNA being disrupted structurally; a powerful mouse model to study the functional importance of the segment of DNA being disrupted.

Lipodystrophy
A clinical condition of either complete or partial loss of fat depot. Many specific syndromes are genetically determined, but it can also be drug-induced, most notably those caused by protease inhibitors commonly used for the treatment of AIDS.

Lipotoxicity
Dysfunction of nonadipose tissues as a result of fat accumulation.

Diabetes mellitus, a chronic progressive metabolic disorder of glucose homeostasis, is a major cause of morbidity and mortality. It is broadly categorized into type 1 and type 2, each with distinctly different pathogenesis. Diabetes mellitus is extremely prevalent, with type 2 making up nearly 90% of the cases and the incidence continues to increase at an alarming rate, in parallel with aging and the increasing prevalence of obesity worldwide. Although the pathogenesis of these two major categories is different, they share similar long-term complications as a result of inadequate glycemic control. It is well established that poor glycemic control is the major causative factor in the development of microvascular complications, affecting primarily the kidney, eyes, and peripheral nerves. Recent data suggest that hyperglycemia also plays a critical role in the development of macrovascular complications. To date, a complete normalization of glucose metabolic disturbances in diabetes remains an elusive goal. Even excellent glycemic control evades a large number of affected subjects. However, advances in our understanding of the pathogenesis of the disorder at cellular and molecular levels over the past several decades had led to significant improvements in treatment outcomes.

Insulin resistance is highly prevalent, and particularly so in Western societies. It can be found in a number of metabolic disorders, including the metabolic syndrome, impaired glucose tolerance, and in its extreme form, type 2 diabetes. In the current paradigm, insulin resistance plays a central role in the predisposition of an individual to the development of type 2 diabetes. Initially, compensatory overproduction of insulin by pancreatic β-cells enables an individual to sustain normal blood glucose levels. Over time, progressive β-cell dysfunction and loss of cell mass leads to relative insulin deficiency and diabetes ensues. Obesity not only directly causes insulin resistance, but also contributes to the progressive failure of β-cells through a number of mechanisms, including free fatty acid–induced β-cell apoptosis. Such maladaptive responses are also associated with an increased ectopic deposition of fat to nonadipose tissues, in turn, aggravating the preexisting insulin resistance.

The development of clinical diabetes is a result of combined genetic susceptibility and environmental stressors. With only a handful of exceptions, diabetes is a polygenic disorder. Identification of candidate genes may provide a more refined framework for either better understanding of the pathophysiological processes and/or for identifying novel targets for pharmacological interventions.

1 Introduction

Diabetes mellitus is a chronic progressive disease caused by inherited and/or acquired deficiency in the production of insulin, or by the ineffectiveness of the insulin produced, resulting in elevated blood glucose levels. Diabetes is broadly categorized as type 1, in which the pancreas fails to produce insulin, and type 2, in which the body fails to respond properly to the action of insulin. Diabetes is extremely prevalent. Recent data suggest that approximately 150 million people worldwide have diabetes mellitus and amongst them, approximately 90% have type 2. This number will increase to at least 300 million by 2025 (WHO 2003). Much of this increase will occur in developing countries due to aging, unhealthy lifestyle, and the increasing epidemic of obesity, all predisposing factors for type 2 diabetes. Although the two types of diabetes differ fundamentally in their pathogenesis, they share common features in the clinical complications including those directly attributable to chronic hyperglycemia. With regard to pathogenesis, this review will focus on insulin resistance, the cardinal pathogenic factor for the development of type 2 diabetes mellitus.

Insulin resistance of the peripheral tissue predates the development of frank diabetes. This has been documented in a variety of conditions, including obesity, as well as in individuals with impaired fasting glucose. To maintain glucose homeostasis, the pancreatic β-cells overproduce insulin, leading to hyperinsulinemia. As the disease progresses, β-cell decompensation develops and frank diabetes ensues.

1.1 Gene–Environment Interactions

Type 2 diabetes is increasingly being recognized as a polygenic disorder. Nevertheless, environmental factors are considered responsible for the alarming rise in diabetes incidence in all age groups including the young since no major change in the gene pool has occurred. Furthermore, the success of intervention trials in preventing the progression to diabetes in vulnerable populations by reducing the glycemic load through acarbose (Study to Prevent Non-Insulin-Dependent Diabetes Mellitus, or STOP-NIDDM trial) or modest lifestyle change of low saturated fat weight-loss diets with regular exercise (Finnish Diabetes Prevention Study and the US Diabetic Prevention Program) has emphasized the role of environmental factors, especially when combined. Current thinking on the genesis of diabetes links genes and environment through insulin resistance. A progression is seen where insulin resistance increases over time, promoted by adiposity, especially intra-abdominal fat. Insulin levels rise and then decline over time as blood glucose levels rise due to pancreatic insufficiency. Genetic factors influencing insulin resistance, body fat distribution and β-cell failure are obviously key factors in this scenario. Interestingly, more than 80% of subjects with diabetes are obese. Observational studies support the notion that diet high in glycemic load increases the risk of development of type 2 diabetes, whereas a high fiber diet reduces the risk. Current data also support the notion that the quality of dietary fat significantly influences the development of insulin resistance and diabetes. Physical activities also confer beneficial effects on insulin resistance and improvement of glycemic control. A single bout of aerobic exercise can significantly reduce plasma glucose levels. Likewise, endurance exercise training also improves insulin resistance and glycemic control.

From the genetic standpoint, the age-adjusted increase in diabetes and obesity in the last 100 years in western nations cannot be attributed to a change in the gene pool. Rather, it is likely, in large measure, to be related to changes in diet and lifestyle. From the evolutionary perspective, our genetic makeup is ill-adapted to life in the twenty-first century. Such a lifestyle is typified by lack of need to exercise and a continuous abundance of food.

Furthermore, the nature of the food has changed. For example, traditional starchy foods that are digested more slowly, such as beans, peas, lentils, pumpernickel breads, and parboiled rice, have been replaced in the modern diet by white bread, potatoes, and glutinous white rice that tend to taste sweeter due to salivary amylolytic digestion in the mouth. As a result, lower glycemic index foods have tended to be exchanged for higher glycemic index foods. This dietary change has been even more marked in traditional cultures where contemporary carbohydrate foods have replaced very low glycemic index traditional foods in the diets, Pima Indians and Australian Aborigines of the groups that have seen some of the greatest increases in diabetes incidence. The thrifty genome has therefore been challenged by twenty-first century lifestyles. Much of the difficulty in this situation of identifying the single gene or group of genes responsible is due to the fact that the whole genome evolved to deal with a radically different lifestyle. The genetic cause of our present epidemic is therefore polygenic.

2 Peripheral Insulin Resistance

Clearance of blood glucose by target tissues is mediated by a family of facilitative transporters (GLUTs). While genes for 11 GLUT isoforms have been identified, only 7 have been shown to transport sugars (GLUTs 1–5, 8, 9). GLUT1 is expressed in erythrocytes and in vascular endothelial cells in the brain, whereas GLUT3 is expressed in neurons. GLUT1 and GLUT3 act in concert to mediate glucose uptake past the blood–brain barrier into the neurons. GLUT2 is expressed widely in a number of tissues including liver, intestine, kidney, and pancreatic β-cells. These transporters participate as part of the glucose sensing system in the β-cells and as the transport system in the intestinal cells for the absorption of dietary glucose. GLUT5 is a fructose transporter, GLUT8 appears to be important in blastocyst development, and GLUT9 is expressed in the brain and leukocytes. GLUT4, found predominantly in skeletal muscle and adipose tissues, is quantitatively the most important insulin-mediated transporter in the clearance of blood glucose. Insulin-mediated rapid cell surface translocation of the preformed and intracellularly sequestered GLUT4 remains the most investigated hypothesis for the mechanism of how insulin mobilizes glucose uptake.

2.1 Insulin Signaling

The binding of insulin to its heterotetrameric receptor (Fig. 1) results in intracellular autophosphorylation of the β-subunit and increases tyrosine kinase activity. These activated insulin receptors modulate the activities of their target effector proteins through phosphorylating a number of proximal intermediate proteins including members of the insulin receptor substrate family (IRS1-4), Shc, Cbl, and so on. An important example of this sequential activation is the tyrosine phosphorylation of IRS proteins resulting in the availability of docking sites of p85, the regulatory subunit of the type 1A phosphatidylinositol 3-kinase (PI3K). PI3K has long been recognized to play a critical role in mediating insulin-stimulated GLUT4 translocation. Activation of PI3K leads to the formation of 3′ phosphoinositide, which in turn, sequentially phosphorylates and activates phophoinositidyl-dependent protein kinase (PDK1) and PKB (also

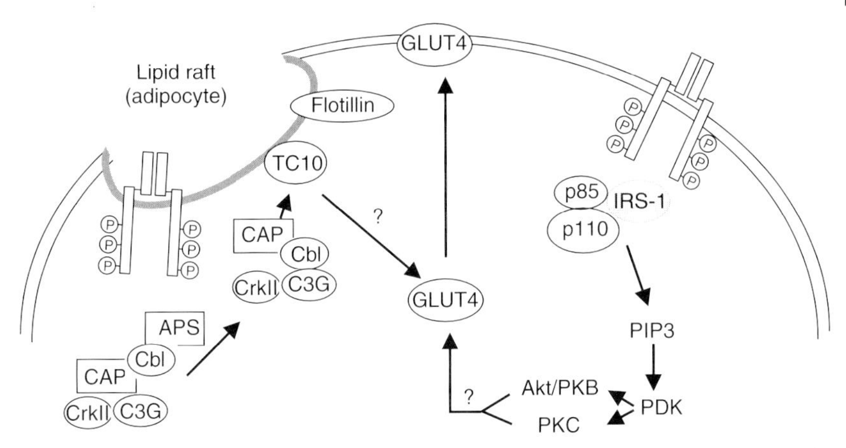

Fig. 1 Signaling cascades in insulin receptor mediated GLUT4 translocation – cross talk between IRS-mediated pathways and lipid raft-associated pathways.

known as Akt). In addition, protein kinase Cζ (PKCζ) has also been shown to participate in insulin-stimulated GLUT4 translocation. However, data to date have not identified a dominant role for either pathway.

Several lines of evidence strongly suggest that activation of PI3K in the generation of phosphatidylinositol 3,4,5-triphosphate is necessary but not sufficient for mediating the insulin-stimulated GLUT4 translocation. Novel insights recently developed were built on the observations of the comparative studies on the proximal insulin receptor tyrosine–phosphorylated substrates. It was found that in adipocytes, only insulin can induce tyrosine phosphorylation of Cbl. Two accessory proteins, APS and Cbl-associated adaptor protein (CAP), appear to act in concert. The former probably enables the insulin receptor to tyrosine phosophorylate Cbl and the latter, through its ability to interact with the caveolar protein flotillin, recruits the complex to the lipid raft in adipocytes. During this process, a small adaptor protein CrkII is also recruited to the lipid raft through its association with the phosphorylated Cbl. CrkII, in turn, interacts with the guanylnucleotide exchange factor, C3G, through one of the two tandem SH3 domains. Putting it together, insulin induces tyrosine phosphorylation of Cbl and recruits the APS/Cbl/CAP/CrkII/C3G complex to the lipid raft at the plasma membrane. Subsequent screens for small GTP binding proteins lead to the discovery of TC10, a member of the Rho family, which is instrumental in the mediation of insulin-induced, PI3K-independent stimulation of GLUT4 translocation. It therefore appears that the TC10 pathway functions in parallel with PI3K to stimulate fully GLUT4 translocation in response to insulin and these two major components are spatially separated and distinctly compartmentalized. The discovery of this novel lipid raft–associated pathway has provided a new framework for unraveling the complexity of insulin-mediated GLUT4 translocation.

2.2
Exercise and Glucose Uptake by Skeletal Muscle

Glucose can also be taken up by skeletal muscles by a number of factors independent of insulin, including muscle contractions, hypoxia, nitric oxide, and bradykinin. Increased GLUT4 translocation has been demonstrated in mediating the increased glucose uptake for each of the above factors. In spite of its great therapeutic potential, the cellular and molecular signaling mechanism of the contraction-induced glucose uptake by skeletal muscles remains to be fully elucidated. Many candidates, including glycogen, adenosine, nitric oxide (NO), PKC, and 5′ AMP-activated protein kinase (AMPK) have been studied.

It has been shown that glycogen content of skeletal muscle modulates glucose uptake during contractions. Physiologically, glycogen-store depletion after exercise requires repletion to restore contractile functions. If carbohydrate supplementation is inadequate after exercise, glycogen is stored only incompletely. On the other hand, with the availability of carbohydrates, the restoration mechanism may indeed result in supercompensated state, rebuilding glycogen levels to beyond that of the resting levels. Studies showed that surface membrane GLUT4 protein content after contraction was inversely related to the initial glycogen levels in fast-twitch fibres. This modulator effect of glycogen has also been reported in humans. These authors reported a near doubling of plasma glucose clearance rate during one hour of exercise at 70% of maximum oxygen consumption when muscle glycogen content was 185 μmol gm^{-1} versus 800 μmol gm^{-1}. On the contrary, the stimulatory effect of insulin on glucose transport/GLUT4 recruitment was decreased when glycogen was supercompensated. The mechanism behind acute enhanced insulin action after a bout of exercise remains poorly understood. Three to four hours post exercise (when most of the acute effects of exercise ought to have subsided), insulin's effects on glucose uptake and glycogen synthase activity are increased but there was no evidence of increase in insulin-signaling intermediaries. It was suggested that this acute effect of insulin post exercise might be mediated by pathways downstream of Akt and glycogen synthase kinase 3 (GSK3). In isolated muscles, inhibition of protein synthesis with cycloheximide did not prevent muscle contraction–induced increases in insulin sensitivity, suggestive of a protein synthesis independent pathway. The molecular mechanisms underlying the modulatory effects of glycogen on insulin action are not known. Current *in vivo* data point to the possibilities of (1) increased insulin-induced activation of Akt and (2) increased AMPK activities in association with a low glycogen content. However, it is intriguing to note that glycogen depletion is not a necessary condition to observe AMPK activation with 5-aminoimidazole-4-carboxamide ribonucleotide (AICAR) and further research is required to elucidate the precise role of glycogen in postexercise insulin action.

2.2.1 **Skeletal Muscle Insulin Resistance in Diabetics**

It has been well established that skeletal muscle is the major site of insulin-stimulated glucose uptake in the whole body glucose homeostasis. *In vivo* studies consistently showed significant impairment in postprandial insulin mediated glucose uptake in both type 2 diabetic patients and in glucose tolerant first-degree relatives. Tracer studies further localized the defect to the nonoxidative pathways for glucose metabolism. *Ex vivo* assays quantifying insulin-stimulated glucose uptake on abdominal muscle surgical biopsy specimens were developed to show a marked impairment in glucose transport in morbidly obese patients with or without diabetes. In concordance with this finding, muscle biopsy specimens similarly obtained from lean to moderately obese type 2 diabetic subjects revealed a significant reduction in insulin-stimulated transport of 3-O-methylglucose, a non-metabolizable glucose analog, at both physiological and pharmacological concentrations of insulin. Similar defects in insulin-mediated glucose transport have also been reported in rodent models of insulin resistance. The defects of glucose transport observed in diabetics on the basis of *in vitro* and *ex vivo* studies correlate exceedingly well with those of *in vivo* analyses.

In recent years, the technique of open biopsy of the vastus lateralis muscle under local anesthetics has made *ex vivo* specimens more assessable, and therefore, has permitted more systematic analyses. Using an exofacial *bis*-mannose photolabelling technique, Lund et al. were able to show that insulin stimulation of the muscle strip leads to an increase in the cell surface GLUT4 protein level. Since this surface level of GLUT4 is strongly correlated with the 3-O-methylglucose uptake and because of the fact that the GLUT4 turnover rate is not affected by insulin, it is concluded that the major mechanism of glucose transport in human skeletal muscle is GLUT4 translocation to the plasma membrane.

In type 2 diabetic individuals, insulin resistance is not generally associated with a reduction in skeletal muscle GLUT4 levels, but the insulin-induced translocation of GLUT4 is markedly impaired. In insulin-resistant subjects, insulin stimulation of muscle IRS-1 tyrosine phosphorylation and of IRS-1 immunoprecipitable PI3K activity is significantly less compared to insulin-sensitive controls. A reduction in insulin-stimulated Akt/PKB kinase activity in skeletal muscle from insulin-resistant type 2 diabetic individuals has also been reported, albeit not consistently. The role of Akt signaling in the impairment of glucose transport seen in type 2 diabetic individuals remains unclear. There have been a number of reports showing a reduction in insulin-stimulated glycogen synthase activity in diabetic subjects. Insulin-stimulated mitogen-activated protein kinase (MAPK) phosphorylation is not significantly altered in muscles from type 2 diabetic individuals. It appears that skeletal muscle insulin resistance in type 2 diabetic individuals is largely a result of defects in the proximal insulin-signaling steps. Interestingly, a survey of the literature on rodent models of insulin resistance, especially those data based on Zucker *fa/fa* rats, yielded remarkably similar conclusions regarding the molecular defects in mediating the insulin-stimulated glucose transport.

2.2.2 Exercise as Therapeutic Intervention in Insulin Resistance

Although our current state of knowledge in the impact of exercise on glucose homeostasis at the molecular level remains to be fully explored, many lines of evidence point to exercise as a highly desirable form of intervention in the treatment and/or prevention of type 2 diabetes among susceptible individuals. The most compelling evidence comes from the published results on the Diabetes Prevention Program trial and the Finnish Diabetes Prevention Study. Rigorous lifestyle changes, including a minimum of 150 min of physical activity per week to achieve at least a 7% reduction in body weight remains the most efficacious way of preventing the development of type 2 diabetes in susceptible subjects.

2.2.3 Acute Exercise and Insulin Resistance

It is well established that in type 2 diabetes patients, significant glucose lowering can be achieved during an acute, single bout of physical activity. A concomitant reduction in plasma insulin level has also been observed. Mechanistic studies in both type 2 diabetic subjects and in rodent models yielded very similar conclusions. Immediately after a bout of exercise (45–60 min of cycling at 60–70% of V_{max} in the human study), Kennedy et al. reported a 74% increase in plasma membrane GLUT4 protein in vastus lateralis muscle, an improvement comparable to the postexercise increase seen in nondiabetic subjects. This improvement, on the other hand, was not related to a change in GLUT4 protein expression. Similar findings were reported by King et al. in the obese Zucker rat model. These findings are consistent with the observations that contractile-induced skeletal muscle glucose transport remains intact both in animal models of insulin resistance and in type 2 diabetic subjects.

2.2.4 Chronic Exercise and Insulin Resistance

The potential beneficial effect of endurance exercise training in glucose metabolism and insulin resistance has been extensively evaluated. To date, both rodent models and human studies yielded positive results in the effect of exercise training on glucose metabolism. In rodent models, prolonged exercise consistently results in improvement in whole body glucose tolerance, whole body glucose disposal, insulin-stimulated glucose transport, GLUT4 translocation, and GLUT4 protein expression. In insulin-resistant and type 2 diabetics, endurance exercise training leads to improvements in glucose tolerance and whole body insulin-mediated glucose disposal.

The mechanisms underlying the beneficial effects of endurance training on glucose metabolism in insulin-resistant rodent models, insulin-resistant, and type 2 diabetic subjects remain poorly understood. In obese Zucker rats, exercise training results in long-term adaptive changes including upregulation of GLUT4 protein expression, increase GLUT4 translocation, and cell surface GLUT4 after insulin stimulation. However, the impact of exercise on the insulin receptor signaling pathway is less certain. Data pertaining to the effect of exercise on skeletal muscle insulin signaling in diabetic subjects are equally sparse. A better understanding of these molecular determinants will greatly enhance our ability to develop more effective therapeutic exercise strategies to combat insulin resistance and prevent onset of type 2 diabetes.

2.3 Adipocytes, Obesity, and Insulin Resistance

It has long been well recognized that obesity is a strong risk factor for the development of insulin resistance and diabetes. Likewise, up to 80% of type 2 diabetic subjects are either overweight or obese. This strong correlation is also applicable to many genetic isolates with extraordinary high prevalence of type 2 diabetic subjects including the Pima Indians and the Sandy Lake Oji-Cree. While insulin action is key to virtually all facets of adipocyte differentiation, maturation, and physiologic functions, excess fat mass promotes insulin resistance.

2.3.1 Insulin and Adipocytes

Insulin is a key regulator of nearly all aspects of fat-cell biology including the promotion of the differentiation of preadipocytes to adipocytes. In mature adipocytes, insulin action simultaneously promotes lipogenesis and inhibits lipolysis. Meanwhile, insulin stimulates the uptake of glucose in insulin-sensitive tissues via activation of the glucose transporter and the uptake of fatty acids from circulating triglycerides (TG) in very low density lipoproteins and chylomicrons through increasing the lipoprotein lipase activity. In addition, insulin has been known to be mitogenic to the adipose tissues as well as promoting coordinated gene expressions through transcription factors of either the sterol response element-binding protein 1 (SREBP1) variants or the forkhead protein family.

In adipocytes, insulin also acts through binding to and activating the insulin receptor with a subsequent activation of PI3K, the latter being a necessary but not a sufficient condition for the induction of the Glut transporters for glucose uptake. The non-PI3K factor in adipocyte insulin signaling remains to be fully elucidated. Previous studies implicated both the Akt/PKB pathway and the PKCλ/ζ pathway, but the exact role of each in mediating glucose transport remains controversial. As discussed earlier, the lipid raft–associated TC10 pathway is also implicated in the non-PI3K signaling. In the case of adipocytes, it is further suggested that this TC10 pathway interacts with contical actin, the adipocyte-specific unique actin cytoskeleton organization, in the modulation of GLUT4 translocation.

It has been suggested that pathways that mediate the various metabolic effects of insulin may diverge downstream of the PI3K step, providing a partial explanation for the differential sensitivity of different pathways to the action of the same ambient concentration of insulin. For example, a significantly lower concentration of insulin is adequate to inhibit adipocyte lipolysis but not sufficient to reverse the hyperglycemia through increased glucose uptake. In insulin-resistant and diabetic states, adipose tissue shares similar functional defects in insulin signaling including impaired binding of insulin to the receptors, phosphorylation of the receptors, and the downstream IRS. Meanwhile, adipose tissues also manifest tissue-specific defects that are distinct from those of skeletal muscles, including a selective reduction in IRS-1 expression resulting in the recruitment of IRS-2 as the main docking protein for PI3K. In skeletal muscles, both IRS-1 and IRS-2 protein levels are normal and the signaling defect was caused by a reduction in PI3K activity in association with both IRS-1 and -2.

2.3.2 Adipose Tissue in Glucose Homeostasis and Energy Storage

The quantitative importance of adipose tissue in the regulation glucose disposal has been continuously revised. On the basis of earlier metabolic studies, it has been held that insulin-mediated glucose uptake by adipose tissue, at least in lean individuals, is relatively insignificant when compared with skeletal muscle. More recent studies in humans as well as in transgenic animal models seem to suggest the contrary. Selective overexpression of the *GLUT4* gene in adipose tissue resulted in a significant increase in whole body insulin sensitivity. A selective disruption of the *GLUT4* gene in adipose tissue led to a degree of insulin resistance comparable to that seen in the skeletal muscle-specific *glut4* knockout mice.

In humans, treatment with the $\beta 3$ selective agonist CL316,243 increases glucose uptake selectively in fat, which is also associated with an increase in whole body glucose uptake and suppression of hepatic glucose production. While these observations collectively attest to the possibility that direct insulin-mediated uptake of glucose by adipose tissue may parallel that of skeletal muscle quantitatively, one needs to allow for contributions of other nondirect mediators.

Adipose tissue has long been viewed as a major energy depot. Excess energy intake is stored in adipocytes in the form of TGs. Insulin is a key activator of not only the action of lipoprotein lipase in the uptake of lipoprotein-derived fatty acids but also inhibits the activity of hormone-sensitive lipase to prevent lipolysis. In addition, insulin also acts directly on the insulin-responsive sterol response element–binding protein 1 (*SREBP1*) gene whose gene products, SREBP1a and SREBP1c serve as transcription factors in regulating the expressions of a series of genes including those involved in promoting endogenous lipogenesis. Insulin-deficient and insulin-resistant states including obesity are associated with enhanced lipolysis of the adipose tissue, resulting in an elevated level of circulating free fatty acids (FFA). Increased adipose lipolysis may result in a variety of dysmetabolic consequences including peripheral insulin resistance, hepatic insulin resistance, increased hepatic TG overproduction, and aggravation of pancreatic β-cell insulin secretory function. More recently, elevated FFA concentration has also been shown to result in myocardial dysfunction. Collectively, this FFA-induced multiorgan dysfunction has been dubbed lipotoxicity. The unifying concept behind such otherwise unrelated organ dysfunction is that of excessive intracellular accumulation of TGs as a consequence of nutrient excess and altered energy disposal. In this paradigm, leptin action seems to be the key switch in the proper channeling of the metabolic fate of the excess FA flux commonly seen in insulin-resistant states and obesity.

2.3.3 Lipotoxicity

Caloric intake in excess of expenditure results in an increase in adiposity and associated insulin resistance and hyperinsulinemia. Insulin resistance at the adipose tissue level results in increased lipolysis but the compensatory hyperinsulinemia may impact directly on lipid metabolism in nonadipose tissues. *SREBP1*, a gene encoding a lipogenic transcription factor, remains insulin-responsive in the insulin-resistant milieu and has been shown to be upregulated in a variety of nonadipose tissues and may play an important role in endogenous lipogenesis observed in these tissues. Excess circulating FFAs are capable of entering nonadipose tissues in the body in a fatty acid binding protein independent manner. Under normal circumstances, such excess free FA is efficiently channeled to the β-oxidation pathway and the excess energy dissipated as heat. However, in the absence of leptin action (either due to leptin deficiency or leptin resistance), nonoxidative metabolism of FFAs prevail leading to tissue steatosis, an accumulation of TGs in the nonadipose tissues. There is also evidence suggesting that high levels of FFAs and their subsequent metabolism increase the consumption of antioxidants and cause oxidative damage. In the mouse model of leptin deficiency (*ob/ob*), leptin receptor deficiency (*db/db*), and a rat model of leptin unresponsiveness (*fa/fa*), marked obesity is associated with a generalized form of steatosis, namely, significant accumulation of TG in skeletal muscle, cardiac muscle, pancreatic beta cells, kidneys, and liver. In a murine model of congenital lipodystrophy, a paucity of adipose tissue also results in severe insulin resistance and steatosis in multiple nonadipose tissue depots. More intriguingly, the metabolic defect can be reversed by either treatment with leptin or with white fat transplantation. On the other hand, a similar transplantation experiment with white adipose without leptin fails to correct the metabolic abnormalities in another model for lipoatrophy, attesting to the critical importance of leptin in the reversal of insulin resistance and diabetes. In this particular model, a paucity of adipose tissue results in an obligate shunting of excess energy to nonadipose tissues. A concomitant lack of leptin thus promotes nonoxidative disposal of FA in nonadipose tissues, resulting in TG accumulation.

Studies on the relevance of this lipotoxicity model in humans have entered the literature. Treatment of human subjects with lipodystrophy-associated diabetes and insulin resistance with recombinant leptin has shown very encouraging beneficial effects. In the first study, 8 of the 9 lipodystrophic female subjects had diabetes, low plasma leptin levels, and fatty livers. A twice-daily leptin injection for 4 months resulted in a mean 1.9% reduction in HgA1C, a 6% reduction in fasting plasma TGs, and a 28% reduction in liver volume. In the latter study, 3 patients with severe lipodystrophy, diabetes, hepatic steatosis, and marked leptin deficiency underwent a chronic leptin-replacement regime, which resulted in a marked reversal in insulin resistance, hepatic steatosis, and hypertriglyceridemia.

2.3.4 Visceral Obesityff

Numerous lines of evidence suggest that visceral fat depots are more important in the promotion of dysmetabolic status and coronary risk than subcutaneous fat. In the Quebec Cardiovascular Study, visceral obesity was characterized by a clustering of a metabolic triad, namely, hyperinsulinemia, hyperapolipoprotein B, and small dense LDL, and was associated with a 20-fold increase in the risk of coronary heart disease over 5 years. Similar findings have also been reported in other cohorts.

Since insulin resistance and dyslipidemia are more closely associated with visceral fat mass than subcutaneous fat, a variety of biochemical and metabolic factors have been explored in explaining such differences. Visceral fat is characterized by enhanced lipolysis, resulting in an increase in FFA flux, especially into the portal circulation. Kissebah et al. showed that, *in vitro*, palmitate exposure caused a dose-dependent reduction in cell surface receptor binding of insulin in isolated hepatocytes, suggestive of a reduction in hepatic extraction of insulin. Differences in gene regulation in the different fat depots have also been reported. Collectively, these studies suggest that the deep abdominal peritoneal fat depot is less effective in insulin-mediated glucose uptake, expresses relatively less leptin, and more PPARγ. Plasminogen activator inhibitor-1 (PAI-1) is produced by adipocytes and its plasma level is strongly correlated with the degree of visceral adiposity but not subcutaneous fat in humans. Circulating PAI-1 has been implicated to play a role in thrombus formation and continues to be investigated as an emerging coronary risk factor.

2.3.5 Effect of Adipocyte Size on Insulin Resistance

Several lines of experimental evidence suggest that adipocyte size influences the impact of adipose tissues on the development of diabetes. Enlarged adipocyte size appears to be a better clinical predictor of diabetes independent of insulin resistance. Paolisso et al. showed that the predictive value of plasma FFA concentrations on diabetes can be accounted for entirely by the adipocyte size. A large fat cell size is predictive of the development of type 2 diabetes in Pima Indians independent of age, sex, percent body fat, and fat distribution.

Though intriguing, the biological basis of such association remains speculative. Early stage obesity, secondary to overnutrition, is characterized by overrepresentation by large adipocytes (hypertrophic obesity). Such large adipocytes might be more resistant to differentiation upon further nutrient overload. As a result, continued caloric intake will progress into an accumulation of fat in nonadipose tissues including muscle, liver, and pancreatic β-cells, resulting in insulin resistance and in case of β-cells, increased apoptosis, and loss of β-cell mass ensue. Additional putative contributing factors include a disproportionally high rate of secretion of TNFα and release of FFA by the large adipocytes, aggravating insulin resistance.

Adipose differentiation is a complex, highly regulated process. The ability for a given adipose depot to recruit preadipocyte differentiation and proliferation influences the cell number and size of the fat cells. Of the various genes and hormones that regulate and modulate the fat cell differentiation machinery, three transcription factors play particularly important roles in our understanding of adipogenesis, namely, peroxisome proliferator activator receptor γ (PPARγ), CAAT enhancer binding proteins (C/EBP)s, and adducin 1 (ADD1)/SREBP1.

PPARγ, a member of the PPAR subfamily of the nuclear hormone receptor superfamily, is expressed at high levels in adipose tissue. Two splice variants, PPARγ-1 and PPARγ-2, have been identified and the former has been shown to be expressed at low levels in a variety of other tissues. This receptor is induced very early in adipocyte differentiation and has been identified as a regulator of many fat-specific genes and a *master* regulator that can trigger the entire adipogenic program. PPARγ forms heterodimers with the retinoic acid receptor α (RARα), which is recognized by the DR-1 sequence in a variety of fat-specific genes including aP2. The key regulatory role of PPARγ in adipogenesis is best exemplified by the efficient way in which *PPARγ* gene transfected fibroblasts can be induced to differentiate into adipocytes.

C/EBPα is induced relatively late in the adipocyte differentiation program. Similar to PPARγ, C/EBPα also binds to many adipose-specific genes. At a physiologic level of expression, C/EBPα is a relatively weak inducer of adipogenesis. However, when coexpressed with PPARγ, C/EBPα can become extremely potent in the induction of adipocyte differentiation.

ADD1/SREBP1 is a basic helix–loop–helix protein independently cloned as an early gene induced in adipogenesis and as a sterol response element–binding protein involved in lipogenesis. SREBP1 and its close homolog SREBP2 are products of two different genes, but the former is expressed as two splice variants SREBP1a and SREBP1c. The primary protein products of each gene are large transmembrane proteins inserted in the endoplasmic reticulum membrane in a hairpin fashion. Two sequential proteolytic steps initiated by SREBP cleavage-activating protein (SCAP) result in the release of the terminal portion of the protein into the nucleus to impart its transcriptional activity. It has been shown that ectopic expression of

SREBP1 in fibroblasts under conditions favorable for adipogenesis augment PPARγ-induced differentiation into adipocytes.

Recent studies have added new insights into the once elusive mechanism by which thiazolidenediones (TZD), a class of synthetic PPARγ ligands already approved for clinical use, ameliorate insulin resistance. Treatment of obese female *fa/fa* Zucker rats with pioglitazone for up to 28 days decreased the insulin resistant and hyperlipidemic states and caused a shift in body adiposity. In the TZD-treated rats, there was a depot-specific increase in the number of smaller fat cells due to the appearance of new fat cells and the shrinkage and/or disappearance of the existing large mature fat cells. This is consistent with the notion that treatment of rats with a PPARγ agonist might have induced the formation of new small adipocytes that may have better lipid storage capacity as well as being more insulin-sensitive, leading to a lower lipolytic activity. These changes may in turn result in a reduction in ectopic, nonadipose TG accumulation. In the case of the skeletal muscles, this may result in a significant degree of improvement in skeletal muscle insulin sensitivity. In humans, adipocyte restructuring after treatment with a TZD (troglitazone or pioglitazone) was shown to increase total body adiposity, but there was a shift from visceral fat to the subcutaneous fat in conjunction with a reduction in insulin resistance. In this case, it appears that an increase in adipose tissue mass after treatment with a TZD due to a redistribution from the large visceral fat to the smaller subcutaneous fat may be critical in the amelioration of insulin resistance.

2.3.6 Adipose Tissue as Endocrine Organ

Recent studies in adipocyte biology have expanded our conventional view of adipose tissue being primarily a fat depot for storage of energy in the form of TGs. In this traditional paradigm, insulin action on the adipocytes plays the key role of stimulating glucose uptake and lipogenesis and simultaneously inhibits lipolysis, promoting fat storage. Defects in fuel partitioning including increased adipose mass and/or increased lipolysis with increased circulating free fatty acids would result in insulin resistance, dyslipidemia, and β-cell dysfunction. We have now accrued increasing evidence that adipose tissues are programmed to secrete a large variety of substances that may act as signaling molecules in a hormonal, paracrine, and/or autocrine fashion. The secretory factors identified to date span a large variety including enzymes (lipoprotein lipase and adipsin), cytokines (tumor necrosis factor α (TNFα), interleukin 6 (IL6)), hormones (adiponectin, leptin, resistin, acylation stimulation protein, and corticosteroids), growth factors (vascular endothelial growth factor), and thrombogenesis modulators (plasminogen activator inhibitor-1 (PAI-1)). These substances are differentially regulated and are functionally diverse. Amongst them, leptin, adiponectin, TNFα, and resistin are most studied for their roles in energy homeostasis, insulin signaling, and glucose/lipid metabolism. The rapid pace of new research findings in this area does not permit an updated discussion within the scope of this paper. Readers are referred to more updated research papers and reviews.

2.3.7 Leptin

The leptin gene was first cloned when the *circulating factor* in the obese *ob/ob* mice was sought. The gene product, leptin, is a 16-kDa protein synthesized and secreted mainly by the white adipose tissue. The *ob/ob* mice, now known to be leptin deficient, develop severe hyperphagia, reduced metabolic rate, massive obesity, and a variety of endocrinal abnormalities including infertility. Leptin acts through binding with the leptin receptor in the central nervous system with its main function in regulating food intake. Defective leptin receptors have been identified as the cause of obesity, insulin resistance, and diabetes in the *db/db* mice. In humans, monogenic causes of leptin deficiency–related obesity have been described but are extremely rare. The large majority of obese humans have an elevated level of circulating leptin, suggesting that *leptin resistance* is a key metabolic defect in obesity. Not surprisingly, administration of leptin to obese humans led to minimal benefit in the reversal of excess body weight and insulin resistance. On the contrary, in the relatively uncommon cases of subjects with partial lipodystrophy, affected subjects have been found to be hypoleptinemic, and leptin treatment resulted in a reversal of the metabolic abnormalities.

The leptin receptor was first isolated from mouse choroid plexus by expression cloning. This receptor is normally expressed at high levels in hypothalamic neurons and in other cell types, including T cells and vascular endothelial cells. There are several isoforms of the leptin receptor with the long form being predominantly expressed in the hypothalamus. Binding and activation of the leptin receptor results in downstream alterations of gene expression through the JAK/STAT pathway. One functional implication of this signaling is its possible linkage with the metabolic action of protein tyrosine phosphatase 1B (PTP1B). In addition to playing a role in the inactivation of insulin signaling through dephosphorylation of the insulin receptor, PTP1B also can dephosphorylate the leptin receptor–associated tyrosine kinase JAK. Leptin-deficient mice made PTP1B deficient were found to have much reduced weight gain and an increased metabolic rate, supportive of the notion that the signaling pathway cross talk is biologically significant at the whole organism level.

Leptin expression is regulated by a number of factors. Positive modulators include intracellular glucosamine (a product of glucose flux), insulin, TNFα, glucocorticoids, interleukin-1, and so on. It is intriguing to note that in cultured adipocytes, leptin mRNA expression is correlated with the lipid content and adipocyte size, but the underlying mechanism for this adaptive change remains unknown. Negative modulators include PPARγ activation, catecholamines, cAMP agonists, β-adrenergic receptor agonists, and thyroid hormones. In addition to the extensively characterized action of leptin in the regulation of energy balance, glucose homeostasis and insulin signaling, leptin has also been implicated in other biological functions, including immune activities by stimulating proliferation of T lymphocytes and in maturation of reproductive organs and in gestation. Future studies are needed to fully understand the biological function of this adipocyte-derived hormone and its role as a therapeutic in the treatment of obesity and diabetes.

2.3.8 Adiponectin (Acrp30/AdipQ)

Adiponectin was originally identified as a highly expressed gene during adipocyte differentiation. The gene was expressed exclusively by adipocytes, and the secreted protein contains an amino-terminal collagenous domain and a carboxy-terminal globular domain, the latter bearing significant homology with collagens X and VIII and complement C1q. Plasma level of adiponectin is inversely associated with

BMI in humans and is reduced in a number of murine models of insulin resistance. Furthermore, adiponectin level is lower in CHD patients compared to subjects without CHD, even after adjustment for BMI. Several lines of evidence suggest that adiponectin may have direct biological action as an insulin sensitizer by reducing hepatic glucose output as well as reducing skeletal TG content. Treatment of mice with adiponectin induced a large and sustained weight loss in a high fat/high sucrose–fed model mouse. Recent data also suggest that adiponectin confers antiatherogenic actions. Data available to date suggest the adiponectin may be a promising therapeutic candidate for the prevention of diabetes and CHD.

3
Pancreatic β-cell Dysfunction

In pancreatic β-cells, ambient fatty acids modulate the glucose-induced insulin secretion. The potency of fatty acids to promote glucose-induced insulin secretion varies with chain length and the degree of unsaturation. Acute lowering of plasma fatty acid levels is associated with a decreased insulin response to glucose. However, longer exposure of islets to FFA results in a reduction in insulin secretion in the hyperglycemic milieu. This effect of differential impact of acute versus chronic exposure of plasma FFA on insulin secretion was also observed in humans. These observations are consistent with the original Randle hypothesis based on studies in cardiac and skeletal muscle. An elevation of FFA would promote oxidation of fatty acids, which in turn inhibits pyruvate dehydrogenase activity, resulting in an accumulation of cytosolic glucose-6-phosphate with associated inhibition of hexokinase for glucose uptake. Since glycolysis is the pathway for glucose metabolism in β-cells, this FA-induced alteration in glucose uptake may play an important role in the development of β-cell secretory dysfunction.

3.1
Effect of FFA on β-cell Apoptosis

Pancreatic islets in diabetics are characterized by a reduced β-cell mass. This morphologic change is postulated to be largely a result of an accelerated β-cell apoptosis during the progression from prediabetic state to diabetes. The notion of a progressive reduction in β-cell mass caused by accelerated β-cell apoptosis as part of the natural history of the development of type 2 diabetes is supported by autopsy studies on obese and diabetic subjects when compared with lean, nondiabetic controls. Excess accumulation of TGs in β-cells, similar to many other nonadipose tissues, is a cardinal feature of the insulin-resistant state. The progressive TG accumulation has been attributed to both an elevated plasma level of FFA as substrate and a shift toward nonoxidative metabolic pathway in conjunction with an increase in endogenous lipogenesis. *In vitro* data strongly suggest that the modest increase in TG content in islets is causative of the compensatory islet cell hyperplasia and hyperinsulinemia. However, further increase in islet TG eventually leads to decompensation, triggering an accelerated apoptosis and precipitous loss of β-cell mass. In conjunction with the progression of the disease, cellular content of ceramide accumulation causes β-cell decompensation by inducing cellular apoptosis. As such, it has been suggested that FFA-induced β-cell apoptosis may, at least in part, be due to an increased supply of substrate for the synthesis of ceramide. Inhibition of ceramide synthetase with fumonisin B1 can inhibit FFA-induced apoptosis. In parallel, increased plasma FFA is also associated with an induction of iNOS, promoting nitric oxide (NO)-induced apoptosis.

The mechanism of FFA-induced β-cell apoptosis remains incompletely understood. Recent studies suggest that the PKB pathway may play a key role. Selective transgene expression of a constitutively active form of PKB in β-cells resulted in markedly increased β-cell mass largely through preserving cell survival. This may be explained on the basis that PKB phosphorylation prevents the nuclear translocation of a number of its downstream substrates including forkhead box (Fox) transcription factors, glycogen synthase kinase-3α/β (GSK3$\alpha\beta$), bcl-2 associated death agonist (BAD), caspase 9, and murine double minute 2 (MDM2). Nuclear translocation of such factors would promote proapoptotic transactivation. *In vitro* experiments using a pancreatic β-cell line INS-1 showed that not only can FFA induce apoptosis by inhibition of PKB activation but also can inhibit the IGF-1-induced PKB phosphorylation.

Recent studies suggest that leptin can effectively protect the islet cells by promoting β-oxidative metabolism of surplus FFAs and reducing lipogenesis, especially during periods of overnutrition. Intriguingly, all fatty acids are not equally capable of inducing apoptosis. Saturated fatty acids are highly effective in the induction of apoptosis while unsaturated fatty acids are not. Human β-cell culture studies revealed that when mixing the two types of fatty acids, the unsaturated FAs are capable of preventing saturated FA-induced apoptosis.

Functional characterizations of glucagon-like peptides (GLPs) over the recent years have led to the discovery that GLP-1 may protect pancreatic β-cells from apoptosis. GLP-1 confers its biological actions through binding to the GLP1-receptor, a G-protein coupled receptor expressed in pancreatic β-cells. Not only does GLP-1 stimulate insulin secretion, it also exerts a number of other antidiabetic actions including inhibition of glucagon secretion, reduction of food intake and regulation of cell proliferation, differentiation, and apoptosis. Several lines of experimental evidence support the notion that the GLP-1 receptor is the key mediator of the antiapoptotic properties of GLP-1. In Zucker diabetic rats and in *db/db* mice, short-term infusion of GLP-1 was associated with a marked increase in β-cell mass and the development of new islet clusters. *In vivo* treatment of mice with a long-acting GLP-1 analog NN2211 and *in vivo* treatment of exendin-4, a GLP-1 receptor agonist, both led to a reduction in β-cell apoptosis. A more detailed mechanistic understanding of how GLP-1 regulates cell proliferation and apoptosis is potentially useful in the development of GLP-1 based therapeutics for the treatment or prevention of diabetes.

4
Insulin Resistance and the Liver

Hepatic glucose overproduction is a cardinal feature of type 2 diabetes mellitus. In addition, hypertriglyceridemia in association with low levels of high density lipoprotein cholesterol (HDL-C) levels, elevated levels of apolipoprotein (apo) B100, and less buoyant (small dense) low-density lipoproteins (LDL) are frequently observed in subjects with diabetes and insulin resistance. Recent studies continue to unravel the complex underlying mechanisms both in the cause of hepatic insulin resistance and its metabolic sequelae.

An elevated blood level of FFA has been shown to promote skeletal muscle insulin resistance and β-cell dysfunction. The mechanisms by which elevated levels of FFA contribute to hepatic glucose overproduction are less well known. A recent study by Massillon et al. showed that the glucose-6-phosphatase gene can be induced by an increased level of FFA by Intralipid infusion on a background of pancreatic clamping of the FFA levels. In a somewhat more physiological model, Oakes et al. examined the effect of high-fat feeding in rats on the development of insulin resistance. Compared to chow fed rats, high fat feeding significantly impaired insulin sensitivity, resulting in hyperinsulinemia. While the compensatory increase in insulin level was able to suppress FFA release, it did not normalize the intracellular availability of FA in skeletal muscle. In the liver, glycogen synthesis was reduced because of the down regulation of glycogen synthase. Glucogenesis was enhanced in conjunction with increased FFA oxidation. However, an acute blockade of FFA oxidation with etomoxir reduced gluconeogenesis and yet, increased glucose output was unaffected. This is likely because of an alternate pathway of a sustained increase in the glucose-6-phosphatase/glucose kinase activity ratio, favoring increased glucose output. In humans, a progressive increase in FFA load through Intralipid infusion resulted in an increasing percent contribution of gluconeogenesis to the observed endogenous glucose overproduction. The relative importance of glycogenolysis and gluconeogeneis in FFA-induced hepatic glucose overproduction has been studied in healthy human subjects. Physiologically, insulin suppresses hepatic glucose output more by inhibiting glycogenolysis than by gluconeogenesis. Elevated plasma FFA levels attenuated the suppression of glucose production by interfering with insulin suppression of glycogenolysis. More recently, Lam et al. demonstrated that cell membrane translocation of PKC-δ and increased FFA oxidation may also be major contributors to the FFA-induced hepatic glucose overproduction.

FFAs have also been proposed as the link between intra-abdominal fat and peripheral insulin resistance. Obesity is associated with higher fasting plasma FFA concentrations. Intra-abdominal fat drains via the portal vein directly to the liver. At high concentrations, it has been shown to interfere with hepatic insulin binding and clearance, which even at low FFA concentrations, is only 50% efficient. It is suggested that increased insulin levels in the peripheral circulation secondary to reduced hepatic clearance downregulate peripheral glucose receptors in muscle and in the long term lead to progressive insulin resistance.

4.1 Hepatic Insulin Resistance and Dyslipidemia

Insulin resistance and diabetes are commonly associated with plasma hypertriglyceridemia, elevated apoB100, and a reduced level of HDL-C. While reduced lipoprotein lipase in the light of insulin resistance may have contributed to the accumulation of very low-density lipoproteins (VLDL) in this dyslipidemic state, a hepatic overproduction of VLDL seems to be of primary importance. Many metabolic factors influence the rate of secretion of VLDL (production rate). Since the backbone apoB100 is constitutively synthesized in great excess and each VLDL particle contains one apoB100 molecule, the rate of apoB100 (hence VLDL) entering the circulation is governed by a number of factors in the VLDL synthetic machinery, including the translocation efficiency, lipidation of apoB with TGs, and the rate of degradation of the protein. Grossly, it appears that it is the lipids available in the ER lumen that dictate the amount of apoB eventually secreted.

Lipidation of apoB with TGs is largely mediated by the microsomal transfer protein (MTP). In addition to participating in the assembly and maturation of the VLDL particles, MTP also plays a role in stimulating apoB production through inhibition of ubiquitination, a key step in apoB degradation. Hepatic expression of MTP has been shown to be modulated by insulin through interaction of an insulin response element in its promoter. In addition, response elements to other transcription factors including adaptor protein complex (AP-1), hepatic nuclear factor 1 (HNF-1), and hepatic nuclear factor 4 (HNF-4) have also been identified. MTP mRNA expression is also regulated by hepatic sterol content and is thought to be mediated by SREBPs.

The cytosolic content of TGs, the major neutral lipid in the mature VLDL, correlates with the rate of VLDL secretion. Increased availability of cytosolic TG, either through de novo synthesis, or through esterification of FFA derived from the circulation, stimulates the VLDL production and secretion. Recently, a series of studies performed on a hamster model of insulin resistance through fructose feeding have advanced our understanding of the molecular mechanisms of insulin resistance–associated hepatic VLDL overproduction. In this model of oral fructose-induced insulin resistance, the authors observed an increase in intracellular apoB stability, and elevated levels of MTP in association with enhanced VLDL particle assembly. Impaired hepatic insulin signaling was evidenced by a reduced tyrosine phosphorylation of the insulin receptor and IRS-1. In addition, elevated protein mass and activity of PTP-1B, and a reduced level of PI3K activity were observed. The level of ER-60, a cysteine protease, was significantly reduced, consistent with the observed increase in apoB stability.

4.2 The Role of SREBP in Hepatic Insulin Resistance

SREBP1a and 1c are two variants of the same *SREBP1* gene, each using a different transcription start site. SREBP1a is a potent transcription activator of all SREBP responsive genes, including those involved in both endogenous cholesterol and fatty acid and TG synthesis. SREBP1c, on the other hand, primarily transactivates the enzymes involved in lipogenesis. In adult liver, SREBP1c is the predominant SREBP-1 isoform. The selective coordinated transduction by SREBP1c of the fatty acid program of synthesis has been demonstrated *in vivo* in a transgenic mouse model, which overexpresses the truncated version of SREBP1c that terminates prior to the membrane attachment domain. In mice, fasting markedly reduces the amount of SREBP1c in the nuclei. There is a corresponding reduction in the mRNA abundance of the SREBP1c target genes, that is, those involved in *de novo* lipogenesis. On refeeding a high carbohydrate/low fat diet, Horton et al. reported a markedly exaggerated overabundance of SREBP1c (increased 4- to 5-fold over nonfasting baseline) in association with a parallel increase in hepatic mRNA levels of the lipogenic genes.

Recent data on a rodent model showed that the insulin-induced hepatic lipogenesis is mediated by modulating SREBP1c at the transcriptional level. By comparing two distinct murine models of severe insulin resistance, namely, the *ob/ob* (leptin-deficient) mice and the adipose-specific SREBP transgenic (lipoatrophic) mice, marked hyperinsulinemia differentially caused a significant reduction in IRS-2 mRNA but continued to stimulate SREBP1c, resulting in hepatic lipogenesis and VLDL overproduction. A more recent study further defined the role SREBP1c in the obesity–insulin resistance connection. Targeted disruption of SREBP1 in the *ob/ob* mice by cross breeding the two knockout strains selectively ameliorated the enhanced lipogenesis seen in the *ob/ob* mice without any significant alterations in fat mass and insulin resistance.

Diets rich in polyunsaturated fatty acids (PUFA) including 18 : 2 (ω-6), 20 : 5, and 22 : 6 (ω-3) have been shown to significantly lower plasma lipid levels. In a rodent model, dietary PUFA suppresses hepatic lipogenesis by reducing the nuclear content of mature SREBP1. PUFA reduces the nuclear abundance of SREBP1 by first acutely inhibiting the proteolysis of SREBP1 followed by a reduction in hepatic SREBP1 mRNA and precursor protein content. More recently, Yoshikawa et al. also showed that the suppressive effects of PUFA on SREBP1-induced lipogenesis is mediated through inhibition of the binding of the liver X receptor (LXR) to LXR response element (LRE) on the SREBP1 gene. Our understanding of the molecular mechanism of the interaction between fatty acids and transcription factors in lipid and glucose metabolism may continue to define more specific targets for treatment of hepatic insulin resistance and its associated deleterious phenotypes.

5 Insights from Gene-targeted Mouse Models

Transgenic and knockout mouse technology has provided very powerful tools for dissecting the effects of individual genes involved in insulin signaling and effector pathways. The outcomes of these studies have been extensively reviewed and will be summarized here.

5.1 Insulin Receptor (IR) Knockout Mice

The insulin receptor gene has been knocked out in mice both globally and in a tissue-specific manner. The homozygous IR knockout mice (IR−/−) appear normal at birth but rapidly develop severe metabolic phenotypes of insulin resistance at the time of suckling. These include severe hyperglycemia, ketoacidosis, hypertriglyceridemia, and steatohepatitis. Hepatic content of glycogen was reduced. Muscle was partially atrophic and the amount of adipose tissue was reduced. The mice generally died within one week of birth. On the contrary, the phenotype of the heterozygous knockout mice (IR+/−) was relatively mild. One report showed that the mice developed hyperinsulinemia at 4 to 6 months of age but only about 10% eventually developed diabetes. The insulin receptor gene has been disrupted selectively in a number of tissues, which collectively provided great insight into glucose homeostasis.

5.2 β-cell-specific IR Knockout Mice (βIRKO)

By using the cre-loxP mediated conditional gene inactivation technique, the murine IR gene was disrupted selectively in pancreatic β-cells. The βIRKO mice were normoglycemic and began to develop significant hyperinsulinemia by 6 months of age. However, insulin secretion in response to a glucose load injected peritoneally was impaired, representing a specific impaired first-phase secretory response to glucose and not to arginine at early age. A chronic progressive impairment of glucose tolerance to an intraperitoneal glucose load was detected as early as 2 months of age. The same glucose tolerance test unmasked a reduced level of insulin levels at 4 months of age. This selective loss of first-phase secretory response is remarkably similar to the early stage of development of diabetes in humans. Furthermore, this secretory defect was observed in spite of a quantitative preservation of β-cell GLUT2 transporter. Although it is unlikely that acute first-phase secretory defects observed in humans is caused by specific insulin receptor gene defects, these data are suggestive of β-cell insulin signaling playing an important role.

5.3
Muscle-specific IR Knockout Mice (MIRKO)

A cre-loxP system was also used to generate muscle-specific IR knockout mice. Skeletal muscle IR expression was reduced by more than 95% in conjunction with a comparable loss of insulin-stimulated IR and insulin receptor substrate-1 (IRS-1) phosphorylation. At a whole body level, these mice showed a selective increase in fat mass, plasma TG, and FFA levels, but their blood glucose levels, insulin levels, and glucose tolerance were normal.

Resting uptake of 2-deoxyglucose (2DG) by skeletal muscle is normal in the MIRKO mice. Likewise, exercise-induced 2DG uptake with and without insulin is also normal. Glycogen synthase activity is also normal in both the resting state and with exercise but not after insulin stimulation.

To further evaluate the role of muscle IR in the development of type 2 diabetes, the authors cross-bred the MIRKO into the heterozygous IR knockout mice IR+/−, and the resultant mice of interest are designated MIRKO-50%. These mice begin to develop hyperinsulinemia by 3 months. At 6 months, hyperinsulinemia becomes more severe and many animals showed hyperglycemia in a fed state.

The relatively mild glycemia phenotype in the MIRKO mice was addressed by Kim et al. In the absence of IR in skeletal muscles, there was a redistribution of the substrate to liver and adipose tissue, resulting in an expanded fat mass, hypertriglyceridemia, and elevated FFA levels, features of the prediabetes syndrome. Such intriguing findings challenge the conventional view that skeletal muscle insulin resistance is the main driving force in the development of type 2 diabetes.

5.4
Liver-specific IR Knockout Mice (LIRKO)

The liver-specific IR knockout mice develop severe insulin resistance and impaired glucose tolerance. There is a loss of inhibition of gluconeogenesis and hepatic glucose overproduction despite hyperinsulinemia, the latter was the result of a combined increase in insulin secretion and reduced clearance. The hepatic glucose overproduction seen in the LIRKO mice is also not responsive to exogenously administered insulin, bringing up the concept that hepatic glucose overproduction can continue even in the presence of hyperglycemia in the LIRKO mice and that glucose autoregulation of hepatic glucose production does not occur in the absence of insulin signaling in liver. In this model, the severe hyperinsulinemia had led to a simultaneous differential regulation of a number of insulin-responsive genes. While phosphoenolpyruvate carboxykinase (*PEPCK*) gene expression was found to be upregulated in the LIRKO mouse, the expression of glucokinase and liver pyruvate kinase were coordinately downregulated. Likewise, the transcription of the *SREBP1c* gene continues to be positively responsive to insulin. The loss of IR in the LIRKO mouse liver results in a failure of the insulin receptor through divergent pathways that regulate gene expression, which in turn, contribute to the severe metabolic phenotypes. More interestingly, it is the analyses of these coordinated changes in gene expression that led to the identification of cis-acting DNA regulatory elements (insulin response element), which mediate insulin's effect.

5.5 Fat-specific IR Knockout Mice (FIRKO)

Fat-specific IR knockout mice were recently reported. These animals showed a sustained reduction in total fat mass and are protected from age-related glucose intolerance despite consuming higher amount of calories. There was also a reduction in total body TG content and a polarized distribution of fat cell size. In addition, these mice enjoyed a significantly prolonged life span as compared to wild-type mice. More intriguing is the observation that the phenotype of these mice is quite similar to those of calorie-restricted mice, leading the authors to suggest that it is the leanness that is the key determinant to longevity, not caloric restriction *per se*.

6 Hyperglycemic Complications

6.1 Clinical Impact – Microvascular Complications

The clinical spectrum of diabetic microvascular complications has been extensively reviewed. Two landmark clinical trials over the past decade, the Diabetes Control and Complications Trial (DCCT) for type 1 diabetic subjects and United Kingdom Prospective Diabetes Study (UKPDS) for type 2 diabetic subjects, have provided compelling evidence that hyperglycemia is the primary causative factor in the development and progression of microvascular complications, namely, retinopathy, neuropathy, and nephropathy. In these trials, excellent glycemic control has been demonstrated unequivocally to retard the progression of the microvascular complications (DCCT 1993, UKPDS 1998). Diabetic microvascular disease in retina, glomerulus, and vasa nervorum share similar pathogenic changes. At the onset of hyperglycemia, elevated intracellular glucose concentrations directly impact on vascular tone regulators, including reduced bioavailablity of nitric oxide (NO), increased production of vasoconstrictors including endothelin-1 and angiotensin II, and elaboration of permeability factors such as vascular endothelial growth factor (VEGF). Microvascular cell loss and the compensatory endothelial cell proliferation and qualitative and quantitative changes in extracellular matrix predispose to increased vascular permeability. Excessive matrix accumulation promoted by transforming growth factor beta 1 (TGFβ1) and deposition of periodic acid Schiff (PAS) positive plasma proteins also contribute to microvascular occlusions, leading to local ischemic changes.

Our understanding of how hyperglycemia acts as the common denominator for the development of such an array of microvascular complications has recently been conceptualized by establishing a linkage between four major metabolic pathways: increased polyol pathway, increased advance glycation end-product (AGE) formation, activation of PKC, and increased hexosamine flux. A unifying hypothesis in linking these four pathways has recently been put forth. Evidence available to date suggests that the intracellular excess glucose–induced mitochondrial superoxide (O_2^-) overproduction may be the key in the simultaneous activation of all of these pathways (Fig. 2). In this hypothesis, excess mitochondrial O_2^- production partially inhibits the glycolytic enzyme GAPDH, which in turns diverts its upstream glycolytic metabolites from glycolysis into pathways of glucose overutilization. Accumulation of cellular glucose would result in the formation of sorbitol and fructose through the polyol pathway. Excess fructose-6-phosphate is metabolized into glucosamine. Glyceraldehyde-3-phosphate in excess would provide a substrate for the increased formation of diacylglycerol for PKC activation as well as increased formation of advanced glycation end (AGE) products.

6.1.1 Intracellular Oxidative Stress

An alternate pathway by which hyperglycemia may confer oxidative stress is the overproduction of superoxide by the mitochondrial electron-transport chain. In the mitochondria, the generation of ATP is coupled to the electron-transport chain, through which molecular oxygen (O_2) may be reduced to water by electron donors NADH and $FADH_2$ from the Krebs cycle. Protons are pumped through the inner membrane of the mitochondria, sustaining an electrochemical gradient and electrons flow through intermediates

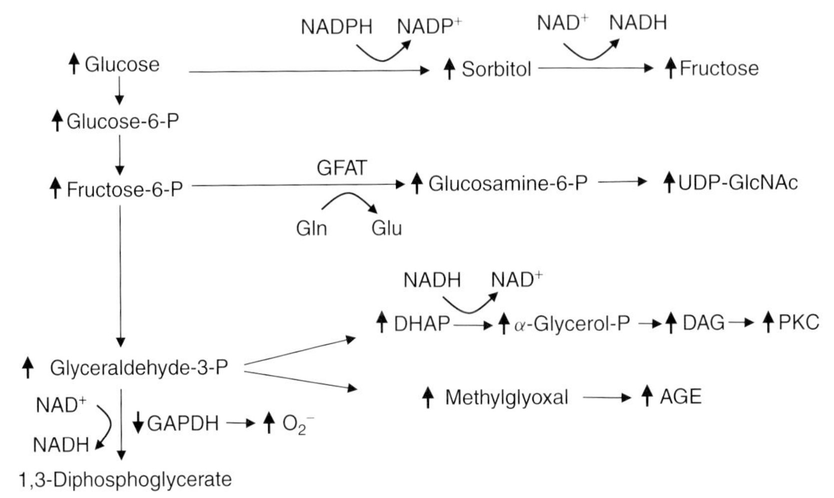

Fig. 2 Impact of elevated cellular glucose levels on multiple metabolic pathways in the promotion of diabetic complications.

such as ubisemiquinone. It has been shown that high glucose availability results in increased production of electron donors, which in turns raises the proton gradient. This, in turn, prolongs the lifetime of the electron-transport intermediates. When this number goes beyond a certain threshold, the intermediate will redirect its reactivity, greatly increasing the conversion of molecular oxygen to superoxide anion (O_2^-). This pathway of hyperglycemia-induced mitochondrial O_2^- overproduction can be attenuated by overexpression of mitochondrial superoxide dismutase (MnSOD) and uncoupling protein-1 (UCP-1).

6.1.2 The Polyol Pathway Flux

Aldose reductase is the rate-limiting enzyme mediating the reduction of glucose and other carbonyl compounds. The reduction is NADPH-dependent and has a low K_m for glucose and with elevated glucose, aldose reductase converts glucose to sorbitol, which in turn can be converted to fructose. These reactions cause simultaneous consumption of NADPH and NAD^+ respectively, leading to a reduction of cellular concentration of NADPH and increased ratio of NAD/NAD^+. A cellular depletion of NADPH would result in a reduction in the regeneration of the reduced form of glutathione (GSH), which in turn aggravates the intracellular oxidative stress. On the other hand, an increased $NADH/NAD^+$ ratio would inhibit the enzyme glyceraldehyde-3 phosphate dehydrogenase (GAPDH), leading to an accumulation of triose phosphate. The downstream effect of this includes an increased formation of methylglyoxal, a precursor of AGEs and diacylglycerol (DAG), the latter leading to the activation of protein kinase C (PKC).

6.1.3 Advanced Glycation End Products

Intracellular glucose at elevated concentration may also be metabolized through nonenzymatic pathways to several reactive dicarbonyls, which in turn react with amino groups of intracellular and extracellular proteins to form AGEs. Intracellular auto-oxidation of glucose leads to the formation of glyoxal. Decomposition of the Amadori product has been shown to generate 3-deoxyglucosone and fragmentation of glyceraldehyde-3-phosphate and dihydroxyacetone phosphate will form methylglyoxal. The two reactive dicarbonyls glyoxal and methylglyoxal may be detoxified by the glyoxalase system. The biological action of intracellular AGEs precursors may damage cells through a number of mechanisms. AGEs' modification of intracellular proteins results in alteration of both the structure and function of cellular proteins. AGEs' formation on extracellular matrix interferes with both matrix–matrix and cell–matrix interactions. AGEs-modified plasma proteins are capable of binding to AGEs receptors (RAGEs) on a variety of cell types including endothelial cells, mesangial cells, and macrophages. In the latter case, binding of AGEs to RAGEs in macrophages may induce production of reactive oxygen species.

6.1.4 Activation of Protein Kinase C

More than 10 isoforms of protein kinase C (PKC) have been described and nine of them are known to be activated by diacylglycerol (DAG). Elevated intracellular levels of glucose promote the *de novo* synthesis of DAG through the glycolytic intermediate dihydroxyacetone. Increased DAG primarily activates the two isoforms PKC-β and -δ, which explains the activation of a large array of downstream cellular and molecular processes responsible for the pathogenesis of a variety of hyperglycemia-induced complications, most notably, those related to the microvascular diabetic complications.

PKC-β activation has been shown to mediate abnormal blood flow in both retinal and renal vasculatures, possibly through reduced NO production and/or an increased activity of endothelin-1, one of the most potent vasoconstrictors. In mesangial cells, PKC activation induces endothelin-1 stimulated MAP-kinase activity, which may contribute to diabetic nephropathy. PKC activation has also been invoked in hyperglycemia-induced expression of VEGF in smooth muscle cells, which may account for the observed increase in vascular permeability and angiogenesis in chronic hyperglycemia. Increased microvascular matrix accumulation in long standing hyperglycemia may also be attributable to PKC activation. The increase in PAI-1 expression in the hyperglycemic state is a predisposing factor for increased thrombogenicity and the increased NFκB activity may account for the increased expression of a wide array of proinflammatory genes. Lastly, hyperglycemia has been shown to be associated with an increased activity of the vascular NADPH oxidase system and PKC activation as a mediator has also been implicated.

6.1.5 The Hexosamine Pathway

Intracellular accumulation of fructose-6-phosphate of the glycolytic pathway may be shunted to the hexosamine pathway to produce glucosamine. The rate-limiting enzyme for this pathway, glutamine:fructose-6-phosphate amidotransferase (GFAT), converts fructose-6-phosphate to glucosamine phosphate, which will subsequently be converted to UDP-*N*-acetylglucosamine. The pathological significance of this pathway can be related to the hyperglycemia-induced upregulation of the transcription of TGF-α, TGF-β1, and PAI-1. Several lines of evidence suggest that the hexosamine-mediated transcription upregulation of *PAI-1* gene in a number of tissues including vascular cells and mesangial cells is through Sp1. Glycosylation modification of the transcription factor Sp1 may also contribute to the enhancement in its transcription activation activity. In addition to PAI-1, hyperglycemia-induced O-acetylglucosaminylation is implicated in the modulation of the activity and function of a number of other cellular proteins including eNOS modification at the Akt site and a number of PKC isoforms, promoting diabetes-related complications.

6.1.6 The Effect of Hyperglycemia on Insulin Signaling

The direct effect of hyperglycemia on insulin signaling has recently been reviewed. A 4-h incubation of rat entensor digitorum longus (EDL) muscle *ex vivo* with 25 mM glucose resulted in a 50% reduction in insulin-stimulated glucose incorporation into glycogen and a significant reduction in the insulin-stimulated phosphorylation and activation of Akt/PKB. The insulin receptor tyrosine phosphorylation, PI3K activation, and MAP-kinase activation were not affected. Furthermore, the same laboratory also demonstrated previously, both *in vivo* and *in vitro*, that acute hyperglycemia causes cellular accumulation of malonyl coA, an inhibitor of palmitoyl transferase. The resultant accumulation of palmitoyl-coA could be a source of cellular ceramide. Interestingly, incubating the EDL muscle with C2-ceramide also resulted in similar signaling defects, raising the speculation that ceramide may play a role in mediating this signaling defect.

In cultured human vascular endothelial cells, incubation of high glucose has resulted in a reduced activation of Akt/PKB in conjunction with an increase in apoptosis, a known downstream effect of Akt/PKB. Concomitantly, cellular concentration of diacyl glyceral (DAG) and the PKC activity were both elevated.

6.1.7 The Effect of Hyperglycemia on β-cell Function

It has been well accepted that it is the development of β-cell dysfunction on the background of progressive insulin resistance that precipitates the development of frank type 2 diabetes. The early clinical defect of β-cell dysfunction, a loss of the first-phase insulin secretion in response to glucose stimulation, precedes fasting hyperglycemia by up to 5 years. Initially, hypersecretion of insulin is able to compensate for the insulin resistance, maintaining euglycemia. It is the progressive loss of β-cell mass, in part through accelerated apoptosis, and other progressive dysfunction that precipitates the decompensation, and fasting hyperglycemia ensues. Recent studies examined the potential mechanism underlying glucose-induced β-cell apoptosis. In light of the fact that β-cells have the highest level of expression of *O*-linked monosaccharide *N*-acetylglucosamine (*O*-GlcNAc) transferase (OGT), recent studies suggest that hyperglycemia may cause β-cell apoptosis through first being metabolized through the hexosamine pathway, resulting in an excessive rate of *O*-GlcNAc modification of a variety of proteins and eventually accelerated apoptosis.

Several lines of evidence suggest that hyperglycemia *per se* in turn further aggravates β-cell dysfunction as well as insulin resistance. Persistent hyperglycemia has been shown to desensitize the β-cells to glucose stimulation. The mechanisms include a reduction in the intracellular concentration of signaling molecules, for example, malonyl coA, which increases fatty acid oxidation, and desensitization of the ATP-dependent potassium channels. Meanwhile, hyperglycemia also contributes to the aggravation of peripheral insulin resistance through downregulation of the glucose transporter. This sets up a vicious cycle and the worsening of hyperglycemia will in turn aggravate β-cell function.

6.1.8 The Effect of Hyperglycemia on Oxidative Stress

Oxidant stress continues to be considered to play a critical role in atherogenesis despite the disappointing results of antioxidant trials (HOPE, BHF/MRC). Reactive oxygen species (ROS) generally consist of a number of free radicals, including superoxide anion (O_2^-) and hydroxyl radicals (OH) and their downstream reactive intermediates including peroxynitrite radicals ($ONOO^-$) and hydrogen peroxide (H_2O_2). The free radicals are highly reactive chemically and are able to initiate chain reactions. Superoxide anion radicals are generated by a number of enzyme systems in the vessel walls, including NAD(P)H oxidase, xanthine oxidase, lipoxygenase, cycloxygenase, mitochondrial oxidative phosphorylation, and so on. In defense, a family of superoxide dismutases (SOD) has evolved to transform O_2^- into hydrogen peroxide (H_2O_2). However, the H_2O_2 can, through either a Haber–Weiss reaction or the Fenton reaction, in turn lead to the formation of highly reactive hydroxide (OH^-) and hydroxyl radicals (OH). Alternatively, two endogenous defense systems, catalase and glutathione peroxidase, are able to reduce

H_2O_2. The former reduces to water and molecular oxygen and the latter, using glutathione as cosubstrate, is able to reduce H_2O_2 and lipid peroxides to water and lipid alcohols respectively. In the vessel wall, O_2^- can rapidly inactivate nitric oxide (NO) by reacting with it to produce a highly reactive species, peroxinitrite ($ONOO^-$). Therefore, O_2^- produced in excess not only provides a source for continued production of reactive oxidative species to induce tissue and lipid peroxidation, it also attenuates the potentially beneficial vasodilatory effect of NO in the vessel wall.

NAD(P)H oxidase was initially described in phagocytes as an enzyme complex capable of producing bursts of O_2^- upon stimulation for bacterial killing. The neutrophil NADH oxidase is an enzyme complex consisting of a membrane-bound cytochrome complex b558, made up of two main subunits gp91phox and p22phox and three cytosolic components, p47phox, p67phox, and rac-1. In its quiescent state, the enzyme system is inactive, but upon activation, the cytosolic components are recruited to associate with the membranous cytochrome complex. NADPH, FAD, and heme are bound to their binding sites in the gp91phox subunit and the electron transfer from NADPH to O_2 ensues, resulting in a rapid burst of O_2^- production. In recent years, nonphagocytic NAD(P)H oxidase homologs have been identified in a variety of cell types including endothelial cells, smooth muscle cells, and adventitial fibroblasts in the vessel wall. The relative abundance of each of the components in vascular cells varies with cell type. Specifically, vascular smooth muscle cells contain only trace amounts of gp91phox and p67phox. On the other hand, nox-1, a member of a family of gp91phox-like homologs, is abundant in vascular smooth muscle cells. In endothelial cells and fibroblasts, all five components are present. In both endothelial cells and smooth muscle cells, p22phox and distinct gp91phox homologs appear to be functionally important in the generation of O_2^-.

6.2 Clinical Impact – Macrovascular Complications

6.2.1 Relevance of NADPH Oxidase in Atherogenesis

Several lines of evidence suggest that NADPH oxidases play important roles in the development of atherosclerosis. First, lesion macrophage NADPH oxidase, together with 12/15 lipoxygenase and myeloperoxidase may contribute to ongoing oxidative modification of LDL, perpetuating the progression of the atherosclerotic lesion. Intracellular release of reactive oxygen species (ROS) from the NAD(P)H oxidase system may mediate the action of growth factors and cytokines, contributing to vascular hypertrophy and inflammation. Data from studies of animal models also showed consistent results. Rabbits fed on a long-term high-cholesterol diet show increased vascular O_2^- production and endothelial dysfunction. The potential role of NADPH oxidase in early atherogenesis is further supported by the observation that oscillatory shear stress, but not laminar shear stress, causes a sustained increase in NADPH oxidase–derived O_2^-. These findings may, in part, explain the propensity for atherosclerotic lesions to form at bifurcations. In addition to the flow perturbation, a number of growth factors and cytokines known to contribute to atherogenesis, including IL-6, angiotensin II, and TNF-α, have been shown to activate the NADPH oxdiases. In turn, the NADPH oxidase–derived O_2^- has been shown to activate the downstream intracellular signaling pathways including p38MAPK (mitogenesis) and Akt/PKB (cell survival kinase) proinflammatory markers including vascular adhesion molecules and MCP-1. Vascular smooth muscle cell proliferation and migration, two important features of atherogenesis, have also been shown to be upregulated by NADPH oxidase–derived O_2^- production.

The direct effect of glucose on NADPH oxidase activity has recently been investigated. By using cultured bovine aortic smooth muscle cells and endothelial cells, Inoguchi et al. demonstrated that incubation of the cells with a high concentration of glucose for 72 h resulted in a PKC-dependent increase in superoxide production. Park et al. recently reported that high glucose concentration (25 mM) directly upregulates endothelin-1 mRNA in bovine retinal endothelial cells in a PKC-β and -δ dependent manner. In human umbilical vascular endothelial cells (HUVEC), endothelin-1 increases the production of O_2^- through upregulation of gp91phox mRNA in a dose-dependent manner and the activation is mediated mainly by the type B endothelin receptor. We recently demonstrated that a 1-h exposure of the freshly isolated C57Bl/6 mouse aorta to 30 mM of glucose increased the aortic O_2^- production by 4 fold, as determined by the lucigenin-derived chemiluminescence method.

6.2.2 The Effect of Hyperglycemia on Thrombosis

It has long been recognized that diabetes mellitus is associated with a hypercoagulable state. Patients with diabetes are significantly more prone to die of a thrombotic event. In concordance with the observed increased thrombotic event rates in diabetic patients, clinical indicators of abnormal coagulation, clotting, fibrinolytic factors, and platelet functions have been studied extensively and have been comprehensively reviewed recently. Virtually all the parameters considered have been found to be abnormal in diabetic patients. On the other hand, the relative contribution of elevated blood glucose on these abnormalities is less well studied. Limited studies are available in exploring the effect of acute hyperglycemia on healthy volunteers. It has been reported that factor VII activity, factor VIII activity, and tissue factor pathway inhibitor (TFPI) can be induced acutely through glucose infusion. Shechter et al. reported a positive correlation between glucose level and platelet-dependent thrombosis in diabetic patients with coronary heart disease.

Elevated plasma levels of plasminogen activator inhibitor type-1 (PAI-1) is one of the established markers of the metabolic syndrome. In addition, increased plasma levels of PAI-1 have also been reported in a number of acute prothrombotic states, including myocardial infarction, deep vein thrombosis, and disseminated intravascular coagulation. PAI-1 is expressed in a variety of tissues and may mediate tissue-specific thrombosis. Obese human subjects also have an elevated level of PAI-1 and the main source of circulating PAI-1 is visceral adipose tissue. In addition, *PAI-1* is also expressed in vascular endothelial cells and smooth muscle cells, and is regulated by a number of external stimuli including insulin. In insulin-resistant states, both *in vivo* and *in vitro* experiments suggest that PAI-1 gene expression continues to be responsive to insulin. Thus, glucose transport and *PAI-1* gene expression are mediated by different pathways.

Recent studies continue to support a positive relationship between elevated glucose and PAI-1 expression. Gabriely et al. showed that hyperglycemia induces the gene expression of PAI-1 in adipose tissue and is mediated by the activation of the hexosamine pathway. Hyperglycemia has also been shown to induce *PAI-1* expression in vascular endothelial cells, which is at least mediated by the activation of the MAP kinase and PKC pathways.

6.2.3 The Effect of Hyperglycemia on Endothelial Dysfunction

The potential causative role of hyperglycemia in atherogenesis continues to draw a great deal of attention. In the UKPDS trial, Stratton et al. reported a positive association between the updated HgA1C and the incidence of the cardiovascular events. For a 1% reduction in the HgA1C, there was an associated 14% reduction in the incidence of myocardial infarction (8–21%, $p < 0.0001$). The underlying mechanism for this rather modest association remains incompletely understood. In addition to the indirect effect through increased PAI-1 levels, the potential impact of hyperglycemia on endothelial function has been extensively explored but will only be briefly reviewed here.

Endothelial dysfunction is a hallmark of atherosclerosis and may predate the appearance of atherosclerotic plaques. Several lines of evidence suggest that elevated levels of glucose directly impact on endothelial function. In a recent dog model study, Gross et al. showed that acute hyperglycemia-induced by intravenous infusion of D-glucose resulted in the development of endothelial dysfunction as assessed by coronary blood flow responses to acetylcholine, and this effect was mediated by an increased production of ROS. In a cultured endothelial cell model, exposure of bovine aortic endothelial cells to a high concentration of glucose led to a reduction in endothelial NO synthase mRNA expression and glycogen synthase activity. Elevated glucose resulted in reduction in the eNOS gene expression, glycogen synthetase activity, ERK 1,2 signaling, p38, Akt expression, and Cu/Zn superoxide-dismutase activity. Uncoupling of eNOS has also been described in diabetic subjects. Tetrahydrobiopterin (BH_4), a cofactor for endothelial nitric oxide synthase, becomes oxidized in the hyperglycemic milieu. As a result, instead of producing NO to react with O_2^- and yield $ONOO^-$, the uncoupled eNOS is capable of producing more O_2^-. Lowering the availability of BH4 in diabetics has been shown to be causative of inducing eNOS uncoupling. This reaction further results in increased production of O_2^- but a reduced availability of nitric oxide.

7 Genetic Approaches to New Genes – Candidate Genes and Susceptibility Genes

It has been well accepted that type 2 diabetes mellitus is a complex metabolic disease with strong genetic influences. Although monogenic causes of insulin resistance and diabetes such as MODY (maturity-onset diabetes of the young) have been identified, the overwhelming majority of the cases are polygenic in nature. Identification of such genes and their functional roles in the development of insulin resistance and diabetes may provide important insights for the identification of novel targets for therapeutic interventions. The candidate gene approach has been informative in identifying a number of monogenic causes of insulin resistance, albeit only accounting for a very small proportion of all type 2 diabetes. In addition, this approach also provided insights into the role of common genetic variants of a few additional genes.

7.1 Monogenic Syndromes of Insulin Resistance and Diabetes Mellitus

7.1.1 Insulin Receptor Mutations

Over 50 different mutations in the insulin receptor gene have been described. In addition to severe insulin resistance, these affected subjects are often found to have other physical abnormalities including abnormal facies, intrauterine growth retardation, acanthosis nigricans, and so on. The phenotype of these mutations is typically transmitted in an autosomal recessive fashion, but the obligate heterozygous parents and some of the other heterozygous relatives often develop milder forms of insulin resistance comparable to those seen in type 2 diabetic subjects. Functional characterizations of these naturally occurring mutations have provided important insights into the structure–function relationships in various domains of the receptor.

7.1.2 Insulin Gene Mutations

Rare mutations in the insulin gene (*INS*) can cause hyperinsulinemia and insulin resistance. Mechanisms of impaired signaling include defective binding to the insulin receptor and changes in proinsulin leading to inefficient cleavage of proinsulin. On the other hand, a number of *INS* mutations and polymorphisms have been detected with no associated susceptibility to diabetes.

7.1.3 Familial Partial Lipodystrophy (FPLD)

Familial partial lipodystrophy (FPLD) of the Dunnigan type (also called *lipoatrophy*), is a rare form of insulin resistance associated with a characteristic form of regional loss of subcutaneous fat tissue after puberty. Other clinical phenotypes include insulin resistance, hypertriglyceridemia, glucose intolerance/diabetes, and severe hypoleptinemia. The causative defective gene was recently identified as the *LMNA*, lamin A/C gene, and the syndrome is transmitted in an autosomal dominant fashion. Several mutations of the *LMNA* gene have been identified to cause FPLD. The *LMNA* gene codes for two splice variants lamin A and C that heterodimerize, and the gene products are nuclear envelope proteins. The mechanism by which *LMNA* gene mutations cause FPLD remains unknown. In light of the remarkable similarities in the metabolic abnormalities when compared to the adipose-specific *SREBP1* transgenic mice, a murine model of generalized lipoatrophy, it has been speculated that a physical interaction between SREBP and the nuclear envelope might be essential for insulin signaling. It is of interest to note that mutations in other regions of the same *LMNA* gene can cause Emery–Dreifuss muscular dystrophy and a cardiomyopathy. In addition to mutations in *LMNA* gene, two mutations in the *PPARγ* gene have also been reported to be associated with the FPLD syndrome. Glucose transporters GLUT1, GLUT2, and GLUT4 are encoded by the *SLC2A1, SLC2A2,* and *SLC2A4* genes respectively. The *SCL2A1* gene is expressed ubiquitously in a constitutive manner but is particularly abundant in the brain. Rare mutations of the *SLC2A1* gene cause seizure disorders due to ineffective transport of glucose across the blood–brain barrier. The *SCL2A2* gene is expressed primarily in the liver and pancreas. Rare mutations of this gene have been described with the phenotype to include impaired glucose uptake and glycogen accumulation in the liver and kidneys. *SCL2A4* encodes for GLUT4 in myocytes and adipocytes and plays a crucial role in glucose uptake by these tissues for utilization. Surprisingly, there are no single gene mutations or common polymorphisms that have been consistently demonstrated to influence diabetes susceptibility.

7.1.4 Insulin Receptor Substrates

Being located at the proximal arm of the insulin-signaling cascade, insulin receptor substrates genes (*IRS*) are natural candidates for insulin resistance and diabetes. A rare mutation in the *IRS1* gene has been detected in a subject with type 2 diabetes. On the other hand, common polymorphisms have been reported to be associated with type 2 diabetes in a number of population studies.

7.1.5 Maturity Onset Diabetes of the Young (MODY)

This is a relatively rare group of monogenic causes of early-onset type 2 diabetes (~2–5% of all type 2 diabetes cases) transmitted in an autosomal dominant manner. The affected are typically not obese and they develop diabetes as early as in childhood, but most by the third decade. All but one type of MODY, namely, MODY2, are caused by mutation of specific transcription factors and the typical primary metabolic defects are characterized by glucose-stimulated insulin secretion rather than insulin resistance. MODY2 results from mutations in the glucokinase (*GCK*) gene, encoding for glucose kinase in the glycolytic pathway. In pancreatic β-cells, a defective glycolytic pathway results in an impairment of signaling for glucose-induced insulin secretion. Genes identified to be responsible for the other MODY syndromes through either positional cloning or candidate gene approaches include a variety of transcriptional factors. Once again, defective β-cell insulin secretion is the common primary defect. Genes mutated to cause the other MODY types include hepatic nuclear factor 4α (*HNF4*α) gene in MODY1, hepatic nuclear factor 1α (*HNF1*α) gene for MODY3, insulin promoting factor 1 (*IPF1*) for MODY4, and transcription factor 2 gene (*TCF2*) for MODY5. The heterogeneity in the genes responsible for different MODY syndromes is reflected by the heterogeneity in clinical phenotypes.

Mutation in the *HNF1*α gene can cause MODY3 and represents the most commonly described MODY subtype. Hyperglycemia in MODY3 is typically severe and is frequently associated with microvascular complications. It is of interest to note that most mutations leading to MODY3 are dominant negative because mice heterozygous for the null mutation in the murine *HNF1*α gene have a normal phenotype. Data available to date suggest that only mutations located in the transactivation domain of HNF1α have a dominant negative effect.

MODY1 is caused by mutations in the transcription factor *HNF4*α, but are much less prevalent than MODY3. The mutations result in defective binding to the target DNA sequences. HNF4α is a member of the steroid/thyroid hormone receptor superfamily and an upstream regulator of HNF1α expression. Its activity is in turn modulated by a number of metabolic factors including long-chain fatty acids. The binding may result in either the activation or inhibition of HNF4α transcriptional activity as a function of chain length degree of saturation, and therefore, makes this transcription factor a potential therapeutic target for dietary fatty acid manipulation.

A deletion mutation in the transcription factor insulin promoter factor-1 (*IPF1*) gene is found to cosegregate with a large MODY kindred. Subjects heterozygous for the mutation develop diabetes of variable severity and one homozygote for the mutated gene developed pancreatic agenesis at birth and suffered both endocrine and exocrine pancreatic deficiency. The importance of *IPF-1* in pancreatic development was confirmed by the pancreatic agenesis phenotype seen in the embryo of the *ipf-1* knockout mice.

Transcription factor 2 (TCF2)/HNF1β mutations have been described in subjects with early-onset diabetes consistent with MODY. In addition, these subjects also developed severe nondiabetic kidney disease and Mullerian aplasia.

7.2 Common Genetic Variants of Other Candidate Genes

Common variants in a number of functional candidate genes including β2- and β3-adrenergic receptor genes, PPARγ, uncoupling protein 1 gene, and so on, have been studied extensively. Two such genes, β3-adrenergic receptor (*ADRB3*) and *PPARγ*, will be briefly reviewed here.

7.2.1 **ADRB3**

ADRB3 is a G-protein coupled receptor expressed largely in visceral adipose tissue. Activation by β-agonists results in cellular accumulation of cAMP that will enhance lipolysis and thermogenesis. A Trp64Arg in the *ADRB3* gene polymorphism was first discovered in Pima Indians, and homozygotes have an increased tendency to develop type 2 diabetes. A follow-up study in a Finn cohort found a positive association between this polymorphism and abdominal obesity and insulin resistance, generating further interest in exploring the potential clinical utility of this genetic variant. However, an increasing number of studies reporting negative associations dampened the pursuit of this candidate gene.

7.2.2 **PPARγ**

PPARγ gene encodes for two splice variants, PPARγ1 and PPARγ2. The former is expressed in a variety of tissues including muscle and adipose, but PPARγ2 is expressed virtually exclusively in adipose tissue. To date, two rare forms of dominant negative PPARγ mutations have been implicated as monogenic causes of insulin resistance and diabetes but they appear to cause lipodystrophy. Of the numerous common variants described for PARγ2, the Pro12Ala PPARγ polymorphism has received a great deal of attention. While the association between this polymorphism with diabetes and obesity has been variable, the association between this polymorphism and insulin resistance has been more consistent. Several populations of different ethnicity showed enhanced insulin sensitivity amongst those carrying the Ala12 allele.

7.3 Genome Scanning and Positional Cloning Strategies

The genome-wide linkage approach permits a systematic search throughout the genome for susceptibility genes in complex diseases such as diabetes without any prior knowledge of the functions of the genes. This technique complements the candidate gene approach because it is increasingly clear that type 2 diabetes susceptibility genes are likely to encode for proteins of no obvious functional links to current established pathways.

Many genome scans have been reported and the results summarized in a number of recent reviews.

The *NIDDM1* locus on chromosome 2q was identified in Mexican Americans. Mahtani et al. reported a linkage at a locus near *MODY3* on chromosome 12 in Finnish families with the major phenotype being an insulin secretion defect. The identification of the calpain 10 (*CAPN10*) gene, which encodes the intracellular nonlysosomal calcium-activated cysteine protease calpain-10 as the gene involved on the *NIDDM1* locus, is a successful example of the positional cloning approach. The identification of calpain-10 holds promise in discovering novel biochemical pathways in glucose homeostasis and insulin resistance.

By using both positional cloning and a candidate gene approach, Hegele et al. have identified, in the Oj-Cree of Sandy Lake in Northern Ontario, a novel mutation in the *HNF1A* gene (*HNF1A Gly319Ser*), which is strongly linked to early-onset type 2 diabetes. The homozygotes and heterozygotes for this mutation are expected to have odd ratios of 4 and 2 in the development of type 2 diabetes, respectively. It is intriguing to observe that despite the same gene as in MODY3 being mutated, the clinical features of subjects with this *HNF1A Gly319Ser* mutation bear little resemblance to the MODY3 phenotypes.

8
Summary and Conclusions

Type 2 diabetes mellitus is an emerging epidemic worldwide, with the incidence rising in parallel with a continuing rise in the prevalence of obesity. With only a handful of exceptions, type 2 diabetes is a polygenic disorder with the clinical phenotype of insulin resistance and frank diabetes brought on by environmental factors. Treatment modalities have originally been focusing on the normalization of hyperglycemia by lifestyle modification and antidiabetic pharmacological agents, singly or in combination. More recently, rigorous lifestyle measures as well as a number of glucose lowering agents have been found to be effective in the prevention of type 2 diabetes. However, much remains to be done in our understanding of the known players in the pathogenesis of insulin resistance, β-cell dysfunction, and diabetes at the cellular and molecular levels. To significantly advance our understanding of this disorder, a multidisciplinary approach is essential.

In the area of insulin signaling, new players continue to be discovered and our understanding of the critical steps in signaling and the interactions between them and external environmental factors will prove to be informative for new therapeutic targets. The field of cross talk between specific signaling pathways and their interactions with cell membrane microdomains represent an exciting area of future research. A systematic dissection of the signaling pathways by the generation of knockout and transgenic mice will also offer novel insights into the functional roles of these individual players at the whole mammalian organism level. The genome projects will soon make available an enormous amount of DNA sequence data from various species, including humans, and will facilitate discovery of novel genes involved in insulin resistance and diabetes. Our evolving view of the adipose tissue as not just a passive fat depot but rather as being a dynamic and active secretory organ, together with a better understanding of biological functions and how they interact with external modulating factors, will be critical for us in developing effective strategies to combat increasing adiposity and their metabolic and functional sequelae. The biological actions of adiponectin and its interactions with other adipocyte-derived hormones and cytokines appear to be an area critical to our understanding of how obesity may contribute to insulin resistance. In addition, partitioning of energy as mediated by distribution and redistribution of body fat may hold the key for our future understanding of links between obesity, insulin resistance, and diabetes and for better design of therapeutics. The discovery of glucagon-like peptides (GLPs) and their respective receptors holds great promise to effectively control hyperglycemia and to reestablish the β-cell proliferation/apoptosis balance with the ultimate goal of preserving β-cell mass.

Uncontrolled hyperglycemia continues to be the main threat for the development of microvascular disease and our currently available therapeutics frequently fail to completely normalize glycemia or to maintain control. Tracking the pathophysiology of high glucose–induced long-term diabetic microvascular complications to the function at subcellular organelle function will facilitate better preventive measures. Likewise, the role of hyperglycemia in the development of macrovascular complications continues to stimulate great interest. Current understanding of the role of elevated blood glucose, especially postprandial hyperglycemia on arterial wall function, including endothelial function, oxidative stress, and the interaction with lipoproteins, may lead to the identification of novel targets for atherogenesis in diabetes.

Acknowledgment

The author wishes to thank Dr. David Jenkins for his contribution and his critical review of the manuscript. D. Ng is a recipient of a Canadian Institute for Health Research New Investigator Career Scholarship.

See also Adipocytes; Diabetes Insipidus, Molecular Biology of; Immunology; Innate Immunity.

Bibliography

Books and Reviews

Adeli, K., Taghibiglou, C., Van Iderstine, S.C., Lewis, G.F. (2002) Mechanisms of hepatic very low-density lipoprotein overproduction in insulin resistance, *Trends Cardiovasc. Med.* **11**, 170–176.

Brownlee, M. (2002) Biochemistry and molecular cell biology of diabetic complications, *Nature* **414**, 813–820.

Busch, C.P., Hegele, R.A. (2001) Genetic determinants of type 2 diabetes mellitus, *Clin. Genet.* **60**, 243–254.

Chisolm, G.M., Steinberg, D. (2000) The oxidative modification hypothesis of atherogenesis: an overview, *Free Radical Biol. Med.* **28**, 1815–1826.

Friedman, J., Halaas, J.L. (1998) Leptin and the regulation of body weight in mammals, *Nature* **395**, 763–770.

Horton, J.D., Goldstein, J.L., Brown, M.S. (2002) SERBPs: activators of the complete program of cholesterol and fatty acid synthesis in the liver, *J. Clin. Invest.* **109**, 1125–1131.

Kahn, B.B., Flier, J.S. (2000) Obesity and insulin resistance, *J. Clin. Invest.* **106**, 473–481.

Kahn, A.H., Pessin, J.E. (2002) Insulin regulation of glucose uptake: a complex interplay of intracellular signalling pathways, *Diabetologia* **45**, 1475–1483.

Mauvais-Jarvis, F., Kulkarni, R.N., Kahn, C.R. (2002) Knockout models are useful tools to dissect the pathophysiology and genetics of insulin resistance, *Clin. Endocrinol.* **57**, 1–9.

Sheetz, M.J., King, G.L. (2002) Molecular understanding of hyperglycemia's adverse effects for diabetic complications, *JAMA* **288**, 2579–2588.

Sorescu, D., Szocs, K., Griendling, K.K. (2001) NAD(P)H oxidases and their relevance ot atherosclerosis, *Trends Cardiovasc. Med.* **11**, 124–131.

Unger, R.H., Orci, L. (2001) Diseases of liporegulation: new perspective on obesity and related disorders, *FASEB J.* **15**, 312–321.

Primary Literature

Accili, D., Drago, J., Lee, E.J., Johnson, M.D., Cool, M.H., Salvatore, P., Asico, L.D., Jose, P.A., Taylor, S.I., Westphal, H. (1996) Early neonatal death in mice homozygous for a null allele of the insulin receptor gene, *Nat. Genet.* **12**, 106–109.

Agarwal, A.K., Garg, A. (2002) A novel heterozygous mutation in peroxisome proliferator-activated receptor-gamma gene in a patient with familial partial lipodystrophy, *J. Clin. Endocrinol. Metab.* **87**, 408–411.

Aguilar-Bryan, L., Bryan, J. (1999) Molecular biology of adenosine triphosphate-sensitive potassium channels, *Endocr. Rev.* **20**, 101–135.

Ahima, R.S., Dushay, J., Flier, S.N., Prabakaran, D., Flier, J.S. (1997) Leptin accelerates the onset of puberty in normal female mice, *J. Clin. Invest.* **99**, 391–395.

Amer, P. (1995) Differences in lipolysis between human subcutaneous and omental adipose tissues, *Ann. Med.* **27**, 435–438.

Andreasson, K., Galuska, D., Thorne, A., Sonnenfeld, T., Wallberg-Henriksson, H. (1991) Decreased insulin-stimulated 3-0-methylglucose transport in in vitro incubated muscle strips from type II diabetic subjects, *Acta Physiol. Scand.* **142**, 255–260.

Azevedo, J.L. Jr., Carey, J.O., Pories, W.J., Morris, P.G., Dohm, G.L. (1995) Hypoxia stimulates glucose transport in insulin-resistant human skeletal muscle, *Diabetes* **44**, 695–698.

Babior, B.M. (1999) NADPH oxidase: an update, *Blood* **93**, 1464–1498.

Balon, T.W., Nadler, J.L. (1996) Nitric oxide mediates skeletal glucose transport, *Am. J. Physiol.* **270**(6 Pt 1), E1058–E1059.

Balon, T.W., Nadler, J.L. (1997) Evidence that nitric oxide increases glucose transport in skeletal muscle, *J. Appl. Physiol.* **82**, 359–363.

Barroso, I., Gurnell, M., Crowley, V.E., Agostini, M., Schwabe, J.W., Soos, M.A., Maslen, G.L., Williams, T.D., Lewis, H., Schafer, A.J., Chatterjee, V.K., O'Rahilly, S. (1999) Dominant negative mutations in human PPARgamma associated with severe insulin resistance, diabetes mellitus and hypertension, *Nature* **402**(6764), 880–883.

Berg, A.H., Combs, T.P., Du, X., Brownlee, M., Scherer, P.E. (2001) The adipocyte-secreted protein Acrp30 enhances hepatic insulin action, *Nat. Med.* **7**, 947–953.

Bjorntorp, P. (1990) "Portal" adipose tissue as a generator of risk factors for cardiovascular disease and diabetes, *Arteriosclerosis* **10**, 493–496.

Bluher, M., Kahn, B.B., Kahn, C.R. (2003) Extended longevity in mice lacking the insulin receptor in adipose tissue, *Science* **299**(5606), 572–574.

Bluher, M., Michael, M.D., Peroni, O.D., Ueki, K., Carter, N., Kahn, B.B., Kahn, C.R. (2002) Adipose tissue selective insulin receptor knockout protects against obesity and obesity-related glucose intolerance, *Dev. Cell* **3**, 25–38.

Boden, G., Cheung, P., Stein, T.P., Kresge, K., Mozzoli, M. (2002) FFA cause hepatic insulin resistance by inhibiting insulin suppression of glycogenolysis, *Am. J. Physiol. Endocrinol. Metab.* **283**, E12–E19.

Bonadonna, R.C., Del Prato, S., Bonora, E., Saccomani, M.P., Gulli, G., Natali, A., Frascerra, S., Pecori, N., Ferrannini, E., Bier, D., Cobelli, C., DeFronzo, R.A. (1996) Roles of glucose transport and glucose phosphorylation in muscle insulin resistance of NIDDM, *Diabetes* **45**, 915–925.

Brozinick, J.T. Jr., Etgen, G.J. Jr., Yaspelkis, B.B. III, Ivy, J.L. (1992) Contraction-activated glucose uptake is normal in insulin-resistant muscle of the obese Zucker rat, *J. Appl. Physiol.* **73**, 382–387.

Brozinick, J.T. Jr., Etgen, G.J. Jr., Yaspelkis, B.B. III, Kang, H.Y., Ivy, J.L. (1993) Effects of exercise training on muscle GLUT-4 protein content and translocation in obese Zucker rats, *Am. J. Physiol.* **265**(3 Pt 1), E419–E427.

Brun, R.P., Tontonoz, P., Forman, B.M., Ellis, R., Chen, J., Evans, R.M., Spiegelman, B.M. (1996) Differential activation of adipogenesis by multiple PPAR isoforms, *Genes Dev.* **10**, 974–984.

Bruning, J.C., Michael, M.D., Winnay, J.N., Hayashi, T., Horsch, D., Accili, D., Goodyear, L.J., Kahn, C.R. (1998) A muscle-specific insulin receptor knockout exhibits features of the metabolic syndrome of NIDDM without altering glucose tolerance, *Mol. Cell* **2**, 559–569.

Bruning, J.C., Winnay, J., Bonner-Weir, S., Taylor, S.I., Accili, D., Kahn, C.R. (1997) Development of a novel polygenic model of NIDDM in mice heterozygous for IR and IRS-1 null alleles, *Cell* **88**, 561–572.

Butler, A.E., Janson, J., Bonner-Weir, S., Ritzel, R., Rizza, R.A., Butler, P.C. (2003) Beta-cell deficit and increased beta-cell apoptosis in humans with type 2 diabetes, *Diabetes* **52**, 102–110.

Calles-Escandon, J., Garcia-Rubi, E., Mirza, S., Mortensen, A. (1999) Type 2 diabetes: one disease, multiple cardiovascular risk factors, *Coron. Artery Dis.* **10**, 23–30.

Carayannopoulos, M.O., Chi, M.M., Cui, Y., Pingsterhaus, J.M., McKnight, R.A., Mueckler, M., Devaskar, S.U., Moley, K.H. (2000) GLUT8 is a glucose transporter responsible for insulin-stimulated glucose uptake in the blastocyst, *Proc. Natl. Acad. Sci. U.S.A.* **97**, 7313–7318.

Carr, M.E. (2001) Diabetes mellitus: A hypercoagulable state, *J. Diabetes Complications* **15**, 44–54.

Carpentier, A., Mittelman, S.D., Lamarche, B., Bergman, R.N., Giacca, A., Lewis, G.F. (1999) Acute enhancement of insulin secretion by FFA in humans is lost with prolonged FFA elevation, *Am. J. Physiol. Endocrinol. Metab.* **276**, E1055–E1066.

Casanueva, F.F., Dieguez, C. (1999) Neuroendocrine regulation and actions of leptin, *Front. Neuroendocrinol.* **20**, 317–363.

Cerielo, A., Motz, E., Cavarape, A., Lizzio, S., Russo, A., Quatraro, A., Giugliano, D. (1997) Hyperglycemia counterbalances the antihypertensive effect of glutathione in diabetic patients: evidence linking hypertension and glycemia through the oxidative stress in diabetes mellitus, *J. Diabetes Complications* **11**, 250–255.

Chehab, F.F., Qiu, J., Mounzih, K., Ewart-Toland, A., Ogus, S. (2002) Leptin and reproduction, *Nutr. Rev.* **60**(10 Pt 2), S39–S46.

Cheng, A., Uetani, N., Simoncic, P.D., Chaubey, V.P., Lee-Loy, A., McGlade, C.J., Kennedy, B.P., Tremblay, M.L. (2002) Attenuation of leptin action and regulation of obesity by protein tyrosine phosphatase 1B, *Dev. Cell* **2**, 497–503.

Chiang, S.H., Baumann, C.A., Kanzaki, M., Thurmond, D.C., Watson, R.T., Neudauer, C.L., Macara, I.G., Pessin, J.E., Saltiel, A.R. (2001) Insulin-stimulated GLUT4 translocation requires the CAP-dependent activation of TC10, *Nature* **410**(6831), 944–948.

Chiasson, J.L., Josse, R.G., Gomis, R., Hanefeld, M., Karasik, A., Laakso, M. (2002) Acarbose for prevention of type 2 diabetes mellitus: the STOP-NIDDM randomised trial, *Lancet* **359**, 2072–2077.

Cline, G.W., Petersen, K.F., Krssak, M., Shen, J., Hundal, R.S., Trajanoski, Z., Inzucchi, S., Dresner, A., Rothman, D.L., Shulman, G.I. (1999) Impaired glucose transport as a cause of decreased insulin-stimulated muscle glycogen synthesis in type 2 diabetes, *N. Engl. J. Med.* **341**, 240–246.

Cnop, C.M., Hannaert, J.C., Hoorens, A., Eizirik, D.L., Pipeleers, D.G. (2001) Inverse relationship between cytotoxicity of free fatty acids in pancreatic islet cells and cellular triglyceride accumulation, *Diabetes* **50**, 1771–1777.

Colombo, C., Cutson, J.J., Yamauchi, T., Vinson, C., Kadowaki, T., Gavrilova, O., Reitman, M.L. (2002) Transplantation of adipose tissue lacking leptin is unable to reverse the metabolic abnormalities associated with lipoatrophy, *Diabetes* **51**, 2727–2733.

Craven, P.A., Studer, R.K., DeRubertis, F.R. (1994) Impaired nitric oxide-dependent cyclic guanosine monophosphate generation in glomeruli from diabetic rats. Evidence for protein kinase C-mediated suppression of the cholinergic response, *J. Clin. Invest.* **93**, 311–320.

Czech, M.P., Corvera, S. (1999) Signaling mechanisms that regulate glucose transport, *J. Biol. Chem.* **274**, 1865–1868.

De Keulenaer, G.W., Chappell, D.C., Ishizaka, N., Nerem, R.M., Alexander, R.W., Griendling, K.K. (1998) Oscillatory and steady laminar shear stress differentially affect human endothelial redox state: role of a superoxide-producing NADH oxidase, *Circ. Res.* **82**, 1094–1101.

de Souza, C.J., Eckhardt, M., Gagen, K., Dong, M., Chen, W., Laurent, D., Burkey, B.F. (2001) Effects of pioglitazone on adipose tissue remodeling within the setting of obesity and insulin resistance, *Diabetes* **50**, 1863–1871.

de Souza, C.J., Hirshman, M.F., Horton, E.S. (1997) CL-316,243, a beta3-specific adrenoceptor agonist, enhances insulin-stimulated glucose disposal in nonobese rats, *Diabetes* **46**, 1257–1263.

Deeb, S.S., Fajas, L., Nemoto, M., Pihlajamaki, J., Mykkanen, L., Kuusisto, J., Laakso, M., Fujimoto, W., Auwerx, J. (1998) A Pro12Ala substitution in PPARgamma2 associated with decreased receptor activity, lower body mass index and improved insulin sensitivity, *Nat. Genet.* **20**, 284–287.

DeFronzo, R.A., Gunnarsson, R., Bjorkman, O., Olsson, M., Wahren, J. (1985) Effects of insulin on peripheral and splanchnic glucose metabolism in noninsulin-dependent (type II) diabetes mellitus, *J. Clin. Invest.* **76**, 149–155.

DeFronzo, R.A., Jacot, E., Jequier, E., Maeder, E., Wahren, J., Felber, J.P. (1981) The effect of insulin on the disposal of intravenous glucose. Results from indirect calorimetry and hepatic and femoral venous catheterization, *Diabetes* **30**, 1000–1007.

Dela, F., Larsen, J.J., Mikines, K.J., Ploug, T., Petersen, L.N., Galbo, H. (1995) Insulin-stimulated muscle glucose clearance in patients with NIDDM. Effects of one-legged physical training, *Diabetes* **44**, 1010–1020.

Derave, W., Lund, S., Holman, G.D., Wojtaszewski, J., Pedersen, O., Richter, E.A. (1999) Contraction-stimulated muscle glucose transport and GLUT-4 surface content are dependent on glycogen content, *Am. J. Physiol.* **277**(6 Pt 1), E1103–E1110.

Despres, J.P. (2001) Health consequences of visceral obesity, *Ann. Med.* **33**, 534–541.

Diabetes Control and Complications Trial (DCCT), The Diabetes Control and Complications Trial Research Group. (1993) The effect of intensive treatment of diabetes on the development and progression of long-term complications in insulin-dependent diabetes mellitus., *N. Engl. J. Med.* **329**, 977–986.

Dobbins, R.L., Chester, M.W., Daniels, M.B., McGarry, J.D., Stein, D.T. (1998) Circulating fatty acids are essential for efficient glucose-stimulated insulin secretion after prolonged fasting in humans, *Diabetes* **47**(10), 1613–1618.

Doege, H., Bocianski, A., Scheepers, A., Axer, H., Eckel, J., Joost, H.G., Schurmann, A. (2001) Characterization of human glucose transporter (GLUT) 11 (encoded by SLC2A11), a novel sugar-transport facilitator specifically expressed in heart and skeletal muscle, *Biochem. J.* **359**(Pt 2), 443–449.

Dolan, P.L., Tapscott, E.B., Dorton, P.J., Dohm, G.L. (1993) Contractile activity restores insulin responsiveness in skeletal muscle of obese Zucker rats, *Biochem. J.* **289**(Pt 2), 423–426.

Dohm, G.L., Tapscott, E.B., Pories, W.J., Dabbs, D.J., Flickinger, E.G., Meelheim, D., Fushiki, T., Atkinson, S.M., Elton, C.W., Caro, J.F. (1988) An in vitro human muscle preparation suitable for metabolic studies. Decreased insulin stimulation of glucose transport in muscle from morbidly obese and diabetic subjects, *J. Clin. Invest.* **82**, 486–494.

Drucker, D.J. (2003) Glucagon-like peptides: regulators of cell proliferation, differentiation, and apoptosis, *Mol. Endocrinol.* **17**, 161–171.

Du, X.L., Edelstein, D., Dimmeler, S., Ju, Q., Sui, C., Brownlee, M. (2001) Hyperglycemia inhibits endothelial nitric oxide synthase activity by posttranslational modification at the Akt site, *J. Clin. Invest.* **108**, 1341–1348.

Du, X.L., Edelstein, D., Rossetti, L., Fantus, I.G., Goldberg, H., Ziyadeh, F., Wu, J., Brownlee, M. (2000) Hyperglycemia-induced mitochondrial superoxide overproduction activates the hexosamine pathway and induces plasminogen activator inhibitor-1 expression by increasing Sp1 glycosylation, *Proc. Natl. Acad. Sci. U.S.A.* **97**, 12222–12226.

Duerrschmidt, N., Wippich, N., Goettsch, W., Broemme, H.J., Morawietz, H. (2000) Endothelin-1 induces NAD(P)H oxidase in human endothelial cells, *Biochem. Biophys. Res. Commun.* **269**, 713–717.

Eitel, K., Staiger, H., Brendel, M.D., Brandhorst, D., Bretzel, R.G., Haring, H.U., Kellerer, M. (2002) Different role of saturated and unsaturated fatty acids in beta-cell apoptosis, *Biochem. Biophys. Res. Commun.* **299**, 853–856.

Eriksson, J., Koranyi, L., Bourey, R., Schalin-Jantti, C., Widen, E., Mueckler, M., Permutt, A.M., Groop, L.C. (1992) Insulin resistance in type 2 (non-insulin-dependent) diabetic patients and their relatives is not associated with a defect in the expression of the insulin-responsive glucose transporter (GLUT-4) gene in human skeletal muscle, *Diabetologia* **35**, 143–147.

Etgen, G.J. Jr., Jensen, J., Wilson, C.M., Hunt, D.G., Cushman, S.W., Ivy, J.L. (1997) Exercise training reverses insulin resistance in muscle by enhanced recruitment of GLUT-4 to the cell surface, *Am. J. Physiol.* **272**(5 Pt 1), E864–E869.

Farilla, L., Hui, H., Bertolotto, C., Kang, E., Bulotta, A., Di, Mario, U., Perfetti, R. (2002) Glucagon-like peptide-1 promotes islet cell growth and inhibits apoptosis in Zucker diabetic rats, *Endocrinology* **43**, 4397–4408.

Fisher, J.S., Gao, J., Han, D.H., Holloszy, J.O., Nolte, L.A. (2002) Activation of AMP kinase enhances sensitivity of muscle glucose transport to insulin, *Am. J. Physiol. Endocrinol. Metab.* **282**, E18–E23.

Foretz, M., Guichard, C., Ferre, P., Foufelle, F. (1999) Sterol regulatory element binding protein-1c is a major mediator of insulin action on the hepatic expression of glucokinase and lipogenesis-related genes, *Proc. Natl. Acad. Sci. U.S.A.* **96**, 12737–12742.

Froguel, P., Velho, G. (2001) Genetic determinants of type 2 diabetes, *Recent Prog. Horm. Res.* **56**, 91–105.

Fruebis, J., Tsao, T.S., Javorschi, S., Ebbets-Reed, D., Erickson, M.R., Yen, F.T., Bihain, B.E., Lodish, H.F. (2001) Proteolytic cleavage product of 30-kDa adipocyte complement-related protein increases fatty acid oxidation in muscle and causes weight loss in mice, *Proc. Natl. Acad. Sci. U.S.A.* **98**, 2005–2010.

Furuta, H., Furuta, M., Sanke, T., Ekawa, K., Hanabusa, T., Nishi, M., Sasaki, H., Nanjo, K. (2002) Nonsense and missense mutations in the human hepatocyte nuclear factor-1 beta gene (TCF2) and their relation to type 2 diabetes in Japanese, *J. Clin. Endocrinol. Metab.* **87**, 3859–3863.

Gabriely, I., Yang, X.M., Cases, J.A., Ma, X.H., Rossetti, L., Barzilai, N. (2002) Hyperglycemia induces PAI-1 gene expression in adipose tissue by activation of the hexosamine biosynthetic pathway, *Atherosclerosis* **160**, 115–122.

Goldberg, H.J., Scholey, J., Fantus, I.G. (2000) Glucosamine activates the plasminogen activator inhibitor 1 gene promoter through Sp1 DNA binding sites in glomerular mesangial cells, *Diabetes* **49**, 863–871.

Gagnon, J., Mauriege, P., Roy, S., Sjostrom, D., Chagnon, Y.C., Dionne, F.T., Oppert, J.M., Perusse, L., Sjostrom, L., Bouchard, C. (1996) The Trp64Arg mutation of the beta3 adrenergic receptor gene has no effect on obesity phenotypes in the Quebec Family Study and Swedish Obese Subjects cohorts, *J. Clin. Invest.* **98**, 2086–2093.

Gautier, J.F., Mourier, A., de Kerviler, E., Tarentola, A., Bigard, A.X., Villette, J.M., Guezennec, C.Y., Cathelineau, G. (1998) Evaluation of abdominal fat distribution in noninsulin-dependent diabetes mellitus: relationship to insulin resistance, *J. Clin. Endocrinol. Metab.* **83**, 1306–1311.

Gavrilova, O., Marcus-Samuels, B., Graham, D., Kim, J.K., Shulman, G.I., Castle, A.L., Vinson, C., Eckhaus, M., Reitman, M.L. (2000) Surgical implantation of adipose tissue reverses diabetes in lipoatrophic mice, *J. Clin. Invest.* **105**, 271–278.

Ghosh, S., Langefeld, C.D., Ally, D., Watanabe, R.M., Hauser, E.R., Magnuson, V.L., Nylund, S.J., Valle, T., Eriksson, J., Bergman, R.N., Tuomilehto, J., Collins, F.S., Boehnke, M. (1999) The W64R variant of the beta3-adrenergic receptor is not associated with type II diabetes or obesity in a large Finnish sample, *Diabetologia* **42**, 238–244.

Ghilardi, N., Ziegler, S., Wiestner, A., Stoffel, R., Heim, M.H., Skoda, R.C. (1996) Defective STAT signaling by the leptin receptor in diabetic mice, *Proc. Natl. Acad. Sci. U.S.A.* **93**, 6231–6235.

Glogowski, E.A., Tsiani, E., Zhou, X., Fantus, I.G., Whiteside, C. (1999) High glucose alters the response of mesangial cell protein kinase C isoforms to endothelin-1, *Kidney Int.* **55**, 486–499.

Gnudi, L., Shepherd, P.R., Kahn, B.B. (1996) Over-expression of GLUT4 selectively in adipose tissue in transgenic mice: implications for nutrient partitioning, *Proc. Nutr. Soc.* **55**(1B), 191–199.

Goodyear, L.J., Giorgino, F., Sherman, L.A., Carey, J., Smith, R.J., Dohm, G.L. (1995) Insulin receptor phosphorylation, insulin receptor substrate-1 phosphorylation, and phosphatidylinositol 3-kinase activity are decreased in intact skeletal muscle strips from obese subjects, *J. Clin. Invest.* **95**, 2195–2204.

Gorlach, A., Brandes, R.P., Nguyen, K., Amidi, M., Dehghani, F., Busse, R. (2000) A gp91phox containing NADPH oxidase selectively expressed in endothelial cells is a major source of oxygen radical generation in the arterial wall, *Circ. Res.* **87**, 26–32.

Gould, G.W., Holman, G.D. (1993) The glucose transporter family: structure, function and tissue-specific expression, *Biochem. J.* **295**, 329–341.

Griendling, K.K., Sorescu, D., Ushio-Fukai, M. (2000a) NAD(P)H oxidase: role in cardiovascular biology and disease, *Circ. Res.* **86**, 494–501.

Griendling, K.K., Sorescu, D., Lassegue, B., Ushio-Fukai, M. (2000b) Modulation of protein kinase activity and gene expression by reactive oxygen species and their role in vascular physiology and pathophysiology, *Arterioscler. Thromb. Vasc. Biol.* **20**, 2175–2183.

Gross, E.R., LaDisa, J.F., Weihrauch, D., Olson, L.E., Kress, T.T., Hettrick, D.A., Pagel, P.S., Warltier, D.C., Kersten, J.R. (2003) Reactive oxygen species modulate coronary wall shear stress and endothelial function during hyperglycemia, *Am. J. Physiol. Heart Circ. Physiol.* **284**, 1552–1559 Jan 23 [epub ahead of print].

Guilherme, A., Czech, M.P. (1998) Stimulation of IRS-1-associated phosphatidylinositol 3-kinase and Akt/protein kinase B but not glucose transport by beta1-integrin signaling in rat adipocytes, *J. Biol. Chem.* **273**, 33119–33122.

Hanis, C.L., Boerwinkle, E., Chakraborty, R., Ellsworth, D.L., Concannon, P., Stirling, B., Morrison, V.A., Wapelhorst, B., Spielman, R.S., Gogolin-Ewens, K.J., Shepard, J.M., et al. (1996) A genome-wide search for human non-insulin-dependent (type 2) diabetes genes reveals a major susceptibility locus on chromosome 2, *Nat. Genet.* **13**, 161–166.

Hara, K., Okada, T., Tobe, K., Yasuda, K., Mori, Y., Kadowaki, H., Hagura, R., Akanuma, Y., Kimura, S., Ito, C., Kadowaki, T. (2000) The Pro12Ala polymorphism in PPAR gamma2 may confer resistance to type 2 diabetes, *Biochem. Biophys. Res. Commun.* **271**, 212–216.

Heart Protection Study Collaborative Group. (2002) MRC/BHF heart protection study of antioxidant vitamin supplementation in 20,536 high-risk individuals: a randomised placebo-controlled trial, *Lancet* **360**, 23–33.

Hegele, R.A. (1995) Dyslipidaemia and coronary heart disease: nature vs nurture, *Br. J. Hosp. Med.* **54**, 143–146.

Hegele, R.A., Cao, H., Anderson, C.M., Hramiak, I.M. (2000) Heterogeneity of nuclear lamin A mutations in Dunnigan-type familial partial lipodystrophy, *J. Clin. Endocrinol. Metab.* **85**, 3431–3435.

Hegele, R.A., Cao, H., Frankowski, C., Mathews, S.T., Leff, T. (2002) PPARG F388L, a transactivation-deficient mutant, in familial partial lipodystrophy, *Diabetes* **51**, 3586–3590.

Hegele, R.A., Cao, H., Harris, S.B., Hanley, A.J., Zinman, B. (1999) The hepatic nuclear factor-1alpha G319S variant is associated with early-onset type 2 diabetes in Canadian Oji-Cree, *J. Clin. Endocrinol. Metab.* **84**, 1077–1082.

Henriksen, E.J. (2002) Effects of acute exercise and exercise training on insulin resistance, *J. Appl. Physiol.* **93**, 788–796.

Herman, W.H., Fajans, S.S., Ortiz, F.J., Smith, M.J., Sturis, J., Bell, G.I., Polonsky, K.S., Halter, J.B. (1994) Abnormal insulin secretion, not insulin resistance, is the genetic or primary defect of MODY in the RW pedigree, *Diabetes* **43**, 40–46.

Hertz, R., Magenheim, J., Berman, I., Bar-Tana, J. (1998) Fatty acyl-CoA thioesters are ligands of hepatic nuclear factor-4alpha, *Nature* **392**(6675), 512–516.

Holloszy, J.O., Kohrt, W.M., Hansen, P.A. (1998) The regulation of carbohydrate and fat metabolism during and after exercise, *Front. Biosci.* **3**, D1011–D1027.

Horikawa, Y., Iwasaki, N., Hara, M., Furuta, H., Hinokio, Y., Cockburn, B.N., Lindner, T., Yamagata, K., Ogata, M., Tomonaga, O., Kuroki, H., Kasahara, T., Iwamoto, Y., Bell, G.I. (1997) Mutation in hepatocyte nuclear factor-1 beta gene (TCF2) associated with MODY, *Nat. Genet.* **17**, 384–385.

Horikawa, Y., Oda, N., Cox, N.J., Li, X., Orho-Melander, M., Hara, M., Hinokio, Y., Lindner, T.H., Mashima, H., Schwarz, P.E., del Bosque-Plata, L., Horikawa, Y., Oda, Y., Yoshiuchi, I., Colilla, S., Polonsky, K.S., Wei, S., Concannon, P., Iwasaki, N., Schulze, J., Baier, L.J., Bogardus, C., Groop, L., Boerwinkle, E., Hanis, C.L., Bell, G.I. (2000) Genetic variation in the gene encoding calpain-10 is associated with type 2 diabetes mellitus, *Nat. Genet.* **26**, 163–175.

Horton, J.D., Bashmakov, Y., Shimomura, I., Shimano, H. (1998) Regulation of sterol regulatory element binding proteins in livers of fasted and refed mice, *Proc. Natl. Acad.Sci. U.S.A.* **95**, 5987–5992.

Hu, E., Tontonoz, P., Spiegelman, B.M. (1995) Transdifferentiation of myoblasts by the adipogenic transcription factors PPARγ and C/EBPα, *Proc. Natl. Acad. Sci.* **92**, 9856–9860.

Hughes, V.A., Fiatarone, M.A., Fielding, R.A., Kahn, B.B., Ferrara, C.M., Shepherd, P., Fisher, E.C., Wolfe, R.R., Elahi, D., Evans, W.J. (1993) Exercise increases muscle GLUT-4 levels and insulin action in subjects with impaired glucose tolerance, *Am. J. Physiol.* **264**(6 Pt 1), E855–E862.

Ido, Y., Carling, D., Ruderman, N. (2002) Hyperglycemia-induced apoptosis in human umbilical vein endothelial cells: inhibition by the AMP-activated protein kinase activation, *Diabetes* **51**, 159–167.

Inoguchi, T., Li, P., Umeda, F., Yu, H.Y., Kakimoto, M., Imamura, M., Aoki, T., Etoh, T., Hashimoto, T., Naruse, M., Sano, H., Utsumi, H., Nawata, H. (2000) High glucose level and free fatty acid stimulate reactive oxygen species production through protein kinase C–dependent activation of NAD(P)H oxidase in cultured vascular cells, *Diabetes* **49**, 1939–1945.

Inoue, N., Ramasamy, S., Fukai, T., Nerem, R.M., Harrison, D.G. (1996) Shear stress modulates expression of Cu/Zn superoxide dismutase in human aortic endothelial cells, *Circ. Res.* **79**, 32–37.

Isakoff, S.J., Taha, C., Rose, E., Marcusohn, J., Klip, A. (1995) The inability of phosphatidylinositol 3-kinase activation to stimulate GLUT4 translocation indicates additional signaling pathways are required for insulin-stimulated glucose uptake, *Proc. Natl. Acad. Sci. U.S.A.* **92**, 10247–10251.

Ishii, H., Jirousek, M.R., Koya, D., Takagi, C., Xia, P., Clermont, A., Bursell, S.E., Kern, T.S., Ballas, L.M., Heath, W.F., Stramm, L.E., Feener, E.P., King, G.L. (1996) Amelioration of vascular dysfunctions in diabetic rats by an oral PKC beta inhibitor, *Science* **272**(5262), 728–731.

Isomaa, B., Henricsson, M., Lehto, M., Forsblom, C., Karanko, S., Sarelin, L., Haggblom, M., Groop, L. (1998) Chronic diabetic complications in patients with MODY3 diabetes, *Diabetologia* **41**, 467–473.

Koch, M., Rett, K., Maerker, E., Volk, A., Haist, K., Deninger, M., Renn, W., Haring, H.U. (1999) The PPARgamma2 amino acid polymorphism Pro 12 Ala is prevalent in offspring of Type II diabetic patients and is associated to increased insulin sensitivity in a subgroup of obese subjects, *Diabetologia* **42**, 758–762.

Jacob, S., Stumvoll, M., Becker, R., Koch, M., Nielsen, M., Loblein, K., Maerker, E., Volk, A., Renn, W., Balletshofer, B., Machicao, F., Rett, K., Haring, H.U. (2000) The PPARgamma2 polymorphism pro12Ala is associated with better insulin sensitivity in the offspring of type 2 diabetic patients, *Horm. Metab. Res.* **32**, 413–416.

Jenkins, D.J., Thorne, M.J., Wolever, T.M., Jenkins, A.L., Rao, A.V., Thompson, L.U. (1987) The effect of starch-protein interaction in wheat on the glycemic response and rate of in vitro digestion, *Am. J. Clin. Nutr.* **45**, 946–951.

Jenkins, D.J., Wolever, T.M., Taylor, R.H., Ghafari, H., Jenkins, A.L., Barker, H., Jenkins, M.J. (1980) Rate of digestion of foods and postprandial glycaemia in normal and diabetic subjects, *Br. Med J.* **281**, 14–17.

Jiang, T., Sweeney, G., Rudolf, M.T., Klip, A., Traynor-Kaplan, A., Tsien, R.Y. (1998) Membrane-permeant esters of phosphatidylinositol 3,4,5-trisphosphate, *J. Biol. Chem.* **273**, 11017–11024.

Jonsson, J., Carlsson, L., Edlund, T., Edlund, H. (1994) Insulin-promoter-factor 1 is required for pancreas development in mice, *Nature* **371**(6498), 606–609.

Joshi, R.L., Lamothe, B., Cordonnier, N., Mesbah, K., Monthioux, E., Jami, J., Bucchini, D. (1996) Targeted disruption of the insulin receptor gene in the mouse results in neonatal lethality, *EMBO J.* **15**, 1542–1547.

Kadonaga, J.T., Courey, A.J., Ladika, J., Tjian, R. (1988) Distinct regions of Sp1 modulate DNA binding and transcriptional activation, *Science* **242**, 1566–1570.

Kadowaki, T., Hara, K., Kubota, N., Tobe, K., Terauchi, Y., Yamauchi, T., Eto, K., Kadowaki, H., Noda, M., Hagura, R., Akanuma, Y. (2002) The role of PPARgamma in high-fat diet-induced obesity and insulin resistance, *J. Diabetes Complications* **16**, 41–45.

Kaur, J., Singh, P., Sowers, J.R. (2002) Effects of glycemic control and other determinants on vascular disease in type 2 diabetes, *Am. J. Med.* **113**Suppl. 6A, 12S–22S.

Kawai, T., Takei, I., Oguma, Y., Ohashi, N., Tokui, M., Oguchi, S., Katsukawa, F., Hirose, H., Shimada, A., Watanabe, K., Saruta, T. (1999) Effects of troglitazone on fat distribution in the treatment of male type 2 diabetes, *Metabolism* **48**, 1102–1107.

Kawanaka, K., Han, D.H., Nolte, L.A., Hansen, P.A., Nakatani, A., Holloszy, J.O.

(1999) Decreased insulin-stimulated GLUT-4 translocation in glycogen-supercompensated muscles of exercised rats, *Am. J. Physiol.* **276**(5 Pt 1), E907–E912.

Kawanaka, K., Nolte, L.A., Han, D.H., Hansen, P.A., Holloszy, J.O. (2000) Mechanisms underlying impaired GLUT-4 translocation in glycogen-supercompensated muscles of exercised rats, *Am. J. Physiol. Endocrinol. Metab.* **279**, E1311–E1318.

Kennedy, J.W., Hirshman, M.F., Gervino, E.V., Ocel, J.V., Forse, R.A., Hoenig, S.J., Aronson, D., Goodyear, L.J., Horton, E.S. (1999) Acute exercise induces GLUT4 translocation in skeletal muscle of normal human subjects and subjects with type 2 diabetes, *Diabetes* **48**, 1192–1197.

Kim, J.K., Michael, M.D., Previs, S.F., Peroni, O.D., Mauvais-Jarvis, F., Neschen, S., Kahn, B.B., Kahn, C.R., Shulman, G.I. (2000) Redistribution of substrates to adipose tissue promotes obesity in mice with selective insulin resistance in muscle, *J. Clin. Invest.* **105**, 1791–1797.

Kim, Y.B., Nikoulina, S.E., Ciaraldi, T.P., Henry, R.R., Kahn, B.B. (1999) Normal insulin-dependent activation of Akt/protein kinase B, with diminished activation of phosphoinositide 3-kinase, in muscle in type 2 diabetes, *J. Clin. Invest.* **104**, 733–741.

Kim, J.B., Spiegelman, B.M. (1996) ADD1/SREBP1 promotes adipocyte differentiation and gene expression linked to fatty acid metabolism, *Genes Dev.* **10**, 1096–1107.

King, P.A., Betts, J.J., Horton, E.D., Horton, E.S. (1993) Exercise, unlike insulin, promotes glucose transporter translocation in obese Zucker rat muscle, *Am. J. Physiol.* **265**(2 Pt 2), R447–R452.

Kissebah, A.H., Peiris, A.N. (1989) Biology of regional body fat distribution: relationship to non-insulin-dependent diabetes mellitus, *Diabetes Metab. Rev* **5**, 83–109.

Kitamura, T., Ogawa, W., Sakaue, H., Hino, Y., Kuroda, S., Takata, M., Matsumoto, M., Maeda, T., Konishi, H., Kikkawa, U., Kasuga, M. (1998) Requirement for activation of the serine-threonine kinase Akt (protein kinase B) in insulin stimulation of protein synthesis but not of glucose transport, *Mol. Cell. Biol.* **18**, 3708–3717.

Knowler, W.C., Barrett-Connor, E., Fowler, S.E., Hamman, R.F., Lachin, J.M., Walker, E.A., Nathan, D.M. (2002) Diabetes Prevention Program Research Group. Reduction in the incidence of type 2 diabetes with lifestyle intervention or metformin, *N. Engl. J. Med.* **346**, 393–403.

Kobayashi, H., Nakamura, T., Miyaoka, K., Nishida, M., Funahashi, T., Yamashita, S., Matsuzawa, Y. (2001) Visceral fat accumulation contributes to insulin resistance, small-sized low-density lipoprotein, and progression of coronary artery disease in middle-aged non-obese Japanese men, *Jpn. Circ. J.* **65**, 193–199.

Korshunov, S.S., Skulachev, V.P., Starkov, A.A. (1997) High protonic potential actuates a mechanism of production of reactive oxygen species in mitochondria, *FEBS Lett.* **416**, 15–18.

Kotani, K., Ogawa, W., Matsumoto, M., Kitamura, T., Sakaue, H., Hino, Y., Miyake, K., Sano, W., Akimoto, K., Ohno, S., Kasuga, M. (1998) Requirement of atypical protein kinase clambda for insulin stimulation of glucose uptake but not for Akt activation in 3T3-L1 adipocytes, *Mol. Cell. Biol.* **18**, 6971–6982.

Kraegen, E.W., James, D.E., Storlien, L.H., Burleigh, K.M., Chisholm, D.J. (1986) In vivo insulin resistance in individual peripheral tissues of the high fat fed rat: assessment by euglycaemic clamp plus deoxyglucose administration, *Diabetologia* **29**, 192–198.

Krook, A., Roth, R.A., Jiang, X.J., Zierath, J.R., Wallberg-Henriksson, H. (1998) Insulin-stimulated Akt kinase activity is reduced in skeletal muscle from NIDDM subjects, *Diabetes* **47**, 1281–1286.

Krook, A., Whitehead, J.P., Dobson, S.P., Griffiths, M.R., Ouwens, M., Baker, C., Hayward, A.C., Sen, S.K., Maassen, J.A., Siddle, K., Tavare, J.M., O'Rahilly, S. (1997) Two naturally occurring insulin receptor tyrosine kinase domain mutants provide evidence that phosphoinositide 3-kinase activation alone is not sufficient for the mediation of insulin's metabolic and mitogenic effects, *J. Biol. Chem.* **272**, 30208–30214.

Krook, A., Bjornholm, M., Galuska, D., Jiang, X.J., Fahlman, R., Myers, M.G. Jr., Wallberg-Henriksson, H., Zierath, J.R. (2000) Characterization of signal transduction and glucose transport in skeletal muscle from type 2 diabetic patients, *Diabetes* **49**, 284–292.

Ku, D.N., Giddens, D.P., Zarins, C.K., Glagov, S. (1985) Pulsatile flow and atherosclerosis in the human carotid bifurcation. Positive correlation between plaque location and low oscillating shear stress, *Arteriosclerosis* **5**, 293–302.

Kulkarni, R.N., Bruning, J.C., Winnay, J.N., Postic, C., Magnuson, M.A., Kahn, C.R. (1999) Tissue-specific knockout of the insulin receptor in pancreatic beta cells creates an insulin secretory defect similar to that in type 2 diabetes, *Cell* **96**, 329–339.

Lam, T.K., Yoshii, H., Haber, C.A., Bogdanovic, E., Lam, L., Fantus, I.G., Giacca, A. (2002) Free fatty acid induced hepatic insulin resistance: a potential role for protein kinase C-delta, *Am. J. Physiol. Endocrinol. Metab.* **283**, E682–E691.

Lambeth, J.D., Cheng, G., Arnold, R.S., Edens, W.A. (2000) Novel homologs of gp91phox, *Trends Biochem. Sci.* **25**, 459–461.

Larsen, J.J., Dela, F., Kjaer, M., Galbo, H. (1997) The effect of moderate exercise on postprandial glucose homeostasis in NIDDM patients, *Diabetologia* **40**, 447–453.

Lebovitz, H.E. (2000) Pathogenesis of type 2 diabetes, *Drug Benefit Trends* **12**Suppl. A, 8–16.

Lee, Y., Hirose, H., Zhou, Y.T., Esser, V., McGarry, J.D., Unger, R.H. (1997) Increased lipogenic capacity of the islets of obese rats: a role in the pathogenesis of NIDDM, *Diabetes* **46**, 408–413.

Lefebvre, A.M., Laville, M., Vega, N., Riou, J.P., van Gaal, L., Auwerx, J., Vidal, H. (1998) Depot-specific differences in adipose tissue gene expression in lean and obese subjects, *Diabetes* **47**, 98–103.

Lewis, G.F. (1997) Fatty acid regulation of very low density lipoprotein production, *Curr. Opin. Lipidol.* **8**, 146–153.

Li, Y., Hansotia, T., Yusta, B., Ris, F., Halban, P.A., Drucker, D.J. (2003) Glucagon-like peptide-1 receptor signaling modulates beta cell apoptosis, *J. Biol. Chem.* **278**, 471–478.

Lisinski, I., Schurmann, A., Joost, H.G., Cushman, S.W., Al-Hasani, H. (2001) Targeting of GLUT6 (formerly GLUT9) and GLUT8 in rat adipose cells, *Biochem. J.* **358**, 517–522.

Listenberger, L.L., Ory, D.S., Schaffer, J.E. (2001) Palmitate-induced apoptosis can occur through a ceramide-independent pathway, *J. Biol. Chem.* **276**, 14890–14895.

Liu, K., Paterson, A.J., Chin, E., Kudlow, J.E. (2000) Glucose stimulates protein modification by O-linked GlcNAc in pancreatic beta cells: linkage of O-linked GlcNAc to beta cell death, *Proc. Natl. Acad. Sci. U.S.A.* **97**, 2820–2825.

Lord, G.M., Matarese, G., Howard, J.K., Baker, R.J., Bloom, S.R., Lechler, R.I. (1998) Leptin modulates the T-cell immune response and reverses starvation-induced immunosuppression, *Nature* **394**, 897–901.

Lund, S., Holman, G.D., Zierath, J.R., Rincon, J., Nolte, L.A., Clark, A.E., Schmitz, O., Pedersen, O., Wallberg-Henriksson, H. (1997) Effect of insulin on GLUT4 cell surface content and turnover rate in human skeletal muscle as measured by the exofacial bis-mannose photolabeling technique, *Diabetes* **46**, 1965–1969.

MacDonald, P.E., El-Kholy, W., Riedel, M.J., Salapatek, A.M., Light, P.E., Wheeler, M.B. (2002) The multiple actions of GLP-1 on the process of glucose-stimulated insulin secretion, *Diabetes* **51**Suppl. 3, S434–S442.

Maedler, K., Spinas, G.A., Dyntar, D., Moritz, W., Kaiser, N., Donath, M.Y. (2001) Distinct effects of saturated and monounsaturated fatty acids on beta-cell turnover and function, *Diabetes* **50**, 69–76.

Maffei, M., Fei, H., Lee, G.H., Dani, C., Leroy, P., Zhang, Y., Proenca, R., Negrel, R., Ailhaud, G., Friedman, J.M. (1995) Increased expression in adipocytes of ob RNA in mice with lesions of the hypothalamus and with mutations at the db locus, *Proc. Natl. Acad. Sci. U.S.A.* **92**, 6957–6960.

Mahtani, M.M., Widen, E., Lehto, M., Thomas, J., McCarthy, M., Brayer, J., Bryant, B., Chan, G., Daly, M., Forsblom, C., Kanninen, T., Kirby, A., Kruglyak, L., Munnelly, K. et al. (1996) Mapping of a gene for type 2 diabetes associated with an insulin secretion defect by a genome scan in Finnish families, *Nat. Genet.* **14**, 90–94.

Malik, N.M., Carter, N.D., Murray, J.F., Scaramuzzi, R.J., Wilson, C.A., Stock, M.J. (2001) Leptin requirement for conception, implantation, and gestation in the mouse, *Endocrinology* **142**, 5198–5202.

Mammarella, S., Romano, F., Di Valerio, A., Creati, B., Esposito, D.L., Palmirotta, R., Capani, F., Vitullo, P., Volpe, G., Battista, P., Della Loggia, F., Mariani-Costantini, R., Cama, A. (2000) Interaction between the G1057D variant of IRS-2 and overweight in the pathogenesis of type 2 diabetes, *Hum. Mol. Genet.* **9**, 2517–2521.

Marette, A., Mauriege, P., Marcotte, B., Atgie, C., Bouchard, C., Theriault, G., Bukowiecki, L.J.,

Marceau, P., Biron, S., Nadeau, A., Despres, J.P. (1997) Regional variation in adipose tissue insulin action and GLUT4 glucose transporter expression in severely obese premenopausal women, *Diabetologia* **40**, 590–598.

Massillon, D., Barzilai, N., Hawkins, M., Prus-Wertheimer, D., Rossetti, L. (1997) Induction of hepatic glucose-6-phosphatase gene expression by lipid infusion, *Diabetes* **46**, 153–157.

Mathers, J. (2002) Choosing your carbohydrates to prevent diabetes, *Br. J. Nutr.* **88**, 107–108.

Mathias, S., Pena, L.A., Kolesnick, R.N. (1998) Signal transduction of stress via ceramide, *Biochem. J.* **335**(Pt 3), 465–480.

Matsuzawa, Y., Funahashi, T., Nakamura, T. (1999) Molecular mechanism of metabolic syndrome X: contribution of adipocytokines adipocyte-derived bioactive substances, *Ann. N.Y. Acad. Sci.* **892**, 146–154.

Michael, M.D., Kulkarni, R.N., Postic, C., Previs, S.F., Shulman, G.I., Magnuson, M.A., Kahn, C.R. (2000) Loss of insulin signaling in hepatocytes leads to severe insulin resistance and progressive hepatic dysfunction, *Mol. Cell* **6**, 87–97.

Miller, F.J. Jr., Gutterman, D.D., Rios, C.D., Heistad, D.D., Davidson, B.L. (1998) Superoxide production in vascular smooth muscle contributes to oxidative stress and impaired relaxation in atherosclerosis, *Circ. Res.* **82**, 1298–1305.

Mohamed-Ali, V., Pinkney, J.H., Coppack, S.W. (1998) Adipose tissue as an endocrine and paracrine organ, *Int. J. Obes. Relat. Metab. Disord.* **22**, 1145–1158.

Moldovan, L., Moldovan, N.I., Sohn, R.H., Parikh, S.A., Goldschmidt-Clermont, P.J. (2000) Redox changes of cultured endothelial cells and actin dynamics, *Circ. Res.* **86**, 549–557.

Montague, C.T., O'Rahilly, S. (2000) The perils of portliness: causes and consequences of visceral adiposity, *Diabetes* **49**, 883–888.

Montague, C.T., Prins, J.B., Sanders, L., Zhang, J., Sewter, C.P., Digby, J., Byrne, C.D., O'Rahilly, S. (1998) Depot-related gene expression in human subcutaneous and omental adipocytes, *Diabetes* **47**, 1384–1391.

Mori, Y., Murakawa, Y., Okada, K., Horikoshi, H., Yokoyama, J., Tajima, N., Ikeda, Y. (1999) Effect of troglitazone on body fat distribution in type 2 diabetic patients, *Diabetes Care* **22**, 908–912.

Musi, N., Fujii, N., Hirshman, M.F., Ekberg, I., Froberg, S., Ljungqvist, O., Thorell, A., Goodyear, L.J. (2001) AMP-activated protein kinase (AMPK) is activated in muscle of subjects with type 2 diabetes during exercise, *Diabetes* **50**, 921–927.

Nesher, R., Karl, I.E., Kipnis, D.M. (1985) Dissociation of effects of insulin and contraction on glucose transport in rat epitrochlearis muscle, *Am. J. Physiol.* **249**(3 Pt 1), C226–C232.

Ng, D., Xuan W.L. (2002) Unpublished data.

Nishigori, H., Yamada, S., Kohama, T., Tomura, H., Sho, K., Horikawa, Y., Bell, G.I., Takeuchi, T., Takeda, J. (1998) Frameshift mutation, A263fsinsGG, in the hepatocyte nuclear factor-1beta gene associated with diabetes and renal dysfunction, *Diabetes* **47**, 1354–1355.

Noyman, I., Marikovsky, M., Sasson, S., Stark, A.H., Bernath, K., Seger, R., Madar, Z. (2002) Hyperglycemia reduces nitric oxide synthase and glycogen synthase activity in endothelial cells, *Nitric Oxide* **7**, 187–193.

Oakes, N.D., Cooney, G.J., Camilleri, S., Chisholm, D.J., Kraegen, E.W. (1997) Mechanisms of liver and muscle insulin resistance induced by chronic high-fat feeding, *Diabetes* **46**, 1768–1774.

Okamoto, Y., Kihara, S., Ouchi, N., Nishida, M., Arita, Y., Kumada, M., Ohashi, K., Sakai, N., Shimomura, I., Kobayashi, H., Terasaka, N., Inaba, T., Funahashi, T., Matsuzawa, Y. (2002) Adiponectin reduces atherosclerosis in apolipoprotein E-deficient mice, *Circulation* **106**, 2767–2770.

O'Brien, R.M., Granner, D.K. (1996) Regulation of gene expression by insulin, *Physiol. Rev.* **76**, 1109–1161.

Opara, E.C., Garfinkel, M., Hubbard, V.S., Burch, W.M., Akwari, O.E. (1994) Effect of fatty acids on insulin release: role of chain length and degree of unsaturation, *Am. J. Physiol.* **266**(4 Pt 1), E635–E639.

Oral, E.A., Simha, V., Ruiz, E., Andewelt, A., Premkumar, A., Snell, P., Wagner, A.J., DePaoli, A.M., Reitman, M.L., Taylor, S.I., Gorden, P., Garg, A. (2002) Leptin-replacement therapy for lipodystrophy, *N. Engl. J. Med.* **346**, 570–578.

Ouchi, N., Kihara, S., Arita, Y., Nishida, M., Matsuyama, A., Okamoto, Y., Ishigami, M., Kuriyama, H., Kishida, K., Nishizawa, H., Hotta, K. et al. (2001) Adipocyte-derived plasma protein, adiponectin, suppresses lipid accumulation and class A scavenger receptor expression in human monocyte-derived macrophages, *Circulation* **103**, 1057–1063.

Paolisso, G., Tataranni, P.A., Foley, J.E., Bogardus, C., Howard, B.V., Ravussin, E. (1995) A high concentration of fasting plasma non-esterified fatty acids is a risk factor for the development of NIDDM, *Diabetologia* **38**, 1213–1217.

Park, J.Y., Takahara, N., Gabriele, A., Chou, E., Naruse,K., Suzuma, K., Yamauchi, T., Ha, S.W., Meier, M., Rhodes, C.J., King, G.L. (2000) Induction of endothelin-1 expression by glucose: an effect of protein kinase C activation, *Diabetes* **49**, 1239–1248.

Perez-Martin, A., Raynaud, E., Mercier, J. (2001) Insulin resistance and associated metabolic abnormalities in muscle: effects of exercise, *Obes. Rev.* **2**, 47–59.

Petersen, K.F., Oral, E.A., Dufour, S., Befroy, D., Ariyan, C., Yu, C., Cline, G.W., DePaoli, A.M., Taylor, S.I., Gorden, P., Shulman, G.I. (2002) Leptin reverses insulin resistance and hepatic steatosis in patients with severe lipodystrophy, *J. Clin. Invest.* **109**, 1345–1350.

Prentki, M., Corkey, B.E. (1996) Are the β-cell signaling molecules rmalonyl-CoA and cytosolic long-chain acyl-CoA implicated in multiple tissue defects of obesity and NIDDM?, *Diabetes* **45**, 273–283.

Rahier, J., Boebbles, R.M., Henquin, J.C. (1983) Cellular composition of the human diabetic pancreas, *Diabetologia* **24**, 366–371.

Randle, P.J., Garland, P.B., Hales, C.N. (1963) The glucose fatty acid cycle: its role in insulin sensitivity and the metabolic disturbances of diabetes mellitus, *Lancet* **1**, 785–789.

Rao, A.K., Chouhan, V., Chen, X., Sun, L., Boden, G. (1999) Activation of the tissue factor pathway of blood coagulation during prolonged hyperglycemia in young healthy men, *Diabetes* **48**, 1156–1161.

Ravussin, E., Smith, S.R. (2002) Increased fat intake, impaired fat oxidation, and failure of fat cell proliferation result in ectopic fat storage, insulin resistance, and type 2 diabetes mellitus, *Ann. N.Y. Acad. Sci* **967**, 363–378.

Richelsen, B., Pedersen, S.B., Moller-Pedersen, T., Bak, J.F. (1991) Regional differences in triglyceride breakdown in human adipose tissue: effects of catecholamines, insulin, and prostaglandin E2, *Metabolism* **40**, 990–996.

Richter, E.A., Wojtaszewski, J.F., Kristiansen, S., Daugaard, J.R., Nielsen, J.N., Derave, W., Kiens, B. (2001) Regulation of muscle glucose transport during exercise, *Int. J. Sport Nutr. Exerc. Metab.* **11**(Suppl.) S71–S77.

Rivellese, A.A., de Natale, C., Lilli, S. (2002) Type of dietary fat and insulin resistance, *Ann. N.Y. Acad. Sci.* **967**, 329–335.

Roden, M., Stingl, H., Chandramouli, V., Schumann, W.C., Hofer, A., Landau, B.R., Nowotny, P., Waldhausl, W., Shulman, G.I. (2000) Effects of free fatty acid elevation on postabsorptive endogenous glucose production and gluconeogenesis in humans, *Diabetes* **49**, 701–707.

Rondinone, C.M., Wang, L.M., Lonnroth, P., Wesslau, C., Pierce, J.H., Smith, U. (1997) Insulin receptor substrate (IRS) 1 is reduced and IRS-2 is the main docking protein for phosphatidylinositol 3-kinase in adipocytes from subjects with non-insulin-dependent diabetes mellitus, *Proc. Natl. Acad. Sci. U.S.A.* **94**, 4171–4175.

Ruderman, N.B., Saha, A.K., Vavvas, D., Witters, L.A. (1999) Malonyl-CoA, fuel sensing, and insulin resistance, *Am. J. Physiol.* **276**(1 Pt 1), E1–E18.

Ryder, J.W., Yang, J., Galuska, D., Rincon, J., Bjornholm, M., Krook, A., Lund, S., Pedersen, O., Wallberg-Henriksson, H., Zierath, J.R., Holman, G.D. (2000) Use of a novel impermeable biotinylated photolabeling reagent to assess insulin- and hypoxia-stimulated cell surface GLUT4 content in skeletal muscle from type 2 diabetic patients, *Diabetes* **49**, 647–654.

Saengsirisuwan, V., Kinnick, T.R., Schmit, M.B., Henriksen, E.J. (2001) Interactions of exercise training and lipoic acid on skeletal muscle glucose transport in obese Zucker rats, *J. Appl. Physiol.* **91**, 145–153.

Saha, A.K., Laybutt, D.R., Dean, D., Vavvas, D., Sebokova, E., Ellis, B., Klimes, I., Kraegen, E.W., Shafrir, E., Ruderman, N.B. (1999) Cytosolic citrate and malonyl-CoA regulation in rat muscle in vivo, *Am. J. Physiol.* **276**(6 Pt 1), E1030–E1037.

Saha, A.K., Vavvas, D., Kurowski, T.G., Apazidis, A., Witters, L.A., Shafrir, E., Ruderman, N.B. (1997) Malonyl-CoA regulation in skeletal muscle: its link to cell citrate and the glucose-fatty acid cycle, *Am. J. Physiol.* **272**(4 Pt 1), E641–E648.

Salmeron, J., Ascherio, A., Rimm, E.B., Colditz, G.A. et al. (1997) Dietary fiber, glycemic load, and risk of NIDDM in men, *Diabetes Care* **20**, 545–550.

Samad, F., Pandey, M., Bell, P.A., Loskutoff, D.J. (2000) Insulin continues to induce plasminogen activator inhibitor 1 gene expression in insulin-resistant mice and adipocytes, *Mol. Med.* **6**, 680–692.

Samad, F., Yamamoto, K., Loskutoff, D.J. (1996) Distribution and regulation of plasminogen activator inhibitor-1 in murine adipose tissue in vivo. Induction by tumor necrosis factor-alpha and lipopolysaccharide, *J. Clin. Invest.* **97**, 37–46.

Sato, R., Miyamoto, W., Inoue, J., Terada, T., Imanaka, T., Maeda, M. (1999) Sterol regulatory element-binding protein negatively regulates microsomal triglyceride transfer protein gene transcription, *J. Biol. Chem.* **274**, 24714–24720.

Schneider, B.S., Faust, I.M., Hemmes, R., Hirsch, J. (1981) Effects of altered adipose tissue morphology on plasma insulin levels in the rat, *Am. J. Physiol.* **240**, E358–E362.

Shechter, M., Merz, C.N., Paul-Labrador, M.J., Kaul, S. (2000) Blood glucose and platelet-dependent thrombosis in patients with coronary artery disease, *J. Am. Coll. Cardiol.* **35**, 300–307.

Shepherd, P.R., Gnudi, L., Tozzo, E., Yang, H., Leach, F., Kahn, B.B. (1993) Adipose cell hyperplasia and enhanced glucose disposal in transgenic mice overexpressing GLUT4 selectively in adipose tissue, *J. Biol. Chem.* **268**, 22243–22246.

Shepherd, P.R., Withers, D.J., Siddle, K. (1998) Phosphoinositide 3-kinase: the key switch mechanism in insulin signalling, *Biochem. J.* **333**(Pt 3), 471–490.

Sherman, W.M., Friedman, J.E., Gao, J.P., Reed, M.J., Elton, C.W., Dohm, G.L. (1993) Glycemia and exercise training alter glucose transport and GLUT4 in the Zucker rat, *Med. Sci. Sports Exerc.* **25**, 341–348.

Shimabukuro, M., Zhou, Y.-T., Levi, M., Unger, R.H. (1998) Fatty acid-induced β cell apoptosis: A link between obesity and diabetes, *Proc. Natl. Acad. Sci. U.S.A.* **95**, 2498–2502.

Shimano, H., Horton, J.D., Shimomura, I., Hammer, R.E., Brown, M.S., Goldstein, J.L. (1997) Isoform 1c of sterol regulatory element binding protein is less active than isoform 1a in livers of transgenic mice and in cultured cells, *J. Clin. Invest.* **99**, 846–854.

Shimomura, I., Bashmakov, Y., Ikemoto, S., Horton, J.D., Brown, M.S., Goldstein, J.L. (1999) Insulin selectively increases SREBP-1c mRNA in the livers of rats with streptozotocin-induced diabetes, *Proc. Natl. Acad. Sci. U.S.A.* **96**, 13656–13661.

Shimomura, I., Hammer, R.E., Ikemoto, S., Brown, M.S., Goldstein, J.L. (1999) Leptin reverses insulin resistance and diabetes mellitus in mice with congenital lipodystrophy, *Nature* **401**(6748), 73–76.

Shimomura, I., Hammer, R.E., Richardson, J.A., Ikemoto, S., Bashmakov, Y., Goldstein, J.L., Brown, M.S. (1998) Insulin resistance and diabetes mellitus in transgenic mice expressing nuclear SREBP-1c in adipose tissue: model for congenital generalized lipodystrophy, *Genes Dev.* **12**, 3182–3194.

Shimomura, I., Matsuda, M., Hammer, R.E., Bashmakov, Y., Brown, M.S., Goldstein, J.L. (2000) Decreased IRS-2 and increased SREBP-1c lead to mixed insulin resistance and sensitivity in livers of lipodystrophic and ob/ob mice, *Mol. Cell* **6**, 77–86.

Shulman, G.I., Rothman, D.L., Jue, T., Stein, P., DeFronzo, R.A., Shulman, R.G. (1990) Quantitation of muscle glycogen synthesis in normal subjects and subjects with non-insulin-dependent diabetes by 13C nuclear magnetic resonance spectroscopy, *N. Engl. J. Med.* **322**, 223–228.

Skyler, J.S. (1998) Insulin therapy in type 2 diabetes mellitus, in: DeFronzo R.A (Ed.) *Current Therapy of Diabetes Mellitus*, Mosby-Year Book, St. Louis, AMO, pp. 108–116.

Smith, U. (1985) Regional differences in adipocyte metabolism and possible consequences in vivo, *Int. J. Obes.* **9**Suppl. 1, 145–148.

Sobel, B.E. (2002) Effects of glycemic control and other determinants on vascular disease in type 2 diabetes, *Am. J. Med.* **113**Suppl. 6A, 12S–22S.

Song, X.M., Kawano, Y., Krook, A., Ryder, J.W., Efendic, S., Roth, R.A., Wallberg-Henriksson, H., Zierath, J.R. (1999) Muscle fiber type-specific defects in insulin signal transduction to glucose transport in diabetic GK rats, *Diabetes* **48**, 664–670.

Spiegel, S., Merrill, A.H. Jr. (1996) Sphingolipid metabolism and cell growth regulation., *FASEB J.* **10**, 1388–1397.

Stratton, I.M., Adler, A.I., Neil, H.A., Matthews, D.R., Manley, S.E., Cull, C.A., Hadden, D., Turner, R.C., Holman, R.R. (2000) Association of glycaemia with macrovascular and microvascular complications of type 2 diabetes (UKPDS 35): prospective observational study, *BMJ* **321**(7258), 405–412.

Staubs, P.A., Nelson, J.G., Reichart, D.R., Olefsky, J.M. (1998) Platelet-derived growth factor inhibits insulin stimulation of insulin receptor substrate-1-associated phosphatidylinositol 3-kinase in 3T3-L1 adipocytes without affecting glucose transport, *J. Biol. Chem.* **273**, 25139–25147.

Steppan, C., Lazar, M. (2002) Resistin and obesity-associated insulin resistance, *Trends Endcrinol. Metab.* **13**, 18–23.

Stoffel, M., Duncan, S.A. (1997) The maturity-onset diabetes of the young (MODY1) transcription factor HNF4alpha regulates expression of genes required for glucose transport and metabolism, *Proc. Natl. Acad. Sci. U.S.A.* **94**, 13209–13214.

Sundaresan, M., Yu, Z.X., Ferrans, V.J., Irani, K., Finkel, T. (1995) Requirement for generation of H2O2 for platelet-derived growth factor signal transduction, *Science* **270**, 296–299.

Suzuki, M., Akimoto, K., Hattori, Y. (2002) Glucose upregulates plasminogen activator inhibitor-1 gene expression in vascular smooth muscle cells, *Life Sci.* **72**, 59–66.

Thorburn, A.W., Brand, J.C., O'Dea, K., Spargo, R.M., Truswell, A.S. (1987) Plasma glucose and insulin responses to starchy foods in Australian aborigines: a population now at high risk of diabetes, *Am. J. Clin. Nutr.* **46**, 282–285.

Thorburn, A.W., Gumbiner, B., Bulacan, F., Wallace, P., Henry, R.R. (1990) Intracellular glucose oxidation and glycogen synthase activity are reduced in non-insulin-dependent (type II) diabetes independent of impaired glucose uptake, *J. Clin. Invest.* **85**, 522–529.

Thornalley, P.J. (1990) The glyoxalase system: new developments towards functional characterization of a metabolic pathway fundamental to biological life, *Biochem. J.* **269**, 1–11.

Tomas, E., Lin, Y.S., Dagher, Z., Saha, A., Luo, Z., Ido, Y., Ruderman, N.B. (2002) Hyperglycemia and insulin resistance: possible mechanisms, *Ann. N.Y. Acad. Sci.* **967**, 43–51.

Tontonoz, P., Hu, E., Spiegelman, B.M. (1994) Stimulation of adipogenesis in fibroblasts by PPARγ2, a lipid-activated transcription factor, *Cell* **79**, 1147–1156.

Topper, J.N., Cai, J., Falb, D., Gimbrone, M.A. Jr. (1996) Identification of vascular endothelial genes differentially responsive to fluid mechanical stimuli: cyclooxygenase-2, manganese superoxide dismutase, and endothelial cell nitric oxide synthase are selectively up-regulated by steady laminar shear stress, *Proc. Natl. Acad. Sci. U.S.A.* **93**, 10417–10422.

Tozzo, E., Gnudi, L., Kahn, B.B. (1997) Amelioration of insulin resistance in streptozotocin diabetic mice by transgenic overexpression of GLUT4 driven by an adipose-specific promoter, *Endocrinology* **138**(4), 1604–1611.

Trent, D.F., Fletcher, D.J., May, J.M., Bonner-Weir, S., Weir, G.C. (1984) Abnormal islet and adipocyte function in young B-cell-deficient rats with near-normoglycemia, *Diabetes* **33**, 170–175.

Tuomilehto, J., Lindstrom, J., Eriksson, J.G., Valle, T.T., Hamalainen, H., Ilanne-Parikka, P., Keinanen-Kiukaanniemi, S., Laakso, M., Louheranta, A., Rastas, M., Salminen, V., Uusitupa, M. (2001) Finnish Diabetes Prevention Study Group. Prevention of type 2 diabetes mellitus by changes in lifestyle among subjects with impaired glucose tolerance, *N. Engl. J. Med.* **344**, 1343–1350.

Tuttle, R.L., Gill, N.S., Pugh, W., Lee, J.P., Koeberlein, B., Furth, E.E., Polonsky, K.S., Naji, A., Birnbaum, M.J. (2001) Regulation of pancreatic beta-cell growth and survival by the serine/threonine protein kinase Akt1/PKBalpha, *Nat. Med.* **7**, 1133–1137.

Unger, R.H. (1995) Lipotoxicity in the pathogenesis of obesity-dependent NIDDM. Genetic and clinical implications, *Diabetes* **44**, 863–870.

Unger, R.H. (2001) Lipotoxicity of beta-cells in obesity and in other causes of fatty acid spillover, *Diabetes* **50**Suppl. 1, S118–S121.

United Kingdom Prospective Diabetes Study (UKPDS), UK Prospective Diabetes Study (UKPDS) Group. (1998) Effect of intensive blood-glucose control with metformin on complications in overweight patients with type 2 diabetes (UKPDS 34), *Lancet* **352**, 854–865.

Ushio-Fukai, M., Zafari, A.M., Fukui, T., Ishizaka, N., Griendling, K.K. (1996) p22phox is a critical component of the superoxide-generating NADH/NADPH oxidase system and regulates angiotensin II-induced hypertrophy in vascular smooth muscle cells, *J. Biol. Chem.* **271**, 23317–23321.

Vaxillaire, M., Abderrahmani, A., Boutin, P., Bailleul, B., Froguel, P., Yaniv, M., Pontoglio, M. (1999) Anatomy of a homeoprotein revealed by the analysis of human MODY3 mutations, *J. Biol. Chem.* **274**, 35639–35646.

Velho, G., Froguel, P. (1998) Genetic, metabolic and clinical characteristics of maturity onset diabetes of the young, *Eur. J. Endocrinol.* **138**, 233–239.

Vinik, A., Bell, G. (1988) Mutant insulin syndromes, *Horm. Metab. Res.* **20**, 1–10.

Wallberg-Henriksson, H. (1992) Exercise and diabetes mellitus, *Exerc. Sport Sci. Rev.* **20**, 339–368.

Wang, Q., Somwar, R., Bilan, P.J., Liu, Z., Jin, J., Woodgett, J.R., Klip, A. (1999) Protein kinase B/Akt participates in GLUT4 translocation by insulin in L6 myoblasts, *Mol. Cell. Biol.* **19**, 4008–4018.

Watson, R.T., Pessin, J.E. (2001a) Intracellular organization of insulin signaling and GLUT4 translocation, *Recent Prog. Horm. Res.* **56**, 175–193.

Watson, R.T., Pessin, J.E. (2001b) Subcellular compartmentalization and trafficking of the insulin-responsive glucose transporter, GLUT4, *Exp. Cell Res.* **271**, 75–83.

Watson, R.T., Shigematsu, S., Chiang, S.H., Mora, S., Kanzaki, M., Macara, I.G., Saltiel, A.R., Pessin, J.E. (2001c) Lipid raft microdomain compartmentalization of TC10 is required for insulin signaling and GLUT4 translocation, *J. Cell Biol.* **154**, 829–840.

Wells-Knecht, K.J., Zyzak, D.V., Litchfield, J.E., Thorpe, S.R., Baynes, J.W. (1995) Mechanism of autoxidative glycosylation: identification of glyoxal and arabinose as intermediates in the autoxidative modification of proteins by glucose, *Biochemistry* **34**, 3702–3709.

Weyer, C., Foley, J.E., Bogardus, C., Tataranni, P.A., Pratley, R.E. (2000) Enlarged subcutaneous abdominal adipocyte size, but not obesity itself, predicts type II diabetes independent of insulin resistance, *Diabetologia* **43**, 1498–1506.

Widen, E., Lehto, M., Kanninen, T., Walston, J., Shuldiner, A.R., Groop, L.C. (1995) Association of a polymorphism in the beta 3-adrenergic-receptor gene with features of the insulin resistance syndrome in Finns, *N. Engl. J. Med.* **333**, 348–351.

Williams, B., Gallacher, B., Patel, H., Orme, C. (1997) Glucose-induced protein kinase C activation regulates vascular permeability factor mRNA expression and peptide production by human vascular smooth muscle cells in vitro, *Diabetes* **46**, 1497–1503.

Wojtaszewski, J.F., Hansen, B.F., Gade, J., Kiens, B., Markuns, J.F., Goodyear, L.J., Richter, E.A. (2000) Insulin signaling and insulin sensitivity after exercise in human skeletal muscle, *Diabetes* **49**, 325–331.

Wojtaszewski, J.F., Higaki, Y., Hirshman, M.F., Michael, M.D., Dufresne, S.D., Kahn, C.R., Goodyear, L.J. (1999) Exercise modulates postreceptor insulin signaling and glucose transport in muscle-specific insulin receptor knockout mice, *J. Clin. Invest.* **104**, 1257–1264.

Wojtaszewski, J.F., Jorgensen, S.B., Hellsten, Y., Hardie, D.G., Richter, E.A. (2002a) Glycogen-dependent effects of 5-aminoimidazole-4-carboxamide (AICA)-riboside on AMP-activated protein kinase and glycogen synthase activities in rat skeletal muscle, *Diabetes* **51**, 284–292.

Wojtaszewski, J.F., Nielsen, J.N., Richter, E.A. (2002b) Invited Review: Effect of acute exercise on insulin signaling and action in humans, *J. Appl. Physiol.* **93**, 384–392.

WHO Fact Sheets. www.who.int/ncd/dia/.

Wrede, C.E., Dickson, L.M., Lingohr, M.K., Briaud, I., Rhodes, C.J. (2002) Protein kinase B/Akt prevents fatty acid-induced apoptosis in pancreatic beta -cells (INS-1), *J. Biol. Chem.* **277**, 49676–49684.

Xu, J., Nakamura, M.T., Cho, H.P., Clarke, S.D. (1999) Sterol regulatory element binding protein-1 expression is suppressed by dietary polyunsaturated fatty acids. A mechanism for the coordinate suppression of lipogenic genes by polyunsaturated fats, *J. Biol. Chem.* **274**, 23577–23583.

Xu, J., Teran-Garcia, M., Park, J.H., Nakamura, M.T., Clarke, S.D. (2001) Polyunsaturated fatty acids suppress hepatic sterol regulatory element-binding protein-1 expression by accelerating transcript decay, *J. Biol. Chem.* **276**, 9800–9807.

Yahagi, N., Shimano, H., Hasty, A.H., Matsuzaka, T., Ide, T., Yoshikawa, T. et al. (2002) Absence of sterol regulatory element-binding protein-1 (SREBP-1) ameliorates fatty livers but not obesity or insulin resistance in Lep(ob)/Lep(ob) mice, *J. Biol. Chem.* **277**, 19353–19357.

Yamauchi, T., Kamon, J., Waki, H., Terauchi, Y., Kubota, N., Hara, K., Mori, Y., Ide, T., Murakami, K., Tsuboyama-Kasaoka, N., Ezaki, O.

et al. (2001) The fat-derived hormone adiponectin reverses insulin resistance associated with both lipoatrophy and obesity, *Nat. Med.* **7**, 941–946.

Yen, C.J., Beamer, B.A., Negri, C., Silver, K., Brown, K.A., Yarnall, D.P., Burns, D.K., Roth, J., Shuldiner, A.R. (1997) Molecular scanning of the human peroxisome proliferator activated receptor gamma (hPPAR gamma) gene in diabetic Caucasians: identification of a Pro12Ala PPAR gamma 2 missense mutation, *Biochem. Biophys. Res. Commun.* **241**, 270–274.

Yerneni, K.K., Bai, W., Khan, B.V., Medford, R.M., Natarajan, R. (1999) Hyperglycemia-induced activation of nuclear transcription factor kappaB in vascular smooth muscle cells, *Diabetes* **48**, 855–864.

Yokoyama, C., Wang, X., Briggs, M.R., Admon, A., Wu, J., Hua, X., Goldstein, J.L., Brown, M.S. (1993) SREBP-1, a basic-helix-loop-helix-leucine zipper protein that controls transcription of the low density lipoprotein receptor gene, *Cell* **75**, 187–197.

Yoshikawa, T., Shimano, H., Yahagi, N., Ide, T., Amemiya-Kudo, M., Matsuzaka, T. et al. (2002) Polyunsaturated fatty acids suppress sterol regulatory element-binding protein 1c promoter activity by inhibition of liver X receptor (LXR) binding to LXR response elements, *J. Biol. Chem.* **277**, 1705–1711.

Yusuf, S., Sleight, P., Pogue, J., Bosch, J., Davies, R., Dagenais, G. (2000) Effects of an angiotensin-converting-enzyme inhibitor, ramipril, on cardiovascular events in high-risk patients. The Heart Outcomes Prevention Evaluation Study Investigators, *N. Engl. J. Med.* **342**, 145–153.

Zhang, Y., Proenca, R., Maffei, M., Barone, M., Leopold, L., Friedman, J.M. (1994) Positional cloning of the mouse obese gene and its human homologue, *Nature* **372**(6505), 425–432.

Zhou, Y.P., Grill, V.E. (1995) Palmitate-induced beta cell insensitivity to glucose is coupled to decreased pyruvate dehydrogenase activity and enhanced kinase activity in rat pancreatic islets, *Diabetes* **44**, 194–199.

Zierath, J.R., He, L., Guma, A., Odegoard Wahlstrom, E., Klip, A., Wallberg-Henriksson, H. (1996) Insulin action on glucose transport and plasma membrane GLUT4 content in skeletal muscle from patients with NIDDM, *Diabetologia* **39**, 1180–1189.

Zierath, J.R., Krook, A., Wallberg-Henriksson, H. (2000) Insulin action and insulin resistance in human skeletal muscle, *Diabetologia* **43**, 821–835.

Zobolotny, J.M., Bence-Hanulee, K.K., Stricker-Krongrad, A. et al. (2002) PTP1B regulates leptin signal transduction in vivo, *Dev. Cell* **2**, 385–387.

Zou, M.H., Shi, C., Cohen, R.A. (2002) Oxidation of the zinc-thiolate complex and uncoupling of endothelial nitric oxide synthase by peroxynitrite. *J. Clin. Invest.* **109**, 817–826.

Intracellular Fatty Acid Binding Proteins and Fatty Acid Transport

Judith Storch[1] *and Lindsay McDermott*[2]
[1] *Rutgers University, New Brunswick, NJ*
[2] *University of Glasgow, Glasgow, UK*

Keywords

Cell Transfections
This is a method whereby a biologically active protein is introduced (chemically or electrically) to cultured mammalian cells. Live cells are then assayed to determine the effects of the introduced protein. Results from such experiments may reveal how the protein behaves within the cell.

Encyclopedia of Molecular Cell Biology and Molecular Medicine, 2nd Edition. Volume 7
Edited by Robert A. Meyers.

ISBN: 3-527-30549-1

Fatty Acids
This is a long carbon chain carboxylic acid that condenses with glycerol to form fat. The carbon chain can vary in length and number and positioning of doulble bonds.

Knockout Mouse
This is a scientific model used to explore the role that a particular gene may play in humans. Within the knockout mouse, the gene of interest is removed using artifical methods; all cells within the resulting mutant mouse contain the same mutation. The appearance, biochemical characteristics, and behavior of the knockout mouse can provide some indication of the gene's normal role in the mouse.

Lipid-binding Proteins
These are small, soluble intracellular proteins capable of non–covalently binding fatty acids and other small hydrophobic ligands. It is believed to participate in lipid transport.

Lipid Transport
This is the movement of lipids from the site of synthesis to their site of utilization.

Long-chain fatty acids (FA) are required by cells as membrane phospholipid constituents, metabolic substrates, precursors for signaling molecules, and mediators of gene expression. They are in constant flux and need to enter and leave cells rapidly and, presumably, in a regulated manner. The relatively low aqueous solubility of fatty acids would strongly suggest that specific and efficient mechanisms must exist for their intracellular transport. High levels of fatty acid–binding proteins (FABPs) are found within cells, and, although it has been shown that these proteins noncovalently bind fatty acids with high affinity, their true *in vivo* functions have remained elusive. This chapter focuses on recent findings assessing the transport function of FABPs, and on data supporting putative mechanisms by which FABPs may be involved in cellular FA uptake, efflux, and intracellular transport.

1 Equilibrium Binding of Fatty Acids to FABPs

There are 12 members of the mammalian FABP family, each with specific tissue expression and, with the exception of the retinol and retinoic acid–binding proteins, each is named after the first tissue of isolation (Table 1). In the tissues where they are found, FABPs are typically expressed abundantly, at reported levels of up to 13% cytosolic protein and, as their name suggests, are capable of noncovalently binding long-chain fatty acids with high affinity and a 1 : 1 molar stoichiometry. Ileal lipid-binding protein (I-LBP) and liver fatty acid–binding protein (L-FABP) prove exceptions to this rule: both can bind more bulky, hydrophobic ligands such as lysophospholipids, bile acids, eicosanoids,

Tab. 1 Members of the family of mammalian intracellular fatty acid–binding proteins.

Name	*Occurrence*
E-FABP (K-FABP)	Epidermis, adipose, mammary tissue, tongue epithelia, testis
H-FABP	Heart muscle, cardiac and skeletal muscle, brain, mammary gland, kidney, adrenals, ovaries, testis
B-FABP	Brain, central nervous system
M-FABP	Peripheral nervous system
A-FABP	Adipose, macrophages
I-FABP	Small intestine
I-LBP	Small intestine (distal)
L-FABP	Liver, small intestine
CRABP-I	Brain, skin, testis
CRABP-II	Epidermis, adrenal
CRBP-I	Liver, kidney, testis, lung
CRBP-II	Small intestine

and some drugs, I-LBP does not bind fatty acids, and L-FABP is capable of binding two fatty acids simultaneously. A recent NMR spectroscopic study showed two distinct binding environments for these FA, and suggested that binding of the first fatty acid precedes and may facilitate binding of the second.

Equilibrium binding studies have been used in an attempt to elucidate the functional characteristics of each FABP type. Most recently, the ligand-binding specificity of eight human FABPs (heart, liver, intestine, adipocyte, myelin, epidermal, brain fatty acid–binding protein, and ileal lipid-binding protein) were directly compared by Zimmerman et al. using the Lipidex assay. By determining the equilibrium distribution of FAs between the resin and FABP, binding affinities were measured and, with the exception of I-LBP, ranged from 0.2 to 4.01 μM. The results obtained showed that the proteins have a lower affinity for palmitic acid than for oleic and arachidonic acids. Contrary to these findings and using the ADIFAB (acrylodated intestinal fatty acid–binding protein) method, Richieri and colleagues showed that human FABPs from brain, heart, intestine, liver, and myelin pertained little or no selectivity for a particular FA, and obtained K_d values ranging from 2 to 400 nM. ADIFAB consists of intestinal fatty acid–binding protein covalently modified with the fluorescent acrylodan group, which exhibits a marked red-shift in fluorescence emission maximum upon fatty acid binding, enabling unbound concentrations of FAs to be accurately measured. A further study using isothermal titration calorimetry (ITC) to determine the binding of FAs to human L-FABP yielded a stearate K_d value approximately 100 times larger (weaker binding) than that determined by Richieri et al., and concluded, again in contrast to Richieri et al., that L-FABP preferentially binds unsaturated relative to saturated FAs. It is possible, however, that the poor solubility of stearate, as well as the high fatty acid concentrations necessary for the ITC injection method, resulted in a lower stearate concentration in the reaction vessel, thus producing lower values for heat change, and therefore uncertain K_d values.

In spite of these discrepancies over absolute K_d values, it is clear from these and earlier *in vitro* binding experiments that FABPs bind long-chain saturated and unsaturated FA. Indeed, NMR and X-ray crystallographic structures of holo FABPs reveal the position of the bound FA within the individual protein structures. In some structures, including intestinal FABP (I-FABP) and adipocyte FABP (A-FABP), the fatty acid adopts a bent conformation, while in others, such as heart FABP (H-FABP), it adopts a U-shaped conformation.

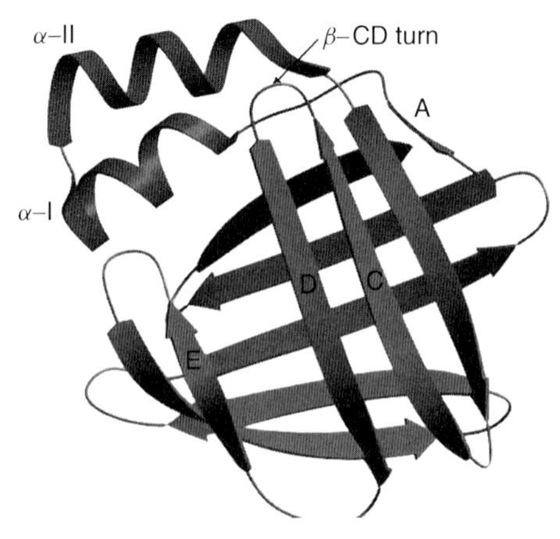

Fig. 1 The three-dimensional structure of a representative cytoplasmic fatty acid–binding protein. The protein comprises a flattened β-barrel capped by two short α-helices. The helices and closely positioned α-turns are believed to behave as a portal for ligand entry and exit.

Notwithstanding a wide variance in primary sequence, all members of the FABP family consist of a β-barrel structure capped by two α-helices; the latter is believed to behave as a portal for ligand entry and release (Fig. 1). This hypothesis was supported recently by fluorescence-based experiments comparing A-FABP and a triple mutant (V32G, F57G, K58G), designed to enlarge the putative portal opening by reducing the size of portal amino acids. By comparing analinonaphthalene sulfonic acid (ANS) and oleate-binding affinities and ANS-binding rates, it was found that enlargement of the A-FABP portal region increased ligand accessibility into the cavity with only modest effects on ligand-binding affinity, suggesting that dynamic fluctuations in this region regulate cavity access. Indeed, the solution structure of apo I-FABP was shown to exhibit a higher degree of mobility in this portal area in comparison with that of the ligand-bound I-FABP complex relative to other domains in these proteins. In an attempt to better understand the mechanism by which FAs bind to and dissociate from the binding cavities of FABPs, Richieri et al. constructed 31 single amino acid mutants within the portal region and in the region of the gap between the βD- and βE-strands of I-FABP, and determined binding affinities and rate constants for FA binding. Together with experiments examining these parameters as a function of ionic strength, it was suggested that the FA initially binds through an electrostatic interaction to Arg56 on the surface of the protein, before inserting into the binding cavity, with a reversal of these steps for the dissociation reaction.

The structural and biochemical studies of FA binding to FABPs do not directly demonstrate a ligand transport function for the proteins, however, they nevertheless, are consistent with their participation in such processes. It has been suggested that upon binding to FABP, the FAs traverse the aqueous cytoplasm in a more energetically favorable manner owing to increased aqueous solubility of FAs, in that lateral diffusion of FABP-bound FAs proceeds more rapidly than that of membrane-bound FAs.

2
In vitro Fatty Acid–Transfer Properties of FABPs

The mechanisms by which members of the FABP family transfer FAs to membranes has been examined in a series of *in vitro* experiments. These studies examined the rate of anthroyloxy-labeled fatty acid (AOFA) movement from FABPs to model acceptor vesicles containing a nonexchangeable fluorescence quencher, using a fluorescence resonance energy transfer assay.

Transfer of fluorescent FA analogues from adipocyte, heart, intestinal, epidermal, brain, and myelin FABPs (A-, H-, I-, E-, B-, and M-FABPs), and of retinol from cellular retinal–binding protein I (CRBP-I), to membrane targets appears to involve collision of the holo protein with both zwitterionic and anionic membranes, with more effective collisional interactions occurring with the latter. FA transfer rates were directly proportional to acceptor membrane phospholipid concentration, and were modulated by changes in the acceptor vesicle charge, implying that FABP and membrane come into direct physical contact, with ligand transfer occurring during the collision. Further studies suggested that positively charged (lysine) residues on the protein surface are involved in this ligand transfer process and are likely to participate in the formation of effective FABP:membrane complexes that involve electrostatic interactions. Site-directed mutagenesis of A-FABP and H-FABP demonstrated that lysines in the helical cap domain are important for establishing these ionic interactions. In particular, lysine residues on αI, αII, and the βCD-turn of both proteins, and possibly the βA-strand of A-FABP, but not H-FABP, were shown to be directly involved in these charge:charge interactions. Indeed, by removing I-FABPs two α-helices, fatty acid transfer no longer occurred via a collisional mechanism. Furthermore, creation of chimeric proteins of A- and H-FABPs revealed that the αII-helix is important in determining the absolute fatty acid transfer rates, while the αI-helix appears to be particularly important in regulating protein sensitivity to the negative charge of membranes. The FABP α-helical domain, part of the same region believed to enable FA entry and exit, is therefore extremely important not only for the direct interaction with membrane phospholipids during ligand transfer to membranes but also in regulating FA transfer rates.

The direct interaction of the FABPs with membranes was further assessed using fluorescence-based assays and direct physical measurements and, as suggested by the transfer experiments, the α-helical domain and surface lysine residues therein proved particularly important for membrane association. Preincubation of anionic vesicles with I-FABP prevented the subsequent binding of the peripheral membrane protein cytochrome c, suggesting that the I-FABP was membrane-bound. In contrast, helix-less I-FABP demonstrated 80% less efficiency in preventing cytochrome c binding than intact I-FABP. Further direct support for I-FABP:membrane interactions was provided by surface pressure measurements, Brewster angle microscopy, and infrared reflection-absorption spectroscopy (IRRAS), which revealed that I-FABP interacted with 1,2-dimyristoyl phosphatidic acid monolayers to a stronger extent than its helix-less variant. IRRAS studies also showed I-FABP to induce a stronger conformational ordering of the lipid acyl chains than helix-less I-FABP. The interaction of A-FABP with vesicles was also directly measured using FTIR spectroscopy, and it was found that A-FABP interacts much more strongly with acidic than zwitterionic membranes, and that neutralization of A-FABP positive surface charges by acetylation considerably weakens its interactions with negatively charged vesicles. As for I-FABP, these data supported observations gained from the cytochrome c–binding assay, whereby A-FABP, but not acetylated A-FABP, was able to prevent subsequent cytochrome c binding to model anionic membranes. Notably, in all these studies, the degree of membrane interaction correlated directly with the rate of fatty acid transfer, indicating that FABP: membrane interactions are functionally related to their fatty acid transport properties. While the primary mode of FABP:membrane interaction appears to involve the establishment of ion pairs between positive charges on the protein surface and negative membrane charges, recent mutagenesis studies have suggested that hydrophobic interactions between I-FABP helix II residues and membrane phospholipids may also play a role in establishing the collisional interactions that promote FA transfer. Indeed, it has been noted that those FABPs displaying a collisional mechanism for fatty acid transfer, possess a conserved, solvent-exposed, bulky hydrophobic side chain located on αII, namely, a phenylalanine, leucine, isoleucine, or methionine. L-FABP on the other hand, displaying a diffusional mechanism for fatty acid transfer, possesses a Glu in this position. The aforementioned results for I-FABP thus suggest that exposed hydrophobic residues could be involved in membrane association, thereby rendering a subsequent interaction with a receptor protein more efficient and permitting the exchange of ligand without its entry into an aqueous phase.

L-FABP and CRBP-II are the only members of the FABP family that were found not to transfer FA, or retinol in the case of CRBP-II, via a collisional mechanism. L-FABP transfers FAs to membranes almost 50-fold slower than members of the family, exhibiting a collisional FA transfer mechanism. Despite overall structural similarity, this particular FABP appears to transfer its ligand via aqueous diffusion, a mechanism that does not involve direct protein–membrane contact. In support of this mechanism, the rate of FA transfer from L-FABP to membranes was modulated by neither the concentration of acceptor membranes nor their composition; however, changes in the ionic strength of the buffer directly affected the transfer rates, indicating that the transfer rate is regulated by aqueous solubility of the fatty acid. Recently, it was demonstrated that the alpha-helical region of L-FABP is responsible for its diffusional mechanism of fatty acid transfer to membranes. This was deduced by the creation of a pair of chimeric proteins, one containing the ligand-binding domain of I-FABP and the alpha-helical region of L-FABP (alphaLbetaIFABP) and the other containing the ligand-binding pocket of L-FABP and the helical domain of I-FABP (alphaIbetaLFABP). Transfer rates from the chimeric proteins compared to those of the wild type indicated that the slower rate of FA transfer observed for L-FABP relative to that of I-FABP is, in part, determined by the helical domain of the proteins. In addition, absolute FA transfer from alphaLbetaIFABP to membranes occurred by aqueous diffusion whereas FA transfer from alphaIbetaLFABP occurred via protein–membrane collisional interactions. Like L-FABP, CRBP-II also displayed a diffusional mechanism, and indeed proved ineffective in preventing cytochrome c from binding to phospholipid vesicles. Davies et al., however, were able to observe L-FABP binding to anionic phospholipid vesicles using the fluorescent probe DAUDA, an undecanoate (11 : 0) derivative. Upon binding to FABP, the fluorescence maximum of DAUDA becomes blue- shifted and exhibits a substantial increase in intensity. By mixing a preformed DAUDA:L-FABP complex with anionic vesicles, an immediate decrease in DAUDA fluorescence occurred, indicating a release of DAUDA from the protein. As this did not occur with zwitterionic vesicles, the results suggested that interaction of L-FABP with the anionic membrane interface induces a rapid conformational change, resulting in a reduced affinity of DAUDA for the protein. The nature of this interaction was suggested to involve both electrostatic and nonpolar forces. It is important to note, however, that these L-FABP–membrane interactions were observed at very low ionic strength assay conditions, but were not found at physiologic ionic strength.

From these experiments, it was hypothesized that those FABPs exhibiting a collisional mechanism for FA transfer are most likely to be involved in the targeted transfer of FAs, whereby the proteins interact either with specific membrane lipids and/or membrane protein domains to transfer their fatty acid ligand. Recently, it was also discovered that transfer of anthroyloxylated fatty acids (AOFAs) from phospholipid membranes to I-FABP (in the opposite direction from that initially examined) also occurs via a collisional mechanism. It is possible then that FABPs may utilize membrane–protein interactions not only for the acquisition of ligand but also for their delivery. By displaying a diffusional mechanism for fatty acid transfer whereby no direct protein–membrane interaction occurs, it is likely that L-FABP and CRBP-II may function in the capacity of cytosolic fatty acid or retinol reservoirs. Nonetheless, just as the apparent membrane interactions may be surrogates for FABP–protein interactions, the absence of apparent L-FABP or CRBP-II–membrane interactions does not preclude protein–protein interactions; however, the nature of any such interactions is likely to be different from that of the collisional FABPs.

It is important to note that much of the work examining the mechanisms of FA transfer to model membranes has involved the use of rodent FABPs. With numerous links between FABPs and several chronic diseases in humans becoming ever clearer, a direct comparison of fatty acid transfer properties of rodent and human proteins demonstrated that the rates and mechanisms of FA transfer from L-, I-, A- and H-FABPs observed for rodent proteins were the same for their human counterparts. An additional approach used to examine the transport function of the FABPs is fluorescence recovery after photobleaching (FRAP) for solution conditions. In such studies, the effective diffusion (Deff) of the fluorescent probe *N*-(7-nitro-2,1,3-benzoxadiazol-4-yl)-stearate (NBDS) is evaluated in individual cells or in so-called model cytosol. Using this approach, it has been found that the rate of NBDS movement correlates directly with the intracellular level of L-FABP in HepG2 cells and hepatocytes, the level of I-FABP in embryonic stem cells, the total FABP level (L-FABP + I-FABP) in rat enterocytes isolated from different intestinal segments, and the L-FABP concentration in solutions prepared to resemble cytosol with intracellular membranes. The mechanism for these effects likely involves FABPs acting to limit fatty acid partitioning into immobile membranes, thereby increasing the rate of movement of the fatty acid. Additionally, specific effects of FABPs in cells were also suggested by results in which permeabilized HepG2 cells were used to generate various cytosolic compositions by incubation with different protein-containing solutions; in these studies, albumin was found to be only 4-fold more effective than equal concentrations of L-FABP in increasing the Deff of NBDS, despite the fact that its fatty acid–binding capacity is of far greater magnitude greater than that of L-FABP, as is its FA-binding affinity.

3 Transfection Studies of FABP Function

To assess FABP function within a more physiological milieu, FABP genes have been transfected into model cell cultures and subsequent changes in FA uptake and metabolism examined.

A series of experiments examining L-cell fibroblasts or embryonic stem cells transfected with L-FABP and/or I-FABP appeared to suggest a role for L-FABP in cellular FA uptake; when NBDS or the fluorescent fatty acid analog *cis*-parinaric acid (cPnA) were added to cells expressing L-FABP and compared with control cells, or cells expressing I-FABP, a 2-fold increase in fluorescence intensity was observed. However, care should be taken in the interpretation of these results given that the quantum yield for cPnA binding to L-FABP differs by a similar degree to that of cPnA binding to I-FABP. Indeed, in contrast to these results, it was reported more recently that expression of L-FABP in transfected L-cell fibroblasts did not reduce the uptake of ^{3}H palmitic acid, its oxidation, or induction of lipid accumulation, although uptake of ^{3}H phytanic acid was found to be reduced. Thus, apparently similar studies provide divergent results. More recently, nevertheless, Wolfrum et al. used peroxisome proliferators to increase L-FABP levels in HepG2 cells, and antisense L-FABP to decrease L-FABP mRNA expression, and the net oleate uptake was shown to correlate directly with the L-FABP content of the cells. When I-FABP was expressed at 2-fold higher concentrations in L-cells, it was found that cPnA uptake was lower relative to cells with lesser I-FABP levels. In contrast, the decreased level of I-FABP expression in differentiated relative to undifferentiated embryonic stem cells was also correlated with a decrease in fatty acid uptake. Results of experiments involving overexpression of Ala^{54} and Thr^{54}-I-FABP (two I-FABP forms created by a single base pair alteration in the human I-FABP gene) in Caco-2 intestinal cells must also be viewed carefully. A 2-fold increase in net fatty acid uptake was obtained from Thr^{54}-I-FABP-transfected cells and approximately 5-fold more triacylglycerol was secreted into the basolateral medium relative to Ala^{54}-I-FABP-transfected cells. In differentiated cells of both lines, the endogenous levels of L-FABP were decreased relative to control cells, although L-FABP levels, nevertheless, remained 2- to 3-fold higher than levels of I-FABP. Moreover, it has subsequently been demonstrated that, in contrast to earlier indications, Caco-2 cells do in fact express I-FABP. A recent publication, nevertheless, reported that parent Caco-2 cells as well as mock-transfected cells fail to express detectable levels of I-FABP mRNA or protein at any stage of differentiation; upon transfecting cells with I-FABP, radiolabeled oleic acid was used to monitor fatty acid metabolism, and it was deduced that I-FABP expression in intestinal cells leads to reduced triacylglycerol secretion. Clearly, the precise function of I-FABP in the enterocyte remains uncertain. Given the additional variable of high levels of expression of L-FABP in this cell type as well, an understanding of the functional properties of enterocyte FABPs, in particular, remains a challenge.

CHO cells transfected with A-FABP showed a 1.5- to 2-fold increase in the net uptake of oleate and, interestingly, expression of equivalent levels of a mutated form of A-FABP with reduced fatty acid binding failed to produce a change in uptake. In contrast, expression of A-FABP in L6 myoblasts did not alter the net fatty acid uptake, although the degree of differentiation in these cells differed from that of control cells. Recently, electroporation of A-FABP into 3T3-L1 preadipocytes was used as an alternative technique for modifying the level of cellular FABP directly. Incorporation of A-FABP was found to result in an increase in the initial rate of palmitate uptake relative to that of control cells, supporting an A-FABP-mediated transport function.

An early report suggested an increase in net fatty acid uptake in cells overexpressing H-FABP, whereas overexpression of H-FABP in L6 myoblasts was reported not to alter uptake. Expression of CRBP-I in Caco-2 cells increased net retinol uptake by about 2-fold; in this clone, a large decrease in endogenous Caco-2 expression of CRBP-II was also found. Results from experiments involving CRABPI knockout ABI cell lines suggested that CRABPI functions to regulate the intracellular concentrations of retinoic acid and to maintain high levels of oxidized retinoic acid metabolites such as 4-oxoretinoic acid within cells.

Thus, some studies have yielded conflicting results, although some have provided strong evidence for an intracellular FABP-mediated transport function. A general concern with stable transfections in cultured cell lines is that parallel alterations may be occurring due to clonal variability, and secondary changes due to the altered expression of a specific gene, in this case an FABP, may also occur. The latter issue is not as serious as the former, as it at least points to the potential involvement of the FABP in a cellular process, if not to its precise role at the molecular level. It would seem of great importance for transfection studies, especially, to demonstrate a dose-dependent functional response to FABP expression. Thus far, only the studies of Wolfrum et al. have been so rigorously performed. The use of direct protein transfer techniques, including streptolysis, electroporation, and lipid- and peptide-based protein transfer reagents, avoid entirely the issue of clonal variability, and in large part the concern about secondary changes in cellular processes. These approaches have not yet been widely applied to the FABPs.

4
Cellular Fatty Acid Transport via FABP–Protein Interactions

As noted above, *in vitro* studies suggested a potential for FABPs to act as targeting proteins, conveying their ligands to particular domains on organellar membranes, and/or to specific protein receptors. Recently, a number of protein–protein interactions involving FABPs have indeed been reported, suggesting that fatty acids may be transported around the cell in a regulated manner.

Using yeast two-hybrid assays, an interaction between A-FABP and hormone-sensitive lipase (HSL) was discovered and further confirmed with experiments including GST-pulldowns and coimmunoprecipitation of the HSL:A-FABP complex. The A-FABP interaction domain of HSL was found to reside in the N-terminal portion of the protein, whereas the catalytic domain is known to be localized to the

C-terminus. Recent studies showed that HSL residues His194 and Glu199 appear to be critical for interactions with A-FABP. Furthermore, it has been shown that A- and E-FABPs, but not I-or L-FABPs, bind to HSL in a 1 : 1 molar stoichiometry only in the presence of oleate. However, A-, E-, L-, H-, and I-FABPs all stimulated HSL activity to an equivalent and relatively modest extent, approximately two-fold. A point mutant of A-FABP which does not display FA binding did not cause this modest stimulation. These results suggest that the effect of the FABPs on HSL is likely via binding of fatty acid but that a protein–protein interaction is not required for the FABP effect. It appears then that binding and activation of HSL by FABPs are separate and distinct functions. Taken as a whole, results from these experiments suggest that A-FABP may function to traffic fatty acids away from the triglyceride droplet after hydrolysis by HSL, thereby promoting further lipolysis by diminishing end-product inhibition. It is still unclear exactly where A-FABP is taking the fatty acid. However, given the evidence of an association between H-FABP and the cytoplasmic domain of the putative transmembrane fatty acid transporter CD36/FAT in milk-fat globule membranes, as determined by gel filtration and coimmunoprecipitation, it is possible that A-FABP transports its bound fatty acid to the plasma membrane and, via an interaction with CD36/FAT, promotes the efflux of the fatty acid out of the cell (Fig. 2). Alternatively, in the adipocyte or in other cell types, an FABP could transport its bound fatty acid to internal sites for re-esterification.

Again using the yeast two-hybrid system, an interaction between L-FABP and the lipid-activated transcription factor

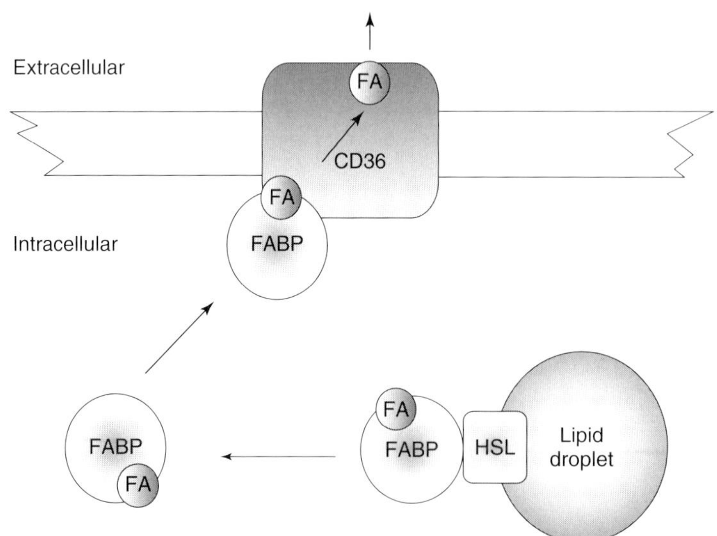

Fig. 2 Fatty acid–binding proteins and fatty acid efflux. It is possible that upon hydrolysis of lipid by hormone sensitive lipase (HSL) or other lipases, the resulting fatty acid (FA) is bound by fatty acid–binding protein (FABP) and transported through the cytoplasm to CD36/FAT, whereupon the fatty acid is off-loaded for efflux out of the cell. Adipocyte FABP-HSL and heart FABP-CD36 complexes have been identified.

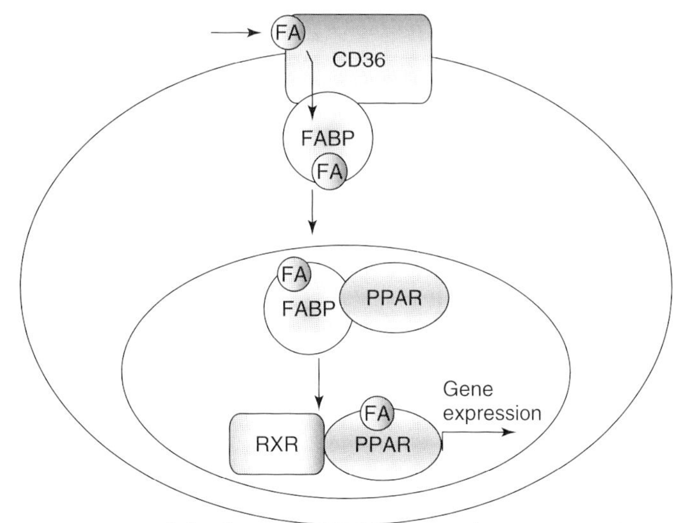

Fig. 3 Fatty acid–binding proteins and gene regulation. Fatty acids (FA) taken up by CD36/FAT may then be bound in the intracellular space by fatty acid–binding proteins (FABP) for transport through the cytoplasm to the nucleus. Upon entering the nucleus, the FABP:FA complex binds to PPAR, the FA becomes PPAR-bound, and gene expression is initiated. CD36 has been shown to play a role in fatty acid uptake, FABPCD36 complexes have been identified, and an association between liver FABP and PPARa has been shown. There is also evidence to suggest that adipocyte and epidermal FABPs may be partially localized to the nucleus.

peroxisome proliferator activated receptor α (PPARα) was found. This was further assessed by pull-down assays and immunoprecipitation, and was shown to be independent of ligand binding. PPARα is believed to be a nuclear target for fatty acids and initiates gene expression of enzymes involved in lipid metabolism. Such an association, therefore, suggests that L-FABP serves to directly traffic its fatty acid ligand, gained possibly from an interaction with CD36/FAT and/or other transmembrane transporters, to the nucleus and, thereby, directly functions in the regulation of gene expression (Fig. 3). Indeed, recent results indicate that A- and E-FABPs localize in the nuclei of 3T3-LI adipocytes as well as the cytoplasm, suggesting that these two FABPs may also exert their action at the level of the nucleus.

An interaction between the extracellular domain of CD36 and the S100A8/S100A9: arachidonic acid complex has also been identified in endothelial cells, implying a role for this protein in transcellular eicosanoid metabolism. Given that the coordinate regulation of gene expression and similar abundance of cytoplasmic FABPs and membrane FA transporters have been repeatedly demonstrated, it is highly possible that a concerted, vectorial mechanism for FA transport exists.

5
Insights into FABP Function from Null Mice

Several mouse models null for different FABPs have been created by targeted gene disruption. These are providing important support for the transport functions of these proteins, as well as new insights into additional potential functions. Of the four FABP knockouts described thus far, the H-FABP-null mouse provides perhaps the clearest example of functional consequences. Mice lacking expression of H-FABP displayed a substantial decrease in long-chain fatty acid uptake into the heart. Further studies in cardiac myocytes isolated from wild-type and H-FABP$^{-/-}$ animals demonstrated that defective fatty acid uptake and oxidation appears to be the underlying cause of the phenotype observed in the whole animal. Interestingly, the CD36/FAT-null mouse displays a very similar FA uptake phenotype, again suggesting a concerted action by the two types of FA transport proteins in cellular FA trafficking. In the H-FABP-null animals, as well as in the CD36 nulls, physiological compensation for the decreased FA uptake appears to occur by an increase in glucose uptake and oxidation in the heart, and not by compensatory increases in other members of the FABP family. H-FABP is also expressed in the mammary gland; however, the H-FABP$^{-/-}$ mice were reported to have no overt phenotype in this tissue.

Mice null for I-FABP were also reported to show no apparent compensation with other FABPs. The I-FABP$^{-/-}$ mice developed hyperinsulinemia that was independent of body weight gain, an unusual dissociation. The I-FABP knockout mice gained more weight and had higher levels of serum triglycerides. This could indicate an involvement of the protein in lipid absorption, metabolism, or secretion, which requires further investigation. An intriguing finding in this model was that the effects on weight and serum triglycerides were observed only in male mice. This gender dependency suggests a previously unexplored interaction between I-FABP and sex hormones.

In contrast to the H-FABP and I-FABP nulls, mice null for the ap2 gene, which encodes A-FABP, showed a dramatic increase in expression of another FABP, keratinocyte FABP (K-FABP), in adipose tissue, perhaps accounting for the absence of a dramatic phenotype in the animals. Initial investigations of low fat–fed mice showed few differences between wild-type and A-FABP$^{-/-}$ animals; however, feeding a high-fat diet resulted in lower levels of plasma insulin and reduced adipocyte mRNA levels of tumor necrosis factor α (TNFα) relative to wild-type mice; the absence of hyperinsulinemia appeared to occur despite the presence of high-fat diet-induced obesity. Conversely, however, it was found that younger aP2$^{-/-}$ mice, despite maintaining lower glucose levels, did in fact develop hyperinsulinemia on a high-fat diet; the plasma insulin levels were directly correlated with the degree of adiposity in both wild-type and A-FABP$^{-/-}$ mice. Further, adipocyte TNFα secretion was not reduced relative to wild-type mice. These results indicate that a dissociation between the development of obesity and the development of hyperinsulinemia, a hallmark of obesity, is not apparent in the A-FABP$^{-/-}$ mice under all circumstances.

The aP2$^{-/-}$ mice have also been reported to have modest decreases in lipolysis in some, but not all, investigations, as well as a small increase in basal levels of *de novo* fatty acid synthesis. Interestingly, the pancreatic insulin secretory response to β-adrenergic stimulation was suppressed in aP2$^{-/-}$ mice, and when aP2 deficiency was introduced into the genetically obese ob/ob mouse, animals lacking A-FABP showed decreased pancreatic insulin secretion as well as improved glucose tolerance. Thus, despite the discrepancies and despite the fact that the mechanisms of the null phenotypic changes are not clear, the collective results indicate that further explorations of the role of A-FABP in fatty acid flux and systemic lipid and glucose metabolism are warranted. A recent report of dramatic increases in serum levels and adipose expression of a bone morphogenic protein in the A-FABP-null mice may point to a heretofore unrecognized interaction between fatty acid metabolism and bone development. At this point, however, the nature of any such association remains to be explored.

A potentially critical involvement for A-FABP in the development of diet-induced atherosclerosis has been recently revealed by studies of A-FABP$^{-/-}$ mice crossed with mice deficient in apoE, the latter being a well-established model of dietary atherosclerosis. ApoE$^{-/-}$ animals develop severe coronary arterial occlusion on a high-fat diet; in dramatic contrast, the A-FABP$^{-/-}$ /apoE$^{-/-}$ mice developed only trivial lesions, strongly indicating a role for A-FABP in the accumulation of lipid-rich foam cells in the arterial intima. As A-FABP is highly expressed not only in adipose tissue but also in macrophages, the results suggested that it plays a critical role in the development of hypercholesterolemia-induced atherosclerosis, likely at the level of the macrophage. Indeed, recent bone marrow transplantation studies showed that macrophage-expressed A-FABP, rather than adipocyte A-FABP, was likely to be primarily involved in the development of dietary atherosclerosis. K-FABP was also found to be expressed in macrophages; however, unlike the compensatory upregulation of expression observed in adipose tissue of the ap2$^{-/-}$ mice, macrophage K-FABP expression remained unchanged. The mechanisms for the apparently proatherosclerotic effects of macrophage A-FABP do not appear to be primarily related to cholesterol ester accumulation, as this was altered to a modest extent, if at all, in macrophages from the A-FABP$^{-/-}$ /apoE$^{-/-}$ mice. Interestingly, levels of interleukin 6, interleukin 1β, macrophage inflammatory proteins 1α and 1β, and macrophage chemoattractant protein 1 were decreased in the apoE/A-FABP-deficient macrophages relative to apoE$^{-/-}$, indicating a role for A-FABP in inflammatory cytokine and chemokine expression.

Recently, the generation of a skin-type FABP (K- or E-FABP)-null mouse (E-FABP$^{-/-}$) was reported. Here, too, it appears that compensatory changes in another FABP are found: mice lacking E-FABP showed increased expression of H-FABP in skin. A dramatic phenotype was not observed in the E-FABP$^{-/-}$ mice; however, changes in the rate, although not the extent of transepidermal water loss, were observed. The mechanisms by which the FABPs may participate in the water barrier function of skin are not yet known. Recently, the construction of transgenic mice overexpressing the *FABP5* gene encoding E-FABP in adipocytes was reported. In adipocytes from FABP5 transgenic mice, the total FABPprotein level was found to increase 150% as compared to the wild type due to a 10-fold increase in the level of E-FABP and an unanticipated 2-fold downregulation of the A-FABP. These results suggested that there is a positive relationship between lipolysis and the total level of FABP, but not between lipolysis and a specific FABP type.

L-FABP knockout mice displayed a dramatically reduced cytosolic fatty acid–binding capacity in the mouse liver. Moreover, a dramatic shift in liver lipid distribution in favor of cholesterol, cholesterol esters, and phospholipids was observed. L-FABP, sterol carrier protein 2

(SCP-2) and SCP-x may contribute to this change in liver lipid distribution; levels of SCP-2 (a known cholesterol-binding protein) increased by approximately 75% while levels of SCP-x decreased in L-FABP null mice. As yet, however, the precise mechanism and functional consequences of the shift in hepatocellular lipid distribution in L-FABP null mice are unknown. It is also worth noting that in an independently generated L-FABP$^{-/-}$ mouse, no changes in SCP-2 levels were detected. In other studies the role of L-FABP in peroxisomal oxidation and metabolism of branched-chain fatty acids was investigated using cultured primary hepatocytes isolated from livers of L-FABP null and wild-type mice. L-FABP gene ablation reduced maximal, but not initial, uptake of phytanic acid (the most common branched-chain fatty acid), and inhibited phytanic acid peroxisomal oxidation and microsomal esterification. L-FABP gene-ablated hepatocytes were also found to contain decreased levels of free fatty acids and triglyceride. L-FABP null mice have been further used to test the hypothesis that L-FABP limits the availability of long-chain fatty acids for oxidation and for PPARα, a fatty acid–binding transcription factor that determines the capacity of hepatic fatty acid oxidation. Indeed, it was found that the mechanisms whereby L-FABP affects fatty acid oxidation may vary with physiological condition; under fasting conditions, hepatic L-FABP appears to contribute to hepatic LCFA oxidation and ketogenesis by a nontranscriptional mechanism, whereas L-FABP appears to activate ketogenic gene expression in fed mice. Independently of these experiments Newberry et al. generated mice with a targeted deletion of the endogenous L-FABP gene and have characterized their response to alterations in hepatic fatty acid flux following prolonged fasting. The resulting data point to an inducible defect in fatty acid utilization in fasted L-FABP null mice that involves targeting of substrate for use in triglyceride metabolism.

6 Perspectives

It has often been suggested that the expression of more than one type of FABP in a single tissue or even a single cell type, is a strong predictor that FABPs perform functions other than or in addition to bulk binding and transport. Recent studies of Widstrom et al., using displacement of the fluorescent probe ANS, showed that H-FABP can bind arachidonic and linoleic acid metabolites, which serve as components of cell signaling cascades. If the FABPs do in fact act to target as well as bind FAs, as the data reviewed here suggest, then there must be specific signals for protein trafficking. An interesting feature of A-FABP is that Phe57 on the β-CD turn rotates by more than 90° into the binding cavity of the apoprotein relative to the holoprotein. As discussed earlier, differences in apo-FABP versus holo-FABP tertiary structures are especially notable in the helix-turn-helix portal domain. Differences in the regulation of fluorescent FA transfer from I-FABP to membranes relative to transfer in the opposite direction further suggests differential trafficking properties of apo- and holo-FABPs. Thus, structural changes in apo- versus holo-FABPs might act to alter interactions with other cellular structures, prevent competition between apo- and holo-proteins for interaction with the same receptor, act as a signal for a ligand-bound protein, or change FABP affinity for a membrane. Evidently, there is much to be learned about the precise mechanisms by which FABPs participate in the uptake, efflux, and intracellular transport and metabolism of their small hydrophobic ligands.

See also Annexins; Membrane Traffic: Vesicle Budding and Fusion; Membrane Transport.

Bibliography

Books and Reviews

Glatz, J.F.C., Luiken, J.J.F.P., van Bilsen, M., van der Vusse, G.J. (2002) Cellular lipid binding proteins as facilitators and regulators of lipid metabolism, *Mol. Cell. Biochem.* **239**, 3–7.

Hamilton, J.A. (2002) Fatty acids bind to proteins: the inside story from protein structures, *Prostaglandins Leukot. Essent. Fatty Acids* **67**, 65–72.

Hamilton, J.A. (2004) Fatty acid interactions with proteins: what X-ray crystal and NMR solution structures tell us, *Prog. Lipid Res.* **43**, 177–199.

Hanhoff, T., Lucke, C., Spener, F. (2002) Insights into binding of fatty acids by fatty acid binding proteins, *Mol. Cell. Biochem.* **239**, 45–54.

Van der Vusse, G.J., van Bilsen, M., Glatz, J.F.C., Hasselbaink, D.M., Luiken, J.J.F.P. (2002) Critical steps in cellular fatty acid uptake and utilization, *Mol. Cell. Biochem.* **239**, 35–43.

Zimmerman, A.W., Veerkamp, J.H. (2002) New insights into the structure and function of fatty acid binding proteins, *Cell. Mol. Life Sci.* **59**, 1096–1116.

Primary Literature

Abumrad, N., Coburn, C., Ibrahimi, A. (1999) Membrane proteins implicated in long-chain fatty acid uptake by mammalian cells: CD36, FATP and FABPm, *Biochim. Biophys. Acta* **1441**, 4–13.

Alpers, D.H., Strauss, A.W., Ockner, R.K., Bass, N.M., Gordon, J.I. (1984) Cloning of a cDNA encoding rat intestinal fatty acid binding protein, *Proc. Natl. Acad. Sci. U. S. A.* **81**, 313–317.

Atshaves, B.P., Foxworth, W.B., Frolov, A., Roths, J.B., Kier, A.B., Oetama, B.K., et al. (1998) Cellular differentiation and I-FABP protein expression modulate fatty acid uptake and diffusion, *Am. J. Physiol.* **274**, C633–C644.

Atshaves, B.P., McIntosh, A.M., Lyuksyutova, O.I., Zipfel, W., Webb, W.W., Schroeder, F. (2004) Liver fatty-acid binding protein gene ablation inhibits branched-chain fatty acid metabolism in cultured primary hepatocytes, *J. Biol. Chem.* **279**, 30954–30965.

Atshaves, B.P., Storey, S.M., Petrescu, A., Greenberg, C.C., Lyuksyutova, O.I., Smith, R., Schroeder, F. (2002) Expression of fatty acid binding proteins inhibits lipid accumulation and alters toxicity in L-cell fibroblasts, *Am. J. Physiol. Cell Physiol.* **283**, 688–703.

Baier, L.J., Bogardus, C., Sacchettini, J.C. (1996) A polymorphism in the human intestinal fatty acid binding protein alters fatty acid transport across Caco-2 cells, *J. Biol. Chem.* **271**, 10892–10896.

Balendiran, G.K., Schnutgen, F., Scapin, G., Borchers, T., Xhong, N., Lim, K., et al. (2000) Crystal structure and thermodynamic analysis of human brain fatty acid- binding protein, *J. Biol. Chem.* **275**, 27045–27054.

Bernlohr, D.A., Angus, C.W., Lane, M.D., Bolanowski, M.A., Kelly, T.J. Jr. (1984) Expression of specific mRNAs during adipose differentiation: identification of an mRNA encoding a homologue of myelin P2 protein, *Proc. Natl. Acad. Sci. U. S. A.* **81**, 5468–5472.

Binas, B., Danneberg, H., McWhir, J., Mullins, L., Clark, A.J. (1999) Requirement for the heart-type fatty acid binding protein in cardiac fatty acid utilization, *FASEB J.* **13**, 805–812.

Chen, A.C., Yu, K., Lane, M.A., Gudas, L.J. (2003) Homozygous deletion of the CRABPI gene in AB1 embryonic stem cells results in increased CRABPII gene expression and decreased intracellular retinoic acid concentration, *Arch. Biochem. Biophys.* **411**, 159–173.

Claffey, K.P., Crisman, T.S., Ruiz-Opazo, N., Brecher, P. (1988) Expression of rat heart fatty acid binding protein cDNA in rat L6 myoblasts, *FASEB J.*, **2**, A1783.

Clark, A.J., Neil, C., Gusterson, B., McWhir, J., Binas, B. (2000) Deletion of the gene encoding H-FABP/MDGI has no overt effects in the mammary gland, *Transgenic Res.* **9**, 439–444.

Coburn, C.T., Knapp, F.F., Febbraio, M., Beets, A.L., Silverstein, R.L., Abumrad, N.A. Jr.

(2000) Defective uptake and utilization of long chain fatty acids in muscle and adipose tissues of CD36 knockout mice, *J. Biol. Chem.* **275**, 32523–32529.

Coe, N.R., Bernlohr, D.A. (1998) Physiological properties and functions of intracellular fatty acid-binding proteins, *Biochim. Biophys. Acta* **1391**, 287–306.

Coe, N.R., Simpson, M.A., Bernlohr, D.A. (1999) Targeted disruption of the adipocyte lipid-binding protein (aP2 protein) gene impairs fat cell lipolysis and increases cellular fatty acid levels, *J. Lipid Res.* **40**, 967–972.

Corsico, B., Liou, H.L., Storch, J. (2004) The alpha-helical domain of liver fatty acid bindingprotein is responsible for the diffusion-mediated transfer of fatty acids to phospholipid membranes, *Biochemistry* **43**, 3600–3607.

Corsico, B., Cistola, D.P., Frieden, C., Storch, J. (1998) The helical domain of intestinal fatty acid binding protein is critical for collisional transfer of fatty acids to phospholipid membranes, *Proc. Natl. Acad. Sci. U. S. A.* **95**, 12174–12178.

Darimont, C., Gradoux, N., Cumin, F., Baum, H.P., De Pover, A. (1998) Differential regulation of intestinal and liver fatty acid-binding proteins in human intestinal cell line (Caco-2): role of collagen, *Exp. Cell Res.* **244**, 441–447.

Davies, J.K., Thumser, A.E., Wilton, D.C. (1999) Binding of recombinant rat liver fatty acid-binding protein to small anionic phospholipid vesicles results in ligand release: a model for interfacial binding and fatty acid targeting, *Biochemistry* **38**, 16932–16940.

Eads, J., Sacchettini, J.C., Kromminga, A., Gordon, J.I. (1993) Escherichia coli-derived rat intestinal fatty acid binding protein with bound myristate at 1.5 A resolution and I-FABPArg106→Gln with bound oleate at 1.74 A resolution, *J. Biol. Chem.* **268**, 26375–26385.

Ellinghaus, P., Wolfrum, C., Assmann, G., Spener, F., Seedorf, U. (1999) Phytanic acid activates the peroxisome proliferator-activated receptor alpha (PPARalpha) in sterol carrier protein 2-/ sterol carrier protein x-deficient mice, *J. Biol. Chem.* **274**, 2766–2772.

Erol, E., Kumar, L.S., Cline, G.W., Shulman, G.I., Kelly, D.P., Binas, B. (2004) Liver fatty acid binding protein is required for high rates of hepatic fatty acid oxidation but not for the action of PPARalpha in fasting mice, *FASEB J.* **18**, 347–349.

Feng, L., Hatten, M.E., Heintz, N. (1994) Brain lipid-binding protein (BLBP): a novel signaling system in the developing mammalian CNS, *Neuron* **12**, 895–908.

Fraser, H., Coles, S.M., Woodford, J.K., Frolov, A.A., Murphy, E.J., Schroeder, F., Bernlohr, D.A., Grund, V. (1997) Fatty acid uptake in diabetic rat adipocytes, *Mol. Cell. Biochem.* **167**, 51–60.

Gedde-Dahl, A., Kulseth, M.A., Ranheim, T., Drevon, C.A., Rustan, A.C. (2002) Reduced secretion of triacylglycerol in Caco-2 cells transfected with intestinal fatty acid binding protein, *Lipids* **37**, 61–68.

Gericke, A., Smith, E.R., Moore, D.J., Mendelsohn, R., Storch, J. (1997) Adipocyte fatty acid-binding protein: interaction with phospholipid membranes and thermal stability studied by FTIR spectroscopy, *Biochemistry* **36**, 8311–8317.

Giguere, V., Lyn, S., Yip, P., Siu, C.H., Amin, S. (1990) Molecular cloning of cDNA encoding a second cellular retinoic acid- binding protein, *Proc. Natl. Acad. Sci. U. S. A.* **87**, 6233–6237.

Glatz, J.F., van der Vusse, G.J. (1996) Cellular fatty acid-binding proteins: their function and physiological significance, *Prog. Lipid Res.* **35**, 243–282.

Gordon, J.I., Alpers, D.H., Ockner, R.K., Strauss, A.W. (1983) The nucleotide sequence of rat liver fatty acid binding protein mRNA, *J. Biol. Chem.* **258**, 3356–3363.

Helledie, T., Antonius, M., Sorensen, R.V., Hertzel, A.V., Bernlohr, D.A., Kolvraa, S., et al. (2000) Lipid-binding proteins modulate ligand-dependent trans-activation by peroxisome proliferator-activated receptors and localize to the nucleus as well as the cytoplasm, *J. Lipid Res.* **41**, 1740–1751.

Herr, F.M., Aronson, J., Storch, J. (1996) Role of portal region lysine residues in electrostatic interactions between heart fatty acid binding protein and phospholipid membranes, *Biochemistry* **35**, 1296–1303.

Herr, F.M., Matarese, V., Bernlohr, D.A., Storch, J. (1995) Surface lysine residues modulate the collisional transfer of fatty acid from adipocyte fatty acid binding protein to membranes, *Biochemistry* **34**, 11840–11845.

Herr, F.M., Li, E., Weinberg, R.B., Cook, V.R., Storch, J. (1999) Differential mechanisms of retinoid transfer from cellular retinol-binding proteins types I and II to phospholipid membranes, *J. Biol. Chem.* **274**, 9556–9563.

Hertzel, A.V., Bennaars_eiden, A., Bernlohr, D.A. (2002) Increased lipolysis on transgenic animals overexpression the epithelial fatty acid binding protein in adipose cells, *J. Lipid Res.* **43**, 2105–2111.

Hodsdon, M.E., Cistola, D.P. (1997) Discrete backbone disorder in the nuclear magnetic resonance structure of apo intestinal fatty acid-binding protein: implications for the mechanism of ligand entry, *Biochemistry* **36**, 1450–1460.

Hotamisligil, G.S., Johnson, R.S., Distel, R.J., Ellis, R., Papaioannou, V.E., Spiegelman, B.M. (1996) Uncoupling of obesity from insulin resistance through a targeted mutation in aP2, the adipocyte fatty acid binding protein, *Science* **274**, 1377–1379.

Hsu, K.T., Storch, J. (1996) Fatty acid transfer from liver and intestinal fatty acid-binding proteins to membranes occurs by different mechanisms, *J. Biol. Chem.* **271**, 13317–13323.

Issemann, I., Green, S. (1990) Activation of a member of the steroid hormone receptor superfamily by peroxisome proliferators, *Nature* **347**, 645–650.

Jefferson, J.R., Powell, D.M., Rymaszewski, Z., Kukowska-Latallo, J., Lowe, J.B., Schroeder, F. (1990) Altered membrane structure in transfected mouse L-cell fibroblasts expressing rat liver fatty acid-binding protein, *J. Biol. Chem.* **265**, 11062–11068.

Jefferson, J.R., Slotte, J.P., Nemecz, G., Pastuszyn, A., Scallen, T.J., Schroeder, F. (1991) Intracellular sterol distribution in transfected mouse L-cell fibroblasts expressing rat liver fatty acid-binding protein, *J. Biol. Chem.* **266**, 5486–5496.

Jenkins, A.E., Hockenberry, J.A., Nguyen, T., Bernlohr, D.A. (2002) Testing of the portal hypothesis: analysis of a V32G, F57G, K58G mutant of the fatty acid binding protein of the murine adipocyte, *Biochemistry* **41**, 2022–2027.

Jenkins-Krutchen, A.E., Bennaars-Eiden, A., Ross, J.R., Shen, W.J., Kraemer, F.B., Bernlohr, D.A. (2003) Fatty acid-binding protein-hormone-sensitive lipase interaction. Fatty acid dependence on binding, *J. Biol. Chem.* **278**, 47636–47643.

Kennedy, M.W., Beauchamp, J. (2000) Sticky-finger interaction sites on cytosolic lipid-binding proteins? *Cell. Mol. Life Sci.* **57**, 1379–1387.

Kerkhoff, C., Sorg, C., Tandon, N.N., Nacken, W. (2001) Interaction of S100A8/S100A9-arachidonic acid complexes with the scavenger receptor CD36 may facilitate fatty acid uptake by endothelial cells, *Biochemistry* **40**, 241–248.

Kim, H.K., Storch, J. (1992a) Free fatty acid transfer from rat liver fatty acid-binding protein to phospholipid vesicles. Effect of ligand and solution properties, *J. Biol. Chem.* **267**, 77–82.

Kim, H.K., Storch, J. (1992b) Mechanism of free fatty acid transfer from rat heart fatty acid-binding protein to phospholipid membranes. Evidence for a collisional process, *J. Biol. Chem.* **267**, 20051–20056.

Kodukula, S., Corsico, B., Storch, J. (2002) Role of Polar Residues in the Helical Domain of Enterocyte Fatty Acid Binding Proteins (FABP) in the mechanism of Fatty Acid (FA) Transfer to Phospholipid (PL) Membranes, *FASEB J.*

Krey, G., Braissant, O., L'Horset, F., Kalkhoven, E., Perroud, M., Parker, M.G., Wahli, W. (1997) Fatty acids, eicosanoids, and hypolipidemic agents identified as ligands of peroxisome proliferator-activated receptors by coactivator- dependent receptor ligand assay, *Mol. Endocrinol.* **11**, 779–791.

Krieg, P., Feil, S., Furstenberger, G., Bowden, G.T. (1993) Tumor-specific overexpression of a novel keratinocyte lipid-binding protein. Identification and characterization of a cloned sequence activated during multistage carcinogenesis in mouse skin, *J. Biol. Chem.* **268**, 17362–17369.

Layne, M.D., Patel, A., Chen, Y.H., Rebel, V.I., Carvajal, I.M., Pellacani, A., et al. (2001) Role of macrophage-expressed adipocyte fatty acid binding protein in the development of accelerated atherosclerosis in hypercholesterolemic mice, *FASEB J.* **15**, 2733–2735.

Le Beyec, J., Delers, F., Jourdant, F., Schreider, C., Chambaz, J., Cardot, P., Pincon-Raymond, M. (1997) A complete epithelial organization of Caco-2 cells induces I-FABP and potentializes apolipoprotein gene expression, *Exp. Cell Res.* **236**, 311–320.

Levin, M.S. (1993) Cellular retinol-binding proteins are determinants of retinol uptake and metabolism in stably transfected Caco-2 cells, *J. Biol. Chem.* **268**, 8267–8276.

Li, E., Demmer, L.A., Sweetser, D.A., Ong, D.E., Gordon, J.I. (1986) Rat cellular retinol-binding protein II: use of a cloned cDNA to define its primary structure, tissue-specific expression, and developmental regulation, *Proc. Natl. Acad. Sci. U. S. A.* **83**, 5779–5783.

Liou, H.L., Storch, J. (2001a) Role of surface lysine residues of adipocyte fatty acid-binding protein in fatty acid transfer to phospholipid vesicles, *Biochemistry* **40**, 6475–6485.

Liou, H.L., Storch, J. (2001b) The role of adipocyte fatty acid binding protein (AFABP) in fatty acud uptake in 3T3-LI cells: altering cellular AFABP levels by electroporation [abstract], *FASEB J.* **15**, A1090.

Liou, H.L., Kahn, P.C., Storch, J. (2002) Role of the helical domain in fatty acid transfer from adipocyte and heart fatty acid-binding proteins to membranes: analysis of chimeric proteins, *J. Biol. Chem.* **277**, 1806–1815.

Luxon, B.A. (1996) Inhibition of binding to fatty acid binding protein reduces the intracellular transport of fatty acids, *Am. J. Physiol.* **271**, G113–G120.

Luxon, B.A., Weisiger, R.A. (1993) Sex differences in intracellular fatty acid transport: role of cytoplasmic binding proteins, *Am. J. Physiol.* **265**, G831–G841.

Luxon, B.A., Milliano, M.T. (1997) Cytoplasmic codiffusion of fatty acids is not specific for fatty acid binding protein, *Am. J. Physiol.* **273**, C859–C867.

Luxon, B.A., Milliano, M.T. (1999) Cytoplasmic transport of fatty acids in rat enterocytes: role of binding to fatty acid-binding protein, *Am. J. Physiol.* **277**, G361–G366.

Makowski, L., Boord, J.B., Maeda, K., Babaev, V.R., Uysal, K.T., Morgan, M.A., et al. (2001) Lack of macrophage fatty-acid-binding protein aP2 protects mice deficient in apolipoprotein E against atherosclerosis, *Nat. Med.* **7**, 699–705.

Martin, G.G., Danneberg, H., Kumar, L.S., Atshaves, B.P., Erol, E., Bader, M., Schroeder, F., Binas, B. (2003) Decreased liver fatty acid binding capacity and altered liver lipid distribution in mice lacking the liver fatty acid-binding protein gene, *J. Biol. Chem.* **278**, 21429–21438.

McCullagh, K.J., Juel, C., O'Brien, M., Bonen, A. (1996) Chronic muscle stimulation increases lactate transport in rat skeletal muscle, *Mol. Cell. Biochem.* **156**, 51–57.

Murphy, E.J. (1998) L-FABP and I-FABP expression increase NBD-stearate uptake and cytoplasmic diffusion in L cells, *Am. J. Physiol.* **275**, G244–G249.

Murphy, E.J., Prows, D.R., Jefferson, J.R., Schroeder, F. (1996) Liver fatty acid-binding protein expression in transfected fibroblasts stimulates fatty acid uptake and metabolism, *Biochim. Biophys. Acta* **1301**, 191–198.

Nemecz, G., Hubbell, T., Jefferson, J.R., Lowe, J.B., Schroeder, F. (1991) Interaction of fatty acids with recombinant rat intestinal and liver fatty acid-binding proteins, *Arch. Biochem. Biophys.* **286**, 300–309.

Newberry, E.P., Xie, Y., Kennedy, S., Han, X., Buhman, K.K., Luo, J., Gross, R.W., Davidson, N.O. (2003) Decreased hepatic triglyceride accumulation and altered fatty acid uptake in mice with deletion of the liver fatty-acid binding protein gene, *J. Biol. Chem.* **278**, 51664–51672.

Owada, Y., Takano, H., Yamanaka, H., Kobayashi, H., Sugitani, Y., Tomioka, Y., et al. (2002) Altered water barrier function in epidermal-type fatty acid binding protein-deficient mice, *J. Invest. Dermatol.* **118**, 430–435.

Pelsers, M.M., Lutgerink, J.T., Nieuwenhoven, F.A., Tandon, N.N., van der Vusse, G.J., Arends, J.W., et al. (1999) A sensitive immunoassay for rat fatty acid translocase (CD36) using phage antibodies selected on cell transfectants: abundant presence of fatty acid translocase/CD36 in cardiac and red skeletal muscle and up- regulation in diabetes, *Biochem. J.* **337**(3), 407–414.

Pelton, P.D., Zhou, L., Demarest, K.T., Burris, T.P. (1999) PPARgamma activation induces the expression of the adipocyte fatty acid binding protein gene in human monocytes, *Biochem. Biophys. Res. Commun.* **261**, 456–458.

Prinsen, C.F., Veerkamp, J.H. (1998) Transfection of L6 myoblasts with adipocyte fatty acid-binding protein cDNA does not affect fatty acid uptake but disturbs lipid metabolism and fusion, *Biochem. J.* **329**(2), 265–273.

Prows, D.R., Schroeder, F. (1997) Metallothionein-IIA promoter induction alters rat intestinal fatty acid binding protein expression, fatty acid uptake, and lipid metabolism in transfected L-cells, *Arch. Biochem. Biophys.* **340**, 135–143.

Prows, D.R., Murphy, E.J., Schroeder, F. (1995) Intestinal and liver fatty acid binding proteins differentially affect fatty acid uptake and esterification in L-cells, *Lipids* **30**, 907–910.

Repa, J.J., Mangelsdorf, D.J. (2000) The role of orphan nuclear receptors in the regulation of cholesterol homeostasis, *Annu. Rev. Cell Dev. Biol.* **16**, 459–481.

Richieri, G.V., Ogata, R.T., Kleinfeld, A.M. (1994) Equilibrium constants for the binding of fatty acids with fatty acid- binding proteins from adipocyte, intestine, heart, and liver measured with the fluorescent probe ADIFAB, *J. Biol. Chem.* **269**, 23918–23930.

Richieri, G.V., Low, P.J., Ogata, R.T., Kleinfeld, A.M. (1999) Binding kinetics of engineered mutants provide insight about the pathway for entering and exiting the intestinal fatty acid binding protein, *Biochemistry* **38**, 5888–5895.

Richieri, G.V., Ogata, R.T., Zimmerman, A.W., Veerkamp, J.H., Kleinfeld, A.M. (2000) Fatty acid binding proteins from different tissues show distinct patterns of fatty acid interactions, *Biochemistry* **39**, 7197–7204.

Rolf, B., Oudenampsen-Kruger, E., Borchers, T., Faergeman, N.J., Knudsen, J., Lezius, A., Spener, F. (1995) Analysis of the ligand binding properties of recombinant bovine liver- type fatty acid binding protein, *Biochim. Biophys. Acta* **1259**, 245–253.

Sacchettini, J.C., Said, B., Schulz, H., Gordon, J.I. (1986) Rat heart fatty acid-binding protein is highly homologous to the murine adipocyte 422 protein and the P2 protein of peripheral nerve myelin, *J. Biol. Chem.* **261**, 8218–8223.

Schaap, F.G., Binas, B., Danneberg, H., van der Vusse, G.J., Glatz, J.F. (1999) Impaired long-chain fatty acid utilization by cardiac myocytes isolated from mice lacking the heart-type fatty acid binding protein gene, *Circ. Res.* **85**, 329–337.

Scheja, L., Makowski, L., Uysal, K.T., Wiesbrock, S.M., Shimshek, D.R., Meyers, D.S., et al. (1999) Altered insulin secretion associated with reduced lipolytic efficiency in aP2−/− mice, *Diabetes* **48**, 1987–1994.

Schroeder, F., Jefferson, J.R., Powell, D., Incerpi, S., Woodford, J.K., Colles, S.M., et al. (1993) Expression of rat L-FABP in mouse fibroblasts: role in fat absorption, *Mol. Cell. Biochem.* **123**, 73–83.

Sha, R.S., Kane, C.D., Xu, Z., Banaszak, L.J., Bernlohr, D.A. (1993) Modulation of ligand binding affinity of the adipocyte lipid-binding protein by selective mutation. Analysis in vitro and in situ, *J. Biol. Chem* **268**, 7885–7892.

Shaughnessy, S., Smith, E.R., Kodukula, S., Storch, J., Fried, S.K. (2000) Adipocyte metabolism in adipocyte fatty acid binding protein knockout mice (aP2−/−) after short-term high-fat feeding: functional compensation by the keratinocyte [correction of keratinocyte] fatty acid binding protein, *Diabetes* **49**, 904–911.

Shen, W.J., Sridhar, K., Bernlohr, D.A., Kraemer, F.B. (1999) Interaction of rat hormone-sensitive lipase with adipocyte lipid- binding protein, *Proc. Natl. Acad. Sci. U. S. A.* **96**, 5528–5532.

Shen, W.J., Liang, Y., Hong, R., Patel, S., Natu, V., Sridhar, K., et al. (2001) Characterization of the functional interaction of adipocyte lipid- binding protein with hormone-sensitive lipase, *J. Biol. Chem.* **276**, 49443–49448.

Smith, E.R., Storch, J. (1999) The adipocyte fatty acid-binding protein binds to membranes by electrostatic interactions, *J. Biol. Chem.* **274**, 35325–35330.

Spitsberg, V.L., Matitashvili, E., Gorewit, R.C. (1995) Association and coexpression of fatty-acid-binding protein and glycoprotein CD36 in the bovine mammary gland, *Eur. J. Biochem.* **230**, 872–878.

Storch, J., Veerkamp, J.H., Hsu, K.T. (2002) Similar mechanisms of fatty acid transfer from human and rodent fatty acid-binding proteins to membranes:Liver, intestine, heart, muscle and adipose tissue FABPs, *Mol. Cell. Biochem.* **239**, 25–33.

Sundelin, J., Laurent, B.C., Anundi, H., Tragardh, L., Larhammar, D., Bjorck, L., et al. (1985) Amino acid sequence homologies between rabbit, rat, and human serum retinol-binding proteins, *J. Biol. Chem.* **260**, 6472–6480.

Thompson, J., Winter, N., Terwey, D., Bratt, J., Banaszak, L. (1997) The crystal structure of the liver fatty acid-binding protein. A complex with two bound oleates, *J. Biol. Chem.* **272**, 7140–7150.

Thumser, A.E., Storch, J. (2000) Liver and intestinal fatty acid-binding proteins obtain fatty acids from phospholipid membranes by different mechanisms, *J. Lipid Res.* **41**, 647–656.

Thumser, A.E., Wilton, D.C. (1996) The binding of cholesterol and bile salts to recombinant rat liver fatty acid-binding protein, *Biochem. J.* **320**(3), 729–733.

Thumser, A.E., Tsai, J., Storch, J. (2001) Collision-mediated transfer of long-chain fatty acids by neural tissue fatty acid-binding proteins (FABP): studies with fluorescent analogs, *J. Mol. Neurosci.* **16**, 143–150.

Uysal, K.T., Scheja, L., Wiesbrock, S.M., Bonner-Weir, S., Hotamisligil, G.S. (2000) Improved glucose and lipid metabolism in genetically obese mice lacking aP2, *Endocrinology* **141**, 3388–3396.

van der Lee, K.A., Vork, M.M., De Vries, J.E., Willemsen, P.H., Glatz, J.F., Reneman, R.S., et al. (2000) Long-chain fatty acid-induced changes in gene expression in neonatal cardiac myocytes, *J. Lipid Res..* **41**, 41–47.

Van Nieuwenhoven, F.A., Willemsen, P.H., van der Vusse, G.J., Glatz, J.F. (1999) Co-expression in rat heart and skeletal muscle of four genes coding for proteins implicated in long-chain fatty acid uptake, *Int. J. Biochem. Cell Biol.* **31**, 489–498.

Walz, D.A., Wider, M.D., Snow, J.W., Dass, C., Desiderio, D.M. (1988) The complete amino acid sequence of porcine gastrotropin, an ileal protein which stimulates gastric acid and pepsinogen secretion, *J. Biol. Chem.* **263**, 14189–14195.

Wang, H., He, Y., Kroenke, C.D., Kodukula, S., Storch, J., Palmer, A.G., Stark, R.E. (2002) Titration and exchange studies of liver fatty acid-binding protein with (13)C-labeled long-chain fatty acids, *Biochemistry* **41**, 5453–5461.

Weise, M.J., Hsieh, D., Hoffman, P.M., Powers, J.M., Brostoff, S.W. (1980) Bovine peripheral nervous system myelin P2 protein: chemical and immunological characterization of the cyanogen bromide peptides, *J. Neurochem.* **35**, 393–399.

Widstrom, R.L., Norris, A.W., Spector, A.A. (2001) Binding of cytochrome P450 monooxygenase and lipoxygenase pathway products by heart fatty acid-binding protein, *Biochemistry* **40**, 1070–1076.

Wilkinson, T.C., Wilton, D.C. (1986) Studies on fatty acid-binding proteins. The detection and quantification of the protein from rat liver by using a fluorescent fatty acid analogue, *Biochem. J.* **238**, 419–424.

Witthuhn, B.A., Bernlohr, D.A. (2001) Upregulation of bone morphogenetic protein GDF-3/Vgr-2 expression in adipose tissue of FABP4/aP2 null mice, *Cytokine* **14**, 129–135.

Wolfrum, C., Borchers, T., Sacchettini, J.C., Spener, F. (2000) Binding of fatty acids and peroxisome proliferators to orthologous fatty acid binding proteins from human, murine, and bovine liver, *Biochemistry* **39**, 1469–1474.

Wolfrum, C., Borrmann, C.M., Borchers, T., Spener, F. (2001) Fatty acids and hypolipidemic drugs regulate peroxisome proliferator-activated receptors alpha- and gamma-mediated gene expression via liver fatty acid binding protein: a signaling path to the nucleus, *Proc. Natl. Acad. Sci. U. S. A.* **98**, 2323–2328.

Wolfrum, C., Buhlmann, C., Rolf, B., Borchers, T., Spener, F. (1999a) Variation of liver-type fatty acid binding protein content in the human hepatoma cell line HepG2 by peroxisome proliferators and antisense RNA affects the rate of fatty acid uptake, *Biochim. Biophys. Acta* **1437**, 194–201.

Wolfrum, C., Ellinghaus, P., Fobker, M., Seedorf, U., Assmann, G., Borchers, T., Spener, F. (1999b) Phytanic acid is ligand and transcriptional activator of murine liver fatty acid binding protein, *J. Lipid Res.* **40**, 708–714.

Wootan, M.G., Storch, J. (1994) Regulation of fluorescent fatty acid transfer from adipocyte and heart fatty acid binding proteins by acceptor membrane lipid composition and structure, *J. Biol. Chem.* **269**, 10517–10523.

Wu, F., Corsico, B., Flach, C.R., Cistola, D.P., Storch, J., Mendelsohn, R. (2001) Deletion of the helical motif in the intestinal fatty acid-binding protein reduces its interactions with membrane monolayers: Brewster angle microscopy, IR reflection-absorption spectroscopy, and surface pressure studies, *Biochemistry* **40**, 1976–1983.

Xu, Z., Bernlohr, D.A., Banaszak, L.J. (1993) The adipocyte lipid-binding protein at 1.6-A resolution. Crystal structures of the apoprotein and with bound saturated and unsaturated fatty acids, *J. Biol. Chem.* **268**, 7874–7884.

Young, A.C., Scapin, G., Kromminga, A., Patel, S.B., Veerkamp, J.H., Sacchettini, J.C. (1994) Structural studies on human muscle fatty acid binding protein at 1.4 A resolution: binding interactions with three C18 fatty acids, *Structure* **2**, 523–534.

Zimmerman, A.W., van Moerkerk, H.T., and Veerkamp, J.H. (2001) Ligand specificity and conformational stability of human fatty acid-binding proteins, *Int. J. Biochem. Cell Biol..* **33**, 865–876.

Intracellular Fatty Acid Binding Proteins in Metabolic Regulation

John M. Stewart
Mount Allison University, Sackville, New Brunswick, Canada

Encyclopedia of Molecular Cell Biology and Molecular Medicine, 2nd Edition. Volume 7
Edited by Robert A. Meyers.

ISBN: 3-527-30549-1

Keywords

Allosteric
Literally "other space." The action of a modifying molecule on a protein at a site other than the usual site at which the protein performs its function.

Anabolic
Those reactions in biological systems that synthesize more complex molecules from simpler molecules, usually requiring energy.

Binding Site
A cavity formed by the folding of the protein backbone with the ability to bind a specific molecule in a discrete three-dimensional relationship to the bulk of the protein structure.

Catabolism
Those reactions in biological systems that break down complex molecules to simpler molecules, usually releasing energy that is trapped and converted to ATP.

Dissociation Constant
The dissociation constant (K_d) is a measure of the equilibrium between protein-bound and unbound ligand. K_d is also the concentration of ligand that occupies half of the protein binding sites in the system. A measure of the strength of ligand binding; a smaller K_d value represents stronger binding than a larger K_d.

Enzyme
A protein synthesized by the cell that behaves as a catalyst to a chemical reaction. Almost all chemical reactions in a biological system are catalyzed by a separate and specific enzyme. Enzyme names usually end in "-ase."

Fatty Acids
A class of biological molecules that consist of long, hydrophobic carbon chains with a single carboxylic acid functional group; poorly soluble in water and, hence, hydrophobic.

Gluconeogenesis
The synthesis of "new" glucose from smaller molecules such as pyruvate. In some senses, a reverse of the results of glycolysis.

Glycation
The chemical reaction between a carbohydrate and an amine group of a protein to form a covalent C=N bond (a Schiff base). The surface structure and operation of the protein is usually altered by this reaction.

Glycolysis
The series of 10 biochemical, enzyme-catalyzed reactions that converts glucose into pyruvate and is located in the cytoplasm of the cell.

Hyperlipidemia
A concentration of lipids larger than consistent with normal levels.

Inhibition Constant
The concentration of an enzyme inhibitor that reduces the enzyme rate by one half (K_i).

Isoform
In proteins, a variant of a protein coded for by a different gene and contains one, or more, amino acid substitutions.

Isosteric
Literally, "same space." The action of a modifying molecule on a protein at a site, which is the same as that where the protein function occurs.

Ketone Bodies
When carbohydrate is curtailed, as in fasting, starvation, or diabetes, fatty acids are converted to a class of small, four-carbon molecules called *ketone bodies*, usually in the liver. Ketone bodies are more soluble and are transported in the circulation to muscles where they are converted to energy. One of the ketone bodies decomposes spontaneously to produce acetone that is exhaled.

Krebs Cycle
A central, cyclic series of eight biochemical, enzyme-catalyzed reactions taking place in the mitochondria that further degrade carbon resources from carbohydrates, fatty acid oxidation, and protein break down to carbon dioxide, extracting the chemical energy inherent in these structures. This is also known as the *citric acid cycle*.

Ligand
A molecule that binds to a specific protein binding site. Here, fatty acids are ligands to fatty acid binding proteins.

Metabolism
A combination of all catabolic and anabolic reactions.

Metabolic Flux
The rate at which a specific metabolic pathway, or a segment of that pathway, processes metabolites through the system. For example, the flux of glucose through glycolysis

would be the rate at which the combined reactions of glycolysis convert glucose to pyruvate.

Monomeric
A protein with only one subunit; opposite of multimeric, where a number of individual protein molecules form a complex.

Null mice
Laboratory mice in which a gene has been deleted.

Nuclear Hormone Receptors
A large family of proteins in the nucleus that bind various ligands, associate with other transcription factor proteins to form dimers, and then target expression sequences on genes, activating their transcription.

β-Oxidation
A series of enzyme-catalyzed reactions that occur mainly in the mitochondria that converts fatty acids into two carbon units for entrance into the Krebs Cycle.

Phosphorylation
The addition of a phosphate group, usually from ATP, to a simple molecule or to a protein through an ester bond formation. Protein phosphorylation is usually a switch to turn the protein activity on or off.

Peroxisome
A subcellular organelle or compartment.

Polyunsaturated Fatty Acids
Long-chain fatty acids that contain two or more carbon–carbon double bonds.

Protein
A polymer of amino acid linked by a peptide bond. The sequence of the 20 types of amino acids used in protein synthesis is dictated by the gene sequence.

Long-chain fatty acids are paradoxical: they are fuel, structural components, signaling molecules, and, at elevated levels, are toxic. Because of this nature of fatty acids, control of their concentration and movement within cells is important. Central to the biological and biochemical role of fatty acids is a family of proteins referred to as the *fatty acid binding proteins* (FABPs). After early discoveries that the FABPs were the intracellular transport system for fatty acids, their other potential roles, spurred on by the discovery that there were many types of FABPs, have continued to mount. This ancient family of proteins emerged before the divergence of the vertebrate and invertebrate lines (more than one billion years ago) in response to the necessity of manipulating energy-rich fatty acids. A number of ancillary functions seem to have accrued to this diverse protein family that place them in a central position of energy metabolism, fuel sensing, and monitoring the concentrations of fatty acids derived from the diet.

1 Introduction

1.1 Fatty Acid Binding Proteins

Fatty acid binding proteins (FABPs) are a family of small intracellular proteins with a molecular weight between 14 000 and 15 000 (about 133 amino acids). With intracellular concentrations ranging from 200 to 400 μM, they can represent between 3 and 15% of soluble protein inside cells, depending upon the animal species and the tissue. For example, in humans, heart cells' FABP represents between 3 and 5% of soluble protein, while in the flight muscle cells of migrating Western Sandpipers, it can be 20% of soluble cellular protein. While the protein family has been observed in all animals (vertebrates and invertebrates), there is great three-dimensional structural similarity. The molecule is compact (about 250 × 350 × 400 nm) and, although generally referred to as a *β-clam motif* with one "side" pinched inward, has a barrel shape. Liver- and muscle-type FABP are shown in Fig. 1. All strands are hydrogen-bonded to the preceding and succeeding strands with only one exception, known as the *gap*. This rift in the continuous surface of the barrel (facing the reader in Fig. 1) allows flexibility of the barrel structure without significant disruption of the H-bonded strands around the rest of its circumference. At the "top" of the gap region (the structure is usually oriented with the helix-turn-helix motif at the top and back) is the portal region where the base of the helix-turn-helix and right turns of two *β*-strands are proximal. It is through the portal region that the ligand probably enters and exits the internal binding site. The internal volume to accommodate ligands has been determined and can vary from about 250 to 440 A^3, with the largest observed for liver FABP that accommodates two, rather than the usual single, fatty acid molecule.

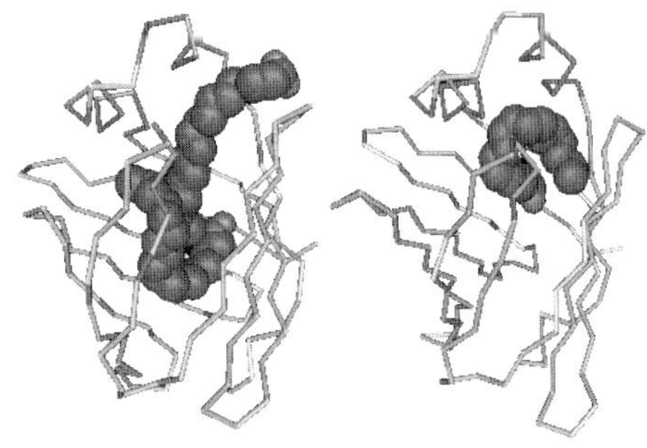

Fig. 1 The peptide backbone structures of rat liver FABP (left panel) with both binding sites filled with oleate and human muscle FABP (right panel) with its single binding site occupied. The images were derived from Cn3D v 4.0 from NCBI with accessions codes 1HMS (muscle protein) and 1LFO (liver protein).

1.2 General Introduction to Fatty Acid Binding Protein Function

FABPs have functions arising from their ability to bind long-chain fatty acids (FAs) and to interact with either the membranous or aqueous fraction of the cell (predominantly the cytoplasm). The potential of interactions between FABPs and other proteins has been supported by evidence that FABPs can alter enzyme kinetics, interact directly with both hormone-sensitive lipase and a cell membrane fatty acid transport protein (CD36), and with peroxisome proliferator–activated receptors (PPAR) in the nucleus that are involved in turning on the expression of various genes. Studies with null mice in which the gene for heart FABP was ablated reflected not only the role of the protein in upregulating genes of FA metabolism but also the compensatory activation of the glycolytic system and apparent reduction of the usual controls of glycolysis.

FABPs can also bind and transport signaling molecules such as bilirubin, prostaglandin E1, and lipoxygenase metabolites of the fatty acid arachidonic acid, although some signaling lipids such as sleep-inducing oleamide in the central nervous system do not bind to brain FABPs. Further, FABPs may protect polyunsaturated fatty acids from free radical damage (peroxidation) through preferential binding of them while greater partitioning of oxidation products of polyunsaturated fatty acids into FABPs, when compared with unilamellar vesicles, may indicate protection of membranes from these monohydroxy fatty acids. Thus, FABPs are involved in intracellular signaling, cell development, and membrane protection, and it appears that they are more than intracellular ferries of poorly soluble hydrophobic molecules.

The idea is developed herein of an interaction between carbohydrate and fatty acid energy metabolism (also known as *fuel sensing* or *fuel preference*) that involves FABPs. Evidence is presented to support the contention that interactions between nonlipid metabolites and FABPs can modulate the partitioning of long-chain FAs into FABP binding sites. Additionally, experimental results showing the potential for reciprocal interaction between FAs and enzymes of glycolysis will be presented. Finally, different FABP types (muscle and liver) respond differently to nonlipid metabolites. The suggested cross talk between the two fuel streams in energy metabolism is centered on substrate cycles involving the enzyme pairs hexokinase-glucose 6-phosphatase and phosphofructokinase-fructose 1,6-phosphatase in muscle and glucokinase-glucose 6-phosphatase in liver. Differential involvement of different FABP types with separate metabolic loci may provide some insight into why there are multiple intracellular fatty acid binding proteins but only a single major extracellular fatty acid binding protein.

1.3 Established Interactions between Carbohydrate and Fatty Acid-based Energy Production

The exchange of rudimentary status information (predominantly concentration) between carbohydrate and FA catabolisms by substrate level control and inhibition/activation of participating enzymes is well established. The reciprocal interaction between FA levels and glucose use, referred to as the *Randle cycle*, suggests that modulation of glycolytic flux depends upon the mitochondrial production of citrate which, when transported to the cytoplasm, inhibits the major flux generator of this pathway, phosphofructokinase-I (PFK-I). In effect, an elevated citrate concentration inhibits PFK-I and slows the flow of carbohydrate through this point in the pathway. The other major effect of FA is to activate a pyruvate dehydrogenase kinase that phosphorylates mitochondrial pyruvate dehydrogenase and decreases its activity. This curtails the movement of pyruvate into mitochondrial pathways (Krebs Cycle), and an overall decrease is observed in glucose oxidation. This accounts well for the commonly observed decrease of glucose oxidation when lipid levels are increased.

The control of mitochondrial oxidation of fatty acids (β-oxidation) is vested in gating entry of FAs into the mitochondrial matrix through coordinated actions of carnitine palmitoyltransferase I (CPT I), palmitoylcarnitine translocase, and matrical carnitine palmitoyltransferase II. A powerful inhibitor of CPT I and part of the fuel sensing system is malonyl-CoA, the first committed metabolite in FA synthesis and itself a product of cytoplasmic citrate. CPT I is also under phosphorylation control in liver, where phosphorylation by cAMP-protein kinase increases the CPT I activity by 30 to 80%. Thus, an increase in mitochondrially produced citrate (from carbohydrate or other sources) increases its partition to the cytoplasm. An increased cytosolic concentration of citrate inhibits glucose processing at PFK-I and shifts the carbon resources it represents into the pathway that synthesizes fatty acids. The first metabolite along this anabolic pathway is malonyl-CoA that "shuts down" entry of fatty acid carbon into the mitochondria where it is oxidized.

1.4
The Involvement of FABP in Metabolism: Working Hypothesis

An additional level of lipidic control of glucose phosphorylation in muscle has recently been reported: CoA esters of long-chain FAs (in rat and human skeletal muscle) and long-chain FAs (in bovine heart) inhibit hexokinase I. This enzyme, by adding a phosphate group to glucose, traps glucose in the cell. There are many older reports that long-chain FAs inhibit a number of other enzymes, but uncertainty arises from those reports since unrealistically large FA concentrations were employed, and detergental effects cannot be ruled out. One careful kinetic study from that pre-FABP era was the inhibitory effect of palmitate on muscle PFK-I with an inhibition constant (K_i) of 25 μM, although the physical binding of FAs to PFK-I has not been studied. Thus, if FAs can influence enzymes of glycolysis (and elsewhere, see below), an involvement of FABP in delivering the FA to the sites of action, presumably through protein:protein interaction, is strongly implicated since such fatty acids have vanishingly small solubility in the aqueous cytosol. Other areas where long-chain FAs have influence, such as the ADP/ATP translocase-porin-hexokinase complex of mitochondria, might also be examined productively.

If FABPs modulate FA metabolism above the level of the genetic machinery, to what signals do they respond? Clearly, any productive modulation of FABP function must act in response to the instantaneous status of the cell/tissue. Some years ago, we suggested that the metabolic status of the cell might be part of the signaling mechanism influencing FA binding characteristics of various FABPs. Starting with this idea, we began to search for metabolites or conditions within the cell that might influence the FABP operation through modulation of formation of the protein:FA complex. This would probably be an allosteric modulation (in the strictest sense of the term, i.e. not at the fatty acid binding site) since competitive interaction, such as observed with ferri- and ferroheme, would be a binary system (on/off) and not optimally adaptive to a dynamic intracellular milieu. It must be pointed out that, in a strictly thermodynamic sense (i.e. mass action), the effect would result from repartitioning effects. Although it might seem unusual that a small, monomeric protein with a relatively small surface area that changed little upon binding FAs would display allosteric properties usually observed with large, multimeric enzyme systems, such regulation has been demonstrated for other small proteins: cytochrome *c* has a saturable binding site for ATP that regulates the electron transport system and hence oxygen consumption and ATP production. If FABPs are involved in the productive modulation of lipid metabolism and in targeting of FABPs, for example, to the mitochondria, there must be an exchange of information between the binding of FAs and the metabolite status of the cell.

1.5 Criteria for Physiological Relevance of Metabolite Modulation of Fatty Acid Binding to FABP

In the search for possible modulators of FABPs, we compared the degree of binding of reporter ligands (radioactively labeled or fluorescent fatty acids) in the presence and absence of physiological concentrations of metabolites. The assay conditions used in such studies are particularly important since FA binding is effected by ionic strength, pH, and by temperature, and it is vital to mimic the intracellular conditions as closely as possible. Statistically significant changes in binding of reporter ligands to FABP must occur at modulator concentrations encountered under physiological conditions. We examined potential modulators at concentrations found in unstressed tissue since metabolic flux is normally curtailed compared to the maximal flux attainable.

A second level criterion we employ is that the equilibrium dissociation constant (the apparent K_d) of any modulator must also fall within physiological concentration ranges to be part of a normal regulatory network. Finally, interactions that influence FA binding should be consistent with what is known about particular metabolic pathways and physiological states.

1.6 Liver FABP

As an initial test of the working hypothesis, we examined the effect of glycolytic metabolites on the binding of both ^{14}C-labeled oleate and the fluorescent probe *cis*-parinarate (cPnA) to rat L-FABP [27] at 37 °C. To do this, we set the concentration of reporter molecules at the value of the K_d of the ligand, where binding is most sensitive to change. After establishing the FABP:FA complex, we challenged it with test metabolites. We reported that, of the early glycolytic intermediates, the most effective modulators of the ability of L-FABP to bind FAs were glucose and glucose-6-phosphate (G6P). Both of these metabolites increase the ability of FABP to bind long-chain FAs 30 and 40% for 6 mM glucose and 0.25 mM G6P respectively. An alternate binding assay using the fluorescent probe cPnA provided the same result. In examining the effect of increasing concentrations of these two metabolites on oleate and cPnA binding, the concentrations required to produce a 50% increase in the observed FA binding change were 6 mM and 0.2 mM respectively, within the normal physiological range of glucose and G6P in rat liver. As part of this study, we also examined glucose-1-phosphate (G1P), phosphate ion, fructose-6-phosphate (F6P), fructose-1,6-bisphosphate (F-1,6-P_2), and fructose-2,6-bisphosphate (F-2,6-P_2), phosphoenol pyruvate (PEP), ATP, ADP, AMP and cAMP, NAD^+ and NADH, $NADP^+$ and NADPH, and acetyl-CoA, none of which had an effect on FA binding to rat L-FABP.

A recent study included an examination of the effect of glucose and G6P on [1-^{14}C]oleate binding to human H-FABP and L-FABP. While there was no effect of glucose or G6P on oleate binding as compared to that on the control, the human muscle-type FABP was affected and was similar (although smaller in magnitude) to the effects on bovine H-FABP, as described below.

We extended this study to L-FABP of another rodent species, Richardson's ground squirrel (*Spermophilus richardsonii*), arguing that such a mechanism may be more generally observed. In this system, glucose and G6P also increased the binding of cPnA to the FABP at 37 °C compared to that of the control by 20.6 ± 0.5% and 41.4 ± 2.2% respectively ($n = 4; p < 0.05$), when examined under the same conditions, as reported before, for rat and using *cis*-parinarate as a fluorescent probe.

In summary, Glc and G6P increased the ability of L-FABP to bind FAs in three mammalian systems and to different degrees. Since the number of ligand molecules bound to the protein was not likely to increase (2 : 1 binding stoichiometry for L-FABP), the effect was probably centered on changes in the binding strength for fatty acid. It should be noted that the effect of glucose on liver-type FABP to glucose is exhibited in frogs (*Rana pipens* and *Rana sylvatica*) and turtle (*Chrysemus picta belli*). The similarity of response in mammals, amphibians, and reptiles suggests a more general phenomenon. We are presently determining amino acid sequences of these proteins and carrying out topological mapping of three-dimensional structures in an attempt to identify surface features of the protein that would best provide information for a phase of site-directed mutational studies that could create the glucose effect or eliminate it.

Because of different experimental protocols and few reports of work with pure proteins, it is difficult to find a general metabolic scenario into which the L-FABP work can be fit. However, recent examination of the effect of oleate on the glycolytic/gluconeogenic balance and glycogen systems of fasted rat hepatocytes is consistent with a role for FABP. In the presence of S4048, an inhibitor of endoplasmic reticulum G6P translocase, oleate increased intracellular G6P 3-fold and glycogen production 1.5-fold. Further, oleate did not have an inhibitory effect on glucokinase (the liver form of hexokinase), but stimulated G6P phosphatase flux (part of the putative hexokinase/G6P phosphatase substrate cycle) and decreased overall lactate production. The increase in G6P was attributed to increased gluconeogenesis. Thus, it appears that long-chain fatty acids, while interacting with early glycolytic enzymes, foster gluconeogenesis in liver. Presumably, an increase in G6P or glucose could increase the partitioning of

FAs to the FABP and behave as a communicative link to lipid status.

Other studies in our laboratory with rat L-FABP have examined the role of the ketone body, D-3-hydroxybutyrate, where, at normal physiological levels (0.14 mM), there was only a small increase in the binding of FA (10%). At levels mirroring diabetic or fasting/starvation hepatic levels (about 7 mM), there was a 45.5% increase in FA binding. The effect was saturable with an inhibition constant, K_i (3-hydroxybutyrate), of 4.9 mM, which is not physiological under euglycemic or nonstarvation conditions. Increased FA binding with elevated 3-hydroxybutyrate (when carbohydrate supply is curtailed as in diabetes or fasting/starvation) could reverse a tendency to gluconeogenesis and modify delivery of available FAs.

1.7 Heart/Muscle FABP

For heart tissue, which has different hexokinase (HK) isoforms and a different FABP type from liver (H-FABP or M-FABP), the situation appears equally complex (note: muscle-type and heart-type FABPs are the same protein and will be referred to as H-FABP here). We showed that bovine heart hexokinase (type I) is inhibited by long-chain fatty acids in a manner that correlates positively with chain length and degree of saturation, but is activated by FAs containing less than 12 carbons. Hexokinase also has a saturable binding site for cPnA and oleate with dissociation constant (K_d) values of 3 μM and 1.3 μM respectively (by fluorescence competition assay), while the short and medium-chain FAs that activated HK did not bind to this site. In other words, long-chain fatty acids bind strongly to hexokinase in a discrete binding site to inhibit the trapping of blood borne glucose within the cell. We can now report work on the FABP purified from bovine heart and for heart-type FABP from the spadefoot toad (*Scaphiopus couchii*).

Figure 2 shows the effects of a number of early glycolytic intermediates on FA binding to the FABP Bovine H-FABP was insensitive to glucose and G6P, as reported for human H-FABP and with toad heart FABP. There was no significant effect on the binding of FAs by physiological concentrations of F6P, but there was significant inhibition of FA binding by $F16P_2$. G1P and P_i did not have any effect on the binding of the fluorescent probe. In contrast to mammalian L-FABP, FA binding by bovine H-FABP was unchanged by 3-hydroxybutyrate.

Unlike hepatic tissue, muscle and heart tissue do not actively produce glucose or ketone bodies. Heart and red skeletal muscles import carbon resources (carbohydrates, fatty acids, and ketone bodies) in response to energy needs and are predominantly aerobic. This essentially unidirectional resource flow is reflected in the poise of muscle metabolic pathways, where the major control points are PFK-I, pyruvate dehydrogenase, and mitochondrial import of FAs. As discussed above, both long-chain FAs and their CoA derivatives inhibit glucose phosphorylation in heart and skeletal muscle. The focus of the interactions in muscle tissue appears to be PFK-I. Long-chain FAs inhibit muscle PFK-I ($K_i = 25$ μM) and $F\text{-}1,6\text{-}P_2$ (at 0.03 mM) inhibits FA binding, establishing the possibility of a reciprocal interaction. Elevated FA could decrease glycolytic flux through inhibiting actions on both HK and PFK-I, while decreased $F\text{-}1,6\text{-}P_2$ could lift resting state inhibition of FA binding to this FABP type.

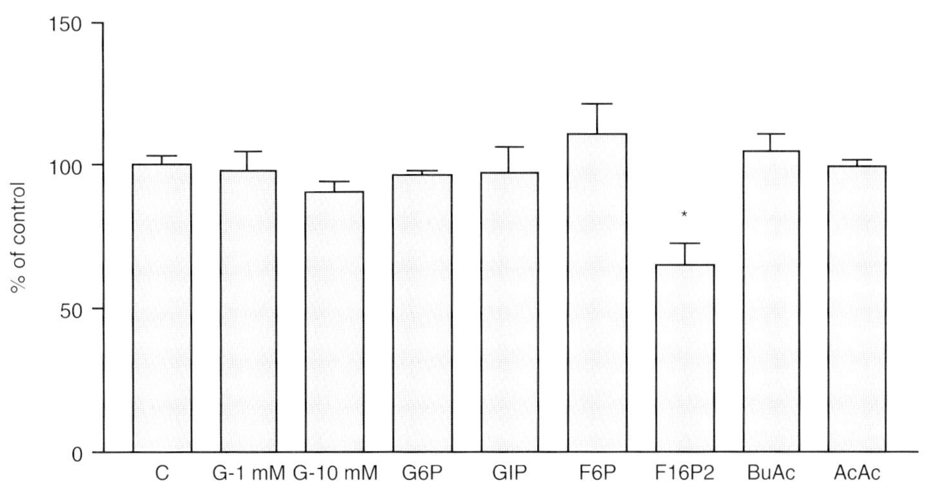

Fig. 2 The effect of various metabolites on the binding of *cis*-parinaric acid (1 μM) to bovine heart FABP compared with the control (no nonlipid metabolite). The assays were carried out at 37 °C, in 50 mM sodium phosphate buffer, pH 7. The assays contained 20 μg of pure proteins per milliliter. The concentrations of metabolites not indicated were G6P, 0.5 mM; F6P, 0.1 mM; F1,6P2, 0.03 mM; 3-hydroxybutyrate (ButAc), 0.1 mM; acetoacetate (AcAc), 0.02 mM. The asterisk indicates statistically significant differences from the control at the 95% confidence limit. The experiments were replicated three to five times with the data representing the means and the error bars the standard deviation.

2 Potential of Formation of Schiff Bases: Nonenzymatic Glycation of FABPs

Carbohydrates that undergo ring opening and formation of a carbonyl group can form Schiff bases with primary amines such as lysine residues that are prevalent in FABP and which appear involved in the collisional mechanism proposed for loading and unloading muscle-type FABPs. Since many of the glycolytic intermediates that can ring open to produce the carbonyl function group influence binding of FAs to FABPs (G6P, glucose, F-1,6-P_2) while G1P and F-2,6-P_2 (which are locked in the closed hemiacetal and hemiketal rings) do not affect FA binding, it is possible that formation of Schiff bases with protein amino groups may cause the observed altered sequestration of FAs. To examine this possibility, we incubated a number of purified FABPs and bovine serum albumin with fluorescently tagged glucose (2-(*N*-(7-nitrobenz-2-oxa-1,3-diazol-4-yl)amino)-2-deoxyglucose; 2-NBDG) at room temperature and in 50 mM sodium phosphate, pH 7.0. We monitored the proteins for indications of covalent fluorescent tagging that would result from Schiff base formation. Periodically, aliquots of the incubation mixtures were removed, and 2-NBDG was separated from protein with a small size-exclusion column. The protein fraction was examined spectrofluorometrically to detect fluorescent tagging (excitation at 473 nm, emission at 532 nm). Figure 3 shows the effect of incubating a number of FABPs and BSA with the fluorescent tag for three days. Only after 48 h of incubation were proteins tagged with the fluorescent

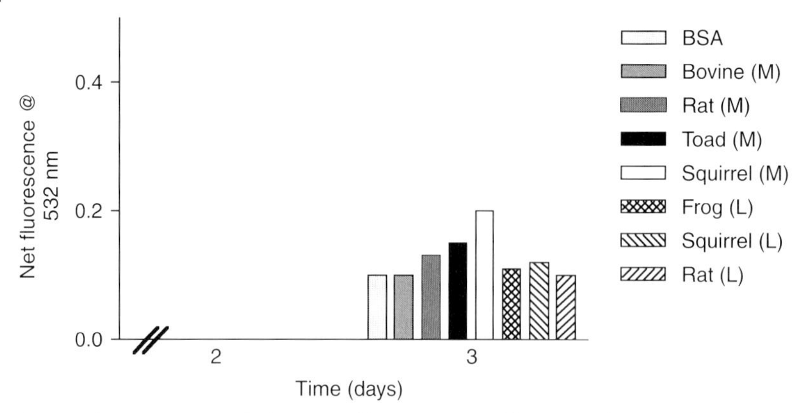

Fig. 3 The time course of glycation of bovine serum albumin and assorted FABPs (30 μg/ml) with D-glucose analog NBDG (80 μM) in 10 mM sodium phosphate buffer, pH 7.0, and 1% (w/v) sodium azide at 25 °C. Over the time course, as proteins (after separation from small molecules with size exclusion) became glycated, they became fluorescently tagged (473-nm excitation, 532-nm emission). A fluorescence value of 0.5 represents 1-nmol fluorescent tag.

carbohydrate derivative at detectable levels. The slowness of the nonenzymatic glycations we observed was consistent with nonenzymic glycation rates of human serum albumin (days) using other fluorometric methods. Thus, it is unlikely that Schiff base formation was significant during the time required for making the measurements of the effects of metabolites on FA binding (minutes).

3 Theoretical Effects and Implications of Reciprocal Cross Talk

3.1 How Much Fatty Acid Would Be Available to Interact with Hexokinase?

The similarity of dissociation constants of HK and FABP for oleate and cPnA leads to an important question surrounding such mutual interactions. Given the large cellular concentration of FABP, is there enough unbound FA to interact with HK? To model the amount of unbound FA in the presence of two different binding proteins requires the concentration value of both proteins, the total FA concentration, and the equilibrium dissociation constants (K_d) of both binding proteins to be known. The intracellular concentration of FABP has been reported to be 85 μM, while some estimates provide ranges of 200 to 400 μM. An intermediary value of 250 μM muscle-type FABP was used in the following model.

Determining the molar concentration of HK in heart tissue is more difficult since it is not directly reported in the literature. We can estimate from the maximal tissue activity (V_{max}) and the measured catalytic rate constant (k_{cat}) of the enzyme a reasonable enzyme concentration since $V_{max} = k_{cat}[E]_T$, where $[E]_T$ is the total enzyme. The k_{cat} of type I HK is 64 s^{-1}, while the maximal activity in mammalian heart has been reported between 5 and 10 μmol min^{-1} g^{-1} tissue for mammals. A value of 8 μmol min^{-1} g^{-1} tissue (porcine heart) was used here.

Finally, the estimate must convert the tissue weight–based activities to concentrations: the assumption of 75% water content will suffice. Thus, with $V_{max} = 8\ \mu\text{mol min}^{-1}\ \text{g}^{-1}$ and $k_{cat} = 64\ \text{s}^{-1}$, we can estimate a minimum $[\text{E}]_\text{T} = 3\ \mu\text{M}$. Using a K_d of 1 μM for FABP and 1.3 μM for HK, we can model equilibrium concentrations of unbound FA and degree of saturation of both protein systems at different initial FA concentrations. Figure 4(a and b) shows the result of this very simple equilibrium model.

The amount of free FA in the simple model in which FABP is present is enough to saturate HK, as indicated by Fig. 4(a) that shows the binding isotherms of FABP at various free FA concentrations, and

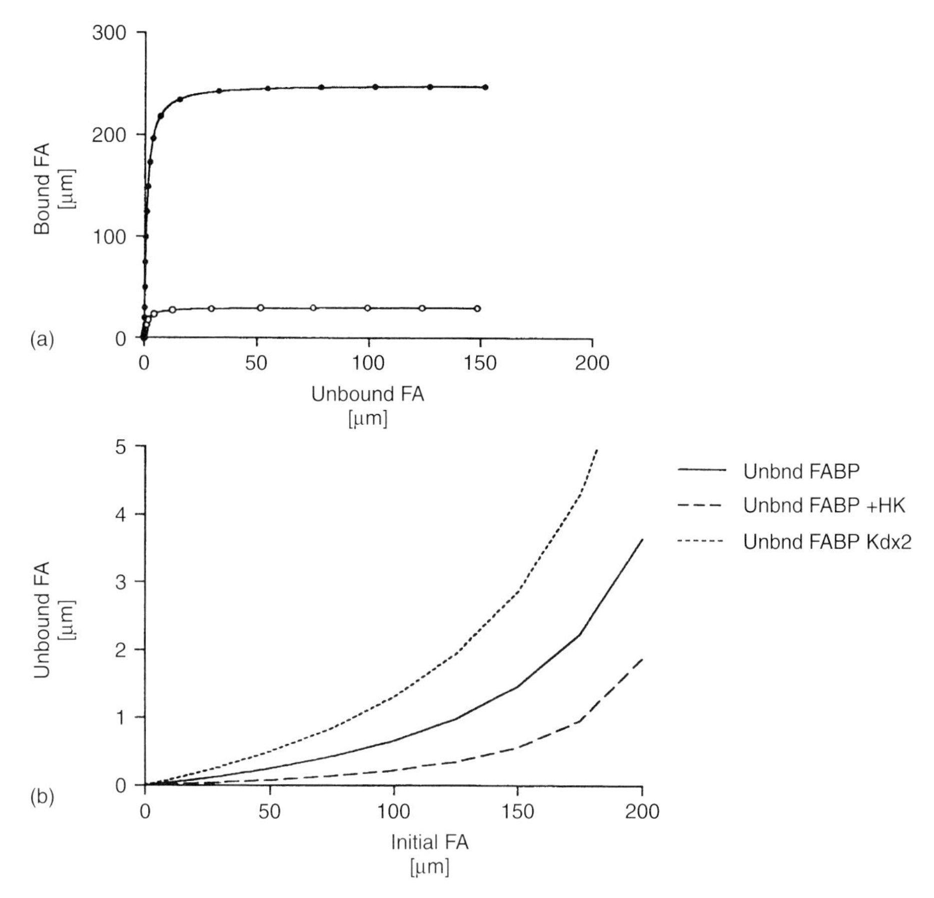

Fig. 4 (a) The degree of saturation of a muscle-type FABP and hexokinase versus the amount of unbound fatty acid (oleate) in the model system. The saturation curve of hexokinase (lower curve) was generated using the concentration remaining after FABP equilibrated its binding site with ligand. The hexokinase binding curve was magnified by 10 to be observable at this scale. (b) The residual concentration of unbound fatty acid at various initial fatty acid concentrations in the presence of FABP and of FABP plus hexokinase. The upper line indicates the effect on the amount of unbound fatty acids if binding strength of FABP was decreased by a factor of two (i.e. K_d increased twofold).

the binding curve of HK using the FA remaining unbound in the presence of FABP. Figure 4(b) shows the estimated concentration of unbound FA in the presence of FABP, of FABP plus HK, and the effect on the amount of unbound FA when the K_d value of the FABP for oleate is doubled (i.e. binding is weakened), as is seen in toad muscle FABP in the presence of F-1,6-P_2. Thus, even with FABP present, providing very low concentrations of unbound FAs, unbound FA could be sequestered on HK.

4 Difference in Modulation of Fatty Acids and between Different Types of FABP

The data suggest that liver FABPs and muscle types of FABP respond different nonlipid metabolites. While the liver types increase the degree of FA binding in the presence of glucose and G6P, with an activation of binding also elicited by ketone bodies, muscle type does not respond to glucose or G6P, is inhibited by F-1,6-P_2, and does not respond to ketone body. If FABPs are involved in fuel selection, the different metabolic poise of these two tissues would demand that they operate differently. The differential response may offer some part of an explanation of the presence of different FABPs in different tissues and perhaps multiple forms in some tissues.

5 Where Else Should We Look?

5.1 Other Enzymes Influenced by Fatty Acids

The influence of long-chain FAs on enzyme kinetics, specifically hexokinase, and PFK-I have been noted above. FAs also have an effect on the myoglobin system and bind strongly to cytochrome *c* altering their functions in the cell. As part of an examination of potential sites of interference during hyperlipidemia, we examined a battery of enzymes to see if elevated FA concentrations altered their reaction kinetics. We examined the enzymes of glycolysis (HK, PFK, PK, lactate dehydrogenase, aldolase, triosephosphate isomerase, pentose phosphate cycle (G6P dehydrogenase), the citric acid cycle (citrate synthase, malate dehydrogenase, and α-ketoglutarate dehydrogenase), and β-oxidation (CPT-I)). Additionally, we examined the effects of fatty acids on the enzymes that degrade reactive oxygen species (catalase, superoxide dismutase). Table 1 shows the results of incubation of enzymes of major metabolic systems with oleate. In these studies, all enzymes were obtained from commercial sources, and assays were established that provided maximal specific activities. We compared the maximal enzyme activities over a range of FA concentrations with the highest being 0.2 mM. The enzymes listed above, and not in Table 1, showed no effect upon incubation with long-chain fatty acids.

The clearest conclusion from this survey is that most enzymes are not sensitive to concentrations of FAs. Those enzymes that are sensitive to FAs are those that have controlling functions on various metabolic pathways and indicate that there is a closer connection between lipid and carbohydrate metabolism than has generally been appreciated.

5.2 The PPAR System

The peroxisomal proliferator-activated receptors (PPAR) are lipid-binding proteins,

Tab. 1 The effect of oleate on the specific activity of selected enzymes and proteins. The values are mean specific activity ± standard deviation and the number of trials are given in parentheses.

Enzyme	*No oleate*	*0.2 mM oleate*
	[μmol min^{-1} mg^{-1} protein]	*[μmol min^{-1} mg^{-1} protein]*
Glycolysis		
Hexokinase (bovine heart)	2.74 ± 0.10 (3)	0.55 ± 0.03 (3)[b]
Phosphofructokinase (rabbit muscle)	18.6 ± 0.9 (3)	0.32 ± 0.07 (3)[b]
Pyruvate kinase (rabbit muscle)	9.8 ± 1.6 (6)	10.2 ± 1.6 (3)
Lactate dehydrogenase (rabbit muscle)	1.8 ± 0.1 (5)	1.7 ± 0.01 (5)
Krebs cycle		
Citrate synthase (porcine heart)	54.8 ± 2.5 (3)	51.3 ± 3.1 (3)
Pentose phosphate cycle		
G-6-P dehydrogenase (yeast)	26.6 ± 0.7 (3)	0.20 ± 0.04 (3)[b]
G-6-P dehydrogenase (bovine adrenals)	0.67 ± 0.04 (6)	0.55 ± 0.20 (6)
Electron transport		
Ferrocytochrome *c* oxidation[a]	0.050 ± 0.005 (4)	0.100 ± 0.015 (4)[b]
Reactive oxygen		
Catalase (bovine liver)	11.2 ± 0.4 (6)	11.4 ± 0.3 (5)
Superoxide dismutase (SOD) (bovine erythrocyte)		
L-DOPA oxidation-no SOD	0.250 ± 0.002 (3)	0.260 ± 0.005 (3)
L-DOPA oxidation-10 mg SOD	0.13 ± 0.1 (3)	0.14 ± 0.01 (3)

[a] Cytochrome *c* autoxidation was measured with Hansatech DW2/2 oxygen electrode at pH 7, 37 °C, in 50 mM sodium phosphate buffer with an initial [cytochrome *c*] = 0.25 mM. Units are μmol O_2 min^{-1} mg^{-1} protein.
[b] Indicates values that were significantly different ($p < 0.05$).

members of the nuclear hormone receptors family, and transcription factors that initiate expression of genes after forming a complex with the retinoid X receptor proteins (RXR). The PPAR:RXR complex binds to specific expression elements of the gene. PPARs (types α and γ interact with FA-loaded FABPs in the nucleus, and a fatty acid is transferred to the PPAR. It is the PPAR:FA complex than can then bind to the RXR. Fatty acid binding proteins along with bound fatty acids are translocated into the nucleus where they interact with the PPARs. FABPs without bound FAs are excluded from the nuclear matrix. Conceivably, alteration of the binding strength of FABPs for their fatty acids would influence the amount of import of the loaded FABP and perhaps the ease at which the FA is transferred to a PPAR. If, in their interaction with nonlipid metabolites, FABPs mediate a potential interaction between glycolytic status and the expression of FA metabolic enzymes and proteins, they are more than transport barges for poorly soluble fatty acids.

6 Summary

The binding of FA to FABP can be modulated by nonlipid metabolites.

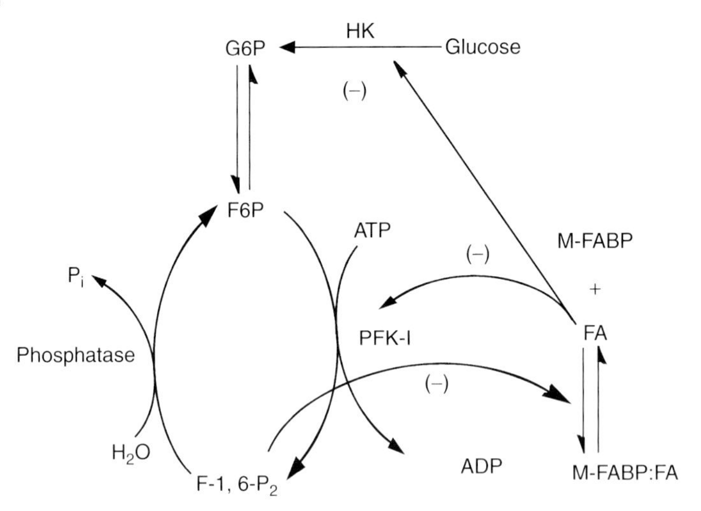

Fig. 5 A model of the interaction scheme between early glycolysis and FABP in mammalian muscle or heart. The bracketed negative signs indicate inhibition of either enzyme activity or binding to FABP. The values of the inhibition constants are given.

Figure 5 summarizes one model (muscle FABP/PFK-I) that forms the basis of present endeavors. There is a differential response of L- and M-FABP to metabolites. These differential responses may offer insight into why there are different FABP types expressed in the same tissue. While the ability of FAs to bind to cytoplasmic enzymes may alter the concentrations of unbound FAs (usually sequestered in membranes), the reverse, where FABPs can influence enzyme substrates, must be evaluated. A consequence of FABP interactions with glycolytic intermediates may influence the concentration of available glucose in the case of hexokinase and F-1,6-P_2 in the case of PFK-I. Significant fractions of cytoplasmic glycolytic substrates could be associated with FABPs. As in conservation of critical, thermodynamic characteristics of fatty acids binding to FABPs across species, the effects of metabolites on FA binding to FABPs being retained across diverse species support this possible new role for FABPs.

Acknowledgments

Many thanks to John A. Blakely, my long-time research associate, for discussions, reading of the manuscript, and technical expertise, and to my colleagues Janet M. Storey and Kenneth B. Storey, Carleton University, for stimulating discussions over the years. The work of a number of students has been essential in this work, notably (in no particular order), Steven Wiseman, Jaime F. Snow (nee Claude), Stephen J. Crozier, Daniel Jardine, Dr. Monica Henry, Philip Karpowicz, Dr Kevin Larade, Shauna Flavelle, Megan Miller, Ashley Woods (glycation work), Luke E. Gauthier, Nathalie Boudreau, Dr Justin MacDonald, Ted Wright, Vanessa Dewling, Christopher Martin, Jonathan D. Schinkel, R. James Allan, and Dr Frank O'Dea. The National Sciences and Engineering Research Council of Canada, the Heart and Stroke Foundation of New Brunswick and Mount Allison University must be acknowledged for the funding that allowed the research to proceed.

See also Biological Regulation by Protein Phosphorylation; Lipid and Lipoprotein Metabolism; Metabolic Basis of Cellular Energy.

Bibliography

Books and Reviews

Alkadi, H.A., Fox, K.A. (2004) Do we need additional markers of myocyte necrosis: the potential value of heart fatty-acid-binding protein? *QJM* **97**, 187–198.

Besnard, P., Niot, I., Poirier, H., Clement, L., Bernard, A. (2002) New insights into the fatty acids-binding protein (FABP) family in the small intestine, *Mol. Cell. Biochem.* **239**, 139–147.

Duplus, E., Forest, C. (2002) Is there a single mechanism for fatty acid regulation of gene transcription, *Biochem. Pharmacol.* **64**, 893–901.

Duttaroy, A.K., Spener F. (Eds.) (2003) *Cellular Proteins and their Fatty Acids in Health and Disease*, Wiley-VCH GmbH & Co. KgaA, Weinheim.

Glatz, J.F., Bonen, A., Luiken, J.J. (2002) Exercise and insulin increase muscle fatty acid uptake by recruiting putative fatty acid transporters to the sarcolemma, *Curr. Opin. Clin. Nutr. Metab. Care* **5**, 365–370.

Hanhoff, T., Lucke, C., Spener, F. (2002) Insights into binding of fatty acids by fatty acid binding proteins, *Mol. Cell. Biochem.* **239**, 45–54.

Laakso, M. (2004) Gene variants, insulin resistance, and dyslipidaemia, *Curr. Opin. Lipidol.* **15**, 115–120.

Londraville, R.L. (1996) Intracellular fatty acid-binding proteins: putting lower vertebrates in perspective, *Braz. J. Med. Biol. Res.* **29**, 707–720.

Randle, P.J. (1998) Regulatory interactions between lipids and carbohydrates: the glucose fatty acid cycle after 35 years, *Diabetes Metab. Rev.* **14**, 263–283.

Scopes, R.K. (1987) *Protein Purification: Principles and Practice*, 2nd edition, Springer-Verlag, New York.

Stewart, J.M. (2000) The cytoplasmic fatty-acid-binding proteins: thirty years and counting, *Cell. Mol. Life Sci.* **57**, 1345–1358.

Stremmel, W., Pohl, L., Ring, A., Herrmann, T. (2001) A new concept of cellular uptake and intracellular trafficking of long-chain fatty acids, *Lipids* **36**, 981–989.

Van der Vusse, G.J., van Bilsen, M., Glatz, J.F., Hasselbaink, D.M., Luiken, J.J. (2002) Critical steps in cellular fatty acid uptake and utilization, *Mol. Cell. Biochem.* **239**, 9–15.

Weiss, E.P., Brown, M.D., Shuldiner, A.R., Hagberg, J.M. (2002) Physiol, *Genomics* **10**, 145–157.

Primary Literature

Binas, B., Danneberg, H., McWhir, J., Mullins, L., Clark, A.J. (1999) Requirement for the heart-type fatty acid binding protein in cardiac fatty acid utilization, *FASEB* **13**, 805–812.

Chu, R., Lim, H., Brumfield, L., Liu, H., Herring, C., Ulintz, P., Reddy, J.K., Davison, M. (2004) Protein profiling of mouse livers with peroxisome proliferators-activated receptor alpha activation, *Mol. Cell. Biol.* **24**, 6288–6297.

Craig, D.B., Wallace, C.J. (1993) ATP binding to cytochrome c diminishes electron flow in the mitochondrial respiratory pathway, *Protein Sci.* **2**, 966–976.

Dutta-Roy, A.K., Gopalswamy, N., Trulzsch, D.V. (1987) Prostaglandin E1 binds to Z protein of rat liver, *Eur. J. Biochem.* **162**, 615–619.

Ek, B.A., Cistola, D.P., Hamilton, J.A., Kaduce, T.L., Spector, A.A. (1997) Fatty acid binding proteins reduce 15-lipoxygenase-induced oxygenation of linoleic acid and arachidonic acid, *Biochim. Biophys. Acta* **17**, 75–85.

Ek-von Mentzer, B.A., Zhang, G., Hamilton, J.A. (2001) Binding of 13-HODE and 15-HETE to phospholipid bilayers, albumin and intracellular fatty acid binding proteins. Implications for transmembrane and intracellular transport and for protection from lipid peroxidation, *J. Biol. Chem.* **276**, 15575–15580.

Fang, T.-Y., Alechina, O., Aleshin, A.E., Fromm, H.J., Honzatko, R.B. (1998) Identification of a phosphate regulatory site and a low affinity binding site for glucose 6-phosphate in the N-terminal half of human brain hexokinase, *J. Biol. Chem.* **273**, 19548–19553.

Font, B., Vial, C., Gautheron, D.C. (1975) Intracellular and submitochondrial localization of pig heart hexokinase, *FEBS Letts.* **56**, 24–29.

Glatz, J.F., Vork, M.M., Cistola, D.P., van der Vusse, G.J. (1993) Cytoplasmic fatty acid binding protein: significance for intracellular transport of fatty acids and putative role on signal transduction pathways, *Prostaglandins Leukot. Essent. Fatty Acids* **48**, 33–41.

Gustafson, L.A., Neeft, M., Reijngoud, D.-J., Kuipers, F., Sauerwein, H.P., Romijn, J.A., Herling, A.W., Burger, H.-J., Meijer, A.J. (2001) Fatty acid and amino acid modulation of glucose cycling in isolated rat hepatocytes, *Biochem. J.* **358**, 665–671.

Harano, Y., Kashiwagi, A., Kojima, H., Suzuki, M., Hashimoto, T., Shigeta, Y. (1985) Phosphorylation of carnitine palmitoyltransferase and activation by glucagon in isolated rat hepatocytes, *FEBS Letts.* **188**, 267–272.

Jia, Y., Qi, C., Kashireddi, P., Surapureddi, S., Zhu, Y.J., Rao, M.S., Le Roith, D., Chambon, P., Gonzalez, F.J., Reddy, J.K. (2004) Transcription coactivator PBP, the peroxisome proliferators-activated receptor (PPAR)-binding protein, is required for PPARalpha-regulated gene expression in liver, *J. Biol. Chem.* **279**, 24427–24434.

Jolly, C.A., Hubbell, T., Behnke, W.D., Schroeder, F. (1997) Fatty acid binding protein: stimulation of microsomal phosphatidic acid formation, *Arch. Biochem. Biophys.* **341**, 112–121.

Kaikaus, R.M., Bass, N.M., Ockner, R.K. (1990) Functions of fatty acid binding proteins, *Experientia* **46**, 617–630.

Kerner, J., Hoppel, C. (2000) Radiochemical malonyl-CoA decarboxylase assay: activity and subcellular distribution in heart and skeletal muscle, *Biochim. Biophys. Acta* **1486**, 1–17.

Kerner, J., Distler, A.M., Minkler, P.E., Parland, W., Peterman, S.M., Hoppel, C.L. (2004) Phosphorylation of rat liver mitochondrial carnitine palmitoyltransferase-I (CPT-I): effect on the enzyme's kinetic properties, *J. Biol. Chem.* **279**, 41104–41113.

Kim, H.K., Storch, J. (1992) Mechanism of free fatty acid transfer of rat heart fatty acid-binding protein to phospholipids membranes. Evidence for a collisional process, *J. Biol. Chem.* **267**, 20051–20056.

Liou, H.-L., Storch, J. (2001) Role of surface lysine residues of adipocyte fatty acid-binding protein in fatty acid transfer to phospholipids vesicles, *Biochemistry* **40**, 6475–6485.

Nemecz, G., Hubbell, T., Jefferson, J.R., Lowe, J.B., Schroeder, F. (1991) Interaction of fatty acids with recombinant rat intestinal and liver fatty acid-binding proteins, *Arch. Biochem. Biophys.* **286**, 300–309.

Qi, C., Zhu, Y., Reddy, J.K. (2000) Peroxisome proliferators-activated receptors, coactivators, and downstream targets, *Cell Biochem. Biophys.* **32**, 187–204.

Ramadoss, C.S., Uyeda, K., Johnston, J.M. (1976) Studies on the fatty acid inactivation of phosphofructokinase, *J. Biol. Chem.* **251**, 98–107.

Randle, P.J., Priestman, D.A., Mistry, S.C., Halsall, A. (1994) Glucose fatty acid interactions and the regulation of glucose disposal, *J. Cell. Biochem.* **55**(Suppl.), 1–11.

Richieri, G.V., Ogata, R.T., Zimmerman, A.W., Veerkamp, J.H., Kleinfeld, A.M. (2000) Fatty acid binding proteins from different tissues show distinct patterns of fatty acid interactions, *Biochemistry* **39**, 7197–7204.

Ruderman, N.B., Saha, K.A., Vavvas, D., Witters, L.A. (1999) Malonyl-CoA, fuel sensing, and insulin resistance, *Am. J. Physiol. Endocrinol. Metab.* **276**, E1–E18.

Schaap, F.G., Binas, B., Danneberg, H., van der Vusse, G.J., Glatz, J.F. (1999) Impaired long-chain fatty acid utilization by cardiac myocytes isolated from mice lacking the heart-type fatty acid binding protein gene, *Circ. Res.* **85**, 329–337.

Shen, W.-J., Sridhar, K., Bernlohr, D.A., Kraemer, F.B. (1999) Interaction of rat hormone-sensitive lipase with adipocyte lipid-binding protein, *Proc. Natl. Acad. Sci. U.S.A.* **96**, 5528–5532.

Shen, W.J., Liang, Y., Hong., R., Patel, S., Natu, V., Sridhar, K., Jenkins, A., Bernlohr, D.A., Kraemer, F.B. (2001) Characterization of the functional interaction of adipocyte lipid-binding protein with hormone-sensitive lipase, *J. Biol. Chem.* **276**, 49443–49448.

Stewart, J.M. (1990) Free fatty acids enhance the oxidation of oxymyoglobin and inhibit the peroxidase activity of metmyoglobin, *Biochem. Cell Biol.* **68**, 1096–1102.

Stewart, J.M., Blakely, J.A. (2000) Long chain fatty acids inhibit and medium chain fatty acids activate mammalian cardiac hexokinase, *Biochim. Biophys. Acta Mol. Cell. Biol. Lipids* **1484**, 278–286.

Stewart, J.M., Blakely, J.A., Johnson, M.D. (2000) The interaction of ferrocytochrome *c* with long-chain fatty acids and their CoA and carnitine esters, *Biochem. Cell Biol.* **78**, 675–681.

Stewart, J.M., Dewling, V.F., Wright, T.G. (1998) Fatty acid binding to rat liver fatty acid-binding protein is modulated by early glycolytic intermediates, *Biochim. Biophys. Acta* **1391**, 1–6.

Stewart, J.M., English, T.E., Storey, K.B. (1998) Comparisons of the effects of temperature on the liver fatty acid binding proteins from hibernator and nonhibernator mammals, *Biochem. Cell Biol.* **76**, 593–599.

Stewart, J.M., Blakely, J.A., Boudreau, N.M., Storey, K.B. (2001) A comparison of oleamide in the brains of hibernating and non-hibernating Richardson's ground squirrel (*Spermophilus richardsonii*) and its inability to bind to brain fatty acid binding protein, *J. Therm. Biol.* **27**, 309–315.

Stewart, J.M., Claude, J.F., MacDonald, J.A., Storey, K.B. (2000) The muscle fatty acid binding protein of spadefoot toad (*Scaphiopus couchii*), *Comp. Biochem. Physiol. B* **125**, 347–357.

Stewart, J.M., Slysz, G.W., Pritting, M., Muller-Eberhard, U. (1996) Ferriheme and ferroheme are isosteric inhibitors of fatty acid binding to rat liver fatty acid binding protein, *Biochem. Cell Biol.* **74**, 249–255.

Spitsberg, V.L., Matitashvili, E., Gorewit, R.C. (1995) Association and coexpression of fatty-acid-binding protein and glycoprotein CD36 in the bovine mammary gland, *Eur. J. Biochem.* **230**, 872–878.

Storch, J., Thumser, A.E.A. (2000) The fatty acid transport function of fatty acid-binding proteins, *Biochem Biophys. Acta* **1486**, 28–44.

Thompson, A.L., Cooney, G.J. (2000) Acyl-CoA inhibition of hexokinase in rat and human skeletal muscle is a potential mechanism of lipid-induced insulin resistance, *Diabetes* **49**, 1761–1765.

Thumser, A.E., Tsai, J., Storch, J. (2001) Collision-mediated transfer of long-chain fatty acids by neural tissue fatty binding proteins (FABP): studies with fluorescent analogs, *J. Mol. Neurosci.* **16**, 143–150.

Tuominen, E.K., Zhu, K., Wallace, C.J., Clark-Lewis, I., Craig, D.B., Rytomaa, M. (2001) ATP induces a conformation change in lipid-bound cytochrome c, *J. Biol. Chem.* **276**, 19356–19362.

Veerkamp, J.H., Zimmerman, A.W.J. (2001) Fatty acid-binding proteins of nervous tissue, *J. Mol. Neurosci.* **16**, 133–142.

Veerkamp, J.H., van Moerkerk, T.B., Zimmerman, A.W. (2000) Effect of fatty acid-binding proteins on intermembrane fatty acid transport studies on different types and mutant proteins, *Eur. J. Biochem.* **267**, 5959–5966.

Vork, M.M., Glatz, J.F.C., van der Vusse, G.J. (1993) Significance of cytoplasmic fatty acid-binding protein for the ischemic heart, *J. Theor. Biol.* **160**, 207–222.

Wieckowski, M.R., Brdiczka, D., Wojtczak, L. (2000) Long-chain fatty acids promote opening of the reconstituted mitochondrial permeability transition pore, *FEBS Letts.* **484**, 61–64.

Williamson, D.H., Brosnan, J.T. (1994) Concentrations of metabolites in animal tissues, *Methods Enzyme Anal.* **4**, 2266–2302.

Wolfrum, C., Borrmann, C.M., Borchers, T., Spener, F. (2001) Fatty acids and hypolipidemic drugs regulate peroxisome proliferator-activated receptors alpha- and gamma-mediated gene expression via liver fatty acid binding protein: a signaling path to the nucleus, *Proc. Natl. Acad. Sci. U.S.A.* **98**, 2323–2328.

Zoellner, H., Hou, J.Y., Hochgrebe, T., Poljak, A., Duncan, M.W., Golding, J., Henderson, T., Lynch, G. (2001) Fluorometric and mass spectrometric analysis of nonenzymatic glycosylated albumin, *Biophys. Res. Commun.* **284**, 83–89.

Intracellular Signaling in Cancer

Chittam U. Thakore[1], *Brian D. Lehmann*[1], *James A. McCubrey*[2,3], *and David M. Terrian*[1,3]
[1]*The Brody School of Medicine at East Carolina University, Anatomy & Cell Biology, Greenville, NC, USA*
[2]*The Brody School of Medicine at East Carolina University, Microbiology & Immunology, Greenville, NC, USA*
[3]*Leo Jenkins Cancer Center, Greenville, NC, USA*

Encyclopedia of Molecular Cell Biology and Molecular Medicine, 2nd Edition. Volume 7
Edited by Robert A. Meyers.

ISBN: 3-527-30549-1

Keywords

Cancer Gene
A mutated gene that has been causally implicated in oncogenesis (i.e. conferring a clonal growth advantage) by at least two independent reports confirming somatic/germline mutations in primary patient specimens.

Chromosomal Translocation
Genomic rearrangements that create chimeric genes; the most common mutation in cancer genes, often resulting in fusion proteins that include an oncogenic kinase catalytic domain and regulatory elements of an unrelated protein.

Human Kinome
A catalog of the 518 protein kinase genes encoded by the human genome and viewable at www.kinase.com.

Kinase Domain
A conserved sequence of amino acids composed of 12 subdomains and responsible for the catalytic transfer of the γ-phosphate of ATP to protein substrates.

Neoplasm
Any new growth of tissue that is both uncontrolled and progressive.

Abbreviations

ALCL	anaplastic large-cell lymphoma
ALK	anaplastic lymphoma kinase
ALL	acute lymphoblastic leukemia
ARG	Abl-related gene
ATM	ataxia telangiectasia mutated kinase
CDK	cyclin-dependent kinase

CML	chronic myelogenous leukemia
EGFR	epidermal growth factor receptor
ERK	extracellular signal regulated kinase
FGFR	fibroblast growth factor receptor
FLT3	FMS-like tyrosine kinase 3
GDNF	glial cell line–derived neurotrophic factor
GOF	gain-of-function
GRB	growth factor receptor bound
JAK	Janus kinase
JNK	Jun-amino-terminal kinase
JNKK	JNK kinase
LCK	T-lymphocyte protein–tyrosine kinase
MAP	mitogen-activated protein
MEK	MAP kinase–kinase
NSCLC	non–small-cell lung cancer
NTRK	neurotrophin tyrosine receptor kinase
PDGFR	platelet-derived growth factor receptor
PDK1	3-phosphoinositide-dependent protein kinase-1
PI3K	phosphoinositide 3-kinase
PIM	proviral integration of the moloney murine leukemia virus
PLCγ	phospholipase C γ
PTEN	phosphatase and tensin homolog
PTK	protein–tyrosine kinase
RET	rearranged in transformation
RTK	receptor protein–tyrosine kinase
SCF	stem cell growth factor
SGK	serum glucocorticoid regulated kinase
SH	Src homology domain
STAT	signal transducers and activators of transcription
STRAD	STE20-related adapter
STK11	serine–threonine kinase 11
TCR	T-cell receptor
TPR	translocated promoter region

Cancer genes that encode protein kinases are the most commonly mutated class of genes to be causally linked to the progressive outgrowth of malignant neoplasms. Protein kinases operate as important molecular switches that impinge, directly or indirectly, upon an integrated signaling circuitry that often becomes genetically reprogrammed, through successive mutations, to transmit many of the growth-stimulating and survival signals that govern the transformation of normal human cells into a cancerous cell mass. Of the 291 genes recently included in a comprehensive catalog of known human cancer genes, 9.3% (27 distinct genes) encode kinase domain sequences. This number far exceeds a random prediction of 6.3

genes (2%) and reinforces the hypothesis that a vast majority of human cancer cells carry genetic defects in the growth-signaling circuitry that enable these cells to grow autonomously and evade cellular senescence and death. First, we provide a general overview of a prototypical growth-signaling circuit and a detailed analysis of the canonical Ras/Raf/MEK/ERK pathway, because it has a central role in cancer cell signaling. This analysis is followed by a discussion of each of the kinases known to be mutated in cancer; we will describe the domain organization and normal function(s) of the protein, its oncogenic alterations in cancer and how this alters its activity, and identify the tumor/cancer types in which these changes are most common.

1 Overview of the Prototypical Growth-signaling Circuit

1.1 Signaling Pathways in Normal Human Cells

Protein phosphorylation is the most common posttranslational protein modification in mammalian cells and is a fundamental mechanism for the direct or indirect control of physiological processes of crucial importance to cell growth and survival. Protein phosphorylation is mediated by a superfamily of protein kinases comprising ~2% of the human genome, and a diverse collection of unrelated substrate proteins, comprising more than 10% of the genome. Collectively, these proteins control the responsiveness of a cell to its environment by coordinating complex functions such as normal embryonic development, cell division, differentiation, adhesion, motility, and metabolic activity.

All eukaryotic protein kinases have conserved a ~250 to 300 amino acid "kinase domain," composed of 12 subdomains and 12 relatively (>95%) invariant residues. The kinase domain uses the γ-phosphate of ATP or GTP to generate phosphate monoesters using two types of acceptors: (a) protein phenolic groups of tyrosine amino acids (i.e. the protein-tyrosine kinases, PTK) and (b) protein alcohol groups on serine/threonine amino acids (i.e. the protein Ser/Thr kinases). A third class, called dual specificity protein kinases, are able to phosphorylate either substrate (both Tyr and Ser/Thr). The kinase domain folds into a two-lobed three-dimensional structure separated by a deep hydrophobic cleft. When in a catalytically active conformation, the kinase domain is able to perform three interdependent functions: (a) binding and orienting the ATP/GTP phosphate donor as a complex with a divalent cation (Mg^{2+} or $Mn2^{2+}$), (b) binding and orienting the protein substrate, and (c) transferring the γ-phosphate from ATP/GTP to the acceptor hydroxyl residue of the protein substrate. The latter reaction occurs within a subdomain referred to as the *catalytic loop*. In a normal cell, the catalytic activity of most protein kinases is constrained by an intrasteric inhibition at both the ATP/GTP and substrate peptide binding sites within the kinase domain. This suppression of catalytic activity is often achieved when a "pseudosubstrate" sequence, located outside of the kinase domain and containing the consensus recognition sequence of a true peptide substrate lacking the phosphoryl acceptor, interacts with the kinase domain and thereby blocks or distorts the active site. This form of autoinhibition is common among both the PTK and Ser/Thr kinase families and serves as just one example of the multiple layers of regulatory constraint normally exerted on a protein kinase in order to maintain a balance between the rate of cell-cycle progression and cell death. Other important examples of mechanisms capable of relieving the autoinhibition of a protein kinase include the role of protein–protein interactions (both homo- and hetero-dimerization), and activating ligands and/or second messengers. Limiting the intensity/duration of changes in protein phosphorylation are protein phosphatases and a diverse array of mechanisms controlling the biogenesis and half-life of the protein kinase.

Our understanding of how multiple protein kinases normally function within a signaling pathway has been derived largely from studies focused on delineating immediate upstream and downstream interactions through gain- and loss-of-function manipulations, arranging these interactions into orderly and conditional cascades of molecular events, and ultimately linking cell surface receptors to an effector that is capable of eliciting a biological response. Analyses of such linear pathways have provided valuable insights into the interactions between components, both within and between certain pathways.

Many of the genes that cause cancer encode receptor PTKs (RTK), a subclass of transmembrane-spanning receptor that responds to certain extracellular cues (ligands) with a disinhibition of their intrinsic PTK activity. The cytoplasmic Ser/Thr protein kinases Akt and Raf function intracellularly as important points for the convergence of many of the signals emanating from these upstream oncogenic RTKs. A

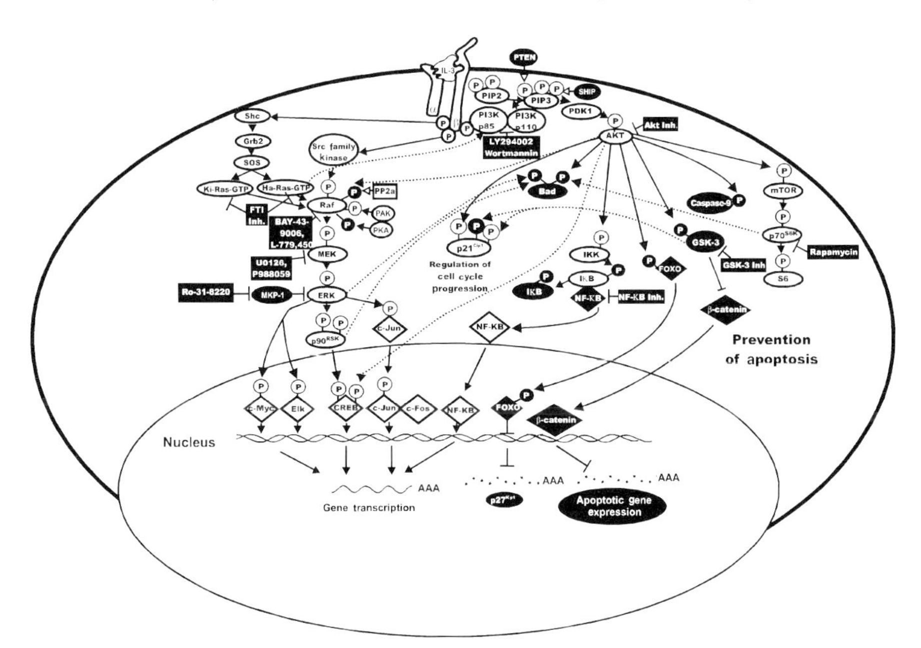

Fig. 1 Interactions between Raf/MEK/ERK and PI3K/Akt pathways and sites of action of small molecular weight signal transduction pathway inhibitors.

typical graphic representation of such a signaling network is shown in Figure 1, encompassing Akt, Raf, and several interactive signaling pathways regulating the phosphorylation of its downstream effectors, such as c-Myc, Elk, CREB, NF-kappa B, and Forkhead. These phosphorylated proteins, in turn, mediate the effects of Akt and Raf on gene transcription, and on the proliferation, growth, differentiation, and survival of the cell.

1.2
Genetic Rewiring of Signaling Circuits in Cancer Cells

Most PTKs that have thus far been implicated in cancer undergo mutations that result in a loss of their normal autoinhibition of catalytic activity. In certain signaling "cascades," such as the mitogen-activated protein (MAP) kinase pathway (Fig. 1), this loss of repression would be sufficient to trigger the constitutive activation of all downstream components, since the activity of MAP kinases is dependent on activation of the upstream kinase. However, the regulation of kinase activity is generally more complex, making it more difficult to predict how the oncogenic deregulation of a given kinase may contribute to cellular transformation. For example, the activation of 3-phosphoinositide-dependent protein kinase-1 (PDK1) is not sufficient to induce the phosphorylation of certain of its downstream substrates (e.g. p70 ribosomal S6 kinase). Rather, these substrate proteins must first undergo conformational rearrangements that render their PDK1 phosphorylation sites accessible to PDK1. Such modes of substrate-directed regulation are not frequently integrated into a typical signaling diagram. A traditional view of signaling networks provides a valuable and empirical foundation for generating meaningful hypotheses pertaining to the potential downstream effects that a constitutively active (mutant) oncogenic kinase might have within a network such as that shown in Figure 1. However, these predictions are often intuitive and can be somewhat simplistic when applied to the downstream effectors of a protein kinase such as PDK1.

Systems biology and computational modeling have now entered the field of cell biology and are changing the way scientists think about signaling networks in complex biological systems. A goal of systems biology is to understand, from a global perspective, the constellation of biological behaviors that can emerge from single/multiple mutations within the components of a cell's extensive network of signaling modules. This approach to interrogating cells, together with high-throughput genomic and proteomic screening approaches, promises to lead to the identification of unexpected "cluster groups" of genes/proteins of similar or related functions that are mutually influenced by experimentally defined mutational events. Though this approach presents a formidable challenge, it constitutes a major new frontier in the study of signaling in cancer, and has gained a significant degree of traction with the recent identification of a limited and well-characterized set of cancer genes encoding signaling molecules.

1.3
Identification of Activating Mutations in Cancer Genes Encoding Signaling Molecules

Of the 291 genes recently included in a comprehensive catalog of known human cancer genes, 27 distinct genes encode kinase domain sequences (see Cancer Genome Project: http://www.sanger.ac.uk/CGP/). With the identification of each of these mutated kinases, it will now be possible to apply the principles of systems biology to a global analysis of how the introduction of each mutant kinase alters cellular behavior, and to exploit genomic and proteomic strategies to extract responsive cluster groups. With such an expansive inventory of new data, computational modeling can then begin to evaluate whether a specific combination of mutations ("hits") would be capable of transforming a normal cell into a cancer cell. The oncogenic themes likely to arise from such an effort can be expected to vary among cancers and according to the sequence in which genetic mutations are introduced within a given cell.

The success of the epidermal growth factor receptor (EGFR) kinase inhibitor gefitinib (Iressa®) in the treatment of non–small-cell lung cancer (NSCLC) and the ABL kinase inhibitor imatinib (Gleevec®) in treating chronic myeloid leukemia (CML) demonstrate the potential for targeting specific activating mutations in protein kinases that promote proliferative signals in cancer cells. However, it is also now apparent that the effectiveness of such treatments will be limited, and will be critically dependent on identifying the dominant genetic lesion responsible for activating a given kinase in each malignancy, and thereafter developing drugs to target this Achilles' heel. For example, only about 10 to 19% of patients with NSCLC respond to treatment with Iressa®, although a majority of these patients express the EGFR. However, a subset of patients, representing ~10% of lung cancers with tumors positive for mutations clustered around the ATP-binding pocket of the tyrosine kinase domain, respond dramatically to Iressa®. Therefore, the patient's response to therapy will be dependent on correctly matching a given kinase inhibitor to the causal oncogenic event in each case. As a prelude to this emerging effort, this article will describe each of the 27 protein kinases known to be capable of functioning as dominant human cancer genes in turn. We will describe the domain organization and normal function(s) of the protein, its oncogenic alterations in cancer and how this alters its activity, and identify the tumor/cancer types in which these changes are most common. First, however, we begin with an analysis of the canonical Ras/Raf/MEK/ERK pathway that is of such importance in cancer cell signaling.

2 Ras/Raf/Mek/Erk Pathway

The Ras/Raf/MEK/ERK pathway is a central signal transduction pathway, which transmits signals from multiple cell surface receptors to transcription factors in the nucleus. This pathway is frequently referred to as the MAP kinase pathway because it can be stimulated by mitogens, cytokines, and growth factors. The pathway can be activated by Ras stimulating the membrane translocation of Raf. This pathway also interacts with many different signal transduction pathways including phosphatidylinositol 3-kinase (PI3K)/Akt and Jak/STAT (See below).

2.1 Ras

Ras is a small GTP-binding protein, which is the common upstream molecule of several signaling pathways including Raf/MEK/ERK, PI3K/Akt, and RalEGF/Ral. The switch regions of the Ras proteins are in part responsible for the switch between the inactive and active states of the protein. Switching between these states has been associated with conformational changes in the switch regions, which allows the binding of GTPase activating proteins and guanine nucleotide exchange factors. When Ras is active, GTP is bound, whereas when Ras is inactive GDP is bound. GTPases inactivate the Ras proteins while guanine nucleotide exchange factors activate the Ras proteins by stimulating the removal of phosphate from GTP or addition of GTP, respectively.

Different mutation frequencies have been observed between *Ras* genes in human cancer. There are three different Ras family members, Ha-Ras, Ki-Ras and N-Ras. The Ras proteins show varying abilities to activate the Raf/MEK/ERK and PI3K/Akt cascades; Ki-Ras has been associated with Raf/MEK/ERK while Ha-Ras is associated with PI3K/Akt activation. Amplification of *Ras* proto-oncogenes and activating mutations that lead to the expression of constitutively active Ras proteins are observed in approximately 30% of all human cancers. The effects of Ras on proliferation and tumorigenesis have been documented in immortal cell lines. However, antiproliferative responses of oncogenic Ras have also been observed in non-transformed fibroblasts, primary rat Schwann cells, and primary fibroblast cells of human and murine origins. This premature G_1 arrest and subsequent senescence is dependent on the Raf/MEK/ERK pathway, and has been shown to be mediated by many cell-cycle inhibitory molecules including $p15^{Ink4b}/p16^{Ink4a}$ and $p21^{Cip1}$.

2.2 Raf

The Raf protein family consists of A-Raf, B-Raf, and Raf-1, which are involved in the regulation of proliferation, differentiation, and apoptosis, induced after cytokine stimulation. The Raf proteins have three distinct functional domains: CR1, CR2, and CR3. The CR1 domain is necessary for Ras binding and subsequent activation. The CR2 domain is a regulatory domain. CR3 is the kinase domain. Deletion of the CR1 and CR2 domains produces a constitutively active Raf protein due, in part, to the removal of phosphorylation sites, which serve to negatively regulate the kinase in the CR2 domain.

The *A-Raf* gene is located on the X chromosome in humans and produces a 68-kDa protein. The highest level of A-Raf expression in the adult is in the urogenital tract. A-Raf deletion results in postnatal lethality, attributed to neurological and gastrointestinal defects. The role of A-Raf has remained elusive. A-Raf is the weakest Raf kinase in terms of activation of ERK activity. ERK activation occurs in A-Raf knock-out mice indicating that other Raf isoforms can compensate for this deficiency. Some studies have indicated that A-Raf has an important role in stimulating the growth of hematopoietic cells. Moreover, we have observed that overexpression of activated A-Raf will abrogate the cytokine-dependence of hematopoietic cells.

The *B-Raf* gene is located on chromosome 7q34. B-Raf encodes a 94-kDa protein. The highest expression of B-Raf is in the testes and neuronal tissue. The *B-Raf* gene produces splice variants, the physiological roles of which have not yet been elucidated. The loss of B-Raf expression in mice results in intrauterine death between days 10.5 and 12.5. The B-Raf knock-out mouse embryo displays enlarged blood vessels and increased apoptosis of differentiated endothelial cells. This is an indication that Raf kinases can regulate apoptosis. B-Raf and Raf-1 activity are negatively regulated by Akt phosphorylation. Akt also has other effects on the presentation of apoptosis, and will be discussed in more detail later. In contrast, other studies have shown that overexpression of B-Raf in Rat-1 cells results in decreased apoptosis because of inhibition of caspase activity. Historically, B-Raf is the strongest Raf in terms of induction of MEK activity, as determined by *in vitro* kinase assays. B-Raf activation is different from either Raf-1 or A-Raf activation; it requires the phosphorylation of only one regulatory residue, whereas activation of Raf-1 and A-Raf require two phosphorylation events. B-Raf is also regulated by Akt and the serum glucocorticoid regulated kinase (SGK). Recently it has been shown that B-Raf may be required in the activation of Raf-1, and that B-Raf and Raf-1 may be induced at different times following stimulation. Moreover, the three different Raf proteins may have different subcellular pools, and Raf-1 and A-Raf may be localized, in some cases, to the mitochondrion. The roles of B-Raf in human neoplasia will be discussed later.

The *Raf-1* gene is located on chromosome 3p25 and produces a 74-kDa protein. The *Raf-1* proto-oncogene was the first *Raf* gene to be cloned. It is the cellular homolog of *v-Raf* contained in the naturally occurring acute retrovirus 3611-MSV. *Raf-1* is ubiquitously expressed in adult tissues with highest expression in muscle, cerebellum, and fetal brain. It is the most studied Raf isoform. A dominant-negative version of Raf-1 inhibits Ha-Ras-induced transformation and tumor formation. Raf-1 has important roles in apoptosis as it phosphorylates and inactivates Bad. Raf-1 phosphorylates and co-immunoprecipitates with Bcl-2 and also regulates Bag and Bad expression, in BCR/ABL expressing cells. Recently, Raf-1 has been proposed to have roles independent of MEK/ERK that are often involved in regulating cell-cycle progression and apoptosis. These new roles for Raf-1 will be discussed below.

The activation of the three Raf isoforms is complex and not totally understood. Ras is known to activate B-Raf independently of Src activity while Raf-1 and A-Raf require an Src-like kinase for complete activation. An Src-like kinase phosphorylates Tyr340 and Tyr341 on Raf-1 and Tyr301 and Tyr302 on A-Raf, which are not present in B-Raf. B-Raf contains aspartic acid at the corresponding residues. Aspartic acid residues are negatively charged and are believed to confer a constitutively active configuration at this site. B-Raf is constitutively phosphorylated on Ser445 (a site equivalent to Ser338 in Raf-1), which is phosphorylated as a result of Ras activation. The Ras binding domains of B-Raf and Raf-1 have a greater affinity for Ras than A-Raf. This suggests that A-Raf may have a primary activator other than Ras. Phosphorylation of the activation loop of Raf-1 and B-Raf is necessary, but not sufficient, for activation. The deletion or mutation of inhibitory phosphorylation sites present at certain residues by site-directed mutagenesis activates Raf proteins. Akt also phosphorylates Raf-1 at Ser259, which has been associated with inhibition of Raf-1 activity.

Several protein kinases are known to regulate Raf-1 activity. cAMP-dependent protein kinase inhibits Raf-1. cAMP activates PKA, resulting in phosphorylation of Raf-1 on Ser43 and Ser621, and inhibiting Raf-1 activity. This contrasts with B-Raf activation in response to cAMP-dependent protein kinase. Raf-1 is activated through $p21^{Cip1}$-associated protein. Activation of conventional protein kinase C isoforms (α, β, and γ) stimulates Raf-1 activity in a baculovirus system. Protein kinase Cε is associated with c-N-Ras and Raf-1 and is necessary for activation of Raf-1 by phorbol 12myristate 13acetate in fibroblasts. RKIP, a Raf-1-interacting protein, inhibits Raf-1 activation of MEK1. 14-3-3 proteins are known to bind Raf. The interaction of 14-3-3 proteins with phosphorylated Ser259 and Ser621 inhibits Raf-1. Protein phosphatase 2A can remove the phosphate on Ser259, allowing Raf to become disassociated from 14-3-3. This allows Raf-1 to assume a conformation in which it can be phosophorylated and activated by other kinases including $p21^{Cip1}$-associated protein and Src family kinases.

Raf-1 was originally characterized as phosphorylating MEK1, but B-Raf is the primary MEK1 activator in bovine brain extracts, NIH 3T3 cells, PC12 cells and other cells. Raf-dependent activation of p90 ribosomal S6 kinase ($p90^{Rsk}$) and NF-κB are activated by an Raf-1 mutant, which no longer binds or phosphorylates MEK1. This suggests that Raf-1 has physiological substrates other than MEK1. Recently, the possible roles of Raf-1 in the Raf/MEK/ERK signal transduction pathway have become a matter of controversy due to the discovery that B-Raf is the more potent activator of MEK and that many of the "functions" of Raf-1 still persist in Raf-1 knock-out mice, but not in B-Raf knock-out mice. Raf-1 has been postulated to have important roles in cell-cycle progression, activation of the p53 and NF-κB transcription factors, and the prevention of apoptosis. Raf-1 has been postulated to have nonenzymatic functions and serve as a docking protein. Raf-1 has been proposed to have important functions at the mitochondrial membrane. Interestingly, it has been shown that Bcr-Abl (see Sect. 3.1.2) may interact with Raf-1 to alter the distribution of Bag at the mitochondrial membrane and hence regulate (prevent) apoptosis in hematopoietic cells.

Recently, the mechanism of B-Raf regulation has been more intensively investigated. B-Raf activation may occur through the GTP-binding protein Rap-1, which is in turn activated by RA-guanine nucleotide exchange factor-1 and Src. cAMP-dependent protein kinase may activate Src in some cells, resulting in Rap-1 activation. There are three regulatory residues on B-Raf – Ser444, Thr598, and Ser601. The Ser/Thr kinases that phosphorylate these residues are not known. Activated B-Raf can interact with Ras, which in turn results in the activation of Raf-1. B-Raf can activate MEK, which results in the activation of ERK and downstream kinases and transcription factors. Rheb and 14-3-3 proteins can bind and inhibit B-Raf activity. B-Raf can also be negatively affected by phosphorylation by either SGK or Akt. Akt has been shown to phosphorylate B-Raf on three residues – Ser364, Ser428, and Thr439 (see below). However, Akt may have greater affinity for other substrates besides B-Raf. SGK has been proposed to be a more relevant kinase for regulating B-Raf through phosphorylation of Ser364. B-Raf may also activate Raf-1 *via* Ras-GTP. While Rap-1 activates B-Raf, it may also serve to inactivate Raf-1. Thus the pathways for activation and inactivation of B-Raf and Raf-1 are complex, and may at times appear contradictory.

2.3 Mek

MEK proteins are the primary downstream targets of Raf. The *MEK* family of genes consists of five genes: *MEK1, MEK2, MEK3, MEK4, and MEK5*. This family of dual specificity kinases has both Ser/Thr and Tyr kinase activity. The structure of MEK consists of an amino-terminal negative regulatory domain and a carboxy-terminal MAP kinase-binding domain that is necessary for binding and activation of ERKs. Deletion of the regulatory MEK1 domain results in constitutive MEK1 and ERK activation.

MEK1 is a 393 amino acid protein with a molecular weight of 44 kDa. MEK1 is modestly expressed in embryonic development and is elevated in adult tissue, with the highest levels detected in brain tissue. Knock out of functional MEK1 results in lethality due to placental vascularization problems. Mice with dominant-negative MEK1 have also been generated; these mice were viable although defects in T-cell development occurred. MEK1 requires phosphorylation of Ser218 and Ser222 for activation. The substitution of these residues with aspartate or glutamic acid led to an increase in activity and foci formation in NIH3T3 cells. Mutated MEK1 constructs demonstrate that the activation of ERK does not require a functional MEK1 kinase domain. Replacement of the amino-terminus of MEK1 with the hormone-binding domain of the estrogen receptor produces a construct with kinase activity responsive to the presence of estrogen analogs. This construct is an invaluable tool in research into MEK1 signaling. Studies with this construct demonstrated that activated MEK1 could abrogate the cytokine dependency of certain hematopoietic cells. Constitutive activity of MEK1 in primary cell culture promotes senescence and induces p53 and $p16^{INK4a}$, whereas the opposite was observed in immortalized cells or cells lacking either p53 or $p16^{INK4a}$. Constitutive activity of MEK1 inhibits NF-κB transcription, by negatively regulating $p38^{MAPK}$ activity. Pharmaceutical companies have developed inhibitors of MEK. The two most widely used are U0126 and PD98059, because they are commercially available. These two inhibitors have IC_{50}s of 70 nM and 2 μM, respectively. PD98059 inhibits activation, while U0126 inhibits activity. Thus, these two inhibitors have a noncompetitive mechanism of inhibiting ERK activity.

2.4 Erk

The main physiological substrates of MEK are the members of the ERK (extracellular signal regulated kinase), or MAP kinase, family of genes. The ERK family consists of four distinct groups of kinases: ERK, Jun amino-terminal kinases (JNK1/2/3), $p38^{MAPK}$ (p38 $\alpha/\beta/\gamma/\delta$), and ERK5. In addition, there are ERK3, ERK4, ERK6, ERK7, and ERK8 kinases, which are related to ERK1 and ERK2, but have different modes of activation, with biochemical roles that are not as well characterized. The ERK 1 and 2 proteins are the most studied with regards to Raf signaling in hematopoietic cells.

ERK1 and *ERK2* encode 42 and 44 kDa proteins, respectively. These proteins were originally isolated by their ability to phosphorylate microtubule-associated protein 2. ERKs are activated through dual phosphorylation of threonine and tyrosine residues by MEK1 kinases. This dual phosphorylation activates ERK; however, ERK activity is downregulated by serine/threonine phosphatases, tyrosine phosphatases, or dual specificity phosphatases. ERK dimerization occurs subsequent to phosphorylation. This dimerization maintains the activation of ERK and promotes nuclear localization of the protein. Activated ERK preferentially phosphorylates serine/threonine residues preceding a proline.

Mice lacking ERK1 have defective T-cell development, similar to transgenic mice containing a dominant-negative MEK1. Recently, different roles have been suggested for ERK1 and ERK2. While these two proteins are often detected at similar levels, their expression patterns may change. ERK1 has been postulated to inhibit ERK2 expression. Increased ERK2 activity is associated with cell proliferation. Interestingly, ERK1 has been associated with cognitive brain functions including learning.

2.5 Roles of the Ras/Raf/Mek/Erk Pathway in Neoplasia

Mutations at three different Ras codons (12, 13, and 61) convert Ras into a constitutively active protein. These point mutations can be induced by environmental mutagens. Given the high level of mutations that have been detected at Ras, this pathway has always been considered a key target for therapeutic intervention. Approximately 30% of human cancers have Ras mutations. Ras mutations are frequently observed in certain hematopoietic malignancies including myelodysplastic syndromes, juvenile myelomonocytic leukemia, and acute myeloid leukemia. Ras mutations are often one step in tumor progression and mutations at other genes (chromosomal translocations, gene amplification, and tumor suppressor gene inactivation) have to occur for a complete malignant phenotype to be manifested. Pharmaceutical companies have developed many farnesyl transferase inhibitors, which suppress the farnesylation of Ras, preventing it from localizing to the cell membrane.

As stated previously, there are three different *Ras* genes – *Ki-Ras, Ha-Ras,* and *N-Ras.* The biochemical differences between these Ras proteins have remained elusive. *Ki-Ras* mutations have been more frequently detected in human neoplasia than *Ha-Ras* mutations. Ras has been shown to activate both the Raf/MEK/ERK and the PI3K/Akt pathways. Thus, mutations at Ras should theoretically activate both pathways simultaneously. Ras mutations have a key role in malignant transformation as both of these pathways can prevent apoptosis as well as regulate cell-cycle progression. Recently it was shown that there is specificity in terms of the ability of Ki-Ras and Ha-Ras to induce the Raf/MEK/ERK and PI3K/Akt pathways. Ki-Ras preferentially activates the Raf/MEK/ERK pathway, while Ha-Ras preferentially activates the PI3K/Akt pathway. Therefore, if Ras inhibitors could be developed which would specifically inhibit one particular Ras protein, it might be possible to inhibit one of the downstream pathways. This might, under certain circumstances, be advantageous. Furthermore, decreases in ERK expression may affect differentiation responses. Thus in certain tumors, one might desire to inhibit the effects that Ras has on the PI3K/Akt pathway as opposed to the effects it has on Raf. Targeting of Ha-Ras as opposed to Ki-Ras might inhibit apoptosis suppression by Ha-Ras, but not the effects Ki-Ras has on cell-cycle progression and differentiation.

Overexpression of the Raf/MEK/ERK cascade has been frequently observed in human neoplasia. A prime consequence of this activation may be the increased expression of growth factors, which can potentially further activate this cascade by an autocrine loop. Many cytokine and growth factor genes contain transcription factor binding sites, which are bound by transcription factors (Ets, Elk, Jun, Fos, CREB) that are often activated by the

Raf/MEK/ERK cascade. Identification of the mechanisms responsible for activation of this cascade has remained elusive. Genetic mutations at Raf, MEK, or ERK were thought to be relatively rare in human neoplasia. For many years, numerous scientists were of the opinion that the activation of the Raf/MEK/ERK cascade was mainly due to mutations at Ras, and hence studies aimed at elucidating the mechanisms of Ras activation were launched.

While mutations at the *Raf* gene in human neoplasia have been detected, they have not until recently gained the clinical importance that Ras mutations readily achieved. Because of more innovative, high-throughput DNA sequencing, scientists have recently discovered that the *B-Raf* gene is frequently mutated in certain cancers including hematopoietic tumors. Approximately 60% of the melanomas surveyed in one study were observed to have mutations at *B-Raf* (see Sect. 3.2.1). This result provides relevance to the investigation of signal transduction pathways because, by understanding how *B-Raf* is activated, one Ras-dependent and one Ras-independent event, scientists could predict why a single missense mutation in *B-Raf* permitted ligand-independent activation, whereas in similar mutation events it could not be predicted whether the mutation events would result in either Raf-1 or A-Raf activation, as they requires multiple activation events. Interestingly, in 22 tumors (melanomas, colorectal, and NSCLC tumors) that were examined, there were 10 with mutations at *B-Raf* and 10 with mutations at Ras. Out of these two tumors had mutations at *B-Raf* and Ras; two did not have mutations at either gene. Thus *B-Raf* transformation does not appear to require Ras, and many tumors had mutations at one or the other, but not both, genes. Recently, it has been suggested that B-Raf mutations occur during the process of progression of the tumor, and not during the establishment of the tumor. This was suggested by analysis of *B-Raf* mutations in different developmental stages of melanomas.

Raf inhibitors have been developed and some are being used in clinical trials. Raf-1 has at least 13 regulatory phosphorylation residues. This makes inhibition of Raf activity difficult, as certain phosphorylation events stimulate Raf activity while others inhibit and promote Raf association with 14-3-3 proteins, rendering it inactive though present in the cytoplasm. Certain Raf inhibitors were developed which inhibit the Raf kinase activity as determined by assays with purified Raf proteins and substrates (e.g. MEK). Some Raf inhibitors may affect a single Raf isoform (e.g. Raf-1), others may affect Raf proteins, which are more similar (Raf-1 and A-Raf), while other Raf inhibitors may affect all three Raf proteins (Raf-1, A-Raf and B-Raf). We have observed that the L-779,450 inhibitor suppresses the effects of A-Raf and Raf-1 more than those of *B-Raf*. Knowledge of the particular *Raf* gene that is mutated or overexpressed in certain tumors may provide critical information for planning treatment for the patient. Inhibition of certain *Raf* genes might prove to be beneficial, while inhibition of other *Raf* genes under certain circumstances might prove detrimental. Thus, the development of unique Raf inhibitors may prove useful in human cancer therapy.

Activation of Raf is complex and requires numerous phosphorylation and dephosphorylation events. Prevention of Raf activation by targeting kinases (e.g. Src or Akt) and phosphatases involved in Raf activation may be a mechanism to inhibit/regulate Raf activity. It is worth noting that some of these kinases normally inhibit Raf activation (e.g. Akt). A major limitation of this approach would be the specificities of these kinases and phosphatases. Inhibiting these kinases/phosphatases could result in activation or inactivation of other proteins and would have other effects on cell physiology.

Dimerization of Raf proteins is critical for their activity. We often think of a single Raf protein carrying out its biochemical activity. However, Raf proteins dimerize with themselves and other Raf isoforms to become active. Drugs such as coumermycin, which inhibit Raf dimerization, and others such as geldanmycin, which prevent interaction of Raf with 14-3-3 proteins, suppress Raf activity. Various Raf isoforms may dimerize and result in chimeric molecules, which may have different biochemical activities. Little is known about the heterodimerization of Raf proteins.

Downstream of Raf lies MEK. Currently, it is believed that MEK1 is not frequently mutated in human cancer. However, aberrant expression of MEK1 has been detected in many different types of cancer, and mutated forms of MEK1 will transform fibroblast, hematopoietic, and other cell types. Useful inhibitors to MEK have been developed, which display high degrees of specificity. The successful development of MEK inhibitors may be due to the relatively few phosphorylation sites on MEK involved in activation/inactivation. MEK inhibitors are in clinical trials.

Downstream of MEK lies ERK. To our knowledge no small-molecular-weight ERK inhibitors have been developed yet. However, inhibitors to ERK could prove very useful as ERK can phosphorylate many targets (Rsk, c-Myc, Elk, etc.), which have growth-promoting effects. There are at least 2 ERK molecules regulated by the Raf/MEK/ERK cascade, ERK1 and ERK2. Little is known about the different *in vivo* targets of ERK1 and ERK2. However, ERK2 has been postulated to have pro-proliferative effects while ERK1 has antiproliferative effects. Development of specific inhibitors to ERK1 and ERK2 might eventually prove useful in the treatment of certain diseases.

The MAP kinase phosphatase-1 (*DUSP1*) removes the phosphates from ERK. MAP kinase phosphatase-1 is mutated in certain tumors and could be considered to be a tumor suppressor gene. An inhibitor to this phosphatase has been developed (Ro-31-8220).

3 Protein Kinases Encoded by Cancer Genes

3.1 Protein-tyrosine Kinase Family

3.1.1 Receptor Protein-tyrosine Kinase Subfamily

More than half (55%) of the protein kinases included in the current census of human cancer genes are RTKs. Somatic chromosomal translocations that result in the fusion of a given tyrosine kinase domain of one gene with the regulatory region of another unrelated gene are most common. In many instances, a number of different genes will form chimeric transcripts with the same *RTK* gene. Generally, the fusion proteins encoded by these translocated cancer genes phosphorylate their protein substrates in a ligand-independent manner. Such mutations are often found in lymphomas, leukemia, epithelial neoplasms, and solid tumors.

ALK The anaplastic lymphoma kinase (*ALK*) gene, located on human chromosome 2p23, is expressed only in the small intestine, testis, and brain, and encodes a Type Ia membrane protein (carboxy-terminal kinase domain oriented intracellularly) with a glycosylated amino-terminal extracellular region that binds the ligands pleiotrophin and midkine. Ligand-binding specificity is dependent on two cysteine-rich clusters that are adjacent to the major ligand-binding domain. The extracellular domain of ALK also contains a low-density lipoprotein-A module, suggesting that ALK may also interact with low-density lipoproteins. Other motifs present in the extracellular domain include a glycine-rich patch and MAM domain (Meprin/A-5/protein-tyrosine phosphatase Mu), which may play a role in mediating certain cell–cell interactions. Like other RTKs, discussed below, tyrosine residues in the intracellular domain of ALK regulate its intrinsic kinase activity, and provide phosphorylation-dependent recruitment sites for the Src homology 2 (SH-2) domains of various adapter proteins and enzymes that propagate intracellular signaling cascades. For ALK, these include the insulin receptor substrate-1, Shc protein, and phospholipase C-γ. The adapter protein Shc forms complexes with GRB2 (growth factor receptor-bound protein 2) and the Son of Sevenless protein to activate the Ras/Raf/MEK/ERK signaling cascade (Fig. 1). Although ALK-deficient mice do not display any marked developmental defects, the *drosophila* homolog (i.e. *DAlk*) is required for normal development of the gut musculature. A common feature among all ALK chimeric proteins is the presence of an oligomerization domain in the partner protein and oligomerization-mediated kinase activation. Although the intracellular localization of these fusion proteins are predominantly determined by the amino-terminal fusion partner, all fusion proteins result in aberrant activation of multiple downstream signaling cascades that are responsible for cellular transformation. Approximately half the non-Hodgkin's lymphomas classified as anaplastic large-cell lymphomas (ALCL) are associated with a common somatic chromosomal translocation, t(2;5)(p23;q35). This chimeric gene encodes an oncogenic fusion protein comprised of the complete cytosolic and kinase domains of ALK joined to the amino-terminal 117 amino acid residues of the nucleolar phosphoprotein nucleophosmin. The transforming activity of the fusion protein (nucleophosmin-ALK) requires the activation of its kinase function, a result of the oligomerization mediated by the nucleophosmin segment. Although less common, other ALK fusion partners include TPM3, TFGxl, TFGs, ATIC, CLTC, MSN, TPM4, MyH9, RANBP2, ALO17, and CARS. Constitutive ALK signaling activates pathways that either enhance proliferation or target the PI3K/Akt survival pathway. Proliferation is enhanced by activation of the MAP kinase pathway through Ras or phospholipase C-γ (PLC-γ). PLC-γ-mediated hydrolysis of membrane phospholipids produces localized increases in inositol triphosphate and diacyglycerol, which, in turn, stimulate the release of Ca^{2+} from intracellular stores and the activity of the Ser/Thr protein kinase C. Although the precise nature of the interaction between PI3K and ALK is unknown, a potential complex includes nucleophosmin-ALK and the regulatory p85 subunit of PI3K with GAB2, Shc, and/or CrkL, an interaction recently identified in the human Karpass 299 cell line.

EGFRs The EGFR subfamily of RTKs encompasses four closely related receptors, EGFR (also known as ErbB-1 or HER1),

ErbB-2 (HER2), ErbB-3 (HER3), and ErbB-4 (HER4), which drive proliferation, differentiation, survival, and cell motility. Of these, *EGFR* and *ErbB-2*, located on chromosomes 7p12 and 17q21, respectively, are best characterized for their role in human cancers. EGFRs are comprised of an extracellular domain (containing two receptor ligand-binding domains and a furin-like domain), a transmembrane region, and the cytoplasmic domain (containing the catalytic kinase domain). EGFR signaling is normally triggered by the binding of its natural ligand, such as EGF, betacellulin, heparin-binding EGF, and transforming growth factor-α. These EGFR ligands are all membrane-bound proteins containing one or more EGF-like motifs that interact with the receptor. Members of the EGFR subfamily are capable of forming as many as 10 different homodimers and heterodimers when occupied by an appropriate ligand and the affinity of each EGFR combination differs for each ligand. For example, an EGFR homodimer binds EGF, betacellulin, and heparin-binding EGF with about the same affinity, whereas an ErbB2:3 heterodimer preferentially binds the heregulin β isoform versus the heregulin α isoform. Receptor dimerization initiates the transphosphorylation of the PTK activation loop, which is required for optimal kinase function. Multiple tyrosine residues in the cytoplasmic tail are also phosphorylated, permitting the recruitment of adapter signaling proteins. Adapter proteins containing SH-2 domains recognize and bind phosphorylated Tyr residues and feed into signaling pathways including the Ras/Raf/MEK/ERK, PI3K/Akt, and PLCγ/PKC pathways. The result is the activation of proliferative and cell-survival signals. Potential biological responses are complex and diverse, depending upon receptor composition, ligand identity, and duration of signaling. The *EGFR (ErbB-1)* and *ErbB-2* genes are commonly amplified, rearranged, or mutated in many human cancers, including head and neck cancer, NSCLC, breast cancer, and gliomas. For example, EGFR is overexpressed in 40to 80% of NSCLCs. As discussed above (see Sect. 1.3), however, only a subset of these patients responds to the EGFR kinase inhibitor Iressa®. Therefore, overexpression by itself is not sufficient to predict responses, and the evidence suggests that activating mutations are also required. In NSCLC, these activating mutations are somatic and clustered around the ATP-binding pocket of the EGFR catalytic domain. All mutations are heterozygous and consist either of small, in-frame deletions, or amino acid substitutions within exons 17 to 19. Importantly, these mutations lead to increased responsiveness to EGF and also confer susceptibility to the inhibitor Iressa®. In contrast, rearrangements within the extracellular domain of EGFR are common in gliomas and colorectal cancer, and are associated with kinase activation. A number of monoclonal antibodies targeting the extracellular ligand-binding region (e.g. cetuximab) have been developed to specifically block such mutant EGFRs. The *erbB-2* gene is amplified in 20 to 30% of breast and ovarian cancers. ErbB-2 does not directly bind to ligands and yet its overexpression promotes oncogenesis through decreases in attenuation pathways and signal potentiation. ErbB-2 amplifies signaling indirectly by increasing the ligand affinity of its heterodimeric binding partners (i.e. EGFR, ErbB-3, and ErbB-4) and prolonging the surface expression of these RTKs by slowing their rate of internalization. In addition, ErbB-2 enhances EGFR recycling and reduces lysosomal targeting.

FGFRs The fibroblast growth factor receptor (FGFR) family is comprised of four high-affinity RTKs (FGFR1-4) that share several structural features in common, including two or three immunoglobulin-like domains and an acidic box in their amino-terminal extracellular domains, a single transmembrane domain, and an intracellular PTK domain split by a short (14 amino acid) insertion. While three of these RTKs have now been classified as cancer genes (*fgfr1–3*), the fourth (*fgfr4*) is also known to be overexpressed in mammary and ovarian carcinomas. The chromosomal locations of the *fgfr1-fgfr3* genes are 8p11.2, 10q26, and 4p16.3, respectively. Distinct FGFR mutations have been shown to mediate a variety of autosomal disorders of bone growth and embryonic development and a discussion of each would be beyond the scope of the present article. However, it is important to note that the majority of mutant FGFRs identified to date possess elevated PTK activity relative to their wild-type counterparts, resulting in ligand-independent activation. While mutations in the extracellular domain of FGFRs have been associated with dwarfism, Pfeiffer syndrome, and other developmental abnormalities, the most common mutations associated with human cancer involve a subset of acquired (somatic) translocations that target chromosome band 8p11 (*fgfr1*), and will serve as an instructive example of how the constitutive activation of FGFRs might contribute to oncogenesis. Crystal structures of an inactive FGFR1 reveal that the activation loop normally occludes substrate tyrosine binding to the catalytic site. In solution, equilibrium exists between this inactive conformation (called *cis* inhibition) and the active loop conformation. The family of fibroblastic growth factors (FGFs) capable of activating the FGFRs includes at least 24 members sharing 30 to 50% amino acid sequence homology, and are designated FGF-1 through FGF-24. Certain FGF family members are considered to be oncogenic (e.g. FGF-3, FGF-4, FGF-5, and FGF-6) and all have been implicated in the control of a variety of biological processes. In response to ligand occupancy, FGFRs undergo a ligand-induced conformational shift that promotes dimerization and *trans*phosphorylation of tyrosine residues in the activation loops of the binding partner. This leads to a movement of the activation loop away from the catalytic site that permits substrate access, while *trans*phosphorylation provides binding sites for recruitment of signal transducers and activators of transcription (STATs) *via* their SH-2 domains. In acute myelogenous leukemia, chromosomal rearrangements lead to a fusion between the ZNF198 mRNA at exon 17 and FGFR1 at exon 9, containing the kinase domain. *ZNF198* is a widely expressed gene that encodes a 155-kDa protein with five Zn finger–related motifs (known as MYM domains) that are retained in the ZNF198-FGFR1 chimeric protein. ZNF198-FGFR1 has constitutive PTK activity and is thought to rely on STAT5 activation in the transformation of hemopoietic cells. FGFR1 phosphorylation of STAT5 promotes dimerization, and dimeric STATs are exported to the nucleus, where these latent transcription factors associate with the promoters of early response genes such as *c-fos*, *c-jun*, and *c-myc* and initiate transcription of cell-cycle regulatory genes. FGFRs are mitogenic for endothelial cells and play a significant role in angiogenesis by stimulating the production and release of agents that break down basement membranes.

FLT3 FMS-like tyrosine kinase 3 (FLT3) is a member of the platelet-derived growth factor receptor (PDGFR) family and is located on human chromosome 13q12. Four of the five members of the PDGFR gene family are regarded as authentic cancer genes. These include KIT and the PDGRs type A (PDGFRA) and PDGFR type B, in addition to FLT3. FLT3 contains five extracellular ligand-binding immunoglobulin-like (Ig) domains and a cytoplasmic split tyrosine kinase motif. FLT3 is expressed in hematopoietic cells, placenta, gonads, and brain. The ligand for FLT3 is a hematopoietic growth factor termed FL (Flt3 ligand). The *fl* gene encodes a transmembrane protein, but alternative splicing can result in the generation of a soluble form of the FLT3 ligand. Flt3 binds both the membrane-bound and soluble forms of FL. FL binding activates tyrosine kinase activity and stimulates proliferation. Although FL does not efficiently induce proliferation of normal cells by itself, it does reinforce signaling *via* other growth factor receptors. FLT3 deficient mice develop normally, although they possess specific deficiencies in primitive B-lymphoid progenitors, and $FLT^{-/-}$ stem cells are impaired in their ability to reconstitute T cells and myeloid cells when transplanted into bone marrow. Thus, FLT3 signaling is critical for normal development of stem cells and B cells. FLT3 signaling is mediated by phosphorylation-dependent interactions between its carboxy-terminal cytoplasmic tail and enzymes and adapter proteins, such as the regulatory p85 subunit of PI3K, RAS/GTPase activating protein, PLC-γ, Shc, and GRB2. The activity of FLT3 is normally autoinhibited by intramolecular interactions between the cytosolic juxtamembrane domain and kinase domain, preventing ATP and substrate from binding to the activation loop. Although FLT3 is not normally expressed in mature hematopoietic cells, somatic mutants of the *flt3* gene are frequently expressed in hematological malignancies. Examples include FLT3 expression in 70% of acute myelogenous leukemias, about 30% of acute lymphoblastic leukemias (ALL), and some CML cases in lymphoid blast crisis. Although a number of different activating mutations of the *flt3* gene have been documented, two types are by far the most common. Small tandem duplications of amino acids within the juxtamembrane domain and point mutations within the activation loop occur in 24 and 7% of acute myelogenous leukemia patients, respectively. Although duplications within the juxtamembrane domain are highly variable, each is thought to move this cytosolic domain away from the activation loop, triggering kinase activity. Point mutations of the nucleotides encoding Asp835, within the FLT3 activation loop, similarly disrupt the autoinhibition of kinase activity. Since FLT3 mutations tend to be associated with a poor prognosis in leukemia, it is thought that mutant FLT3 may cooperate with other leukemia oncogenes to confer a more aggressive phenotype of this disease.

KIT The proto-oncogene *c-kit*, located on human chromosome 4q12, encodes the stem cell growth factor (SCF) receptor KIT/SCFR, a human homolog of the *v-kit* Hardy-Zuckerman 4 feline sarcoma viral oncogene. The domain organization of KIT has been conserved by all members of the PDGFR family of glycosylated, single-pass transmembrane proteins containing five extracellular immunoglobulin-like domains and a split intracellular PTK domain. The cytosolic juxtamembrane region appears to provide a crucial level of KIT autoinhibition, as mutations within this region are often associated with constitutive, ligand-independent, kinase activity.

The importance of this receptor in normal hematopoiesis and melanogenesis was established by the finding that naturally occurring germline loss-of-function mutations result in developmental defects in both processes. In a normal cell, KIT activation stimulates the proliferation and differentiation of cells such as hematopoietic stem cells, mast cells, and germ cells through the canonical RTK pathway; that is, SCF binding initiates KIT dimerization and the transphosphorylation of the dimeric KIT subunits. These phosphotyrosine residues enable the docking of SH-2 domain–containing signaling molecules including GRB2, SHP-1 (a protein-tyrosine phosphatase expressed in hematopoietic cells), and members of the Src family of cytosolic tyrosine kinases. The dephosphorylation of KIT by SHP-1 functionally limits the activity of this RTK and the half-life of the receptor is limited by the adapter protein APS, which promotes ubiquitin-mediated degradation of KIT through the recruitment of E3 ubiquitin ligase c-Cbl. Thus, both SHP-1 and APS may function as tumor suppressors for KIT-induced malignancies. Downstream of KIT, PI3K-dependent activation of Akt and the phosphorylation/inactivation of the proapoptotic protein Bad appear to be crucial mediators of KIT-induced cellular transformation. More than 30 gain-of-function (GOF) mutations in *KIT* result in a constitutively active kinase and have now been identified in a variety of different human malignancies. These mutations can generally be found in one of two regions. Deletions of a few codons within the autoinhibitory juxtamembrane region, encoded by exon 11, are observed in gastrointestinal stromal tumors and single amino acid changes in exon 17, in the carboxy-terminal half of the kinase domain, are associated with mast cell leukemia and to a lesser extent with T-cell lymphoma and acute myelogenous leukemia. Other GOF point mutations have been identified in exon 13, encoding the amino-terminal half of the KIT kinase domain. In addition to these somatic mutations, GOF mutations of the *KIT* gene have also been associated with germline tumors such as testicular seminomas and ovarian dysgerminomas. Experimentally, the co-expression of KIT and its ligand SCF has been shown to enhance the autocrine growth of cells and tumorigenesis. Approximately 80% of the patients with activating *KIT* mutations in exon 11 respond to the EGFR inhibitor Gleevec®.

MET The human *Met* (HGF receptor) gene is located on chromosome 7q21-q31 and is expressed in a variety of organs and cell types. Following its glycosylation, the MET precursor protein (190 kDa) is cleaved into a 50 kDa α-chain and 140-kDa β-chain that are linked *via* disulfide bonds. After membrane insertion of the β-subunit, these heterodimers form a mature receptor consisting of an extracellular α-subunit joined to a single-pass membrane protein (i.e. β-subunit) with a large ectodomain and intracellular PTK domain. Stimulation of MET by its natural ligand, hepatocyte growth factor/scatter factor, can lead to a broad range of responses in normal cells, such as increased proliferation, scattering, enhanced motility, and angiogenesis. The downstream effects of MET activation include the stimulation of both the PI3K and MAP kinase pathways (see Fig. 1). A series of autophosphorylation events, including phosphorylation within the β-chain activation loop at Tyr1234 and Tyr1235, and a substrate docking site at Tyr1249 and Tyr1356, stimulate MET catalytic activity and facilitate the docking of SH-2 domain- and

MET-binding domain (MBD)-containing proteins including Src, GRB2, PI3K, and PLC-γ. Proteosomal degradation of MET is mediated through interactions between the multisubstrate docking site of MET and c-Cbl. A score of MET-activating mutations, several of which are located in the PTK domain of the β-subunit, have been identified in various solid tumors and metastatic carcinomas. Of particular interest are various germline as well as somatic missense mutations within exons 16 to 19 of the *Met* gene. These mutations commonly generate amino acid substitutions within the kinase domain. For example, the N1118Y substitution impacts the highly conserved ATP-binding site while Y1253D affects the regulatory site on MET in squamous cell cancer. MET overexpression has been observed in various tumor specimens, including NSCLC, mesothelioma, pancreatic cancer, osteosarcoma, ovarian carcinoma, as well as metastatic lesions. Studies of small cell lung carcinoma have revealed that interactions between PI3K and MET sustain survival and motility. Direct involvement of MET in the process of oncogenesis is implicated by the presence of germline as well as somatic missense mutations, which characterize hereditary papillary renal carcinoma.

NTRK1 and NTRK3 Neurotrophins are a family of four secreted and growth-promoting polypeptides known to be widely distributed in both neural and non-neural tissues – that is, nerve growth factor, brain-derived neurotrophic factor, and neurotrophins 3 and 4 (NT3 and NT4). Three homologous neurotrophin tyrosine receptor kinases (NTRK1–3) are responsive to one or more of the neurotrophins and both NTRK1 and NTRK3 are known to form oncogenic fusion proteins in soft tissue tumors. The human *Nrtk1* and *Nrtk3* genes are located on chromosomes 1q21-q22 and 15q25, respectively. While nerve growth factor is the preferred ligand of NTRK1, NTRK3 binds neurotrophin 3 with high affinity. NTRK receptors share a common structural organization of their amino-terminal extracellular domains, which includes three leucine-rich 24-residue motifs flanked by two cysteine clusters and a pair of juxtamembrane C2-type immunoglobulin-like domains. The major ligand-binding domains are localized to the membrane-proximal immunoglobulin-like domains, which are also required to inhibit spontaneous dimerization and activation of the receptors in the absence of ligand. The intracellular carboxy-terminus of these receptors contains a tyrosine kinase domain plus several Tyr-containing motifs. These motifs are involved in regulating kinase activity and also in the phosphorylation-dependent recruitment of several small G proteins (e.g. Ras, Rap-1, and members of the Cdc-42/Rac/Rho family) and intermediates in the MAP kinase, PI3K, and PLCγ signaling cascades, ultimately resulting in activation of gene expression in a cell-dependent manner. Chromosomal translocations that form oncogenic fusion proteins containing the kinase domains of either NTRK1 or NTRK3, have been identified in various carcinomas. In the case of NTRK1, oncogenic activity may be conferred by a chromosomal translocation involving the 5′-dimerization domain of a non-muscle tropomyosin gene and the 3′-kinase domain of *Ntrk1*, giving rise to a constitutively active PTK. In other cases, rearrangements have been identified involving either the 5′-*TPR* (translocated promoter region) from the *Met* oncogene, or the 5′-region of *TFG* (*trk* fused gene). The *TPR-Ntrk1* translocation forms a fusion protein composed of the amino-terminal portion of TPR and the carboxy-terminal tyrosine kinase domain of NTRK1. Transgenic mice that express human TPR-NTRK1 in the thyroid develop follicular hyperplasia and papillary carcinoma. Chromosomal translocations involving *Ntrk3* often involve the breakpoint regions t(12;15)(p13;q25) and result in a chimera consisting of the NTRK3 kinase domain and the dimerization domain of the ETV6/Tel transcription factor. The oncogenic fusion proteins produced by any of the rearrangements discussed above exhibit three common characteristics that are absent from NTRK1 or NTRK3; ubiquitous cytosolic expression, ligand-independent dimerization *via* coiled-coil domain interactions, and constitutive kinase activity. These fusions have been implicated in the genesis of human colon carcinoma and papillary thyroid carcinoma in the case of NTRK1, and in secretory breast carcinoma and congenital fibrosarcoma in the case of NTRK3.

PDGFRA and PDGFRB PDGFRA and PDGFRB are encoded by genes located in human chromosomes 4q11-q13 and 5q31-q32, respectively. PDGFRA exhibits a pronounced structural and sequence similarity with members of the PDGFR family, including PDGFRB. The main distinction between PDGFRA and PDGFRB is exemplified by their ligand preferences, with PDGFRB selectively interacting with PDGF homodimers containing B type polypeptide chains (PDGF BB) while PDGFRA interacts more promiscuously with the PDGF AA, BB, and AB dimers. In addition, these receptors exhibit temporally distinct and tissue-specific activities in the mammalian embryo and adult. For example, PDGFRA expression in the nervous system coincides with the development of glial cells, which includes processes of glial cell migration and proliferation. The functional PDGFRA consists of an extracellular domain containing five immunoglobulin-like domains punctuated by eight N-linked glycosylation sites, a transmembrane domain, and two cytoplasmic kinase domains (domains 1 and 2). The extracellular immunoglobulin-like domains are essential for the proper folding of the PDGFRA protein, thereby enabling receptor–ligand interactions. Ligand engagement induces receptor dimerization and autophosphorylation, which generates phosphorylation-dependent recruitment sites for adapter proteins and enzymes capable of activating the PI3K and MAP kinase pathways (e.g. Src, PI3K, PLC-γ, the Crk group of adapter proteins, and SHP-2). PDGFRA is also capable of stimulating c-Jun N-terminal kinase (JNK1) to promote apoptosis, and inhibit cell-cycle progression *via* the cyclin-dependent kinase (CDK) inhibitor p21^{Cip1}. Signaling through the PDGFRA also stimulates membrane ruffling, cytoskeletal rearrangements, and migration in a cell-type-specific manner. PDGFRA activity is limited by ligand-induced Src activation and receptor trafficking/internalization. In cancer, a potentially activating deletion of an 81 amino acid stretch within the fifth immunoglobulin-like domain of the extracellular stalk of PDGFRA has been described. This deletion mutant is found amplified in glioblastoma, where it is likely to give rise to an alternatively folded receptor capable of sustaining PDGFRA catalytic activity in the absence of ligand. Recently, a transforming *PDGFRA* deletion mutant lacking immunoglobulin-like domain-encoding exons 8 and 9 has also been identified in gliomas. This mutant protein is capable of autophosphorylation and, therefore, activation in the absence of PDGF. In addition, mutations and deletions within the PDKFRA activation loop (exon 18) and juxtamembrane region (exon 12), that have been detected in gastrointestinal pacemaker cell tumors (i.e. "gastrointestinal stromal tumors"), are also capable of producing ligand-independent activation. PDGFRA and its ligands have also been reported to facilitate tumorigenesis in the absence of genetic mutations, by establishing autocrine signaling loops, as in the case of medulloblastomas where autocrine stimulation promotes metastasis.

RET The glial cell line–derived neurotrophic factor (GDNF) family, consisting of GDNF, neurturin, artemin, and persephin, all transmit signals promoting neuronal survival, proliferation, and differentiation *via* the transmembrane receptor tyrosine kinase RET (rearranged in transformation) and its co-receptors GFRα-1 to -4. RET is normally expressed throughout the nervous system, within neural crest derived cells, and the urogenital system, and is required for neuronal and kidney development. The GFRα1 to 4 co-receptors are glycosyl-phosphatidylinositol-anchored proteins that selectively recognize and bind certain members of the GDNF family and associate with RET kinase to trigger intracellular signaling. For instance, GDNF preferentially binds to the GFRα1-RET heterodimer. The *RET* gene has been mapped to human chromosome 10q11.2 and is known to give rise to a transcript which is alternatively spliced to yield the short, middle, and long isoforms of RET, each containing an extracellular domain with a cadherin-related motif and a cysteine-rich region. When bound to GDNF, GFRα1 dimerizes with RET, and stimulates transphosphorylation of specific Tyr residues within the cytoplasmic domain of RET. The three RET isoforms have a number of Tyr residues in common, including those residing within the kinase domain and carboxy-terminus, though the long RET isoform contains an additional carboxy-terminal Tyr1096 residue. Once phosphorylated, these residues serve as scaffolding sites for signaling complexes. The conserved Tyr residues bind GRB10, PLCγ, and Shc, while Tyr1096 binds GRB2 specifically and links RET activation to the cell proliferation and survival pathways controlled by the Ras/Raf/MEK/ERK and PI3K/AKT pathways respectively. Underscoring the importance of Tyr1062, are studies that demonstrate that a point mutation within this codon significantly decreases the transforming activity of the RET protein. RET-mediated transformation results from various activating germline mutations in early-onset multiple endocrine neoplasia types 2A and 2B and late-onset familial medullary thyroid carcinoma. In addition, genetic rearrangements have been identified in papillary thyroid carcinoma. Chromosomal translocations involving the intracellular domain of *RET* have given rise to 10 known fusion proteins (RET/PTC1 -9 and RET/ELKS) prevalent in papillary thyroid carcinoma in a geographically distinct manner. These include translocations between *RET* and *H4*, *RIα*, *ELE1*, *RFG5*, *hTIF*, *RFG7*, *kinectin*, *RFG9*, and *ELKS*. The most common rearrangements, RET/PTC1 and RET/PTC3, involve inversions of the *H4* inv(10)(q11.2q21) and *ELE1* genes and account for 90% of all rearrangements in papillary thyroid carcinoma. Studies suggest that RET/PTC becomes ectopically localized to the cytoplasm where it activates STAT3 by phosphorylation of Tyr705. The constitutively active STAT3 promotes cellular transformation *via* stimulation of *cyclin-D1, D2, D3*, and *bcl-XL* transcription. RET/PTC has also been implicated in the transcriptional upregulation of the osteopontin gene, resulting in an increase in the proliferative and invasive activity of cells that respond to autocrine stimulation by osteopontin. While RET/PTC rearrangements predominate in papillary thyroid carcinoma, germline mutations within *RET* are responsible for the onset of multiple endocrine neoplasia types 2A and 2B, and familial medullary thyroid carcinoma. Point mutations or base pair duplications of the *RET* gene have been reported in patients with multiple endocrine neoplasia 2A and familial medullary thyroid carcinoma. Missense mutations within *RET* have also been described in patients with multiple endocrine neoplasia 2A or 2B. Point mutations observed in multiple endocrine neoplasia 2A and familial medullary thyroid carcinoma, target the extracellular cysteine residues that normally contribute to intramolecular disulfide bonds, and promote RET dimerization. Formation of RET homodimers leads to its ligand-independent activation. On the other hand, RET mutations in multiple endocrine neoplasia 2B do not induce dimerization; rather, mutations target the catalytic core and may alter substrate specificity by inducing conformational changes in the region containing Tyr905, a conserved residue important for catalysis.

3.1.2 Cytoplasmic Protein-tyrosine Kinase Subfamily

ABL1/ABL2 The *ABL1* gene (chromosomal band 9q34.1), along with the closely related *ABL2* or *ARG* (Abl-related gene; chromosomal band 1q24-q25), belongs to the Abelson family of genes that encode prototypic non-receptor tyrosine kinases involved in various cellular processes. The *ABL1* gene product, c-Abl is a ubiquitously expressed protein that colocalizes with actin-based cytoskeletal structures in the cytoplasm, and binds chromatin when imported to the nucleus. These intermolecular interactions are mediated by the carboxy-terminal actin- and DNA-binding domains of c-Abl. The amino-terminus of c-Abl contains an SH-3 domain and an SH-2 domain. Autoinhibition of c-Abl is achieved by a novel variation on a familiar theme, and the details of this mechanism only recently became apparent when the crystal structure of the protein became available for examination. As in the case of the closely related Src kinases, the amino-terminus of the protein functions as an autoinhibitory "Cap" region preventing substrate docking with the kinase domain. However, this is not achieved by the binding of its SH-2 domain to a phosphotyrosine residue, as in other Src-like kinases. Rather, an amino-terminal myristate is responsible for binding to a hydrophobic pocket in the kinase domain that stabilizes an autoinhibited conformation. It is this distinction in the autoinhibitory mechanisms employed by c-Abl as compared with those of the other Src kinases that explains the ability of the drug imatinib (Gleevec®) to selectively inhibit the catalytic activity of c-Abl, but not that of c-Src. Targeted disruption of c-Abl in mice is characterized by lymphopenia, failure to thrive, and neonatal mortality. In the cytoplasm, c-Abl participates in the propagation of mitogenic signals via the Ras/Raf/MEK/ERK and PI3K/Akt pathways. This signaling is negatively regulated by the interactions of c-Abl with PSTPIP1, a PEST-type protein Tyr phosphatase that dephosphorylates c-Abl. c-Abl signaling in the nucleus results in a suppression of cell growth *via* the c-Abl-dependent transactivation of various proapoptotic genes. During the early G1 phase of the cell cycle, this nuclear activity of c-Abl is repressed due to its sequestration by the pocket protein Rb. Following an oncogenic and reciprocal chromosomal translocation, control of c-Abl activity is lost in cells, and this generates what has been termed the *Philadelphia chromosome*. This mutation is present in most cases of CML and many cases of ALL. Although similar translocations occur in CML and ALL, their c-Abl products are distinct, producing 210 kDa and 180 to 185 kDa proteins respectively. Both involve a t(9;22) (q34, q11) translocation of the *ABL* gene to the center of the *BCR* gene, producing a fused transcript that translates into a chimeric Bcr-Abl protein that is constitutively active. C-Abl activity is sustained by the loss of its autoinhibitory (myristoylated) amino-terminal tail and the dimerization triggered by the presence of coiled-coil domains within the amino-terminus of Bcr. Bcr-Abl dimerization, in turn, allows for the transphosphorylation of the fusion protein and activation of oncogenic signaling. The formation of Bcr-Abl homo-oligomers and hetero-oligomers with Bcr further promotes oncogenesis by localizing the protein in the cytosol and preventing DNA binding. The use of the small molecular tyrosine kinase inhibitor Gleevec® in the treatment of CML patients who are Bcr-Abl-positive has met with success, but drug resistance occurs frequently. Mutations of residues in or near the ATP-binding pocket of c-Abl prevent Imatinib binding and account for a majority of drug-resistant cases. The homologous *ARG* gene is involved in a similar translocation, but with a different fusion partner, Ets variant gene 6 (*ETV6/Tel*). ETV6/Tel is a member of the Ets family of transcription factors and is frequently found to be rearranged with a variety of translocation partners in human leukemias. This fusion protein contains the complete SH-3, SH-2, and PTK domains of ARG along with the helix–loop–helix oligomerization domain of ETV6/Tel. The translocation t(1;12)(q25;p13) causes malignancy in adult patients with acute myelogenous leukemia.

JAK-2 Janus kinase 2 (JAK-2) is a receptor-associated PTK essential for propagating certain cytokine induced signals. Located on human chromosome 9p24, the *JAK-2* gene encodes five JH domains, which are regions common to all members of the JAK kinase family and are sites of interaction between the protein kinase and its associated cytokine receptor. An amino-terminal JH-1 domain contains the kinase domain responsible for the phosphorylation of its preferred substrate; the Tyr, neutral, basic amino acid motif. Carboxy-terminal to the catalytic domain is a regulatory pseudokinase domain, referred to as the JH-2 domain. Normally, JAK-2 activity is cytokine-dependent. Interleukin-3 (IL-3), IL-5, and/or IL-6 interactions with cytokine receptors initiate receptor dimerization and the activating transphosphorylation of the constitutively associated JAK-2 proteins. Once active, JAK-2 phosphorylates Tyr residues on the carboxy-terminal region of cytokine receptors, providing binding sites for recruitment of STATs *via* their SH-2 domains. This in turn permits JAK-2 to phosphorylate STAT proteins, thereby promoting the disassociation of STATs from cytokine receptors, and their dimerization. When exported to the nucleus, dimeric STATs associate with the promoters of early response genes such as *c-fos*, *c-jun*, and *c-myc* and initiate transcription of cell-cycle regulatory genes such as *cyclin D1* and pro-survival genes such as *bcl-XL*. Consequently, the constitutive activation of JAK-2 significantly enhances cell proliferation and survival. Deregulation of JAK-2 activity arises as a result of a somatic chromosomal translocation between the dimerization domain of the ETS-like transcription factor Tel and JAK-2. The resulting fusion protein contains the amino-terminal protein dimerization domain of Tel and carboxy-terminal JAK-2 catalytic domain (JH-1), and is able to form homodimers capable of transphosphorylating its binding partner, thus leading to constitutive activation. The Tel-JAK-2 fusion protein is ectopically localized within cells and may undergo unusual interactions with members of the early gene induction pathway in a cytokine-independent manner, thereby supporting cellular transformation. The presence of the *Tel-JAK-2* translocation and resulting transformation of hematopoietic cells underscore the importance of JAK-2 signaling in hematopoiesis. Examples of the Tel-JAK-2 driven transformation include cases of T-cell childhood acute lymphocytic leukemia, acute myeloid leukemia, acute lymphocytic leukemia, and atypical chronic myeloid leukemia. The finding that JAK-2 inhibition by a specific kinase blocker (i.e. AG-490) induces apoptosis of leukemic cells further supports the direct linkage between constitutive JAK-2 activation and malignant transformation. It is of note that while neither JAK-1 nor JAK-3 meets the criteria for inclusion in the cancer genome, both these cytoplasmic PTKs also signal to the nucleus *via* STATs and are overexpressed in various leukemias.

LCK The T-lymphocyte protein-tyrosine kinase (*LCK*) is a member of the Src-tyrosine kinase family and is located on the short arm of human chromosome 1 (1p34). LCK expression predominates in T cells, where its activity is crucial for normal T-cell development and activation. LCK is closely associated with the T-cell receptors (TCR) CD4 and CD8. LCK is a multidomain protein, typical of the Src kinases, including SH-3, SH-2, tyrosine kinase domains, and a unique domain responsible, among other things, for the membrane anchoring of LCK. During the process of acquiring adaptive immunity, T-cell activation is initiated by the binding of CD4 and CD8 to antigen-presenting cells *via* type I and type II glycoproteins of the major histocompatibility complex. The proximity of LCK to these activated TCRs allows for the transduction of signals responsible for the T-cell response, which includes the production of IL-2 and inositide (1,4,5) triphosphate. LCK activation requires autophosphorylation of Tyr394 and dephosphorylation of Tyr505, in the carboxy-terminal tail. The protein Tyr phosphatase CD45 regulates LCK activity in a bimodal fashion, not only by mediating Tyr505 dephosphorylation (enhanced activity) but also by dephosphorylation of Tyr394 (suppressed activity). LCK promotes T-cell activation by phosphorylating immunoreceptor Tyr-based activation motifs present within TCR complexes. When phosphorylated, these motifs enable recruitment of SH-2 domain–containing proteins (e.g. ZAP70) to the TCR. In addition, LCK phosphorylates the membrane–cytoskeleton linker protein ezrin to promote T-cell activation. In T cells, the LKC-dependent phosphorylation of the linker protein Shc also initiates mitogenic signaling *via* the Ras/Raf/MEK/ERK pathway. Because of the critical position occupied by LCK in the TCR signaling pathway, its necessarily transient activity is highly regulated through ubiquitination under the control of the Cbl ubiquitin ligase. LCK mutations resulting in constitutive autophosphorylation and/or deregulated catalytic function have been described in T-cell leukemia, where they contribute to the oncogenic potential of this kinase. Insertion of the tripeptide QKP between the cytosolic SH-2 and kinase domains may also enhance catalytic function by disrupting the normal repressive interaction of its SH-2 domain with its catalytic site. Three other recorded mutations, resulting in a substitution within the region critical for interaction between LCK and CD4 and two alternative substitutions within the kinase domain, may result in a GOF for LCK. In addition to these mutations, a chromosomal translocation (t(1;7)(p34;q34)) has been observed in T-cell acute lymphoblastic leukemia, resulting in an oncogenic fusion protein (i.e. βTCR:LCK) that contains the βTCR enhancer region and the 5′ promoter region of LCK. In the case of the βTCR:LCK fusion, a GOF would result from an alleviation of transcriptional repression by the βTCR enhancer region.

3.2 Protein-serine/Threonine Kinase Family

3.2.1 Tyrosine Kinase-like Subfamily

BRAF BRAF is discussed above in Sect. 2.2.

3.2.2 Homologs of Yeast Sterile 7, Sterile 11, Sterile 20 Kinase Subfamily

JNKK The c-Jun terminal kinase–kinase (*JNKK*) gene is located on human chromosome 17p11. This oncogene encodes a dual specificity kinase catalyzing Thr180 and Tyr182 phosphorylation of p38 MAP kinase and JNK1, 2, and 3 in response to environmental stress, including changes in cellular osmolarity. The functional consequences of JNKK signaling are cell-type-, environment-, and stimulus-specific. Regulation of JNKK occurs upstream of the kinase as a result of the ubiquitin-mediated degradation of MAPK kinase–kinase (MEKK1). Ubiquitination prevents phosphorylation of JNKK by MEKK1 and abrogates downstream activation of JNK. JNK activation requires stable complex formation between JNKK and JNK. The amino-terminal region of JNKK facilitates this interaction as it codes for a conserved MAP kinase-docking site (D-domain). JNKK is able to modulate JNK signaling to AP-1 transcription factors by competitive sequestration of JNK *via* D-domain interactions. In a classic response to UV irradiation, JNKK mediated activation of JNK results in apoptosis through a p53-dependent mechanism. Downregulation of JNKK or its downstream signaling components provides cells with a growth advantage. Moreover, a loss of JNKK expression has been observed in various cancers, including prostate cancer. An overall *JNKK* mutation rate of ~5% has been observed in various types of tumors. The mutations include two nonsense mutations, five missense mutations, and one splice site mutation. Not surprisingly, these mutations generally serve to inactivate JNKK signaling. In keeping with its tumor suppressor classification, the *JNKK* gene is a known target for homozygous deletions in pancreatic adenocarcinomas, biliary adenocarcinomas, and breast carcinomas.

3.2.3 Calcium/Calmodulin-dependent Protein Kinase Subfamily

PIM-1 A common site for proviral integration of the moloney murine leukemia virus (PIM) in mice encodes the *PIM-1* gene. The Ser/Thr kinase PIM-1 is the product of this gene, which is located on the p12 segment of human chromosome 6. In addition to PIM-1, PIM-2 and PIM-3 make up the PIM family of kinases and may serve redundant functions as observed in PIM-1-deficient mice, where compensation for the loss of *PIM-1* was evident. PIM-1 functions to promote hematopoiesis in response to growth factors such as IL-2 and IL-3 and its expression is confined, throughout fetal development, to hematopoietic sites within the liver and spleen, as well as within the thymus, lymph-nodes, and testis. High expression of PIM-1 has also been observed in myeloid cell and B-cell lines. PIM-1 has been characterized as a 33-kDa cytoplasmic protein, though recent studies have revealed that its nuclear localization is crucial for its antiapoptotic function in Burkitt's lymphoma. Upregulation of PIM-1 has been implicated in BCR/ABL mediated leukemogenesis where it acts downstream of STAT5 to confer protection from apoptosis and promote cell-cycle progression. PIM-1 overexpression has also been shown to positively influence a variety of human leukemias, though overexpression alone is not sufficient for cellular transformation, and cooperation between PIM-1 and c-myc is often observed during this process. The identification of a breakpoint region on chromosome 6 has raised the possibility of a translocation between *PIM-1* and the proto-oncogene *c-ABL* (t(6;9)(p21; q33)) in some cases of myeloid leukemia. In response to cytokines and various mitogens, *PIM-1* expression is induced *via* the JAK/STAT5 and MAP kinase pathways and its activity may be modulated through a series of autophosphorylation events. With the exception of Etk, which promotes the activation of PIM-1 *via* phosphorylation of Tyr218, upstream kinase regulators of PIM-1 have not been identified. Recent studies show that phosphorylation of the nuclear mitotic apparatus protein by PIM-1 facilitates interactions between this protein, the heterochromatic protein-1β, and the spindle proteins dynein and dynactin, which are crucial for mitosis. Furthermore, PIM-1 alleviates transcriptional repression by directly phosphorylating HPI, a member of the transcriptional repression machinery, thereby regulating chromatin structure. Other putative substrates of PIM-1 include p100, a co-activator of the transcription factor c-myb, the cell-cycle phosphatase Cdc25A, and the cyclin-dependent kinase inhibitor $p21^{cip1}$. It is evident that PIM-1 regulates the activity of many proteins that have the potential to promote tumor growth and, as expected, its expression is tightly regulated *via* transcriptional and translational mechanisms. The presence of dyad symmetry elements enables the *PIM-1* gene to form stem loop structures, and to block transcriptional elongation in the absence of appropriate stimuli. Another regulatory element is present in the form of an A/U rich region within the 3′-unstranslated region of PIM-1 mRNA. This region promotes RNA instability, which can be overcome in the presence of certain cytokines and mitogens. Translation of the PIM-1 message is regulated by the presence of a 5′-untranslated region containing an internal ribosomal entry site. PIM-1 activity may also be regulated *via* autophosphorylation, while its turnover rate is dependent upon ubiquitination. In the event of deregulated PIM-1 expression, oncogenesis is most likely facilitated by the PIM-1-dependent stimulation of cell-cycle progression and mitosis.

STK11 Serine–threonine kinase 11 (STK11), also known as LKB1, is a crucial regulator of cellular metabolism and polarity. This cytoplasmic kinase is ubiquitously expressed in human tissues and exhibits catalytic activity only in the presence of its binding partner STRAD (STE20-related adapter). As STRAD lacks intrinsic kinase activity, the association of the pseudokinase domain of STRAD with the kinase domain of STK11 is postulated to result in intermolecular interactions capable of stimulating the STK11 catalytic core. In the cytoplasm, the association of STK11 with STRAD gives rise to a series of STK11 autophosphorylation events and results in the phosphorylation of STRAD. Induction of cell polarity by STK11 may result from its association with the Ser/Thr kinase PAR1. This leads to destabilization and polarized localization of microtubules. STK11 also plays a role in axis induction during development by stimulating the Wnt signaling pathway. The tumor suppressor activity of STK11 is mediated by the ataxia-telangiectasia mutated kinase (ATM, see below). In response to the cellular stress caused by ATP depletion, STK11 directly phosphorylates Thr172 of the AMP-activated kinase. This results in the downregulation of ATP-consuming processes, such as the translational activity mediated by the mammalian target of rapamycin, and upregulation of ATP-producing processes such as glucose uptake. Therefore, the loss of STK11 expression, observed in various carcinomas, offers cells a distinct growth advantage by removing the inhibition normally placed on protein synthesis and mitogenesis. The tumor suppression activity of STK11 makes it a target for inactivating mutations such as those observed in patients with the hereditary Peutz-Jeghers syndrome. Indeed, the *STK11* gene was mapped to region p13.3 of human chromosome 19 – a site susceptible to germline mutations in patients with Peutz-Jeghers syndrome. In addition to the formation of benign intestinal hamartomatus polyps, individuals with Peutz-Jeghers syndrome exhibit a predisposition toward gastrointestinal, pancreatic, ovarian, testicular, uterine, and breast carcinomas. Studies have identified various deviations of the wild-type STK11 transcript. These are caused by translational frameshift insertions or deletions, nonsense mutations, missense mutations, and disruption of STK11 mRNA splicing through exonal splice site mutations. The functional consequence of these mutations is the generation of an inactive STK11 protein. Germinal *STK11* mutations are often accompanied by somatic mutations. Somatic mutations leading to homozygous deletion of *STK11* or *STK11* truncations have also been reported in patients with non-Peutz-Jeghers syndrome related pancreatic cancer, mucinous minimal deviation adenocarcinoma of the uterine cervix, lung adenocarcinoma, and malignant melanoma. In general, *STK11* mutations affect the biological activity of STK11 by one of two common means. Large truncations or targeted mutations of invariant residues within the STK11 kinase domain destroy its ability to bind ATP and catalyze phosphotransfer. The SL26 mutant present in some cases of Peutz-Jeghers syndrome does not affect kinase activity; rather it promotes the loss of cytoplasmic retention, which is accompanied by nuclear localization of STK11. It is hypothesized that nuclear localization results from the inability of the SL26 mutant to interact with its cytoplasmic substrates and thereby remain in the cytoplasm. Nuclear localization also leads to silencing of STK11 signaling because of the lack of substrates and activating factors. Paradoxically, owing to a marked increase in the AMP/ATP ratio, cells with loss of STK11 expression are rendered more susceptible to cell death caused by agents such as the

AMP analog AICAR. This response may represent a unique therapeutic opportunity for those patients whose carcinomas exhibit a loss of STK11 activity.

3.2.4 Containing Pka, Pkg, Pkc Subfamilies

AKT2 The *AKT2* gene, located at 9q13.1-q13.2, produces a phosphoinositide-dependent Ser/Thr kinase, crucial for regulating cellular events including apoptosis, differentiation, proliferation, and metabolism. The gene product of AKT2, also referred to as protein kinase B beta (PKB-β), is a broadly expressed protein containing a single 3-phosphoinositide-binding pleckstrin homology (PH) domain at its amino terminal. Adjacent to this pleckstrin homology domain is the kinase domain, followed by a carboxy-terminal regulatory domain. The AKT isoforms (AKT1 and AKT2) function downstream of various RTKs, and the biogenesis of an active AKT requires the sequential recruitment of the inactive protein to the plasma membrane followed by the transphosphorylation of a conserved Thr residue within the activation loop (Thr308 in AKT1 and Thr306 in AKT2) and Ser residue within a carboxy-terminal hydrophobic motif (Ser473 in AKT1 and Ser474 in AKT2). AKT is recruited to the plasma membrane through the binding of its amino-terminal PH domain to phosphatidylinositol-3, 4, 5-triphosphate (PIP_3), a lipid product of PI3K. This protein--lipid interaction is tightly controlled by the lipid phosphatase *PTEN* (phosphatase and tensin homolog), which is also a cancer (tumor) suppressor gene that undergoes both somatic and germline inactivating mutations. By anchoring AKT to the membrane lipid bilayer, PIP_3 alters the conformation of the AKT kinase domain and exposes the activation loop phosphorylation site to PDK1. Following this "priming" step, AKT is phosphorylated on its hydrophobic motif by a yet-to-be-identified protein Ser/Thr kinase. Integrin-linked kinase was shown to be essential for this final step in the maturation of AKT, but whether this kinase is directly responsible for AKT phosphorylation remains unresolved. The authentic AKT hydrophobic motif kinase may be the DNA-dependent protein kinase. Good evidence for this has recently been presented, but has not yet been confirmed. Once activated, AKT is capable of enhancing cellular proliferation and the generation of survival signals through some of the pathways and effector molecules included in Figure. 1. *AKT2* was included in the census of cancer genes because of its amplification in ovarian and pancreatic cancers. The region that is amplified in ovarian carcinomas (19q13.1-q13.2) spans the entire *AKT2* gene. However, no activating mutations have been documented for this kinase in cancer and, therefore, tumors in which *AKT2* has been amplified may not be responsive to small-molecule inhibitors of kinase activity. Depending on the genetic profile presented by a given patient, other oncogenes might contribute to the transforming activity of AKT2. For example, the coexistence of an *AKT2* amplification and loss-of-function mutation in *PTEN* would allow an enhancement of AKT2 signaling in the presence of endogenous growth factors. The human prostate epithelial cancer cell line LNCaP harbors such a *PTEN* mutation and possesses constitutive AKT kinase activity. Another interesting candidate in ovarian cancer is *SEI-1*. *SEI-1* is located in a genetic region of chromosome 19q13.1-q13.2, neighboring *AKT2*. This genetic region is coamplified along with *AKT2* in a subset of ovarian cancer cell lines. The *SEI-1* gene encodes a CDK4-binding protein, which renders the activity of cyclin D/CDK4 complexes resistant to the inhibitory effect of $p16^{INK4a}$. Given the proliferation-related functions of AKT2 and SEI-1, it is possible that both gene amplifications at chromosome 19 have the potential to advance the growth of certain ovarian tumors.

3.2.5 Containing Cdk, Mapk, Gsk3, Clk Subfamilies

CDK4 and the INK4 inhibitors The human cyclin-dependent kinase 4 (*CDK4*) gene is located at chromosome 12q14 and encodes a small (303 amino acid) cell-cycle regulatory protein Ser/Thr kinase essential for coordinating the cell's progression through the early growth (G1) phase of the cell cycle. CDK4 is activated by the binding of D-type cyclins (which impart basal activity to the kinase) and, by phosphorylation (which fully activates the kinase). When active, CDK4 interferes with the tumor suppressor activity of the retinoblastoma protein Rb and its homologs. CDK4 is inactivated by the binding of the INK4 family of CDK inhibitors. G1 progression depends on extracellular mitogenic signals that upregulate cyclin D expression, committing the cell to another round of cell division. INK4 proteins bind monomeric CDK4, preventing its stable interaction with cyclin D. They also bind the cyclin D – CDK4 complex, forming an inactive ternary complex. There are four INK4 proteins ($P16^{INK4a}$, $P15^{INK4b}$, $P18^{INK4c}$, and $P19^{INK4d}$), all of which possess ankyrin repeats involved in binding CDK4. The binding of these proteins substantially alters the conformation of the CDK4 ATP-binding site and renders the kinase inactive. Mutations leading to a loss of control over the CDK4/INK4/Rb pathway are among the most common in human cancer. CDK4 is often amplified and overexpressed in glioblastoma (50%), uterine cervical carcinoma (26%), breast carcinoma (16%), and osteosarcoma (16%). In addition, a germline arginine-to-cysteine substitution at codon 24 has been reported in some cases of melanoma. This mutation contributes to malignant transformation by preventing the interaction between $p16^{INK4a}$ and CDK4, but is rarely encountered in other human neoplasms. More common are mutations in the binding region of the *INK4* genes, lowering the affinity of the INK4 proteins for CDK4. In the case of $p16^{INK4a}$, such mutations occur frequently in leukemia (58%), bladder carcinoma (50%), glioma (35%), nasopharyngeal carcinoma (35%), and pancreatic cancer (21%). Promoter methylation is also a common mechanism of inactivation of the *INK4* genes in cancer.

3.2.6 Phosphatidylinositol Kinase (Pik)-Related Subfamily

ATM The human tumor suppressor gene ataxia-telangiectasia mutated (*ATM*) is located on chromosome 11q22-q23 and is frequently mutated in a rare autosomal recessive disorder causing neuronal degeneration, immunologic deficiency, radiosensitivity, and cancer predisposition. Over one-third of ataxia-telangiectasia patients develop lymphoid cancers including non-Hodgkin's lymphoma, Hodgkin's lymphoma, and other leukemias. The *ATM* gene encodes a large (3056 amino acid) nuclear protein that is expressed in testis, spleen, and thymus of adult mice and plays a critical role in maintaining genetic stability in response to DNA damage. ATM is a member of the recently identified PIK-related subfamily that phosphorylates Ser/Thr residues, rather than lipids. All members contain a conserved carboxy-terminal catalytic domain, referred to as the PI-kinase domain, and a novel amino-terminal FAT (FRAP, ATM, and TRRAP) domain that possibly regulates the conformation of the PI-kinase domain. Structurally, ATM is composed of two regions referred to as the *head* and an *arm*. Three-dimensional reconstructions of ATM bound to DNA suggest that the kinase uses its arm to clamp around the double helix. Whether this protein–DNA interaction is required for the ATM-dependent repair of double-strand breaks is not known. ATM activation in response to DNA double-strand breaks is rapidly triggered following ionizing radiation functions to arrest the cell-cycle progression of damaged cells in G1, S, or G2. The G1/S checkpoint is activated through the induction of p53. ATM positively regulates the p53 pathway by increasing the half-life of the p53 protein, either by directly phosphorylating Ser15, or indirectly through activation of other kinases such as Chk2 and c-Abl. ATM, Chk2, and c-Abl are all capable of phosphorylating p53, thus stabilizing the later protein by interfering with its binding to the p53 regulatory protein, murine double minute 2. In addition to regulating the G1/S checkpoint, ATM engages the S and G2/M checkpoints in response to ionizing radiation. It triggers the G2/M checkpoint by Chk2-mediated phosphorylation of Ser216 on Cdc25C, thereby promoting the binding of 14-3-3 proteins. The Cdc25C/14-3-3 complex is then exported from the nucleus, preventing Cdc25C from activating the nuclear Cdc2/cyclinB complex required for mitosis. In a similar fashion, ATM controls the S-phase checkpoint through the Chk2-dependent phosphorylation of Cdc25A Ser123. This leads to Cdc25A degradation. Most of the ATM mutations associated with cancer are derived from large deletions, causing protein truncations that inactivate the kinase. Other inactivating mutations are represented by missense mutations (e.g. L1420F) or small in-frame deletions/insertions. This loss of ATM function is thought to predispose to cancer by increasing genetic instability.

Acknowledgments

National Institutes of Health Grant (R01 CA98195 to JAM) and the DOD Prostate Cancer Research Program (DAMD 17-02-1-0053 to DMT) supported this work.

See also Cancer Stem Cells; Cell Signaling During Primitive Hematopoiesis; Cellular Interactions; Epigenetic Mechanisms in Tumorigenesis; Homeodomain Proteins; Mutagenesis, Malignancy and Genome Instability; Oncology, Molecular.

Bibliography

Books and Reviews

Blume-Jensen, P., Hunter, T. (2001) Oncogenic kinase signaling, *Nature* **411**, 355–365.

Futreal, P.A., Coin, L., Marshall, M., Down, T., Hubbard, T., Wooster, R., Rahman, N., Stratton, M.R. (2004) A census of human cancer genes, *Nature Rev.* **4**, 177–183.

Hanks, S.K., Hunter, T. (1995) Protein kinases 6. the eukaryotic protein kinase superfamily: kinase (catalytic) domain structure and classification, *FASEB J.* **9**, 576–596.

Hubbard, S.R., Till, J.H. (2000) Protein tyrosine kinase structure and function, *Annu. Rev. Biochem.* **69**, 373–398.

Steelman, L.S., Pohnert, S.C., Shelton, J.G., Franklin, R.A., Bertrand, F.E., McCubrey, J.A. (2004) JAK/STAT, Raf/MEK/ERK, PI3K/Akt and BCR-ABL in cell cycle progression and leukemogenesis, *Leukemia* **18**, 189–218.

Terrian, D.M., (Ed.), (2003) *Cancer Cell Signaling*, Humana Press, Totowa.

Primary Literature

Agnes, F., Shamoon, B., Dina, C., Rosnet, O., Birnbaum, D., Galibert, F. (1994) Genomic structure of the downstream part of the human FLT3 gene: exon/intron structure conservation among genes encoding receptor tyrosine kinases (RTK) of subclass III, *Gene* **145**, 283–288.

Alessi, D.R., Andjelkovic, M., Caudwell, B., Cron, P., Morrice, N., Cohen, P., Hemmings, B.A. (1996) Mechanism of activation of protein kinase B by insulin and IGF-1, *EMBO J.* **15**, 6541–6551.

Bai, R.Y., Ouyang, T., Miething, C., Morris, S.W., Peschel, C., Duyster, J. (2000) Nucleophosmin-anaplastic lymphoma kinase associated with anaplastic large-cell lymphoma activates the phosphatidylinositol 3-kinase/Akt antiapoptotic signaling pathway, *Blood* **96**, 4319–4327.

Bischof, D., Pulford, K., Mason, D.Y., Morris, S.W. (1997) Role of the nucleophosmin (NPM) portion of the non-Hodgkin's lymphoma-associated NPM-anaplastic lymphoma kinase fusion protein in oncogenesis, *Mol. Cell. Biol.* **17**, 2312–2325.

Blume-Jensen, P., Janknecht, R., Hunter, T. (1998) The kit receptor promotes cell survival via activation of PI 3-kinase and subsequent Akt-mediated phosphorylation of Bad on Ser136, *Curr. Biol.* **8**, 779–782.

Bosch, E., Cherwinski, H., Peterson, D., McMahon, M. (1997) Mutations of critical amino acids affect the biological and biochemical properties of oncogenic A-Raf and Raf-1, *Oncogene* **15**, 1021–1033.

Bossotti, R., Isacchi, A., Sonnhammer, E.L.L. (2000) FAT: a novel domain in PIK-related kinases, *TIBS* **25**, 225–227.

Brown, V.L., Harwood, C.A., Crook, T., Cronin, J.G., Kelsell, D.P., Proby, C.M. (2004) p16INK4a and p14ARF tumor suppressor genes are commonly inactivated in cutaneous squamous cell carcinoma, *J. Invest. Dermatol.* **122**, 1284–1292.

Carow, C.E., Levenstein, M., Kaufmann, S.H., Chen, J., Amin, S., Rockwell, P., Witte, L., Borowitz, M.J., Civin, C.I., Small, D. (1996) Expression of the hematopoietic growth factor receptor FLT3 (STK-1/Flk2) in human leukemias, *Blood* **87**, 1089–1096.

Cazzaniga, G., Tosi, S., Aloisi, A., Giudici, G., Daniotti, M., Pioltelli, P., Kearne, L., Biondi, A. (1999) The tyrosine kinase abl-related gene ARG is fused to ETV6 in an AML-M4Eo patient with a t(1;12)(q25;p13): molecular cloning of both reciprocal transcripts, *Blood* **94**, 4370–4373.

Chang, F., Lee, J.T., Navolanic, P.M., Steelman, J.G., Blalock, W.L., Franklin, R.A., McCubrey, J.A. (2003) Involvement of PI3K/Akt pathway in cell cycle progression, apoptosis, and neoplastic transformation: a target for cancer chemotherapy, *Leukemia* **17**, 590–603.

Chang, F., Steelman, L.S., Lee, J.T., Shelton, J.G., Navolanic, P.M., Blalock, W.L., Franklin, R.A., McCubrey, J.A. (2003) Signal transduction mediated by the Ras/Raf/MEK/ERK pathway from cytokine receptors to transcription factors: potential targeting for therapeutic intervention, *Leukemia* **17**, 1263–1293.

Clark, S.S., McLaughlin, J., Crist, W.M., Champlin, R., Witte, O.N. (1987) Unique forms of the abl tyrosine kinase distinguish Ph1-positive CML from Ph1-positive ALL, *Science* **235**, 85–88.

Cong, F., Spencer, S., Cote, J.F., Wu, Y., Tremblay, M.L., Lasky, L.A., Gogg, S.P. (2000) Cytoskeletal protein PSTPIP1 directs the PEST-type protein tyrosine phosphatase to the c-Abl kinase to mediate Abl dephosphorylation, *Mol. Cell.* **6**, 1413–1423.

Crews, C.M., Alessandrini, A., Erikson, R.L. (1992) The primary structure of MEK, a protein kinase that phosphorylates the ERK gene product, *Science* **258**, 478–480.

Cunningham, D., Humblet, Y., Siena, S., Khayat, D., Bleiberg, H., Santoro, A., Bets, D., Mueser, M., Harstrick, A., Verslype, C., Chau, I., Van Cutsem, E. (2004) Cetuximab monotherapy and cetuximab plus irinotecan in irinotecan-refractory metastatic colorectal cancer, *N. Engl. J. Med.* **351**, 337–345.

de Klein, A., van Kessel, A.G., Grosveld, G., Bartram, C.R., Hagemeijer, A., Bootsma, D., Spurr, N.K., Heisterkamp, N., Groffen, J., Stephenson, J.R. (1982) A cellular oncogene is translocated to the Philadelphia chromosome in chronic myelocytic leukemia, *Nature* **300**, 765–767.

Dong, J., Phelps, R.G., Qiao, R., Yao, S., Benard, O., Ronai, Z., Aaronson, S.A. (2003) B Raf oncogenic mutations correlate with progression rather than initiation of human melanoma, *Cancer Res.* **63**, 3883–3886.

Druker, B.J. (2004) Molecularly targeted therapy: have the floodgates opened? *The Oncologist* **9**, 357–360.

Druker, B.J., Talpaz, M., Testa, D.J., Peng, B., Buchdunger, E., Ford, J.M., Lydon, N.B., Kantarjian, H., Capdeville, R., Ohno-Jones, S., Sawyers, C.L. (2001) Efficacy and safety of a specific inhibitor of the BCR-ABL tyrosine kinase in chronic myeloid leukemia, *N. Engl. J. Med.* **344**, 1031–1037.

Eley, G., Frederick, L., Wang, X.Y., Smith, D.I., James, C.D. (1998) 3′ end structure and rearrangements of EGFR in glioblastomas, *Genes Chromosomes Cancer* **23**, 248–254.

Falck, J., Mailand, N., Syljuasen, R.G., Bartek, J., Lukas, J. (2001) The ATM-Chk2-Cdc25A checkpoint pathway guards against radioresistant DNA synthesis, *Nature* **410**, 766–767.

Feng, J., Park, J., Cron, P., Hess, D., Hemmings, B.A. (2004) Identification of a PKB/Akt hydrophobic motif Ser-473 kinase as DNA-dependent protein kinase, *J. Biol. Chem.*: **279**, 41189–41196.

Gatti, R.A., Berkel, I., Boder, E., Braedt, G., Charmley, P., Concannon, P., Ersoy, F., Foroud, T., Jaspers, N.G.J., Lange, K. et al. (1988) Localization of an ataxia-telangiectasia gene to chromosome 11q22–23, *Nature* **336**, 577–580.

Gilliland, D.G., Griffin, J.D. (2002) The roles of FLT3 in hematopoiesis and leukemia, *Blood* **100**, 1532–1542.

Goldman, J.M., Melo, J.V. (2003) Chronic myeloid leukemia – advances in biology and new approaches to treatment, *N. Engl. J. Med.* **349**, 1451–1464.

Golub, T.R., Goga, A., Barker, G.F., Afar, D.E., McLaughlin, J., Bohlander, S.K., Rowley, J.D., Witte, O.N., Gilliland, D.G. (1996) Oligomerization of the ABL tyrosine kinase by the Ets protein TEL in human leukemia, *Mol. Cell Biol.* **16**, 4107–4116.

Haeder, M., Rotsch, M., Bepler, G., Hennig, C., Havemann, K., Heimann, B., Moelling, K. (1988) Epidermal growth factor receptor expression in human lung cancer cell lines, *Cancer Res.* **48**, 1132–1136.

Hall, M., Peters, G. (1996) Genetic alterations of cyclins, cyclin-dependent kinases, and Cdk inhibitors in human cancer, *Adv. Cancer Res.* **68**, 67–108.

Hantschel, O., Superti-Furga, G. (2004) Regulation of the c-Abl and Bcr-Abl tyrosine kinases, *Nat. Rev. Mol. Cell. Biol.* **5**, 33–44.

Heath, C., Cross, N.C. (2004) Critical role of STAT5 activation in transformation mediated by ZNF198-FGFR1, *J. Biol. Chem.* **279**, 6666–6673.

Huang, W., Kessler, D.S., Erikson, R.L. (1995) Biochemical and biological analysis of Mek1 phosphorylation site mutants, *Mol. Biol. Cell* **6**, 237–245.

Iijima, Y., Okuda, K., Tojo, A., Tri, N.K., Setoyama, M., Sakaki, Y., Asano, S., Tokunaga, K., Kruh, G.D., Sato, Y. (2002) Transformation of Ba/F3 cells and Rat-1 cells by ETV6/ARG, *Oncogene* **21**, 4374–4383.

Iwahara, T., Fujimoto, J., Wen, D., Cupples, R., Bucay, N., Arakawa, T., Mori, S., Ratzkin, B., Yamamoto, T. (1997) Molecular characterization of ALK, a receptor tyrosine kinase expressed specifically in the nervous system, *Oncogene* **14**, 439–449.

Jeffrey, P.D., Tong, L., Pavletich, N.P. (2000) Structural basis of inhibition of CDK-cyclin complexes by INK4 inhibitors, *Genes Dev.* **14**, 3115–3125.

Johnson, D.E., Williams, L.T. (1993) Structural and functional diversity in the FGF receptor multigene family, *Adv. Cancer Res.* **60**, 1–41.

Keung, Y.K., Beaty, M., Steward, W., Jackle, B., Pettnati, M. (2002) Chronic myelocytic leukemia with eosinophilia, t(9;12)(q34;p13), and ETV6-ABL gene rearrangement: case report and review of the literature, *Cancer Genet. Cytogenet.* **138**, 139–142.

Khosravi, R., Maya, R., Gottlieb, T., Oren, M., Shiloh, Y., Shkedy, D. (1999) Rapid ATM-dependent phosphorylation of MDM2 precedes p53 accumulation in response to DNA damage, *Proc. Natl. Acad. Sci. U.S.A.* **96**, 14973–14977.

King, C.R., Kraus, M.H., Aaronson, S.A. (1985) Amplification of a novel v-erbB-related gene in a human mammary carcinoma, *Science* **229**, 974–976.

Kulkarni, S., Reiter, A., Smedley, D., Goldman, J.M., Cross, N.C.P. (1999) The genomic structure of ZNF198 and location of breakpoints in the t(8;13) myeloproliferative syndrome, *Genomics* **55**, 118–121.

Lacronique, V., Boureux, A., Monni, R., Dumon, S., Mauchauffe, M., Mayeux, P., Gouilleux, F., Berger, R., Gisselbrecht, S., Ghysdael, J., Bernard, O.A. (2000) Transforming properties of chimeric TEL-JAK proteins in Ba/F3 cells, *Blood* **95**, 2076–2083.

Leonard, J.H., Kearsley, J.H., Chenevix-Trench, G., Hayward, N.K. (1991) Analysis of gene amplification in head-and-neck squamous-cell carcinoma, *Int. J. Cancer* **48**, 511–515.

Llorca, O., Rivera-Calzada, A., Grantham, J., Willison, K.R. (2003) Electron microscopy and 3D reconstruction reveal that human ATM kinase uses an arm-like domain to clamp around double-stranded DNA, *Oncogene* **22**, 3867–3874.

Lyman, S.D., Stocking, K., Davison, B., Fletcher, F., Johnson, L., Escobar, S. (1995) Structural analysis of human and murine flt3 ligand genomic loci, *Oncogene* **11**, 1165–1172.

Lynch, T.J., Bell, D.W., Sordella, R. et al. (2004) Activating mutations in the epidermal growth factor receptor underlying responsiveness of non-small-cell lung cancer to Gefitinib, *N. Engl. J. Med.* **350**, 2129–2038.

Ma, P.C., Maulik, G., Christensen, J., Salgia, R. (2003) c-Met: structure, functions and potential for therapeutic inhibition, *Cancer Metastasis Rev.* **22**, 309–325.

Marais, R., Light, Y., Paterson, H.F., Mason, C.S., Marshall, C.J. (1997) Differential regulation of Raf-1, A-Raf, and B-Raf by oncogenic ras and tyrosine kinases, *J. Biol. Chem.* **272**, 4378–4383.

Maroe, N., Rottapel, R., Rosnet, O., Marchetto, S., Lavezzi, C., Mannoni, P., Birnbaum, D., Dubreuil, P. (1993) Biochemical characterization and analysis of the transforming potential of the FLT3/FLK2 receptor tyrosine kinase, *Oncogene* **8**, 909–918.

Matsuoka, S., Huang, M., Elledge, S.J. (1998) Linkage of ATM to cell cycle regulation by the Chk2 protein kinase, *Science* **282**, 1893–1897.

McWhirter, J.R., Galasso, D.L., Wang, J.Y. (1993) A coiled-coil oligomerization domain of Bcr is essential for the transforming function of Bcr-Abl oncoproteins, *Mol. Cell. Biol.* **13**, 7587–7595.

Meydan, N., Grunberger, T., Dadi, H., Shahar, M., Arpaia, E., Lapidot, Z., Leeder, J.S., Freedman, M., Cohen, A., Gazit, A., Levitzki, A., Roifman, C.M. (1996) Inhibition of acute lymphoblastic leukemia by a Jak-2 inhibitor, *Nature* **379**, 645–648.

Mora, A., Komander, D., van Aalten, D.M.F., Alessi, D.R. (2004) PDK1, the master regulator of AGC kinase signal transduction, *Semin. Cell Dev. Biol.* **15**, 161–170.

Morris, S.W., Kirstein, M.N., Valentine, M.B., Dittmer, K.G., Shapiro, D.N., Saltman, D.L., Look, A.T. (1994) Fusion of a kinase gene, ALK, to a nucleolar protein gene, NPM, in non-Hodgkin's lymphoma, *Science* **263**, 1281–1284.

Muenke, M., Schell, U., Hehr, A., Robin, N.H., Losken, H.W., Schinzel, A., Pulleyn, L.J., Rutland, P., Reardon, W., Malcolm, S., Winter, R.M. (1994) A common mutation in the fibroblast growth factor receptor 1 gene in Pfeiffer syndrome, *Nat. Genet.* **8**, 269–274.

Neilson, K.M., Friesel, R. (1996) Ligand-independent activation of fibroblast growth factor receptors by point mutations in the extracellular, transmembrane, and kinase domains, *J. Biol. Chem.* **271**, 25049–25057.

Passos-Bueno, M.R., Wilcox, W.R., Jabs, E.W., Sertie, A.L., Alonso, L.G., Kitoh, H. (1999) Clinical spectrum of fibroblast growth factor receptor mutations, *Hum. Mutat.* **14**, 115–125.

Pluk, H., Dorey, K., Superti-Furga, G. (2002) Autoinhibition of c-Abl, *Cell* **108**, 247–259.

Roche-Lestienne, C., Soenen-Cornu, V., Gradel=Duflos, N., Lai, J.L., Phillip, N., Facon, T., Fenaux, P., Preudhomme, C. (2002) Several types of the Abl gene can be found in chronic myeloid leukemia patients resistant to STI571, and they can pre-exist to the onset of treatment, *Blood* **100**, 1014–1018.

Russell, J.P., Powell, D.J., Cunnane, M., Greco, A., Portella, G., Santoro, M., Fusco, A., Rothstein, J.L. (2000) The TRK-T1 fusion protein induces neoplastic transformation of thyroid epithelium, *Oncogene* **19**, 5729–5735.

Saltzman, A., Stone, M., Franks, C., Searfoss, G., Munro, R., Jaye, M., Ivashchenko, Y. (1998) Cloning and characterization of human Jak-2 kinase: high mRNA expression in immune cells and muscle tissue, *Biochem. Biophys. Res. Commun.* **246**, 627–633.

Shtivelman, E., Lifshitz, B., Gale, R.P., Canaani, E. (1985) Fused transcript of abl and bcr genes in chronic myelogenous leukemia, *Nature* **315**, 550–554.

Sommer, S.S., Jiang, Z., Feng, J., Buzin, C.H., Zheng, J., Longmate, J., Jung, M., Moulds, J., Dritschilo, A. (2003) ATM missense mutations are frequent in patients with breast cancer, *Cancer Genet. Cytogenet.* **145**, 115–120.

Sozeri, O., Vollmer, K., Liyanage, M., Frith, D., Kour, G., Mark, G.E., Stabel, S. 3rd (1992) Activation of the c-Raf protein kinase by protein kinase C phosphorylation, *Oncogene* **7**, 2259–2262.

Tang, T.C.-M., Sham, J.S.T., Xie, D., Fang, Y., Huo, K.-K., Wu, Q.-L., Guan, X.-Y. (2002) Identification of a candidate oncogene *SEI-1* within a minimal amplified region at 19q13.1 in ovarian cancer cell lines, *Cancer Res.* **62**, 7157–7161.

Thompson, F.H., Nelson, M.A., Trent, J.M., Guan, X.Y., Liu, Y., Yang, J.M., Emerson, J., Adair, L., Wymer, J., Balfour, C., Massey, K., Weinstein, R., Alberts, D.S., Taetle, R. (1996) Amplification of 19q13.1-q13.2 sequences in ovarian cancer. G-band, FISH, and molecular studies, *Cancer Genet. Cytogenet.* **87**, 55–62.

Thorstenson, Y.R., Roxas, A., Kroiss, R., Jenkins, M.A., Yu, K.M., Bachrich, T., Muhr, D., Wayne, T.L., Chu, G., Davis, R.W., Wagner, T.M., Oefner, P.J. (2003) Contributions of ATM mutations to familial breast and ovarian cancer, *Cancer Res.* **63**, 3325–3333.

Tybulewicz, V.L., Crawford, C.E., Jackson, P.K., Bronson, R.T., Mulligan, R.C. (1991) Neonatal lethality and lymphopenia in mice with a homozygous disruption of the c-abl proto-oncogene, *Cell* **65**, 1153–1163.

Van Etten, R.A., Jackson, P., Baltimore, D. (1989) The mouse type IV c-abl gene product is a nuclear protein, and activation of transforming ability is associated with cytoplasmic localization, *Cell* **58**, 669–678.

Weber, C.K., Slupsky, J.R., Herrmann, C., Schuler, M., Rapp, U.R., Block, C. (2000) Mitogenic signaling of Ras is regulated by differential interaction with Raf isozymes, *Oncogene* **19**, 169–176.

Wolfel, T., Hauer, M., Schneider, J., Serrano, M., Wolfel., C., Klehmann-Hieb, E., De Plaen, E., Hankeln, T., Meyer, K.H., Beach, D. (1995) A p16INK4a-insensitive CDK4 mutant targeted by cytolytic T lymphocytes in a human melanoma, *Science* **269**, 1281–1284.

Wong, A.J., Ruppert, J.M., Bigner, S.H., Grzeschik, C.H., Humphrey, P.A., Bigner, D.S., Vogelstein, B. (1992) Structural alterations of the epidermal growth factor receptor gene in human gliomas, *Proc. Natl. Acad. Sci. USA* **89**, 2965–2969.

Zhang, B.H., Guan, K.L. (2000) Activation of B-Raf kinase requires phosphorylation of the conserved residues Thr598 and Ser601, *EMBO J.* **19**, 5429–5439.

Zimmermann, S., Moelling, K. (1999) Phosphorylation and regulation of Raf by Akt (protein kinase B), *Science* **286**, 1741–1744.

Zuo, L., Weger, J., Yang, Q., Goldstein, A.M., Tucker, M.A., Walker, G.J., Hayward, N., Dracopoli, N.C. (1996) Germline mutations in the p16INK4a binding domain of CDK4 in familial melanoma, *Nat. Genet.* **12**, 97–99.

Ionizing Radiation Damage to DNA

Clemens von Sonntag
Leibniz-Institut für Oberflächenmodifizierung (IOM), Leipzig, Germany

Keywords

Cross-Link
Covalent bond between two macromolecular moieties, usually formed when two macromolecular free radicals recombine.

DNA Strand Breakage
Event induced by certain kinds of chemical damage, through free-radical processes involving the DNA molecule but also through the action of enzymes that try to repair the DNA damage.

Free Radical
Chemical species that is highly reactive because it possesses an unpaired electron.

Encyclopedia of Molecular Cell Biology and Molecular Medicine, 2nd Edition. Volume 7
Edited by Robert A. Meyers.

ISBN: 3-527-30549-1

Heteroatom
Atom in a molecule of an organic compound that is neither a carbon nor a hydrogen atom.

Ionizing Radiation
Energetic photons (X rays, γ-rays) and energetic particles (β-rays, fast electrons from electron accelerators, positrons, α-rays and other fast-moving ions, neutrons) that cause multiple ionization events along their trajectory through matter.

Peroxyl Radical
Generated by addition of dioxygen to a free radical.

Pulse Radiolysis
Technique for studying the kinetics of the chemical reactions of short-lived intermediates: for example, the aqueous solution of a substrate is exposed to a short (nanoseconds-to-microseconds range) burst of high-energy electrons from an electron accelerator.

Radical Ion
Free radical with a positive or negative charge.

Ionizing radiation represents an important risk factor to the living organism. It has always been part of the natural environment, hut it is also now a technological phenomenon, lending added urgency to the study of its effects on living matter. It has been established that the most sensitive target in the living cell is the DNA, which undergoes radiation damage through free-radical processes. This effect has two practical aspects beyond its purely scientific interest: to assess the radiation risk with a view to minimize it, perhaps by chemical means, and to apply ionizing radiation selectively for therapeutic purposes (e.g. in cancer treatment). These goals will not be optimally achieved without adequate knowledge of the chemical processes that ionizing radiation sets in train in DNA.

1 Energy Absorption

Absorption of ionizing radiation by matter leads to the formation of radical cations and electrons (reaction 1) and electronically excited molecules (reaction 2).

$$\mathrm{M} + \text{ionizing radiation} \longrightarrow \mathrm{M}^{\bullet+} + \mathrm{e}^- \quad (1)$$

$$\mathrm{M} + \text{ionizing radiation} \longrightarrow \mathrm{M}* \quad (2)$$

Energy deposition is inhomogeneous. In the case of sparsely ionizing radiation (e.g. γ-rays or high-energy electrons), the average distance between two regions of energy deposition is about 200 Å in matter of about unit density (i.e. in aqueous media). With densely ionizing radiation (e.g. α-particles) these regions, which are called *spurs* and are centered on the location of a primary ionization event, strongly overlap to form a continuous ionization track. The rate of energy deposition is proportional to the electron density. In the living cell, about 70% of the energy is absorbed by the water and about 30% by the organic matter and other solutes.

The radiolysis of water leads to the formation of $^{\bullet}OH$ radicals, solvated electrons, and $H^{\bullet}$ atoms. The water radical cation readily transfers a proton to the neighboring water molecules, thereby yielding an $^{\bullet}OH$ radical (reaction 3), and the electron becomes solvated (reaction 4). Electronically excited water may decompose into $^{\bullet}OH$ and $H^{\bullet}$ (reaction 5).

$$H_2O^{\bullet +} \longrightarrow {}^{\bullet}OH + H^+ \quad (3)$$

$$e^- + nH_2O \longrightarrow e^-_{aq} \quad (4)$$

$$H_2O* \longrightarrow H^{\bullet} + {}^{\bullet}OH \quad (5)$$

For sparsely ionizing radiation, the radiation-chemical yields, or G values, (The G value is expressed in molecules per 100 eV, which corresponds to 1.037×10^{-7} mol J^{-1}.) are $G(^{\bullet}OH) = 2.8 \times 10^{-7}$ mol J^{-1}, $G(e^-_{aq}) = 2.7 \times 10^{-7}$ mol J^{-1}, $G(H^{\bullet}) = 0.6 \times 10^{-7}$ mol J^{-1}. In the spurs, some radical combination occurs yielding the "molecular products" H_2 and H_2O_2 ($G(H_2) = 0.45 \times 10^{-7}$ mol J^{-1}, $G(H_2O_2) = 0.8 \times 10^{-7}$ mol J^{-1}). The water-derived radicals are all highly reactive. Hence, their formation and reactions cause an important part, possibly the major part, of the radiation damage to the living cell.

2
Radiation-induced DNA Lesions

Ionizing radiation is absorbed by the various cell components with practically equal probability. This is in contrast with UV radiation, which is predominantly absorbed by the cellular nucleic acids. While the mechanism of inactivation in the two cases (UV vs ionizing radiation) is quite different, DNA again is by far the most sensitive target to damage induced by ionizing radiation. Subionization UV radiation mainly causes base dimerizations (e.g. formation of thymine dimers at 260 nm) with barely any strand breakage. Ionizing radiation produces very little of this type of damage but causes, characteristically, DNA strand breakage. Ionizing radiation base damage is largely due to free-radical reactions, including those caused by the $^{\bullet}OH$ radicals generated by the radiation absorbed in the aqueous medium surrounding the DNA. In this way, a considerable number of different kinds of damage arise whose products one may group into the following categories:

- Altered bases
- Altered sugar moiety
  - DNA strand break
  - Release of unaltered bases
- Cross-links
  - DNA-DNA
  - DNA-protein

Apart from phosphate-ester bond cleavage, which constitutes a strand break, the phosphate moiety as such is not modified. It is important to realize that the cluster-type energy deposition of the ionizing radiation will cause some of the damaging events also to appear in clusters; that is, there is a considerable likelihood that, especially in double-stranded DNA, one damaged site will be close to another, or even to several others. These "clustered lesions" have also been termed *locally multiply damaged sites* (LMDS). They may involve damaged bases and one or two single-strand breaks. Two opposite single-strand breaks will result in a DNA double-strand break.

In the living cell, the DNA is surrounded not only by proteins (e.g. histones in eukaryotic cells) but also by a high concentration of low molecular weight organic material, among others the thiol glutathione. These react very efficiently (i.e. at practically diffusion-controlled rates) with the water-derived radicals ($^{\bullet}$OH, e^-_{aq}, H$^{\bullet}$) formed in the aqueous surrounding of DNA. Hence, these are effectively scavenged, and so the DNA is largely protected against, say, $^{\bullet}$OH-attack. Only $^{\bullet}$OH radicals generated in the close vicinity of DNA stand some chance of reacting with and hence doing damage to the genetic material.

3
Reaction of DNA Radicals

Damage by ionizing radiation, induced either by the direct effect or by water radicals, will result in DNA radicals; these are the precursors of the final (nonradical) damage at the product stage. The thiol glutathione (GSH), which is present in living cells at comparatively high concentrations (approaching 10^{-2} mol dm^{-3}) can react with the DNA radicals (R$^{\bullet}$) by H-transfer (reaction 6).

$$R^{\bullet} + GSH \longrightarrow RH + GS^{\bullet} \qquad (6)$$

This reaction is thermodynamically slightly favored because the S−H bond is relatively weak. Recently, it has been realized that the peptide C−H bond in proteins is even somewhat lower, but it has not yet been established whether proteins attached to DNA (e.g. histones) reduce $^{\bullet}$OH-induced damage not only by scavenging $^{\bullet}$OH radicals but also by H-donation to a DNA radical.

In competition with this H-donation reaction, dioxygen can add to the DNA radicals, thereby forming the corresponding peroxyl radicals (reaction 7).

$$R^{\bullet} + O_2 \longrightarrow ROO^{\bullet} \qquad (7)$$

These peroxyl radicals give rise to the final products. Some of the peroxyl radicals are capable of eliminating $HO_2^{\bullet}/O_2^{\bullet-}$, but the major part will decay bimolecularly (reaction 8).

$$2ROO^{\bullet} \longrightarrow \text{products} \qquad (8)$$

Low-molecular–weight peroxyl radicals often terminate at dose to diffusion-controlled rates. In high-molecular–weight material such as DNA, the diffusion of radical-bearing segments is considerably restricted. Hence, they may persist for relatively long times and may undergo reactions other than reaction (8) or the elimination of $HO_2^{\bullet}/O_2^{\bullet-}$.

The reaction of the primary DNA radicals with glutathione is usually termed *chemical repair*. However, a repair in the true sense is achieved only if the radical site to be repaired has been created by H-abstraction (e.g. reaction 9).

$$RH + {}^{\bullet}OH \longrightarrow R^{\bullet} + H_2O \qquad (9)$$

In DNA, such sites are present only at the sugar moiety or at the methyl group of thymine. For the most part, $^{\bullet}$OH reacts by addition to a C=C or C=N bond (e.g. reaction 10). Subsequent "repair" by H-donation leads to the formation of a hydrate (reaction 11) rather than the original molecule.

$$R_2C{=}CR_2 + {}^{\bullet}OH \longrightarrow HOCR_2{-}CR_2^{\bullet} \quad (10)$$

$$HOCR_2{-}CR_2^{\bullet} + GSH \longrightarrow HOCR_2{-}CHR_2 + GS^{\bullet} \quad (11)$$

Apparently, the cellular enzymatic repair system can cope with this kind of damage much better than the kind resulting from peroxyl radical reactions (cf the "oxygen effect" in radiobiology).

4
Model Studies

Although the chemical nature of some kinds of base damage has been studied *in vivo*, most of our knowledge of radiation-induced DNA damage is derived from model studies. Electron spin resonance studies for the identification of radicals have mainly been carried out in the solid state at low temperatures. In contrast, practically all product and kinetic (pulse radiolysis) studies have been done in dilute aqueous solutions. Thus, they reflect mostly the indirect effect.

In the direct effect, radical cations (and electrons) are the species primarily produced (reaction 1). To mimic this reaction and to study the fate of radical cations in aqueous solutions, photoionization with a laser (monophotonic at $\lambda = 193$ nm or biphotonic at $\lambda = 248$ nm) or electron transfer to photoexcited quinones (Q) were used with some advantage (e.g. reaction 12).

$$Q* + \text{nucleobase} \longrightarrow Q^{\bullet -} + \text{nucleobase}^{\bullet +} \quad (12)$$

The nucleobase radical cations are strong acids. Quick deprotonation ensues at a heteroatom, but in thymine, deprotonation can eventually materialize at the exocyclic methyl group carbon, in competition with a nucleophilic addition of water to the carbon 6 position. Although after deprotonation the heteroatom-centered radical predominates, the radical cation in near-neutral media is always present at low "equilibrium" concentrations. In contrast to the situation at the heteroatom, deprotonation at the methyl position is practically irreversible under these conditions.

Nucleobase cations are strong oxidants (for a sugar-derived radical cation see below), and in DNA those derived from thymine, cytosine, and adenine may oxidize a neighboring guanine (G), the nucleobase with the lowest reduction potential. Hole transfer through DNA is now a well-documented process. The ultimate sinks of the hole are GG doublets and, even better, GGG triplets whose reduction potentials are substantially lower than that of a single G. Mechanistically, this hole transfer can be described by an incoherent hopping process involving quantum-mechanical channels.

The electrons formed in the ionization process are readily scavenged by the nucleobases (a diffusion-controlled reaction). The radical anions thus formed are strong bases. Hence, it is not surprising that those derived from adenine, guanine, and cytosine are protonated by water on the submicrosecond timescale. The thymine radical anion has a pK_b value of about 7.0, and therefore is much longer lived than the other radical anions.

Most of these radical anions (and to a lesser extent their heteroatom-protonated forms) have pronounced reducing properties; that is, they are capable of retransferring the electron to an oxidant.

However, they are metastable species with respect to ultimate conversion into carbon-protonated intermediates (e.g. in thymine, the final protonation site is C-6). These radicals no longer have reducing properties.

This sequence of events also has a bearing on the reactions of the radical anions with dioxygen. For example, the thymine radical anion and its heteroatom-protonated form react with dioxygen by forming $O_2^{\bullet -}/HO_2^{\bullet}$ thereby regenerating the nucleobase, while in the reaction of dioxygen with the C-6-protonated radical anion, the thymine molecule is destroyed.

Besides the formation of nucleobase radical cations (on account of the direct effect) and radical anions (by e^- attachment), one must consider the reactions of the $^{\bullet}OH$ radical as a major contributor to radiation-induced DNA damage. It mostly adds to the double bonds of the nucleobases, but it also abstracts H atoms from the sugar moiety and the methyl group in thymine.

For the investigation of $^{\bullet}OH$-induced reactions in isolation, in radiation-chemical experiments it is standard practice to convert e_{aq}^- (from reaction 4) into $^{\bullet}OH$ by saturating the solution to be irradiated with nitrous oxide (reaction 13).

$$e_{aq}^- + N_2O \longrightarrow {}^{\bullet}OH + N_2 + OH^- \qquad (13)$$

As a result, the radical species now consist of 90% $^{\bullet}OH$ and 10% $H^{\bullet}$; that is, the observed reactions and their products are dominated by the effects of the $^{\bullet}OH$ radical. In the pyrimidines, $^{\bullet}OH$-addition to the C-5 position yields a reducing radical, while an addition to the C-6 position yields a radical with oxidizing properties. These properties can be defined using suitable redox probes. The pulse radiolysis technique has allowed the characterization of these radicals and the determination of their yields. In the case of the purines, the redox titration technique is not as straightforward, and the assignment of the sites of $^{\bullet}OH$-addition is complicated by rapid ring-opening and water elimination reactions.

Our present knowledge of pyrimidine free-radical chemistry (in particular, uracil and thymine, their methyl derivatives as well as their nucleosides) is much more extensive than that of the purines. It is obvious from the published data that only a fraction of the primary purine $^{\bullet}OH$-adduct radicals have shown up in the form of products. Considerable effort will be required to bring the purine (and cytosine) free-radical chemistry to a satisfactory level of understanding. This situation is also reflected in the determination of the base product yields from irradiated DNA, where a considerable deficit in the product yields (related to the primary $^{\bullet}OH$ radical yield) is observed.

5
DNA Strand Breakage and Cross-linking

Solvated electrons do not cause DNA strand breakage, but $^{\bullet}OH$ radicals do. In competition with addition to the nucleobases, they also abstract H atoms from the sugar moiety. This has two possible consequences: in the subsequent reactions a strand break is induced and an unaltered base is released, or base release occurs without giving rise to strand breakage. Hence, base release always predominates somewhat over strand breakage. A number of altered sugars that are related to these processes have been identified, both in the absence and in the presence of dioxygen. On the basis of detailed model studies, the reactions and their kinetics leading to strand breakage in the absence of dioxygen are fairly well understood. The primary step is the abstraction of the H atom at C-4′:

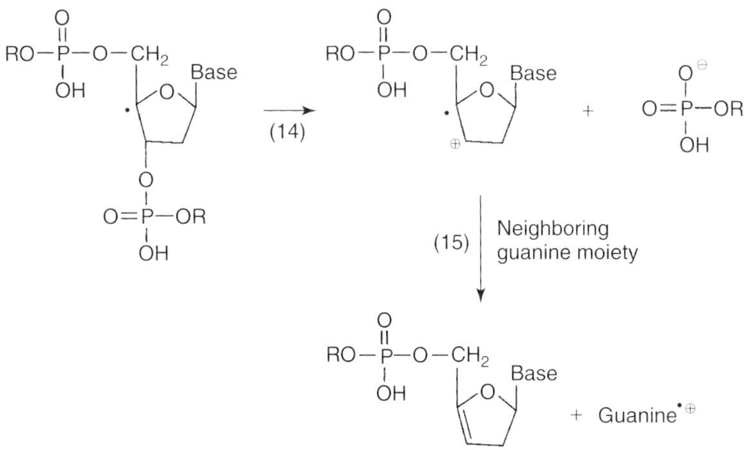

(14)

$O=P(O^{\ominus})(OH)-OR$

(15) Neighboring guanine moiety

+ $Guanine^{\bullet\oplus}$

This radical then eliminates a neighboring phosphate (linked to a fragment of the DNA strand) (reaction 14), leaving behind a radical cation. This radical cation has oxidizing properties and can oxidize a neighboring guanine moiety (reaction 15) with the consequence of hole migration through DNA (see above). In competition, the sugar radical cation reacts with water, either at the position that has eliminated the phosphate, or at C-4′. In the former case, the other phosphate function may be eliminated by the same mechanism; in double-stranded DNA, this sequence of events produces a clean gap in the affected strand (the end groups of the two fragments are phosphate groups) with the loss of some information because of the disappearance of the damaged nucleoside. In the latter case, the base is also lost, but additionally at the end of one of the fragment strands, a damaged sugar remains linked to the phosphate group. Enzymatically speaking, this is a "dirty" end group. Details of the mechanism of DNA strand breakage under conditions of oxygenation are less well understood, but some of the relevant sugar lesions have been detected. Model systems (ribose 5-phosphate) indicate that under these conditions C-5′ should be an additional site of attack and, in analogy, one would expect (no experimental evidence yet) the C-3′ peroxyl radical also to be a potential precursor for strand breakage.

So far, sugar damage has been discussed only in terms of $^{\bullet}OH$ radicals attacking this moiety. A contribution of the direct effect (ionization of the sugar moiety and the phosphate groups) must also be considered, but experimental evidence is not yet available. However, there is another interesting aspect. In polynucleotides such as poly(U) and poly(C), there is convincing evidence that base radicals are the major precursors of the sugar radicals that lead to strand breakage and the release of an unmodified base at the site of the damaged sugar. It is less clear whether such a radical transfer from the base to the sugar moiety can also occur in DNA.

In mammalian cells, DNA double-strand breaks are observed alongside single-strand breaks approximately in the ratio of 1 : 25. This poses the question of how these double-strand breaks are formed. It has been argued here that they result from clustered lesions. In the literature, an additional one-hit route has been suggested that involves a radical transfer from the already broken strand to the sugar moiety of the opposite strand, followed by breakage of this strand.

Carbon-centered radicals are known to add to the C=C bonds of nucleobases. Such reactions, as well as radical–radical combination reactions involving macroradicals, in principle allow the formation of DNA–protein and DNA–DNA cross-links. In special cases, such products have been observed with biological material, albeit in yields considerably lower than DNA double-strand breaks.

See also Free Radicals in Biochemistry and Medicine; Mutagenesis, Malignancy and Genome Instability; Nucleic Acids (DNA) Damage and Repair.

Bibliography

Books and Reviews

Dizdaroglu, M. (1991) Chemical determination of free radical-induced damage to DNA, *Free Radical Biol. Med.* **10**, 225–242.

Hüttermann, J. (1991) Radical ions and their reactions in DNA and its constituents, in: Lund, A., Shiotani, M. (Eds.) *Radical Ionic Systems*, Kluwer, Dordrecht, pp. 435–462.

Michael, B.D., Held, K.D., Harrop, H.A. (1983) Biological Aspects of DNA Radioprotection, in: Nygaard, O., Simic, M.G. (Eds.) *Radioprotectors and Anticarcinogens*, Academic Press, New York, pp. 325–338.

Teebor, G.W., Boorstein, R.J., Cadet, J. (1988) The repairability of oxidative free radical mediated damage to DNA: A review, *Int. J. Radiat. Biol.* **54**, 131–150.

von Sonntag, C. (1987) *The Chemical Basis of Radiation Biology*, Taylor & Francis, London.

von Sonntag, C., Hagen, U., Schön-Bopp, A., Schulte-Frohlinde, D. (1981) Radiation-induced strand breaks in DNA: chemical and enzymatic analysis of end groups and mechanistic aspects, *Adv. Radiat. Biol.* **9**, 109–142.

Ward, J.F. (1988) DNA damage produced by ionizing radiation in mammalian cells: identities, mechanisms of formation, and repairability, *Prog. Nucleic Acid Res. Mol. Biol.* **35**, 95–125.

Primary Literature

Giese, B., Amaudrut, J., Köhler, A., Spormann, M., Wessely, S. (2001) Direct observation of hole transfer through DNA by hopping between adenine bases and by tunnelling, *Nature* **412**, 318–320.

Schulte-Frohlinde, D., Simic, M.G., Görner, H. (1990) Laser-induced strand break formation in DNA and polynucleotides, *Photochem. Photobiol.* **52**, 1137–1151.

Labeling, Biophysical

Gertz I. Likhtenshtein
Ben-Gurion University of the Negev, Beer-Sheva, Israel

Encyclopedia of Molecular Cell Biology and Molecular Medicine, 2nd Edition. Volume 7
Edited by Robert A. Meyers.

ISBN: 3-527-30549-1

Keywords

Dipole–dipole Interaction
Arises from Coulomb or magnetic fields mutually induced by electrostatic or magnetic dipoles respectively.

Exchange Interaction
Originates from inter- or intramolecular interaction between electrons due to their quantum mechanical delocalization (e.g. overlap of orbitals populated with the electrons).

Label
A compound that binds covalently to an object under study and whose properties enable monitoring by appropriate physical methods.

Probe
A compound that either adds on non-covalently or diffuses freely in the medium.

Spin
The intrinsic angular momentum of an unpaired electron that induces magnetic momentum.

In solving problems of molecular biology, molecular biophysics and biochemistry of proteins, nuclear acids, and biomembranes, it is necessary to investigate molecular dynamics properties of biological objects and spatial disposition of their individual parts. One must also know the depth of immersion of definite centers in a biological matrix, that is, the availability of enzyme sites to substrates, distance of electron tunneling between a donor and an acceptor group, position of a biophysical label in a membrane and in a protein globule, distribution of the electrostatic field around molecules, and so on. The biophysical labeling approach consists, in principle, of the modification of chosen sites of an object of interest by special compounds (labels and probes) whose properties make it possible to trace the state of the biological matrix by appropriate physical methods. Four types of compound are most often used as labels and probes: (1) centers with unpaired electrons, spins (stable nitroxide radicals and paramagnetic metal complexes); (2) fluorescent and phosphorescent chromophores; (3) electron-scattering groups of heavy atoms (e.g. polygold or polymercury compounds); and (4) Mössbauer atoms (e.g. ^{57}Fe), which give the nuclear γ-resonance spectra. Three types of problems can be solved by means of labeling technique on a molecular level: (1) recording of conformational changes in proteins, enzymes, membranes, and other structures; (2) the investigation of the microrelief, micropolarity, and intra- and intermolecular mobility of biological objects; and (3) the determination of the distance between chosen parts of systems of interest. The labeling approach makes it possible to work in whole native cellular and subcellular structures in aqueous, diluted solutions. The measurements are fast and widely available.

1 Principles

1.1 Spin Labeling

1.1.1 General

In the overwhelming majority of stable nitroxide radicals (NRs) used as spin labels and probes, the nitroxide group is stabilized by a substituent on the α-carbon atoms; methyl groups, for example (Fig. 1). Spin labels generally include fragments of already known, widely accepted specific reagents for functional groups of proteins, lipids, nucleic acids, and other biological compounds. The nitroxide group with unpaired electron produces simple, well resolved, and comparatively easily treated electron spin resonance (ESR) spectra that are sensitive to the molecular motion of the nitroxide band and its interactions with other paramagnetic materials. In a constant magnetic field of strength H_0, electron spins orient themselves in two directions, that is, along the field and against the field (the Zeeman Effect). Superposition of a perpendicular electromagnetic field of frequency ν in the microwave region results in reorientation of the spins and resonance absorption of the microwave energy occurs. The condition of the resonance is the equality

$$h\nu = g\beta H_0, \qquad (1)$$

where β is the Bohr magneton and g is the g-factor that is characteristic of magnetic momentum of the nitroxide.

The unpaired electron of nitroxide is delocalized over the oxygen and nitrogen. In the spherical symmetrical S state, the delocalized electron undergoes a contact interaction with the spin of the nitrogen nucleus that causes (for ^{14}N) isotropic hyperfine splitting (A_{iso}) into three components corresponding to orientation of the nuclear spin with magnetic quantum number $m = 1$ along ($m = -1$), against ($m = +1$), and perpendicular to ($m = 0$) the constant magnetic field, correspondingly. The A_{iso} is proportional to the electron spin density on the nuclear spin.

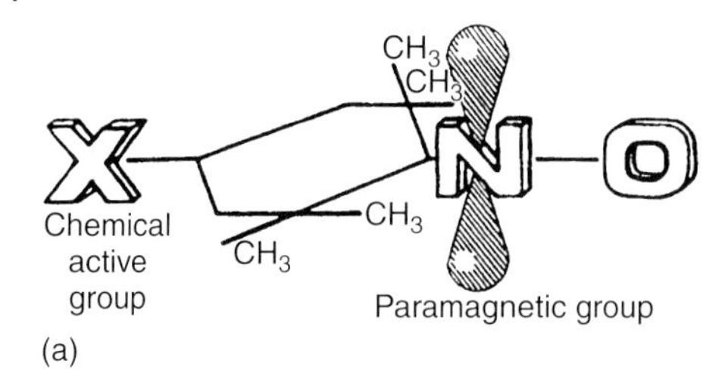

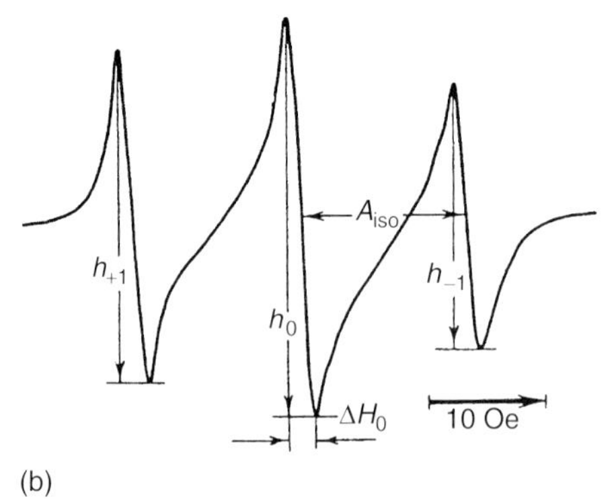

Fig. 1 (a) Schematic presentation for a nitroxide spin label, and (b) schematic spectrum of a nitroxide in solution.

Another type of electron–nuclear spin interaction is associated with localization of the electron on the p-orbital of nitrogen, and therefore, depends on the angle between the direction of the axes of the p-orbital and the constant magnetic field. Such an interaction gives rise to anisotropic hyperfine splitting (A_{aniso}).

1.1.2 Rotational Diffusion of Nitroxide Label

The nitroxides are asymmetric particles (Fig. 1a) characterized by three values of the g-factor (g_x, g_y, g_z) and A_{aniso} (A_x, A_y, A_z). In the immobilized state (frozen or supercooled solutions), the ESR spectra are superpositions of the *microspectra* of an enormous number of radicals with different orientations. The motion of the radicals results in averaging anisotropic effects. According the theory, a value of the time of the nitroxide isotropic rotational diffusion by one radian (correlation time, τ_c) in the region $\tau_c = 1 \times 10^{-10} \div 2 \times 10^{-9}$ s can be calculated from ESR spectra by using the formula,

$$\tau_c = \frac{2.7 \times 10^{-10}}{\left\{\left[\left(\frac{I_0}{I_-}\right)^{1/2} - 1\right] \Delta H_0\right\}} \quad (2)$$

where ΔH_0 is the width of the central component of the spectrum in milliteslas (mT) and I_0 and I_{-1} are the intensities of components with $m = 0$ and $m = -1$ respectively (Fig. 1b). Special methods were developed for the measurement $\tau_c = 10^{-4} \div 10^{-8}$ s.

1.1.3 Nitroxides as Dielectric, pH, Redox, and Imaging Probes

The magnetic parameters of NRs, the g-factor, and the hyperfine splittings A_{iso} and A_{aniso} are sensitive to the properties of the medium (e.g. micropolarity and ability to form hydrogen bonds with the $N - \dot{O}$ group). The physical reason for such dependence is that the contribution of the resonance structure on the right

$$N - \dot{O} \longleftrightarrow \dot{N}^{+} - O^{-}$$

causes an increase in the spin density of unpaired electrons on the nitrogen atom and therefore increases the A_{iso} and A_{aniso} values. For example, $A_{iso} = 1.4 + 0.064\ (\varepsilon_0 - 1)/(\varepsilon_0 + 1)$, where ε_0 is the dielectric constant and A_{iso} is given in millitesla. Protonation of the $N - \dot{O}$ group leads to the value of $\Delta A_{iso} = 0.57$ mT. At protonation of other groups of the nitroxide rings, the value ΔA_{iso} can reach 0.35 mT. The g-factor is also reported to be sensitive to the medium.

NRs are relatively stable toward oxidation, but they can be readily reduced to the corresponding hydroxylamines. The redox potential of a nitroxide (0.2–0.3 eV) is high enough to oxidize such biological compounds as ascorbic acids, semiquinones, superoxide radical, and Fe^{+2}. Therefore, NRs can serve as redox probes.

1.1.4 Double-labeling Technique. Spin Label–Spin Probe Method: Spin-oxymetry

Under certain circumstances, the ESR spectra of paramagnetic centers will respond suitably to the approach of other centers. Two types of spin–spin interactions are distinguished: (1) dipole–dipole interaction associated with induction by the magnetic dipole of one paramagnetic group of a local magnetic field at the site of another paramagnetic center; and (2) exchange interaction caused by an overlap of the orbitals of unpaired electrons.

For a pair of dipoles, the value of magnetic interaction is

$$\Delta H_{dd} = \frac{\mu_1 \mu_2}{r^3} - \frac{(\mu_1 r)(\mu_2 r)}{r^5} \qquad (3)$$

where μ_1 and μ_2 are magnetic moments and r is the distance between the spins.

The degree of overlap of orbitals with unpaired electrons is quantitatively characterized by the value of the exchange integral I. The value of I is about 10^{13} s^{-1} at the van der Waals distance and decreases exponentially with increasing distance between the spins. The exchange interaction usually prevails over the dipole–dipole interaction at $r < 0.8$–1.0 nm and therefore is suitable for the study of closely disposed paramagnetic materials.

Dipole–dipole interaction is applicable to systems with remote centers. For example, the distance between the pairs of nitroxide spin labels up to 3.0 to 3.5 nm can be calculated by formula:

$$r = 1.6(\Delta H_{dd})^{-1/3}, \qquad (4)$$

where ΔH_{dd} (mT) is the broadening of the ESR line.

Recently, many modifications of pulse ESR have been designed that allowed the improvement of the distance-measurement accuracy and to expand range of distance available for ESR spectroscopy. The principal advantage of such pulse methods is the direct determination of spin-relaxation parameters that, in turn, are directly related to spin–spin interactions, depending on distances.

The investigation of the paramagnetic contribution of ions to the value of the spin–lattice relaxation rate spin label has resulted in the extension of the distance that can be measured up to 6 nm. When a nitroxide spin label encounters a spin probe, that is, a chemically inert paramagnetic species (e.g. ferricyanide or diphenylchromium) that diffuses freely in solution, the magnitude of broadening of the label ESR line ΔH (mT) is related to the rate constant of the exchange interaction k_{ex} (M^{-1}s^{-1}) by the following equation:

$$\Delta H = 6.52 \times 10^{-9} k_{ex} C \qquad (5)$$

where C is the molar spin probe concentration. Measurement of ESR spectra parameters allows one to determine the value of k_{ex} in the range of 10^6 to 10^{10} M^{-1} s^{-1}. For a particular pair of paramagnetic species, the value of k_{ex} is proportional to the collision frequency, and therefore, depends on chemical structure, microviscosity, steric hindrances, concentration of O_2, and distribution of electrostatic charges in the region of encounters with the use nitroxide probes of different electrostatic charges. Hence, after preliminary empirical calibration, the spin label–spin probe method based on examining k_{ex} can be used for experimental study of these factors in biological objects.

1.1.5 New ESR Techniques

In high-frequency ESR spectroscopy (HF ESR) that generates strong magnetic fields (up to 9 T), the increased magnetic field leads to increasing spectral sensitivity to motion dynamics sensitivity in a slow-motion regime. Recently, significant progress in HF ESR has been achieved with the use of millimeter-wave quasi-optic technique, permitting the construction of a 9-T, 250-GHz (1.2 mm) spectrometer.

The 2D ESR technique relies on selective irradiation of a particular resonance line with a microwave frequency field and observation of the resulting effects in the rest of the spectrum. This method provides a 2D display of the homogeneous linescape across an inhomogeneous ESR spectrum. This approach allows direct study of dynamic processes (rotational and translational diffusion, electron transfer) and static dipole and exchange spin–spin interactions. The possibility of the two-dimensional approach has been significantly extended with the use Fourier transform electron spin resonance (2D FT ESR) and electron spin–echo techniques. Multiple quantum ESR recently developed for measuring distances between spins (r) longer than 1.2 nm is based upon double-quantum coherence pulsed ESR methods. Introducing an extensive cycling of a four-pulse sequence allows the determination of dipole–dipole splitting in the homogeneous ESR spectrum. The latter is directly connected to the r value. The electron nuclear double resonance (ENDOR) method provides direct structural information about a paramagnetic molecule orientation and conformation. Another method that is important for structure assignment is the electron–nuclear–nuclear triple resonance spectroscopy, which is an extension of the ENDOR method.

Among applications of new ESR techniques based on continuous wave illumination (CW ESR), the following ones can be pointed out: (1) site-directed spin labeling; (2) rigid incorporation of a spin label in proteins backbone alpha chain; (3) using fast Fourier transform deconvolution; (4) ENDOR of labeled enzyme-active centers; (5) pairwise interaction spin–spin interaction on a solely tumbling macromolecule; (6) and measurements of depth of immersion and location of paramagnetic centers.

1.2 Luminescent Labeling

1.2.1 General

The main parameters of both fluorescence and phosphorescence are their intensity I, the line shape and frequency of the maxima of spectra ν_{max}, the quantum yield ϕ, and the lifetime of the excited state τ^*. A significant difference between fluorescence and phosphorescence is that the lifetimes of the excited state during fluorescence ($\tau^*_f = 10^{-8} - 10^{-10}$ s) are much shorter than phosphorescence lifetimes ($\tau^*_{ph} = 10^{-3} - 10^2$ s). Another distinguishing feature of a phosphorescent molecule is the formation of the triplet, a paramagnetic state after excitation, and such a chromophore may serve as a triplet label.

1.2.2 Rotational Diffusion of Luminescent Labels

Chromophores that are being used as luminescent labels usually have asymmetric adsorption and emission transition moments and are sensitive to the polarization of incident light. If the rotation time τ_R of the chromophore is comparable to or shorter than its excited lifetime τ^*, the emitted light will be depolarized (Fig. 2). A characteristic parameter of depolarization

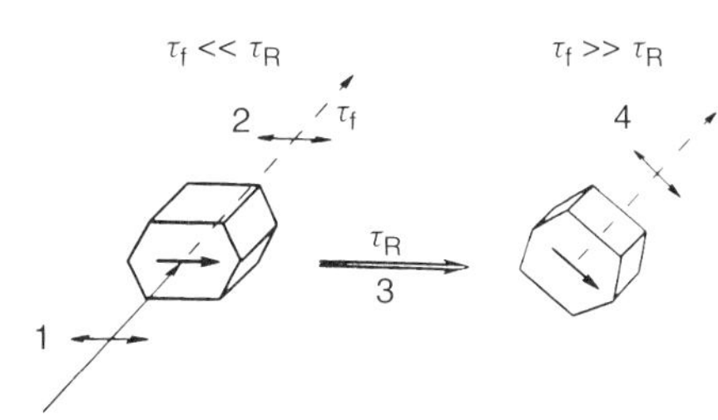

Fig. 2 Schematic representation of the dependence of fluorescent depolarization of a chromophore on the rotational correlation time τ_R and the excited state lifetime τ_f: 1, irradiation by polarized incident light; 2, polarized fluorescence from *immobile* chromophore; 3, rotation of the chromophore; 4, depolarized fluorescence from mobile chromophore.

is the anisotropy,

$$r = \frac{I_{\parallel} - I_{\perp}}{I_{\parallel} + 2I_{\perp}}, \tag{6}$$

where $I_{\parallel}$ and $I_{\perp}$ are the intensities of emitted light passing through the polarizer with the electric axis directed parallel or perpendicular, respectively, to the polarization of the incident light. For a spherical molecule,

$$\frac{r_0}{r} = 1 + \frac{\tau^*}{\tau_R} = 1 + \frac{3k_B T\tau^*}{4\pi b^3 \eta} \tag{7}$$

where r_0 is the anisotropy for a chromophore without rotation, β is the radius of rotation, and η is the viscosity of the medium. Measurement of the dependence of r/r_0 on T/η allows determination of the value of τ_R if τ^* is known. The value of τ_R can be measured directly from the change in the anisotropy after pulse radiation of the sample under study. The aforementioned technique is concerned with rotational diffusion of protein molecules in solution or lipids in membranes with $\tau_R = 10^{-8} - 10^{-9}$ s (fluorescence) and macromolecules in membranes $\tau_R = 10^{-5} - 10^{-6}$ s (phosphorescence).

1.2.3 Molecular Dynamics and Micropolarity of the Media

The excitation of a chromophore is accompanied by a change in the electric dipole moment of the molecule. This involves a change in the energy of interaction with the surrounding dipole molecules. If the characteristic dipole rotational correlation time τ_R in the medium is much longer than the lifetime of the fluorescence or phosphorescence τ^*, the dipoles have no chance to follow the change in the light-induced electric field, and the transition from the *unsolvated* excited state to the ground state takes place. In another limiting case, where $\tau_R \ll t^*$, the interaction between the dipoles and excited molecule lowers the energy of the system, and emission is effected from the *solvated* level. The relaxation can be followed by time-resolved spectroscopy, measuring the kinetic value of the shift of the maximum of the fluorescent or phosphorescent spectra $\nu_{max}(t)$ (relaxation shift):

$$\nu_{max}(t) - \nu_{max}(\infty) = [\nu_{max}(0) - \nu_{max}(\infty)] \times \exp\left(-\frac{t}{\tau_r}\right), \tag{8}$$

where the indexes t, ∞, and 0 are related to the ν_{max} of the time-resolved emission spectrum at a given moment t, $t \to \infty$ and $t = 0$. The value of τ_R can be also estimated by measuring the temperature dependence of the ν_{max} of steady state emission spectra. The value of the relaxation shift depends on the electrostatic dipole moment of the excited chromophore and on the parameters of polarity (in the simplest case, dielectric constant) of the medium. Hence, the latter can be estimated for biological objects (e.g. active centers of enzymes, membranes) after preliminary calibration, using model systems with known properties.

1.2.4 Resonance Energy Transfer Between Chromophores; Quenching of Fluorescence; Fluorescent Recovery after Photobleaching (FRAP)

The electrostatic dipole–dipole interaction between chromophores permits the nonradiative transfer of energy from an excited singlet state of a molecule (a donor, D) to an unexcited singlet state of another molecule (an acceptor, A) via an inductive resonance mechanism. According the theory developed by Förster, the efficiency of the transfer:

$$E_T = 1 - \frac{\phi}{\phi_0} = \frac{R_0^6}{R_0^6 + R^6} \quad (9)$$

where ϕ and ϕ_0 are the quantum yields of the donor fluorescence with and without the acceptor; R is the distance between D and A, and R_0 is the critical distance at which the probability of energy transfer is equal to that of donor excited state deactivation:

$$R_0 = \frac{162K^2}{\pi^5 n^4} \int_0^{\infty} \frac{f_D(\nu)\varepsilon_A(\nu)\,d\nu}{\nu^4}, \quad (10)$$

where the n is refractive index of the medium, $\int$ is the spectrum overlap of the donor fluorescence $f_{D(u)}$ and the acceptor adsorption $\varepsilon_{A(\upsilon)}$, and K^2 is the orientation factor of the dipole pair. For random reorientation of the donor and acceptor transition moments, $K^2 = 2/3$. Measurement of E makes it possible to estimate distances between the centers by 1.0 to 8.0 nm. Lanthanide (chelates of terbium or europium) based fluorescence resonance energy transfer (FRET) is a recent modification of the technique with a number of technical advantages, yet relies on the same fundamental mechanism. The advantages are (1) the value of the critical radius is 5.0 to 7.0 nm; (2) since the donor's emission is unpolarized, the efficiency of energy transfer depends primarily on the distance between donor and acceptor, and not on their relative orientation.

The phenomenon of quenching excited singlet states has been used in the investigation of collisions between fluorescent labels and quenching molecules. The process can be described by the Stern–Volmer equation:

$$\phi_0 \phi = 1 + K_q[Q] = 1 + \tau^* k_q[Q], \quad (11)$$

where K_q is the Stern–Volmer constant and k_q is the quenching rate constant. The range of the k_q values available for the fluorescence quenching technique is 10^8 to 10^9 $M^{-1}s^{-1}$. Molecules with polar groups and heavy atoms, many aromatic molecules, and electron acceptors are efficient quenchers, with $k_q = (3\text{–}6) \times 10^9$ $M^{-1}s^{-1}$.

The relatively slow diffusion of fluorescent probes and labeled macromolecules is studied by Fluorescent Recovery after Photobleaching (FRAP) techniques, which are based on photobleaching of small portions of the objects under study by a laser pulse, followed by monitoring of fluorescence recovery as a result of diffusion from an adjacent portion of the system.

1.2.5 Triplet, Photochrome, Triplet–photochrome, and Spin-triplet-photochrome Labeling

The excited triplet state has two unpaired electrons and can be considered a specific biradical, triplet label. Therefore, the processes of deactivation of the triplet state can be accelerated owing to exchange spin–spin interaction with other paramagnetic molecule by means of triplet–triplet (T–T) energy transfer or intercrossing. In particular, the experimental values of k_{TT}, the rate constant of the triplet state deactivation, can be written as follows:

$$k_{TT} = 10^{15} \exp(-26R), \qquad (12)$$

where R (nm) is the distance between the triplet label and a quenching molecule separated by *nonconducting* media (e.g. molecules or groups with saturated chemical bonds) and the result is given in reciprocal seconds. Dynamic exchange interactions upon collisions between exchanging centers in solution are very effective. The long lifetime of a triplet label in the excited state (τ^*_{ph} is in the range 10^{-1}–10^{-6} s) offers possibilities for measuring the rates of such slow dynamic processes as collisions of proteins in solution and diffusion of small molecules in rigid membranes. Nitroxide spin probes, complexes of paramagnetic metals, biological molecules (say, quinones, vitamins, etc.), and O_2 have proved to be effective quenchers of triplet labels.

The phenomenon of photochromism involves light-induced reversible transitions of a chromophore (A) to another form (B); for example, the photoisomerization a fluorescent *trans*-stilbene derivative to the non-fluorescent *cis*-stilbene. The latter includes the rotation of phenyl groups as a necessary step in the transformation. In a viscous medium and in systems with strong steric hindrance, the rotation can be the rate-determining stage; therefore, the experimental rate constant of the photoisomerization k_{is} of the photochromic label can be a characteristic of the microviscosity of an object of interest.

Photochrome labeling allows one to investigate relaxation processes in biological systems using sensitive and widely available fluorescence techniques. An approach was developed to determine the rate constant of very rare collisions, including those between macromolecules in rigid media such as biological membranes. The photochrome–triplet labeling approach is based on monitoring the rate at which the photochromic stilbene label *cis–trans* isomerization is sensitized by the triplet–triplet energy transfer between this label and a triplet label (say, Erythrosine). The relatively long lifetime of the excited triplet state of the triplet label and nonreversible character of the photoisomerization make it possible to *integrate* information on the energy transfer by accumulating the photoreaction product, a *trans*-stilbene derivative. The characteristic time of the photochrome–triplet labeling method is about 1 s. These unique characteristics of the approach allow one to determine the rate constant of very rare collisions, including those between macromolecules in such rigid media as proteins in biological membranes. An additional step in the cascade reaction scheme is the quenching of the sensitizer triplet state with relatively low-concentration radicals (Fig. 3). This technique combines the three types of biophysical probes: stilbene photochrome probe, triplet probe, and stable nitroxide-radical spin probe, which depresses the sensitizer exited triplet state.

1.2.6 Dual Fluorophore-spin Labeling: High-sensitivity Redox Probes, Spin and Nitric Oxide Traps

The quantitative characterization of quenching and redox processes is based upon the use of two molecular subfunctionalities (a fluorescent chromophore and a stable nitroxide radical) tethered together

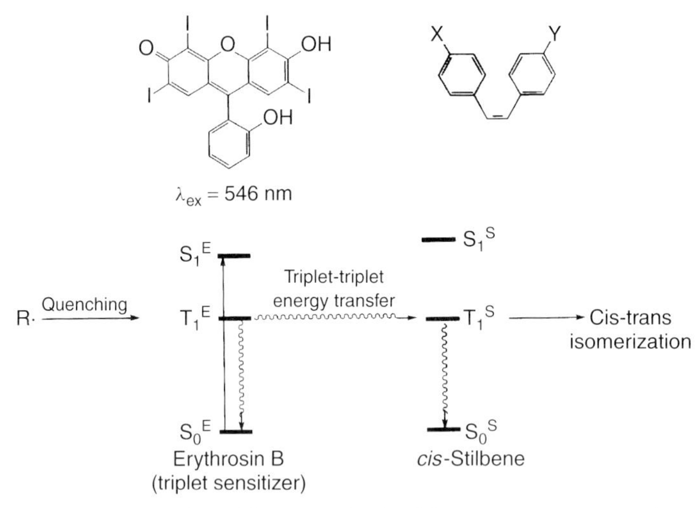

Fig. 3 Representation of the energy levels of the cascade reactants and a competition between the $T_1^E \rightarrow T_1^S$ and $T_1^E \rightarrow S_0^S$ processes.

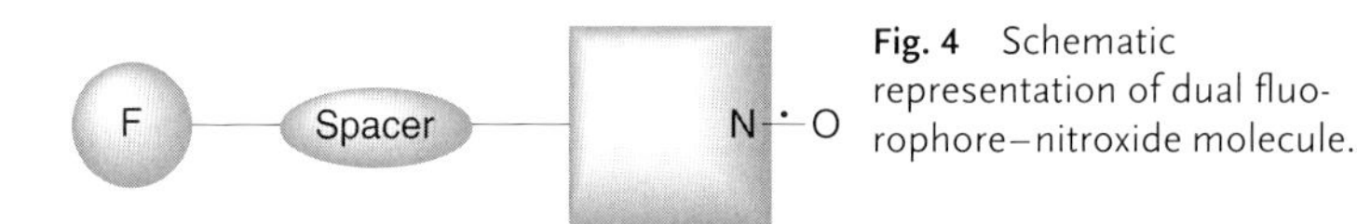

Fig. 4 Schematic representation of dual fluorophore–nitroxide molecule.

by a spacer. A typical double probe consists of fluorophore and nitroxide fragments connected by functional groups (a spacer) (Fig. 4). In the dual molecule, the nitroxide is a strong intramolecular quencher of the fluorescence from the chromophore fragment. Chemical or photoreduction of this fragment to a hydroxylamine derivative, oxidation of the nitroxide fragment, or addition of an active radical yields the fluorescence increase and the parallel decay of the fragment ESR signal. A reducing substrate (e.g. antioxidant analyte, semiquinone and superoxide radicals, nitric oxide in the nanomolar concentration scale) reduces the nitroxide function resulting in a decay of the nitroxide ESR signal plus concomitant enhancement of the chromophore fluorescence. Organic synthetic chemistry allows playing with the chemical structure of dual molecules of different fluorescence, ESR, and redox properties with a variety of bridges (spacers) tethering the fluorophore and nitroxide fragments. It was demonstrated that variation of the chemical structure of nitroxide moieties allows measuring of the concentration of nitric oxide, superoxide, and antioxidants (ascorbic and uric acids, flavons) in biological liquids (human blood plasma, fruits juices), that is to say, establishing the reducing status of these objects.

1.2.7 New Fluorescence Technique

In confocal microscopy, a tiny open volume element of about 0.2 fL is created by a focused laser beam. When a fluorescent molecule enters the confocal volume, it becomes excited. In this technique, the space distribution of a fluorescent fluorophore and the fluctuation of fluorescence intensity are analyzed. Recent developments in fluorescence spectroscopy and microscopy have made it possible to detect and image single molecules. These techniques allow the conducting of spectroscopic measurements for studying chemical and biological species and their interaction with the environment. Single molecular measurement offers time resolution to monitor dynamic processes such as translation, orientation, and enzymatic reactions on a timescale from milliseconds to ten seconds. Fluorescent quantum efficiencies greater than 0.1 can be studied with this technique. The techniques that have evolved to the level of single molecule sensitivity at room temperature are flow cytometry, confocal fluorescence correlation spectroscopy, and microdroplet technique.

Two-photon molecular excitation is performed by very high local intensity provided by tight focusing in a laser-scanning microscopy. This technique is combined with the temporal concentration of femtosecond pulsed lasers that produce a stream of pulses with a pulse duration of about 100 fs at a repetition rate of about 80 MHz. Advantages of the two-photon laser spectroscopy are as follows: high resolution, tolerance of infrared light by biological objects, different selection rule, and vibronic coupling. In home dual-color cross-correlation fluorescence spectroscopy experiments, a sample containing two fluorophores, with different emissions in each molecule, is irradiated with two lasers (or with one laser) to perform simultaneous excitation of the fluorophores. This technique in combination with confocal laser microscopy allows the separation of a microscopic volume with two different fluorophores from a volume with only one fluorophore, and therefore, the monitoring of dissociation of the dual-labeled molecules or association of two single-labeled molecules.

With the development of the near-infrared (NIR) laser diodes, the synthesis of fluorescent dyes with excitation and emission in the NIR wavelength regime has accelerated in the past decade for microscopy application. Fluorescence lifetime imaging is one of the techniques that attract much attention. Fluorescence lifetimes possess the benefit of being independent of local intensity, concentration and photobleaching of the fluorophore, and scattering in complex structure. This

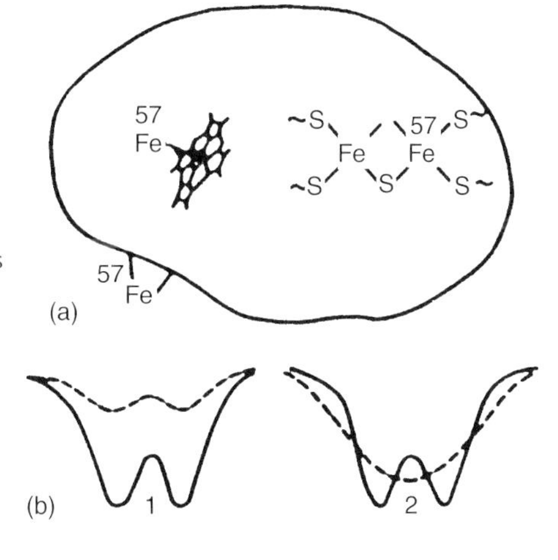

Fig. 5 The Mössbauer labeling methods. (a) Possible methods of introducing labels (^{57}Fe) into proteins. (b) Expected changes in Mössbauer spectral parameters for labels mobilities with $\tau_c = 10^{-7}$ s and amplitude of 0.01–0.02 nm: solid lines, absence of such a mobility; dashed lines, show presence. Curve 1, vibrational motion (f'decreases; line width does not change); curve 2, diffusional motion (line width increases, f'does not change).

technique in combination with FRET allows the investigation of interactions of proteins and nucleic acids within a cell.

1.3 Miscellaneous Labeling Methods

1.3.1 Mössbauer Labels

The Mössbauer method is based on the phenomenon of nuclear gamma resonance (NGR), for example, the transition of a nucleus (e.g. ^{57}Fe) from the ground state to the excited state upon absorption of a γ-quantum emitted by ^{57}Co. The dependence of the intensity the emitting and absorbing nuclei is indicated in the so-called *Mössbauer spectrum* (Fig. 5). Parameters of the NGR spectra are extremely sensitive to the amplitude x of the Mössbauer atom oscillation and the diffusion coefficient D, and therefore, to the dynamics of an object of interest. For example, for ^{57}Fe:

$$f' = 10^{-2300\langle x\rangle^2} \quad (13)$$

and when the amplitude increases by 0.05 nm, the f' value decreases by a factor of 10^5. Mössbauer atoms can be introduced into various portions of biological structures, say active centers of Fe-containing enzymes, via chemical reactions, absorption, or biosyntheses. In a complimentary technique of the Rayleigh scattering of Mössbauer radiation, the radiation is incident upon the sample. The sample does not need to be labeled with ^{57}Fe; nevertheless, the analysis of the radiation gives the mean square displacement of the scattering atoms averaging over all the atoms of the sample.

1.3.2 Nuclear Magnetic Resonance Probes

Probes modified specifically with isotopes (^{2}H, ^{13}C, ^{15}N, ^{19}F, ^{17}O, ^{28}Na, etc.) are characterized by distinct and easily interpreted NMR spectra. Such modifications can be exemplified by the introduction of ^{13}C into the C=O group of lysine, ^{15}N into the amide group of glycine, and ^{2}H into specific positions of lipid probes. The most widely used NMR probes that are commonly combined with spin labeling are deuterated probes. Isotopes ^{13}C and ^{15}N are the most suitable for studying the molecular dynamics of biological molecules.

1.3.3 Electron and X-ray Scattering Labels

The relative intensity of electron scattering and X-ray scattering (I/I_0) increases, other conditions being equal, with atomic number (Z) of the component atom: $I/I_0 \sim Z^2$. A number of polynuclear complexes and clusters of heavy metals are used nowadays as electron-scattering labels: colloid particles and clusters of gold, polymercury mercarbides, and compounds of palladium and iodine. Analysis of electron micrographs and X-ray scattering of the labeled biological objects allows one to investigate the location of specifically labeled portions.

2 Applications

2.1 Enzymes and Proteins

2.1.1 Active Sites of Enzymes

Monitoring the mobility of spin and luminescence labels and measuring the magnetic and electrostatic interactions between them can be used to study many phenomena, including the chemical nature and mutual disposition of modified catalytic and binding chemical groups, the size and form of a cleft of enzyme-active centers, the distribution of electrostatic charges around the centers, the location of specific zones on an enzyme complex, and the rotation of protein globules.

Serine proteases are used as a *proving ground* for various versions of the spin-labeling method. The principal catalytic groups of the α-chymotrypsin active center, namely, Ser195, His57, and Met192, were modified by nitroxide spin labels. The ESR spectra of spin-labeled preparations were highly sensitive to fine changes in the structure of the active centers at the interactions with the substrate analogs. When nitroxide derivatives of sulfamides having bridges of varying lengths between the aromatic groups and nitroxide fragment were included in the enzyme-active center, the mobility of such a fragment could be drastically enhanced by using a probe longer than 1.45 nm. Physical labeling methods were applied to determine the mutual arrangement of Fe_nS_n clusters, the ATP center of nitrogenase and the location of these centers on the enzyme macromolecules, and for estimation of distances between the nucleotide-binding centers of H^+ ATP synthase (4.0–5.0 nm).

The labeling method was used to probe the molecular nature of the binding pocket of a G protein–coupled receptor and the events immediately following the binding and activation. The substance P peptide, a potent agonist for the neurokinin-1 receptor, has been modified with a nitroxide spin probe specifically attached at Lys-3. The EPR spectrum of the bound peptide indicates that the Lys-3 portion of the agonist is highly flexible. A slight increase in the mobility of the bound peptide in the presence of a non-hydrolyzable analog of GTP was detected, indicative of the alternate conformational states described for this class of receptor. *S*-adenosylmethionine (AdoMet) synthetase (I) of *Escherichia coli* is one of numerous enzymes that have a flexible polypeptide loop that moves from gate access to the active site in a motion that is closely coupled to catalysis. Site-directed spin labeling was used to introduce nitroxide groups at 2 positions in the loop to illuminate how the motion of the loop is affected by substrate binding. The results suggest that the motion of the loop may be an intrinsic property of the protein and not be strictly ligand-modulated.

The confocal fluorescence coincidence analysis has been employed for a rapid homogeneous assay for restriction endonuclease *Eco*RI. This methodology has been improved by the application of two-photon excited, dual-color, cross-correlation spectroscopy on the level of single diffusing molecules. A double-strand of DNA was labeled with Rhodamin green and Texas red. The kinetics of the enzymatic cleavage of the labeled DNA by restriction endonuclease was monitored by this new technique. The single-molecule fluorescence technique was used for the study of differences in the chemical reactivity of individual molecules of enzymes. Enzyme molecules are presented at very low concentration (7.6×10^{-17} to 1.5×10^{-16} M) in a narrow capillary and each discrete molecule produces a discrete zone. The kinetics of synthesis of a fluorescent product, 2′-(2-benzothiazol)-6′-hydrobenzthiazol, produced by single alkaline phosphatase molecule has been investigated.

Rhodanase was labeled at its catalytic site with the phosphorescence probe eosin isothiocyanide. The accessibility of molecules to the probe was determined by phosphorescence lifetime-quenching studies. The use of a novel pulsed ESR technique for distance measurement based on the detection of double-quantum coherence (DQC), which yields high quality dipolar spectra, significantly extended the range of measurable distances in proteins using nitroxide spin-labels. Eight T4 lysozyme (T4L) mutants, doubly labeled with methanethiosulfonate spin-label (MTSSL), have been studied using DQC-ESR at 9 and 17 GHz. The distances span the range from 2.0 to 4.7 nm.

The physical basis of an approach rests upon the effect of the local electrostatic potential upon dynamic interactions at encounters with charged quenching molecules resulting in fluorescence (phosphorescence), or between a stable radical, for example, nitroxide, and another charged paramagnetic species. In such cases, the relaxation parameters, that is, the lifetime of the fluorescence (phosphorescence) chromophore or spin–spin and spin–lattice relaxation rates of paramagnetic species are dependent upon the frequency of encounters, and therefore, on local electrostatic fields. Electron-carrier horse heat cytochrome c and dioxygen-carrier sperm-whale myoglobin served as models for determining local electrostatic charges in the vicinity of paramagnetic active sites of metalloenzymes and metalloproteins. Calculations of local charge Z_x in the vicinity of a paramagnetic particle (such as the active site of a metalloprotein or a spin label) colliding with a nitroxide or metallocomplex with known charge Z_p can be carried out with the use of the Debye equation.

Nitroxide spin molecules are convenient probes for the ENDOR application for solving some problems of enzyme catalysis. From ENDOR studies of molecular structures and conformations, several spin-labeled amino acid derivatives incorporated into enzyme-active sites have been reported. For instance, the spin-labeled transition-state analog in the α-chymotripsin reaction, *N*-(2,2,5,5-tetramethyl-1-oxypyrrolinyl)-L-phenylalaninal, has been synthesized. The conformation of the acyl- moiety of the substrate analog in the active site of the reaction intermediate differs significantly from that of the free substrate in solution.

2.1.2 Conformational Changes and Molecular Dynamics

Labeling methods were applied for detecting conformational changes, including large-scale denaturation transitions, predenaturing phenomena, allosteric effects (transmitted from one enzyme subunit to another), conformational changes in enzyme structures other than quaternary (transglobular allosteric transition), and changes in the state of various segments of the active centers of enzymes due to the action of specific reagents on the adjacent segments.

Allosteric effects were first reported by McConnell and associates in studies of horse hemoglobin with the β-SH group modified by a spin label 1.3 to 1.4 nm from the heme active center. Transglobular transitions were studied in a number of proteins, including lysozyme, myoglobin, aspartate aminotransferase, and myosin.

Biophysical labeling methods provide a unique possibility for monitoring local intermolecular dynamics properties in a wide range of correlation times (τ_c is in the range 10^2 to 10^{-10} s). They enable the investigation of the dynamics of different parts of protein globules in the vicinity of the spin, fluorescence, phosphorescence, and Mössbauer labels under various conditions (temperatures 30–330 K, water content, viscosity, substrates, inhibitor additions, etc.). The experiments revealed the following tendency: appropriate physical methods detect the mobility of labels starting from the temperature of liquid nitrogen; the lower the value of the characteristic frequency of the method, the lower the temperature at which the label mobility can be recorded. Thus, the mobility recorded results from a gradual softening of the protein–water matrix rather than from an individual phase transition.

At physiological ambient conditions, all labeling methods indicate the mobility of labels and surrounding media in the nanosecond temporal regions. A correlation between the temperature dependencies of nonharmonic parameters of Mössbauer atoms in heme, the heme spectra Soret band Gaussian broadening, non-elastic neutron scattering, and the molecular dynamics simulation in myoglobin was demonstrated. According to these data, the anharmonic intramolecular mobility of proteins increases dramatically at $T > 200$ K. This conclusion was confirmed by low-temperature experiments with the use of the fluorescence dynamic Stokes shift, neutron scattering, and molecular dynamics simulation in other proteins, superoxide dismutase, lysozyme, elastase, bacteriorhodopsin, and RNAse. It is necessary to stress that the anharmonic nanosecond dynamics of the proteins recorded with labeling methods only appear if the water content of the sample being studied exceeds a critical value by about 10 to 25%. According to experimental data on Mössbauer spectroscopy, at ambient temperatures, the myoglobin heme group exhibits the unharmonic nanosecond motion with $\langle \Delta x \rangle_{nh} > 0.02$ nm. The flexibility of the cavity of the myoglobin active site is evidenced by the mobility of a spin probe, a derivative of isocyanate attached to the heme group in the single crystal.

The mobility of a single deuterated tryptophane located in a loop of the active site of triosephosphate isomerase has been investigated by solid-state deuterium NMR and solution state 19F NMR. The rate of the loop's opening and closing was detected using samples of the enzyme in the presence and in the absence of a substrate analog DL-glycerol 3-phosphate at temperatures ranging from -15 to $+45\,^{\circ}$C. It was shown that the rate of the loop's opening and closing is of order 10^4 s^{-1}. Hinge-bending and substrate-induced conformational transitions in T4 lysozyme in solution were confirmed in a study by site-directed labeling. Both single and pairs of nitroxide spin labels were introduced into different domains of the protein followed by monitoring distances between the labels by ESR technique. In the absence of a substrate, the results are consistent with a hinge-bending motion, which opens the active site cleft. When substrate binding takes place, the relative domain movement occurs.

Simulation studies of typical intramolecular energy transfer experiments reveal that both static and dynamic conformational distribution information can thus be obtained at a single temperature and viscosity. This method was used for the investigation of the refolding transition of *E. coli* adenylate kinase (AK) by monitoring the refolding kinetics of a selected 20-residue helical segment in the CORE domain of the protein.

The local and global dynamics of the *Sulfolobus solfataricus* β-glycosidase were studied by ESR and time-resolved fluorescence techniques. For EPR investigations, the protein was covalently modified by the maleimido nitroxide spin label, which is specific for cysteine -SH groups, at position 344 and at position 101, where Ser101 was changed into a cysteine by site-directed mutagenesis. The labeled proteins underwent temperature perturbation in the range 290 to 335 K. The general dynamic information was deduced from the analysis of the fluorescence emission decay of the tryptophanyl residues that are present in each region of the protein structure.

2.2 Membranes

Specific modification of individual lipid and protein parts of biological and model membranes by whole sets of labels and probes allows the biophysical labeling method to be extremely effective in investigating membrane dynamics and microstructure. Monitoring of the rotational diffusion of nitroxide, fluorescent, and phosphorescent derivatives of lipids and proteins, and their collisions, enables one to determine thickness, the microviscosity profile of membranes, membrane heterogeneity, and phase transitions, as well as to discover phenomena related to the lateral diffusion of lipid and protein associations in membranes. Spin and fluorescence labeling approaches have indicated significant similarities between the structural and dynamic properties of model bilayer liposomes and the lipid phase of such biological membranes such as erythrocytes, the sarcoplasic reticulum, and microsomes. A number of specific problems associated with the intrinsic inhomogeneity of cells, biochemical and biophysical processes (including redox ones), and cell organization can be readily solved by means of labeling approaches. In differentiated cells (ovarian cells, spermatozoa), the FRAP method indicated the presence of protein domains and of different microcompartments in various parts of the cells that differed with respect to diffusion coefficients and mobile fractions. The values of intercellular pH monitored by a fluorescent probe (fluorescein after enzymatic deacetylation) were found to differ in various compartments of the cells.

The distribution of LW peptide between coexisting ordered and disordered lipid domains was probed by measuring the LW Trp fluorescence quenched by a nitroxide-labeled phospholipid that is concentrated in disordered lipid domains. Strong quenching of the Trp fluorescence (relative to quenching in model membranes lacking domains) showed that LW peptide was concentrated in quencher-rich disordered domains and was largely excluded from ordered domains. Nitroxide spin label studies with high-field/high-frequency ESR and two-dimensional Fourier transform ESR enable one to accurately detect distances in biomolecules, unravel the details of the complex dynamics in proteins, characterize the dynamic structure of membrane domains, and discriminate between bulk lipids and boundary lipids that coat transmembrane peptides or proteins.

The EPR spectroscopic characterization of a recently developed magnetically oriented spin-labeled model membrane system was reported. The oriented membrane system is composed of a mixture of a bilayer-forming phospholipid and a short-chain phospholipid that breaks up the extended bilayers into bilayered micelles or bicelles that are highly hydrated (approx. 75% aq.). Paramagnetic lanthanide ions (Tm^{3+}) were added to align the bicelles such that the bilayer normal is collinear with the direction of the static magnetic field. The nitroxide spin probe 3β-doxyl-5α-cholestane (cholestane) was used to demonstrate the effects of macroscopic bilayer alignment through the measurement of orientational dependent hyperfine splittings that were close to A_{yy}. The EPR signals of cholestane inserted into oriented and randomly dispersed DMPC-rich bilayers have been investigated over the temperature range of 298 to 348 K.

2D-ELDOR spectroscopy has been employed to study the dynamic structure of the liquid-ordered (L_O) phase versus that of the liquid-crystal (L_C) phase in multibilayer phospholipid vesicles without (L_C) and with (L_O) cholesterol, using end-chain and headgroup labels and spin-labeled cholestane. Fluorescence correlation spectroscopy was used for investigation of the size and the size distribution of liposomes loaded with a fluorescence probe.

2.3 Nucleic Acids

The following methods have been developed for modification of nucleic acids: (1) covalent attachment of spin and fluorescent labels to functional groups of the nucleotide bases; (2) covalent attachment of labels to hydroxyl groups on the ribose glucosyl rings; (3) alkylation or acylation of the SH group of the 4-thiouracil or the NH_2 group on the aminoacyl residue of tRNA; (4) enzymatic synthesis of nucleic acids with the use of a labeled base; and (5) intercalation of spin and fluorescent probes between bases in the double-helix portion of nucleic acids, and complexation with paramagnetic (Mn^{+2}) or Mössbauer (^{57}Fe) atoms. The extent of the immobilization of the labels and probes and the spin–spin interactions between them are determined by the structure of the nucleic acid, the length and flexibility of the label, and ambient conditions (pH, temperature, ionic strength, destructive forces, and degrees of binding of various compounds). The thermal- and pH-induced transitions of nucleic acids at their melting temperatures are accompanied by a drastic increase in the mobility of the attached spin labels. The ESR spectra of spin-labeled preparations of nucleic acids are sensitive to fine conformational changes as well.

The dynamics of iron nuclei in the condensates obtained by interaction of Fe^{III} with DNA, (DNA monomer), has been investigated by ^{57}Fe Mössbauer spectroscopy. Functions of the parameters isomer shift and nuclear quadrupole splitting in temperature ranges 20 to 260 K were employed to investigate the dynamics of Fe nuclei and showed linear trends in the temperature ranges 20 to 150 and 150 to 260 K, respectively, the latter with larger slopes. Site-directed spin labeling utilizes site-specific attachment of a stable nitroxide radical to probe the structure and dynamics of nucleic acids. 4-thiouridine base was introduced at each of six different positions in a 23-nucleotide RNA molecule. The 4-thiouridine derivatives were subsequently modified with one of three methanethiosulfonate nitroxide reagents to introduce a spin label at specific sites. The motion is similar to that found for a structurally related probe on helical sites in proteins, suggesting a similar mode of motion. The nitroxide spin-labels are covalently linked to a deoxyuridine residue using either a monoacetylene or diacetylene tether. A detailed analysis of the EPR spectrum of duplex DNA in solution, spin-labeled using the diacetylene tether, demonstrates that the motion of the nitroxide can be modeled in terms of this independent uniaxial rotation together with motion of the DNA, which is consistent with the global tumbling of the duplex.

New genome sequence information is rapidly increasing the number of nucleic acid (NA) targets of use for characterizing and treating diseases. Detection of these targets by fluorescence-based assays is often limited by fluorescence background from unincorporated or unbound probes that are present in large excess over the target. To solve this problem, energy transfer–based probes have been developed and used to reduce the fluorescence from unbound probes. The quenching approach of a two-color NA assay with a correlated, two-color, single-molecule fluorescence detection was employed. A method for generating fluorescent probes from RNA or DNA samples was described. In this method, an amine-modified base was added to the ends of the primers so that a dye could be introduced into each DNA and onto the end of each product. This allows labeling of a very small amount of total RNA in such a way that reproducible experiments were performed. It was shown that FRET between fluorescein labeled to poly(C) and an intercalator agent takes place when single-stranded poly(C) hybridizes with poly(I).

The following application of the labeling approach can be also pointed out: (1) protein-induced DNA blending studied by labeling a 30-base pair strand of DNA, one with europium donor chelate and the other with the acceptor dye Cy5; (2) detection of the association between actin and the protein dystropin inside a muscle cell; (3) time-resolved identification of individual label mononucleotides in water; (4) DNA sequences on the single molecule level; (5) detection of individual tumor marker molecules in neat human.

2.4 Cells and Organs

Fluorescence lifetime imaging microscopy using multiphoton excitation techniques is now finding an important place in the imaging of protein–protein interactions and intracellular physiology. An example of a typical application of the system was provided in which the fluorescence resonance energy transfer between a donor–acceptor pair of fluorescent proteins within a cellular specimen is measured. The recent development of automated fluorescence imaging systems has enabled fluorescence microscopy to be used for the purposes of compounds screening. This technique has found various applications, including screening for the effect of kinase inhibitors on the cytoskeleton, agonist-stimulated receptor internalization, and protein phosphorylation and acetylation.

L-band (low-frequency) EPR imaging instrumentation has been developed in order to elucidate spatially defined differences in tissue metabolism and the redox status. For example, 3D images of nitroxyl compounds in a rat tail and isolated rat heart have been reported, which reflect the local metabolic and oxygenation status. Very recently, a new magnetic resonance imaging method (MRI) was presented that can "visualize" free radical generation in animals by means of MRI. The advantages of MRI, the combination of MRI and EPR, and PEDRI (proton electron double resonance imaging), and recent progress of these methods were demonstrated.

The kinetics of reductions of the radical R*, 5-dimethylaminonaphthalene-1-sulfonyl-4-amino-2,2,6,6-tetramethyl-1-piperidine-oxyl, by blood and its components were studied using the EPR technique. The results demonstrate that: (1) the erythrocytes catalyze the redaction of R* by ascorbate; (2) the rate of redaction of the radical is high though it does not penetrate the cells; and (3) in human erythrocytes, there is an efficient electron transfer route through the cell membrane. The study points out that R* is a suitable spin label for measuring the reduction kinetics and antioxidant capacity in blood as expressed by reduction by ascorbate. The nitroxides are reduced to corresponding hydroxylamine in cellular incubations as well as *in vivo* incubations. *Tissue hypoxia* is known to significantly enhance the rate of the nitroxide reduction.

Fluorescence *in situ* hybridization (FISH)-based techniques were used in the clinical detection of genetic alterations in tumor cells. Fluorescent DNA probes also enabled screening for very subtle chromosomal changes. A growing number of FISH-based cytogenetic tests are employed in clinical laboratories to support a physician's diagnoses of the causes and the course of a disease. FISH-based analyses have been applied very successfully to the analysis of single cells and have demonstrated the existence of cell clones of different chromosomal make-up within human tumors. An immunofluorescence assay was developed to identify proteins specifically binding to oligonucleoside phosphorodithioate (ODN) aptamers from a bead-bound ODN library. It was shown that high-resolution localization of genes on chromosomes is possible with specially correlated introduction of fluorescent DNA probes (fluorescence *in situ* hybridization). This method allows the mapping of cloned genes to a locus defined by a chromosome band.

2.5 Miscellaneous

2.5.1 Biologically Active Compounds, Biological Analysis, Pharmokinetics

Syntheses and applications have been reported for many types of spin- and fluorescent-labeled analogs of amino acids, steroids, phospho-organic compounds, alkaloids and terpenes, nucleotides and their fragments, coenzymes (vitamins B_{12} and B_6, protohemin, NADH, etc.), and haptens. Advances in the synthetic chemistry of nitroxide analogs of medical preparations such as morphine, local anesthetics, anticancer drugs (analogs of thiophosphoamin, rubromycin, adriamycin, nitroarylaziridnes, nitrosourea, diethyleneimineurethanic acids, and anthracycline drugs) have provided the basis for further development in basic and applied biochemical and pharmacokinetic investigations.

Some of the most promising applications of fluorescence and spin labeling appear to entail the analysis of biologically active compounds, metal ions, biopolymers, and enzymes. Changes in fluorescence, rotational diffusion parameters, dipole–dipole electrostatic and spin–spin interactions under contact between physical probes, and the compound being tested allow one to work out an appropriate analytical procedure in each particular case. A number of procedures featuring fluorescent and spin-labeled chelating reagents have been developed for quantitative assays of Ca^{2+}, Ni^{2+}, Cu^{2+}, Zn^{2+}, Co^{2+}, and Na^{+} ions, and for determination of enzymatic activity. The ability of chromophores (propidium, acridine orange, etc.) to change the intensity of fluorescence upon binding with nucleic acids has been widely used in analytical biochemistry. Lanthanide fluorescence chelates have been successfully developed as the fluorescence labels for highly sensitive bioassays. The applications of lanthanide fluorescence labels for time-resolved fluoroimmunoassay (TR-FIA), DNA hybridization assay, cell activity assay, and fluorescence imaging microscopy have been widely investigated.

Nitroxide probes are used as *in vivo* imaging reagent and oximetry probes. The pharmacokinetics and spatial distribution of the nitroxide in tumor tissue were followed and compared with those in normal tissue. The tumor tissues showed significant heterogeneity in the nitroxide distribution and higher reducing rate compared to the normal tissue.

3 New Trends in Biophysical Labeling

The following main trends in development and application biophysical labeling may be pointed out:

- Synthesis of fluorescent and phosphorescent dyes with excitation and emission in the NIR wavelength regime for microscopy and analysis application.
- Syntheses and application of dual fluorophore–nitroxide probes of different redox potential and fluorophore fragment nature.
- Site-directed incorporation of spin and luminescent label in protein, nuclear acids followed by the investigation with the use modern physical methods.
- Growing employment of recently developed ESR techniques (high-frequency ESR spectroscopy, Fourier transform 2D ESR, ELDOR electron spin-echo techniques, ELDOR and ENDOR, and multiple-quantum ESR) and fluorescence methods such as confocal microscopy, single molecular measurement, multiphoton molecular excitation, fluorescence lifetimes imaging, cascade systems, etc.
- Expanding objects under investigation including cells, organs, and organisms in prospect.
- Wide application in biomedical research and medical diagnostics.

Bibliography

Books and Reviews

Berliner, L. (Ed.) (1998) *Spin Labeling. The Next Millennium*, Academic Press, New York, p. 14.

Berliner, L., Eaton, S., Eaton, G. (Eds.) (2000) *Distance Measurements in Biological Systems by ESR*, Kluwer Academic Publishers, Dordrecht.

Blake, R.A. (2001) Cellular screening assays using fluorescence microscopy, *Curr. Opin. Pharmacol.* **1**, 533–539.

Blank, A., Dunnam, C.R., Borbat, P.P., Freed, J.H. (2003) High resolution electron spin resonance microscopy, *J. Magn. Reson.* **165**, 116–127.

Borbat, P.P., Costa-Filho, A.J., Earle, K.A., Moscicki, J.K., Freed, J.H. (2001) Electron spin resonance in studies of membranes and proteins, *Science* **291**, 266–269.

Kocherginsky, N., Swartz, H.M. *Nitroxide Spin Labels*, CRC Press, Boca Raton, FL, 1995, pp. 114–172.

Krinichnyi, V.I. (1994) *2-mm Wave Band ESR Spectroscopy of Condensed Systems*, CRC Press, Boca Raton, FL.

Krishnan, R.V., Masuda, A., Centonze, V.E., Herman, B. (2003) Quantitative imaging of protein-protein interactions by multiphoton fluorescence lifetime imaging microscopy using a streak camera, *J. Biomed. Opt.* **8**, 362–367.

Lakowicz, J.R. (1983) *Principles of Fluorescence Spectroscopy*, 2nd edition, Plenum Press, New York.

Lakowicz, J.R., Gryczynski, I., Shen, Y., Malicka, J., D'Auria, S., Gryczynski, Z. (2002) Fluorescence spectral engineering – biophysical and biomedical applications, *Springer Ser. Fluoresc.* **2**, 43–68.

Likhtenshtein, G.I. (1976) *Spin Labeling Method in Molecular Biology*, Wiley-Interscience, New York.

Likhtenshtein, G.I. (1993) *Biophysical Labeling Methods in Molecular Biology*, Cambridge University Press, Cambridge, New York.

Likhtenshtein, G.I. (1996) Spin and fluorescence immunoassay in solution, in: Lefkovits, I., Nezlin, R. (Eds.) *Immunology Methods Manual*, Pergamon Press, London, pp. 540–550.

Likhtenshtein, G.I. (2003) *New Trends in Enzyme Catalysis and Mimicking Chemical Reactions*,

Kluwer Academic Publishers, Dordrecht, New York.

Likhtenshtein, G.I., Febrario, F., Nucci, R. (2000) Intramolecular dynamics and conformational transitions in proteins studied by biophysical labeling methods. Common and specific features of proteins from thermophylic microorganisms, *Spectrochem. Acta Part A: Biomol. Spectrosc.* **56**, 2011–2031.

Papper, V., Likhtenshtein, G.I. (2001) Substituted stilbenes: A new view on well-known systems: new application in chemistry and biophysics, *J. Photochem. Photobiol., A: Chem.* **140**, 39–52.

Parak, F., Reinish, L. (1986) Mössbauer effect in the study of structure dynamics, *Methods Enzymol.* **131**, 568–607.

Ratner, V., Kahana, E., Eichler, M., Haas, E. (2002) A general strategy for site-specific double labeling of globular proteins for kinetic FRET studies, *Bioconjugate Chem.* **13**, 1163–1170.

Rettig, W., Stremel, B., Schrader, S., Seifer, H. (Eds.) (1999) *Applied Fluorescence in Chemistry, Biology and Medicine*, Springer-Verlag, Berlin.

Selvin, P.R. (1999) Luminescent lanthanide chelates for improved resonance energy transfer and application to biology, in: Rettig, W., Stremel, B., Schrader, S., Seifer, H. (Eds.) *Applied Fluorescence in Chemistry, Biology and Medicine*, Springer-Verlag, Berlin, pp. 457–487.

Swartz, H.M. (2002) Measuring real levels of oxygen in vivo: opportunities and challenges, *Biochem. Soc. Trans.* **30**, 248–252.

Trommer, W.E., Vogel, P.D. (1992) Photoaffinity spin labeling, in: Zhdanov, R.I. (Ed.) *Bioactive Spin Labels*, Springer, Berlin, pp. 405–427.

Vo-Dinh, T. (2003) life time-based imaging, in: Herman, P., Lin, H.-J., Lakowicz, J.R. (Eds.) *Biomedical Photonics Handbook 9/1–9/30*, CRC Press LLC, Boca Raton, FL.

Weier, H.-U.G., Greulich-Bode, K.M., Ito, Y., Lersch, R.A., Fung, J. (2002) FISH in cancer diagnosis and prognostication: From cause to course of disease, *Expert Rev. Mol. Diagn.* **2**, 109–119.

Xiang, C.C., Brownstein, M.J. (2003) Preparing fluorescent probes for microarray studies, *Methods Mol. Biol.* **224**, 55–60.

Yuan, J., Wang, G., Matsumoto, K. (2001) Lanthanide fluorescence chelates and their applications in bioassays, *Trends Inorg. Chem.* **7**, 109–117.

Primary Literature

Boehmer, M., Enderlein, J. (2003) Fluorescence spectroscopy of single molecules under ambient conditions: methodology and technology, *Chem. Phys. Chem.* **4**, 792–808.

Borbat, P.P., Mchaourab, H.S., Freed, J.H. (2002) Protein structure determination using long-distance constraints from double-quantum coherence ESR: study of T4 lysozyme, *J. Am. Chem. Soc.* **124**, 5304–5314.

Costa-Filho, A.J., Shimoyama, Y., Freed, J.H. (2003) A 2D ELDOR study of the liquid ordered phase in multilamellar vesicle membranes, *Biophys. J.* **84**, 2619–2633.

Fujii, H., Berliner, L.J. (2003) In vivo detection and visualization of reactive oxygen free radicals by EPR and MRI, *Biomed. Res. Trace Elem.* **14**, 6–10.

Hustedt, E.J., Kirchner, J.J., Spaltenstein, A., Hopkins, P.B., Robinson, B.H. (1995) Monitoring DNA dynamics using spin-labels with different Independent mobilities, *Biochemistry* **34**, 4369–4375.

Krupyanskii, Y.u.F., Esin, S.V., Eshenko, G.V., Mikhailyuk, M.G. (2002) Equilibrium fluctuations in lysozyme and myoglobin, *Hyperfine Interact.* **141/142**, 273–277.

Mangels, M.L., Cardon, T.B., Harper, A.C., Howard, K.P., Lorigan, G.A. (2000) Spectroscopic characterization of spin-labeled magnetically oriented phospholipid bilayers by EPR spectroscopy, *J. Am. Chem. Soc.* **122**, 7052–7058.

McNulty, J.C., Thompson, D.A., Carrasco, M.R., Millhauser, G.L. (2002) Dap-SL: a new site-directed nitroxide spin labeling approach for determining structure and motions in synthesized peptides and proteins, *FEBS Lett.* **529**, 243–248.

Nakazumi, H., Colyer, C.L., Kaihara, K., Yagi, S., Hyodo, H. (2003) Red luminescent squarylium dyes for noncovalent HSA labeling, *Chem. Lett.* **32**, 804,805.

Nolan, R.L., Cai, H., Nolan, J.P., Goodwin, P.M. (2003) A simple quenching method for fluorescence background reduction and its application to the direct, quantitative detection of specific mRNA, *Anal. Chem.* **75**, 6236–6243.

Qin, P.Z., Hideg, K., Feigon, J., Hubbell, W.L. (2000) Monitoring RNA base structure and dynamics using site-directed spin labeling, *Biochemistry* **42**, 6772–6783.

Saphier, O., Silberstein, T., Shames, A.I., Likhtenshtein, G.I., Maimon, E., Mankuta, D., Mazor, M., Katz, M., Meyerstein, D., Meyerstein, N. (2003) The reduction of a nitroxide spin label as a probe of human blood antioxidant properties, *Free Radical Res.* **37**, 301–308.

Schunemann, V., Winkler, H. (2000) Structure and dynamics of biomolecules studied by Mossbauer spectroscopy, *Rep. Prog. Phys.* **63**, 263–353.

Shafer, A.M., Bennett, V.J., Kim, P., Voss, J.C. (2003) Probing the binding pocket and endocytosis of a G protein-coupled receptor in live cells reported by a spin-labeled substance P agonist, *J. Biol. Chem.* **278**, 34203–33421.

Talavera, E.M., Bermejo, R., Crovetto, L., Orte, A., Alvarez-Pez, J.M. (2003) Fluorescence energy transfer between fluorescein label and DNA intercalators to detect nucleic acids hybridization in homogeneous media, *Appl. Spectrosc.* **57**, 208–215.

Taylor, J.C., Markham, G.D. (2003) Conformational dynamics of the active site loop of S-adenosylmethionine synthetase illuminated by site-directed spin labeling, *Arch. Biochem. Biophys.* **415**, 164–171.

Trotta, A., Barbieri Paulsen, A., Silvestri, A., Ruisi, G., Assunta Girasolo, M., Barbieri, R. (2002) The dynamics of ^{57}Fe nuclei in FeIII-DNA condensate, *J. Inorg. Biochem.* **88**, 14–18.

Ligase-mediated Gene Detection

Johan Stenberg, Mats Nilsson, and Ulf Landegren
Department of Genetics and Pathology, Uppsala University, Sweden

Keywords

Ligase
An enzyme capable of joining the 5′ and 3′ ends, respectively, of two adjoining nucleic acid molecules, creating a phosphodiester bond.

Encyclopedia of Molecular Cell Biology and Molecular Medicine, 2nd Edition. Volume 7
Edited by Robert A. Meyers.

ISBN: 3-527-30549-1

5′ and 3′
The two ends of a nucleic acid strand are termed the 5′ and the 3′ ends. Enzymatic synthesis proceeds in the 5′ to 3′ direction.

Ligation template
A nucleic acid molecule with a sequence that allows two probe molecules to hybridize adjacent to each other, allowing ligation of the 5′ end of one to the 3′ end of the other to take place.

DNA ligation is a popular mechanism in analyses of gene sequences. The ligation of pairs of DNA probes upon hybridization to a target nucleic acid sequence creates a contiguous DNA sequence that can be used to infer the presence and nature of target molecules. The ligatable DNA strands can also be coupled to other molecules interacting with targets of interest to increase the range of molecules detectable in ligase-based assays.

In ligase-mediated detection of nucleic acid sequences, the requirements for substrate recognition by ligases may be used to distinguish target sequences that differ in even single nucleotide positions. The covalent link that forms between the probe pairs can be used to connect retrievable functions on one probe to detectable groups on the other. Even more usefully, the resulting contiguous DNA strand can be amplified, along with sequence elements that encode the target specificity of the probes. These principles are being applied in a rapidly increasing range of molecular analyses, including high-throughput genotyping, *in situ* detection of nucleic acid sequences, and highly sensitive protein assays.

1 Principles

1.1 Basis of Ligase-mediated Analyses

Specific nucleic acid sequences can be analyzed using hybridization probes that base pair to the target sequence. For enhanced performance, hybridization probes are often used in combination with nucleic acid–specific enzymes in molecular genetic analyses. One example is the polymerase chain reaction (PCR), where two oligonucleotide probes and a DNA polymerase are used to detect a nucleic acid sequence. In ligase-mediated detection reactions, pairs of DNA probes are joined by the action of a ligase. The joining event is used as a measure of the presence and nature of the nucleic acid sequences. This is possible because ligation critically depends on the ability of the probes to base-pair to a template nucleic acid molecule (Fig. 1). This assay mechanism has several valuable properties:

1. The ligation reaction is inherently very specific since it requires the coincident hybridization of two independent probe molecules immediately next to each other on a template. It is therefore

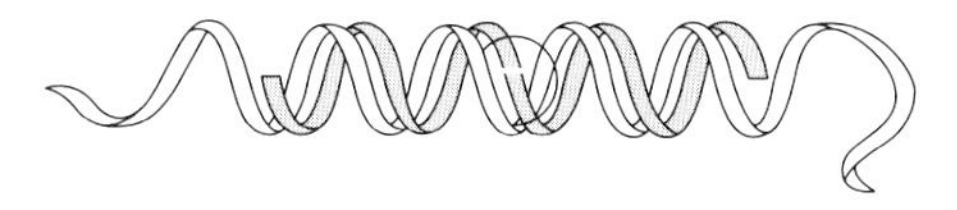

Fig. 1 Oligonucleotide ligation assay. Configuration of two oligonucleotide probes (shaded), hybridizing immediately next to each other on a target nucleic acid strand (white). Where the two probes meet (circled region), they can be connected by a ligase.

unlikely that the probes will join in the absence of the correct template nucleic acid molecule. As a consequence, the assay permits detection of unique nucleic acid sequences in complex populations of DNA molecules, for example, the complete human genome.
2. Variants of sequences can be conveniently distinguished because ligation is strongly inhibited by a mismatch even in a single nucleotide position at the junction of the probes.
3. The act of ligation of probes gives rise to a new DNA sequence not previously present in the reaction. This circumstance can be exploited for detection of the reaction products. For example, detection can be enhanced by exploiting the increased hybridization stability of the ligation product as compared to that of each individual probe. Alternatively, pairs of probes may be used in which one member carries a detectable moiety and the other is bound to a support before or after the ligation assay. Thereby, ligation products can be efficiently isolated and excess probes removed by washes before detection. If the two ligatable probe sequences are located at either end of the same DNA molecule, then the ligation will result in a circularized probe, wound around the target molecule. Finally, ligation products can also be recruited as templates in subsequent nucleic acid-amplification reactions, resulting in highly sensitive detection of target sequences.

Ligation-mediated analysis has undergone a strong development from the characterization of ligase enzymes and the first analytical ligase experiments carried out in the late 1960s and early 1970s, respectively. We will first discuss some fundamental properties of ligation reactions before describing some of the methods that have been developed on the basis of the probe ligation principle.

1.2 Biology of Ligation

Ligases are proteins that serve to join the ends of polynucleotide strands. They use cofactors such as NAD or ATP to supply the energy required to form the phosphodiester bonds. By contrast, the so-called breakage-reunion enzymes, exemplified by the topoisomerases, first break and then reseal DNA strands in a concerted fashion, and do not require an external supply of energy.

In the cell, ligases complete excision-repair reactions in damaged DNA. In addition, they seal interrupted DNA strands during discontinuous DNA replication, and during meiosis they participate in altering the continuity of paternally and maternally inherited DNA sequences through genetic recombination.

The first biochemical characterization of enzymes that join DNA strands appeared in 1967. The isolated ligases soon proved to be of great value as biotechnological tools for the analysis of DNA structure and construction of novel DNA molecules. Using these enzymes, it famously proved possible to assemble recombinant molecules from separate DNA segments *in vitro*, and then clone, propagate, and express these genes in new organisms.

1.3
Properties of Ligases

Several enzymes are available that can join nucleic acid strands in a template-dependent manner. The DNA ligase most commonly used in laboratory work is derived from bacteriophage T4. T4 DNA ligase is a monomeric protein with a molecular weight of 68 kDa. It uses ATP as a cofactor and can join both DNA and RNA strands, provided that these are immediately juxtaposed by being base-paired to complementary strands, which can be either DNA or RNA. Reactions involving RNA strands proceed much less efficiently, however. Ends of double-stranded DNA molecules can be joined by the T4 DNA ligase, even when these have no single-stranded extensions that can hybridize and thus stabilize the interaction between the two ends.

The *E. coli* ligase has a molecular weight of 74 kDa. Unlike the ligases derived from bacteriophages, archaea, and eukaryotes, all of which use ATP as a cofactor, the *E. coli* ligase requires NAD as a cofactor, as is true for other bacterial ligases.

Ligases have also been isolated from thermophilic organisms such as *Thermus thermophilus, Thermus aquaticus,* and *Desulfurolobus ambivalens*. These enzymes operate at high reaction temperatures and survive conditions that denature DNA. At the optimal reaction temperature of these enzymes, oligonucleotides are less likely to interact with mismatched target sequences. Furthermore, the thermostable ligases exhibit no or little tendency to join DNA molecules with base-paired (blunt) ends or with short complementary (sticky) ends. Because of these properties, they offer increased selectivity of detection and convenience in many analytical reactions.

T4 DNA ligase can also be employed to construct nucleic acid strands that branch into two distinct 3′ end sequences by attaching the 5′ phosphate of an oligonucleotide to the 3′ hydroxyl of a nucleotide residue internal to another oligonucleotide. This requires that this second oligonucleotide is constructed such that the internal nucleotide is connected to the downstream sequence by a 2′ to 5′ phosphodiester bond, rather than the regular 3′ to 5′ bond, leaving a free 3′ hydroxyl.

In contrast to the enzymes just mentioned that all require double-stranded substrates, the T4 RNA ligase joins DNA or RNA strands that are not base-paired to a complementary polynucleotide strand. The probability of ligation is increased, however, if the two ends to be ligated are positioned near each other. This enzyme requires ATP as a cofactor.

1.4
Chemistry of Enzymatic Ligation

DNA ligases join DNA strands by using the energy of a pyrophosphate linkage in a nucleotide cofactor to create a new phosphodiester bond, connecting the two strands. This bond is formed between a 5′-phosphate group and a 3′-hydroxyl on two adjacent nucleic acid molecules. The enzyme-catalyzed reaction can be described in four steps (Fig. 2): first, a high-energy enzyme intermediate is formed when an adenylyl group from a cofactor (ATP or NAD) is coupled to the ligase while releasing pyrophosphate

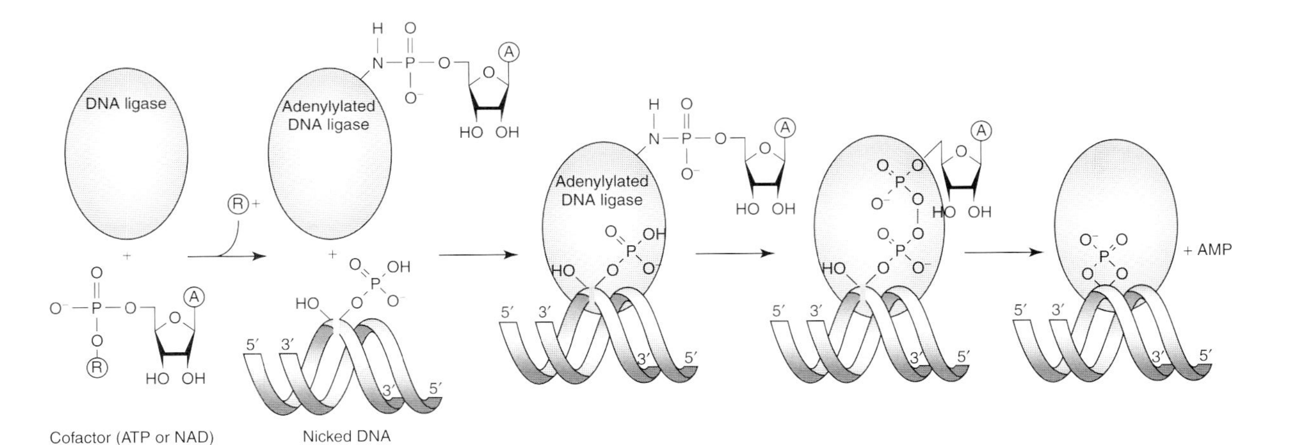

Fig. 2 The chemistry of enzymatic ligation of nucleic acids. The reaction takes place in four reversible reaction steps. In the first step, an adenylyl group from a cofactor, NAD or ATP, is transferred to the ε-amino group of a lysine residue in the DNA ligase enzyme. Thereby the energy required to join two nucleic acid strands is transferred from the cofactor and stored temporarily in the form of a high-energy bond between an adenylyl group and the enzyme. The activated ligase then binds to a nicked site on a DNA duplex. Next, the adenylyl group is transferred to the 5′-phosphate to be connected to the 3′-hydroxyl by ligation. This step recreates a pyrophosphate linkage and preserves the high-energy bond. Finally, the activated 5′ end is attacked by the nearby 3′-hydroxyl group, resulting in the formation of a phosphodiester bond between the 5′ and 3′ ends and releasing AMP. The circled A indicates an adenine group and the circled R represents either a pyrophosphate (in the case of the cofactor ATP) or nicotinamide mononucleotide monophosphate (in reactions in which NAD is a cofactor).

or NMN. The adenylylation occurs at a lysine residue in a conserved motif in the ligase. In the second step, the activated enzyme binds to a nicked site in the substrate. The adenylyl group is then transferred to the 5′-phosphate at the nick, and finally this activated phosphate is attacked by a nearby 3′-hydroxyl group, forming a phosphodiester bond between the two adjoining strands and AMP is released. The enzyme remains bound to the ligation center throughout the ligation process and recognizes the adenylylated phosphate moiety together with the nearby 3′-hydroxyl. Probably, the reactivity of the activated 5′-pyrophosphate is further increased by bridge complexation to the enzyme via a magnesium ion required in the reaction.

1.5 Substrate Requirements in Enzymatic Ligation Reactions

In contrast to the ligation of restriction enzyme–digested DNA fragments, probes in ligation assays generally hybridize stably under assay conditions. The reaction is therefore more similar to the sealing of an interrupted DNA strand in a duplex molecule. However, since the ligation reaction can trap hybridized molecules through the formation of a covalent bond, it is also possible to use short oligonucleotide probes involving just a few hybridizing bases in ligation assays. By contrast, such short probes would exhibit very limited stability in conventional hybridization-based assays.

As mentioned earlier, either RNA or DNA molecules can be used as probes in reactions catalyzed by T4 DNA ligase, although most commonly, DNA oligonucleotides are employed. Enzymatic ligation requires the presence of a phosphate group at the 5′ end of one probe molecule to be joined to a hydroxyl group at a nearby 3′ end. Ligases are very sensitive to mismatches at the nucleotides on either side of the nick to be sealed (Fig. 4). A mismatch at the 3′ end generally is more inhibitory than one at the 5′ end. Mismatches in other positions of the probe-target hybrid can also significantly reduce the efficiency of ligation. The *Tth* DNA ligase has been shown to ligate substrates with a mismatch nine nucleotides away from the nick in the 5′-direction with a lower efficiency than the corresponding matched substrate. Ligases also require a minimal length of base-paired sequences, which differs between enzymes. As an example, T7 DNA ligase is able to join a 5′ hexamer to a nonamer, while *Tth* DNA ligase is not.

1.6 Ribozyme Ligases

RNA molecules have been found that have the capacity to catalyze ligation of two RNA molecules hybridizing in juxtaposition on a complementary nucleic acid molecule. Such ribozyme ligases have been isolated from large pools of combinatorially synthesized RNA sequences by an iterative process of *in vitro* selection for the ability to ligate themselves to another RNA molecule, and amplification of ligation products. Molecules with a ligation capacity greatly increased over background levels were isolated and sequenced. Nonetheless, RNA ligases remain substantially less efficient than the protein enzymes.

1.7 Chemical Ligation

Also, chemical means can be employed to join nucleic acid strands through a

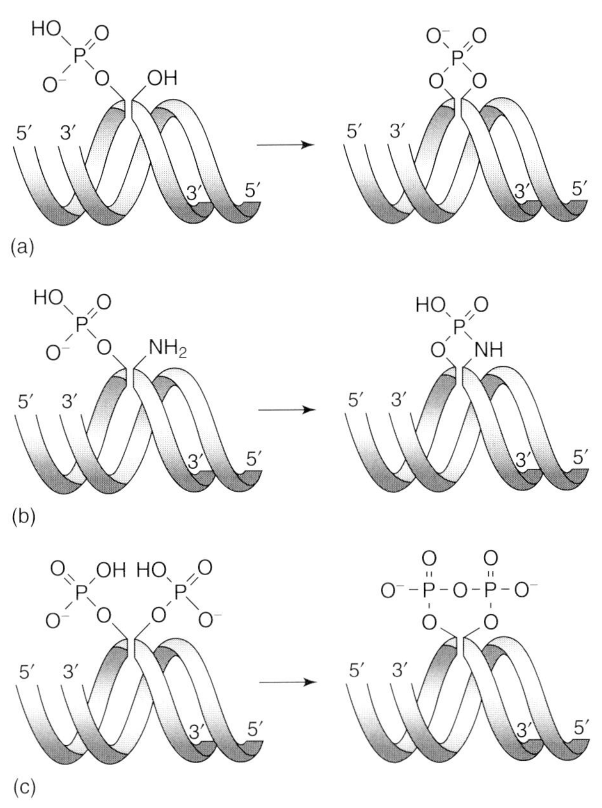

Fig. 3 Chemically induced nucleic acid ligation reactions. Using reagents such as cyanogen bromide or carbodiimide, nucleic acid molecules that are not substrates for the protein enzymes may be joined by chemical ligation. Examples of such reactions are (a) a strand with a 5′-hydroxyl group can be joined to a 3′ phosphate on a nearby strand to form a regular phosphodiester linkage; (b) a 3′ amino function may be joined to a 5′ phosphate, creating a phosphoamide linkage; and (c) two adjoining ends, both bearing phosphate groups, may be joined to form a pyrophosphate linkage.

phosphodiester linkage in a template-dependent manner. Substrate requirements in these reactions may differ from those of enzyme-catalyzed reactions (Fig. 3). For example, cyanogen bromide preferentially participates in the ligation of strands with a 5′-hydroxyl and a 3′-phosphate group, respectively, while protein ligases strictly require a 5′-phosphate group and a 3′-hydroxyl group. It is also possible to modify the ends of nucleic acid molecules so that they can be joined covalently upon hybridization to a template strand, but with no requirement for enzymes or catalysts. For example, upon hybridization of a 3′-phosphorothioate next to a 5′-iodothymidine residue, the two strands will be ligated, forming a phosphorothioate bond.

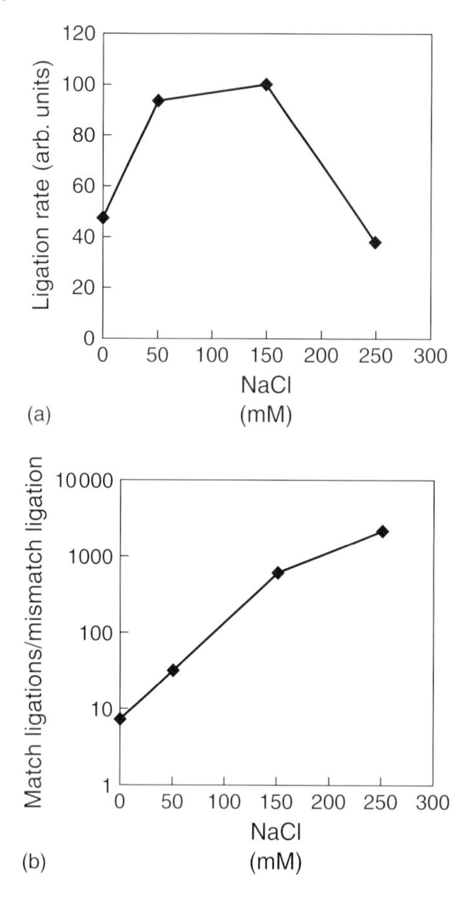

Fig. 4 Ligation at different concentrations of NaCl of two oligonucleotide probes using T4 DNA ligase under limiting conditions. (a) The rate of ligation in the presence of a template oligonucleotide fully matched to the probes. (b) The ratio of the number of ligations on a matched template to the number of ligations on a mismatched at the position opposite the 3′ ultimate nucleotide at the site of ligation.

Also, pairs of oligonucleotides that form an interrupted third strand in a triple helix by binding to the major groove of a DNA duplex may be joined chemically. Since there is no requirement for substrate recognition by an enzyme, chemical ligation is not limited to the joining of natural nucleotide residues; residues containing other sugars or bases may become covalently joined to nearby strands. Moreover, besides phosphodiester bonds, phosphoroamide, phosphorothioate, and pyrophosphate bonds may be created. It is also possible to join fragments that are shorter than what is required by ligase enzymes. Chemical ligation thus extends the repertoire of substrate molecules that may be joined by ligation.

Because of the properties of nucleic acid-ligation reactions discussed above, ligation reactions have proven useful in a number of assays for analysis of target molecules. Next, we will describe some of these assays.

2 Techniques

2.1 Oligonucleotide Joining Reactions for Detection of Nucleic Acids

Ligation of pairs of oligonucleotide probes was used by Khorana and coworkers in the early 1970s to identify target molecules. Since the end of the 1980s, this has become a popular mechanism for genetic analyses, with the demonstration that DNA sequence variants can be efficiently detected and distinguished by comparing the ligation of pairs of probes that are matched at the junction to one but not the other of two target sequence variants. Over a wide range of ligase concentrations, this oligonucleotide ligation assay (OLA) reaction is inhibited by a single mismatched base pair where the two strands to be joined abut. This mechanism is now in routine use in clinical diagnostics.

The amount of signal achieved from an OLA reaction performed on genomic DNA samples is generally not sufficient for unambiguous detection. This can be solved by generating more targets, more ligation events per target, or by amplifying the signal after ligation. More targets can be generated by performing a PCR amplification before ligation. If a thermostable ligase is used, thermal cycling may also be employed during the ligation to allow ligated pairs of probes to dissociate from their targets, giving room for new probes to hybridize and be ligated. In the ligase chain reaction (LCR), these two methods are combined. A second pair of probes, complementary to the first pair, is introduced. The ligation of the first pair of probes generates a template for the second pair and *vice versa*. In this manner, thermal cycling of the ligation reaction results in an exponential increase of ligated probes. The method is associated with a tendency to template-independent probe ligation through blunt-end ligation of the probe pairs. This problem can be avoided by using a gap-fill approach, where a gap of one or a few nucleotides first has to be filled in by the action of a polymerase before ligation can occur. The extension can be performed in an allele-specific manner by designing the 3′ ends of the probes to be extended so that they will be correctly matched to one but not the other of the target sequence variant.

A gap-fill polymerization step may also be included in the OLA test to decrease the number of probes required for an assay. In the regular OLA outlined above, the ability of one locus-specific probe to be joined to two different allele-specific probes is investigated. If the probes instead leave a gap corresponding to the variable position, then combined polymerase extension and ligation reactions can be performed in the presence of either allele-specific nucleotide triphosphate. With this method, allele distinction occurs at two levels; first the right nucleotide has to be introduced, and if a mistake is made, the ligation step will still be inhibited. Also, the number of probes that have to be constructed is decreased, at the cost of performing multiple reactions.

As mentioned, signal amplification can also be performed after ligation. This is most commonly done by introducing primer sequences in the probes so that ligation products can be amplified by PCR. Another amplification method is ligation-initiated Qβ amplification. In this reaction, the bacteriophage Qβ replicase enzyme, which has the capacity to replicate certain RNA molecules and then copy both the original strand and the replication product, is used. With this enzyme, a 10^9-fold amplification can be achieved in 30 min. To use this enzyme in a ligation assay, two RNA probes are constructed in such a way that they form a substrate for the Qβ replicase only after ligation to each other.

Several methods are available for detection of the ligated probes. For example, the members of the probe pair can be modified so that the reaction results in a detectable combination of functions in the ligation product. One of the probes may be biotinylated while the other probe carries a fluorophore, allowing ligation products to be captured on a support and detected. In another method, the remote ends of the ligation probes have complementary sequences carrying two different fluorophores. Upon ligation, the probes

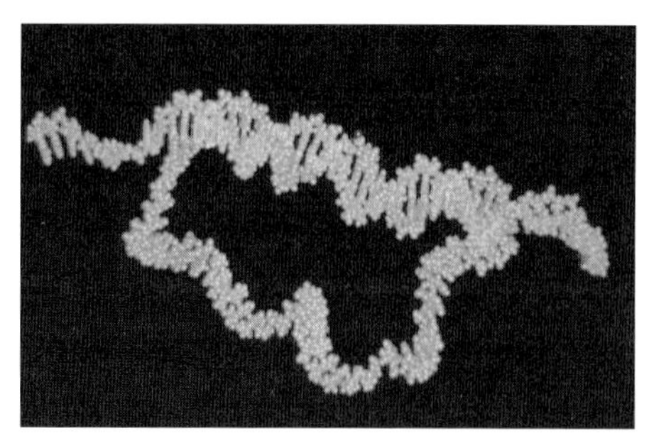

Fig. 5 A padlock probe, designed to be ligated into a circle upon interaction with the proper target sequence. The two ends of the linear probe (yellow) hybridize in juxtaposition on the target sequence (blue). The 5′-phosphate group on one end of the probe about to be joined by ligation to the 3′-hydroxyl at the opposite end is shown in red. Circularized molecules are wound around the target strand and can detach only by strand breakage or by sliding off a nearby end on the target molecule (see color plate p. xxii).

assume a hairpin-loop structure, and exhibit an altered emission spectrum due to the closeness of the two fluorophores. The method has been shown to permit analysis of point mutations in unamplified genomic DNA.

Since any nucleotide mismatches can be distinguished from perfect matches under standard reaction conditions, the oligonucleotide ligation procedure is suitable for automation and can be performed on large series of samples. The throughput of such analyses can be further increased by multiplexing – using many pairs of probes in the same reaction. The detection method must then be adjusted to separate the signals from different pairs of probes. For example, one of the probes of each pair may be size-coded to distinguish different loci by electrophoresis, while the other may be labeled with different fluorophores to identify separate alleles. The method has also been used to measure the copy number of genes. Alternatively, the locus may be represented by a hybridization sequence tag on one probe. Ligation products may be identified after hybridization to DNA microarrays with complements of the tag sequences. With this strategy, it is possible to identify large numbers of genotypes from a single reaction.

2.2
Padlock Probes

In a variant of the OLA technique, the two probes to be joined are in fact the two ends of the same linear probe molecule. Upon hybridization to the target sequence and subsequent ligation, the probe becomes circularized, wound around the target molecule in a padlock-like fashion (Fig. 5). Padlock probes may also be used in a gap-fill reaction to decrease the number of probes required for genotyping by half. If the target has no free ends, then ligated padlock probes will remain linked to the target even after denaturing washes. Thus, in solid-phase assays, unligated probes and any intermolecular ligation products may be efficiently washed away, reducing background signals. The topological link allows padlock probes to be used for localized detection of target DNA or RNA sequences via detectable moieties attached to the probe (Fig. 6).

Besides localized detection, padlock probes also offer particular advantages in multiplex assays. As more probes are combined in one reaction, the risk increases for unwanted ligation between different probes, fortuitously hybridizing next to each other somewhere in the added DNA sample. These ligation products

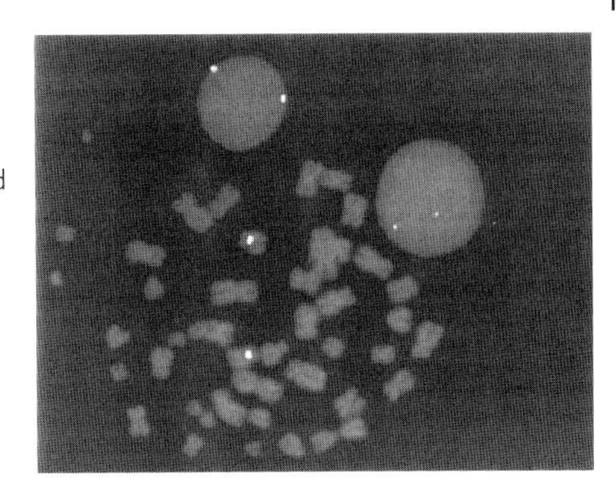

Fig. 6 *In situ* analysis of a repeated DNA sequence, located in the centromeric region of human chromosome 12, using a biotin-modified padlock probe, detected via fluorescein-conjugated streptavidin.

can be ignored, however, since properly ligated padlock probes can be conveniently distinguished from products of inter-probe ligations: Correct ligation of padlock probes results in circular molecules, while ligation of noncognate probe ends results in linear molecules that may be degraded by exonucleolytic treatment, and that fail to template rolling circle replification, as described below.

After ligation, circularized padlock probes may be amplified by PCR across the ligation junction, allowing thousands of different amplification products to be identified by hybridizing to tag microarrays. Circularized padlock probes are also suitable templates for rolling circle replication (RCR) reactions. This reaction is a linear amplification procedure that requires a single primer, and it results in long, single-stranded products that consist of many complements of the DNA circles repeated in tandem. The RCR reaction allows precise quantification and may be monitored in real time by the use of molecular beacons.

The hyper-branched rolling circle amplification (HRCA) reaction is a variant of the RCR procedure that allows a rapid isothermal amplification of circularized probes. Two primers are used; one primer complementary to the circular DNA strand and another one that hybridizes to each complement of the circle in the RCR product. Extension products from the primers along a strand catch up with each other and result in strand displacement of the downstream sequence. This process, in turn, generates new single-stranded sequences where new primers can anneal. The extension and strand displacement process yields a mixture of single- and double-stranded fragments of lengths representing multiples of the starting DNA circle. The amplification reaction is capable of billionfold amplification in a 90-min isothermal incubation. By using distinct primers for allele-specific padlock probes, the HRCA reaction has been used for genotyping.

Padlock probes have also been used to recognize specific target sequences in double-stranded DNA molecules. This can be done by first using PNA (peptide nucleic acid) probes that pry open the DNA duplexes, allowing padlock probes to be attached as the so-called ear ring probes. A similar effect has been achieved with linear probes having midsections that form triplex structures with duplex target DNA molecules. The probes become topologically linked to the target molecules when an oligonucleotide is added that allows the free ends of the probes to be joined. The mechanism has been used to couple targeting peptides or biotin molecules to plasmid molecules.

2.3 Proximity Ligation

Also, proteins and other macromolecules can be detected via DNA ligation reactions using the so-called proximity ligation method. Pairs of binding reagents, such as antibodies or DNA aptamers binding separate determinants on a specific target protein, bring into proximity the 5′ and 3′ ends of oligonucleotides attached to the binding reagents (Fig. 7). By introducing a ligation template, referred to as a connector oligonucleotide, the DNA sequence extensions of proximity probes that have bound the same target molecule can be joined, forming a DNA strand with sequence elements required for amplification, identification, and visualization. By contrast, 5′ and 3′ ends on probes that have failed to bind in pairs to a target molecule rapidly hybridize to one connector oligonucleotide each, preventing any subsequent hybridization to a common connector and ligation. The method can be performed without any washes and offers a sensitivity that is orders of magnitude greater than conventional sandwich ELISA protein detection assays. Both individual and interacting proteins can be demonstrated. The method can also be performed in solid-phase mode for further increased sensitivity of detection, and more recently a variant of the method has been shown to permit sensitive detection of the location of target proteins *in situ*.

2.4 Other Uses of Ligation Reactions

Ligation reactions have also been used to detect bending and twisting of short double-stranded DNA molecules. By studying the efficiency with which a ligase joins the ends of DNA double helices to form circles, nucleotide sequence-dependent or protein-induced bending of the stiff double-stranded DNA molecule may be estimated. In a similar fashion, binding by transcriptional factor complexes can juxtapose separate regulatory gene sequences, allowing otherwise remote DNA fragment ends to be brought together and be ligated.

3 Outlook

Large-scale SNP genotyping studies currently appear to shift from hybridization- and polymerization-based assays to assays that also include a ligation step. Gene expression profiling may well follow the same trend. Furthermore, the proximity

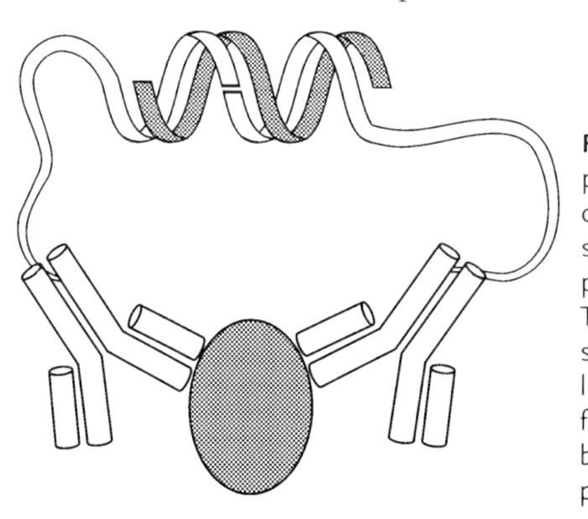

Fig. 7 Proximity ligation. Two proximity probes (white), consisting of antibodies conjugated to oligonucleotides, simultaneously bind to the same target protein molecule (dark shaded). Thereby an oligonucleotide (light shaded) and a ligase, added to template ligation of the two probes, can direct the formation of a DNA strand that can then be amplified and detected to reflect the presence of the target protein.

ligation mechanism can similarly transform large-scale protein expression and interaction studies to a form where DNA strands are created that represent the recognized proteins, and that can be amplified and detected, for example, after hybridization to tag microarrays. The unique event of a ligation reaction connecting two previously separate DNA sequences, or the two ends of individual probe molecules, can be designed to reflect the presence of target DNA, RNA, or protein molecules with excellent specificity, efficiently distinguishing closely similar target molecules. Once formed, the ligation products serve as a class of universally amplifiable, information-carrying molecules. They can therefore be detected with great sensitivity and allow very large numbers of target molecules to be investigated in parallel. In a similar manner, the probe ligation mechanism will also be important to demonstrate the location or the colocation of target molecules in biological specimens. In conclusion, ligation reactions are promising as a general means to investigate large, diverse sets of target molecules, and they may thus serve as the *molecular transistors* that will be required to standardize, miniaturize, and scale-up biological analyses for future highly resolving biological investigations in research, and in a wide range of application areas.

See also Anthology of Human Repetitive DNA; Infectious Disease Testing by LCR; Oligonucleotides.

Bibliography

Books and Reviews

Banér, J., Nilsson, M., Isaksson, A., Mendel-Hartvig, M., Antson, D.O., Landegren, U. (2001) More keys to padlock probes: mechanisms for high-throughput nucleic acid analysis, *Curr. Opin. Biotechnol.* **12**, 11–15.

Demidov, V.V. (2003) Earrings and padlocks for the double helix: topological labeling of duplex DNA, *Trends Biotechnol.* **21**, 148–151.

Doherty, A.J., Dafforn, T.R. (2000) Nick recognition by DNA ligases, *J. Mol. Biol.* **296**, 43–56.

Engler, M.J., Richardson, C.C. (1982) DNA ligases, in: Boyer, P.D. (Ed.) *The Enzymes*, Vol. 15, Academic Press, New York, pp. 3–29.

Gullberg, M., Fredriksson, S., Taussig, M., Jarvius, J., Gustafsdottir, S., Landegren, U. (2003) A sense of closeness: protein detection by proximity ligation, *Curr. Opin. Biotechnol.* **14**, 82–86.

Higgins, N.P., Cozzarelli, N.R. (1979) DNA-joining enzymes: a review, *Methods Enzymol.* **68**, 50–71.

Jarvius, J., Nilsson, M., Landegren, U. (2003) Oligonucleotide ligation assay, *Methods Mol. Biol.* **212**, 215–228.

Nilsson, M., Baner, J., Mendel-Hartvig, M., Dahl, F., Antson, D.O., Gullberg, M., Landegren, U. (2002) Making ends meet in genetic analysis using padlock probes, *Hum. Mutat.* **19**, 410–415.

Primary Literature

Abravaya, K., Carrino, J.J., Muldoon, S., Lee, H.H. (1995) Detection of point mutations with a modified ligase chain reaction (Gap-LCR), *Nucleic Acids Res.* **23**, 675–682.

Alsmadi, O.A., Bornarth, C.J., Song, W., Wisniewski, M., Du, J., Brockman, J.P., Faruqi, A.F., Hosono, S., Sun, Z., Du, Y., Wu, X., Egholm, M., Abarzua, P., Lasken, R.S., Driscoll, M.D. (2003) High accuracy genotyping directly from genomic DNA using a rolling circle amplification based assay, *BMC Genomics* **4**, 21.

Antson, D.O., Isaksson, A., Landegren, U., Nilsson, M. (2000) PCR-generated padlock probes detect single nucleotide variation in genomic DNA, *Nucleic Acids Res.* **28**, E58.

Banér, J., Isaksson, A., Waldenstrom, E., Jarvius, J., Landegren, U., Nilsson, M. (2003) Parallel gene analysis with allele-specific padlock probes and tag microarrays, *Nucleic Acids Res.* **31**, e103.

Banér, J., Nilsson, M., Mendel-Hartvig, M., Landegren, U. (1998) Signal amplification of padlock probes by rolling circle replication, *Nucleic Acids Res.* **26**, 5073–5078.

Barany, F. (1991) Genetic disease detection and DNA amplification using cloned thermostable ligase, *Proc. Natl. Acad. Sci. U.S.A.* **88**, 189–193.

Besmer, P., Miller, R.C. Jr., Caruthers, M.H., Kumar, A., Minamoto, K., Van de Sande, J.H., Sidarova, N., Khorana, H.G. (1972) Studies on polynucleotides. CXVII. Hybridization of polydeoxynucleotides with tyrosine transfer RNA sequences to the r-strand of phi80psu + 3 DNA, *J. Mol. Biol.* **72**, 503–522.

Cherepanov, A.V., de Vries, S. (2002) Dynamic mechanism of nick recognition by DNA ligase, *Eur. J. Biochem.* **269**, 5993–5999.

Crothers, D.M., Drak, J., Kahn, J.D., Levene, S.D. (1992) DNA bending, flexibility, and helical repeat by cyclization kinetics, *Methods Enzymol.* **212**, 3–29.

Dolinnaya, N.G., Tsytovich, A.V., Sergeev, V.N., Oretskaya, T.S., Shabarova, Z.A. (1991) Structural and kinetic aspects of chemical reactions in DNA duplexes. Information on DNA local structure obtained from chemical ligation data, *Nucleic Acids Res.* **19**, 3073–3080.

Ekland, E.H., Szostak, J.W., Bartel, D.P. (1995) Structurally complex and highly active RNA ligases derived from random RNA sequences, *Science* **269**, 364–370.

Fan, J.B., Chen, X., Halushka, M.K., Berno, A., Huang, X., Ryder, T., Lipshutz, R.J., Lockhart, D.J., Chakravarti, A. (2000) Parallel genotyping of human SNPs using generic high-density oligonucleotide tag arrays, *Genome Res.* **10**, 853–860.

Faruqi, A.F., Hosono, S., Driscoll, M.D., Dean, F.B., Alsmadi, O., Bandaru, R., Kumar, G., Grimwade, B., Zong, Q., Sun, Z., Du, Y., Kingsmore, S., Knott, T., Lasken, R.S. (2001) High-throughput genotyping of single nucleotide polymorphisms with rolling circle amplification, *BMC Genomics* **2**, 4.

Fredriksson, S., Gullberg, M., Jarvius, J., Olsson, C., Pietras, K., Gustafsdottir, S.M., Ostman, A., Landegren, U. (2002) Protein detection using proximity-dependent DNA ligation assays, *Nat. Biotechnol.* **20**, 473–477.

Grossman, P.D., Bloch, W., Brinson, E., Chang, C.C., Eggerding, F.A., Fung, S., Iovannisci, D.M., Woo, S., Winn-Deen, E.S., Iovannisci, D.A. (1994) High-density multiplex detection of nucleic acid sequences: oligonucleotide ligation assay and sequence-coded separation, *Nucleic Acids Res.* **22**, 4527–4534.

Hardenbol, P., Baner, J., Jain, M., Nilsson, M., Namsaraev, E.A., Karlin-Neumann, G.A., Fakhrai-Rad, H., Ronaghi, M., Willis, T.D., Landegren, U., Davis, R.W. (2003) Multiplexed genotyping with sequence-tagged molecular inversion probes, *Nat. Biotechnol.* **21**, 673–678.

Hsuih, T.C., Park, Y.N., Zaretsky, C., Wu, F., Tyagi, S., Kramer, F.R., Sperling, R., Zhang, D.Y. (1996) Novel, ligation-dependent PCR assay for detection of hepatitis C in serum, *J. Clin. Microbiol.* **34**, 501–507.

Landegren, U., Kaiser, R., Sanders, J., Hood, L. (1988) A ligase-mediated gene detection technique, *Science* **241**, 1077–1080.

Lizardi, P.M., Huang, X., Zhu, Z., Bray-Ward, P., Thomas, D.C., Ward, D.C. (1998) Mutation detection and single-molecule counting using isothermal rolling-circle amplification, *Nat. Genet.* **19**, 225–232.

Mendel-Hartvig, M., Kumar, A., Landegren, U. (2004) Ligase-mediated construction of branched DNA strands: a novel DNA joining activity catalyzed by T4 DNA ligase, *Nucleic Acids Res.* **32**,, e2.

Nilsson, M., Antson, D.O., Barbany, G., Landegren, U. (2001) RNA-templated DNA ligation for transcript analysis, *Nucleic Acids Res.* **29**, 578–581.

Nilsson, M., Barbany, G., Antson, D.O., Gertow, K., Landegren, U. (2000) Enhanced detection and distinction of RNA by enzymatic probe ligation, *Nat. Biotechnol.* **18**, 791–793.

Nilsson, M., Gullberg, M., Dahl, F., Szuhai, K., Raap, A.K. (2002) Real-time monitoring of rolling-circle amplification using a modified molecular beacon design, *Nucleic Acids Res.* **30**, e66.

Nilsson, M., Krejci, K., Koch, J., Kwiatkowski, M., Gustavsson, P., Landegren, U. (1997) Padlock probes reveal single-nucleotide differences, parent of origin and in situ distribution of centromeric sequences in human chromosomes 13 and 21, *Nat. Genet.* **16**, 252–255.

Nilsson, M., Malmgren, H., Samiotaki, M., Kwiatkowski, M., Chowdhary, B.P., Landegren, U. (1994) Padlock probes: circularizing oligonucleotides for localized DNA detection, *Science* **265**, 2085–2088.

Odell, M., Shuman, S. (1999) Footprinting of Chlorella virus DNA ligase bound at a nick in duplex DNA, *J. Biol. Chem.* **274**, 14032–14039.

Park, Y.N., Abe, K., Li, H., Hsuih, T., Thung, S.N., Zhang, D.Y. (1996) Detection of hepatitis C virus RNA using ligation-dependent polymerase chain reaction in formalin-fixed, paraffin-embedded liver tissues, *Am. J. Pathol.* **149**, 1485–1491.

Pickering, J., Bamford, A., Godbole, V., Briggs, J., Scozzafava, G., Roe, P., Wheeler, C., Ghouze, F., Cuss, S. (2002) Integration of DNA ligation and rolling circle amplification for the homogeneous, end-point detection of single nucleotide polymorphisms, *Nucleic Acids Res.* **30**, e60.

Pritchard, C.E., Southern, E.M. (1997) Effects of base mismatches on joining of short oligodeoxynucleotides by DNA ligases, *Nucleic Acids Res.* **25**, 3403–3407.

Qi, X., Bakht, S., Devos, K.M., Gale, M.D., Osbourn, A. (2001) L-RCA (ligation-rolling circle amplification): a general method for genotyping of single nucleotide polymorphisms (SNPs), *Nucleic Acids Res.* **29**, E116.

Roulon, T., Coulaud, D., Delain, E., Le Cam, E., Helene, C., Escude, C. (2002) Padlock oligonucleotides as a tool for labeling superhelical DNA, *Nucleic Acids Res.* **30**, E12.

Roulon, T., Helene, C., Escude, C. (2002) Coupling of a targeting peptide to plasmid DNA using a new type of padlock oligonucleotide, *Bioconjugate Chem.* **13**, 1134–1139.

Samiotaki, M., Kwiatkowski, M., Parik, J., Landegren, U. (1994) Dual-color detection of DNA sequence variants by ligase-mediated analysis, *Genomics* **20**, 238–242.

Schouten, J.P., McElgunn, C.J., Waaijer, R., Zwijnenburg, D., Diepvens, F., Pals, G. (2002) Relative quantification of 40 nucleic acid sequences by multiplex ligation-dependent probe amplification, *Nucleic Acids Res.* **30**, e57.

Tyagi, S., Landegren, U., Tazi, M., Lizardi, P.M., Kramer, F.R. (1996) Extremely sensitive, background-free gene detection using binary probes and beta replicase, *Proc. Natl. Acad. Sci. U.S.A.* **93**, 5395–5400.

Wabuyele, M.B., Farquar, H., Stryjewski, W., Hammer, R.P., Soper, S.A., Cheng, Y.W., Barany, F. (2003) Approaching real-time molecular diagnostics: single-pair fluorescence resonance energy transfer (spFRET) detection for the analysis of low abundant point mutations in K-ras oncogenes, *J. Am. Chem. Soc.* **125**, 6937–6945.

Weiss, B., Thompson, A., Richardson, C.C. (1968) Enzymatic breakage and joining of deoxyribonucleic acid. VII. Properties of the enzyme-adenylate intermediate in the polynucleotide ligase reaction, *J. Biol. Chem.* **243**, 4556–4563.

Wu, D.Y., Wallace, R.B. (1989) Specificity of the nick-closing activity of bacteriophage T4 DNA ligase, *Gene* **76**, 245–254.

Xu, Y., Karalkar, N.B., Kool, E.T. (2001) Nonenzymatic autoligation in direct three-color detection of RNA and DNA point mutations, *Nat. Biotechnol.* **19**, 148–152.

Yeakley, J.M., Fan, J.B., Doucet, D., Luo, L., Wickham, E., Ye, Z., Chee, M.S., Fu, X.D. (2002) Profiling alternative splicing on fiber-optic arrays, *Nat. Biotechnol.* **20**, 353–358.

Zhang, D.Y., Brandwein, M., Hsuih, T.C., Li, H. (1998) Amplification of target-specific, ligation-dependent circular probe, *Gene* **211**, 277–285.

Lipid and Lipoprotein Metabolism

Clive R. Pullinger and John P. Kane
University of California, San Francisco, CA

Encyclopedia of Molecular Cell Biology and Molecular Medicine, 2nd Edition. Volume 7
Edited by Robert A. Meyers.

ISBN: 3-527-30549-1

Keywords

Apolipoproteins
Proteins associated with the plasma lipoproteins.

Chylomicrons
A class of triglyceride-rich lipoproteins of intestinal origin, isolated by ultracentrifugation at a density of 0.93 g mL^{-1}.

HDL
High-density lipoproteins: density range 1.063 to 1.21 g mL^{-1}.

IDL
Intermediate-density lipoproteins: density range 1.006 to 1.019 g mL^{-1}.

Kringles
Named from their resemblance to Danish pretzels, these triloop peptide domains of 80 to 90 amino acid residues are held rigidly together by three internal disulfide bonds in a 1–6, 2–4, 3–5 pattern. They are found in a number of proteins including apolipoprotein (a), prothrombin, factor XII, plasminogen, tissue plasminogen activator, and urokinase.

LDL
Low-density lipoproteins: density range 1.019 to 1.063 g mL^{-1}.

VLDL
Very low-density lipoproteins, a class of triglyceride-rich lipoproteins: density range 0.93 to 1.006 g mL^{-1}.

Abbreviations

CE	Cholesteryl ester
CETP	cholesteryl-ester transfer protein

LCAT	lecithin:cholesterol acyltransferase
LPL	lipoprotein lipase
LRP	LDL-receptor-like protein
MTP	microsomal triglyceride transfer protein
FFA	free fatty acids
PLTP	phospholipid transfer protein
PPAR	peroxisome proliferator-activated receptor
SREBP	sterol regulatory element binding protein
TM	transmembrane

Lipoprotein metabolism can be subdivided into transport of exogenous lipids, transport of endogenous lipids from the liver to peripheral tissues, and reverse cholesterol transport with eventual excretion of cholesterol and its catabolic end products, bile salts, into the bile. Lipids are transported through the bloodstream as macromolecular complexes called *lipoproteins*, of which there are a number of classes that can be separated by sequential ultracentrifugation. Lipoprotein particles consist of a hydrophobic lipid core that is stabilized by a monolayer of phospholipids, free cholesterol, and apolipoproteins. Nine apolipoproteins are members of a multigene family. Eight of these are clustered at two loci: apoA-I, apoA-IV, apoA-V, and apoC-III on chromosome 11; and apoC-I, apoC-II apoC-IV, and apoE on chromosome 19. The *apoB* gene codes for two major structural proteins, one of which is produced by a novel mRNA editing mechanism.

There are an increasing number of key genes recognized that play important roles in lipid metabolism and transport. Those that code for intracellular proteins involved in cholesterol synthesis, storage, or catabolism include HMGCoA synthase (HMGCS1), HMGCoA reductase (HMGCR), acyl-CoA:cholesterol acyltransferase (ACAT1/SOAT1 and ACAT2/SOAT2), and cholesterol 7α-hydroxylase (CYP7A1). The diacylglycerol O-acyltransferases (DGAT1 and DGAT2) are important in triglyceride synthesis. The microsomal triglyceride transfer protein (MTP) is a major participant in the assembly of very low density lipoproteins (VLDL). There is a family of lipoprotein receptor genes, most notable of which is the LDL receptor. Similarly, there is a family of lipases including lipoprotein lipase (LPL), hepatic triglyceride lipase (LIPC), and endothelial lipase (LIPG). In addition to the apolipoproteins, but also associated with lipoproteins, are a number of gene products that are important for lipid transport. These include lecithin:cholesterol acyltransferase (LCAT), phospholipid transfer protein (PLTP), and cholesterol ester transfer protein (CETP). Key genes involved in reverse cholesterol transport are ATP-binding cassette A1 (ABCA1), scavenger receptor SR-BI/CLA1 (SCARB1), and caveolin (CAV1). Other members of the ABC transporter family are being increasingly seen as major players in lipid transport. For example, ABCA7 and ABCB4 are phospholipid transporters, and ABCB11 is the bile salt export protein.

Over the last decade, a number of nuclear receptors that regulate the genes mentioned above have been identified. These include the sterol regulatory element binding proteins (SREBPs), the liver X receptors (LXRα, NR1H3, and LXRβ, NR1H2), the farnesoid X receptor (FXR, NR1H4), and the peroxisome proliferator-activated receptors (PPARα, PPARγ, and PPARδ).

Recent advances in molecular cell biology and molecular medicine, involving studies of these genes and their products, have significantly increased our understanding of lipoprotein metabolism and atherosclerotic heart disease.

1 Lipid Transport and Disorders

1.1 Exogenous Transport

Exogenous transport, illustrated in Fig. 1, describes the mechanism by which large quantities of dietary triglycerides and cholesterol are moved through the blood for storage. Prior to absorption, triglycerides are hydrolyzed by lipases to fatty acids and β-monoacylglycerol. The lipids are absorbed in the small intestine with the aid of emulsifying bile acids. Triglycerides are resynthesized and much of the cholesterol is esterified in the enterocytes.

1.1.2 Chylomicron Assembly

The hydrophobic triglycerides and cholesterol esters are packaged in the smooth endoplasmic reticulum (ER) of enterocytes as large stable microemulsion particles called *chylomicrons*. The nascent chylomicrons, many over 100 nm in diameter, are exocytosed *via* the Golgi apparatus into the lymphatics, whence they migrate through the thoracic duct into the blood. In addition to a single molecule of the structural protein apoB-48 they contain, primarily, apoA-I, apoA-II, and apoA-IV. As a result of exchange with HDL they lose the A apolipoproteins and gain C apolipoproteins.

The large polypeptide subunit of microsomal triglyceride transfer protein, which has homology to vitellogenin, is an 88-kDa protein the gene for which, *MTP* (chromosome 4q24), is expressed in intestine and liver and is required for lipoprotein assembly. It can be shown, *in vitro*, to enhance the transfer of triglycerides, cholesteryl esters, and phospholipids between phospholipid membranes and to mediate the lipidation of apoB. The activity, present in the microsomal fraction, resides in a heterodimer of MTP with a 58-kDa multifunctionalprotein disulfide isomerase (PDI), which acts as a chaperone that ensures correct protein folding. The MTP promoter is regulated positively by cholesterol and negatively by insulin, the latter *via* the MAPK(erg) cascade. It contains the recognition sequences for AP-1 and for the liver-specific factors HNF-1 and HNF-4. It also contains a modified sterol response element (SRE) at −124 to −116 that binds SREBP. Negative response elements have been found for ethanol and IL-1 at −612 to −142, and −121 to −88, respectively. In a large population-based study, a functional SNP, −493 G/T, was not associated with lipoprotein levels or CAD risk. Other reports have been contradictory, showing either decreased or increased LDL cholesterol in subjects homozygous for the less common −493 T variant. The degree of impact of this SNP is thought to depend on the level of visceral

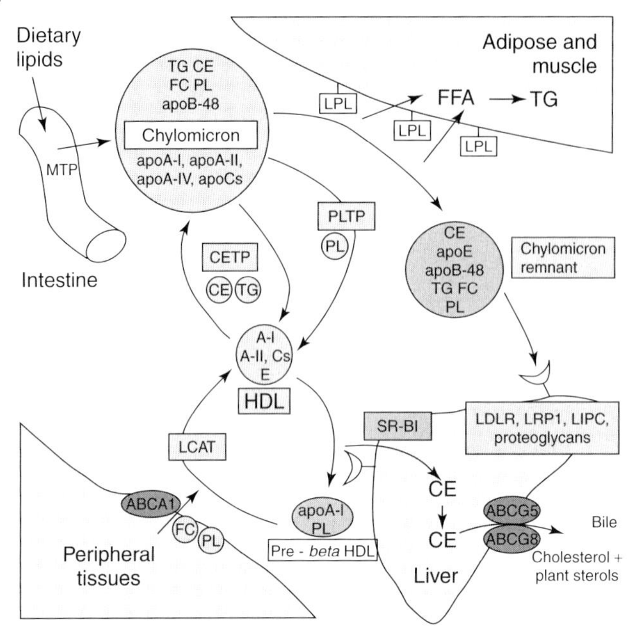

Fig. 1 Exogenous lipid transport (CE, cholesteryl ester; CETP, cholesteryl-ester transfer protein; FC, free cholesterol; LCAT, lecithin:cholesterol acyltransferase; LDLR, LDL receptor; LIPC, hepatic lipase; LPL, lipoprotein lipase; LRP1, LDL receptor-related protein 1; MTP, microsomal triglyceride transfer protein; FFA, free fatty acid; PL; phospholipid; PLTP, phospholipid transfer protein; SR-BI, scavenger receptor class B, 1; TG, Triglyceride.).

adipose tissue, and it may contribute to dyslipidemia in diabetes. The level of MTP mRNA was increased in both intestine and liver when hamsters were fed a high fat diet. The response was more rapid in the intestine.

1.1.3 Chylomicron Catabolism

1.1.4 Lipoprotein Lipase

Most of the triglyceride core of chylomicrons is hydrolyzed by the enzyme lipoprotein lipase (LPL), a triglyceride hydrolase that is synthesized at a number of sites, notably adipose tissue, skeletal muscle, heart, adrenal glands, placenta, and mammary gland. LPL, originally referred to as *clearing factor lipase*, is secreted from parenchymal cells and is transported to the vascular endothelial luminal surface where it attaches, as a noncovalent homodimer, to a high-affinity binding site. The precise nature of this interaction is unknown but is thought to involve cell membrane-anchored heparan sulfate proteoglycans such as perlecan, the glypicans, and the syndecans, and a specific heparin-sensitive binding protein. The hydrolysis of triglyceride releases free fatty acids (FFA) and β-monoacylglycerol. FFA are taken up by adipose tissue where they are re-esterified and stored as triglyceride. In other tissues, such as the heart, FFA can be oxidized as a source of energy. LPL can also facilitate the cellular binding and uptake of lipoproteins by lipoprotein receptors.

The gene for LPL (chromosomal location, 8p21.3) spans 30 kb, is comprised of 10 exons, and codes for a 475-amino acid-residue protein with a 27-residue signal peptide. Two sizes of mRNA, 3.35 and 3.75 kb, are observed. Under different physiological conditions, LPL is regulated hormonally primarily by post-transcriptional control. This control, in the case of insulin, is *via* stabilization of mRNA, and, with thyroid hormone and adrenaline, this is at the translational level. Regulatory cis-acting sequences and transcription factors responsible for basal promoter activity have been identified, as has a promoter region responsible for estrogen suppression of transcription. In adults, tissue-specific suppression of *LPL* gene expression in the liver involves the binding of transcription factor RF-2-LPL to a NF-1-like site in the region −591 to −288 of the transcription initiation site. The role of a peroxisome proliferator response element (PPRE), which binds peroxisome proliferator-activated receptors α or γ (PPARα and PPARγ) is enigmatic. Induction of PPARs with their ligands has been reported to either induce or suppress LPL transcription. A further complication is that changes in LPL mRNA do not always correlate with enzyme activity. An SRE at −90 to −81 that binds SREBP is responsible for the upregulation seen in adipose tissue as a result of cholesterol depletion. In tissues other than adipose or muscle, a positive response to sterols is mediated by an LXRα response element, a DR4 sequence in the first intron at +635 to +650.

The mature enzyme is a 421-residue, 55-kDa, N-linked glycoprotein. Ser132, Asp156 and His241 make up the active site, which is homologous to that in pancreatic, hepatic, and endothelial lipase. All these enzymes have a catalytic site pocket covered by a lid (LPL residues 216 to 239) that opens upon binding of the substrate. These structures play a vital role in determining the substrate specificity of each lipase. Unlike the other lipases in the family, LPL requires apoC-II as a cofactor. Asn43, an N-linked glycosylation site, is essential for catalysis and secretion of LPL. Sites in LPL have been identified that bind apoC-II and lipid. There are two sites rich in arginine and lysine residues, at 279 to 282 and 292 to 304, that bind to heparin.

The rare autosomal recessive disorder, familial lipoprotein lipase deficiency is characterized by diminished LPL activity and chylomicronemia. Numerous mutations in the *LPL* gene have been identified. Many are single-base substitutions; one is a 2-kb insertion and one a 6-kb deletion. Exon 5 appears to be a mutational hot spot, with numerous missense mutations that produce inactive LPL. Homozygosity for G188E or A221del, two of the most common mutations, causes chylomicronemia. Another relatively common mutant, N291S, decreases LPL activity, but is not severe enough on its own to cause chylomicronemia.

Four out of five cases of pregnancy-induced chylomicronemia were shown to be due to partial LPL deficiency caused by mutations in the *LPL* gene. In these subjects, hypertriglyceridemia became much more pronounced during pregnancy, when there is normally a much higher rate of production of VLDL.

Two SNPs in the *LPL* gene, −93 T/G and D9N, are in strong linkage disequilibrium in Caucasians and are associated with differences in the levels of plasma triglyceride. The Asn9 allele is present at a population frequency of 1.5%, with carriers having triglyceride values 25 mg dL^{-1} higher than noncarriers and a higher risk of heart disease. More common is the S447X variant. The 447X allele, with a population frequency as high as 25%, is associated with decreased triglycerides and lower risk of vascular disease.

In transgenic mice expressing a high activity of human LPL in heart, skeletal muscle, and adipose tissue, the plasma levels of VLDL triglyceride were much decreased and HDL2 levels were raised. The clearance of VLDL and the conversion of VLDL to LDL were enhanced. There was no increase in VLDL levels in these animals after sucrose feeding as seen with normal mice. When placed on a high-cholesterol diet, the animals were protected against hypercholesterolemia.

1.1.5 **Chylomicron Remnant Uptake**

The fatty acids released by LPL are taken up by the surrounding tissue, a process facilitated by fatty acid transport proteins. They are stored as triglyceride, or used as fuel. Along with free cholesterol and phospholipid, apoC-II is shed, as the chylomicron particles get progressively smaller. During this process they acquire apoE and cholesterol esters, and lose most of the other apoproteins, except apoB-48, by transfer with HDL. Ultimately, chylomicron remnant particles rich in cholesterol esters are formed. These are rapidly removed at the microvillus surface of hepatocytes by a high-affinity process. Earlier it was thought that a distinct chylomicron remnant receptor existed. Despite much evidence for its existence it eluded purification and cloning. It is now accepted that the LDL receptor can initially bind to chylomicron remnants with apoE acting as the ligand. Remnants bound to the LDL receptor move rapidly to clathrin-coated pits at the base of the microvilli and are endocytosed. Hepatic lipase, which is bound to cell-surface heparan sulfate proteoglycans, is also involved in initial binding to remnants. The space of Disse is rich in apoE, which is also bound to the proteoglycans. The remnant particles accumulate apoE, increasing their affinity for the LDL-receptor-related protein or LRP1 (also referred to as the *α_2-macroglobulin receptor*).

LRP1 (Table 1) is a large multifunctional cell-surface receptor comprised of a large 515-kDa extracellular domain and an 85-kDa membrane-spanning subunit. It is expressed in a number of tissues including liver, adrenal cortex, ovary, fibroblasts, macrophages, and some neurons. In addition to binding to α_2-macroglobulin-protease complexes LRP1 has been shown to bind apoE-enriched βVLDL, Pseudomonas exotoxin A and plasminogen activator-inhibitor complexes. The binding and uptake of these ligands is blocked by a 39-kDa receptor-associated protein (RAP) that copurifies with the receptor. LRP1 binds with high affinity to the C-terminal domain of lipoprotein lipase (residues 378 to 423 of human LPL). Knockout of LRP1 is lethal to the embryo.

Another member of the lipoprotein receptor family, LDL receptor-related protein 2 (LRP2), also referred to as *glycoprotein 330 (gp330)* (Table 1), is expressed by the epithelial cells of a number of tissues. It is found mainly as a cell-surface protein concentrated in clathrin-coated pits and is structurally related both to the LRP1 receptor and to the LDL receptor. Like LRP1, LRP2 binds to RAP, LPL, and apoE-enriched βVLDL. It also binds with high affinity to apoJ. Because LRP2 binds LPL with high affinity, like LRP1, it may be involved in the hepatic clearance of LPL-associated lipoproteins. It is mainly expressed in specialized epithelial cells, particularly in proximal renal tubules. Whatever the precise role of LRP1

Tab. 1 Six human lipoprotein receptor genes.

Gene	*Alternative name*	*HUGO symbol*	*Gene size [kb]*	*Exons*	*Chromosomal location*	*Initial translated protein amino acid residues*	*Mature protein molecular weight [kDa]*
ApoB48 receptor		APOB48R	4.3	4	16p12.1	1088	200
LDL receptor		LDLR	43	18	19p13.2	860	120[a]
LDL receptor-related protein 1	alpha2-macroglobulin receptor	LRP1	92	89	12q13.3	4544	515 and 85
LDL receptor-related protein 2	Glycoprotein 330 or megalin	LRP2	234	79	2q31.1	4655	519
LDL receptor-related protein 8	apoE receptor-2	LRP8	81	19	1p32.3	963	154
VLDL receptor		VLDLR	33	19	9p24.2	873	130

[a] Apparent size is 160 kDa on SDS gels.

and LRP2 in chylomicron remnant removal may be, evidence in mice suggests a greater role for the LDL receptor itself.

A recently recognized receptor expressed on monocytes, macrophages, and endothelial cells is the apolipoprotein B48 receptor (APOB48R), which is capable of binding apoB-48-containing triglyceride-rich lipoproteins. It is possible that this protein plays a role in the formation of foam cells in the artery wall.

1.2 Endogenous Transport

The transport of endogenous lipids, illustrated in Fig. 2, describes the movement through the circulation of lipids synthesized in the liver.

1.2.1 VLDL Assembly

The liver secretes triglyceride-rich lipoproteins analogous to chylomicrons. These nascent VLDL particles are smaller than chylomicrons but have a similar lipid composition. In humans, they contain one copy of apoB-100 as their structural protein as well as apoE and the A and C families of apoproteins. The triglyceride used for their assembly in the liver is derived from chylomicron remnants and from fatty acids released by LPL and taken up by the liver, or from fatty acids synthesized *de novo*. As with chylomicrons, MTP plays a critical role in the assembly of nascent VLDL. Unlike chylomicron secretion, which is maintained only throughout the postprandial period, VLDL production continues, albeit at a reduced rate, even during periods of starvation, when it is

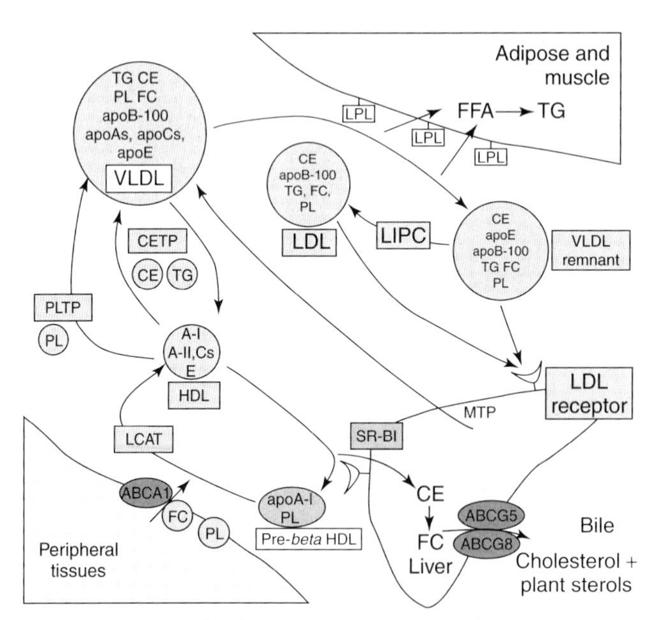

Fig. 2 Endogenous lipid transport (CE, cholesteryl ester; CETP, cholesteryl-ester transfer protein; FC, free cholesterol; LCAT, lecithin:cholesterol acyltransferase; LIPC, hepatic lipase; LPL, lipoprotein lipase; MTP, microsomal triglyceride transfer protein; FFA, free fatty acid; PL, phospholipid; PLTP, phospholipid transfer protein; SR-BI, scavenger receptor class B, 1; TG, Triglyceride.).

an important means of providing muscle fuel.

In the relatively common disorder familial combined hyperlipidemia (see Sect. 1.3.6), overproduction and secretion of VLDL particles by the liver is the cause of the increased level of circulating apoB and triglyceride. The important role of stearoyl-CoA desaturase (SCD), which catalyzes the formation of the monounsaturated fatty acids oleate and palmitoleate, in regulating the supply of triglyceride, phospholipid, and cholesteryl esters for VLDL production, has been recognized.

1.2.2 VLDL Catabolism

Nascent VLDL loses apoA-I and apoE and gains more C apoproteins by a process of exchange with HDL. The triglyceride core of the mature VLDL is hydrolyzed by LPL in a similar manner to chylomicrons, though at a slower rate.

As the hydrolysis of VLDL triglyceride proceeds, transfer with HDL restores apoE and removes most of the C proteins. The resulting VLDL remnants contain a single molecule of apoB-100 and several copies of apoE. About half the remnants are removed by the hepatic LDL receptor, which has high-affinity binding sites for apoE. The remaining remnants lose more triglyceride by transfer to HDL and by the action of hepatic lipase (LIPC), and are converted to LDL particles (Fig. 3) that are rich in cholesterol esters. During this process, the particle decreases in size to about 22 nm in diameter and the apoE and C proteins are lost, leaving apoB-100 as the sole protein.

Hepatic lipase has similar substrate specificity for triglycerides as LPL, but hydrolyses phospholipid two to three times more effectively. It is not activated by apoC-II and does not require Ca^{2+}. Like LPL, its optimum activity is at pH 8.0 to 9.0. Evidence points to a role for hepatic lipase, in addition to LPL, in the production of remnant particles that can be efficiently taken up by the liver. It contributes, *via* an interaction with extracellular proteoglycans, to the process of binding and uptake of remnants by the liver.

The gene for hepatic lipase (chromosome 15q21.3) is 35 kb in length and is composed of 9 exons. The promoter region has tissue-specific glucocorticoid and cAMP response elements. Thyroid hormone has been shown to increase the activity of hepatic lipase by a posttranslational mechanism. Mature hepatic lipase is a glycoprotein (60 kDa) of 476 residues. As a result of alternative splicing, the adrenal gland produces an isoform that is different from the liver enzyme. Four polymorphisms in the LIPC promoter, at −250, −514, −710 and −763, are in almost complete linkage disequilibrium. The rare haplotype, which has a frequency of 0.20, is associated with increased HDL levels, decreased hepatic lipase activity, and more buoyant LDL particles. Hence, LIPC variants play an important role in determining the amount of small dense LDL, which is believed to be associated with increased risk of CAD.

The VLDL receptor (Table 1), a cell-surface receptor that binds with high affinity to VLDL and cholesteryl ester-rich βVLDL, but not to LDL, was first cloned in rabbits. It is expressed in the heart, skeletal muscle, and macrophages, with less expression in lung, kidney, placenta, pancreas, and brain. In contrast to the LDL receptor, to which it shows a very high degree of homology, it is not expressed in the liver. The human VLDL receptor protein is 96% homologous to the rabbit receptor. The human gene gives rise to two forms of the receptor, one with five

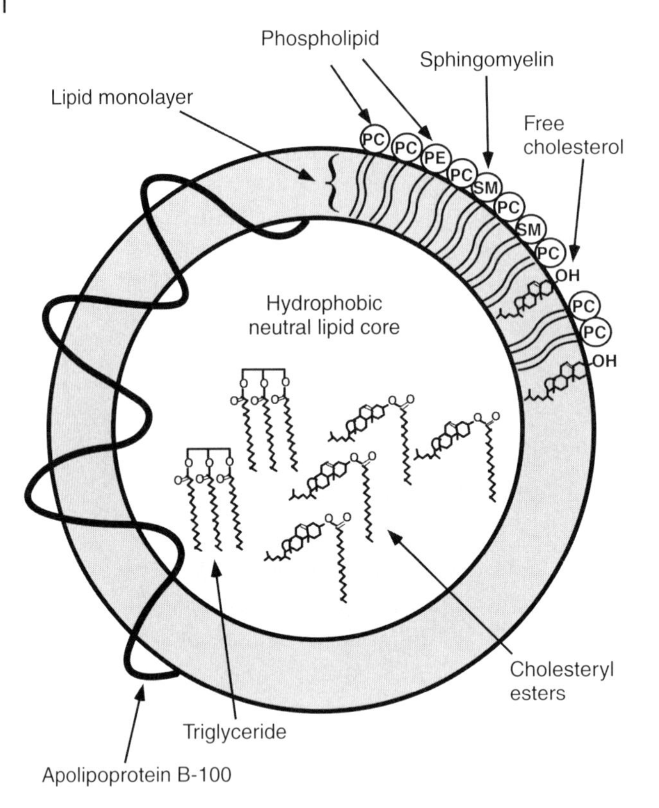

Fig. 3 Structure of low-density lipoprotein (PC, phosphatidylcholine; SM, sphingomyelin; PE, phosphatidylethanolamine.)

domains and one lacking the O-linked sugar domain. It is thought that the VLDL receptor functions in nonhepatic tissues in which fatty acids are actively metabolized. When compared to the LDL receptor gene, the VLDL receptor has an extra exon coding for a ligand binding domain. The VLDL receptor binds to triglyceride-rich lipoproteins *via* apoE, and not apoB, in concert with lipoprotein lipase (LPL), to which it also binds. Like the LDL receptor, it also binds to the receptor-associated protein (RAP). Its role in lipid metabolism has been hard to establish. Studies with VLDL receptor-deficient mice have been difficult to interpret, but it has now been shown that the VLDL receptor does indeed play a role in peripheral triglyceride hydrolysis. Another apoE receptor, homologous to both the LDL receptor and the VLDL receptor, is the low-density lipoprotein receptor-related protein 8 (LRP8) also known as the *apoE receptor-2* (APOER2). It binds with high affinity to apoE-rich βVLDL, and poorly to VLDL and LDL. LRP8 is expressed most highly in the brain and placenta. Both LRP8 and the VLDL receptor have neuronal signaling functions and play important roles during the development of the brain.

The LDL receptor is responsible for the uptake by the liver of VLDL remnants, the binding affinity of which is considerably higher than that of LDL. The ligand on VLDL is apolipoprotein E (apoE).

1.2.3 LDL Catabolism

The liver removes approximately 70% of the LDL particles circulating in plasma, mainly *via* the LDL receptor: the remainder are removed by a nonspecific low-affinity process. On LDL particles the ligand for the receptor is apolipoprotein B (apoB).

The LDL receptor is present on the surfaces of most nucleated cells. The highest concentration of mRNA is in the adrenal gland followed by the corpus luteum of the ovary and then the liver. The half-life of LDLR mRNA in liver is 30 min.

The LDL receptor transcript, a 5.3-kb mRNA, has a long 2.5-kb 3′ UTR, which contains a dinucleotide AT repeat polymorphism. The gene has three 300-bp Alu repeats that may be a recombination hot spot.

The N-terminal 292 residues of LDLR, encoded by exons 1 to 6, make up the ligand binding domain with seven cysteine-rich tandem repeats. Each repeat of approximately 40 residues has three disulfide bonds with a coordinated Ca^{2+} ion. Exons 7 to 14 encode a domain that has a strong homology with the EGF precursor. This 400-residue region, which contains a six-bladed β-propeller motif, is central for endosomal release of the lipoprotein and for receptor recycling to the cell surface. A 50-residue O-linked sugar domain is encoded by exon 15. This is followed by a membrane-spanning domain of 22 residues encoded by exon 16 and the beginning of exon 17. The end of exon 17, and part of exon 18, code for a cytoplasmic C-terminus of 50 residues. This has a signal, NPVY, which is necessary for the incorporation of the receptor into clathrin-coated pits, and subsequent endocytosis. There is another motif between residues 790 and 839, which directs the receptor to the basolateral surface of hepatocytes. The remainder of exon 18 encodes the 3′UTR. At the low pH found in endosomes, the β-propeller region displaces lipoprotein particles bound to the receptor by means of a strong interaction with ligand-binding repeats 4 and 5.

LDL receptor activity is under sterol-dependent regulation, being tightly coupled to the extracellular level of LDL, with most of the regulation at the level of gene transcription. In addition to the major means of regulation by repression of transcription by sterols, the hepatic expression of the LDL receptor is stimulated by thyroid hormone. The proximal promoter of LDL receptor contains a sterol regulatory element (SRE) flanked on either side by Sp1 sites. For a high level of expression, the active form of SREBP acts in synergy with Sp1.

1.3 Disorders of Lipid Transport

1.3.1 Abetalipoproteinemia

The MTP protein and microsomal triglyceride transfer activity are absent in intestinal biopsies from patients with abetalipoproteinemia. In several subjects, this has been shown to be due to homozygous mutations in the gene for the large subunit of MTP. Abetalipoproteinemia is a rare autosomal recessive disease, which is characterized by the absence of apoB-containing lipoproteins in the plasma because of an inability to secrete triglyceride-rich lipoproteins from intestine or liver. The failure to absorb lipids results in a deficiency of the fat-soluble vitamins A, D, E, K and β-carotene. In the absence of dietary vitamin supplementation, this results in neurological and ophthalmological complications during childhood.

1.3.2 Hepatic Lipase Deficiency

Familial hepatic lipase deficiency, a rare condition, is marked by elevated levels of VLDL remnants and by the presence of triglyceride- and phospholipid-rich LDL and HDL. Compound heterozygosity for S267F and T383M is associated with high levels of cholesteryl-ester rich βVLDL in plasma, and premature atherosclerosis.

1.3.3 Familial Hypercholesterolemia

The discovery of the LDL receptor and its subsequent cloning has led to an understanding of the molecular basis of familial hypercholesterolemia (FH). FH is an autosomal codominant disorder characterized by elevated LDL levels, the presence of cholesterol deposits in skin and tendons (xanthomas) and in arteries (atheromas), and premature coronary heart disease (CAD). Clinical disease occurs earlier in men than in women. More extensive xanthomatosis and atherosclerosis are seen in homozygotes, who generally die of myocardial infarction (MI) by age 30. FH is caused by one of several hundred of the known mutations in the LDL receptor gene. The heterozygous frequency is 1 in 500 and the homozygous frequency 1 in a million. This monogenic disorder accounts for 4% of those individuals having type IIa hyperlipidemia as designated by the World Health Organization (those with LDL levels above the 95th percentile and with no other lipoprotein abnormalities).

A number of biochemical studies first led to the recognition that the LDLR receptor defects cause FH. Radiolabeled autologous LDL was found to have a lower fractional catabolic rate (FCR) when injected into FH subjects, when compared with normal individuals. Normally the half-life of LDL is 2.9 days whereas it is 4.7 days in heterozygous patients and 6.6 days in homozygotes. The slightly altered lipid composition of LDL prepared from FH patients did not affect metabolism of the LDL particles when injected into control subjects. Deficient high-affinity specific binding or internalization of LDL was seen in fibroblasts from FH homozygous patients. Kinetic studies indicated that, in addition to lower FCR, there is an increase in the rate of LDL production in FH. This is explained by a reduced rate of clearance of VLDL remnants by the LDL receptor, and a consequential increase in the proportion converted to LDL. The receptor normally clears VLDL remnants much more rapidly than it does LDL because of its higher affinity for apoE than for apoB. In a subset of FH patients the underlying mutation affects only LDL binding, and not that of VLDL remnants, and there is a more modest rise in plasma LDL levels.

Normally the liver takes up 70% of LDL particles through the LDLR, and the remainder through a less well-understood low-affinity nonspecific mechanism. This so-called *scavenger pathway* plays a more dominant role in the removal of LDL in FH. The scavenger pathway operates in macrophages in the artery wall where the cells take up modified LDL particles and can develop into the lipid-laden foam cells that are found in atheromas. Therefore, in FH, the longer residence time of LDL, which leads to more oxidation and other modifications of the LDL particles, is a factor underlying the observed premature CAD. This is especially true of homozygous patients.

A diagnosis of FH has come to be synonymous with LDL receptor gene defects, and many individuals so diagnosed can be shown to have LDL receptor mutations. A few patients, though, do not, and defects at other loci can cause a phenocopy disorder. Where an LDL receptor mutation is known to be present, it has been occasionally observed that other loci can modulate its severity.

Five classes of LDL receptor defect have been established at the molecular level. Class 1 is null alleles caused by different types of mutations that result in the absence of protein synthesis and account for 17% of FH cases. Class 2, comprising the majority of cases (54%), is due to mutations that cause defective transport of the receptor from the ER to the Golgi apparatus, with most mutations occurring in repeats 4 and 5 of the ligand binding domain. Receptor molecules that do not bind normally even though they are found on the cell surface constitute class 3. Most of these are due to in-frame mutations that affect the EGF precursor homology or ligand binding domains. Class 4 mutations are found in the cytoplasmic domain and result in receptors that fail to endocytose due to an inability to cluster in clathrin-coated pits. Class 5 comprises another common set of mutations (22%). Here, binding and endocytosis are normal, but there is a failure to release the LDL particle in the endosomal compartment. Many of these are due to mutations in the β-propeller domain.

The *LDL* receptor gene has been targeted in mice using homologous recombination to produce an FH model. These $Ldlr^{-/-}$ mice have a twofold increase in cholesterol levels due to substantial increases in IDL and LDL, but do not develop substantial atherosclerosis. The addition of cholesterol to the diet accentuates the phenotype. The phenotype in these mice can be temporarily reversed using a recombinant LDL receptor adenovirus.

1.3.4 Familial Ligand-defective Apolipoprotein B-100

As stated above, FH accounts for only 4% of cases with elevated LDL and much effort has been made to identify other genetic loci that are involved, besides the LDL receptor. Autologous LDL, prepared from non-FH hypercholesterolemic patients, was often found to have a low FCR. It was surmised that this could be due to the presence of ligand-defective LDL, and patients were identified whose LDL showed poor binding to LDL receptors on cultured fibroblasts. Because the lipid composition of the LDL was unaffected, it was concluded that an abnormality in apoB due to a genetic defect was responsible. This autosomal codominant disorder is called *familial ligand-defective apoB* (FLDB).

After separation of the two LDL allotypes from one patient with FLDB the particles were shown to have more than 90% reduced affinity for the receptor. In three families, an apoB R3500Q mutation was shown to associate with FLDB. The R3500Q mutation, with a frequency in the general population of 1 in 600, is much more common among patients attending lipid clinics. The hypercholesterolemia seen with R3500Q heterozygotes is milder compared with that seen in FH heterozygotes, and the presence of xanthomas and premature CAD is less frequent. The few homozygotes that have been reported are more severely affected than heterozygotes. Two other FLDB mutations have been reported. The R3531C variant, which is less common than R3500Q, results in milder hypercholesterolemia, consistent with the finding that 3531C LDL particles retain 27% of normal binding affinity for the receptor. FH/FLDB patients with compound heterozygosity for apoB R3531C and LDL receptor P663L are more severely affected than those with the FH mutation alone. The other known mutation, R3500W, is similar in its effects to the R3500Q disorder.

An interesting mechanism has been proposed to explain why the LDL particles are defective in FLDB. The topological arrangement of the apoB polypeptide on the LDL surface, mapped using monoclonal antibodies, displays a "ribbon and bow" structure. Mutations in the region around residues 3500 to 3531 cause conformational changes that disrupt the topology and mask the receptor binding site. A direct interaction between W4369 and R3500 probably helps to preserve the bow arrangement.

1.3.5 Familial Hypobetalipoproteinemia

Individuals with familial hypobetalipoproteinemia (FHB) have apoB mutations that result in the production of C-terminally truncated variants. The length of the truncated protein is monotonically related to the diameter and buoyant density of the lipoprotein formed. Numerous such mutations have been reported, many of which are small deletions. Four mutations that were predicted, B-2, B-9, B-25 and B-29 proteins, were not detected in plasma. B-31 was found only in the HDL density interval, whereas, B-37 was found in HDL, LDL, and VLDL intervals. Those longer than B-46 were found mainly in LDL and VLDL. Three mutants, B-75, B-87 and B-89, had increased binding to the LDL receptor. In one especially interesting case of FHB, a frameshift mutation created a sequence of eight consecutive adenines, caused by a single base pair deletion in exon 26 that predicted the formation of a truncated apoB species. However, this allele was shown to produce full-length apoB-100 in addition to the truncated protein. This was due to an additional adenine being inserted in a portion of the mRNA molecules during transcription, leading to restoration of the correct reading frame.

The phenotype associated with FHB is variable. Simple heterozygotes have low levels of LDL cholesterol but are usually clinically asymptomatic. At its most severe, with the presence of 2 null alleles, FHB is similar to the recessive disorder abetalipoproteinemia (see Sect. 1.3.1 above). Mouse models for FHB have been produced by targeting the *apoB* gene using homologous recombination.

1.3.6 Familial Combined Hyperlipidemia

An independent phenotype termed *familial combined hyperlipidemia* (FCH) was originally described, more than 30 years ago, in families of patients who had survived heart attacks. FCH was originally thought to be a dominant disorder, but more recently it has been shown to obey a multigenic mode of inheritance. It displays high penetrance and is present at a frequency of 1 to 2% in the general population. It is thought to play a role in 15% of cases of premature CAD. Any of three phenotypes can occur, and in any one individual the phenotype can vary over time. Because the genetic basis of FCH is not well understood, it is still identified by phenotype. Elevation of either VLDL or LDL level is seen in some subjects. However, levels of both these lipoproteins are elevated in most individuals with FCH. LDL cholesterol levels are almost always above 100 mg dL^{-1} and IDL is usually raised. High levels of apoB are a hallmark of this disease. Patients also tend to have HDL and LDL particles that are smaller, denser, and more atherogenic. There is much evidence that overproduction of VLDL, by about two-fold, is responsible for the observed dyslipidemia. Such overproduction is thought to be due, in part, to an increase in flux of FFA to the liver. Obesity and insulin resistance, with elevated FFA, are often seen in cases of FCH. The molecular causes of FCH have proved difficult to decipher. Individual linkage studies of FCH families have identified genetic loci with high lod scores, but when data were combined from several studies the associations were considerably diminished, suggesting that FCH is caused by defects at alternate loci and may be multigenic. Thus, FCH is a genetically heterogeneous disorder, and there is evidence to support the involvement of several candidate genes. It has been suggested that the lipoprotein lipase gene plays a modifier role. A number of individuals with FCH have low LPL activity. A major candidate for FCH has been the ApoA-I/ApoC-III/ApoA-IV/ApoA-V locus at chromosome 11q23.3. Some, but not all, linkage studies have identified this locus. Other gene candidates are LCAT (see Sect. 2.3), CETP (see Sect. 2.5), manganese superoxide dismutase (an enzyme involved in lipid oxidation), intestinal fatty acid–binding protein 2 (FABP2), and tumor necrosis factor receptor superfamily, member 1B (TNFRSF1B).

1.3.7 **Autosomal Recessive Hypercholesterolemia**

The molecular basis of the rare disorder, autosomal recessive hypercholesterolemia (ARH), has recently been established. Defects in the LDL receptor and *apoB* genes were ruled out by cosegregation studies. ARH was mapped to chromosome 1p35–p36. This disease was originally referred to as *pseudo-homozygous FH*. The mean plasma cholesterol level of 29 patients was 561 mg dL^{-1}. The probands in two Sardinian families with this rare disorder had clinical symptoms that were analogous to those seen in the homozygous form of FH. One striking difference between ARH and FH was that fibroblasts from ARH patients bound and endocytosed LDL normally even though kinetic studies had shown decreased FCR for LDL. However, the degradation of LDL by monocyte-derived macrophages and EBV-lymphocytes isolated from ARH patients was defective. Normal processing of LDL was seen when the lymphocytes were transfected with the normal *ARH* gene, which codes for a receptor adapter protein. The ARH protein contains a phosphotyrosine binding domain (PTB). These domains bind to NPXY sequences found in the cytoplasmic tails of members of the LDL receptor family. The NPXY sequence targets receptors to clathrin-coated pits. One FH mutation, Y807C (in the *LDLR* gene) which results in defective internalization of the LDL receptor, disrupts the NPXY consensus sequence. Besides binding to the LDL receptor, the ARH protein binds to the receptor adapter protein AP-2, and the clathrin heavy chain. In some signaling proteins, PTB domains are essential for signal transduction. The ARH PTB domain is similar to that found in the two endocytic proteins, numb and Disabled-2 (Dab-2). It is likely that these two proteins, which are expressed in fibroblasts, can substitute for the ARH protein in those cells.

1.3.8 **Autosomal Dominant Hypercholesterolemia**

Besides FH and FLDB a third type of autosomal dominant hypercholesterolemia has been reported. In a large kindred with hypercholesterolemia, in which the LDL receptor and *ApoB* genes had been excluded, linkage was found to chromosome 1q32. The gene responsible for the disorder is *PCSK9*, which codes for the proprotein convertase subtilisin/kexin type 9.

This protein is mainly expressed in liver, kidney, and intestine. The missense mutations detected were thought to cause a gain of function. Its overexpression in mice results in a doubling of total serum cholesterol with no change in HDL. The related protease, PCSK8, cleaves and activates SREBP transcription factors.

1.3.9 Chylomicron Retention Disease and Anderson Disease

An inherited disorder of severe fat malabsorption was reported in 1961 and termed *Anderson disease*. A similar disorder, chylomicron retention disease (CMRD) was reported later, and differed from Anderson disease on the basis of partitioning between membrane-bound and cytosolic compartments of enterocytes, which accumulate large amounts of lipid in both cases. A variant of CMRD is associated with Marinesco–Sjogren syndrome, a neuromuscular disorder. All these diseases are characterized by failure to thrive in infancy, fat-soluble vitamin deficiency, low plasma cholesterol, and the absence of circulating chylomicrons. Hepatic secretion of VLDL appears to be normal. Recently, a coat protein complex, COPII, which facilitates transport from ER to Golgi, has been shown to be required for VLDL and chylomicron secretion. Mutations in the gene *SARA2*, which codes for the protein Sar1B, a GTPase component of COPII, have been shown to underlie Anderson disease and the two forms of CMRD.

1.3.10 Sitosterolemia

Sitosterolemia, also referred to as *phytosterolemia*, is a monogenic disorder that causes elevated levels of LDL. This rare autosomal recessive disease was first described over thirty years ago, and is characterized by highly elevated plasma levels of plant sterols, notably β-sitosterol and campesterol. By 1998, there were only 45 known cases of sitosterolemia. Plant sterols also accumulate in tissues, and are found in atherosclerotic plaque and in xanthomas, where they account for 20% of the sterol content. A characteristic of sitosterolemia is that during the first decade of life patients develop both tendon and tuberous xanthomas. Early CAD and MI are often associated with this disease.

In patients with sitosterolemia, although the absorption of cholesterol is only slightly higher than normal, that of plant sterols is dramatically elevated, from a normal 5% to between 15 and 63%. The underlying molecular basis for sitosterolemia remained a conundrum for decades. Both decreased hepatic *de novo* synthesis of cholesterol and HMGCoA reductase activity were seen. Conversion of cholesterol to bile acids was normal even though cholesterol 7α-hydroxylase activity was lower. There have been contradictory reports as to whether the cholesterol content of liver is affected. Considering that the increased absorption of sterols is insufficient to explain the levels of phytosterols in plasma, an important observation was that overall sterol excretion, particularly of plant sterols, into bile was low.

It has now been determined that mutations in one or the other of the ABC transporter genes *ABCG5* and *ABCG8* cause sitosterolemia. Mutations in the *ABCG5* gene were detected in only some of the patients. An inspection of the *ABCG5* locus at chromosome 2p21 revealed an additional ABC transporter, *ABCG8*. Mutations in the *ABCG8* gene account for the other cases of sitosterolemia. Both *ABCG5* and *ABCG8* are half transporters, each gene has 13 exons and codes for proteins (sterolin-1 and sterolin-2) each with one nucleotide binding fold (NBF) and a single transmembrane domain (TM) comprised of 6 membrane-spanning α-helices. The genes are orientated in a head-to-head arrangement with the transcription start sites 140-bp apart. *ABCG5* and *ABCG8* interact to form a heterodimer in the ER, constituting a functional complex. In all cases, homozygous or compound heterozygous mutations were seen in either *ABCG5* or *ABCG8*. The presence of a single normal copy of each gene rules out the disease.

It seems that *ABCG5* and *ABCG8* have evolved as sterol transporters, which are expressed to the highest degree in liver and small intestine. When coexpressed they are found with apical plasma membrane markers. It is thought that an ER retention motif localizes these proteins in the ER until heterodimerization occurs, the motif is masked, and transportation to the Golgi proceeds. When these genes were overexpressed in mice, the animals had a decreased rate of net absorption of dietary cholesterol and an increased rate of excretion of sterol into bile. Contrary to what is seen in sitosterolemia, hepatic cholesterol synthesis was increased in these mice. With *ABCG5/ABCG8* knockout mice there was a higher fractional absorption of plant sterol from the diet and a high concentration of sitosterol in plasma.

2
Reverse Cholesterol Transport

The hepato-biliary system is the major means by which cholesterol is excreted. Whereas most of the bile salts and free cholesterol in bile are reabsorbed by the ileum and returned to the liver, a fraction is lost in the feces. Some of this is derived from cholesterol on remnant particles or from hepatic synthesis of cholesterol. The rest is from cholesterol returned to the liver from peripheral tissues by an active, specific mechanism. This latter process is termed *reverse cholesterol transport*. This is initiated in the extravascular compartment in cells of extrahepatic tissues, where free cholesterol effluxes with the aid of ATP-binding cassette transporter A1 (ABCA1) and the class B scavenger receptor SR-BI (see Sect. 7.1). In the case of ABCA1, especially in macrophages, an extracellular primary acceptor, pre-*beta*-1 HDL, comprised of phospholipid and two copies of apoA-I, takes up cholesterol. Pre-*beta*-1 HDL has an apparent molecular weight of 67 kDa. Some evidence also indicates that a lipoprotein particle with gamma electrophoretic mobility, containing only apoE, may also play a role. SR-BI facilitates the diffusion of free cholesterol between the plasma membrane and *alpha* HDL particles. Of the total efflux of cholesterol from peripheral tissues, possibly a greater amount is transferred by the SR-BI facilitated process than by the ABCA1 mediated one.

The cholesterol transferred to pre-*beta* HDL and *alpha* HDL is now a substrate for LCAT. The resulting cholesterol esters (CE) accumulate to form a hydrophobic core of larger pre-*beta* HDL particles and ultimately appear in large mature spherical HDL particles of alpha mobility. This CE can be taken up directly by liver *via* the scavenger receptor SR-BI, or transferred to LDL, VLDL, VLDL remnants, or chylomicron remnants by a process that is mediated by CETP.

Several subspecies of HDL are revealed by electrophoresis in nondenaturing gels and by isoelectric focusing. Some of these contain apoA-I but not apoA-II, and are referred to as *LpAI*. Others contain both of these proteins and are referred to as *LpAI/AII*. LpAI and LpAI/AII play metabolically different roles, especially with regard to reverse cholesterol transport. Both are able to promote the efflux of cholesterol from the plasma membrane, but LpAI appears to mediate the translocation of intracellular cholesterol to the cell surface. LpAI/AII is the preferred substrate for hepatic lipase. The level of LpAI in plasma is a better measure of the protective antiatherogenic potential of HDL than LpAI/AII is.

2.1 ATP-Binding Cassette Transporter A1

Phospholipids can move readily ("flip-flop") across the bilayer of certain membranes such as those of the endoplasmic reticulum. However, in the plasma membrane (and also late Golgi, endosomal and bile canalicular membranes) such passive equilibration is restricted and requires an ATP-dependent process involving members of a superfamily of ATP-binding cassette transporter proteins. Unlike the half transporters, ABCG5 and ABCG8, referred to above (Sect. 1.3.10), ABCA1 is a full transporter with 2 TM and 2 NBF domains. ABCA1 is highly expressed in liver where it is involved in the secretion of nascent HDL. In macrophages, particularly in the artery wall, ABCA1 is thought to be important for the removal of a toxic pool of cholesterol, thereby preventing apoptosis.

2.1.1 Tangier Disease

Tangier disease is a rare autosomal disorder, thought originally to be recessive, but now accepted to be codominant, and is characterized by the almost complete absence of HDL. What little HDL is present is pre-*beta* HDL. CE accumulates in many tissues, notably the intestine and reticuloendothelial system, with the accumulation of orange-pigmented lipids in the tonsils and rectal mucosa. Peripheral neuropathies and a syringomyelia-like disorder accompany the disease, together with premature CAD. Fibroblasts cultured from affected patients showed very low efflux of cholesterol to acceptor apoA-I compared to normal cells. Efflux of cholesterol to HDL was little affected while that of phospholipid was greatly decreased. After numerous candidate genes were ruled out, linkage and subsequent studies confirmed that homozygous or compound heterozygous mutations in *ABCA1* underlie all cases of Tangier disease. Heterozygous *ABCA1* mutations account for some cases of primary hypoalphalipoproteinemia, though these are probably less than 10% of cases. The ABCA1 defects that cause Tangier disease appear to result primarily in impaired phospholipid efflux, which leads to secondary impairment of cholesterol efflux.

2.2 Caveolin

All cells actively acquire and efflux cholesterol to the retrieval pathway, to maintain homeostasis in plasma membranes. The process involves movement of cholesterol to the plasma membrane and is probably facilitated by the caveolin proteins coded by three caveolin genes, *CAV1, CAV2,* and *CAV3*. There are two isoforms of caveolin-1, *alpha* and *beta*, because there are two transcription start sites. The human *CAV1* and *CAV2* genes each have 3 exons and are located 18-kb apart at chromosome 7q31.2. The caveolin family of scaffold proteins, notably the 22-kDa caveolin-1, is found in caveolae, which are clathrin-free invaginations of the plasma membrane, rich in free cholesterol and sphingolipids.

Multiple caveolin molecules form highly ordered assemblies within caveolae. Concentrated in the caveolae are components of lipid signaling pathways, transmembrane receptor kinases, protein kinase C, protein kinase A, adenyl cyclase, and MAP signaling intermediates. In addition, the scavenger receptors SR-BI (SCARB1) and CD36 (SCARB3) are also located in caveolae. The caveolin polypeptides, which bind free cholesterol, are embedded in the plasma membrane with both the N- and C-termini in the cytoplasm. Short, approximately 20-residue, motifs in the caveolins bind to and recruit free cholesterol and proteins.

2.3
Lecithin Cholesterol Acyl Transferase

The enzyme LCAT rapidly esterifies the free cholesterol that is transferred to pre-*beta*-1 HDL from plasma membranes. The cholesteryl esters are transferred to HDL with *alpha* electrophoretic mobility. HDL are also acceptors of free cholesterol from other lipoproteins, notably LDL, and this transfer may be LCAT dependent.

The *LCAT* gene (chromosomal location 16q22.1) has six exons. Mature LCAT, secreted by the liver, is a 60-kDa, 416-residue glycoprotein with a hydrophilic C-terminus. There are regions with homology to apoE, LPL, and hepatic lipase (LIPC). Cofactor activation of LCAT has been attributed to apoA-I and apoA-IV.

A number of mutations causing LCAT deficiency have been reported, which result in decreased levels of HDL- and LDL-cholesterol esters. There is also abnormal architecture of virtually all lipoprotein species. These mutations are dispersed in various regions of LCAT, indicating that there are several important domains. Two mutations have been shown to cause cases of Fish-eye disease, which is associated with markedly reduced levels of HDL cholesterol. In one of these cases (T123I), the mutant protein has been shown to be capable of esterifying LDL cholesterol but not HDL cholesterol.

Studies using *in vitro* mutagenesis have shown that, of the four potential N-glycosylation sites, Asn84 is important for full enzymatic activity, though not for intracellular processing, and Asn272 is essential for secretion. The LCAT variant R158C appears to be a natural polymorphism.

2.4
Phospholipid Transfer Protein

Although the liver secretes nascent discoidal HDL, HDL particles are also formed in plasma, with the apolipoproteins and phospholipids being derived from triglyceride-rich lipoproteins (chylomicrons and VLDL). The supply of phospholipid for the formation of HDL particles is mediated by PLTP. Studies on transgenic animals have provided evidence that pre-*beta*-1HDL particles are generated by the activities of PLTP. PLTP-knockout mice have reduced levels of HDL, owing to decreased production and increased catabolism. As well as being important for pre-*beta*-1-HDL formation, PLTP plays an important role in remodeling larger HDL species, by converting smaller HDL_3 particles to larger HDL_2 particles. In addition to PLTP, hepatic lipase too can metabolize HDL_2 to particles of smaller diameter, and plays a role in pre-*beta*-HDL production.

While PLTP is expressed in many tissues, the source for most of the circulating protein is liver and adipose tissue. There is evidence that the *PLTP* gene is up-regulated at the transcriptional level by PPARα, LXR, and possibly by FXR. PLTP is a highly glycosylated protein and has an apparent molecular mass of 81 kDa, with 476 amino acid residues. The human gene (13.3 kb) at chromosome 20q13.12 is composed of 16 exons and is organized in a manner similar to that of the gene for CETP. These two genes appear to have evolved from a common ancestor. PLTP has homology with (in addition to CETP) lipopolysaccharide-binding protein and neutrophil-bactericidal-permeability-increasing protein. PLTP is nonspecific in its transfer capabilities, being able to transfer most phospholipids in addition to diacylglycerol. It has also been shown to promote the exchange transfer of *alpha*-tocopherol (vitamin E) among lipoproteins.

2.5
Cholesteryl Ester Transfer Protein

CETP is responsible for the transfer of the cholesteryl esters formed by LCAT from HDL to the apoB-containing lipoproteins and facilitates the reverse transfer of triglycerides. The HDL triglycerides are subsequently hydrolyzed by hepatic lipase. CETP is an important factor in determining the ultimate size profile of LDL particles. It is a 70-kDa hydrophobic glycoprotein. Most of the circulating CETP is associated with the HDL fraction of plasma. The human gene, on chromosome 16q13, comprises 16 exons spanning 25 kb and codes for a 493-residue prepeptide. It is expressed in the liver, small intestine, adrenal glands, spleen, and adipose tissue. There is evidence for the presence of an alternatively spliced CETP mRNA species, lacking exon 9, in certain tissues and for only limited secretion of the corresponding protein.

Individuals with homozygous familial CETP deficiency have high levels of large, apoE-rich HDL particles and low levels of LDL cholesterol, but do not suffer from premature atherosclerosis. These subjects have markedly decreased catabolism of apoA-I and apoA-II. Two particular mutant alleles, D442G and a splicing defect in intron 14, account for 10% of the variance of HDL in the Japanese population. These alleles are relatively common, with heterozygosities of 7 and 2% respectively. Analysis of the CETP gene promoter revealed eight major haplotypes among Japanese subjects, with one being associated with lower CETP levels and higher HDL levels in plasma.

High-level expression of CETP in transgenic mice (a species in which plasma CETP is normally absent) is associated with an increased rate of catabolism of apoA-I. This leads to a decrease, both in the level of HDL cholesterol and in HDL particle size. These animals develop atherosclerosis more easily than control mice.

In addition to the transfer of HDL cholesterol esters by CETP to other lipoproteins and subsequent uptake by the liver, there are two other minor routes for cholesterol ester clearance. Firstly, apoE-rich HDL particles are endocytosed by hepatic receptors that recognize apoE, and secondly there is selective hepatic uptake of HDL cholesterol esters, but not apoA-I, by SR-BI, a process involving hepatic lipase.

2.6
Endothelial Lipase

Endothelial lipase (LIPG) is a recently recognized member of the lipase gene family, which includes lipoprotein lipase (LPL), hepatic lipase (LIPC), pancreatic lipase (PNLIP), phosphatidylserine-specific phospholipase A1 (PLA1A), and pancreatic lipase-related proteins 1 and 2 (PNLIPRP1 and PNLIPRP2). It has 44 and

41% homology with LPL and hepatic lipase, respectively. In particular, the active site catalytic triad, Ser169, Asp193 and His274, and heparin binding sites are conserved. A notable exception is the lid region responsible for substrate recognition. Unlike these two lipases it does not hydrolyze triglycerides; rather, its preferred substrate is phosphatidylcholine releasing the sn-1 acyl chain. It is expressed in a number of tissues, notably liver, thyroid, smooth muscle, bronchial epithelial cells, macrophages, and placenta. The human gene is at chromosome 18q21.1, and is 31 kb-long with 10 exons. The precise physiological function of endothelial lipase is not clear at present. However, evidence points to a role in HDL metabolism. Endothelial lipase knockout mice (*Lipg*−/−), and mice treated with an antiendothelial lipase antibody, had increased levels of HDL. The overexpression of this lipase in mice led to decreased HDL. Endothelial lipase hydrolyzes the main lipid constituent of HDL, phosphatidylcholine. It is not clear whether this occurs before or after binding of HDL to its receptor, SR-BI, that is, before or after much of the cholesterol content of the HDL particle has been transferred.

3 Cholesterol Metabolism

3.1 Synthesis

Cholesterol is required primarily to maintain the structure and characteristics of cellular membranes. It is also required as a precursor for steroid hormones and bile acids. Much of the cholesterol in cells is obtained from circulating LDL by uptake through the LDL receptor, and by a nonspecific pathway. However, most tissues, in particular, the liver, can synthesize cholesterol.

The isoprenoid biosynthetic pathway, of which cholesterol is the end product, can also supply cells with other compounds such as dolichol and ubiquinone. An isoprenoid intermediate in the cholesterol synthetic pathway, 15-carbon farnesyl pyrophosphate, and the related isoprenoid 20-carbon geranylgeranyl pyrophosphate, are used for the posttranslational prenylation of certain proteins (notably G proteins) at cysteine residues through thioether bonds. Prenylation, essential for membrane localization of a number of proteins, is catalyzed by several prenyltransferases. The gene products that are prenylated include many that are involved in signal transduction pathways. Examples are the ras, Rho, and Rab GTPases, Lamin A and B, and phosphorylase kinases. Cholesterol itself is covalently attached to the N-terminus of Shh, the product of the sonic hedgehog homolog gene (*SHH*) *via* a catalytic activity residing in the C-terminus. The Shh-cholesterol adduct is thereby restricted to zones within the developing cells in which Shh acts.

3.1.1 HMGCoA Synthase and Reductase

The initial precursor for cholesterogenesis is acetyl CoA. After conversion to acetoacetyl CoA, two key regulatory enzymes, hydroxymethylglutaryl-coenzymeA (HMG-CoA) synthase and HMG-CoA reductase are responsible, sequentially, for the formation of mevalonic acid. This is the substrate for the formation, *via* geranyl pyrophosphate and farnesyl pyrophosphate, of squalene. Cyclization of squalene produces the first sterol, lanosterol. This is modified further by a series of enzymes forming cholesterol. The enzymatic reduction of

HMG-CoA to mevalonic acid, which utilizes 2 molecules of NADPH, represents the first and rate-determining step in isoprenoid and cholesterol biosynthesis.

The genes for HMG-CoA synthase, HMG-CoA reductase, farnesyl pyrophosphate synthase, squalene synthase (all regulatory enzymes), and the LDL receptor are coordinately regulated by the amount of cholesterol available to the cell. As well as having unique enhancers, they share common cis-acting sterol regulatory elements (SREs) in their promoters.

The gene for HMG-CoA synthase (HMGCS1) on chromosome 5p12 spans 23 kb and has 9 exons, and the gene for HMG-CoA reductase (HMGCR) on chromosome 5q13.3 spans 24 kb and has 19 exons. Seven membrane-spanning segments, in the 339-residue N-terminus, that anchor the 97-kDa HMG-CoA reductase protein to the smooth ER, act as a sterol-sensing domain, which controls the regulation of enzyme degradation. The catalytic site is located in the 548-residue C-terminus, which projects into the cytosol.

Regulation of both enzymes is *via* negative feedback control by sterol and nonsterol products of mevalonic acid. In addition to transcriptional and translational regulation, HMG-CoA reductase is regulated posttranslationally by phosphorylation of the catalytic domain with AMP-activated protein kinase. This latter mode of cross regulation, which is independent of the feedback control, coordinates isoprenoid synthesis with cellular energy balance.

Several inhibitors of HMG-CoA reductase, namely, the statin class of drugs, cause hepatic induction of LDL receptor expression. This leads to increased clearance of VLDL remnants and LDL, which results in a highly effective lowering of plasma levels of LDL cholesterol. What seemed to be evidence for the existence of an additional gene, coding for a second HMG-CoA reductase isoform expressed in peroxisomes, has been discounted. It is now thought to be merely the result of alternative targeting of the ER enzyme.

3.1.2 Smith–Lemli–Opitz Syndrome

First described in 1964, Smith–Lemli–Opitz Syndrome (SLOS) is a rare developmental disorder characterized by high levels of 7-dehydrocholesterol and low levels of cholesterol in plasma. It is caused by a defect in the gene for the cholesterol biosynthetic pathway enzyme, 7-dehydrocholesterol reductase (DHCR7). This enzyme has a TM sterol-sensing domain similar to that found in HMG-CoA reductase, SREBP cleavage activation protein (SCAP), the Niemann–Pick type C1 protein (NPC-1) and Patched (the Shh receptor). Patients have multiple congenital malformations that vary in clinical severity. The lack of sufficient cholesterol interferes with Shh signaling which is crucial for normal embryogenesis. Congenital disorders less common than SLOS, affecting other sterol modification enzymes, have also been reported.

3.1.3 Acyl CoA: Cholesterol Acyltransferases

Acyl-CoA: cholesterol acyltransferase (ACAT) catalyzes the long-chain fatty acyl esterification of cholesterol and is crucial for the regulation of the level of free cholesterol within cells. Cholesterol esters are used for chylomicron and VLDL production. In steroidogenic tissues, such as the adrenal gland, stored CE is used as the precursor for steroid hormone. Free cholesterol released after uptake of LDL is esterified by ACAT and stored in cytoplasmic lipid droplets. This process is important in the macrophages and smooth muscle cells of the artery wall where it can lead to the formation of CE-loaded foam cells, features of early atherosclerotic lesions. ACAT activity in liver is regulated by the availability of unesterified cholesterol. Hepatic ACAT activity helps to maintain the free cholesterol concentration within the liver and plays a role in regulating the secretion of cholesterol, as a constituent of VLDL, into plasma. Hence, ACAT is involved in the regulation of apoB secretion from the liver. It also plays a role in regulating the efflux of free cholesterol into bile. ACAT activity in the liver is decreased in patients with cholesterol gallstones, and this decrease contributes to the increased availability of free cholesterol for bile secretion.

ACAT is a membrane-bound protein of the endoplasmic reticulum, which initially proved difficult to purify to homogeneity. Two human genes with ACAT activity have been identified. Because the gene for acetoacetyl-CoA thiolase has been given the symbol ACAT, the acyl-CoA: cholesterol acyltransferases have been renamed as *sterol O-acyltransferases*, SOAT1 and SOAT2. The *SOAT1* gene is on chromosome 1q25.2, spans 61 kb, has 16 exons, and codes for a 550-residue protein. SOAT2 is on chromosome 12q13.13, spans 21 kb, has 15 exons, and codes for a 522-residue protein. SOAT2 (ACAT2) was discovered on the basis of high homology with SOAT1 from codon 102 to the C-terminus. The rare phenomenon of trans-splicing appears to be the origin of a minor 4.3 kb ACAT1 mRNA, which has an additional exon from chromosome 7.

There has been confusion regarding the sites of expression of these two genes. *ACAT1* is more widely expressed than *ACAT2*, which is mainly expressed in small intestine at the apical region of the villi and in liver. In humans, ACAT1 activity dominates that of ACAT2 in human liver. However, in ACAT1 knockout mice hepatic CE levels and synthetic rates were normal. The only tissues affected were the adrenals and macrophages. ACAT2 disruption in mice resulted in very low ACAT enzyme activity in liver and intestine. These animals showed resistance to diet-induced hypercholesterolemia and to the formation of cholesterol gallstones because of the decreased intestinal cholesterol absorption. Whatever the relative expression levels of ACAT1 and ACAT2 are in human liver, they are probably different from those in mouse liver.

Both enzymes are membrane bound, each with five TM domains, being found in microsomal cell fractions. It has been proposed that CE that is formed within the ER bilayer by ACAT is processed in one of two possible ways. It can either be directed as lipid droplets into the cytosol or, in cells producing lipoproteins, transferred to apoB in the ER lumen by MTP.

3.2
Catabolism: Bile Acid Metabolism and Disorders

The main bile acids, taurine or glycine conjugates of cholic, deoxycholic, and chenodeoxycholic acids, are synthesized from cholesterol in the liver. This is the primary means by which cholesterol is catabolized, the hepato-biliary system being the main pathway for cholesterol excretion. The first, rate-controlling step of the major pathway of bile acid synthesis, is the formation of 7α-hydroxycholesterol. This is catalyzed by cholesterol 7α-hydroxylase (CYP7A1), a liver-specific, cytochrome P-450-dependent, mixed function oxidase. In humans, the pathway leads to two primary bile acids, cholic acid and chenodeoxycholic acid.

CYP7A1, located in the smooth ER, is subject to feedback regulation at the transcriptional level, with mRNA levels decreased by high levels of bile acids and positively correlated with the availability of cholesterol. *CYP7A1* gene transcription is subject to feedback regulation by bile acids through a bile acid response element (BARE), and is positively correlated with the availability of free cholesterol. CYP7A1 mRNA levels are regulated by thyroid hormone and insulin. There have been numerous studies on the regulation of CYP7A1, and the promoter region has been extensively analyzed. However, there remain a number of unresolved questions, especially with regard to the human gene, and a number of contradictions exist among the studies. It has been emphasized that there are important differences in the types of bile acids among species and the response of CYP7A1 to high-cholesterol diets, especially between humans and widely studied rodents. Nevertheless, there is a current, commonly accepted, model that attempts to explain the regulation of CYP7A1 and bile acid homeostasis, albeit based on rodent data.

The concentration of oxysterols, which are activating ligands for the nuclear receptor LXR-α (NR1H3), is thought to reflect the level of cell cholesterol. High levels of cholesterol, and hence of oxysterols, increase the transcription of CYP7A1, basal transcription being the result of the binding of CPF (or LRH-1, the mouse homolog) encoded by the *NR5A2* gene, and of HNF4-α, to the promoter. There is evidence that HNF-1 binds to and activates the CYP7A1 promoter in humans but not in the rat. A number of bile acids, notably chenodeoxycholic acid (CDCA), are activating ligands for the nuclear receptor FXR (NR1H4). FXR (see Sect. 5.4) is an orphan nuclear hormone receptor expressed within hepatocytes and enterocytes in the terminal ileum. It is thought that FXR regulates CYP7A1 indirectly because no consensus binding site has been identified in the promoter. FXR upregulates the gene for SHP (small heterodimer partner), and the SHP protein binds to and inhibits the activation of CPF (LRH-1) and HNF4-alpha. This model holds up better in rodents than in humans. Unlike its action on the rat cyp7a1 promoter, LXR has only a small effect on the human promoter to which it binds only weakly. It has been reported that FXR only modestly increases SHP transcription in human HepG2 cells. SREBP can activate the human SHP promoter, but not that of the mouse gene. The transcriptional coactivator PGC-1α activates *CYP7A1* gene transcription. *CYP7A1* is repressed by the c-Jun N-terminal kinase pathway, a process that involves fibroblast growth factor 19 (FGF-19), which is regulated by FXR.

The human gene for cholesterol 7α-hydroxylase, which is 10-kb long, lies at chromosome 8q12.1, contains 6 exons, and codes for a 504-residue peptide, which contains sterol and heme binding sites. Overexpression of the gene in hamsters caused a large decline in LDL, and transgenic mice had resistance to diet-induced gallstone formation and to atherosclerosis. There have been conflicting reports as regards the effect on lipid levels in plasma, and the phenotype in general, in knockout mice. In one colony, most homozygous animals died within 18 days of birth unless mothers were supplemented with vitamins and cholic acid was added to the diet. There were no changes in serum lipid levels. The phenotype consisted of fat malabsorption, abnormal lipid excretion, and skin and behavioral abnormalities. A second colony survived on an unsupplemented diet, had elevated serum cholesterol, and had a proatherogenic phenotype. CYP7A1 polymorphisms, notably an SNP at −204 in the promoter, have been associated with differences in LDL-cholesterol levels. Screening of hypercholesterolemic patients, using a differential DNA melting technique, revealed several rare CYP7A1 variants.

Cholesterol 7α-hydroxylase initiates the classical or neutral bile acid synthesis pathway. An additional pathway, the alternative or acidic pathway, does not involve CYP7A1 and is initiated by mitochondrial sterol 27-hydroxylase (CYP27A1). Subsequently, a microsomal enzyme, oxysterol 7α-hydroxylase (CY7B1), introduces a 7α-hydroxyl group. The rest of the pathway is similar to the neutral pathway. However, chenodeoxycholic acid is the main product here, and not cholic acid.

3.2.1 Cholesterol 7α-hydroxylase Deficiency

A number of CYP7A1 coding region variants have been found, one of which is a frameshift mutation (L413fsX414) that results in loss of catalytic activity due to deletion of the heme binding domain of the enzyme, and leads to the disorder *cholesterol 7α-hydroxylase deficiency*. Two homozygous males had highly reduced bile acid excretion, significant hypercholesterolemia and hypertriglyceridemia, and were profoundly resistant to treatment by HMG-CoA reductase inhibitors (statins). They both had early gallstone disease. Liver biopsy and stool bile acid analysis indicated limited upregulation of the alternative bile acid synthetic pathway. The observed doubling of liver cholesterol content, due to low conversion to bile acids, may have resulted in downregulation of hepatic LDL receptors that could account for the raised plasma cholesterol. In terms of the hypercholesterolemia, there appears to be a gene dosage effect, with heterozygotes also affected. Thus, cholesterol 7α-hydroxylase deficiency is an autosomal codominant disorder.

3.2.2 Oxysterol 7α-Hydroxylase Deficiency

A single case of this disorder, associated with lack of synthesis of primary bile acids, was reported in a human infant due to a nonsense mutation in exon 5 of the *CYP7B1* gene. The child had cholestasis and early liver failure, due to accumulation of toxic monhydroxycholenoic acids. CYP7A1 activity was low in this patient, as in normal infants, and therefore was not able to compensate for the absence of CYP7B1. In contrast, *Cyp7b1* knockout mice had normal bile acid metabolism, with plasma cholesterol and triglycerides unchanged.

3.2.3 Cerebrotendinous Xanthomatosis

First reported in 1937, cerebrotendinous xanthomatosis (CTX) is caused by a deficiency of sterol 27-hydroxylase (CYP27A1). It is characterized by a marked reduction in bile acid synthesis, an accumulation of cholesterol and cholestanol in many tissues, and by ataxia, dementia, and premature atherosclerosis. Cyp27a1 knockout mice had even lower bile acid production, were hypertriglyceridemia, but had no CTX-like symptoms.

3.3 Disorders of Intracellular Lipoprotein Degradation

3.3.1 Wolman Disease and Cholesteryl-ester Storage Disease

Degradation of lipoproteins following their uptake by tissues involves lysosomal acid lipase, which catalyzes the hydrolysis of cholesteryl esters and triglycerides. A deficiency in the activity of this enzyme can leads to two disorders, Wolman disease, and cholesteryl ester storage disease (CESD). Both are autosomal recessive disorders caused by mutations in the gene for lysosomal acid lipase (LIPA) on chromosome 10q23.31. The gene has 12 exons and spans 201 kb. Transcription leads, by alternative splicing, to two mRNAs that code for 399- and 401-residue proteins, respectively. Wolman disease is a severe disorder invariably fatal in the first year of life. It is associated with a failure to thrive, multiple gastrointestinal symptoms, including steatorrhea and hepatosplenomegaly, and adrenal calcification. CESD is associated with hepatomegaly, but its symptoms are generally less severe. It sometimes does not manifest in childhood. Owing to hyperlipidemia, severe premature atherosclerosis is common.

3.3.2 Niemann–Pick Disease Type C

Caused by defects in at least two genes, Niemann–Pick Disease Type C, first described in 1958, is a rare autosomal recessive disorder characterized by aberrant intracellular trafficking of endocytosed LDL cholesterol. The cholesterol accumulates in lysosomes. It is often a fatal disease with progressive visceral and neurological symptoms. The human *NPC1* gene spans 55 kb on chromosome 18q11.2 with 25 exons, and encodes a 1278-residue integral membrane protein with a sterol-sensing domain similar to HMG-CoA reductase, SCAP and Patched. The second gene associated with the disorder, NPC2, is on chromosome 14q24.3, spans 13 kb, and codes for a 151-residue protein. This has a lipid-binding motif and functions, together with the NPC1 protein, to facilitate cholesterol transport in the late lysosomal/endosomal pathway.

A gene related to *NPC1* is *NPC1L1* (Niemann–Pick disease, type C1, gene-like 1). The NPC1L1 protein product, highly expressed in enterocyte brush border membranes, plays a poorly understood, but critical, role in the intestinal absorption of cholesterol. It functions in the pathway of cholesterol absorption in which the cholesterol-lowering drug ezetimibe acts. Ezetimibe failed to inhibit the uptake of cholesterol in NPC1L1 knockout mice.

4 Triglyceride and Phospholipid Metabolism

Many triglyceride and phospholipid pathway enzymes are intimately bound to membranes and have proved difficult to purify, sequence, and clone. The steps leading to the formation of diacylglycerol from acyl CoA, *via* phosphatidate, are common to both the triglyceride and phospholipid pathways. Diacylglycerol is synthesized *via* the action of phosphatidic acid phosphatase (PAP). There is a family of PAP enzymes encoded by three genes, *PPAP2A, PPAP2B*, and *PPAP2C*. In each case, isoforms resulting from alternative splicing exist. They have an N-glycosylation site and 6 TM domains, being integral membrane glycoproteins. The PAPs also function in phospholipase D signal transduction.

The last step in the synthesis of phosphatidyl choline, the major phospholipid, is formed by the action of CDP-choline: 1,2-diacylglycerol choline phosphotransferase (or choline/ethanolaminephosphotransferase) an intrinsic, ER and Golgi enzyme. The human gene *CEPT1* located on chromosome 1p13.2 has 9 exons, spans 45 kb, and encodes a 396-residue protein. CEPT also catalyses the formation of phosphatidylethanolamine. First cloned from yeast, the protein has a molecular mass of 46 kDa and seven membrane-spanning helices. CDP-choline itself is formed by the action of CTP:phosphocholine cytidyltransferase (CT). Rat CT has been cloned and has a molecular mass of 42 kDa. It lacks a hydrophobic domain and is attached to membranes by a 58-residue amphipathic α-helix. There are two human CT genes: *PCYT1A* and *PCYT1B*.

In humans, there are two enzymes that catalyze the final step in triglyceride synthesis. The diacylglycerol acyltransferases DGAT1 and DGAT2 utilize fatty acyl-CoA and diacylglycerol as substrates to form triglyceride. The *DGAT1* gene, first recognized because of its homology to ACAT, is at chromosome 8q24.3, has 17 exons, and spans 11 kb. *DGAT2* on chromosome 11q13.3 has 8 exons spanning 33 kb. It is expressed widely, but at high levels in liver and white adipose tissue.

5 Role of Nuclear Hormone Receptors

5.1 Sterol Regulatory Element Binding Proteins

The 5′ flanking region of the LDL receptor (LDLR) gene was found to contain a sterol regulatory element (SRE). In the presence of sterols, this element loses its ability to synergistically enhance transcription *via* two direct repeats that bind to the Sp1 transcription factor. A nuclear protein was identified that binds to the SRE and is termed *sterol regulatory element binding protein* (SREBP). SREs have since been identified in the promoters of other genes involved in sterol and fatty acid synthesis. Three similar transcription factors, SREBP-1a, SREBP-1c, and SREBP-2, all of which are 120-kDa proteins, have been identified, and they all bind to SREs. SREBP-1a, and SREBP-1c differ as a result of separate transcription starts sites in the *SREBF1* gene involving alternative initial exons (human gene: chromosome 17p11.2). The SREBP-2 protein is encoded by the *SREBF2* gene (human gene: chromosome 22q13.2). In mammals, SREBP-1a and SREBP1-c primarily activate the genes of fatty acid synthesis, and SREBP-2 activates those of sterol synthesis. The SREBPs are basic helix–loop–helix leucine zipper (bHLH-zip) transcription factors found in a wide variety of animals. The precursor peptides are attached to the ER membrane with the N-terminal bHLH-zip domain and the C-terminal regulatory domain located in the cytoplasm. The presence of sterols blocks a two-step proteolytic cleavage of SREBP that occurs in sterol-depleted cells when a released N-terminal fragment of the protein enters the nucleus and activates transcription of sterol-regulated genes *via* the SREs. In the presence of sterols, SREBP is bound by its regulatory domain to the SREBP-cleavage-activating protein (SCAP). SCAP has eight membrane-spanning helices in its N-terminus and a C-terminal propeller region composed of multiple WD repeats that binds to SREBP. The SREBP–SCAP complex is retained in the ER by proteins that interact with the sterol-sensing domains of SCAP, and which are encoded by insulin-induced genes 1 and 2 (*INSIG-1* and *INSIG-2*). Other proteins, not yet identified, are thought to retain the Insig proteins in the ER since they do not possess ER retention signals. In the absence of sterols the Insig proteins release SREBP–SCAP, which migrates in COPII (coat protein complex) vesicles to the Golgi apparatus. Here, the initial cleavage of SREBP by *site 1 protease (S1P)* (also called

PCSK8), encoded by the *MBTPS1* gene, occurs in a protein loop in the lumen of the ER. A second cleavage by the metalloproteinase S2P, encoded by the *MBTPS2* gene at site 2 in a transmembrane domain, releases the N-terminal active peptide.

5.2 Peroxisome Proliferator-activated Receptors

The peroxisome proliferator-activated receptor (PPAR) family of transcription factors was discovered in rodents. In rodents, PPARs are activated by chemicals, resulting in larger and more numerous peroxisomes. This is accompanied by increased β-oxidation through activation of the acyl CoA oxidase gene, which contains, in its promoter, a peroxisome proliferator response element (PPRE) consisting of a direct repeat of the motif PuGGTCA, to which PPAR binds as a heterodimer with the 9-cis retinoic receptor (RXR, NR2B1). PPARs play roles in differentiation, cell division, apoptosis, and inflammation. They are also key regulators of lipid and lipoprotein metabolism and of the pathways of gluconeogenesis and glycolysis. There are three human *PPAR* genes, *PPARA* (chromosome 22q13.31), *PPARG* (chromosome 3p25.2), and *PPARD* (chromosome 6p21.31) that encode, respectively, PPARα, PPARγ, and PPARδ. The first, PPARα is expressed at high levels in liver, kidney, muscle, and heart where it activates fatty acid β-oxidation. The second, PPARγ, in contrast, is highly expressed in intestine, adipose, and mammary tissue. It promotes the storage of lipids. The third receptor, PPARδ is more widely expressed than the other two members of the family. There are a number of fatty acid and fatty acid–derived compounds that are ligands for PPARs. In particular, a hydroxyeicosatetraenoic acid (HETE) derivative 8S-HETE and leukotriene B4 are potent activators of PPARα. Natural ligands for PPARγ include prostaglandin J2, 15-HETE and the hydroxyoctadecadienoic acids (HODE) 9-HODE and 13-HODE. The hypolipidemic fibrate drugs and antidiabetic glitazone drugs are synthetic PPAR activators.

5.3 Liver X Receptors

The orphan nuclear receptors LXRα and LXRβ are activated by oxysterols, though the nature of the precise physiological ligand is still unclear. Examples of natural compounds known to activate LXR are 24(S),25-epoxycholesterol and 20(S)-, 22(R)-, 24(S)-, and 27-hydroxycholesterol. The LXRs are key regulators of numerous genes involved in lipid and carbohydrate metabolism, the inflammatory response, and energy homeostasis. They bind as heterodimers with RXR to DR-4 (or LXRE) response elements in the promoters of target genes. These are two hexanucleotide repeats with the consensus (T/A/G)G(G/T)(G/T)(T/C)A, separated by four nucleotides. Lipid transfer and transporter genes that are targets for LXRs include *ABCA1*, *ABCG1*, *ABCG4*, *ABCG5/G8*, *CETP*, and *PLTP*. Gene targets involved in fatty acid synthesis are, *ACC* (acetyl CoA carboxylase), *FAS* (fatty acid synthase), and *SCD* (stearoyl CoA desaturase). Two apolipoprotein genes, *APOC2* and *APOE*, are also upregulated by LXRs as is *SREBP-1c* and *LPL*. The *Cyp7a1* gene in rodents, but not in humans is a target for LXRs. The *LXR* genes themselves are subject to positive feedback autoregulation by LXRs. The human gene for LXRα (*NR1H3*, chromosome 11p11.2) is expressed at high levels in the liver, kidney, adipose tissue, and intestine, whereas that for LXRβ (*NR1H2*, chromosome 19q13.33) is more ubiquitously expressed. In liver, LXRα is induced by insulin, leading to upregulation of lipogenesis and downregulation of gluconeogenesis *via* repression of the rate-limiting enzyme phosphoenolpyruvate carboxykinase gene (*PCK1*).

5.4 Farnesoid X Receptor and Pregnane X Receptor

At high concentrations, farnesol was found to activate an orphan nuclear receptor. Thus, this receptor was named FXR (farnesoid X receptor). Since then FXR (human gene, *NR1H4*, chromosome 12q23.1) has been shown to be a bile acid sensor with the highest activation by chenodeoxycholic acid. Some 25- and 26-hydroxylated bile acid intermediates also have high affinity for FXR. With RXR, it binds as a heterodimer to the bile acid response elements (BARE) of a number of genes involved in bile acid and lipid metabolism. Target genes upregulated by FXR include *ABCC2* (multidrug resistance protein: MRP2), *ABCB11* (canalicular bile salt export pump: BSEP), *FABP6* (fatty acid–binding protein 6, the ileal bile acid-binding protein), *SHP* (see Sect. 3.2 above), *ApoC-II*, and *PLTP*.

A related nuclear receptor, the pregnane X receptor, PXR (human gene, *NR1I2*, chromosome 3q13.33), also binds as a heterodimer with RXR. Natural ligands include lithocholic acid. It activates several cytochrome P450 enzyme genes, notably the cluster of four *CYP3A* family genes, *CYP2B*, *ABCB1 (MDR1)* and *ABCC2 (MRP2)*. These enzymes are involved in drug metabolism, and some of them in lipid synthesis and transport also.

6 The Apolipoproteins

6.1 The Apolipoprotein Multigene Family

Nine apolipoproteins, all of which are exchangeable among lipoproteins, are members of a multigene family. These are *apoA-I, apoA-II, apoA-IV, apoA-V, apoC-I, apoC-II, apoC-III, apoC-IV* and *apoE*. The genes have a similar structure and have evolved from a common ancestor that was similar to apoC-I. Each has 4 exons, except apoA-IV, and apoA-V in which exons 1 and 2 have fused. Within exon 3 there is, in each case, a conserved region of 33 codons. This amphipathic α-helical lipid binding domain is a trimer of 11 amino acids. The fourth exon of each codes for variable numbers of consensus, 22-residue and, in some cases, 11-residue, lipid-binding tandem repeats.

The *apoA-I, apoC-III, apoA-IV* and *apoA-V* genes (Table 2) are clustered in a 48-kb region on chromosome 11 (11q23.3). This cluster is regulated by an enhancer region, necessary for intestinal expression of apoA-I and apoA-III, upstream of the *apoA-III* gene. Transcription factors involved include SP1 and HNF4. In a patient with premature atherosclerosis, a 6-kb inversion affecting the *apoA-I* and *apoC-III* genes resulted in apoA-I and apoC-III deficiency. In another case the entire gene cluster was deleted, resulting in apoA-I, apoC-III and apoA-IV deficiency and very low HDL levels. Population studies using polymorphic markers suggest an association between the A-I/C-III/A-IV/A-V locus and the frequency of hyperlipidemia and atherosclerosis. A case-control study of SNPs at this locus showed significant association between the FCH phenotype and the *ApoA-I* and *ApoA-V* genes.

Tab. 2 The human apolipoprotein genes.

Protein	*Alternative name*	*HUGO symbol*	*Chromosomal location*	*Gene size [kb]*	*Exons*	*Initial translated protein amino acid residues*	*Mature protein molecular weight [kDa]*	*Normal fasting concentration in plasma [mg dL^{-1}]*	*Lipoprotein found on*
ApoA-I		APOA1	11q23.3	1.9	4	267	28	130	HDL, chylomicrons
ApoA-II		APOA2	1q21-q23	1.3	4	100	8.7	40	HDL, VLDL, chylomicrons
ApoA-IV		APOA4	11q23.3	2.6	3	396	46	16	HDL, chylomicrons
ApoA-V		APOA5	11q23.3	2.5	3	363	39	<1	HDL
ApoB		APOB	2p24-p23	45	29	4563 (apoB-100)	549	80	VLDL, IDL, LDL
						2179 (apoB-48)	264	<1	chylomicrons and chylomicron remnants
ApoC-I		APOC1	19q13.2	4.6	4	83	6.6	6	HDL, IDL, VLDL
ApoC-II		APOC2	19q13.2	3.3	4	101	8.9	3	HDL, IDL, VLDL
ApoC-III		APOC3	11q23.3	3.1	4	99	8.9	12	HDL, IDL, VLDL
ApoC-IV		APOC4	19q13.2	3.3	3	127	11.5	<1	VLDL, HDL
ApoD		APOD	3q26.2-qter	15.2	5	189	19–32	10	HDL
ApoE		APOE	19q13.2	3.6	4	317	34	5	HDL, IDL, VLDL
ApoH	beta-2-glycoprotein I	APOH	17q23-qter	17	8	345	50	20	HDL, VLDL, chylomicrons
ApoJ	Clusterin, SP-40/40	CLU	8p21-p12	17	9	427	70	1.4	
ApoLI		APOL1	22q13.1		7			<1	HDL, VLDL
Apo(a)[a]	lipoprotein Lp(a)	LPA	6q26–27	212	39	4548	550	variable[a]	Lp(a)

[a] Data for apo(a) is given for the gene in the public domain genome sequence. The protein size can vary from 300 to 800 kDa. The concentration in human plasma can vary from undetectable to over several hundred mg dL^{-1}.

The genes for apoC-I, apoC-II, apoC-IV, apoE, (Table 2) and an apoC-I pseudogene are clustered on chromosome 19 (19q13.32) spanning 44 kb. The regulation of this cluster is discussed below (Sect. 6.11).

6.2 Apolipoprotein A-I

The major protein component of HDL, apoA-I is an activator of LCAT. The secreted 249-residue propeptide is processed in the plasma to the mature 243-residue (28 kDa) form. A number of rare charge variants have been detected by isoelectric focusing. Two of these, P165R and R173C (apoA-I Milano), in addition to a number of truncated variants, are associated with low levels of HDL. In the case of apoA-I Milano, the decrease in the level of HDL is due to an elevated rate of catabolism of the abnormal apoA-I species. Recently, clinical trials of intravenously infused recombinant apoA-I Milano have shown regression of coronary atherosclerosis. The practicality of this approach *versus* more conventional drug therapy in treating large numbers of patients is questionable. Several mutations, including G26R and L60R, cause amyloidosis. Two others, K107del and P165R, are mildly defective in activating LCAT.

An XmnI RFLP in the 5′flanking region is associated with hypertriglyceridemia in some populations. A possible functional single-base G/A substitution polymorphism in the promoter at nucleotide −75, in one of two inverted repeats, was associated with differences in HDL levels in some populations.

A number of cis-acting elements upstream of the gene have been identified as required for full liver-specific expression of apoA-I. However, while only 256-bp 5′ of the *apoA-I* gene are required for hepatic expression, a construct containing 5.5-kb 5′ and 4-kb 3′ of the human gene was not sufficient for intestinal expression in transgenic mice. This was achieved only when a construct was used that contained both the *apoA-I* and *apoC-III* genes together with 300-bp 5′ to the *apoA-I* gene and 1.4-kb 5′ to the *apoC-III* gene.

ApoA-I production and the level of the mRNA in rat liver are stimulated by thyroid hormone, an effect due both to increased transcription and to stabilization of apoA-I nuclear RNA. HNF-4, a member of the thyroid/steroid hormone receptor superfamily, has been shown to interact with a sequence in the 5′ flanking region and to increase promoter activity. Inhibitory effects have been shown for another DNA-binding protein, ARP-1, at the same site.

The overexpression of human apoA-I in transgenic mice increased the HDL cholesterol levels and prevented early diet-induced atherosclerosis. When these animals were crossed with apoE-deficient mice there was a marked decrease in the development of atherosclerosis, which is a characteristic of the apoE-deficient animals.

Homozygous apoA-I-deficient mice produced by gene targeting had only 20% of the normal level of HDL cholesterol but were generally healthy.

6.3 Apolipoprotein A-II

A secondary protein component of HDL, apoA-II is synthesized and secreted mainly by the liver and intestine. Unlike apoA-I it does not activate LCAT. Like apoA-I, it is regulated both transcriptionally and translationally.

Following cleavage of the 18-residue signal sequence some of the 82-residue propeptide is cleaved prior to secretion, forming the mature 77-residue peptide. The rest is processed in the plasma. ApoA-II contains one free cysteine residue, enabling it to form heterodimers with apoE and apoD, though most of it exists in the form of a homodimer. The physiological function of apoA-II is largely unknown. It may be involved in the modulation of LCAT and HTGL activity.

Numerous studies have explored the regulation of the human gene expression. A diverse set of promoter elements have been identified. An MspI polymorphism at the 3′ end of the gene, in an Alu repeat sequence, has been reported to be associated with differences in the levels of apoA-I and apoA-II in serum.

An increased tendency to develop atherosclerotic lesions was seen in transgenic mice that overexpress human apoA-II. These animals had elevated levels of HDL and the particles were larger with a higher ratio of apoA-II to apoA-I when compared with normal mice.

6.4 Apolipoprotein A-IV

The 46-kDa glycoprotein, ApoA-IV is synthesized mainly in the small intestine and exists in a number of isoforms. The precise role of apoA-IV is unknown, but, like apoA-I, it activates LCAT and may be involved in reverse cholesterol transport. It has been reported to be a satiety factor. ApoA-IV is rapidly catabolized, having a tenfold higher fractional catabolic rate than apoA-I. The rate of synthesis determines the level of apoA-IV in plasma.

Three common polymorphisms exist: N127S, T347S and Q360H. With the latter polymorphism, the fractional catabolic rate of the *His* allele product has been reported to be lower than that of the more common *Gln* allele (frequency 92%). Genotypic variation in the *APOA-1V* gene affects the response of LDL cholesterol to diet.

6.5 Apolipoprotein A-V

Unlike the other apolipoproteins, apoA-V (and apoC-IV) was not discovered first as a protein isolated from plasma. Rather, the presence of apoA-V was revealed in the rat as the most highly upregulated of several novel genes following partial hepatectomy, and as a result of a comparison of the mouse and human genomes. It shows noticeable homology with apoA-I and apoA-IV, and the gene lies close to the cluster on chromosome 11q23 together with the genes for these two apolipoproteins as well as apoC-III. The apoA-V protein was then identified in plasma in the HDL fraction. Overexpression, either as a transgene, or by adenovirus transfection, resulted in decreased levels of plasma triglycerides. In contrast, knockout mice showed a marked increase in triglycerides. Together with the confirmation that functional FXR and PPARα response elements exist in the promoter for apoA-V, this strongly suggests a role in the regulation of triglyceride metabolism in the liver. The upregulation of apoA-V after hepatectomy suggests it has a role in the provision and trafficking of lipid for formation of new cell membranes. The apoA-V sequence contains amphipathic α-helices with an overall 60% α-helical content. Like ApoA-I, it can organize dimyristoylphosphatidylcholine vesicles into discoidal structures with diameters of 15 to 20 nm. However, unlike ApoA-I, it only poorly activates LCAT and does not increase cholesterol efflux from cAMP-stimulated macrophages.

6.6 Apolipoprotein B

In plasma from fasting humans, 90% of circulating apoB is in LDL, 8% in IDL and 2% in VLDL and chylomicrons. A centile nomenclature is used to describe apoB species based on their molecular weights. There are two circulating forms, apoB-48 and apoB-100 (Table 2). C-terminally truncated apoB-48, found on chylomicrons and their remnants, is produced in humans in the intestine. ApoB mRNA (14 kb in length) is edited by an editing complex in the nucleus of intestinal enterocytes in an enzyme-mediated process, such that codon 2153 becomes UAA (a stop codon) in place of CAA (Glu) in the gene sequence. A 26-nucleotide domain, 6662 to 6687, is highly conserved in mammals and necessary for editing to occur. The editing enzyme, Apobec-1, a 27-kDa protein with cytidine deaminase activity, was isolated and cloned from rat intestine. Additional factors are required for the editing to occur. The first to be identified is the Apobec-1 complementation factor (ACF), a 65-kDa protein.

In humans, the liver makes only the apoB-100 found in plasma on VLDL, IDL, and LDL. ApoB mRNA editing is undetectable in liver from humans, pigs, cows, sheep, cats, and monkeys. In liver from rabbits and guinea pigs, editing activity is very low. However, in several species, including dogs, horses, rats, and mice, an appreciable degree of editing has been found. The extent of this hepatic editing may be an important determinant of the level of apoB-containing lipoproteins in the circulation.

There is little acute metabolic regulation of hepatic levels of apoB mRNA. The availability of lipid and the extent of intracellular degradation of apoB are important in regulating secretion rates. The proteasome–ubiquitin pathway is the major means for the degradation of the apoB that fails to be lipidated.

Much larger than the exchangeable apolipoproteins, apoB has a globular N-terminus followed by four alternating amphipathic α-helical and β-sheet domains. This "pentapartite" structure is given as NH_2-$\beta\alpha1$-$\beta1$-$\alpha2$-$\beta2$-$\alpha3$-CO_2H. Through interaction with MTP, the $\beta\alpha1$ domain forms a lipid pocket that promotes lipid accumulation in the early stages of VLDL formation. The $\beta1$ and $\beta2$ hydrophobic surfaces interact with, and organize, the lipoprotein lipid core. The overall structure of native LDL is generally thought to be spherical. ApoB is highly glycosylated *via* 16 N-linked sites, with carbohydrate contributing 35 kDa of the molecular weight of apoB-100. Two regions (residues 1280 to 1320 and 3180 to 3282) are particularly sensitive to cleavage by endoproteases, an indication that they are exposed on the surface of LDL. Thrombin cuts at 1297 and 3249, forming three fragments, T4 (N-terminal), T3 (central) and T2 (C-terminal). The region from residues 3000 to 3600 that flanks the T2/T3 junction contains sequences responsible for binding to the LDL receptor. Receptor-blocking monoclonal antibodies have been mapped to this region. Two positively charged sequences in this region (residues 3147 to 3157 and 3359 to 3367) bind strongly to heparin. The second of these has close homology to the LDL receptor binding domain of apoE.

The tissue-specific regulation of expression of the *apoB* gene is complex with many cis- and trans-acting factors. Regulation of intestinal expression remains largely enigmatic. Studies with transfected cells, *in vitro*, and with transgenic mice have provided an understanding of the liver-specific regulation. Binding sites for a number of nuclear regulatory proteins including ARP-1, EAR-2, EAR-3, HNF-4 and C/EBP have been identified within the promoter region. HNF-4 and C/EBP activate transcription synergistically. ARP-1, EAR-2, and EAR-3 repress transcription. There is a core enhancer in intron 2 that binds HNF-1, C/EBP, and an uncharacterized protein. A weaker enhancer lies in intron 3. In addition to these elements, nuclear matrix association regions (MARs), identified at the 5′ and 3′ ends of the gene, have been shown to play an important role in apoB expression. Tissue-specific transcription in liver and intestine correlates with undermethylation of the 5′ flanking region in these tissues.

There is a hypervariable region, composed of 15-bp AT-rich tandem repeats, located 200-bp downstream of the gene, with at least 14 common alleles ranging from 25 to 52 repeats. Five single-amino-acid substitution polymorphisms, the antigenic group (Ag) series, were first detected in antisera from multiply transfused subjects. Four are in association with unique common apoB polymorphisms. In the case of the fifth, the Ag(x/y) polymorphism, two SNPs in complete linkage disequilibrium are involved. These are BfaI and Eco57I RFLPs, associated with P2712L and N4311S substitutions, respectively. Pro2712/Asn4311 represents the more common Ag(y) epitope. A silent XbaI RFLP in the third base of the codon for Thr2488 is in strong linkage disequilibrium with the Ag(x/y) polymorphism. An association of the XbaI RFLP with differences in lipid levels has been reported in a number of studies, with this polymorphism accounting for possibly 3 to 5% of the variance seen in the amount of LDL cholesterol in the general population.

A common insertion/deletion polymorphism results in either a 27- or 24-residue signal sequence. A rare third allele has 29 residues. A number of studies have examined the association between particular polymorphisms or haplotypes and either lipid levels or incidence of atherosclerotic disease.

6.7 Apolipoprotein C-I

A minor component of HDL, IDL, and VLDL, apoC-I is found mainly in HDL (97%). More than 90% of the circulating apoC-I is of hepatic origin, with moderate expression in skin and trace levels of expression in brain. Little is known concerning its physiological function. It was shown to be a highly effective inhibitor of the binding of apoE-enriched βVLDL to the α_2-macroglobulin receptor/LRP. The expression of the human *apoC-I* gene in transgenic mice resulted in high levels of triglyceride-rich lipoproteins, indicating that apoC-I is involved in the regulation of their metabolism. Evidence in baboons indicates that an N-terminal fragment of apoC-I interacts with apoA-I to inhibit CETP activity. It may also be a cofactor for LCAT. ApoC-I, which has a domain that binds to lipopolysaccharide seems to play a role in infection and inflammation.

A functional apoC-I promoter polymorphism at −317 (a HpaI RFLP), between the *apoC-I* and *apoE* genes, is associated with type III hyperlipidemia *via* the apoE genotype. There is a synonymous SNP in codon 6, and BglI and DraI RFLPs have been reported. Common liver-specific regulation of expression of the *apoC-I* and *apoE* genes is discussed below (Sect. 6.11). ApoC-I deficient mice produced by gene targeting have a moderately elevated level of serum triglycerides. The rate of clearance of remnant lipoprotein particles in these animals is decreased.

6.8 Apolipoprotein C-II

The apolipoprotein apoC-II gene is expressed mainly in the liver, with intestinal cells containing only small amounts of mRNA. It is glycosylated within the cell but is subsequently deglycosylated in the plasma, where it exists mainly in the 8.9 kDa, 79-residue form, although some of it is further processed by removal of the N-terminal 6-amino acids to produce a minor isoform. The only known role of apoC-II is as the requisite cofactor for LPL. The activator and primary LPL binding domains are located in the C-terminal third of apoC-II. Like the other C apolipoproteins, apoC-II is transferred during hydrolysis from chylomicrons and VLDL to HDL. Most of the circulating apoC-II is present in HDL; 10% is in IDL, and 30% is in VLDL and chylomicrons.

The first intron of the *apoC-II* gene contains 4 Alu type sequences and a $(TG)_n(AG)_m$ microsatellite with at least 15 different alleles. The third intron contains a minisatellite composed of 37-bp tandem repeats. There are two common alleles of six and seven repeats with frequencies of 82 and 18%, respectively. AvaII, BanI, BglI, NcoI, PstI, and TaqI, RFLPs have been reported. In human hepatoma cells, apoC-II is upregulated by the nuclear receptor FXR.

Three charge variants of apoC-II, K19T, E38K, and K55Q, all detected by isoelectric focusing, have been reported. The first two are rare, but the K55Q variant is common among individuals of African decent. Whether they are associated with hyperlipidemia is unclear.

6.8.1 Apolipoprotein C-II Deficiency

Autosomal recessive familial chylomicronemia, due to homozygous apoC-II deficiency, has been described in ten instances. It is characterized by fasting chylomicronemia and severe hypertriglyceridemia, and is often accompanied by the presence of xanthomas and pancreatitis. Six cases were due to apoC-II gene mutations that introduced premature stop codons, resulting in either the absence of apoC-II or very low levels of truncated species. One other was due to an intron 2 donor splice defect, and one to an initiation defect M-22V. In one case, a single C insertion between codons 69 and 70 resulted in low levels of a 96-residue frameshifted protein with severely impaired cofactor activity. A major disruption of the *apoC-II* gene, and neighboring *apoC-IV* gene, by a 7.5-kb deletion, has also been reported.

6.9 Apolipoprotein C-III

The glycoprotein, apoC-III is the most abundant protein in VLDL after apoB100, comprising 25 to 30% of the protein mass. Most is found in HDL (60%), 10% in LDL, 10% in IDL and 20% in VLDL. There are three isoforms: apoC-III_{-0}, apoC-III_{-1} and apoC-III_{-2}, with 0, 1, and 2 residues of sialic acid per molecule, respectively. These have different isoelectric points and can be separated by isoelectric focusing.

Little is known about the precise function of apoC-III in lipoprotein metabolism. It inhibits LPL *in vitro*. Although a number of hydrophilic sequences within the peptide interact with LPL, the major inhibitory effect resides in an NH_2 terminal domain. ApoC-III does not compete with apoC-II at the apoC-II activation site on LPL. Glycosylation of apoC-III does not seem to be important for LPL inhibition.

Overproduction of VLDL in rats fed a high-sucrose diet does not alter the rate of apoC-III gene transcription, or the level of mRNA, in liver. Inflammation is linked to a reduction in the level of hepatic apoC-III mRNA and with a decrease in circulating HDL apoC-III content. Thyroid hormone has been reported to improve the efficiency of maturation of the apoC-III mRNA.

Studies using transgenic mice suggest that apoC-III is involved in the apoE-mediated clearance of triglyceride-rich lipoproteins. Overexpression of the human gene in transgenic mice causes elevated levels of VLDL triglyceride with more than double the normal amount of apoC-III per particle, and is accompanied by an increase in circulating nonesterified fatty acids. The VLDL particles have a larger mean diameter and are taken up poorly by lipoprotein receptors. Tissue levels of LPL are normal and it is likely that the hypertriglyceridemia is a result of the low fractional catabolic rate of VLDL and a low rate of clearance of apoB48 remnants in these animals. This low rate is probably due to the combined effect of the increased amount of apoC-III and the decreased amount of apoE, on VLDL. These mice were crossed with those that overexpress apoE to produce animals overexpressing both apoC-III and apoE. These animals had normal levels of triglycerides and chylomicron remnants.

ApoC-III deficient mice, produced by gene targeting, have decreased fasting levels of triglycerides and show an absence of postprandial lipemia with an increased rate of clearance of chylomicrons. An interpretation of these findings is complicated by the fact that the intestinal expression of the neighboring *apoA-I* and *apoA-IV* genes are decreased in these animals.

In a family of Mexican origin an apoC-III Q38K mutant was associated with moderate hypertriglyceridemia. Rare variants, D45N and A23T do not seem to be associated with dyslipidemia. Another mutant, T74A, is not glycosylated. Hyperalphalipoproteinemia with abnormal HDL containing large apoE-rich particles was seen in subjects with an apoC-III K58E mutation. There is an SstI SNP in the 3′ untranslated region of exon 4 and a tetranucleotide repeat polymorphism, with 7 common alleles, in intron 3, at the end of an Alu sequence. The SstI SNP and a polymorphism in a promoter insulin response element are associated with differences in triglyceride levels.

6.10 Apolipoprotein C-IV

That an additional apolipoprotein gene might exist at the 19q13.2 locus along with *apoC-I*, *apoC-II*, and *apoE* was first suggested in a study of the corresponding rat locus. Later, the human *apoC-IV* gene was characterized and shown to be located only 555-bp upstream of *apoC-II*. It lacks a TATA box and is transcribed in liver at a level well below that of the *apoC-II* gene. A higher expression level of apoC-IV in the rabbit has been attributed to the presence of a unique purine-rich tandem repeat sequence in the promoter and 5′ UTR that functions as a GAGA box. The peptide has two amphipathic α-helical regions and has but a limited homology with apoC-I, apoC-II, and apoE. Very low levels of apoC-IV are found in human plasma, mainly in VLDL with some in HDL. Much higher levels of the protein are found in rabbit plasma, where it is primarily associated with VLDL particles. Rabbit apoC-IV was the first to be characterized. The human protein can be O- and N-glycosylated, and exists as a number of sialylation isoforms. When the human protein was expressed in mice there was an associated increase in plasma levels of triglyceride.

6.11 Apolipoprotein E

ApoE mRNA is found in a number of tissues, notably liver, brain, kidney, adrenal gland, macrophages, spleen, and ovary. The sequence between residues 140 and 160 of the mature peptide constitutes the ligand that binds to the LDL receptor. The C-terminal residues 245 to 299 are important for the association of apoE with lipoprotein particles and for tetramer formation. Most of the apoE in the circulation is synthesized in the liver. Much of the newly synthesized apoE, like apoB, is broken down by a degradative pathway. The synthesis and secretion of apoE is reduced by insulin. Half of the apoE present in serum is in HDL, 10% in LDL, 20% in IDL, and 20% in VLDL. It is a glycoprotein with a single O-linked chain at Thr194. The carbohydrate contains one or more sialic acid residues.

There are three common alleles of the apoE gene, $\varepsilon 2$, $\varepsilon 3$, and $\varepsilon 4$, present at frequencies of 9, 77, and 14%, which give rise to E2, E3, and E4 isoforms, respectively. The proteins can be separated by isoelectric focusing, with E4 the most basic, and E2 the most acidic. They differ at residues 112 and 158. E2 is Cys112/Cys158, E3 is Cys112/Arg158 and E4 is Arg112/Arg158. Some of the E3 in plasma exists as a homodimer and some as a heterodimer with apoA-II. The binding affinity of E2 for the LDL receptor is less than 5% that of E3 or E4. This is thought to be due to reorganized salt bridges as a result of the loss of the Arg158 residue. This changes the charge disposition of the surface that is presented to the receptor.

Expression of apoE in liver in transgenic mice was shown to require the presence of a hepatic control region (HCR) located 18-kb and 9-kb downstream of the *apoE* and *apoC-I* genes, respectively. This HCR lies between the *apoC-I* gene and the *apoC-I* pseudogene. Constructs containing a 5-kb apoE promoter but lacking the HCR resulted in expression only in kidney and skin. A second HCR was later located 27-kb downstream of the *apoE* gene. The four genes in the 19q13.2 cluster, *apoC-I*, *apoC-II*, *apoC-IV*, and *apoE*, are coordinately regulated. In liver, this regulation is *via* the two distal 350-bp HCRs (HCR.1 and HCR.2), which show close homology. In macrophages, adipose tissue, brain, and skin regulation is through two homologous 620-bp multienhancer regions (ME.1 and ME.2) situated 3.3-kb and 15.9-kb downstream of the *apoE* gene, respectively. ME.1 and ME.2 contain LXR response elements, and in macrophages the mRNAs for all four genes were upregulated in response to LXR ligands. It has been hypothesized that the induction of these four genes may be important for promoting the efflux of lipid from macrophage foam cells in atherosclerotic lesions.

Transgenic mice that overexpress apoE have a striking decrease in plasma levels of VLDL and LDL due to increased clearance rates of triglyceride-rich lipoproteins.

Homologous recombination was used to target and inactivate the *apoE* gene in mice. These *Apoe*$^{-/-}$ animals have hypercholesterolemia, which is especially severe when they are fed a high fat diet. Plasma levels of VLDL and VLDL remnants, containing mainly apoB-48 rather than apoB-100, are raised. The mice exhibit rapid development of atherosclerotic lesions that are thought to be a result simply of the very high cholesterol levels (400 to 500 mg dL^{-1}) rather than to the atherogenicity of the VLDL and VLDL remnant particles. Mice resulting from crossing *Apoe*$^{-/-}$ mice with those that exclusively produce apoB-100 (*Apob*$^{100/100}$ mice) had lower levels of total cholesterol, similar to the levels in *Ldlr*$^{-/-}$ mice crossed with *Apob*$^{100/100}$ mice. The latter animals had much more severe atherosclerosis due to higher numbers of small LDL.

The abnormal phenotype of *Apoe*$^{-/-}$ mice was to a large extent corrected following bone marrow transplantation from a donor mouse with the normal *apoE* gene, indicating the extent to which apoE synthesized by macrophages can contribute to the pool of lipoprotein apoE.

6.11.1 Dysbetalipoproteinemia and ApoE Deficiency

The 1% of individuals who are homozygous for the $\varepsilon 2$ allele have a tendency to accumulate βVLDL in plasma. Additional unknown genetic or other factors cause 5% of these subjects to develop clinically important dysbetalipoproteinemia with marked accumulation of chylomicron and VLDL remnants. This disorder is referred to as *type III hyperlipidemia*.

Rare isoforms of apoE have been discovered, many of which are defective in binding to the LDL receptor. In four of these cases the mutation results in dominant transmission, and in one case recessive transmission, of type III hyperlipidemia. A double mutant, apoE-$4_{\text{Philadelphia}}$ (E13K and R145C) is associated in the homozygous state with a severe form of type III hyperlipidemia and in the heterozygous state with a moderate form. Transgenic mice have been produced that express the human mutant $\text{apoE}_{\text{Leiden}}$, which contains a seven-amino acid insertion that is a tandem repeat of residues 121 to 127. In these animals there is dominant transmission of hyperlipoproteinemia, characterized by elevated levels of VLDL and LDL.

ApoE deficiency is a rare genetic disorder characterized by less than 1% of normal levels of apoE in plasma. Patients have xanthomas, type III hyperlipidemia and premature atherosclerosis. There is delayed clearance of chylomicron and VLDL remnant particles, leading to elevated levels of cholesterol-rich VLDL and IDL.

6.11.2 Alzheimer Disease and ApoE

The frequency in the population of the $\varepsilon 4$ allele declines with age and that of the $\varepsilon 2$ allele increases. There is much evidence for a highly significant genetic association between the $\varepsilon 4$ allele and late onset Alzheimer's disease (LOAD), both in the familial and the sporadic forms. The effect is proportional to gene dosage, with the $\varepsilon 4/\varepsilon 4$ genotype being a greater risk factor than the $\varepsilon 4/\varepsilon 3$ or $\varepsilon 4/\varepsilon 2$ genotypes. These, in turn, are greater risk factors than the other genotypes ($\varepsilon 3/\varepsilon 3$, $\varepsilon 3/\varepsilon 2$ or $\varepsilon 2/\varepsilon 2$). In families with late-onset Alzheimer's disease, the presence of the $\varepsilon 4/\varepsilon 4$ genotype alone is associated with the disease in almost all of those over 80 years of age. Additional evidence, supporting the link between apoE and Alzheimer's disease, is that the gene locus is in a region of chromosome 19, previously associated with LOAD. The apoE protein has been found in senile plaques and neurofibrillary tangles along with the β-amyloid protein. *In vitro*, both apoE4 and apoE3 bind with high specificity, and irreversibly, to β-amyloid, with apoE residues 244 to 272 being crucial for the interaction. The formation of the complex of β-amyloid with apoE4 is much more rapid than with apoE3.

6.12 Apolipoprotein (a)

Apolipoprotein (a) [apo(a)], synthesized in the liver, is found on a lipoprotein species Lp(a), where it is covalently attached to apoB *via* a disulfide bridge. The two proteins also interact noncovalently. Lp(a) floats in the 1.050 to 1.080 g mL^{-1} density fraction in the ultracentrifuge and can be separated from HDL by gel filtration. The level of Lp(a) within the population varies from less than 0.10 mg dL^{-1} to as much as 500 mg dL^{-1}. Of this variation, 90% is genetically determined, primarily at the *apo(a)* gene locus with a small amount possibly attributed to the *apoE* genotype. One factor that affects the plasma concentration is the size of a particular isoform, with an inverse correlation between the size and the rate of secretion, this rate being determined by the efficiency of posttranslational processing.

High levels of Lp(a) have been shown, in most epidemiological studies, to be an independent risk factor for premature atherosclerosis, but the physiological function of Lp(a) is still essentially unknown, although the high degree of homology of apo(a) with plasminogen appears to be responsible for its inhibitory effect on thrombolysis. Plasminogen and apo(a) compete in binding to fibrin. Recombinant apo(a) binds with high affinity *via* intrachain lysine residues of intact fibrin and to carboxy-terminal lysines on the degraded peptides. To what degree Lp(a) inhibits fibrinolysis physiologically has been questioned. It has been postulated that Lp(a), *via* a reduction in the formation of plasmin, enhances vascular smooth muscle cell proliferation as a result of lower activation of TGFβ.

Thyroid hormone status, besides modulating the plasma concentration of total apoB also affects Lp(a) levels. In the hypothyroid state, there is impaired catabolism. Secretion is suppressed and catabolism elevated in the hyperthyroid state.

It is not clear to what extent Lp(a) is secreted directly from the liver or formed in plasma as a result of association between free apo(a) and LDL. It does not appear to originate from a triglyceride-rich lipoprotein precursor. Contact with LDL is made through only a few kringle repeats with most of the apo(a) extended away from the particle.

The human gene for apo(a), *LPA*, is closely linked to that of plasminogen (*PLG*), the two genes being separated by 38 kb. Two *apo(a)*/plasminogen pseudogenes are present at the same locus. The size of the *apo(a)* gene is highly polymorphic and depends on the number of exons that code for tri-loop peptide structures held together by three internal disulfide bonds and termed *kringles*. Plasminogen has five kringles numbered I to V. Apo(a) has a variable number of kringles, with high homology to plasminogen kringle IV, and a single copy homologous to kringle V. It also has a region that is homologous to the plasminogen protease domain, but it does not have a tissue plasminogen activator cleavage site. Cys4057, in kringle 36 of the published cDNA sequence, one of the kringle IV repeats, forms a disulfide bond with apoB Cys4326. In the noncovalent interaction, apoB lysine 680 plays a critical role. The size polymorphism, which is reflected in the molecular weight range (<300 to >800 kDa) of more than 34 different apo(a) isoforms, can be detected by pulsed-field gel electrophoresis after digestion with KpnI. In addition to these size polymorphisms, 14 SNPs have been reported. There is a common Met to Thr polymorphism in kringle IV type 10. A Trp-to-Arg mutation, also in kringle IV type 10, results in Lp(a) that is defective in binding to the lysine residues of fibrin. A C/T polymorphism in the 5′ untranslated region of the gene is associated with decreased *in vitro* translation.

When fed a high-fat diet, transgenic mice (a species in which Lp(a) is absent) that express human apo(a) had larger areas of lipid-staining atherosclerotic lesions than control animals did. The apo(a) in these animals is not in the form of Lp(a); that is, it is not covalently attached to apoB. However, when these mice were crossed with mice expressing human apoB100, high levels of human Lp(a) were found in the plasma.

6.13 Other Apolipoproteins

6.13.1 Apolipoprotein D

The sialoglycoprotein apoD is related to retinol-binding protein and both are members of the lipocalin superfamily. ApoD shows no homology with the other apolipoproteins. In humans, the gene is expressed in a number of tissues, notably the adrenal glands, cerebellum, spleen, kidney, and pancreas. There is much less expression in the liver and intestine. Cleavage of the 20-residue signal peptide gives rise to the mature form in which the N-terminal Gln is blocked by cyclization. Much of it exists as a 38-kDa heterodimer with apoA-II. Isoelectric focusing detects a number of isoforms with variable numbers of sialic acid residues. It is mainly found in the HDL fraction where it constitutes about 5% of the total protein. It binds to a number of ligands, including progesterone, cholesterol, bilirubin, and arachidonic acid. ApoD possibly plays a role, alongside LCAT and CETP in reverse cholesterol transport. There is increasing evidence that apoD is important for nerve regeneration.

Putative regulatory domains have been reported in the apoD promoter region. Response elements include those for estrogen, progesterone, glucocorticoid, and thyroid hormone. The promoter contains a 26-bp element that could form a Z-DNA structure. MspI and TaqI RFLPs have been studied.

6.13.2 Apolipoprotein H

The physiologic function of apoH has not been determined with any certainty. It is also referred to as *beta-2-glycoprotein I*. In addition to being involved in lipoprotein metabolism, it has also been reported to be involved in coagulation, and is a cofactor for the binding of antiphospholipid autoantibodies to anionic phospholipid. These antibodies are found, for example, in patients with lupus erythematosus. The gene for apoH is alternatively spliced and probably produces a number of distinct polypeptide products.

6.13.3 Apolipoprotein J

An abundant and widely dispersed glycoprotein, ApoJ, also referred to as *clusterin* and *SP-40/40*, is expressed at a high level in the epithelial cells of many tissues. A constituent of the terminal complement complex, apoJ is a multifunctional glycoprotein that inhibits complement-mediated cytolysis. It is found on an unstable subclass of HDL particles that have $\alpha2$ electrophoretic mobility, only a small proportion of which contain apoA-I. ApoJ and apoA-I associate *in vitro*. ApoJ is an inhibitor of the complement membrane-attack complex. ApoJ is stored in platelet granules and released after platelet activation, and it may be involved in wound repair. The other functions of apoJ are thought to include roles in sperm maturation, lipid transport, programmed cell death, and membrane recycling.

6.13.4 Apolipoprotein L Gene Family

ApoL-I, reported to be the trypanosome lytic factor, was discovered on a subpopulation of HDL particles in HDL isolated by selected-affinity immunosorption. The gene for apoL-I is a member of a family of six genes; however, only apoL-I seems to be truly an apolipoprotein. Microarray analysis of prefrontal cortex tissue from schizophrenia subjects showed up-regulation of the *ApoL-I, ApoL-II,* and *ApoL-IV* genes.

7 Scavenger Receptors

7.1 SR-BI/CLA1

The scavenger receptor class B type I, SR-BI, which binds directly to cholesterol, is able to alter the distribution of membrane cholesterol pools, and is localized to cholesterol-rich rafts. It facilitates the selective uptake by the liver of cholesteryl esters from HDL, thereby playing a crucial role in reverse cholesterol transport. SR-BI binds through a C-terminal domain to the multifunctional PDZK1 protein, which may play a scaffolding role. PDZK1 knockout mice have increased serum levels of cholesterol.

Previously, it was thought that initial binding of HDL to the receptor was followed by diffusion of lipid into the plasma membrane down a concentration gradient. It is now believed that, following binding to SR-BI, the complete HDL particle undergoes endocytosis. The cholesteryl esters are hydrolyzed in early endosomes and the HDL particle is then recycled to the plasma membrane and secreted. SR-BI expression correlates with the secretion of cholesterol into bile, SR-BI being expressed in canalicular as well as sinusoidal membranes. Hence, it is thought that SR-BI helps to mediate the secretion of cholesterol into bile, possibly in conjunction with phospholipid transporters such as ABCB4 and sterol transporters ABCG5/G8. In peripheral tissues, SR-BI mediates the diffusion of free cholesterol from the plasma membrane to *alpha* HDL particles. In macrophages, it plays an important antiatherogenic role.

The human gene for SR-BI (also known as *CLAI* and *CD36L1*) is *SCARB1* and is located at chromosome 12q24.31. It spans 86 kbp and has 13 exons. It codes for a 509-residue, 57-kDa, protein, which is N-glycosylated at a number of sites, giving a total molecular weight of 82 to 86 kDa. Both the N- and C-termini are located in the cytoplasm with two transmembrane domains separated by a 408-residue extracellular domain. As a result of alternative splicing, there is a second isoform called *SR-BII,* with a different C-terminus.

7.2 Modified LDL Receptors: MSR1, CD36 and LOX1

Oxidized or chemically modified LDL are taken up by macrophages *via* receptor-mediated endocytosis, independent of the LDL receptor. This may be the mechanism, *in vivo,* for foam cell formation, leading to atherosclerotic lesions. Lipoproteins containing apoB have been isolated from normal and atheromatous arteries.

Initially, from bovine liver, a glycoprotein was isolated that binds to acetylated LDL, oxidized LDL, and endotoxin (bacterial lipopolysaccharide), and was referred to as the *acetylated LDL receptor* or *scavenger receptor*. The ligands for this receptor are numerous and include polyanionic compounds, such as polyribonucleotides. Cloning of the receptor, SR-A, now referred to as the *macrophage scavenger receptor 1 (MSR1)*, revealed the presence of two homologous proteins, the type I and type II class A scavenger receptors. These are produced from a single MSR1 gene by alternative splicing. The human gene at chromosome 8p22 has 10 exons. Isoform type I has 451 residues (50 kDa) and type II 388 residues (38 kDa). They have identical structures except for the C-terminus. The common structures include an NH_2-terminal cytoplasmic tail, a transmembrane region and a collagen-like domain. The active receptors exist as trimers. A third isoform exists, that does not internalize acetyl-LDL and by inhibiting the first two isoforms plays a dominant negative function. Despite increased foam cell formation in cells overexpressing MSR1 *in vitro*, decreased atherosclerosis was observed in MSR1 transgenic mice.

The macrophage receptor, CD36, was found to bind oxidized LDL with high affinity. It belongs to the family of class B scavenger receptors, which includes SR-BI and LIMP-2. CD36 is a multifunctional receptor that is also referred to as the *collagen type I receptor*, and the *thrombospondin receptor*. In phagocytic cells, it is a scavenger receptor that recognizes and removes apoptotic cells. Among the numerous roles attributed to it, CD36 is also thought to function as a fatty acid translocase (FAT) in muscle and adipose tissue. CD36 knockout mice develop hyperlipidemia, and transgenic animals develop hypolipidemia. Supporting the notion that CD36, as a scavenger receptor for oxidized atherogenic lipoproteins, is of major importance in the development of atherosclerosis are data with CD36 knockout mice crossed with apoE-deficient animals. These animals, when put on a high fat diet, developed greatly decreased aortic lesions.

The *CD36* gene has 16 exons, spans 38 kbp, and is located on chromosome 7q21.11. It encodes a 471-amino acid-residue transmembrane protein that is highly glycosylated. Feed-forward upregulation of the *CD36* gene by oxidized LDL is *via* a PPARγ response element in its promoter that is activated by PPARγ ligands such as prostaglandin J2. CD36 is also induced by interleukin-4. During inflammation, it is downregulated *via* TGF-β1 and lipopolysaccharide.

A member of the C-type lectin gene family, the oxidized LDL (lectin-like) receptor 1 (OLR1), also referred to as *LOX1*, has been shown to bind and internalize oxidized LDL. The *OLR1* gene is located on chromosome 12p13.2 with a cluster of natural killer cell receptor genes. Its six exons span 14 kb, and encode a 273-residue protein. It is expressed in a number of tissues, including brain, testis, kidney, and aorta, with the highest levels of mRNA in lung, spinal cord, and placenta. OLR1 is upregulated by a number of cytokines. Case-control studies indicate that polymorphisms at the OLR1 locus are associated with risk of CAD and acute myocardial infarction. There have been conflicting reports on associations with Alzheimer disease.

8 Lipids and Atherosclerosis

Disorders in the transport and metabolism of lipids contribute materially to the development of arteriosclerotic heart disease, stroke, and peripheral vascular disease. The histological hallmark of atherogenesis is the lipid-filled macrophage, which is transformed into a proinflammatory element. Smooth muscle cells can also be transformed into lipid-filled cells that participate in atherogenesis. The accumulation of oxidized lipids leads to the formation of neoantigens and increases oxidative stress, evoking an array of inflammatory mechanisms, ultimately leading to the attenuation of coronary blood flow or its cessation in some regions of the coronary vasculature, resulting in myocardial infarction.

The cell biology of these processes is highly complex, involving many factors that reflect genomic dependency. Rapid progress in the study of the genomic basis of disorders of lipid metabolism and of coronary disease is leading to the investigation of novel genes, thereby opening new avenues for understanding of the mechanisms underlying vascular diseases.

See also Adipocytes; Gene Therapy and Cardiovascular Diseases; Intracellular Fatty Acid Binding Proteins in Metabolic Regulation.

Bibliography

Books and Reviews

Brunzell, J.D., Deeb, S.S. (2001) Familial Lipoprotein Lipase Deficiency, Apo C-II Deficiency, and Hepatic Lipase Deficiency, in: Scriver, C.R., Beaudet, A.L., Sly, W.S., Valle, D. (Eds.) *The Metabolic and Molecular Bases of Inherited Disease*, 8th edition, McGraw-Hill, New York, pp. 2789–2816.

Goldstein, J.L., Hobbs, H.H., Brown, M.S. (2001) Familial Hypercholesterolemia, in: Scriver, C.R., Beaudet, A.L., Sly, W.S., Valle, D. (Eds.) *The Metabolic and Molecular Bases of Inherited Disease*, 8th edition, McGraw-Hill, New York, pp. 2863–2913.

Havel, R.J., Kane, J.P. (2001) Introduction: Structure and Metabolism of Plasma Lipoproteins, in: Scriver, C.R., Beaudet, A.L., Sly, W.S., Valle, D. (Eds.) *The Metabolic and Molecular Bases of Inherited Disease*, 8th edition, McGraw-Hill, New York, pp. 2705–2716.

Kane, J.P., Havel, R.J. (2001) Disorders of the Biogenesis and Secretion of Lipoproteins Containing the B-apolipoproteins, in: Scriver, C.R., Beaudet, A.L., Sly, W.S., Valle, D. (Eds.) *The Metabolic and Molecular Bases of Inherited Disease*, 8th edition, McGraw-Hill, New York, pp. 2717–2752.

Pullinger, C.R., Kane, J.P., Malloy, M.J. (2003) Primary hypercholesterolemia: genetic causes and treatment of five monogenic disorders, *Expert Rev. Cardiovasc. Ther.* **1**, 107–119.

Tall, A.R., Breslow, J.L., Rubin, E.M. (2001) Genetic Disorders Affecting High-Density Lipoprotein Metabolism, in: Scriver, C.R., Beaudet, A.L., Sly, W.S., Valle, D. (Eds.) *The Metabolic and Molecular Bases of Inherited Disease*, 8th edition, McGraw Hill, New York, pp. 2915–2936.

Utermann, G. (2001) Lipoprotein(a), in: Scriver, C.R., Beaudet, A.L., Sly, W.S., Valle, D. (Eds.) *The Metabolic and Molecular Bases of Inherited Disease*, 8th edition, McGraw-Hill, New York, pp. 2753–2787.

Primary Literature

Aalto-Setala, K., Weinstock, P.H., Bisgaier, C.L., Wu, L., Smith, J.D., Breslow, J.L. (1996) Further characterization of the metabolic properties of triglyceride-rich lipoproteins from human and mouse apoC-III transgenic mice, *J. Lipid Res.* **37**, 1802–1811.

Abifadel, M., Varret, M., Rabes, J.P., Allard, D., Ouguerram, K., Devillers, M., Cruaud, C., Benjannet, S., Wickham, L., Erlich, D., Derre, A., Villeger, L., Farnier, M., Beucler, I., Bruckert, E., Chambaz, J.,

Chanu, B., Lecerf, J.M., Luc, G., Moulin, P., Weissenbach, J., Prat, A., Krempf, M., Junien, C., Seidah, N.G., Boileau, C. (2003) Mutations in PCSK9 cause autosomal dominant hypercholesterolemia, *Nat. Genet.* **34**, 154–156.

Adimoolam, S., Jin, L., Grabbe, E., Shieh, J.J., Jonas, A. (1998) Structural and functional properties of two mutants of lecithin-cholesterol acyltransferase (T123I and N228K), *J. Biol. Chem.* **273**, 32561–32567.

Allan, C.M., Walker, D., Segrest, J.P., Taylor, J.M. (1995) Identification and characterization of a new human gene (APOC4) in the apolipoprotein E, C-I, and C-II gene locus, *Genomics* **28**, 291–300.

Altmann, S.W., Davis, H.R., Jr., Zhu, L.J., Yao, X., Hoos, L.M., Tetzloff, G., Iyer, S.P., Maguire, M., Golovko, A., Zeng, M., Wang, L., Murgolo, N., Graziano, M.P. (2004) Niemann-Pick C1 Like 1 protein is critical for intestinal cholesterol absorption, *Science* **303**, 1201–1204.

Anderson, C.M., Townley, R.R., Freemanm, Johansen, P. (1961) Unusual causes of steatorrhea in infancy and childhood, *Med. J. Aust.* (2), 617–622.

Aouizerat, B.E., Allayee, H., Bodnar, J., Krass, K.L., Peltonen, L., de Bruin, T.W., Rotter, J.I., Lusis, A.J. (1999) Novel genes for familial combined hyperlipidemia, *Curr. Opin. Lipidol.* **10**, 113–122.

Beffert, U., Stolt, P.C., Herz, J. (2004) Functions of lipoprotein receptors in neurons, *J. Lipid. Res.* **45**, 403–409.

Berge, K.E., Tian, H., Graf, G.A., Yu, L., Grishin, N.V., Schultz, J., Kwiterovich, P., Shan, B., Barnes, R., Hobbs, H.H. (2000) Accumulation of dietary cholesterol in sitosterolemia caused by mutations in adjacent ABC transporters, [Comment In: *Science* 2000 Dec 1;**290**(5497), 1709–11 UI: 20559920] *Science* **290**, 1771–1775.

Bhattacharyya, A.K., Connor, W.E. (1974) Beta-sitosterolemia and xanthomatosis. A newly described lipid storage disease in two sisters, *J. Clin. Invest.* **53**, 1033–1043.

Bodzioch, M., Orsó, E., Klucken, J., Langmann, T., Böttcher, A., Diederich, W., Drobnik, W., Barlage, S., Büchler, C., Porsch-Ozcürümez, M., Kaminski, W.E., Hahmann, H.W., Oette, K., Rothe, G., Aslanidis, C., Lackner, K.J., Schmitz, G. (1999) The gene encoding ATP-binding cassette transporter 1 is mutated in Tangier disease [see comments], *Nat. Genet.* **22**, 347–351.

Boren, J., Ekstrom, U., Agren, B., Nilsson-Ehle, P., Innerarity, T.L. (2001) The molecular mechanism for the genetic disorder familial defective apolipoprotein B100, *J. Biol. Chem.* **276**, 9214–9218.

Breitling, R., Krisans, S.K. (2002) A second gene for peroxisomal HMG-CoA reductase? A genomic reassessment, *J. Lipid Res.* **43**, 2031–2036.

Briggs, M.R., Yokoyama, C., Wang, X., Brown, M.S., Goldstein, J.L. (1993) Nuclear protein that binds sterol regulatory element of low density lipoprotein receptor promoter. I. Identification of the protein and delineation of its target nucleotide sequence, *J. Biol. Chem.* **268**, 14490–14496.

Brooks-Wilson, A., Marcil, M., Clee, S.M., Zhang, L.H., Roomp, K., van Dam, M., Yu, L., Brewer, C., Collins, J.A., Molhuizen, H.O., Loubser, O., Ouelette, B.F., Fichter, K., Ashbourne-Excoffon, K.J., Sensen, C.W., Scherer, S., Mott, S., Denis, M., Martindale, D., Frohlich, J., Morgan, K., Koop, B., Pimstone, S., Kastelein, J.J., Hayden, M.R. et al (1999) Mutations in ABC1 in Tangier disease and familial high-density lipoprotein deficiency [see comments], *Nat. Genet.* **22**, 336–345.

Brunzell, J.D., Deeb, S.S. (2001) Familial Lipoprotein Lipase Deficiency, Apo C-II Deficiency, and Hepatic Lipase Deficiency, in: Scriver, C.R., Beaudet, A.L., Sly, W.S., Valle, D. (Eds.) *The Metabolic and Molecular Bases of Inherited Disease*, 8th edition, McGraw-Hill, New York, pp. 2789–2816.

Buhman, K.K., Accad, M., Novak, S., Choi, R.S., Wong, J.S., Hamilton, R.L., Turley, S., Farese, R.V. Jr. (2000) Resistance to diet-induced hypercholesterolemia and gallstone formation in ACAT2-deficient mice, *Nat. Med.* **6**, 1341–1347.

Carstea, E.D., Morris, J.A., Coleman, K.G., Loftus, S.K., Zhang, D., Cummings, C., Gu, J., Rosenfeld, M.A., Pavan, W.J., Krizman, D.B., Nagle, J., Polymeropoulos, M.H., Sturley, S.L., Ioannou, Y.A., Higgins, M.E., Comly, M., Cooney, A., Brown, A., Kaneski, C.R., Blanchette-Mackie, E.J., Dwyer, N.K., Neufeld, E.B., Chang, T.Y., Liscum, L., Tagle, D.A. et al (1997) Niemann-Pick C1 disease gene: homology to mediators of cholesterol homeostasis, *Science* **277**, 228–231.

Chang, T.Y., Chang, C.C., Lu, X., Lin, S. (2001) Catalysis of ACAT may be completed within the plane of the membrane: a working hypothesis, *J. Lipid Res.* **42**, 1933–1938.

Chatterton, J.E., Phillips, M.L., Curtiss, L.K., Milne, R., Fruchart, J.C., Schumaker, V.N. (1995) Immunoelectron microscopy of low density lipoproteins yields a ribbon and bow model for the conformation of apolipoprotein B on the lipoprotein surface, *J. Lipid Res.* **36**, 2027–2037.

Chen, W.J., Goldstein, J.L., Brown, M.S. (1990) NPXY, a sequence often found in cytoplasmic tails, is required for coated pit-mediated internalization of the low density lipoprotein receptor, *J. Biol. Chem.* **265**, 3116–3123.

Chester, A., Scott, J., Anant, S., Navaratnam, N. (2000) RNA editing: cytidine to uridine conversion in apolipoprotein B mRNA, *Biochim. Biophys. Acta* **1494**, 1–13.

Chiang, J.Y. (2002) Bile acid regulation of gene expression: roles of nuclear hormone receptors, *Endocr. Rev.* **23**, 443–463.

Chiang, J.Y., Kimmel, R., Stroup, D. (2001) Regulation of cholesterol 7alpha-hydroxylase gene (CYP7A1) transcription by the liver orphan receptor (LXRalpha), *Gene* **262**, 257–265.

Chinetti, G., Fruchart, J.C., Staels, B. (2000) Peroxisome proliferator-activated receptors (PPARs): nuclear receptors at the crossroads between lipid metabolism and inflammation, *Inflamm. Res.* **49**, 497–505.

Cohen, J.C. (1999) Contribution of cholesterol 7alpha-hydroxylase to the regulation of lipoprotein metabolism, *Curr. Opin. Lipidol.* **10**, 303–307.

Cohen, P., Miyazaki, M., Socci, N.D., Hagge-Greenberg, A., Liedtke, W., Soukas, A.A., Sharma, R., Hudgins, L.C., Ntambi, J.M., Friedman, J.M. (2002) Role for stearoyl-CoA desaturase-1 in leptin-mediated weight loss, *Science* **297**, 240–243.

Cohn, J.S., Tremblay, M., Boulet, L., Jacques, H., Davignon, J., Roy, M., Bernier, L. (2003) Plasma concentration and lipoprotein distribution of ApoC-I is dependent on ApoE genotype rather than the Hpa I ApoC-I promoter polymorphism, *Atherosclerosis* **169**, 63–70.

Connelly, P.W., Hegele, R.A. (1998) Hepatic lipase deficiency, *Crit. Rev. Clin. Lab. Sci.* **35**, 547–572.

Couture, P., Otvos, J.D., Cupples, L.A., Lahoz, C., Wilson, P.W., Schaefer, E.J., Ordovas, J.M. (2000) Association of the C-514T polymorphism in the hepatic lipase gene with variations in lipoprotein subclass profiles: The Framingham Offspring Study, *Arterioscler. Thromb. Vasc. Biol.* **20**, 815–822.

Duchateau, P.N., Pullinger, C.R., Orellana, R.E., Kunitake, S.T., Naya-Vigne, J., O'Connor, P.M., Malloy, M.J., Kane, J.P. (1997) Apolipoprotein L, a new human high-density lipoprotein apolipoprotein expressed by the pancreas. Identification, cloning, characterization, and plasma distribution of apolipoprotein L, *J. Biol. Chem.* **272**, 25576–25582.

Dugi, K.A., Dichek, H.L., Santamarina-Fojo, S. (1995) Human hepatic and lipoprotein lipase: the loop covering the catalytic site mediates lipase substrate specificity, *J. Biol. Chem.* **270**, 25396–25401.

Edwards, P.A., Kast, H.R., Anisfeld, A.M. (2002) BAREing it all: the adoption of LXR and FXR and their roles in lipid homeostasis, *J. Lipid Res.* **43**, 2–12.

Endemann, G., Stanton, L.W., Madden, K.S., Bryant, C.M., White, R.T., Protter, A.A. (1993) CD36 is a receptor for oxidized low density lipoprotein, *J. Biol. Chem.* **268**, 11811–11816.

Erickson, S.K., Lear, S.R., Deane, S., Dubrac, S., Huling, S.L., Nguyen, L., Bollineni, J.S., Shefer, S., Hyogo, H., Cohen, D.E., Shneider, B., Sehayek, E., Ananthanarayanan, M., Balasubramaniyan, N., Suchy, F.J., Batta, A.K., Salen, G. (2003) Hypercholesterolemia and changes in lipid and bile acid metabolism in male and female cyp7A1-deficient mice, *J. Lipid Res.* **44**, 1001–1009.

Febbraio, M., Podrez, E.A., Smith, J.D., Hajjar, D.P., Hazen, S.L., Hoff, H.F., Sharma, K., Silverstein, R.L. (2000) Targeted disruption of the class B scavenger receptor CD36 protects against atherosclerotic lesion development in mice, *J. Clin. Invest.* **105**, 1049–1056.

Forman, B.M., Goode, E., Chen, J., Oro, A.E., Bradley, D.J., Perlmann, T., Noonan, D.J., Burka, L.T., McMorris, T., Lamph, W.W. (1995) Identification of a nuclear receptor that is activated by farnesol metabolites, *Cell* **81**, 687–693.

Garcia, C.K., Wilund, K., Arca, M., Zuliani, G., Fellin, R., Maioli, M., Calandra, S., Bertolini, S., Cossu, F., Grishin, N., Barnes, R., Cohen, J.C., Hobbs, H.H. (2001) Autosomal recessive hypercholesterolemia caused by mutations in a putative LDL receptor adaptor protein, *Science* **292**, 1394–1398.

Gillespie, J.G., Hardie, D.G. (1992) Phosphorylation and inactivation of HMG-CoA reductase at the AMP-activated protein kinase site in response to fructose treatment of isolated rat hepatocytes, *FEBS Lett.* **306**, 59–62.

Goldstein, J.L., Schrott, H.G., Hazzard, W.R., Bierman, E.L., Motulsky, A.G. (1973) Hyperlipidemia in coronary heart disease. II. Genetic analysis of lipid levels in 176 families and delineation of a new inherited disorder, combined hyperlipidemia, *J. Clin. Invest.* **52**, 1544–1568.

Graf, G.A., Li, W.P., Gerard, R.D., Gelissen, I., White, A., Cohen, J.C., Hobbs, H.H. (2002) Coexpression of ATP-binding cassette proteins ABCG5 and ABCG8 permits their transport to the apical surface, *J. Clin. Invest.* **110**, 659–669.

Groenendijk, M., Cantor, R.M., de Bruin, T.W., Dallinga-Thie, G.M. (2001) The apoAI-CIII-AIV gene cluster, *Atherosclerosis* **157**, 1–11.

Grundy, S.M., Vega, G.L. (1985) Influence of mevinolin on metabolism of low density lipoproteins in primary moderate hypercholesterolemia, *J. Lipid. Res.* **26**, 1464–1475.

Hall, S., Chu, G., Miller, G., Cruickshank, K., Cooper, J.A., Humphries, S.E., Talmud, P.J. (1997) A common mutation in the lipoprotein lipase gene promoter, −93T/G, is associated with lower plasma triglyceride levels and increased promoter activity in vitro, *Arterioscler. Thromb. Vasc. Biol.* **17**, 1969–1976.

Hegele, R.A., Little, J.A., Vezina, C., Maguire, G.F., Tu, L., Wolever, T.S., Jenkins, D.J., Connelly, P.W. (1993) Hepatic lipase deficiency. Clinical, biochemical, and molecular genetic characteristics, *Arterioscler. Thromb.* **13**, 720–728.

Hirata, K., Dichek, H.L., Cioffi, J.A., Choi, S.Y., Leeper, N.J., Quintana, L., Kronmal, G.S., Cooper, A.D., Quertermous, T. (1999) Cloning of a unique lipase from endothelial cells extends the lipase gene family, *J. Biol. Chem.* **274**, 14170–14175.

Holt, J.A., Luo, G., Billin, A.N., Bisi, J., McNeill, Y.Y., Kozarsky, K.F., Donahee, M., Wang da, Y., Mansfield, T.A., Kliewer, S.A., Goodwin, B., Jones, S.A. (2003) Definition of a novel growth factor-dependent signal cascade for the suppression of bile acid biosynthesis, *Genes Dev.* **17**, 1581–1591.

Innerarity, T.L., Balestra, M.E., Arnold, K.S., Mahley, R.W., Vega, G.L., Grundy, S.M., Young, S.G. (1988) Isolation of defective receptor-binding low density lipoproteins from subjects with familial defective apolipoprotein B-100, *Arteriosclerosis* **8**, 551a.

Issemann, I., Green, S. (1990) Activation of a member of the steroid hormone receptor superfamily by peroxisome proliferators, *Nature* **347**, 645–650.

Janowski, B.A., Willy, P.J., Devi, T.R., Falck, J.R., Mangelsdorf, D.J. (1996) An oxysterol signalling pathway mediated by the nuclear receptor LXR alpha, *Nature* **383**, 728–731.

Jaye, M., Lynch, K.J., Krawiec, J., Marchadier, D., Maugeais, C., Doan, K., South, V., Amin, D., Perrone, M., Rader, D.J. (1999) A novel endothelial-derived lipase that modulates HDL metabolism, *Nat. Genet.* **21**, 424–428.

Jiang, X.C., Bruce, C., Mar, J., Lin, M., Ji, Y., Francone, O.L., Tall, A.R. (1999) Targeted mutation of plasma phospholipid transfer protein gene markedly reduces high-density lipoprotein levels, *J. Clin. Invest.* **103**, 907–914.

Jira, P.E., Waterham, H.R., Wanders, R.J., Smeitink, J.A., Sengers, R.C., Wevers, R.A. (2003) Smith-Lemli-Opitz syndrome and the DHCR7 gene, *Ann. Hum. Genet.* **67**, 269–280.

Jones, B., Jones, E.L., Bonney, S.A., Patel, H.N., Mensenkamp, A.R., Eichenbaum-Voline, S., Rudling, M., Myrdal, U., Annesi, G., Naik, S., Meadows, N., Quattrone, A., Islam, S.A., Naoumova, R.P., Angelin, B., Infante, R., Levy, E., Roy, C.C., Freemont, P.S., Scott, J., Shoulders, C.C. (2003) Mutations in a Sar1 GTPase of COPII vesicles are associated with lipid absorption disorders, *Nat. Genet.* **34**, 29–31.

Kane, J.P., Aouizerat, B.E., Luke, M.M., Shiffman, D., Iakoubova, O., Liu, D., Rowland, C.M., Catanese, J.J., Leong, D.U., Lau, K.F., Louie, J.Z., Tong, C.H., McAllister, L.B., Dabby, L.F., Ports, T.A., Michaels, A.D., Zellner, C., Pullinger, C.R., Malloy, M.J., Devlin, J.J. (2004) Novel Genetic Markers for Structural Coronary Artery Disease, Myocardial Infarction, and Familial Combined Hyperlipidemia: Candidate and Genome Scans of Functional SNPS, in: Matsuzawa, T.K.Y., Nagai, R., Teramoto, T. (Eds.) *Atherosclerosis XIII: Proceedings of the 13th International Symposium*, Elsevier, Kyoto, Japan, pp. 309–312.

Karathanasis, S., Ferris, E., Haddad, I. (1987) DNA inversion with the apolipoproteins

AI/CIII/AIV encoding gene clusters of certain patients with premature atherosclerosis, *Proc. Natl. Acad. Sci. U.S.A.* **84**, 7198–7202.

Kast, H.R., Nguyen, C.M., Sinal, C.J., Jones, S.A., Laffitte, B.A., Reue, K., Gonzalez, F.J., Willson, T.M., Edwards, P.A. (2001) Farnesoid X-activated receptor induces apolipoprotein C-II transcription: a molecular mechanism linking plasma triglyceride levels to bile acids, *Mol. Endocrinol.* **15**, 1720–1728.

Kelley, R.L., Roessler, E., Hennekam, R.C., Feldman, G.L., Kosaki, K., Jones, M.C., Palumbos, J.C., Muenke, M. (1996) Holoprosencephaly in RSH/Smith-Lemli-Opitz syndrome: does abnormal cholesterol metabolism affect the function of Sonic hedgehog? *Am. J. Med. Genet.* **66**, 478–484.

Koschinsky, M.L., Marcovina, S.M. (2004) Structure-function relationships in apolipoprotein(a): insights into lipoprotein(a) assembly and pathogenicity, *Curr. Opin. Lipidol.* **15**, 167–174.

Kuivenhoven, J.A., Pritchard, H., Hill, J., Frohlich, J., Assmann, G., Kastelein, J. (1997) The molecular pathology of lecithin:cholesterol acyltransferase (LCAT) deficiency syndromes, *J. Lipid Res.* **38**, 191–205.

Langer, T., Strober, W., Levy, R.I. (1972) The metabolism of low density lipoprotein in familial type II hyperlipoproteinemia, *J. Clin. Invest.* **51**, 1528–1536.

Lawn, R.M., Wade, D.P., Garvin, M.R., Wang, X., Schwartz, K., Porter, J.G., Seilhamer, J.J., Vaughan, A.M., Oram, J.F. (1999) The Tangier disease gene product ABC1 controls the cellular apolipoprotein-mediated lipid removal pathway [see comments], *J. Clin. Invest.* **104**, R25–R31.

Lee, M.H., Lu, K., Hazard, S., Yu, H., Shulenin, S., Hidaka, H., Kojima, H., Allikmets, R., Sakuma, N., Pegoraro, R., Srivastava, A.K., Salen, G., Dean, M., Patel, S.B. (2001) Identification of a gene, ABCG5, important in the regulation of dietary cholesterol absorption, *Nat. Genet.* **27**, 79–83.

Li, B.L., Li, X.L., Duan, Z.J., Lee, O., Lin, S., Ma, Z.M., Chang, C.C., Yang, X.Y., Park, J.P., Mohandas, T.K., Noll, W., Chan, L., Chang, T.Y. (1999) Human acyl-CoA:cholesterol acyltransferase-1 (ACAT-1) gene organization and evidence that the 4.3-kilobase ACAT-1 mRNA is produced from two different chromosomes, *J. Biol. Chem.* **274**, 11060–11071.

Linton, M.F., Pierotti, V., Young, S.G. (1992) Reading frame restoration with an apolipoprotein B gene frameshift mutation, *Proc. Natl. Acad. Sci. U.S.A.*; **89**, 11431–11435.

Lu, H., Inazu, A., Moriyama, Y., Higashikata, T., Kawashiri, M.A., Yu, W., Huang, Z., Okamura, T., Mabuchi, H. (2003) Haplotype analyses of cholesteryl ester transfer protein gene promoter: a clue to an unsolved mystery of TaqIB polymorphism, *J. Mol. Med.* **81**, 246–255.

Lu, T.T., Makishima, M., Repa, J.J., Schoonjans, K., Kerr, T.A., Auwerx, J., Mangelsdorf, D.J. (2000) Molecular basis for feedback regulation of bile acid synthesis by nuclear receptors, *Mol. Cell.* **6**, 507–515.

Mak, P.A., Laffitte, B.A., Desrumaux, C., Joseph, S.B., Curtiss, L.K., Mangelsdorf, D.J., Tontonoz, P., Edwards, P.A. (2002) Regulated expression of the apolipoprotein E/C-I/C-IV/C-II gene cluster in murine and human macrophages. A critical role for nuclear liver X receptors alpha and beta, *J. Biol. Chem.* **277**, 31900–31908.

Mar, R., Pajukanta, P., Allayee, H., Groenendijk, M., Dallinga-Thie, G., Krauss, R.M., Sinsheimer, J.S., Cantor, R.M., de Bruin, T.W., Lusis, A.J. (2004) Association of the APOLIPOPROTEIN A1/C3/A4/A5 gene cluster with triglyceride levels and LDL particle size in familial combined hyperlipidemia, *Circ. Res.* **94**, 993–999.

Mehta, A., Kinter, M.T., Sherman, N.E., Driscoll, D.M. (2000) Molecular cloning of apobec-1 complementation factor, a novel RNA-binding protein involved in the editing of apolipoprotein B mRNA, *Mol. Cell. Biol.* **20**, 1846–1854.

Mimmack, M.L., Ryan, M., Baba, H., Navarro-Ruiz, J., Iritani, S., Faull, R.L., McKenna, P.J., Jones, P.B., Arai, H., Starkey, M., Emson, P.C., Bahn, S. (2002) Gene expression analysis in schizophrenia: reproducible up-regulation of several members of the apolipoprotein L family located in a high-susceptibility locus for schizophrenia on chromosome 22, *Proc. Natl. Acad. Sci. U. S. A.* **99**, 4680–4685.

Mishra, S.K., Watkins, S.C., Traub, L.M. (2002) The autosomal recessive hypercholesterolemia (ARH) protein interfaces directly with the clathrin-coat machinery, *Proc. Natl. Acad. Sci. U. S. A.* **99**, 16099–16104.

Naureckiene, S., Sleat, D.E., Lackland, H., Fensom, A., Vanier, M.T., Wattiaux, R., Jadot, M., Lobel, P. (2000) Identification of HE1 as the second gene of Niemann-Pick C disease, *Science* **290**, 2298–2301.

Nishimaki-Mogami, T., Une, M., Fujino, T., Sato, Y., Tamehiro, N., Kawahara, Y., Shudo, K., Inoue, K. (2004) Identification of intermediates in the bile acid synthetic pathway as ligands for the farnesoid X receptor, *J. Lipid. Res.* **45**, 1538–1545.

Paszty, C., Maeda, N., Verstuyft, J., Rubin, E.M. (1994) Apolipoprotein AI transgene corrects apolipoprotein E deficiency-induced atherosclerosis in mice, *J. Clin. Invest.* **94**, 899–903.

Pennacchio, L.A., Olivier, M., Hubacek, J.A., Cohen, J.C., Cox, D.R., Fruchart, J.C., Krauss, R.M., Rubin, E.M. (2001) An apolipoprotein influencing triglycerides in humans and mice revealed by comparative sequencing, [Comment In: *Science*. 2001 Oct 5; **294**(5540), 33 UI: 21472613 Comment In: *Science*. 2001 Oct 5; **294**(5540), 33 UI: 21472614] *Science* **294**, 169–173.

Porter, J.A., Young, K.E., Beachy, P.A. (1996) Cholesterol modification of hedgehog signaling proteins in animal development, *Science* **274**, 255–259.

Pullinger, C.R., Malloy, M.J., Shahidi, A.K., Ghassemzadeh, M., Duchateau, P., Villagomez, J., Allaart, J., Kane, J.P. (1997) A novel apolipoprotein C-III variant, apoC-III(Gln38→Lys), associated with moderate hypertriglyceridemia in a large kindred of Mexican origin, *J. Lipid Res.* **38**, 1833–1840.

Pullinger, C.R., Hennessy, L.K., Chatterton, J.E., Liu, W.Q., Love, J.A., Mendel, C.M., Frost, P.H., Malloy, M.J., Schumaker, V.N., Kane, J.P. (1995) Familial ligand-defective apolipoprotein B – Identification of a new mutation that decreases LDL receptor binding affinity, *J. Clin. Invest.* **95**, 1225–1234.

Pullinger, C.R., Eng, C., Salen, G., Shefer, S., Batta, A.K., Erickson, S.K., Verhagen, A., Rivera, C.R., Mulvihill, S.J., Mary, J., Malloy, M.J., Kane, J.P. (2002) Human Cholesterol 7α-Hydroxylase (CYP7A1) Deficiency has a Hypercholesterolemic Phenotype, *J. Clin. Invest.* **110**, 109–117.

Qin, S., Kawano, K., Bruce, C., Lin, M., Bisgaier, C., Tall, A.R., Jiang, X. (2000) Phospholipid transfer protein gene knock-out mice have low-high density lipoprotein levels, due to hypercatabolism, and accumulate apoA-IV-rich lamellar lipoproteins, *J. Lipid Res.* **41**, 269–276.

Rassart, E., Bedirian, A., Do Carmo, S., Guinard, O., Sirois, J., Terrisse, L., Milne, R. (2000) Apolipoprotein D, *Biochim. Biophys. Acta* **1482**, 185–198.

Rawson, R.B. (2003) The SREBP pathway–insights from Insigs and insects, *Nat. Rev. Mol. Cell Biol.* **4**, 631–640.

Remaley, A.T., Schumacher, U.K., Stonik, J.A., Farsi, B.D., Nazih, H., Brewer, H.B. Jr. (1997) Decreased reverse cholesterol transport from Tangier disease fibroblasts. Acceptor specificity and effect of brefeldin on lipid efflux, *Arterioscler. Thromb. Vasc. Biol.* **17**, 1813–1821.

Rose, H.G., Kranz, P., Weinstock, M., Juliano, J., Haft, J.I. (1973) Inheritance of combined hyperlipoproteinemia:evidence for a new lipoprotein phenotype, *Am. J. Med.* **54**, 148–160.

Rudenko, G., Henry, L., Henderson, K., Ichtchenko, K., Brown, M.S., Goldstein, J.L., Deisenhofer, J. (2002) Structure of the LDL receptor extracellular domain at endosomal pH, *Science* **298**, 2353–2358.

Rust, S., Rosier, M., Funke, H., Real, J., Amoura, Z., Piette, J.C., Deleuze, J.F., Brewer, H.B., Duverger, N., Denefle, P., Assmann, G. (1999) Tangier disease is caused by mutations in the gene encoding ATP-binding cassette transporter 1, *Nat. Genet.* **22**, 352–355.

Schonfeld, G. (2003) Familial hypobetalipoproteinemia: a review, *J. Lipid Res.* **44**, 878–883.

Schwarz, M., Lund, E.G., Setchell, K.D.R., Kayden, H.J., Zerwekh, J.E., Björkhem, I., Herz, J., Russell, D.W. (1996) Disruption of cholesterol 7alpha-hydroxylase gene in mice. II. Bile acid deficiency is overcome by induction of oxysterol 7alpha-hydroxylase, *J Biol. Chem.* **271**, 18024–18031.

Segrest, J.P., Jones, M.K., De Loof, H., Dashti, N. (2001) Structure of apolipoprotein B-100 in low density lipoproteins, *J. Lipid Res.* **42**, 1346–1367.

Setchell, K.D., Schwarz, M., O'Connell, N.C., Lund, E.G., Davis, D.L., Lathe, R., Thompson, H.R., Weslie Tyson, R., Sokol, R.J., Russell, D.W. (1998) Identification of a new inborn error in bile acid synthesis: mutation of the oxysterol 7alpha-hydroxylase gene causes severe neonatal liver disease, *J. Clin. Invest.* **102**, 1690–1703.

Shen, P., Howlett, G.J. (1992) Two coding regions closely linked to the rat apolipoprotein E gene: nucleotide sequences of rat apolipoprotein C-I and ECL cDNA, *Arch. Biochem. Biophys.* **297**, 345–353.

Silver, D.L., Tall, A.R. (2001) The cellular biology of scavenger receptor class B type I, *Curr. Opin. Lipidol.* **12**, 497–504.

Simonet, W.S., Bucay, N., Lauer, S.J., Taylor, J.M. (1993) A far-downstream hepatocyte-specific control region directs expression of the linked human apolipoprotein E and C-I genes in transgenic mice, *J. Biol. Chem.* **268**, 8221–8229.

Smith, D.W., Lemli, L., Opitz, J.M. (1964) A newly recognized syndrome of multiple congenital anomalies, *J. Pediatr.* **64**, 210–217.

Smith, J.L., Hardie, I.R., Pillay, S.P., de Jersey, J. (1990) Hepatic acyl-coenzyme A:cholesterol acyltransferase activity is decreased in patients with cholesterol gallstones, *J. Lipid Res.* **31**, 1993–2000.

Smith, J.L., Rangaraj, K., Simpson, R., Maclean, D.J., Nathanson, L.K., Stuart, K.A., Scott, S.P., Ramm, G.A., de Jersey, J. (2004) Quantitative analysis of the expression of ACAT genes in human tissues by real-time PCR, *J. Lipid Res.* **45**, 686–696.

Soria, L.F., Ludwig, E.H., Clarke, H.R.G., Vega, G.L., Grundy, S.M., McCarthy, B.J. (1989) Association between a specific apolipoprotein B mutation and familial defective apolipoprotein B-100, *Proc. Natl. Acad. Sci. U.S.A.* **86**, 587–591.

Steffensen, K.R., Gustafsson, J.A. (2004) Putative metabolic effects of the liver X receptor (LXR), *Diabetes* **53**(1), S36–S42.

Tacken, P.J., Hofker, M.H., Havekes, L.M., van Dijk, K.W. (2001) Living up to a name: the role of the VLDL receptor in lipid metabolism, *Curr. Opin. Lipidol.* **12**, 275–279.

van der Vliet, H.N., Sammels, M.G., Leegwater, A.C., Levels, J.H., Reitsma, P.H., Boers, W., Chamuleau, R.A. (2001) Apolipoprotein A-V: a novel apolipoprotein associated with an early phase of liver regeneration, *J. Biol. Chem.* **276**, 44512–44520.

Vanhamme, L., Paturiaux-Hanocq, F., Poelvoorde, P., Nolan, D.P., Lins, L., Van Den Abbeele, J., Pays, A., Tebabi, P., Van Xong, H., Jacquet, A., Moguilevsky, N., Dieu, M., Kane, J.P., De Baetselier, P., Brasseur, R., Pays, E. (2003) Apolipoprotein L-I is the trypanosome lytic factor of human serum, *Nature* **422**, 83–87.

Veniant, M.M., Withycombe, S., Young, S.G. (2001) Lipoprotein size and atherosclerosis susceptibility in Apoe(−/−) and Ldlr(−/−) mice, *Arterioscler. Thromb. Vasc. Biol.* **21**, 1567–1570.

Venkatesan, S., Cullen, P., Pacy, P., Halliday, D., Scott, J. (1993) Stable isotopes show a direct relation between VLDL apoB overproduction and serum triglyceride levels and indicate a metabolically and biochemically coherent basis for familial combined hyperlipidemia, *Arterioscler. Thromb.* **13**, 1110–1118.

Wetterau, J.R., Combs, K.A., McLean, L.R., Spinner, S.N., Aggerbeck, L.P. (1991) Protein disulfide isomerase appears necessary to maintain the catalytically active structure of the microsomal triglyceride transfer protein, *Biochemistry* **30**, 9728–9735.

Wetterau, J.R., Aggerbeck, L.P., Bouma, M.E., Eisenberg, C., Munck, A., Hermier, M., Schmitz, J., Gay, G., Rader, D.J., Gregg, R.E. (1992) Absence of microsomal triglyceride transfer protein in individuals with abetalipoproteinemia, *Science* **258**, 999–1001.

Yamanaka, S., Zhang, X.Y., Miura, K., Kim, S., Iwao, H. (1998) The human gene encoding the lectin-type oxidized LDL receptor (OLR1) is a novel member of the natural killer gene complex with a unique expression profile, *Genomics* **54**, 191–199.

Yang, T., Espenshade, P.J., Wright, M.E., Yabe, D., Gong, Y., Aebersold, R., Goldstein, J.L., Brown, M.S. (2002) Crucial step in cholesterol homeostasis: sterols promote binding of SCAP to INSIG-1, a membrane protein that facilitates retention of SREBPs in ER, *Cell* **110**, 489–500.

Yokode, M., Pathak, R.K., Hammer, R.E., Brown, M.S., Goldstein, J.L., Anderson, R.G. (1992) Cytoplasmic sequence required for basolateral targeting of LDL receptor in livers of transgenic mice, *J. Cell Biol.* **117**, 39–46.

Zannis, V.I., Kan, H.Y., Kritis, A., Zanni, E., Kardassis, D. (2001) Transcriptional regulation of the human apolipoprotein genes, *Front Biosci.* **6**, D456–D504.

Zuliani, G., Arca, M., Signore, A., Bader, G., Fazio, S., Chianelli, M., Bellosta, S., Campagna, F., Montali, A., Maioli, M., Pacifico, A., Ricci, G., Fellin, R. (1999) Characterization of a new form of inherited hypercholesterolemia: familial recessive hypercholesterolemia, *Arterioscler. Thromb. Vasc. Biol.* **19**, 802–809.

Lipids, Microbial

Colin Ratledge
Department of Biological Sciences, University of Hull, Hull, UK

Encyclopedia of Molecular Cell Biology and Molecular Medicine, 2nd Edition. Volume 7
Edited by Robert A. Meyers.

ISBN: 3-527-30549-1

Keywords

Ceramide
A branched chain lipid based on a long-chain (C_{18}) polyhydroxyamine (= sphingolipid) to which a fatty acid group is attached via the 2-amino group.

Fatty Acids
Long-chain (usually 16 or 18C atoms) aliphatic carboxylic acids normally found esterified to glycerol; see **Triacylglycerol**.

Glycolipid
A sugar (usually a mono- or disaccharide) linked to one or more fatty acyl groups; acts as an emulsifying agent by having both water- and lipid-soluble components.

Lipoglycan
A macromolecular complex where the glucan or glycan (polysaccharide) backbone molecule has a number of fatty acyl substituents. Overall, the molecule retains its carbohydrate properties.

Neutral Lipid
A lipid having no charged group; encompasses mono-, di-, and triacylglycerols plus hydrocarbons, including carotenoids, sterols, and so on.

Phospholipid
Strictly, any lipid containing a phospho group, but usually refers to lipids based on 1-, 2-diacyglycerol-3-phosphate (also known as *phosphatidic acid*); a polar substituent (often choline) may also occur on the phospho group, thereby creating a range of phospholipid types. Phospholipids form the major component of most membranes of cells.

Polar Lipids
A lipid having a polar (charged) group; often taken to be synonymous with phospholipids but will include sphingolipids and other related lipids.

Polyunsaturated Fatty Acids (PUFA)
Fatty acids containing three or more double bonds that are usually methylene interrupted, namely, $-CH=CH-CH_2-CH=CH-$. Polyunsaturated fatty acids can

therefore be designated according to the position of the double bond nearest to the CH_3–terminus: thus, $CH_3-CH_2-CH=CH-$would be known as an ω-3 (or n-3) fatty acid.

Sulfolipid
Any lipid containing a sulfo group, usually $-SO_3-$.

Terpenoid Lipids
Lipids that are based on multiples of the isopentenyl group ($CH_2{:}C(CH_3){\cdot}CH_2{\cdot}CH_2-$) giving rise to sterols, carotenoids, polyprenols, and the side chains of chlorophylls, quinones, and so on.

Triacylglycerol
Member of a class of lipids in which the three alcohol groups of glycerol ($CH_2OH{\cdot}CHOH{\cdot}CH_2OH$) are each esterified with a fatty acid. These, along with phospholipids, are the predominant lipid types in most eukaryotic cells.

Unsaturated Fatty Acids
Fatty acids containing one or more double ($-CH=CH-$) bonds. The position of the double bond is indicated by the first carbon numbering from the carboxylic acid group: $^{1}COOH{\cdot}^{2}CH_2{\cdot}^{3}CH_2{\cdot}^{4}CH_2{\cdot}^{5}CH_2{\cdot}^{6}CH_2{\cdot}^{7}CH_2{\cdot}^{8}CH_2{\cdot}^{9}CH{=}^{10}CH{\cdot}^{11}CH_2{\cdot}^{12}CH{=}$ $^{13}CH{\cdot}^{14}CH_2{\cdot}^{15}CH_2{\cdot}^{16}CH_2{\cdot}^{17}CH_2{\cdot}^{18}CH_3$ is 9, 12-octadecadienoic acid (or linoleic acid). The orientation of the double bond is usually in the *Z* (or cis) configuration though *E* (or trans) bonds are known. The shorthand nomenclature for the acid above is therefore 18 : 2 (*Z*9, *Z*12), abbreviated to 18 : 2 (9, 12) if the stereospecificity is not required to be stated. Sometimes it is convenient just to give the position of the double bond nearest the terminal methyl group because, once this is defined, the position of all the other double bonds, being methylene interrupted, are also defined: thus linoleic acid could be given as 18 : 2 (ω-6) or alternatively as 18 : 2 (n-6). This system is particularly useful for polyunsaturated fatty acids and is used in this chapter. Almost all double bonds in naturally occurring fatty acids are of the *Z*- or cis configuration.

Microorganisms, like all other living cells, contain lipids. These lipids are probably the most diverse of all the major kingdoms, and range from simple fatty acids as occurring in plants and animals, to complex multibranched ether lipids found only in the most primitive of bacteria, the *Archaea*. Lipids are used for both structural and functional activities. The former include the principal component of all membranes – whether it is of the cytoplasm or of the major organelles of the eukaryotic microorganisms: yeasts, fungi, and algae. The functional aspects of lipids include their role as storage reserve materials as well as fulfilling key roles in photosynthetic microorganisms. Both bacteria and eukaryotic microorganisms have the propensity to accumulate during periods of carbon nutrient excess large amounts of lipids, which they can utilize during periods of starvation as sources of carbon, energy, and even water. Present biotechnological applications of microbial lipids are aimed at realizing their potential in the nutritional and health care industries. The molecular biology of lipids, though, is a relatively neglected area but is likely to become a major focus of activity as the exploitability of microbial genes for the improvement of plant oils and fats is recognized.

1 Introduction

Lipids are divided into two major classes: the components based on fatty acids (or fatty acyls) and those based on the terpene, or isoprenoid structure. The former includes all the triacylglycerols (**I**) and acylglycerophospholipids (**II**) that occur throughout all eukaryotic cells, including plants, animals as well as microorganisms except the *Archaea*. The latter class of lipids includes the sterols (**III**) and carotenoid (**IV**) lipids, which are synthesized from a common pathway known as the mevalonate or isoprenoid (**V**) route. The archaeal lipids (see Sect. 2) are also derived from mevalonate. Fig. 1 shows the biosynthetic origins of these two groups.

A further distinction that can be made with lipids is to categorize them as polar or neutral. Polar lipids, which are characteristically the lipids of the cell bilayer membranes, owe their polarity to the presence of a charged group such as ethanolamine or choline, attached to the glycerol group (**II**). These lipids are therefore often referred to as phospholipids, an ambiguous and somewhat imprecise term that merely denotes any lipid containing a phospho group. Other polar lipids that do not contain a phospho group include a number of glycolipids, involving condensation of a fatty acid with a sugar or sugar groups, sulfolipids where a SO_3^- group can be found, peptidolipids involving one or more amino acids, and a number of sphingolipids that are more usually thought of as animal-type lipids, but are found in small but significant amounts in many yeasts and fungi. The entire subject of microbial lipids, including lipids of viruses, has been covered in an extensive two-volume treatise edited by the present author and his colleague, Professor S. G. Wilkinson.

Microorganisms are now divided into three domains: *Archaea*, *Bacteria*, and *Eukarya*. The first two comprise all the prokaryotic organisms that were originally referred to as the *Archaebacteria* and the *Eubacteria*, respectively, and have been combined under the single order of the *Bacteria*. In view of the considerable disparity between the 16S rRNA sequences and the structures of RNA polymerases of these two orders, they are now considered to be distinct evolutionary groups and, therefore, have been given equal ranking. The *Eukarya* domain comprises a group of microorganisms loosely known as the fungi, which are then divided into the yeasts and the molds as useful, but somewhat imprecise terminologies. *Eukarya* also includes the photosynthetic algae. Blue-green algae are now referred to as cyanobacteria and are included in *Bacteria*.

The genetics and molecular biology of microbial lipids is very much in its

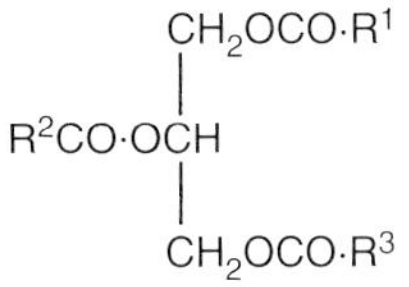

$CH_2OCO{\cdot}R^1$
$R^2CO{\cdot}OCH$
$CH_2OCO{\cdot}R^3$

where R^1CO-, R^2CO- and R^3CO- are fatty acyl groups.

I

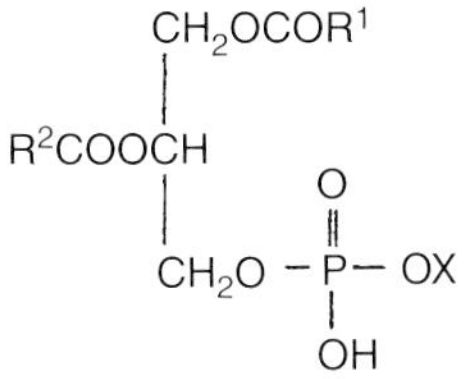

CH_2OCOR^1
R^2COOCH
$CH_2O-P(=O)(OH)-OX$

General structure
R^1CO- and R^2CO- are fatty acyl groups.
X = polar group (choline, serine, ethanolamine etc.)

II

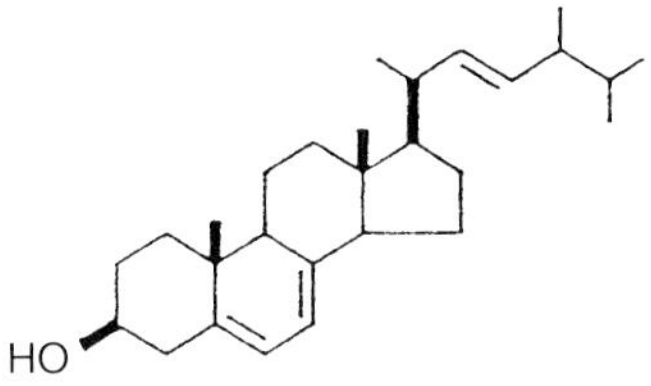

HO

III

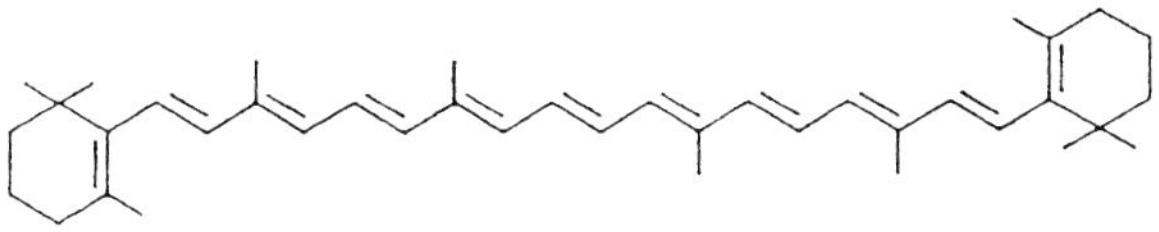

IV

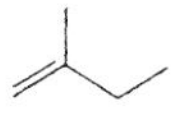

V

infancy and only a few areas have received much attention. The application of genetic techniques have mainly been confined to *Escherichia coli* and *Saccharomyces cerevisiae*, neither of which produces unusual or interesting lipids. Nevertheless, the basis of a genetic understanding of microbial lipids has now been laid. This article is aimed at providing factual information concerning the diversity of lipid molecules found in the different major groups of microorganisms.

2
Archaea (or Archaebacteria)

The distinction between the *Archaea*, the most primitive of the prokaryotic domains, and the *Bacteria* was made principally on the differences in the 16S rRNA sequences. This distinction is also clearly seen in the nature of the lipid components of the two domains. The *Archaea* uniquely contain lipids that are made up of repeating isopentanyl (**VI**) units. The isopentanyl

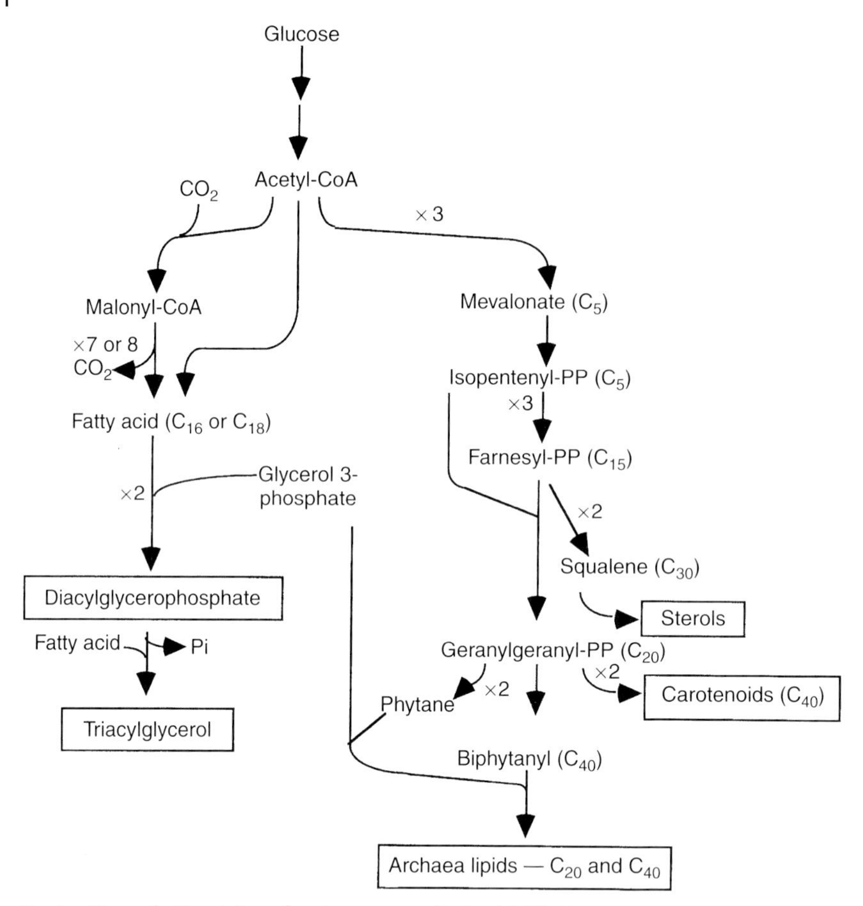

Fig. 1 Biosynthetic origins of major groups of microbial lipids.

unit is derived from the mevalonate pathway of lipid biosynthesis (see Fig. 1), but although this pathway is found in other microorganisms as a source of the terpenoid lipids, in the *Archaea* the entire lipid family are derived from this source.

VI

A further distinction that arises between the archaeal lipids and those of all other organisms is that they are linked to glycerol via ether linkages ($-CH_2-O-CH_2-$) rather than the more common ester link ($-CH_2-O-CO-$). The ether bond is much less susceptible to hydrolysis than the ester linkage and, not surprisingly, then confers considerable stability to the membrane structures of the *Archaea* that are made up with these lipids. The *Archaea* are, indeed, very resistant to extremes of the environment: they comprise the extreme halophiles (*Halobacterium, Halococcus, Natronococcus, Natronobacterium*, and so on), extreme thermophiles (*Thermus, Caldariella, Sulfobolus*, and so on), and organisms that can grow on methane – the methanotrophs or methanogens (*Methanospirillum, Methanosarcina, Methanococcus*, and so on). The somewhat loose term "extremophile" has been given to this group of organisms as a convenient epithet, but it lacks the formal precision of *Archaea* and may often be used to include representatives of the *Bacteria* that show some tolerance to the extremes of the environment.

A further point to note in connection with these glycerol diether lipids (Fig. 2) is that the two isopranoid chains (also called *phytanyl chains*) are attached to the *sn*2- and 3- positions of glycerol rather than the conventional 1- and 2- positions of the diacylglycerols and diacylphosphoglycerols found in all other living cells, including bacteria.

Typical isopranoid ether lipids of *Archaea* are shown in Fig. 2. The lipid given in Fig. 2(a), which is the core lipid of the *Archaea*, is also referred to as *archaeol*, which indicates its association with this domain of microorganisms as well as its status as an alcohol. The lipid given in Fig. 2(c) is known as *caldarchaeol*. As Fig. 2 indicates, not only are there a variety of configurations giving rise to both diethers and tetraether lipids but there are also phosphoglycolipids and sulfolipids, as well as lipoglycans that form part of the cell wall of these microorganisms and are all composed of these phytanyl lipids.

Thus, not only are the lipid membranes of the extremophilic microorganisms composed of highly stable isopranoid ether lipids but so too are many of the lipid components associated with the cell wall.

Figure 3 gives the biosynthesis of the archaeol phospholipids and glycolipids in the extreme halophiles. The C_{20} product of the mevalonate pathway, probably, geranylgeranyl pyrophosphate is considered to condense with dihydroxyacetone and gives a "pre-diether." This becomes the precursor of both the phospholipids and the glycolipids, with the reduced, unsaturated bonds of the geranylgeranyl group being saturated after condensation with the dihydroxyacetone. The biphytanyl (C_{40}) chains are thought to arise by direct head-to-head condensation of two molecules of archaeol (Fig. 2a). The substituted glycerol and phosphoglycero lipids (Figs. 2e and f) are probably synthesized before this condensation reaction; this would include the sugar-containing lipids (glycolipids).

The function of these branched chain ether lipids is to impart stability to the membranes of these bacteria under extremes of temperature, pH, salinity, and so on. Under these extreme conditions, conventional ester-linked lipids would quickly hydrolyze; moreover, unsaturated fatty acyl groups would be quickly oxidized, and thus, "conventional" microorganisms would not, and cannot, survive under these conditions. Since the alkyl chains are completely saturated and furthermore have only short side chains (methyl groups), which allows for close molecular packing, the branched chain ether lipids become liquid crystalline at ambient temperatures. Consequently, their presence in the extreme thermophilic organisms means that these organisms are unable to function until the temperature is at least 70 or even 80 °C. Thus, the high stability of the membranes and cell walls of these organisms means that the organisms cannot grow below temperatures at which conventional microorganisms would quickly die. The presence of pentacyclic rings in the biphytanyl chains (Fig. 2d) ensures that the fluidity of the membranes containing these lipids remains fairly constant as the temperature is increased above 80 °C. Many of these organisms have been isolated from deep-sea thermal vents,

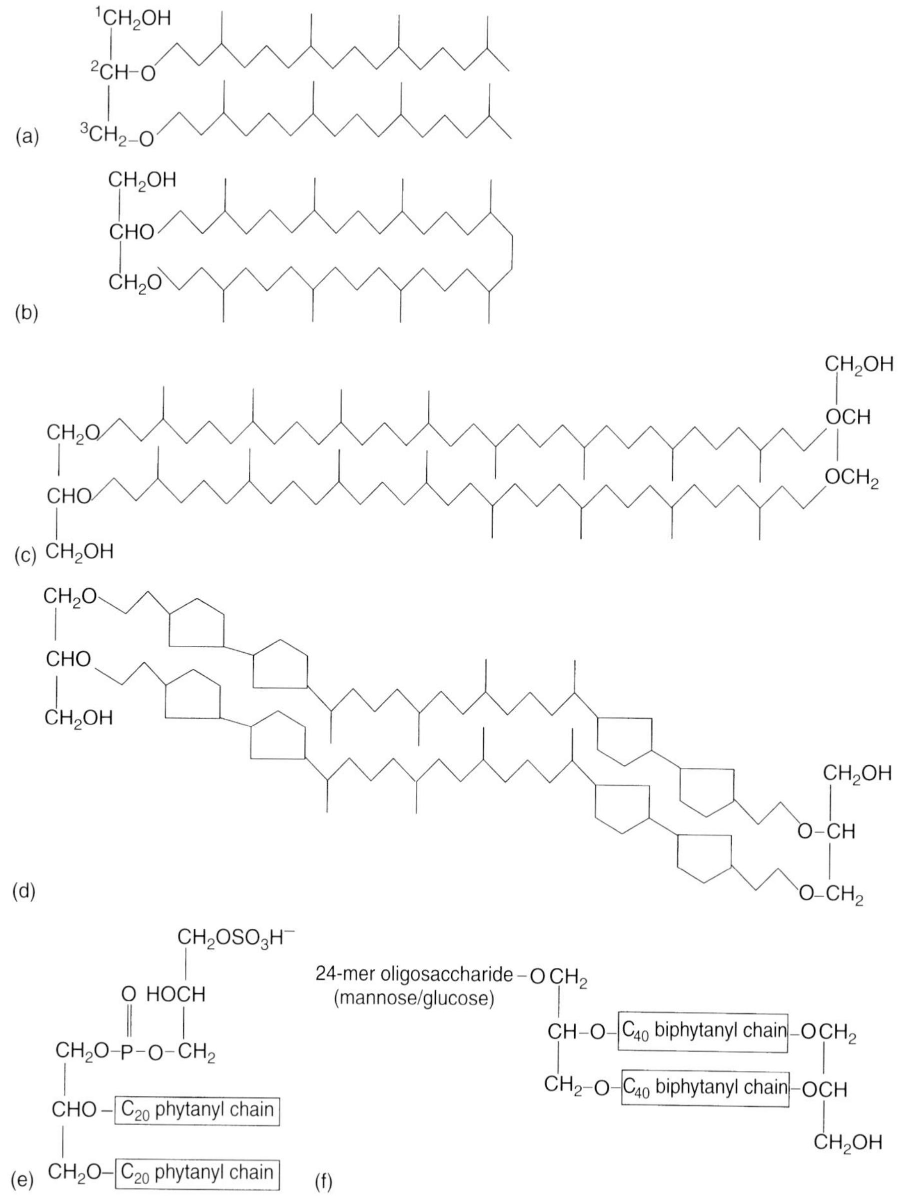

(a) $^{1}CH_2OH$ — ^{2}CH–O — $^{3}CH_2$–O

(b) CH_2OH — CHO — CH_2O

(c) CH_2O — CHO — CH_2OH … CH_2OH — OCH — OCH_2

(d) CH_2O — CHO — CH_2OH … CH_2OH — O–CH — O–CH_2

(e) $CH_2OSO_3H^-$ — O HOCH — CH_2O–P–O–CH_2 — CHO – C_{20} phytanyl chain — CH_2O– C_{20} phytanyl chain

(f) 24-mer oligosaccharide – O CH_2 (mannose/glucose) — CH–O– C_{40} biphytanyl chain –OCH_2 — CH_2–O– C_{40} biphytanyl chain –OCH — CH_2OH

Fig. 2 Isoprenoid lipids of *Archaea:* (a) 2,3-di-*O*-phytanyl*sn*glycerol (archaeol), (b) 2,3-di-*O*-biphytanyl-*sn*-glycerol, (c) glycerol-dialkylglycerol tetraether (caldarchaeol), (d) tetracyclized glycerol dialkylnonitol tetraether (cyclized nonitolcaldarchaeol), (e) phosphoglycerolsulfate dialkylglycerol diether, (f) lipoglycan from *Thermoplasma* mannosyl-glucosyl oligosaccharide of (c).

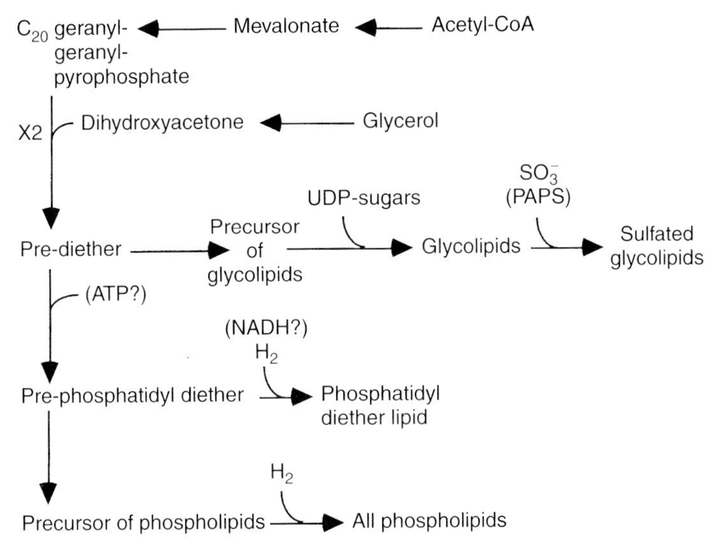

Fig. 3 Proposed pathways for the biosynthesis of the archaeol-containing phospho- and glycolipids in the extreme halophilic *Archaea*. (Adapted from the work of Morris Kates, University of Ottawa, Canada.)

demonstrating that their tolerance to temperatures in excess of 110 °C must be excellent.

Since the taxonomy of *Archaea* is still very much in its infancy, there is a considerable amount of work to be carried out on the structures of many of these organisms. Undoubtedly, some very complex lipids and lipid-containing molecules will be found in the next few years, and these will add to the bewildering array of the di- and tetraether lipids that have already been described.

3 Bacteria

"*Bacteria*" is the term now applied to all bacteria except those now within the *Archaea*. It, therefore, comprises the bulk of "conventional" bacteria that were originally classified as the *Eubacteria*. These organisms synthesize ester-based lipids, usually based on glycerol, from fatty acids that are synthesized via the acetyl-malonyl pathway (see Fig. 1). In most cases, the resultant lipids are used as structural and functional entities in bacteria, being the component of the main phospholipid membranes of the cell. Also, however, they are used in conjunction with polysaccharides and peptides in the construction of the cell wall and other parts of the cell envelope. Bacteria are divided into two major groups, the gram-positive and the gram-negative organisms, reflecting differences in their cell wall structures. The former have a peptide-glycan outer shell acting as the wall, which is then anchored to the inner phospholipid bilayer membrane via lipoteichoic acids (**VII**). In gram-negative bacteria, there is an outer membrane, beyond the peptidoglycan layer, that is a matrix of phospholipids, complex lipopolysaccharides in which the porin uptake proteins are anchored. Lipoproteins are then used to link the outer membrane to the peptidoglycan layer that abuts the inner membrane. Most of the phospholipids are conventional diacylphosphoglycero derivatives (**II**) that are found in all living cells. There are, though, some variations in that ethanolamine, monomethyl-, and dimethylethanolamine may be used instead of choline (= trimethylethanolamine) as the polar basic group attached to the phospho group. Bacteria, apart from the mycobacteria, do not synthesize significant amounts of triacylglycerols (**I**).

The genetics of bacterial lipid metabolism are under active investigation by several major groups worldwide; the group led by John E. Cronan, Jr. (Urbana, IL) is particularly active in this respect.

3.1 Unsaturated Fatty Acids

In many bacteria, unsaturated fatty acids are synthesized by a so-called anaerobic pathway that is not found in other microorganisms. This pathway (Fig. 4) has been mainly studied in *E. coli*. It involves a modification of the existing fatty acid synthetase system (see Fig. 1) where, at the C_{10} level, there occurs an alternative dehydratase enzyme that forms 3*(Z)*-decenoate instead of the 2*(E)*-decenoate. (These terms replace the older descriptors of cis-β, γ-and *trans*-$\alpha\beta$-decenoates, respectively.) The latter intermediate is reduced and leads, via the continuation of fatty acid synthesis, to palmitoyl- and stearoyl-ACP (where ACP is a separable acyl carrier protein in *E. coli* having relative molecular mass of 8847). The 10:1 (*Z*3) intermediate cannot be reduced and is thus successively elongated so that the double bond remains in the molecule and, as it is elongated, the position of the double bond advances down the chain: 10:1 (*Z*3) → 12:1 (*Z*5) → 14:1 (*Z*7) → 16:1 (*Z*9) → 18:1 (*Z*11), which is *cis* (or *Z*)-vaccenic acid. This latter product is thus distinct from oleic acid, 18:1 (*Z*9), found in some bacteria and in all yeasts and molds. The presence of *cis*-vaccenic acid in extracted microbial lipids can be clearly distinguished from oleic acid by good gas chromatographic procedures, especially those using capillary gas chromatography.

The final elongation reaction of the "anaerobic" route is temperature sensitive. The enzyme that condenses 16:1 (*Z*9)-ACP (palmitoleoyl-ACP) with malonyl-ACP has a higher rate of reaction at lower temperatures than at

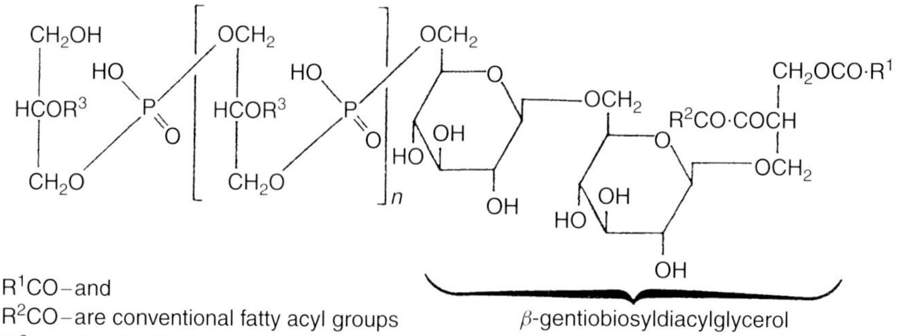

R^1CO–and
R^2CO–are conventional fatty acyl groups
R^3 is mainly H, D-alanyl or *N*-acetylglucosaminyl
n = 30 approx.

β-gentiobiosyldiacylglycerol

VII

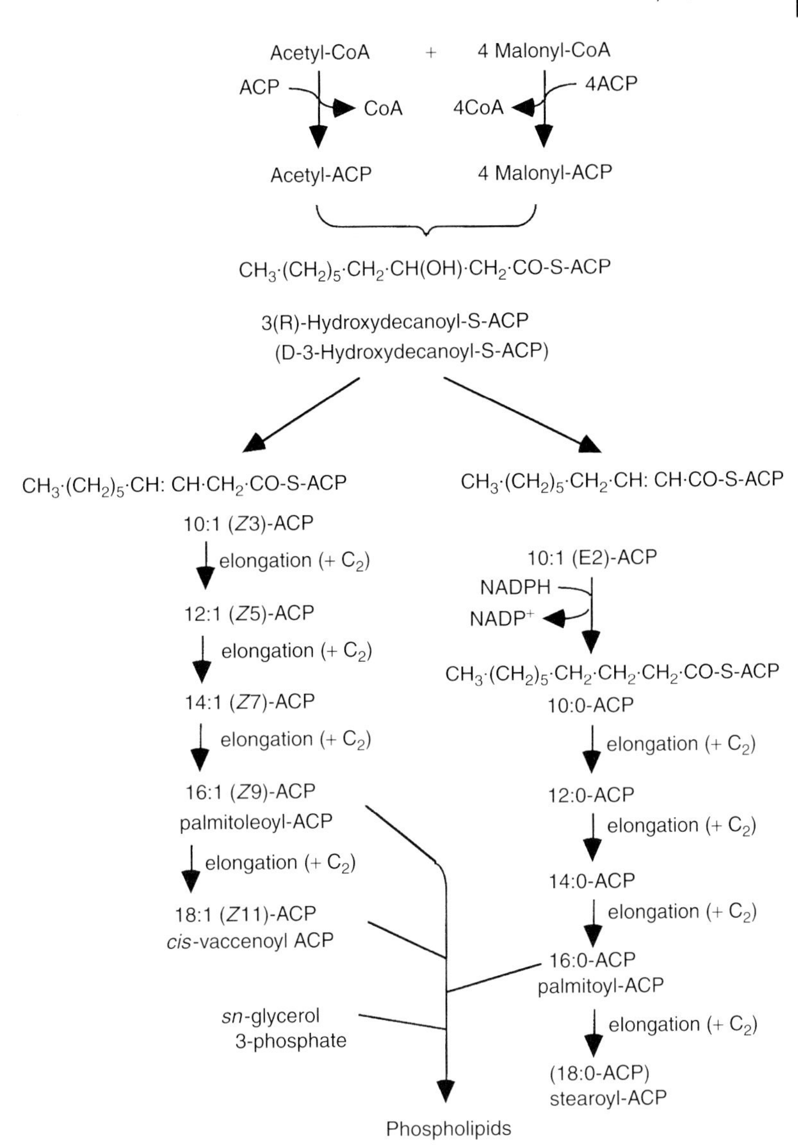

Fig. 4 Biosynthesis of saturated and unsaturated fatty acids by the fatty acid synthetase of *E. coli*. The conventional cyclic pathway of fatty acid biosynthesis condenses acetyl-ACP and malonyl-ACP in a succession of reactions leading to the formation of 3(*R*)-hydroxydecanoyl-*S*-ACP. This can either continue through the cyclic sequence to give the series of saturated fatty acids (right-hand side) or be dehydrated to give a fatty acid with the double bond now in a different position but also of a different isomeric configuration. Since this intermediate, the 3(*Z*)-decenoyl-ACP, now cannot participate in the reductive reaction, the double bond remains as the acyl chain continues to undergo further condensation reactions in the fatty acid synthetase cycle. The final conversion of 16 : 1 (*Z*9)-ACP to 18 : 1 (*Z*11)-ACP is a temperature-controlled reaction in that at low temperatures the condensing enzyme has a higher relative rate of reaction than at higher temperatures (see text).

higher ones. Consequently, at low growth temperatures, more 18:1-ACP is formed than at higher temperatures, and it competes with palmitoyl-ACP for incorporation into phospholipids (see Fig. 4) and results in the formation of a higher proportion of diunsaturated acylphospholipids. This, in turn, ensures a lower temperature for the liquid phase transition of the membrane lipids. This temperature sensitivity of unsaturated fatty acid biosynthesis allows *E. coli* and other bacteria to adapt and to survive at low temperatures by modifying the fatty acyl constituents of the membrane phospholipids, ensuring that they remain as fluid as possible at lower temperatures.

3.2 Polyunsaturated Fatty Acids

Although the vast majority of bacteria do not synthesize polyunsaturated fatty acids because of the lack of specific desaturase enzymes, mycobacteria do synthesize fatty acids with multiple double bonds: for example, phleic acid is a 36 : 5 (4, 8, 12, 16, 20) fatty acid found in *Mycobacterium phlei* and *Mycobacterium smegmatis*. However, the double bonds are not of the methylene-interrupted type as found elsewhere. Such fatty acids, though, have been recognized in a number of bacteria recovered from fish intestines and include eicosapentaenoic acid, 20 : 5 (n-3), and docosahexaenoic acid, 22 : 6 (n-3). Although some interest has been indicated in these bacteria as potential sources of these very long-chain polyunsaturated fatty acids, better sources are various fungi and marine algae (see Sects. 5.3 and 6.2). The bacteria, however, could be useful providers of the genes needed to clone into plants if genetically modified plants are to be created that can synthesize these nutritionally desirable fatty acids.

3.3 Branched Chain Fatty Acids

Branched chain fatty acids with a methyl group at the penultimate or antepenultimate carbon atom (see **VIIIA** and **B**) are synthesized by both gram-positive and gram-negative bacteria. The principal genera among the former group that produce branched chain fatty acids are *Bacillus, Clostridium, Listeria, Micrococcus, Sarcina,* and *Staphylococcus*; of the gram-negatives, *Flavobacterium, Bacteroides, Legionella, Vibrio,* and *Desulphovibrio* are the principal genera, though some species of *Pseudomonas* and *Xanthomonas* also contain these acids. The branched chain fatty acids are synthesized from the derivatives arising from the branched chain amino acids: leucine (or valine) for the iso-fatty acids and isoleucine for the anteiso-fatty acids (see **VIIIA** and **B**).

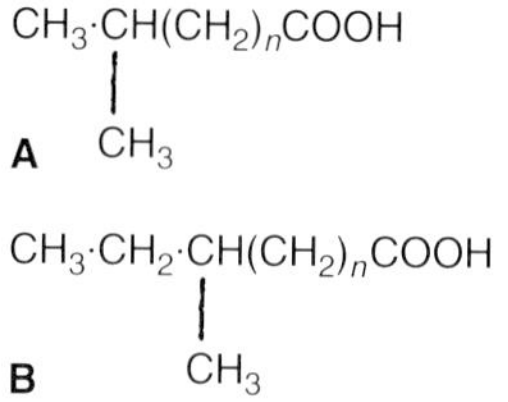

$$CH_3{\cdot}CH(CH_3)(CH_2)_nCOOH$$

A

$$CH_3{\cdot}CH_2{\cdot}CH(CH_3)(CH_2)_nCOOH$$

B

VIII

Fatty acids with a methyl branch in the middle of the chain are also known. The best-known example of this is tuberculostearic acid, 10-methyloctadecanoate (**IX**), which was originally isolated from the tubercle bacillus *Mycobacterium tuberculosis* but has now been found in more than 20 difference species and strains of *Mycobacterium* besides also occurring in a number of other actinomycetes.

A number of other methyl-branched fatty acids also occur in mycobacteria, including small amounts of the iso- and anteiso-fatty acids, as well as multi-methyl-branched fatty acids of which the C_{32} mycocerosic acid 2,4,6,8-tetramethyloctadecanoate is but one of several complex branched chain fatty acids found in this genera (see also Sect. 3.6, Mycolic Acids).

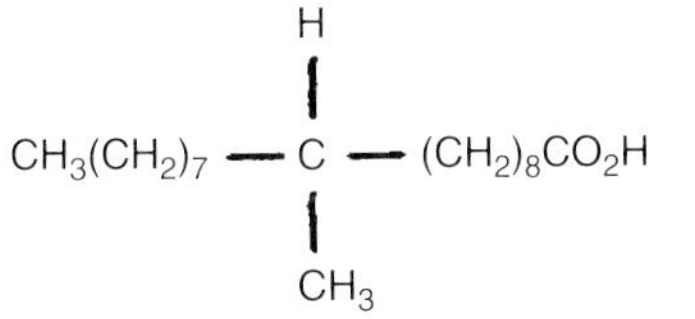

$$CH_3(CH_2)_7-\underset{\underset{CH_3}{|}}{\overset{\overset{H}{|}}{C}}-(CH_2)_8CO_2H$$

IX

3.4 Cyclopropane Fatty Acids

Fatty acids that contain a cyclopropane ring are fairly common in bacterial lipids. The most widely distributed example is lactobacillic acid, Z-11, 12-methyleneoctadecanoic acid (**X**), abbreviated as 19:0 cy. This, as the name would suggest, was originally isolated from *Lactobacillus* spp. but is recognized now in many organisms including *E. coli*, most *Pseudomonas* spp., and some *Rhizobium* and *Clostridium* spp. The corresponding 17 : 0 cy fatty acid also occurs. The presence of these acids can be regarded as indicative of a unique bacterial lipid because they do not occur in eukaryotic microorganisms: a related fatty acid, sterculic acid (an unsaturated 19 : 1 cy fatty acid), though, does occur in some plant lipids.

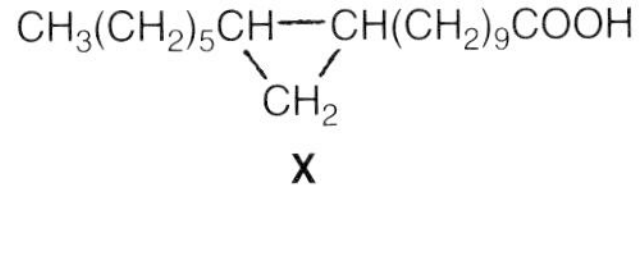

$$CH_3(CH_2)_5CH-CH(CH_2)_9COOH$$
$$\diagdown\ \diagup$$
$$CH_2$$

X

3.5 Hydroxy Fatty Acids

Hydroxy fatty acids are constituents of lipopolysaccharides of the cell envelope of the gram-negative bacteria and have been found in most species that have been studied. 3-Hydroxyalkanoic acids, ranging from C_{11} to C_{21}, have been identified in *E. coli* and species of *Salmonella, Klebsiella, Yersinia, Vibrio, Pseudomonas, Xanthomonas*, and *Rhizobium*. 2-Hydroxy fatty acids also occur to a lesser extent; however, the steric configuration of this hydroxy fatty acid is the *S* (or *L*) form and is thus opposite to the *R* (or *D*) form of the 3-hydroxy fatty acid. This probably indicates separate biosynthetic origins for the two fatty acids. 3-Hydroxy iso-fatty acids may be detected in some species as minor constituents.

The arrangement of the 3-hydroxy fatty acids (of which 3-hydroxy tetradecanoate is the most abundant type) is shown in Fig. 5 as part of the lipid A structure of some enterobacteria. Lipid A is the anchor region of the complex lipopolysaccharide structure in the outer membrane of these organisms. The fatty acids attached to the diglucosamine will then adjoin the periplasmic space.

3.6 Mycolic Acids

Very long chain, hydroxylated, branched chain fatty acids occur in bacteria belonging to the *Mycobacterium, Nocardia, Corynebacterium* group, *Rhodococcus*, and related groups of bacteria. The basic structure is that of a 2-alkyl, 3-hydroxy fatty acid (**XI**). The alkyl side chain may be up to 24 carbon atoms long (as occurs in *M. tuberculosis*); it is shorter in other mycobacteria. In *Corynebacterium* spp. it is only six carbons long. The long-chain group attached

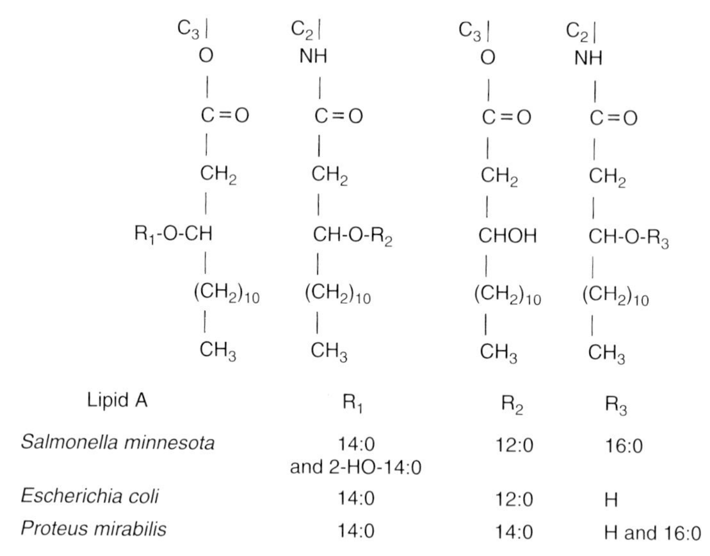

Lipid A	R_1	R_2	R_3
Salmonella minnesota	14:0 and 2-HO-14:0	12:0	16:0
Escherichia coli	14:0	12:0	H
Proteus mirabilis	14:0	14:0	H and 16:0

Fig. 5 Structure of lipid A in three enterobacteria.

$$R{-}CH(OH){-}CH[(CH_2)_nCH_3]{-}CO_2H \qquad n = 5\text{–}23\text{, odd numbers only}$$

XI

$$CH_3{-}(CH_2)_{17}{-}\underset{cis}{CH{=}CH}{-}(CH_2)_{13}{-}\underset{trans}{CH{=}CH}{-}CH(CH_3){-}(CH_2)_{17}{-}CH(OH){-}CH[(CH_2)_{21}{-}CH_3]{-}CO_2CH_3$$

XII

at C-3 of the mycolic acid is up to 61 carbon atoms, making some of the mycolic acids the largest unsubstituted lipids that occur in any organism. A typical mycolic acid, as found in *M. smegmatis* as its methyl ester ($C_{80}H_{156}O_3$), is shown in structure **XII**.

$$HOCH(R)CH_2CO{\cdot}O{-}\left[CH(R)CH_2CO{\cdot}O\right]_n{-}CH(R)CH_2CO{\cdot}OH$$

(n = up to 30 000)

XIII

3.7
Polyesters

Bacteria, apart from some exceptions in the *Mycobacterium* genus, do not synthesize triacylglycerols, which are the usual storage lipid of eukaryotic microorganisms (see Sects. 4.1 and 5.3). Lipid storage, nevertheless, still occurs in bacteria by the biosynthesis of high molecular weight polyesters of β-hydroxybutyrate (PHB) or polyesters of β-hydroxyalkanoate (PHA). Poly-β-hydroxybutyrate ($R = -CH_3$ in **XIII**), which is the *D* (or *R*) isomer, is synthesized by a large number of bacteria; species of *Alcaligenes, Azotobacter, Bacillus, Nocardia, Pseudomonas, Rhizobium, Rhodococcus,* and *Zoogloea,* plus *E. coli,* have been investigated in numerous laboratories. In some bacteria, the more ubiquitous PHB is replaced by a poly-β-hydroxyalkanoate (PHA) in which the side chain (R in **XIII**) may be up to 12 carbon atoms long. PHB has also been found in some fungi.

PHA, which can reach up to 80% of the biomass of *Alcaligenes eutrophus,* and also in a recombinant strain of *E. coli,* can attain molecular sizes of over 1 million Daltons and, moreover, be produced by the bacteria with a combination of side chains (see **XIII**) which then endows the resultant biopolymer with properties that can be designed with a particular function in mind. Nevertheless, in spite of all these advantages, including biodegradability, neither PHB nor PHA has achieved commercial production mainly because the costs involved are about ten times those of producing polyethylene and related chemical polymers.

Some hope for commercial production was considered to be by way of introducing the requisite three key genes for PHB synthesis (see Fig. 6) into a plant such as rapeseed, sunflower, or soybean. Unfortunately, yields of the biopolymer were not as good as in the bacteria and costs of production were again found to be higher than costs of producing conventional plastics. A limited amount of commercial interest only remains in these unique bacterial polymers once considered as the "green" alternative to petrochemically produced plastics.

Since the pathway for PHB biosynthesis requires only three new enzymes (see

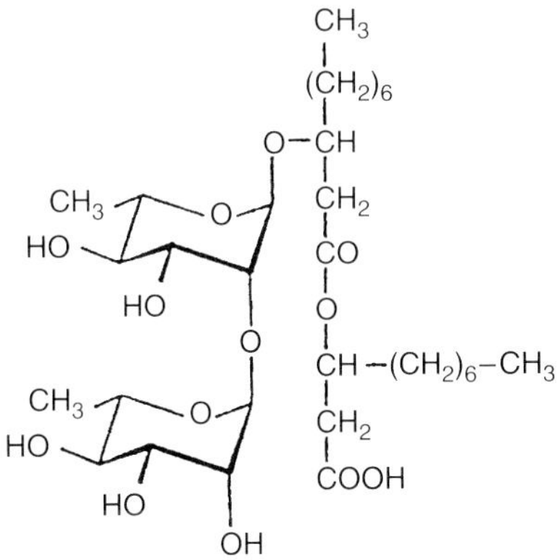

XIV

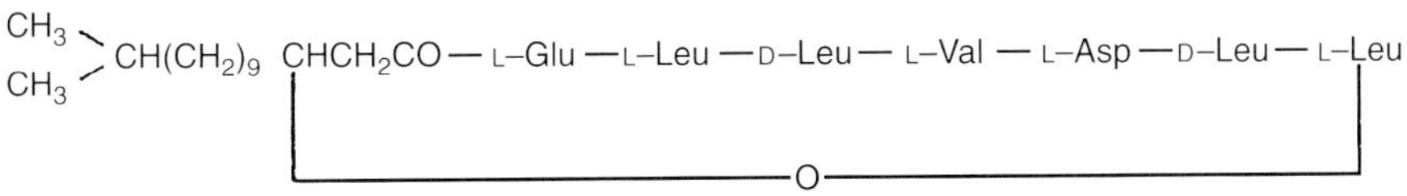

$(CH_3)_2CH(CH_2)_9CHCH_2CO$—L-Glu—L-Leu—D-Leu—L-Val—L-Asp—D-Leu—L-Leu (ring closed via —O— between CH and terminal L-Leu)

XV

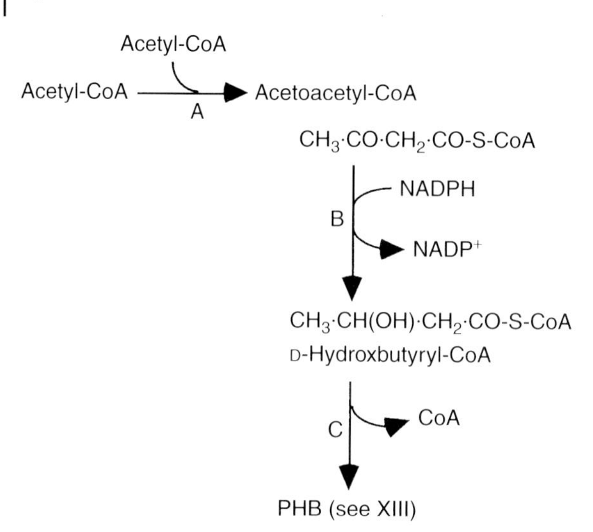

Fig. 6 Biosynthesis of poly-β-hydroxybutyrate in bacteria. Enzymes are 3-ketothiolase (A), acetoacetyl-CoA reductase (B), and PHB synthetase (C).

Fig. 6), prospects of cloning the corresponding three genes into plants has been under consideration. All three genes have been isolated and sequenced. Plants that normally synthesize considerable amounts of oil, thereby already having good supplies of acetyl-CoA, are the obvious recipients of the gene transfer: rapeseed, sunflower, and soybean are all currently receiving attention. However, the cost of producing these materials, even using genetically modified plants, appears to be more than the market is willing to pay for biodegradable plastics.

3.8 Biosurfactant Lipids

Bacteria produce a number of biosurfactants of diverse chemical structures that have the ability to form stable emulsions of oil and water mixtures. Most of the biosurfactants owe their properties to being glycolipids (see **XIV**); that is, they possess both water-soluble and oil-soluble components. Some, however, are lipopeptides (see **XV**). Highest concentrations of these compounds are produced during bacterial growth on oils, either petroleum or plant in origin. Several of these biosurfactants are produced commercially or are undergoing industrial investigation. They include emulsan, which is produced by *Acinecobacter calcoaceticus* RAG-1 and is a polyanionoic, amphipathic, heterolipopolysaccharide of high molecular weight (1 MDa), rhamnolipids (**XIV**) produced by *Pseudomonas* spp., and surfactin (**XV**) produced by *Bacillus subilis*. Other biosurfactants of commercial interest are produced by yeasts and molds (see Sects. 4.4 and 5.4).

HO

XVI

4 Yeasts

4.1 Fatty Acids and Fatty Acyl Lipids

Yeasts are eukaryotic, unicellular fungi that do not form a mycelium and, thus, are differentiated from the filamentous fungi. There are over 600 identified species of yeasts. Yeasts, like all eukaryotic organisms, synthesize their fatty acids via the acetate–malonate route (Fig. 1) and possess specific desaturases for the conversion of saturated fatty acids to unsaturated ones. Most fatty acids are 16 to 18 carbon atoms in length, though small amounts of fatty acids as long as 28 or 30 carbons can be discerned by careful analysis of some species.

Yeasts tend to have a much more limited range of fatty acids than molds: oleic (18 : 1), palmitic (16 : 0), linoleic (18 : 2), and palmitoleic (16 : 1) acids usually account for more than 90% of the total fatty acids. Typical fatty acid profiles of yeasts are shown in Table 1.

Tab. 1 Typical fatty acid profiles of some selected yeasts.

Organism	*Relative amount [% w/w] of major acyl groups*						
	14:0	*16:0*	*16:1*	*18:0*	*18:1*	*18:2*	*18:3*
Ascosporogenous yeasts							
Debaryomyces hansenii	1	20	12	2	28	30	8
Hansenula polymorpha	1	16	2	3	25	30	23
Kluyveromyces marxianus	<1	11	24	5	45	12	3
Lipomyces lipofer[a]	<1	37	4	7	48	3	0
Lipomyces starkeyi[a]	<1	34	6	5	51	3	0
Pichia stipitis	<1	12	5	2	33	40	7
Saccharomyces cerevisiae	1	9	46	2	42	0	0
Schizosaccharomyces pombe	1	4	23	<1	70	0	0
Schwanniomyces occidentalis	<1	16	11	1	42	21	8
Yarrowia lipolytica[a]	3	21	7	7	17	35	0
Zygosaccharomyces rouxii	<1	6	11	3	33	46	0
Basidiosporogenous yeasts							
Leucosporidium scotti	<1	5	2	3	17	34	28
Rhodosporidium toruloides[a]	1	25	<1	13	46	12	2
Sporidiobolus salmonicolor	<1	24	<1	6	36	36	3
Asporogenous yeasts							
Brettanomyces lambicus	2	16	48	0	21	11	0
Candida albicans	<1	20	7	1	22	27	20
Candida boidinii	<1	15	20	3	27	31	0
Candida curvata D[a,b]	<1	29	<1	12	51	6	2
Candida tropicalis	<1	22	5	9	29	26	4
Cryptococcus albidus[a]	<1	12	1	3	73	12	<1
Rhodotorula graminis[a]	1	30	1	12	36	15	4
Trichosporon cutaneum[a]	3	13	0	22	50	13	0
Trichosporon pullulans[a]	<1	15	0	2	57	24	1

[a] Considered to be oleaginous species with oil contents >25% of cell dry weight.
[b] Now reclassified as *Cryptococcus curvatus*.

Fatty acid profiles of all the major genera and of the key species within each genus now have been published. The most systematic and extensive survey was conducted by Lodewyk Kock and his associates in Bloemfontein, South Africa. Although most yeasts (see Table 1) synthesize di- and polyunsaturated fatty acids (18 : 2 and 18 : 3), *Saccharomyces cerevisiae* (baker's and brewer's yeast) does not synthesize unsaturated fatty acids beyond oleic acid (18 : 1). Reports of the occurrence of 18 : 2 and 18 : 3 in *S. cerevisiae* continue to be made, though their presence is attributable to the yeast being grown on a culture medium that contains animal- or plant-derived material bearing traces of these acids. Other yeasts, though, synthesize both 18 : 2 and 18 : 3, which is the α-isomer [i.e. 18 : 3 (9, 12, 15)].

Fatty acids are incorporated into a range of acylglycerophospholipids (**II**) of which phosphatidylcholine, -serine, -ethanolamine, and -inositol are the major ones; diphosphatidylglycerol (cardiolipin) is also common. Many yeasts also synthesize small amounts of sphingolipids (whose function is largely unknown). Unlike bacteria, all yeasts synthesize some triacylglycerols (**I**) along with the functional phospholipids. Triacylglycerols function as a reserve storage product; in some yeasts (about 25 or so species), the amount of triacylglycerol that is stored can be up to 80% of the cell volume. Such yeasts are known as oleaginous species and have been examined as potential sources of oils because the extracted triacylglycerol is similar in composition to several of the commercially produced plant oils.

4.2
Single Cell Oils (SCO)

Microbial lipids that have been considered as sources of edible oils are referred to as *single cell oils* (SCO), as an equivalent term to *single cell proteins,* which describes microorganisms as potential sources of protein or food. In view of the high costs of biotechnological production of such oils, commercial interests have centered on the highest valued oils. For yeasts, this focus has been in the production of a substitute for cocoa butter. (With molds, commercial interest has been directed mainly on the polyunsaturated fatty acids: see Sect. 5.) The main yeast to have been examined was *Candida curvata* (also known as *Apiotrichum curvatum*), now renamed *Cryptococcus curvatus.* To replicate the fatty acid composition of cocoa butter, it was necessary to delete the Δ9-desaturase gene converting stearic acid to oleic acid, thus causing the accumulation of stearic acid up to 40% in the yeast oil (see Table 2). The high content (35%) of stearic acid in cocoa butter gives the product its characteristic hardness at ambient temperatures, with a sharp transition to the molten state at 30 to 32 °C.

The genetic modification of *Candida curvata* was carried out by Henk Smit and colleagues at the Free University of Amsterdam. The yeast has also been explored commercially in New Zealand by Julian Davies (Industrial Research Ltd., formerly the Department of Scientific and Industrial Research of the New Zealand government) using as a low cost substrate the waste lactose from whey arising during cheese manufacture. Even with such a cheap feedstock, however, the biotechnological route has not proved to be commercially competitive against cocoa butter or equivalent fats that are produced from palm oil by fractional recrystallization. Interest in this particular SCO has therefore ceased.

Tab. 2 Cocoa butter fatty acids and the fatty acyl composition of yeast triacylglycerols[a].

Yeast	*Relative amounts [% w/w] of major fatty acyl groups*					
	16:0	*18:0*	*18:1*	*18:2*	*18:3*	*24:0*
WT	17	12	55	8	2	1
WT-NZ[b]	23	23	42	3	1	4
Ufa 33[c]	20	50	6	11	4	4
R22.72	16	43	27	7	1	2
F33.10	24	31	30	6		4
Cocoa butter	23–30	32–37	30–37	2–4		Trace

[a]From *Cryptococcus curvata* [formerly *Candida curvata (Apiotrichum curvatum)*] wild-type (WT) strain, an unsaturated fatty acid auxotrophic mutant (Ufa 33), a revertant mutant (R22.72), and a hybrid derived from Ufa 33 (F33.10) compared to the best results obtained with the wild-type strain grown with limited O_2 supply to diminish desaturase activity.
[b]Wild type grown (in New Zealand) on whey lactose.
[c]Grown with 0.2 g oleic acid L^{-1}.

4.3 Biochemistry of Oleaginicity

The biochemical reason for the relatively sparse distribution of oleaginicity among yeasts (of 600 plus species, only some 25 or so yeasts are known that accumulate >25% lipid) has been identified as the key presence of ATP: citrate lyase (ACL) in the oleaginous species (see Fig. 7). In this scheme, citric acid, which is produced within the mitochondrion, is not further metabolized through the reactions of the tricarboxylic acid cycle because, under the conditions that lead to the accumulation of lipid (exhaustion of N from the culture medium), isocitrate cannot be converted to α-ketoglutarate. This conversion is prevented because adenosine monophosphate (AMP) reaches such a low intracellular concentration under these conditions that isocitrate dehydrogenase, which specifically requires AMP for activity, cannot become fully operational. Thus, when cells have exhausted their supply of N in the medium, the concentration of AMP falls, preventing citrate metabolism; then, provided a supply of carbon (e.g. glucose) is still available, the cells begin to accumulate citrate, which exits the mitochondrion and is cleaved by ACL to yield acetyl–CoA units. The other product from the ACL reaction is oxaloacetate, which is recycled to pyruvate via malic acid; which is decarboxylated to pyruvate by malic enzyme and simultaneously produces the NADPH that is needed for fatty acid biosynthesis. Although all oleaginous yeasts possess ACL, not all possess malic enzyme; at least some species, therefore, must have available an alternative means of generating NADPH and recycling the oxaloacetate. The underlying biochemistry of oleaginicity in yeasts and molds has been largely worked out in the author's laboratory.

The genetics of fatty acid and phospholipid biosynthesis has been examined principally in *S. cerevisiae*, which is not

XVII

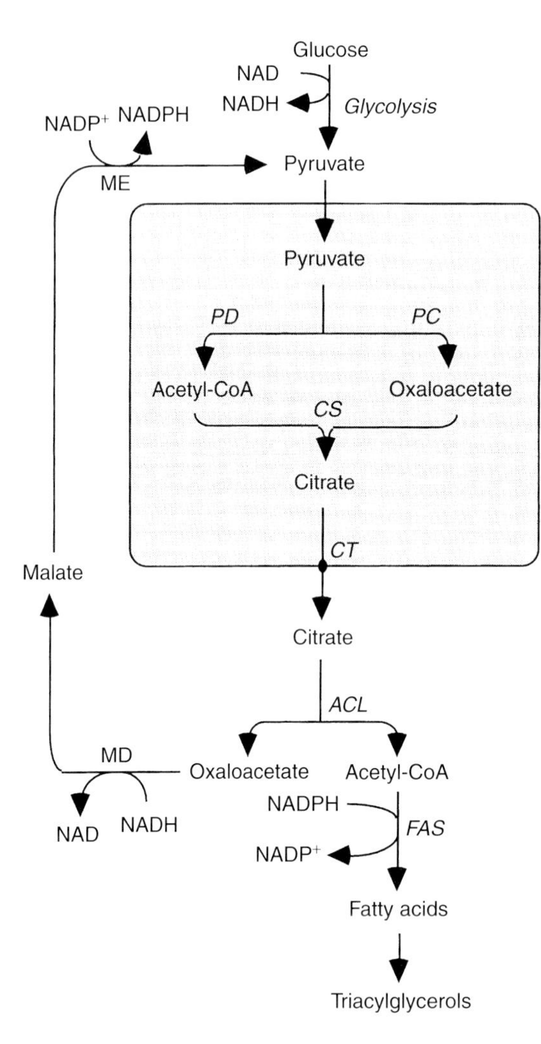

Fig. 7 Mechanism of acetyl–CoA production in oleaginous yeasts and molds leading to the accumulation of single cell oils (triacylglycerols). The metabolism of glucose to pyruvate occurs by *glycolysis* in the cytoplasm; pyruvate is transported into the mitochondrion (shaded box) and is converted to acetyl–CoA and oxaloacetate (by pyruvate dehydrogenase, *PD*, and pyruvate carboxylase, *PC*, respectively), which are synthesized into citrate via citrate synthetase (*CS*). Citrate is translocated via citrate translocase (*CT*) out of the mitochondrion, to be cleaved by ATP: citrate lyase (*ACL*). The acetyl–CoA then is used for fatty acid synthesis (*FAS* = fatty acid synthetase) and oxaloacetate is converted by malate dehydrogenase (*MD*) and malic enzyme (*ME*) back to pyruvate. Malate dehydrogenase uses NADH, which will be provided by the key reactions of glycolysis: *ME* generates the NADPH needed for fatty acid biosynthesis.

an oleaginous species and does not contain ACL activity. Leading investigators of these processes are Susan A. Henry at Cornell University and George M. Carman at Rutgers University, in New Jersey.

4.4 Other Lipids

Yeasts synthesize varying amounts of terpenoid lipids. Sterols, mainly ergosterol (**III**) and zymosterol (**XVI**), occur in all yeasts but usually at no more than 1% of the cell dry weight. However, in some species of *S. cerevisiae*, ergosterol has been reported up to 10%, and, in one case, up to 20% of the cell dry weight. However, this sterol is not of commercial importance.

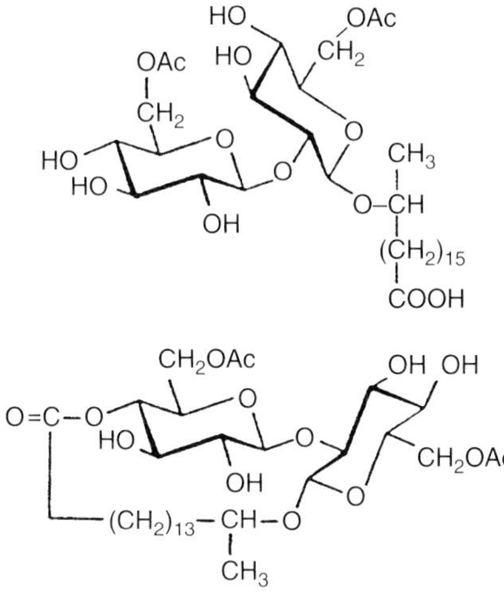

XVIII

Many yeasts do not synthesize carotenoids lipids. These include, obviously, the *Candida* (from *candidus*, Latin for white) yeasts. Various carotenoids, though, are produced in some abundance by the pink and red yeasts: principally, *Sporobolomyces*, *Rhodotorula/Rhodosporidium*, and *Phaffia*. The carotenoid produced by *Phaffia rhodozyma* is astaxanthin (**XVII**), which is now produced commercially as a supplement to fish feed, being a source of red pigmentation for the coloration of salmonid fish.

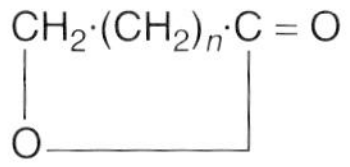

$CH_2 \cdot (CH_2)_n \cdot C=O$ (ring closed through O)

n = 14 or 16

XIX

$CH_3 \cdot (CH_2)_{13}CH(OH)–CH(NH–(CH_2)_{17}–CH_3)–CH_2–OH$

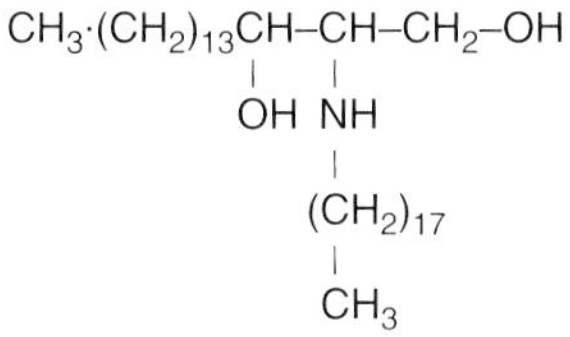

XX

Yeasts, like bacteria, also produce a number of biosurfactants when grown in hydrocarbons or other water-insoluble substrates. The most widely known of these are the sophorolipids (**XVIII**) produced by *Candida bombicola* and related species. Yields may exceed 100 g L^{-1}, but commercial uptake has been very limited because the sophorolipids, though efficacious as emulsificants, have little advantage over the much cheaper, chemically produced surfactants. The key component fatty acyl group of the sophorolipid from *C. bombicola* is a ω- or ω-1-hydroxy fatty acid, which may be lactonized into a macrocyclic lactone (**XIX**) that has some potential use as perfumery intermediate.

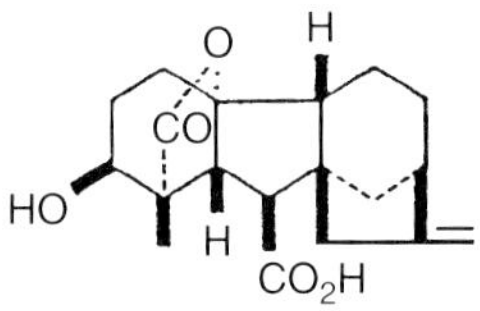

XXI

Although ceramide lipids are usually only minor components of microbial lipids, *N*-stearolyphytosphingosine (**XX**) is now being produced commercially using *Pichia* (formerly *Hansenula*) *ciferri*. The ceramide, after purification, is incorporated into various preparations being used in the cosmetics industry, and is also of some potential in the health care industry for treatment of various skin disorders.

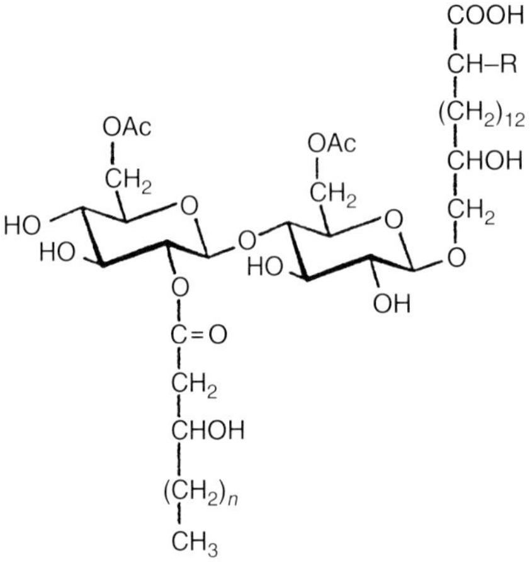

XXII

5
Molds

5.1
Fatty Acids

Some 60 000 individual species of filamentous fungi are known. These microorganisms, which are also referred to as *molds*, synthesize a range of fatty acids, from 12 to 24 carbon atoms in chain length and include polyunsaturated fatty acids up to 22 : 6 (docosahexaenoic acid). Although no systematic survey of mold lipids has been carried out, analysis of most of the easily cultivatable genera has been accomplished. Table 3 gives some of the different ranges of fatty acids that are found. Very little work has been carried out in the genetics of molecular biology of fungal lipids. In a number of species, there may be extensive accumulation of triacylglycerols as storage reserves; oil contents exceeding 70% have been recorded in several instances. The biochemistry of lipid accumulation and of biosynthesis appears to be similar to those processes elucidated with yeasts.

Besides the conventional straight chain fatty acids, some species of the order of phycomycetes produce iso- and anteiso-fatty acids (**VIIIA**, **B**). Ricinoleic acid, 15-hydroxyoleic acid, which is the principal fatty acid found in castor oil, occurs in *Claviceps* species.

5.2
Polyunsaturated Fatty Acids

One of the few differences in fatty acid composition between representatives of the three main orders of molds – the phycomycetes (or lower fungi), the ascomycetes (also known as the fungi imperfecti), and the higher fungi or basidiomycetes – is that the phycomycetes synthesize unsaturated fatty acids of the n-6 series. The ascomycetes and basidiomycetes fungi, in contrast, synthesize n-3 fatty acids. The biosynthetic route to these two series of fatty acids is shown in Fig. 8. These two pathways are similar to those that occur elsewhere: the n-6 route is an "animal"-only route, and the n-3 route is found in plants. Uniquely, some fungi are able to convert arachidonic acid as a member of the n-6 series to eicosapentaenoic acid, which belongs to the n-3 series, by the action of a Δ17-desaturase that is not found in other organisms. The ability of fungi to synthesize fatty acids longer than C_{18} is very restricted, however,

Tab. 3 Typical fatty acid profiles of some selected filamentous fungi.

Organism	*Relative amount [% w/w] of major acyl groups*							
	12:0 and shorter	*14:0*	*16:0*	*18:0*	*18:1*	*18:2*	*18:3*	*Others*
Phycomycetes (lower fungi)								
Conidiobolus nanodes	<1	<1	15	7	26	3	3[a]	20:1, 16% 20:4, 22%
Entomophthora coronata	40	31	9	2	14	2	1[a]	
Mortierella alpina	0	0	14	7	10	6	5[a]	20:3, 6% 20:4, 52%
Mucor hiemalis	0	2	15	10	33	19	19[a]	
Mucor miehei	0	1	23	6	44	15	6[a]	
Pythium ultimum	0	10	16	3	18	17	2[a]	20:1, 3% 20:4, 14% 20:5, 10%
Rhizopus arrhizus	0	19	18	6	22	10	12[a]	
Thraustochytrium aureum	0	2	27	1	35	3	1[a]	20:3, 2% 20:5, 6% 22:5, 6% 22:6, 11%
Ascomycetes (fungi imperfecti)								
Aspergillus niger	2	1	14	8	28	36	9	
Aspergillus terreus	0	2	23	<1	14	40	21	
Claviceps purpurea	0	<1	23	2	19	8	0	12-OH-18:1, 42%
Fusarium moniliforme	0	1	15	11	30	42	1	
Fusarium oxysporum	0	<1	31	12	35	18	1	
Penicillium spinulosum	0	<1	15	7	42	31	1	
Basidiomycetes (higher fungi)								
Agaricus campestris	2	1	15	3	5	66	0	
Tolyposporium ehrenbergii	<1	1	7	5	81	2	0	
Ustilago zeae	0	<1	9	3	54	17	0	20:1, 2% 20:2, 8% 22:0, 1%

[a] γ-linolenic acid, 18:3 (6, 9, 12).

and tends to occur to any appreciable extent only in some representatives of the phycomycetes fungi.

5.3 Single Cell Oils

Commercial interest in fungal oils has centered on the species that produce an abundance of certain polyunsaturated fatty acids. Oils (triacylglycerols) containing such fatty acids are used as dietary supplements, and numerous claims are made for their efficiency in the treatment of disorders as varied as childhood eczema and premenstrual tension, and in the prevention of heart disease. The inclusion of particular polyunsaturated fatty acids into infant milk formulations has also been advocated because these acids, which occur

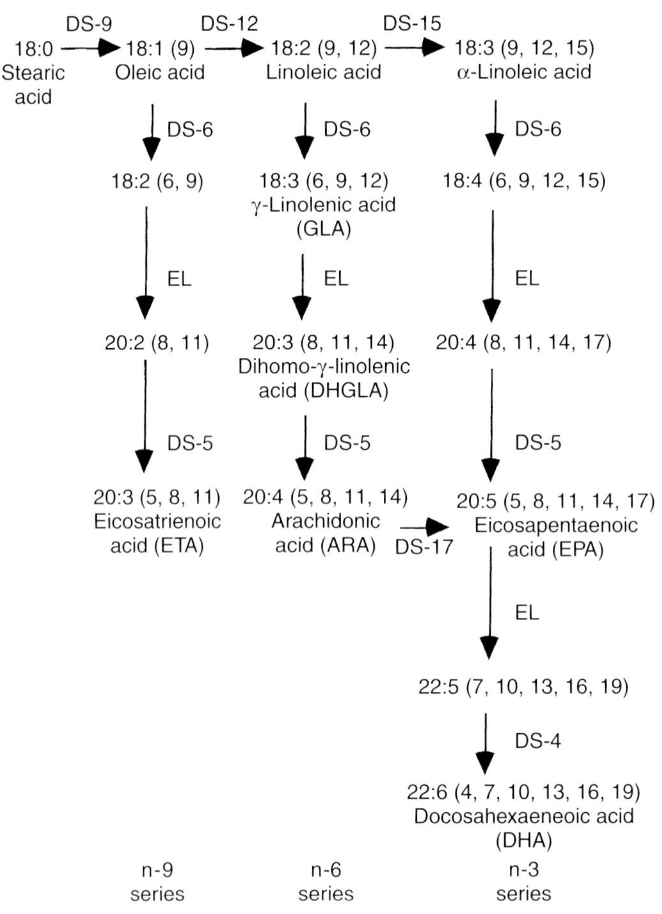

Fig. 8 Pathways of polysaturated fatty acid biosynthesis in fungi: DS-N, desaturase operating at *N*th carbon atom in chain; EL, elongase (acetyl–CoA dependent). The notations n-9, n-6, and n-3 indicate the position of the last double bond that is furthest away from the respective carboxylic acid group.

in human milk, are completely lacking in cow's milk.

Of particular interest are fungi that produce γ-linolenic acid (GLA), 18 : 3 (6, 9, 12), which is currently available only in the seed oil of the evening primrose (*Oenothera bienesis*), borage (*Borago officinalis*), and, as a mixture with the α-isomer, 18 : 3 (9, 12, 15), in *Ribes* spp. (blackcurrant, redcurrant, and so on). The high cost of these plant oils (between $10 and $15 per kg, though prices are very variable and have been both lower and higher in recent years) has made the biotechnological route using species of *Mucor, Mortierella*, and *Rhizobium* particularly attractive. Commercial processes using species of the former two genera were developed in the 1980s and ran into the 1990s, though, neither operates today (2004). In the United Kingdom, a process was operated by J. & E. Sturge Ltd, Yorkshire, using *Mucor circinelloides* grown in fermenters of up to 220 m^3 with the oil being sold under the trade name of Oil of Javanicus. This became the world's first commercially produced single cell oil.

The oil had the highest possible purity and satisfied all regulatory guidelines for its quality. In Japan, a similar process was developed later by Idemitsu Kosan Co. Ltd, Tokyo, using *Mortierella isabellina*. Production of both oils ceased primarily because of the cheaper plant oils that became available, especially borage oil (also known as starflower oil) which had a superior content of GLA over the mold oils (see Table 4 for comparisons between the fungal oils and plant oils containing GLA).

The market for oils rich in GLA now appears to be met by borage and evening primrose oils and both have developed their own niche positions making it difficult for alternative sources of GLA to displace them. This includes the possible production of GLA using a genetically modified plant, such as sunflower, in which the gene coding for the D6-desaturase, that converts the existing linoleic acid (18:2) into GLA (see Fig. 8), has been inserted into the plant. The reluctance of the commercial owners of this technology to develop the GM crop for GLA oil production is perhaps understandable because of the current political pressures (especially in the United Kingdom and other European countries) against such technology and the advent of the cheaper borage oil.

One single cell oil process using a mold continues to be of increasing commercial importance and this is for the production of an oil rich in arachidonic acid (ARA), 20:4 (n-6) – see Fig. 8. This polyunsaturated fatty acid is now recommended for inclusion in infant formulae along with docosahexaenoic acid (DHA) 22:6 (n-3) whose production as another single cell oil is described in Sect. 6.2. The functional roles of ARA and DHA are considered to be in the development of retinal and neural membrane lipids and are thought to be of diminished availability if babies are not being fed upon mother's milk.

Tab. 4 Fatty acid profiles of two commercial fungal oil products compared with evening primrose oil, borage oil, and black currant oil containing γ-linolenic acid.

Oil content, [% w/w]	*Mucor circinelloides*[a] 20	*Mortierella isabellina*[b] ND	*Evening primrose* 18–20	*Borage*[c] 30	*Blackcurrant* 30
Fatty acid	*Relative amount [% w/w]*				
16:0	22–25	27	6–10	9–13	6
18:0	5–8	6	1.5–3.5	3–5	1
18:1	38–41	44	6–12	15–17	10
18:2	10–12	12	65–75	37–41	48
γ-18:3	15–18	8	8–12	19–25	17
α-18:3	0.2		0.2	0.5	13

[a] Production organism used by J. & E. Sturge Ltd., UK
[b] Production organism used by Idemitzu Ltd., Japan; ND, not disclosed but probably about 45–50%.
[c] Also contains 20:1 (4.5%), 22:1 (2.5%), and 24:0 (1.5%); 22:1 is erucic acid.

Supplementation of infant formulae by both these fatty acids has now been recommended in over 60 countries, including the United States and most European countries. Infants receiving these supplements are considered to have improved development of memory and of eyesight.

The organism of choice for ARA production is *Mortierella alpina*. The mold is grown in large-scale fermenters and produces an oil content of about 40% of its biomass with ARA at about 45% of the total fatty acids. Various descriptions of this process have been given. The final oil is diluted with some sunflower oil in order to bring the ARA content down to a standard level of 40%. The oil itself has passed every stringent test and is regarded as entirely safe for ingestion by infants as well as adults, including pregnant females. A profile of the fatty acids of this oil is shown in Table 5.

5.4
Other Lipids

Molds, like yeasts, contain a wide variety of lipids. Some of these lipids are associated with various aspects of fungal differentiation, and certain lipids are found only in association with fungal spores or with particular stages of fungal development. As an example of this, ricinoleic acid (15HO-18 : 1), is found in *Claviceps purpurea*, but only in association with the sclerotial stage of development of this fungus. No 15HO-18 : 1 is found when *C. purpurea* is grown in its mycelial form in submerged culture.

Sterols occur in all molds. Ergosterol (**III**) is the commonest one, but many others (>40) are known.

Carotenoids occur in about 60% of all fungi. β-Carotene (IV) is produced commercially using *Blakeslea trispora* and, though most processes have now ceased owing to the availability of cheaper, chemically produced material, at least one process still continues in eastern Europe.

The group of compounds known as the gibberellins are diterpenes and are biosynthetically derived from a common intermediate of the carotenoid pathway (geranyl geranylpyrophosphate). These are commercially produced as plant growth hormones using *Gibberella fujikuroi*, which

Tab. 5 Fatty acid profiles of single cell oils rich in arachidonic acid and docosahexaenoic acid that are produced commercially by cultivation in large (100 m^3) fermenters.

Oil	*Fatty acid composition (Rel.% w/w)*													
	12:0	*14:0*	*16:0*	*16:1*	*18:0*	*18:1*	*18:2 (n-6)*	*18:3 (n-6)*	*18:3 (n-3)*	*20:3 (n-6)*	*20:4 (n-6)*	*22:5 (n-6)*	*22:6 (n-3)*	*24:0*
ARASCO[a]	–	0.4	8	0	11	14	7	4	–	4	49	–	0	1
DHASCO[b]	4	20	18	2	0.4	15	0.6	–	–	–	–	–	39	–
DHAGold[c]	–	13	29	12	1	1	2	–	3	1	–	12	25	–

[a] Production organism: *Mortierella alpina*.
[b] Production organism: *Crypthecodinium cohnii*.
[c] Production organism: *Schizochytrium* sp. and now (2004) renamed as DHASCO-S.
Note: ARASCO, DHASCO, and DHAGold are registered trade names of Martek Bioscience Corp., MD, United States of America. ARASCO and DHASCO are combined together at a 2 : 1 ratio and used for incorporation into infant formulae under the trade name of Formulaid.

is the telomorphic state of *Fusarium moniliforme*. The principal product is gibberellic acid (**XXI**), though some 80 different variations of this compound are known.

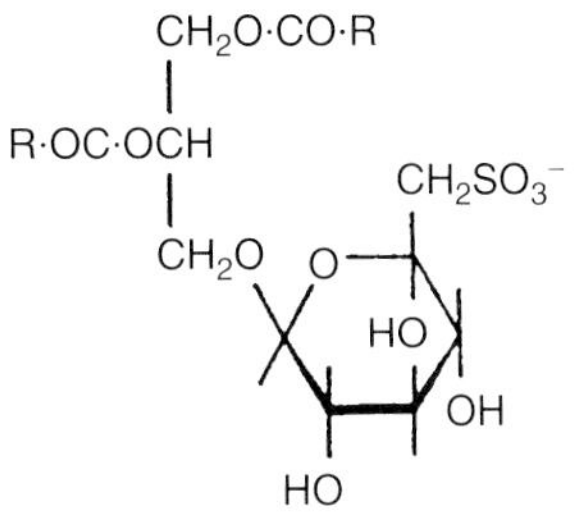

$CH_2O{\cdot}CO{\cdot}R$

$R{\cdot}OC{\cdot}OCH$

CH_2O O $CH_2SO_3^-$

HO OH

HO

XXIII

A number of surfactants are also produced by fungi. Like those produced by other microorganisms, they are effective in creating stable oil–water mixtures. They may, therefore, be functionally useful to microorganisms that inhabit the phyllosphere and, as such, may improve attachment of the fungi (mycelial cells or spores) to plant leaves. A surfactant based on cellobiose from the maize rust fungus *Ustilago maydis* is known as ustilagic acid (**XXII**).

6
Unicellular Algae

The term "algae" encompasses 14 major groups or classes ranging from prokaryotic bacteria through the yellow, green, golden, and brown macroalgae, which includes both the seaweeds and diatoms. Most algae are photosynthetic organisms, though some may grow heterotrophically – that is, using fixed carbon compounds rather than CO_2, from which both carbon and energy may be obtained. When CO_2 is used as the carbon source, sunlight is then usually used, via photosynthesis, as the source of energy.

6.1
Cyanobacterial Lipids

The prokaryotic algae, or cyanobacteria, were originally referred to as the *blue-green algae*. The principal genera are *Anabena, Anacystis, Nostoc, Oscillatoria, Spirulina*, and *Synechococcus*. The lipids of these genera are diverse and, of course, include chlorophyll, which is an essential part of the photosynthetic apparatus. Whereas bacteriochlorophylls are found in the green and purple photosynthetic bacteria (*Rhodospirillaceae, Chromatiaceae,* and *Chlorobiaceae*), plant chlorophyll is found in the cyanobacteria. This then suggests that cyanobacteria are the ancestors of the chloroplasts of plants. The neutral lipids of cyanobacteria include many carotenoid pigments, hydrocarbons, hopanoids, quinones, and, in some species, poly-β-hydroxybutyrate (**XIII**). The principal polar lipids of the cyanobacteria are phosphatidylglycerol and sulfoquinovosyldiacylglycerol (**XXIII**), which are regarded as the sulfolipid characteristic of plants. The fatty acyl substituents of these lipids are "plant-like" rather than "bacteria-like" and include oleic acid (18 : 1), linoleic acid (18 : 2), and either γ- or α-linolenic acid (18 : 3), according to the genus. A number of conjugated lipids also occur that are associated with the cell envelope; these include lipopolysaccharides forming lipid A (see Fig. 5) as part of the cell envelope. The cyanobacterial cells are usually low in total lipid contents (<15%) and, because of the wide diversity of types, are not usually considered commercially exploitable, given the impracticality of purifying individual lipids. However, an exception may be possible with *Spirulina*, which has been researched with some enthusiasm in Israel by Zvi Cohen and his colleagues, at the Ben Gurion University, in the Negev Desert.

Spirulina platensis and *Spirulina maxima* are multicellular, filamentous cyanobacteria that profusely colonize lakes in Africa and Mexico. They have been used as indigenous supplies of food and feed and are of some commercial interest because both have been known for some time to contain γ-linolenic acid. Although the algae grow well in outdoor lagoons in hot, sunny climates and can, indeed, produce in the extracted lipid a GLA content exceeding 20%, the distribution of GLA between neutral lipid (e.g. triacylglycerols), glycolipid, and phospholipid fractions is about 1:9:4. This, coupled with the low ($<$6%) content of lipid in the cells, probably makes it a commercially unattractive source of the fatty acid. With appropriate marketing, however, it could be used to boost the nutritional claims for whole-cell *spirulina* as a dietary supplement.

6.2 Eukaryotic Algae Lipids

6.2.1 Carotenoids

Of the nonfatty acyl lipids, carotenoids have been of some commercial interest. The production of β-carotene (**IV**) by *Dunaliella salina* has been investigated by a number of groups worldwide, with those in Israel, Western Australia, and California being particularly active. Commercialization of this process has now been developed in Western Australia by Aqua-Carotene Ltd., which uses outdoor cultivation ponds to produce the biomass.

Also of increasing commercial interest is another carotenoid, astaxanthin, which is used principally as a colorant for feeding farmed fish. This is produced by *Haematococcus pluvialis*, which can contain over 2% of its biomass as astaxanthin. It is grown in large-scale ponds and is sold as an alga meal by a number of companies. Chemically produced carotenoids, however, remain cheaper than the algal sources but do not have the benefit of being able to claim that they are "natural" ingredients. This is possibly not a deterrent when considering large-scale use where cost is the priority, though, legislation in individual countries may dictate otherwise.

6.2.2 Polyunsaturated Fatty Acids

Several commercial processes now exist for the production of the very long-chain fatty acid, docosahexaenoic acid, 22:6 n-3, (DHA). This fatty acid is now recommended for inclusion in infant formulae along with arachidonic acid, see Sect. 5.3. It is also recommended for dietary supplementation for adults, where it may offer protection against coronary heart diseases. It is preferred, in some cases, to fish oil which, though containing DHA, also contains eicosapentaenoic acid, 20:5 n-3, (EPA) and which may be contraindicated as it can interfere with the uptake and functional role of DHA. Fish oil is not allowed as a supplement for infant formulae in some countries, including the United States, because of its possible contamination with environmental toxins and heavy metals.

The main production organism for DHA is *Crypthecodium cohnii*, which is a unicellular, nonphotosynthetic dinoflagellate. It is grown in large-scale fermenters of up to 100 m^3. The principal company behind the development of the process and the marketing of the oil is Martek Biosciences Corp., Maryland, United States. The oil is known by the trade name of DHASCO and, when mixed with arachidonic acid for incorporation into infant foods, is known as Formulaid. The fatty acid profile of the oil is shown in Table 5, where it can be seen that DHA is the sole polyunsaturated fatty acid thereby making the oil highly desirable as a source of this single PUFA.

Another related process for the production of DHA uses another marine organism known as *Schizochytrium*. The inclusion of this organism as an alga is uncertain as the taxonomic position of the group of organisms, that includes *Schizochytrium* and the related group of thraustochytrid organisms, continues to be researched. *Schizochytrium*, like *Crypthecodinium cohnii*, is nonphotosynthetic and is also grown in large-scale bioreactors using glucose as substrate. The oil produced is slightly different from the oil of *C. cohnii* in that it contains a small but appreciable percentage of docosapentaenoic acid, 22 : 5 n-6. The full profile of the fatty acids are shown in Table 5.

The Schizochytrium process was originally developed by OmegaTech Ltd, Boulder, CO, which is now owned by Martek Biosciences Corp. The oil was originally given the trade name of DHAGold and was used as an animal feed supplement to boost the content of DHA in the final product. It was particularly useful with poultry where it led to eggs being enriched in DHA and thus, could attract a premium price. The oil itself has now received approval as a direct supplement for humans, including infants, and will, in the future, be used for this purpose. Its current trade name is DHASCO-S.

Some interest has continued in the use of photosynthetically grown algae for polyunsaturated fatty acid production. Although the organisms can be grown outdoors using sunlight as energy source and CO_2 as a carbon source, the costs of production remain very high because of the slow growth rate of the cells and the low cell densities that are produced. It may take four to six weeks to generate algal cells at no more than 5 g L^{-1} in an outdoor system, whereas *Schizochytrium*, as discussed earlier, when grown heterotrophically in a stirred fermenter it can reach biomass densities of up to 200 g L^{-1} in about four days. However, there are still several groups who continue to use microalgae for the production of various fatty acids not easily obtainable from other sources. Of particular interest is the possible production of EPA for which no source, as a single PUFA, currently exists apart from very expensive purification from fish oils that contain both EPA and DHA. The demand for an oil rich in EPA as a single PUFA is likely to increase as it has been recommended for use in the treatment of various disorders of the brain, including bipolar disorder and schizophrenia. There have also been suggestions that a high dosage of EPA may help in the treatment of certain cancers though this is highly contentious. For these reasons, there is likely to be considerable interest in obtaining a good supply of this PUFA from whatever source and, currently, photosynthetic algae appear to offer the best prospects.

See also Bacterial Cell Culture Methods; FTIR of Biomolecules; Microbial Development.

Bibliography

Books and Reviews

Cohen, Z. (Ed.) (1999) *Chemicals from Microalgae*, Taylor and Francis, London.

Cohen, Z., Ratledge, C. (Eds.) (2005) *Single Cell Oils – 2*, American Oil Chemists' Society, Champaign, IL.

Cronan, J.E. (2003) Bacterial membrane lipids: where do we stand? *Annu. Rev. Microbiol.* **57**, 203–224.

Horikoshi, K., Grant, W.D. (Eds.) (1998) *Extremophiles: Microbial Life in Extreme Environments*, Wiley Liss, NJ.

Kroes, R., Schaefer, E.J., Squire, R.A., Williams, G.M. (2003) A review of the safety of DHA oil, *Food Chem. Toxicol.* **41**, 1433–1446.

Ratledge, C., Wilkinson, S.G. (Eds.) (1988,1989) *Microbial Lipids*, Vols. 1 and 2. Academic Press, London and New York, 963–726.

Primary Literature

Murphy, D.J. (2001) The biogenesis and functions of lipid bodies in animals, plants and microorganisms, *Prog. Lipid Res.* **40**, 325–438.

Nikaido, H. (2003) Molecular basis of bacterial outer membrane permeability revisited, *Microbiol. Mol. Biol. Rev.* **67**, 593–656.

Ratledge, C., Wynn, J.P. (2001) The biochemistry and molecular biology of lipid accumulation in oleaginous microorganisms, *Adv. Appl. Microbiol.* **51**, 1–51.

Sorger, D., Daum, G. (2003) Triacylglycerol biosynthesis in yeast, *Appl. Microbiol. Biotechnol.* **61**, 289–299.

Wen, Z.-Y., Chen, F. (2003) Heterotrophic production of eicosapentaenoic acid by microalgae, *Biotechnol. Adv.* **21**, 273–294.

Lipoprotein Analysis

Alan T. Remaley[1] *and G. Russell Warnick*[2]
[1] *National Institutes of Health, MD, USA*
[2] *Pacific Biometrics Research Foundation Issaquah, WA, USA*

Encyclopedia of Molecular Cell Biology and Molecular Medicine, 2nd Edition. Volume 7
Edited by Robert A. Meyers.

ISBN: 3-527-30549-1

Keywords

Apolipoprotein
A protein found on lipoprotein particles that functions to maintain the structural integrity of the lipoprotein particle and to modulate lipoprotein metabolism.

Atherosclerosis
Deposition of lipid in the intimal wall of large- to medium-sized arteries that impinges on blood flow and/or predisposes the vessel to thrombosis.

Cholesterol
A 27-carbon sterol that is a common lipid constituent of cell plasma membranes and lipoproteins.

Coronary Artery Disease
The presence of atherosclerosis in the coronary artery vessels of the heart, which can interfere with blood flow.

Lipoprotein
A small complex of lipid and protein that serves as the main carrier of extracellular lipid.

Lipoproteins are submicron- to micron-sized particles that are composed of both lipids and proteins. Lipoproteins are the major extracellular carrier of lipids, such as cholesterol, and thus, they play an important role in the pathogenesis of atherosclerosis. The analysis of lipoproteins is performed in most clinical laboratories to identify patients at risk of coronary heart disease (CHD). Lipoprotein analysis is also essential in monitoring the effectiveness of cholesterol-lowering therapy.

The focus of this chapter will be on methods for lipoprotein analysis that are used for routine diagnostic purposes, but in addition, new promising types of lipoprotein analysis that are still in the realm of basic research will also be highlighted. A brief description of lipid biochemistry and lipoprotein metabolism will be followed by a review of commonly used methods for analyzing lipoproteins. Finally, the chapter will conclude with a summary of how the results of lipoprotein analysis are used in the diagnosis and management of patients with dyslipidemias.

1 Overview of Lipoproteins

1.1 Lipid Biochemistry

Lipids, commonly referred to as *fats*, are composed of mostly carbon and hydrogen (Fig. 1), and are therefore hydrophobic. Two major classes of lipids, based on their relative hydrophobicity, are found on lipoproteins. Phospholipids and cholesterol are classified as amphipathic lipids because, in addition to C-H bonds, they also contain polar groups. In contrast,

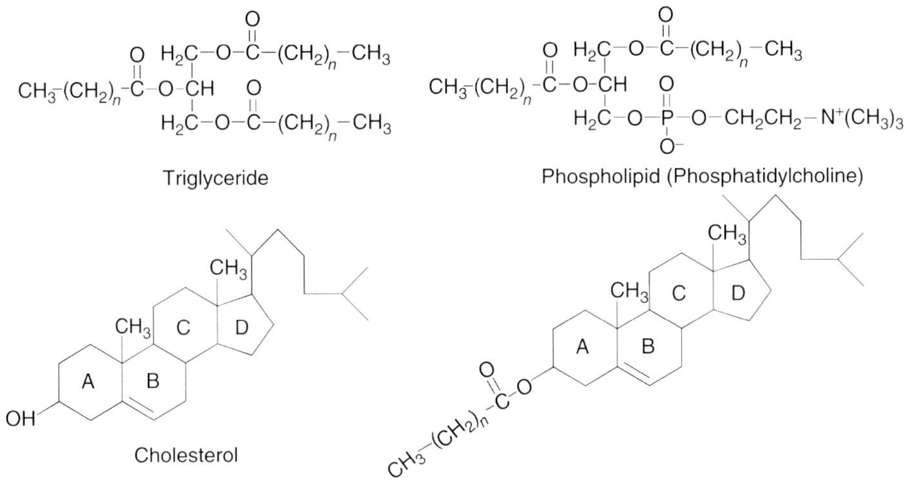

Fig. 1 Chemical structures of lipids found on lipoproteins.

cholesteryl esters and triglycerides do not contain any polar constituents and are therefore more hydrophobic and are classified as neutral lipids.

A prototypical structure of a lipoprotein particle is shown in Fig. 2. Phospholipids and cholesterol form a monolayer on the surface of lipoprotein particles. Phospholipids are oriented so that their charged head groups are facing outward, away from the neutral hydrophobic core. Cholesterol is also oriented so that its polar hydroxyl group on the A ring (Fig. 1) is pointing away from the lipoprotein surface. Cholesteryl esters and triglycerides are found in the core of lipoprotein particles in a disorganized liquid crystalline state.

Except for cholesterol, all of the other major lipids found on lipoprotein have numerous distinct chemical forms (Fig. 1). The different forms of triglyceride can be identified on the basis of the type of fatty acids that they contain. Fatty acids can be classified on the basis of the number of carbon atoms as short-chain (4–6 carbon atoms), medium-chain (8–12 carbon atoms), or long-chain (>12 carbon atoms) fatty acids. Most dietary fatty acids are long chain and contain an even number of carbon atoms. Fatty acids also differ on the basis of the number of unsaturated or double carbon bonds that they contain. Fatty acids with no double bonds are classified as saturated fatty acids, whereas fatty acids with one double bond are monounsaturated and fatty acids with two or more double bonds are polyunsaturated. The different fatty acids are commonly designated by their carbon-chain length and by the number of double bonds. For example, arachidonate, a polyunsaturated fatty acid that has 20 carbon atoms and 4 double bonds is designated as 20:4. The common fatty acids found in human serum triglycerides are palmitate (16:0), oleate (16:1), and stearate (18:0). Triglycerides containing shorter chain fatty acids or unsaturated fatty acids have a lower melting point and are usually liquid at room temperature. Saturated fatty acids are commonly found in food from animal sources, whereas unsaturated fatty acids are abundant in food derived from plants.

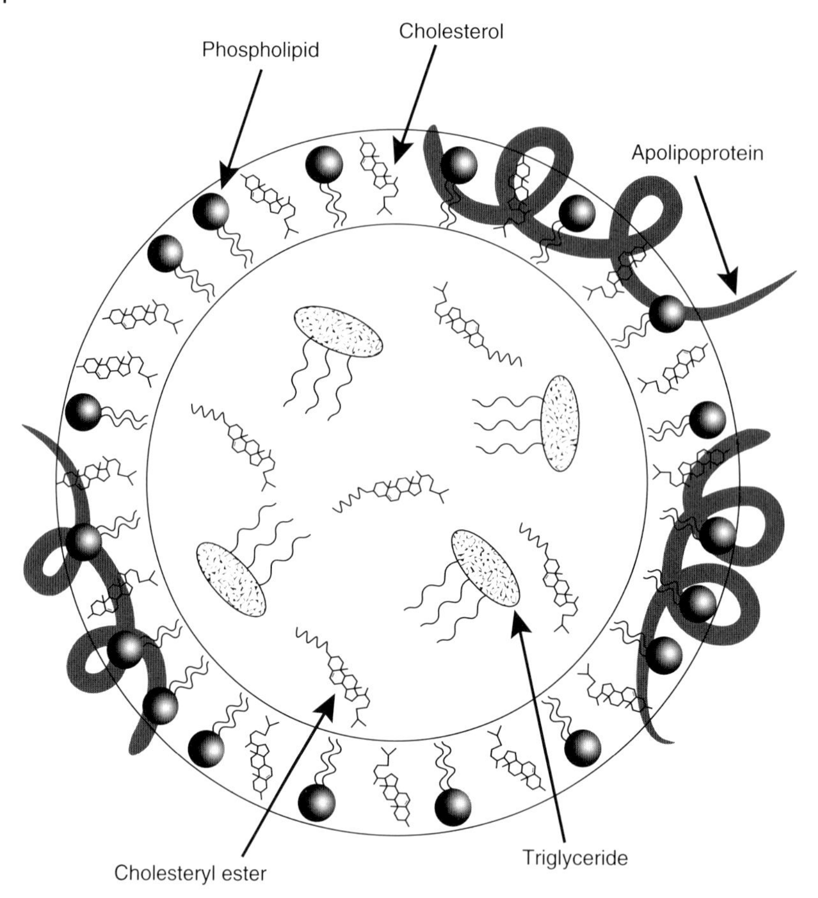

Fig. 2 Model of lipoprotein particle.

Similar to triglycerides, there are different forms of cholesteryl esters, depending on the type of fatty acid that is esterified to the A ring of cholesterol. Cholesteryl linoleate, which is primarily produced by the serum enzyme lecithin: cholesterol acyltransferase (LCAT), and cholesteryl oleate, which is primarily produced by the intracellular enzyme acylcoenzyme A: cholesterol acyltransferase (ACAT), are the two most common forms of cholesteryl esters in humans.

Phospholipids can differ not only by the type of fatty acids that they contain but also by their different head groups (choline, inositol, serine, glycerol, and ethanolamine), which serves as the basis for their nomenclature. Phosphatidylcholine (Fig. 1), which contains a choline head group, is the most common type of phospholipid found on lipoproteins.

1.2 Lipoprotein Structure and Function

Lipoproteins are typically spherical in shape and range in size from 5 to 1200 nm (Table 1). In addition to lipids, lipoproteins also contain proteins, called *apolipoproteins,* which reside on the surface of

Tab. 1 Characteristics of the major human lipoproteins.

Features	*Chylos*	*VLDL*	*LDL*	*HDL*
Density (g mL^{-1})	<0.93	0.93–1.006	1.019–1.063	1.063–1.21
Molecular weight (kDa)	$(0.4–30) \times 10^9$	$(10–80) \times 10^6$	2.75×10^6	$(1.75–3.6) \times 10^5$
Diameter (nm)	80–1200	30–80	18–30	5–12
Total lipid (% weight)	98	89–96	77	50
Triglyceride (% weight)	84	44–60	11	3
Total cholesterol (% weight)	7	16–22	62	19
Site of synthesis	Intestine	Liver	Catabolism of VLDL	Liver, intestine
Electrophoretic mobility	Origin	Pre-beta	Beta	Alpha

Tab. 2 Characteristics of the major human apolipoproteins.

Apolipoprotein	*Molecular weight [kDa]*	*Plasma concentration [mg dL^{-1}]*	*Major lipoprotein location*	*Function*
ApoA-I	28 000	100–200	HDL	Structural, LCAT activator, ABCA1 lipid acceptor
ApoA-II	17 400	20–50	HDL	Structural
ApoA-IV	44 000	10–20	Chylos, VLDL, HDL	Structural
ApoB-100	5.4×10^5	70–125	LDL, VLDL	Structural, LDL receptor ligand
ApoB-48	2.6×10^5	<5	Chylos	Structural, remnant receptor ligand
ApoC-I	6630	5–8	Chylos, VLDL, HDL	Structural
ApoC-II	8900	3–7	Chylos, VLDL, HDL	Structural, LPL cofactor
ApoC-III	9400	10–12	Chylos, VLDL, HDL	Structural, LPL inhibitor
ApoE	34 400	3–15	VLDL, HDL	Structural, LDLreceptor, and remnant receptor ligand
Apo(a)	$(3–7) \times 10^5$	<30	Lp(a)	Unknown

lipoprotein particles (Fig. 2). Apolipoproteins can serve both a structural role in maintaining the integrity of the lipoprotein particle and also a functional role in lipoprotein metabolism. Apolipoproteins contain a structural motif called an *amphipathic helix,* which enables them to bind to lipids. Amphipathic helices are α-helices with approximately half of the helix composed of hydrophilic amino acid residues and the other half composed of hydrophobic amino acid residues. They are oriented on the surface of lipoprotein particles to allow the interaction of the hydrophobic face of the helix with the surface lipids. Apolipoproteins also play a major role in modulating lipoprotein metabolism by acting as ligands for various lipoprotein receptors and by acting as activators and inhibitors of several lipid-modifying enzymes (Table 2).

Several different classification systems based on their varying physical properties can be used for categorizing lipoprotein particles. A density classification system (Chylomicrons (Chylos), very low density lipoproteins (VLDL), intermediate density lipoproteins (IDL), low density lipoproteins (LDL), and high density lipoproteins (HDL)), which was developed on the basis of the original ultracentrifugation system for isolating lipoproteins, is most common (Table 1). The size of the lipoprotein particles is inversely related to their density. Larger lipoprotein particles have relatively more core lipid and less protein and are, therefore, lighter in density.

Three major pathways in lipoprotein metabolism and how the different types of lipoprotein particles are involved in each pathway are shown in Fig. 3. Chylomicrons are the largest and the least-dense lipoprotein particles and contain a single molecule of apoB-48 as well as several other types of apolipoproteins (Table 2). Chylomicrons are produced by the intestine and they participate in the exogenous pathway, which delivers dietary lipids to the liver and to other peripheral tissues. The turbidity of postprandial serum is due to the presence of chylomicrons, which reflect light because of their large size. Once chylomicrons enter the circulation, they are quickly catabolized within a few hours by lipoprotein lipase (LPL) and hepatic lipase (HL), which are on the surface of endothelial cells and hydrolyze triglycerides and phospholipids (Fig. 3). The free fatty acids that are generated are taken up by tissues and are either reesterified and stored in lipid drops or are used as a source of energy. Chylomicrons are transformed by this process into the more dense chylomicron remnants, which are primarily removed by the liver via a chylomicron remnant receptor that binds apoE.

VLDL, IDL, and LDL participate in the endogenous pathway of lipid metabolism, which transports hepatic-derived lipids to peripheral tissues (Fig. 3). VLDL has a single copy of apoB-100; it is produced in the liver and is rich in triglycerides synthesized by the liver. Like chylomicrons, the core lipids of VLDL are acted upon by LPL and HL, and the fatty acids released are taken up by peripheral tissues. As a consequence of the lipolysis, a significant fraction of VLDL is transformed into IDL and ultimately into LDL. The majority of LDL is returned to the liver via the LDL receptor; however, LDL can also be taken up in peripheral tissues by the LDL receptor.

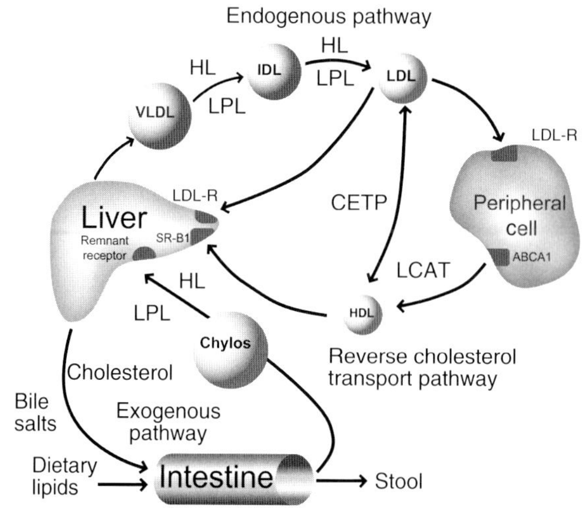

Fig. 3 Diagram of lipoprotein metabolism.

Because of its small size, LDL can also readily infiltrate into the vessel wall and become trapped there by proteoglycans. LDL in the vessel wall is prone to oxidation, which then makes it a better ligand for scavenger receptors on macrophages. Macrophages that take up too much lipid become filled up with intracellular lipid drops and are transformed into foam cells. A collection of foam cells in the vessel walls forms a fatty streak, which is an early precursor of atherosclerotic plaques. Oxidized LDL can also initiate the inflammation of the vessel wall by acting as a chemotactic agent and by inducing the production of various proinflammatory cytokines by macrophages. Another proatherogenic lipoprotein similar to LDL in structure is Lipoprotein (a) (Lp(a)). It contains a single copy of apoB-100, which is covalently linked by a disulfide bond to apo (a), a plasminogen-like protein of unknown function.

High-density lipoprotein is the principal lipoprotein that mediates the reverse cholesterol-transport pathway, delivering the excess cholesterol in peripheral cells back to the liver for excretion (Fig. 3). Because most peripheral cells cannot catabolize cholesterol, the reverse cholesterol-transport pathway is one of the main mechanisms for maintaining cellular cholesterol homeostasis. The principal apolipoprotein on HDL is apoA-I, which is sometimes used as a marker for HDL. In contrast to LDL, HDL has anti-inflammatory properties and is considered to be antiatherogenic because it also mediates the removal of excess cholesterol from peripheral cells. The process of reverse cholesterol transport is initiated when lipid-poor apoA-I interacts with the ABCA1 transporter on the cell surface and extracts excess cholesterol from cells. Cholesterol that is removed from cells is then trapped on HDL after it is transformed into cholesteryl ester by LCAT (Fig. 3). HDL cholesterol is then returned either directly to the liver by the SR-B1 receptor or indirectly after being transferred to LDL by the cholesteryl ester transfer protein (CETP) (Fig. 3).

2
Analysis of Lipoprotein Components

The initial analysis of lipoproteins usually begins with a direct measurement of total cholesterol and triglyceride from serum or plasma, without any prior extraction or fractionation step. Apolipoproteins can also be measured directly on serum or plasma. There are several chemical methods for measuring the various lipids on lipoproteins, which are still used as reference methods by the Centers for Disease Control and Prevention (CDC) and the associated Cholesterol Reference Method Laboratory Network (CRMLN). These chemical methods, however, are complex, laborious, and expensive to perform and have largely been replaced in routine clinical laboratories by automated enzymatic methods.

2.1
Total Cholesterol Assay

A coupled enzyme reaction (Fig. 4), which depends on the specificity of cholesterol oxidase for cholesterol, is the most common enzyme method for measuring total cholesterol. Before measuring cholesterol, cholesteryl ester, which makes up about two-thirds of total cholesterol, is first converted to free cholesterol by cholesteryl esterase. The free cholesterol generated then reacts with cholesterol oxidase, which results in the production of cholestenone

- Cholesteryl ester + H_2O $\xrightarrow{\text{Cholesteryl esterase}}$ Cholesterol + Fatty acid
- Cholesterol + O_2 $\xrightarrow{\text{Cholesterol oxidase}}$ Cholestenone + H_2O_2
- H_2O_2 + Dye $\xrightarrow{\text{Cholesterol oxidase}}$ Color

Fig. 4 Enzymatic assay for cholesterol.

and hydrogen peroxide. The amount of cholesterol in the sample is quantified by measuring the amount of hydrogen peroxide generated by using a horseradish peroxidase–dye system (Fig. 4). The assay is monitored spectrophotometrically at approximately 500 nm and can be performed on serum or plasma without any prior extraction with an organic solvent. Modern reagents include other constituents that eliminate most sources of interference, although specimens with marked elevations of vitamin C or bilirubin and specimens with gross hemolysis can experience interference with the peroxidase catalyzed color reaction. Most commercial assays are developed for specific analyzers and should yield results within 3% of the result obtained by the reference method, with a coefficient of variation of less than 3%. Although it is not frequently performed for diagnostic purposes, free (nonesterified) cholesterol can be simply measured by omitting the cholesterol ester–hydrolase step.

2.2 Triglyceride Assay

There are several different colorimetric enzyme assays for triglycerides, which primarily differ in how they are coupled to an indicator dye. As shown in Fig. 5, all of these methods begin with the enzymatic hydrolysis of fatty acid from the glycerol backbone, with a lipase. In one of the more common methods (Fig. 5), the amount of triglyceride is measured by quantifying the glycerol, using glycerokinase and glycerophosphate oxidase and the same dye-peroxidase system used in the cholesterol assay. One problem with this approach is that the assay will also measure any endogenous free glycerol in a sample. For most samples, this will result in a clinically insignificant positive bias of less than 10 to 20 mg dL^{-1}, but in some disorders, such as diabetes, or in patients receiving total parenteral nutrition, the bias can become clinically significant. It is for this reason

- Triglyceride + H_2O $\xrightarrow{\text{Bacterial lipase}}$ Fatty acid + Glycerol
- Glycerol + ATP $\xrightarrow{\text{Glycerokinase}}$ Glycerophosphate + ADP
- Glycerophosphate + O_2 $\xrightarrow{\text{Glycerophosphate oxidase}}$ Dihydroxyacetone + H_2O_2
- H_2O_2 + Dye $\xrightarrow{\text{Peroxidase}}$ Color

Fig. 5 Enzymatic assay for triglyceride.

that some laboratories compensate for this problem by measuring and correcting for the free glycerol in a blanking procedure or by performing an assay that consumes any free glycerol in the first step, thus eliminating the measurement of any endogenous glycerol later on in the reaction. Alternatively, the manufacturers of some triglyceride assays compensate for this bias by adjusting the value of their calibrators.

2.3 Phospholipid Assay

The measurement of phospholipids has not been shown to have any diagnostic value and is commonly done only in research laboratories. The choline-containing phospholipids, lecithin, lysolecithin, and sphingomyelin, account for at least 95% of total phospholipids in serum and can be measured enzymatically with commercial kits, using phospholipase D, choline-oxidase, and horseradish peroxidase. To measure total phospholipids, independent of the head group, phosphorus is measured chemically after it is released by acid hydrolysis from an extracted sample. The individual types of phospholipids can be measured semiquantitatively after separation by thin-layer chromatography.

2.4 Apolipoprotein Assays

Because of their relatively unique distribution on the different lipoprotein particles (Table 2), apoA-I and apoB-100 in serum are used as an index of the amount of HDL and LDL respectively. Because these two proteins are relatively abundant, they can be measured with specific antibodies by either turbidometric or nephelometric procedures. These assays are now routinely available on standard clinical analyzers and are free from most interferences except for problems related to light scattering from turbid samples with elevated chylomicrons and VLDL. Tests for apoA-I and apoB-100 have been shown to be more precise and accurate than tests for HDL-C and LDL-C. The apolipoproteins have been proposed as being possibly superior for predicting CHD (coronary heart disease) risk, but they have not yet been incorporated into the National Cholesterol Education Program (NCEP) practice guidelines for lipoprotein analysis.

The proatherogenic Lp(a) lipoprotein can also be measured by quantifying the presence of the apo(a) apolipoprotein, using various ELISA or nephelometric methods. There are, however, different size polymorphisms of apo(a), which complicate the immunologic measurement and differ in their correlation with CHD risk. A qualitative assessment of isoform distribution by electrophoresis or by genotyping is also used for characterizing Lp(a) and for estimating CHD risk.

3 Fractionation of Lipoproteins

Numerous methods for fractionation of the various lipoproteins have been developed on the basis of their different physical properties, such as size, density, charge, solubility, and apolipoprotein content. Historically, ultracentrifugation was the primary means of separating the different lipoprotein fractions on the basis of density and is still the foundation of the current lipoprotein classification system (Table 1). The primary reason for fractionating lipoproteins is to provide a measure of their relative abundances. Fractionation is particularly useful for measuring the cholesterol content of LDL (LDL-C) and HDL (HDL-C), which are used in calculating CHD risk. Electrophoresis is the other commonly used method in clinical laboratories for fractionating lipoproteins.

3.1 HDL-C Assays

Until recently, the most common method for measuring HDL-C has been a multistep precipitation procedure, which involved the removal of apoB-containing lipoproteins (chylos, VLDL, IDL, LDL, and Lp(a)) by precipitation and centrifugation, followed by the enzymatic measurement of cholesterol in the supernatant. A wide variety of polyanions, such as heparin, phosphotungstate, and dextran sulfate will bind to positively charged groups on apoB-containing lipoproteins and cause their aggregation and precipitation, in the presence of divalent cations, such as manganese and magnesium. Magnesium has become the divalent cation of choice because of interference with enzymatic lipid assays by manganese. Numerous commercial assays have been developed with the different precipitating reagents. Initially, results from these different assays often yielded discordant results, but this problem has largely been resolved by the ongoing efforts to improve the standardization of lipoprotein assays.

A major limitation of the various HDL-precipitation methods is that specimens with elevated triglycerides often have a positive bias for HDL-C. Because of their low density, specimens with high levels of chylomicrons and triglyceride-rich VLDL will often form aggregates that do not fully sediment upon centrifugation. This leads to a false elevation for HDL-C, because, when cholesterol is measured in the supernatant, not only cholesterol from HDL is measured but cholesterol from any residual aggregates is also measured. This problem can at least be partially solved by predilution of the sample, which facilitates the sedimentation of the aggregates. Although it is laborious, ultracentrifugation or ultrafiltration of the sample to remove any residual aggregates will also eliminate this source of bias.

Unlike the precipitation-based HDL-C tests, most other commonly measured clinical laboratory tests do not require any significant specimen preprocessing. This limitation of the HDL-C-precipitation tests has prompted the recent development of homogenous assays for HDL-C that can be performed directly on serum or plasma, without the physical removal of non-HDL lipoproteins. These assays work by using reagents that chemically mask the non-HDL lipoproteins during the enzymatic measurement of cholesterol associated with HDL. This can be accomplished by using different strategies that involve the use of polymers, detergents, antibodies, or even modified enzymes. These various reagents bind to non-HDL lipoproteins and block their accessibility to cholesterol esterase and cholesterol oxidase. Other homogenous assays selectively deplete non-HDL-C in the first step, thus preventing its measurement later on in the reaction. Several commercial homogenous HDL-C assays are now available on automated instruments, and these assays show relatively good accuracy and typically better precision than precipitation-based assays. Precipitation-based HDL-C assays have now largely been replaced by homogenous HDL-C assays in most routine clinical laboratories, but they still have a role in research laboratories and in specialty lipid laboratories because the homogenous HDL-C assays do not always yield good results in patients with unusual dyslipidemias and/or in patients with severe liver or kidney disease.

3.2 LDL-C Assays

Traditionally, methods for measuring LDL-C have been more difficult to perform than HDL-C assays because there is no convenient precipitation procedure to cleanly separate LDL from the other apoB-containing lipoproteins. As a consequence, the original LDL-C methods involved a procedure commonly referred to as *β-quantification*, which involves both a precipitation and an ultracentrifugation step. In the first step, chylomicrons and VLDL are removed from a sample by ultracentrifugation (105 000 G for 18 h at 10 °C) at the native density of plasma (1.006 g L^{-1}), which results in their flotation at the top of the tube. Chylomicrons and VLDL are then removed with a tube slicer and cholesterol in the infranate, which consists of HDL-C and LDL-C, is measured using an enzymatic assay. HDL-C is then measured either from the infranate or from the original sample after precipitation and removal of LDL. LDL-C is calculated by subtracting HDL-C from the cholesterol value obtained from the infranate. The β-quantification method serves as the basis for the LDL-C reference method, but because of its complexity and the requirement of an ultracentrifuge, the procedure is primarily used by specialty lipid laboratories.

Instead of directly measuring LDL-C, most clinical laboratories calculate it from the results of a fasting sample, using the following Friedewald equation: (LDL-C = Total cholesterol – (HDL-C + triglyceride/5); all units in mg dL^{-1}). In a fasting sample without chylomicrons, cholesterol can be mostly found in LDL, HDL, and VLDL and triglycerides are mostly found in VLDL. The term triglyceride/5 in the equation has been shown to closely approximate VLDL-C because there is about five times more triglyceride than cholesterol in a typical VLDL particle when both values are expressed in units of mg dL^{-1}. LDL-C can, therefore, be simply calculated by subtracting the measured HDL-C and the calculated VLDL-C from the measured total cholesterol. However, the Friedewald equation has been shown to yield inaccurate results in samples that have VLDL particles with abnormal lipid composition. This rarely occurs but can be observed in patients with fasting triglycerides greater than 400 mg dL^{-1}, who have triglyceride-rich VLDL, and in patients with type-III hyperlipidemia, who have cholesterol-rich VLDL. Thus, by the Friedewald equation, LDL-C can be reasonably estimated in most samples by simply measuring total cholesterol, triglyceride, and HDL-C. These three tests and the calculated LDL-C comprise what has now become commonly known as a "lipid panel."

One potential limitation of both the β-quantification procedure and the Friedewald calculation is that they both measure, in addition to LDL-C, the relatively small amounts of cholesterol that are associated with Lp(a) and IDL. In addition, the β-quantification procedure is complex and there have been concerns about the precision and accuracy of the Friedewald equation because of the cumulative effect of errors in the three underlying tests that are used in the calculation. These limitations have led to the development of direct homogenous LDL-C assays. Like the homogenous HDL-C assays, these assays use reagents that selectively block or solubilize and consume the different lipoprotein classes, allowing the direct measurement of LDL-C from an unfractionated sample. Several automated commercial homogenous LDL-C assays are currently available. In general, these assays show good precision and appear to be accurate on normal specimens. Homogenous LDL-C assays, however, have not yet replaced the use of calculated LDL-C in most clinical laboratories for several reasons. First, the homogenous LDL-C assays are relatively new and, in many cases, have not been fully evaluated in patients with the full spectrum of dyslipidemias. Early results, in fact, suggest that the homogenous LDL-C assays are not highly specific in their separation of the various lipoproteins and do not consistently agree with β-quantification and are typically no more reliable than Friedewald calculation. In addition, because the three tests that make up the lipid panel are usually necessary for a complete lipoprotein evaluation, it has not yet been clearly established that performing a separate direct test for LDL-C offers any real advantages in the diagnosis and management of patients with dyslipidemias.

3.3 Electrophoresis of Lipoproteins

Electrophoresis is a relatively convenient method for separating the major classes of lipoproteins. Electrophoresis of lipoproteins can be performed on a variety of matrices, but separation on agarose gels is relatively easy to perform and it readily resolves the major lipoprotein classes on the basis of both size and charge. Some typical electrophoretic patterns of lipoproteins are shown in Fig. 6. HDL shows the fastest migration on agarose gels and migrates to what is called the *α-position*. Chylomicrons barely migrate past the origin of the gel, and LDL typically migrates to the intermediate β-position. VLDL migrates to the pre-β-position between HDL and LDL, but in type-III hyperlipidemia, it usually migrates as a broad band in the β-region. The lipoproteins in agarose gels are usually stained with a dye that stains neutral lipids, such as Oil Red O.

Early on, the electrophoretic separation of lipoproteins was an important diagnostic tool because the original phenotypic classification of dyslipidemias was based on this method. With the discovery of the genetic basis of many of the common dyslipidemias and with the advent of more specific tests, a genetic classification of lipoproteins is now more commonly used. As a consequence, electrophoresis is no longer as valuable diagnostically, but it is still useful for identifying and monitoring patients with rare variants of dyslipidemias. Standard electrophoretic methods are also only semiqualitative, which limits their use. Recently, however, commercial

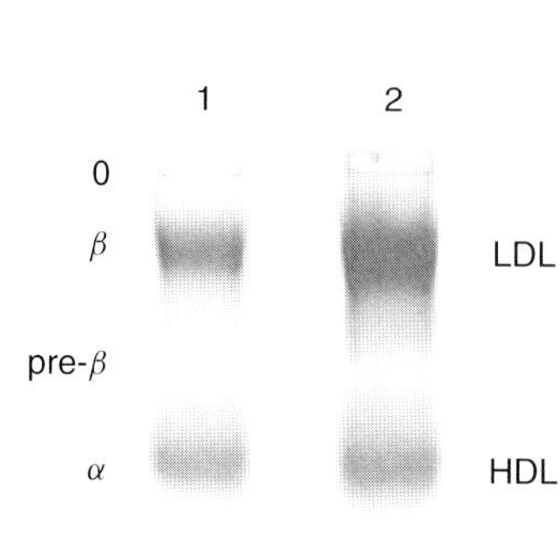

Fig. 6 Agarose gel electrophoresis of a normal subject (Lane 1) and a patient with familial hypercholesterolemia, with an elevated LDL (Lane 2).

automated electrophoretic systems have been developed, which have been shown to be relatively precise and accurate in quantifying the major lipoprotein fractions. There has also been a recent resurgence in the interest of electrophoretic techniques for separating lipoproteins because of their ability to resolve subfractions within each of the major lipoprotein classes.

3.4 Subfractionation of Lipoproteins

Besides the major lipoprotein classes shown in Table 2, it is possible to detect distinct subspecies or subclasses within each class. Most of the methods used in fractionation of the major subclasses of lipoproteins can be modified to improve their resolution for performing subfraction analysis. In addition, nuclear magnetic resonance spectroscopy and density gradient ultracentrifugation in a vertical rotor are particularly useful for performing a comprehensive subfraction analysis of lipoproteins. Nondenaturing gradient gel electrophoresis in polyacrylamide is also a good method for separating the subfractions of lipoproteins by size. When combined with agarose gel electrophoresis, in a two-dimensional gel-electrophoresis system, as many as 10 different species of HDL can be observed.

Subfraction analysis of lipoproteins has largely been performed for basic research applications, but increasingly, subfractions within both LDL and HDL have been shown to potentially have diagnostic importance. A subfraction of LDL called *small dense LDL* has been proposed to be particularly proatherogenic. This form of LDL also appears to be at least partly inheritable. Approximately 25% of the Caucasian population has what is called the *subclass B phenotype* of LDL, which is associated with an increase in small dense LDL and an increased risk for CHD.

In the case of HDL, some forms of HDL have been shown to be particularly efficient in promoting reverse cholesterol transport and may even be a stronger negative predictor of CHD than total HDL. One such form is a small lipid-poor form of HDL, which has a pre-β-type migration on agarose gel electrophoresis. Pre-β-HDL can also be measured by an ELISA assay. Unlike other forms of HDL, pre-β-HDL is discoidal in shape because it lacks a central lipid core of neutral lipids. It consists of a bilayer of phospholipids around which apolipoproteins surround the acyl chains of the phospholipids in a beltlike configuration. Pre-β-HDL has been proposed to be particularly antiatherogenic because it is the form of HDL that initially interacts with cells for removing excess cholesterol by the ABCA1 transporter (Fig. 3).

4 Diagnostic Application of Lipoprotein Analysis

4.1 Standardization of Lipoprotein Analysis

There are three important sources of variability or error that can affect the results of lipoprotein analysis, namely, biological, preanalytical, and analytical variability. There has been great improvement in the last several years in analytical systems for measuring lipids and lipoproteins, and systematic biases among the various lipid and lipoprotein assays have significantly decreased over the last several years. This has occurred because of the partnership between the clinical laboratory community, diagnostic companies, and various governmental regulatory bodies in improving the accuracy of lipoprotein analysis. In order to minimize the misdiagnosis of patients, the National Cholesterol Education Program or NCEP, a National Institute of Health sponsored organization that produces practice guidelines for the diagnosis and treatment of patients for coronary artery disease, has developed accuracy and precision goals for lipoprotein analysis (Table 3). The CDC created the Cholesterol Reference Method Laboratory Network (CRMLN; http://www.cdc.gov/nceh/dls/crmln/crmln.htm), which provides a mechanism for diagnostic companies and routine clinical laboratories to compare the performance of their assays to the reference methods, which are traceable to the definitive assays established by the National Institute of Standards and Technology (NIST). Clinical laboratories are also required, as part of their biannual recertification process, to compare themselves, multiple times each year, to their peers that use the same assays by analyzing unknown control material, through one of the national proficiency testing programs, such as the one offered by the College of American Pathologists. It is also mandatory for clinical laboratories to test their assays daily by measuring control materials. In addition, it is important to use a reagent system with an analyzer for which the assay has been developed, in order to ensure the proper performance of the assay. Because of the underlying errors in total cholesterol and HDL-C assays, it is often particularly difficult, however, for routine clinical laboratories to achieve the NCEP accuracy goals for LDL-C by using the Friedewald equation. Further improvements are considered necessary in the current total cholesterol and HDL-C assays or in direct LDL-C assays to improve the accuracy and precision of LDL-C measurements.

It is important also to consider preanalytical sources as a source of error. It is recommended that a blood sample be collected from subjects in a sitting position because posture can affect the results of lipoprotein analysis, through changes in hemoconcentration. In addition, different types of collection tubes can lead to differences in results. Most clinical laboratories use serum for their analysis, but EDTA plasma is often preferred when lipoproteins are isolated by ultracentrifugation because EDTA helps reduce proteolysis and oxidation of the lipoproteins. EDTA in the collection tube will, however, cause slight dilution of the sample because it increases the osmotic strength of plasma, which causes water from red blood cells to shift to the plasma compartment. Lipoprotein results from EDTA plasma should, therefore, be multiplied by 1.03 to match serum results.

Tab. 3 NCEP analytical performance goals.

	Precision	*Bias*	*Total error*
Total cholesterol:	3% CV	±3%	±8.9%
HDL cholesterol:			
≥42 mg dL^{-1}	4% CV	±5%	±12.8%
<42 mg dL^{-1}	SD < 1.7 mg dL^{-1}		
LDL cholesterol:	4% CV	±4%	±11.8%
Triglycerides:	5% CV	±5%	±14.8%

Other types of anticoagulants also have varying osmotic effects. Fasting (9–12 h) specimens are preferred for lipoprotein analysis because of problems related to measuring LDL-C in the presence of high triglycerides and because of the changes that occur in the lipoprotein levels in the postprandial state.

There is also a large biological variability for lipids and lipoproteins. For total cholesterol, the coefficient of variation for biological variability is approximately 6% so that results from the same individual that are collected only a few days apart can vary by as much as 12%. For triglycerides, the coefficient of biological variability is as large as 20%. Some of the potential sources of biological variability include effects due to acute or subacute illnesses, changes in diet, exercise frequency, body weight, or smoking. It is for this reason that the NCEP recommends that any treatment decisions be based on the average of two to three test results that should be taken at least 2 weeks apart to reduce the effect of any biologic and analytical variability.

4.2 Interpretation of Lipoprotein Assay Results

Adult reference range of lipids and lipoproteins are shown in Table 4. Unlike most other laboratory-test results, these reference ranges do not represent the central 95% interval. For example, approximately 50% of the adult US population has a total cholesterol value greater than the upper limit of 200 mg dL^{-1}. The reference values for lipids and lipoproteins instead represent the ideal test range chosen by expert consensus for reducing the risk of developing cardiovascular disease.

A flowchart of how lipoprotein analysis is used in implementation of the NCEP-III guidelines (http://www.nhlbi.nih.gov/about/ncep) for the diagnosis and treatment of cardiovascular disease is shown in Fig. 7. These guidelines represent the consensus of an expert panel on the scientific literature supporting the treatment of dyslipidemia for the prevention of CHD. It is recommended that a complete lipid

Tab. 4 Adult reference ranges for lipids.

Total cholesterol	140–200 mg dL^{-1}
HDL-C	40–75 mg dL^{-1}
LDL-C	50–130 mg dL^{-1}
Triglyceride	60–150 mg dL^{-1}

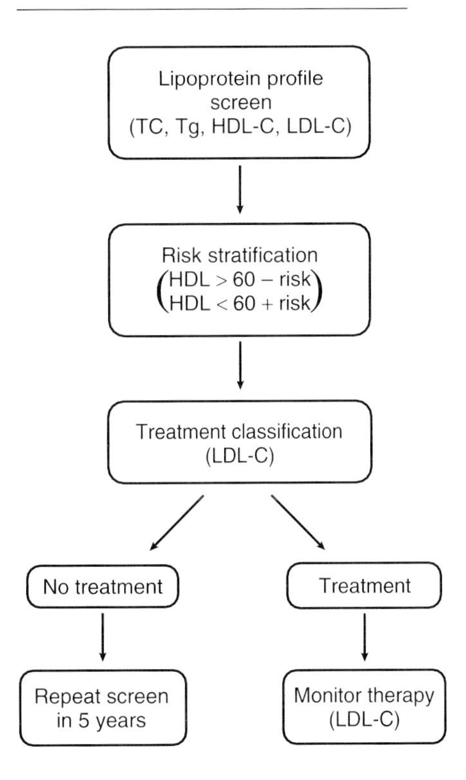

Fig. 7 Flow chart on the use of lipoprotein analysis in NCEP-III guidelines for the diagnosis and treatment of dyslipidemia.

panel test (total cholesterol, triglycerides, HDL-C, and a calculated LDL-C) be performed, as part of the initial screen, on a 9 to 12 h fasting sample. If the total triglycerides are more than 400 mg dL^{-1}, a β quantitation of LDL-C may be necessary or, possibly, a direct LDL-C test. For nonfasting samples, just a total cholesterol and HDL-C test should be performed. If the total cholesterol is less than 200 mg dL^{-1} and the HDL-C is greater than 60 mg dL^{-1} on a nonfasting sample and if the patient is at low risk for CHD, no further analysis is recommended. Each subject is then stratified into four levels of risk for CHD. This stratification is based on previous history of cardiovascular disease and/or its risk equivalent, such as diabetes, as well as other known risk factors for CHD, such as hypertension, smoking, and family history. HDL-C is also used in this risk stratification process. HDL-C greater than 60 mg dL^{-1} is used as a negative risk factor, whereas HDL-C less than 40 mg dL^{-1} is used as a positive risk factor. Each patient is then categorized into a treatment category on the basis of his or her LDL-C. Each level of risk has an ideal upper limit for LDL-C, and the LDL-C limit is set lower for those patients at higher risk. Initially, all patients that have an LDL-C that exceeds the recommended level are counseled to initiate lifestyle changes, such as reducing weight, increasing physical activity, and switching to a low-fat diet. If the lifestyle changes are ineffective in reaching the desirable range of LDL-C and/or if patients have an LDL-C that is beyond a recommended drug-treatment threshold for LDL-C, concurrent treatment with drugs to lower cholesterol is also recommended. The effect of any therapy is monitored, thereafter, by measuring LDL-C. HDL-C and triglycerides may also be important to be monitored, depending on their pretreatment levels and the type of drug therapy used. Other newly developed lipoproteins tests, such as Lp(a), remnant lipoproteins, small dense LDL, and tests for other markers of coronary artery disease, such as CRP and homocysteine, are currently not used in the initial risk stratification of a patient but these can be considered when making treatment decisions. For patients at the lowest risk for CHD, with an LDL-C below 160 mg dL^{-1}, no treatment is recommended and a repeat screen should be performed within 5 years.

5 Perspective

Nearly 50 years have passed since the first methods were developed for separating and analyzing lipoproteins by ultracentrifugation. During this time, great advances have been made in unraveling the various pathways of lipoprotein metabolism and the role of the individual lipoproteins. This has been coupled with the identification and elucidation of many of the genetic causes of dyslipidemias. Of paramount importance have been the epidemiological and animal studies that have revealed the causal association between lipoproteins and coronary heart disease. In many cases, advances in analytical techniques for lipoprotein analysis were pivotal to these discoveries. On the basis of the rapid advances to date in lipoprotein analysis, it is envisioned that there will be many future improvements in these tests, which will lead to even more detailed insights into lipoprotein metabolism and diagnostic tests with even greater clinical utility for preventing coronary heart disease.

See also FTIR of Biomolecules; Gene Therapy and Cardiovascular Diseases; Macromolecules, X-Ray Diffraction of Biological.

Bibliography

Books and Reviews

Albers, J.J., Segrest, J.P. (Eds.) (1986) *Plasma Lipoproteins, Part A: Plasma Lipoproteins: Preparation, Structure, and Molecular Biology. Methods in Enzymology*, Vol. 129. Academic Press, London and New York.

Converse, C.A., Skinner, E.R. (Eds.) (1992) *Lipoprotein Analysis: A practical Approach*. IRL Press, Oxford, New York, and Tokyo.

Li, W.H., Tanimura, M., Luo, C.C., Datta, S., Chan, L. (1988) The apolipoprotein multigene family: biosynthesis, structure, structure-function relationships, and evolution, *J. Lipid Res.* **29**, 245.

Libby, P., Aikawa, M., Schonbeck, U. (2000) Cholesterol and atherosclerosis, *Biochim. Biophys. Acta* **1529**, 299.

Nwokoro, N.A., Wassif, C.A., Porter, F.D. (2001) Genetic disorders of cholesterol biosynthesis in mice and humans, *Mol. Genet. Metab.* **74**, 105.

Pauciullo, P. (2002) Lipoprotein transport and metabolism: a brief update, *Nutr. Metab. Cardiovasc. Dis.* **12**, 90.

Ridker, P.M., Antman, E.M. (1999) Pathogenesis and pathology of coronary heart disease syndromes, *J. Thromb. Thrombolysis* **8**, 167–189.

Rifai, N., Warnick, G.R., Dominiczak, M.H., (Eds.) (2000) *Handbook of Lipoprotein Testing*. AACC Press, Washington, DC.

Segrest, J.P., Albers, J.J. (Eds.) (1986) *Plasma Lipoproteins, Part: B Plasma Lipoproteins: Characterization, Cell Biology, and Metabolism*. Methods in Enzymology, Vol. 128. Academic Press, London and New York.

Primary Literature

Abell, L.L., Levy, B.B., Brody, B.B., Kendall, F.C. (1952) A simplified method for the estimation of total cholesterol in serum and demonstration of its specificity, *J. Biol. Chem.* **195**, 357–66.

Allain, C.C., Poon, L.S., Chan, C.S., et al. (1974) Enzymatic determination of total serum cholesterol, *Clin. Chem.* **20**, 470–5.

Arrese, M.A., Crawford, J.M. (1997) Of plaques and stones: the SR-B1 (scavenger receptor class B, type 1), *Hepatology* **26**, 1072–4.

Bachorik, P.S., Ross, J.W., for the National Cholesterol Education Program Working Group on Lipoprotein Measurement. (1995) National Cholesterol Education Program recommendations for measurement of low-density lipoprotein cholesterol: executive summary, *Clin. Chem.* **41**, 1414–20.

Barrans, A., Jaspard, B., Barbaras, R., et al. (1996) Pre-beta HDL: structure and metabolism, *Biochim. Biophys. Acta* **1300**, 73–85.

Bartlett, G.R. (1959) Phosphorus assay in column chromatography, *J. Biol. Chem.* **234**, 466–8.

Castelli, W.P., Garrison, R.J., Wilson, P.W.F., et al. (1986) Incidence of coronary heart disease and lipoprotein cholesterol levels: the Framingham Study, *JAMA* **256**, 2835–8.

Chapman, M.J., Goldstein, S., Lagrange, D., et al. (1981) A density gradient ultracentrifugal procedure for the isolation of the major lipoprotein classes from human serum, *J. Lipid. Res.* **22**, 339–58.

Cohen, A., Hertz, H.S., Mandel, J., et al. (1980) Total serum cholesterol by isotope dilution/mass spectrometry: a candidate definitive method, *Clin. Chem.* **26**, 854–60.

Cole, T.G. (1990) Glycerol blanking in triglyceride assays: is it necessary? *Clin. Chem.* **36**, 1267–8.

Cooper. G.R., Smith, S.J., Myers, G.L., et al. (1994) Estimating and minimizing effects of biologic sources of variation by relative range when measuring the mean of serum lipids and lipoproteins, *Clin. Chem.* **40**, 227–32.

Dobiasova, M., Frohlich, J.J. (1999) Advances in understanding of the role of lecithin cholesterol acyltransferase (LCAT) in cholesterol transport, *Clin. Chim. Acta* **286**, 257–71.

Eckfeldt, J.H., Copeland, K.R. (1993) Accuracy verification and identification of matrix effects. The College of American Pathologist's protocol, *Arch. Pathol. Lab. Med.* **117**, 381–6.

Erbagci, A.B., Tarakcioglu, M., Aksoy, M., et al. (2002) Diagnostic value of CRP and Lp(a) in coronary heart disease, *Acta Cardiol.* **57**, 197–204.

Fredrickson, D.S., Levy, R.I., Lees, R.S. (1967) Fat transport in lipoproteins: an integrated approach to mechanisms and disorders, *N. Engl. J. Med.* **276**, 148–56.

Friedewald, W.T., Levy, R.I., Fredrickson, D.S. (1972) Estimation of the concentration of low-density lipoprotein cholesterol in plasma, without use of the preparative ultracentrifuge, *Clin. Chem.* **18**, 499–502.

Frohlich, J., Lear, S.A. (2002) Old and new risk factors for atherosclerosis and development of treatment recommendations, *Clin. Exp. Pharmacol. Physiol.* **29**, 838–42.

Genest, J.J., Corbett, H, McNamara J.R, et al. (1988) Effect of hospitalization on high-density lipoprotein cholesterol in patients undergoing elective coronary angiography, *Am. J. Cardiol.* **61**, 998–1000.

Genest, J.J., McNamara, J.R., Salem, D.N., et al. (1991) Prevalence of risk factors in men with premature coronary artery disease, *Am. J. Cardiol.* **67**, 1185–9.

Harris, N., Galpchian, V., Rifai, N. (1996) Three routine methods for measuring high-density lipoprotein cholesterol compared with the reference method, *Clin. Chem.* **42**, 738–43.

Havel, R.J., Eder, H.A., Bragdon, J.H. (1955) The distribution and chemical composition of ultracentrifugally separated lipoproteins in human serum, *J. Clin. Invest.* **34**, 1345–53.

Hevonoja, T., Pentikainen, M.O., Hyvonen, M.T., et al. (2000) Structure of low density lipoprotein (LDL) particles: basis for understanding molecular changes in modified LDL, *Biochim. Biophys. Acta* **1488**, 189–210.

Klotzsch, S.G., McNamara, J.R. (1990) Triglyceride measurements: a review of methods and interferences, *Clin. Chem.* **36**, 1605–13.

Kromhout, D., Menotti, A., Bloemberg, B., et al. (1995) Dietary saturated and trans fatty acids and cholesterol and 25-year mortality from coronary heart disease: the Seven Countries Study, *Prev. Med.* **24**, 308–15.

Kulkarni, K.R., Garber, D.W., Schmidt, C.F., et al. (1992) Analysis of cholesterol in all lipoprotein classes by single vertical ultracentrifugation of fingerstick blood and controlled-dispersion flow analysis, *Clin. Chem.* **38**, 1898–905.

Lamarche, B., Lemieux, I., Despres, J.P. (1999) The small, dense LDL phenotype and the risk of coronary heart disease: epidemiology, patho-physiology and therapeutic aspects, *Diabetes Metab.* **25**, 199–211.

Li, Z., McNamara, J.R., Ordovas, J.M., et al. (1994) Analysis of high-density lipoproteins by a modified gradient gel electrophoresis method, *J. Lipid. Res.* **35**, 1698–711.

Lofland, Jr., H.B. (1964) A semiautomated procedure for the determination of triglycerides in serum, *Anal. Biochem.* **9**, 393–400.

Miremadi, S., Sniderman, A., Frohlich, J. (2002) Can measurement of serum apolipoprotein B replace the lipid profile monitoring of patients with lipoprotein disorders? *Clin. Chem.* **48**, 484–8.

National Cholesterol Education Program (NCEP) Expert Panel on Detection, Evaluation, and Treatment of High Blood Cholesterol in Adults (Adult Treatment Panel III). (2002) Third Report of the National Cholesterol Education Program (NCEP) Expert Panel on detection, evaluation, and treatment of high blood cholesterol in adults (Adult Treatment Panel III) final report, *Circulation* **106**, 3143–421.

Nauck, M., Warnick, G.R., Rifai, N. (2002) Methods for measurement of LDL-cholesterol: a critical assessment of direct measurement by homogeneous assays versus calculation, *Clin. Chem.* **48**, 236–54.

Noble, R.P. (1968) Electrophoretic separation of plasma lipoproteins in agarose gel, *J. Lipid Res.* **9**, 693–700.

Ornish, D., Brown, S.E., Scherwitz, L.W., et al. (1990) Can lifestyle changes reverse coronary heart disease? The lifestyle heart trial, *Lancet* **336**, 129–33.

Rifai, N., Iannotti, E., DeAngelis, K., et al. (1998) Analytical and clinical performance of a homogeneous enzymatic LDL-cholesterol assay compared with the ultracentrifugation-dextran sulfate-Mg^{2+} method, *Clin. Chem.* **44**, 1242–50.

Schaefer, E.J., Lichtenstein, A.H., Lamon-Fava, S., et al. (1995) Lipoproteins, nutrition, aging, and atherosclerosis, *Am. J. Clin. Nutr.* **61**(Suppl. 3), 726S–740S.

Schaefer, E.J., McNamara, J.R., Genest, Jr., J., et al. (1988) Clinical significance of hypertriglyceridemia, *Semin. Thromb. Hemost.* **14**, 142–8.

Schmitz, G., Becker, A., Aslanidis, C. (1996) ACAT/CEH and ACEH/LAL: two key enzymes in hepatic cellular cholesterol homeostasis

and their involvement in genetic disorders, *Z. Gastroenterol.* **34**(Suppl. 3), 68–72.
Shepherd, J., Cobbe, S.M., Ford, I., et al. (1995) Prevention of coronary heart disease with pravastatin in men with hypercholesterolemia: the West of Scotland Coronary Prevention Study, *N. Engl. J. Med.* **333**, 1301–7.
Silver, D.L., Jiang, X.C., Arai, T., et al. (2002) Receptors and lipid transfer proteins in HDL metabolism, *Ann. N. Y. Acad. Sci.* **902**, 103–11.
Steele, B.W., Koehler, D.F., Azar, M.M., et al. (1976) Enzymatic determinations of cholesterol in high-density-lipoprotein fractions prepared by a precipitation technique, *Clin. Chem.* **22**, 98–101.
Stein, E.A., Myers, G.L., for the National Cholesterol Education Program Working Group on Lipoprotein Measurement. (1995) National Cholesterol Education Program recommendations for triglyceride measurement: executive summary, *Clin. Chem.* **41**, 1421–6.
Sugiuchi, H., Irie, T., Uji, Y., et al. (1998) Homogeneous assay for measuring low-density lipoprotein cholesterol in serum with triblock copolymer and a-cyclodextrin sulfate, *Clin. Chem.* **44**, 522–31.
Sugiuchi, H., Uji, Y., Okabe, H., et al. (1995) Direct measurement of high-density lipoprotein cholesterol in serum with polyethylene glycol-modified enzymes and sulfated a-cyclodextrin, *Clin. Chem.* **41**, 717–23.
Sviridov, D., Nestel, P. (2002) Dynamics of reverse cholesterol transport: protection against atherosclerosis, *Atherosclerosis* **161**, 245–54.
Takayama, M., Itoh, S., Nagasaki, T., et al. (1977) A new enzymatic method for determination of serum choline-containing phospholipids, *Clin. Chim. Acta* **79**, 93–8.
Tall, A.R., Costet, P., Wang, N. (2002) Regulation and mechanisms of macrophage cholesterol efflux, *J. Clin. Invest.* **110**, 899–904.
Warnick, G.R., Wood, P.D.J.W., for the National Cholesterol Education Program Working Group on Lipoprotein Measurement. (1995) National Cholesterol Education Program recommendations for measurement of high-density lipoprotein cholesterol: executive summary, *Clin. Chem.* **41**, 1427–33.
Warnick, G.R., Knopp, R.H., Fitzpatrick, V., et al. (1990) Estimating low-density lipoprotein cholesterol by the Friedewald equation is adequate for classifying patients on the basis of nationally recommended cutpoints, *Clin. Chem.* **36**, 15–9.
Warnick, G.R., Nauck, M., Rifai, N. (2001) Evolution of methods for measurement of HDL-cholesterol: from ultracentrifugation to homogeneous assays, *Clin. Chem.* **47**, 1579–96.
Warnick, G.R., Nguyen, T., Bergelin, R.O., et al. (1982) Lipoprotein quantification: an electrophoretic method compared with the Lipid Research Clinics method, *Clin. Chem.* **28**, 2116–20.
Wilson, P.W., D'Agostino, R.B., Levy, D., et al. (1998) Prediction of coronary heart disease using risk factor categories, *Circulation* **97**, 1837–47.
Wu, L.L., Warnick, G.R., Wu, J.T., et al. (1989) A rapid micro-scale procedure for determination of the total lipid profile, *Clin. Chem.* **35**, 1486–91.
Yamagishi, M., Terashima, M., Awano, K., et al. (2000) Morphology of vulnerable coronary plaque: insights from follow-up of patients examined by intravascular ultrasound before an acute coronary syndrome, *J. Am. Coll. Cardiol.* **35**, 106–11.
Zilversmit, D.B. (1995) Atherogenic nature of triglycerides, postprandial lipemia, and triglyceride-rich remnant lipoproteins, *Clin. Chem.* **41**, 153–8.

Liposome Gene Transfection

Nancy Smyth Templeton
Department of Molecular and Cellular Biology, Baylor College of Medicine, Houston, TX, USA

Encyclopedia of Molecular Cell Biology and Molecular Medicine, 2nd Edition. Volume 7
Edited by Robert A. Meyers.

ISBN: 3-527-30549-1

Keywords

Colloidal Suspensions
The density of complexes in a specific volume.

Formulation
Procedures used to make liposomes and complexes.

Gene Therapy
Treatments that are mediated by either the addition of needed genes or the destruction of transcripts encoding unwanted gene products.

Lamellar Structures
Alternation of a lipid bilayer and a water layer.

Liposomes
Bilayers made from lipids.

Nonviral Delivery
Delivery vehicles that are not viruses or any other live organism.

Optimization
Creation of optimal conditions.

Serum Stability
Complexes that do not precipitate in specific percentage of serum.

Systemic Delivery
Delivery in the bloodstream.

Varied results have been obtained using cationic liposomes for *in vivo* delivery. Furthermore, optimization of cationic liposomal complexes for *in vivo* applications is complex involving many diverse components. These components include nucleic acid purification, plasmid design, formulation of the delivery vehicle, administration route and schedule, dosing, detection of gene expression, and others. This chapter will also focus on optimization of the delivery vehicle formulation. These formulation issues include morphology of the complexes, lipids used, flexibility versus rigidity, colloidal suspension, overall charge, serum stability, half-life in circulation, biodistribution, delivery to and dissemination throughout target tissues. Broad assumptions have been frequently made on the basis of the data obtained from focused studies using cationic liposomes. However, these assumptions do not necessarily apply to all delivery vehicles and, most likely, do not apply to many liposomal systems when considering these other key components that influence the results obtained *in vivo*. Optimizing all components of the delivery system is pivotal and will allow broad use of liposomal complexes to treat or cure human diseases or disorders. This chapter will highlight the features of liposomes that contribute to successful delivery, gene expression, and efficacy.

1 Introduction

Many investigators are focused on the production of effective nonviral gene therapeutics and on creating improved delivery systems that mix viral and nonviral vectors. Use of improved liposome formulations for delivery *in vivo* is valuable for gene therapy and would avoid several problems associated with viral delivery. Delivery of nucleic acids using liposomes is promising as a safe and nonimmunogenic approach to gene therapy. Furthermore, gene therapeutics composed of artificial reagents can be standardized and regulated as drugs rather than as biologics. Cationic lipids have been used for efficient delivery of nucleic acids to cells in tissue culture for several years. Much effort has also been directed toward developing cationic liposomes for efficient delivery of nucleic acids in animals and in humans. Most frequently, the formulations that are best to use for transfection of a broad range of cell types in culture are not optimal for achieving efficacy in small and large animal disease models.

Much effort has been devoted to the development of nonviral delivery vehicles due to the numerous disadvantages of viral vectors that have been used for gene therapy. Following viral delivery *in vivo*, immune responses are generated to expressed viral proteins that subsequently kill the target cells required to produce the therapeutic gene product. An innate humoral immune response can be produced to certain viral vectors due to previous exposure to the naturally occurring virus. Random integration of some viral vectors into the host chromosome could occur and cause activation of proto-oncogenes, resulting in tumor formation. Clearance of viral vectors delivered systemically by complement activation can also occur. Viral vectors can be inactivated upon readministration by the humoral immune response. Potential for recombination of the viral vector with DNA sequences in the host chromosome that generates a replication-competent infectious virus also exists. Specific delivery of viral vectors to target cells can be difficult because two distinct steps in engineering viral envelopes or capsids must be achieved. First, the virus envelope or capsid must be changed to inactivate the natural tropism of the virus to enter specific cell types. Then, sequences must be introduced that allow the new viral vector to bind and internalize through a different cell surface receptor. Other disadvantages of viral vectors include the inability to administer certain viral vectors more than once, the high costs for producing large amounts of high-titer viral stocks for use in the clinic, and the limited size of the nucleic acid that can be packaged and used for viral gene therapy. Attempts are being made to overcome the immune responses produced by viral vectors after administration in immune competent animals and in humans, such as the use of gutted adenoviral vectors or encapsulation of viral vectors in liposomes. However, complete elimination of all immune responses to viral vectors may be impossible.

Use of liposomes for gene therapy provides several advantages. A major advantage is the lack of immunogenicity after *in vivo* administration, including systemic injections. Therefore, the nucleic acid-liposome complexes can be readministered without harm to the patient and without compromising the efficacy of the nonviral gene therapeutic. Improved formulations of nucleic acid–liposome complexes can also evade complement inactivation after *in vivo* administration. Nucleic acids of unlimited size can be delivered ranging from single nucleotides to large mammalian artificial chromosomes. Furthermore, different types of nucleic acids can be delivered including plasmid DNA, RNA, oligonucleotides, DNA–RNA chimeras, synthetic ribozymes, antisense molecules, RNAi, viral nucleic acids, and others. Certain cationic formulations can also encapsulate and deliver viruses, proteins or partial proteins with a low isoelectric point (pI), and mixtures of nucleic acids and proteins of any pI. Creation of nonviral vectors for targeted delivery to specific cell types, organs, or tissues is relatively simple. Targeted delivery involves elimination of nonspecific charge interactions with nontarget cells and addition of ligands for binding and internalization through target cell surface receptors. Other advantages of nonviral vectors include the low cost and relative ease in producing nucleic acid-liposome complexes in large scale for use in the clinic. In addition, greater safety for patients is provided using nonviral delivery vehicles due to few or no viral sequences present in the nucleic acids used for delivery, thereby precluding generation of an infectious virus. The disadvantage of nonviral delivery systems had been the low levels of delivery and gene expression produced by "first-generation" complexes. However, recent advances have dramatically improved transfection efficiencies and efficacy of liposomal vectors. Reviews of other *in vivo* delivery systems and improvements using cationic liposomes have been published recently.

Cationic liposome–nucleic acid complexes can be administered via numerous delivery routes *in vivo*. Routes of delivery include direct injection (e.g. intratumoral), intravenous, intraperitoneal, intra-arterial, intrasplenic, mucosal (nasal, vaginal, rectal), intramuscular, subcutaneous, transdermal, intradermal, subretinal, intratracheal, intracranial, and others. Much interest has focused on noninvasive intravenous administration because many investigators believe that this route of delivery is the "holy grail" for the treatment or cure of cancer, cardiovascular, and other inherited or acquired diseases. Particularly for the treatment of metastatic cancer, therapeutics must reach not only the primary tumor but also the distant metastases.

Optimization of cationic liposomal complexes for *in vivo* applications and therapeutics is complex, involving many distinct components. These components include nucleic acid purification, plasmid design, formulation of the delivery vehicle, administration route and schedule, dosing, detection of gene expression, and others. Often I make the analogy of liposome optimization to a functional car. Of course, the engine of the car, analogous to the liposome delivery vehicle, is extremely important. However, if the car does not have wheels, adequate tyres, and so on, the motorist will not be able to drive the vehicle to its destination. This chapter will focus on optimization of these distinct components for use in a variety of *in vivo* applications. Optimizing all components of the delivery system will allow broad use of liposomal complexes to treat or cure human diseases or disorders.

2 Optimization of Cationic Liposome Formulations for Use *in vivo*

Much research has been directed toward the synthesis of new cationic lipids. Some new formulations led to the discovery of more efficient transfection agents for cells in culture. However, their efficiency measured *in vitro* did not correlate with their ability to deliver DNA after administration in animals. Functional properties defined *in vitro* do not assess the stability of the complexes in plasma or their pharmacokinetics and biodistribution, all of which are essential for optimal activity *in vivo*. Colloidal properties of the complexes, in addition to the physicochemical properties of their component lipids, also determine these parameters. In particular, in addition to efficient transfection of target cells, nucleic acid–liposome complexes must be able to traverse tight barriers *in vivo* and penetrate throughout the target tissue to produce efficacy for the treatment of disease. These are not issues for achieving efficient transfection of cells in culture with the exception of polarized tissue culture cells. Therefore, we are not surprised that optimized liposomal delivery vehicles for use *in vivo* may be different than those used for efficient delivery to cells in culture.

In summary, *in vivo* nucleic acid–liposome complexes that produce efficacy in animal models of disease have extended half-life in the circulation, are stable in serum, have broad biodistribution, efficiently encapsulate various sizes of nucleic acids, are targetable to specific organs and cell types, penetrate across tight barriers in several organs, penetrate evenly throughout the target tissue, are optimized for nucleic acid:lipid ratio and colloidal suspension *in vivo*, can be size fractionated to produce a totally homogenous population of complexes prior to injection, and can be repeatedly administered. Recently, we demonstrated efficacy of a robust liposomal delivery system in small and large animal models for lung, breast, head and neck, and pancreatic cancers, and for Hepatitis B and C (Clawson and Templeton, unpublished data). On the basis of efficacy in these animal studies, this liposomal delivery system will be used in upcoming clinical trials to treat these cancers. Our studies have demonstrated broad efficacy in the use of liposomes to treat disease and have dispelled several myths that exist concerning the use of liposomal systems.

3 Liposome Morphology and Effects on Gene Delivery and Expression

Efficient *in vivo* nucleic acid–liposome complexes have unique features including their morphology, mechanisms for crossing the cell membrane and entry into the nucleus, ability to be targeted for delivery to specific cell surface receptors, and ability to penetrate across tight barriers and throughout target tissues. Liposomes have different morphologies based upon their composition and the formulation method. Furthermore, the morphology of complexes can contribute to their ability to deliver nucleic acids *in vivo*. Formulations frequently used for the delivery of nucleic acids are lamellar structures including small unilamellar vesicles (SUVs), multilamellar vesicles (MLVs), or bilamellar invaginated vesicles (BIVs) recently developed in our laboratory (Fig. 1). Several investigators have developed liposomal delivery systems using hexagonal structures; however, they have demonstrated efficiency primarily for the transfection of some cell types in culture and not for *in vivo* delivery. SUVs condense nucleic acids on the surface and form "spaghetti and meatballs" structures. DNA–liposome complexes made using SUVs produce little or no gene expression upon systemic delivery, although these complexes transfect

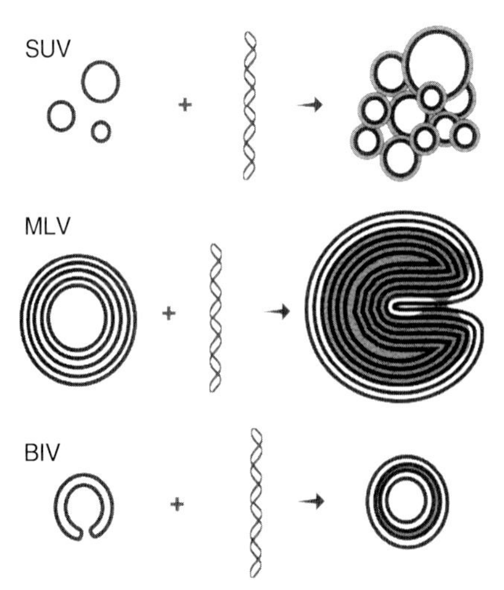

Fig. 1 Diagrams drawn from cryoelectron micrographs of cross sections through vitrified films of various types of liposomes and DNA–liposome complexes. SUVs are small unilamellar vesicles that condense nucleic acids on the surface and produce "spaghetti and meatballs" structures. MLVs are multilamellar vesciles that appear as "Swiss Rolls" after mixing with DNA. BIVs are bilamellar invaginated vesicles produced using a formulation developed in our laboratory. Nucleic acids are efficiently encapsulated between two bilamellar invaginated structures (BIVs).

numerous cell types efficiently *in vitro*. Furthermore, SUV liposome–DNA complexes cannot be targeted efficiently. SUV liposome–DNA complexes also have a short half-life within the circulation, generally about 5 to 10 min. Polyethylene glycol (PEG) has been added to liposome formulations to extend their half-life; however, PEGylation created other problems that have not been resolved. PEG seems to hinder delivery of cationic liposomes into cells due to its sterically hindering ionic interactions, and it interferes with optimal condensation of nucleic acids onto the cationic delivery vehicle. Furthermore, extremely long half-life in the circulation, for example, several days, has caused problems for patients because the bulk of the PEGylated liposomal formulation doxil that encapsulates the cytotoxic agent, doxorubicin, accumulates in the skin, hands, and feet. For example, patients contract mucositis and "Hand and Foot Syndrome" that cause extreme discomfort to the patient. Attempts to add ligands to doxil for delivery to specific cell surface receptors has not resulted in much cell-specific delivery, and the majority of the injected targeted formulation still accumulates in the skin, hands, and feet. Addition of PEG into formulations developed in our laboratory also caused steric hindrance in the bilamellar invaginated structures that did not encapsulate DNA efficiently, and gene expression was substantially diminished.

Some investigators have loaded nucleic acids within SUVs using a variety of methods; however, the bulk of the DNA does not load or stay within the liposomes. Furthermore, most of the processes used for loading nucleic acids within liposomes are extremely time-consuming and not cost effective. Therefore, SUVs are not the ideal liposomes for creating nonviral vehicles for targeted delivery.

Complexes made using MLVs appear as "Swiss rolls" when viewing cross-sections by cryoelectron microscopy. These complexes can become too large for systemic administration or deliver nucleic acids inefficiently into cells due to the inability to "unravel" at the cell surface. Addition of ligands onto MLV liposome–DNA complexes further aggravates these problems. Therefore, MLVs are not useful for the development of targeted delivery of nucleic acids.

Using a formulation developed in our laboratory, nucleic acids are efficiently encapsulated between two bilamellar invaginated vesicles, BIVs. We created these unique structures using 1,2-bis(oleoyloxy)-3-(trimethylammino)propane (DOTAP) and cholesterol (Chol) and a novel formulation procedure. This procedure is different because it includes a brief, low frequency sonication followed by manual extrusion through filters of decreasing pore size. The 0.1 and 0.2 μm filters used are made of aluminum oxide and not polycarbonate that is typically used by other protocols. Aluminum oxide membranes contain more pores per surface area, evenly spaced and sized pores, and pores with straight channels. During the manual extrusion process, the liposomes are passed through each of the four different sized filters only once. This process produces 88% invaginated liposomes. Use of high-frequency sonication and/or mechanical extrusion produces only SUVs.

The BIVs produced condense, unusually large amounts of nucleic acids of any size Fig. 2 or viruses Fig. 3. Furthermore, addition of other DNA condensing agents including polymers is not necessary. For example, condensation of plasmid DNA onto polymers first before encapsulation in the BIVs did not increase condensation or subsequent gene

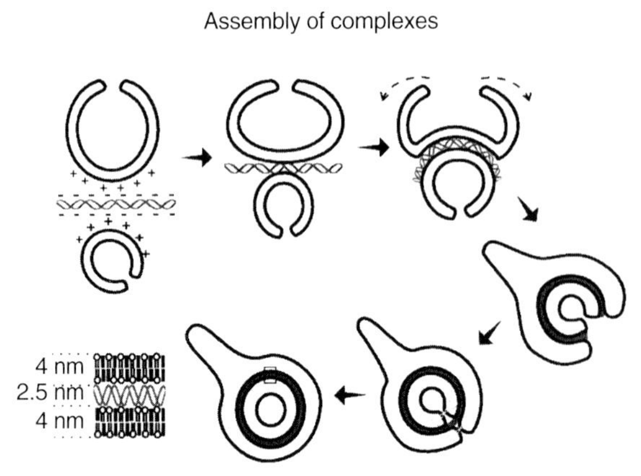

Fig. 2 Proposed model showing cross sections of extruded DOTAP:Chol liposomes (BIVs) interacting with nucleic acids. Nucleic acids adsorb onto a BIV via electrostatic interactions. Attraction of a second BIV to this complex results in further charge neutralization. Expanding electrostatic interactions with nucleic acids cause inversion of the larger BIV and total encapsulation of the nucleic acids. Inversion can occur in these liposomes because of their excess surface area, which allows them to accommodate the stress created by the nucleic acid–lipid interactions. Nucleic acid binding reduces the surface area of the outer leaflet of the bilayer and induces the negative curvature due to lipid ordering and reduction of charge repulsion between cationic lipid headgroups. Condensation of the internalized nucleic acid–lipid sandwich expands the space between the bilayers and may induce membrane fusion to generate the apparently closed structures. The enlarged area shows the arrangement of nucleic acids condensed between two 4 nm bilayers of extruded DOTAP:Chol.

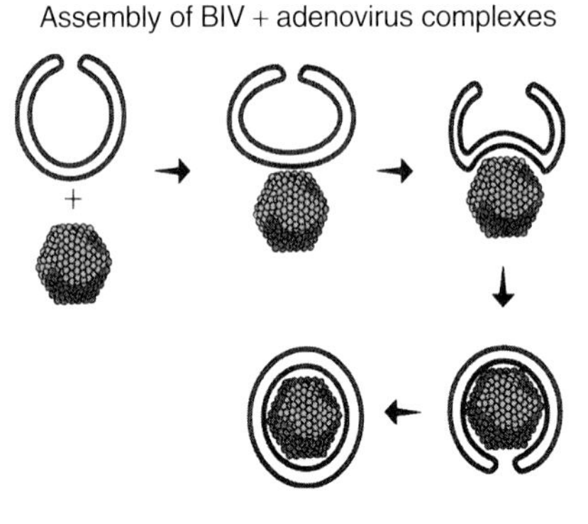

Fig. 3 Proposed model showing cross sections of an extruded DOTAP:Chol liposome (BIV) interacting with adenovirus. Adenovirus interacts with a BIV causing negative curvature and wrapping around the virus particle.

expression after transfection *in vitro* or *in vivo*. Encapsulation of nucleic acids by these BIVs alone is spontaneous and immediate, and, therefore, cost effective, requiring only one step of simple mixing. The extruded DOTAP:Chol–nucleic acid complexes are also large enough so that they are not cleared rapidly by Kupffer cells in the liver and yet extravasate across tight barriers, including the endothelial cell barrier of the lungs in a normal mouse, and diffuse through target organs efficiently. Our recent work demonstrating efficacy for treatment of nonsmall cell lung cancer showed that only BIV DOTAP:Chol-p53 DNA–liposome complexes produced efficacy, and SUV DOTAP:Chol-p53 DNA–liposome complexes produced no efficacy. Therefore, the choice of lipids alone is not sufficient for optimal DNA delivery, and the morphology of the complexes is essential.

4
Optimal Lipids and Liposome Morphology: Effects on Gene Delivery and Expression

Choosing the best cationic lipids and neutral lipids is also essential for producing the optimal *in vivo* formulation. For example, using our novel manual extrusion procedure does not produce BIVs using the cationic lipid dimethyldioctadecylammonium bromide (DDAB). Furthermore, DOTAP is biodegradable, whereas DDAB is not biodegradable. Use of biodegradable lipids is preferred for use in humans. Furthermore, only DOTAP- and not DDAB-containing liposomes produced highly efficient gene expression *in vivo*. DDAB did not produce BIVs and was unable to encapsulate nucleic acids. Apparently, DDAB- and DOTAP-containing SUVs produce similar efficiency of gene delivery *in vivo*; however, these SUVs are not as efficient as BIV DOTAP:Chol. In addition, use of L-α dioleoyl phosphatidylethanolamine (DOPE) as a neutral lipid creates liposomes that cannot wrap or encapsulate nucleic acids. Several investigators have reported efficient transfection of cells in culture using DOPE in liposomal formulations. However, our data showed that formulations consisting of DOPE were not efficient for producing gene expression *in vivo*.

Investigators must also consider the source and lot of certain lipids purchased from companies. For example, different lots of cholesterol from the same vendor can vary dramatically and will affect the formulation of liposomes. Currently, we are using synthetic cholesterol (Sigma, St. Louis, MO). The Food and Drug Administration prefers synthetic cholesterol instead of natural cholesterol that is purified from the wool of sheep for use in producing therapeutics for injection into humans.

Our BIV formulations are also stable for a few years as liquid suspensions. Freeze-dried formulations can also be made that are stable indefinitely even at room temperature. Stability of liposomes and liposomal complexes is also essential, particularly for the commercial development of human therapeutics.

5
Liposome Encapsulation, Flexibility, and Optimal Colloidal Suspensions

A common belief is that artificial vehicles must be 100 nm or smaller to be effective for systemic delivery. However, this belief is most likely true only for large, inflexible delivery vehicles. Blood cells are several microns (up to 7000 nm) in size, and yet have no difficulty circulating in the blood, including through the smallest capillaries. However, sickle cell blood cells, that are rigid, do have problems in the circulation. Therefore, we believe that flexibility is a more important issue than small size. In fact, BIV DNA–liposome complexes in the size range of 200 to 450 nm produced the highest levels of gene expression in all tissues after intravenous injection. Kupffer cells in the liver quickly clear delivery vehicles (including nonviral vectors and viruses) that are not PEGylated and are smaller than 200 nm. Therefore, increased size of liposomal complexes could extend their circulation time, particularly when combined with injection of high colloidal suspensions. BIVs were able to encapsulate nucleic acids and viruses apparently due to the presence of cholesterol in the bilayer (Fig. 4). Whereas, formulations including DOPE instead of cholesterol could not assemble nucleic acids by a "wrapping

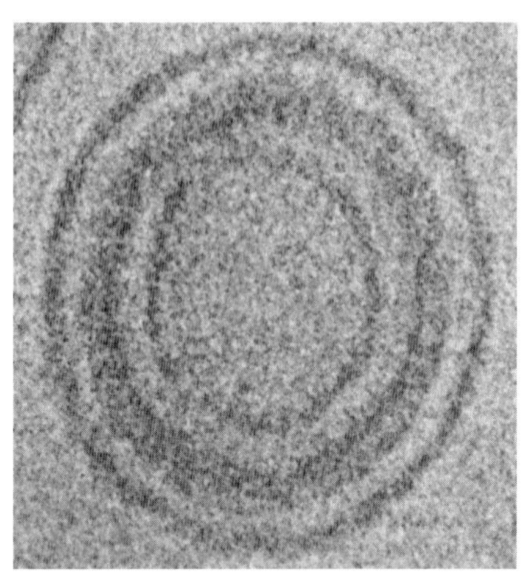

Fig. 4 Cryoelectron micrograph of BIV DOTAP:Chol–DNA–liposome complexes. The plasmid DNA is encapsulated between two BIVs.

Fig. 5 Cryoelectron micrograph of extruded DOTAP:DOPE liposomes complexed to plasmid DNA. Although these liposomes were prepared by the same protocol that produces BIV DOTAP:Chol, these vesicles cannot wrap and encapsulate nucleic acids. The DNA condenses on the surfaces of the liposomes shown.

type" of mechanism (Fig. 5), and produced little gene expression in the lungs and no expression in other tissues after intravenous injections. Because the extruded DOTAP:Chol BIV complexes are flexible and not rigid, are stable in high concentrations of serum, and have extended half-life, they do not have difficulty circulating efficiently in the bloodstream.

We believe that colloidal properties of nucleic acid–liposome complexes also determine the levels of gene expression produced after *in vivo* delivery. These properties include the DNA:lipid ratio that determines the overall charge density of the complexes and the colloidal suspension that is monitored by its turbidity. Complex size and shape, lipid composition and formulation, and encapsulation efficiency of nucleic acids by the liposomes also contribute to the colloidal properties of the complexes. The colloidal properties affect serum stability, protection from nuclease degradation, blood circulation time, and biodistribution of the complexes.

Our *in vivo* transfection data showed that an adequate amount of colloids in suspension was required to produce efficient gene expression in all tissues examined. The colloidal suspension is assessed by measurement of adsorbance at 400 nm using a spectrophotometer optimized to measure turbidity. Our data showed that transfection efficiency in all tissues corresponded to OD400 of the complexes measured prior to intravenous injection.

6 Overall Charge of Complexes and Entry into the Cell

In addition, our delivery system is efficient because the complexes deliver DNA into cells by fusion with the cell membrane and avoid the endocytic pathway (Fig. 6). Cells are negatively charged on the surface, and specific cell types vary in their density of negative charge. These differences in charge density can influence the ability of cells to be transfected. Cationic complexes have nonspecific ionic charge interactions with cell surfaces. Efficient transfection of cells by cationic complexes is, in part, contributed by adequate charge interactions. In addition, recent publications report that certain viruses have a partial positive charge around key subunits of viral proteins on the virus surface responsible for binding to and internalization through target cell surface receptors. Therefore, this partial positive charge is required for virus entry into the cell. Thus, maintenance of adequate positive charge on the surface of targeted liposome complexes is essential for optimal delivery into the cell. Different formulations of liposomes interact with cell surfaces via a variety of mechanisms. Two major pathways for interaction are by endocytosis or by direct fusion with the cell membrane. Preliminary data suggest that nucleic acids delivered *in vitro* and *in vivo* using complexes developed in our lab enter the cell by direct fusion (Fig. 6). Apparently, the bulk of the nucleic acids do not enter endosomes, and, therefore, far more nucleic acid enters the nucleus. Cell transfection by direct fusion produces orders of magnitude increased levels of gene expression and numbers of cells transfected versus cells transfected through the endocytic pathway.

We believe that maintenance of adequate positive charge on the surface of complexes is essential to drive cell entry by direct fusion. Therefore, we create targeted delivery of our complexes *in vivo* without

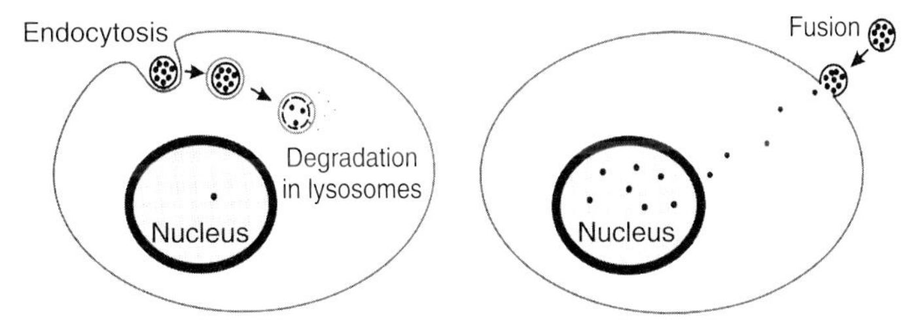

Fig. 6 Mechanisms for cell entry of nucleic acid–liposome complexes. Two major pathways for interaction are by endocytosis or by direct fusion with the cell membrane. Complexes that enter the cell by direct fusion allow delivery of more nucleic acids to the nucleus because the bulk of the nucleic acids do not enter endosomes.

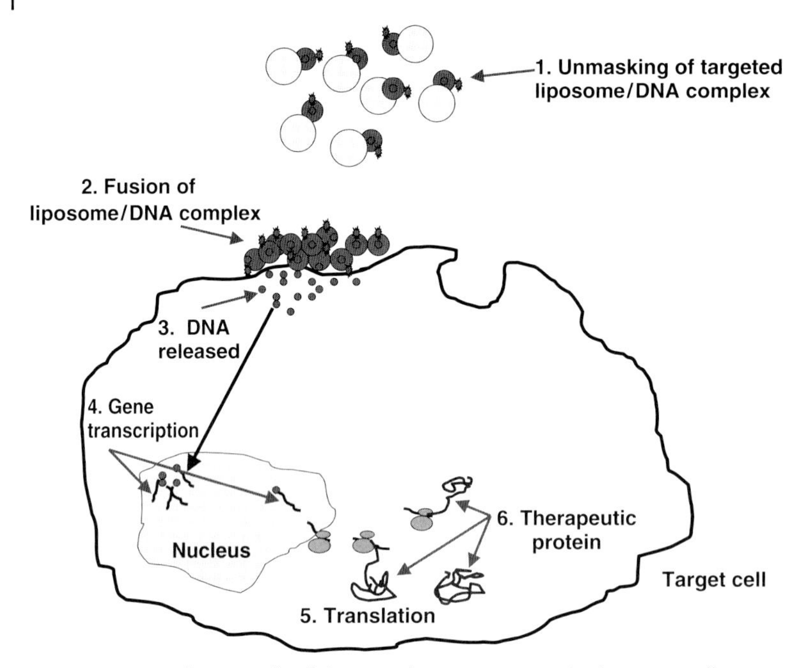

Fig. 7 Optimized strategy for delivery and gene expression in the target cell. Optimization of many steps is required to achieve targeted delivery, deshielding, fusion with the cell membrane, entry of nucleic acids into the cell and to the nucleus, and production of gene expression of a cDNA cloned in a plasmid.

the use of PEG. These ligand-coated complexes also reexpose the overall positive charge of the complexes as they approach the target cells. Through ionic interactions or covalent attachments, we have added monoclonal antibodies, Fab fragments, proteins, partial proteins, peptides, peptide mimetics, small molecules, and drugs to the surface of our complexes after mixing. These ligands efficiently bind to the target cell surface receptor, and maintain entry into the cell by direct fusion. Using novel methods for addition of ligands to the complexes for targeted delivery results in further increased gene expression in the target cells after transfection. Therefore, we design targeted liposomal delivery systems that retain predominant entry into cells by direct fusion versus the endocytic pathway. Figure 7 shows our optimized strategy to achieve targeted delivery, deshielding, fusion with the cell membrane, entry of nucleic acids into the cell and to the nucleus, and production of gene expression of a cDNA cloned in a plasmid.

7 Ligands Used for Targeted Delivery

Using liposomes that encapsulate nucleic acids, ligands can be coated onto the surface of the complexes formed (Fig. 8). We have added monoclonal antibodies, Fab fragments, proteins, partial proteins, peptides, peptide mimetics, small molecules, and drugs to the surface of the complexes after mixing. Ligands are chosen by their ability to efficiently bind to a

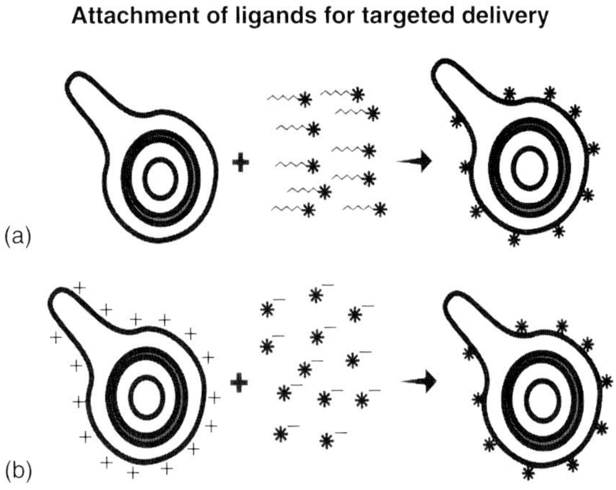

Fig. 8 Cross sections of extruded DOTAP:Chol-nucleic acid–liposome complexes (left) that are coated with ligands (right). Ligands are attached to the surface of preformed nucleic acid–liposome complexes by covalent attachments through "linker lipids" (a) or by ionic interactions (b).

target cell surface receptor while maintaining entry into the cell by direct fusion. Entry into the cell will be discussed further below. The ligands most useful for gene therapeutics in humans will be those that are smallest and possess high affinity for the target receptor. Nonviral systems are desirable because they can be repeatedly administered. Therefore, immune responses may be generated in animals or people upon repeated administration of complexes containing too much ligand or too large a ligand on the surface. These immune responses could cause the targeted therapeutics to be unsafe and/or ineffective for treatments in the clinic.

However, often the "best ligand" is not always available immediately. An investigator could also wait for years for the most appropriate ligand to be generated or produced in the amounts required for large experiments. Our experience shows that much useful information can be generated concerning targeting of a particular cell surface receptor of interest using the "less-than-ideal ligand" *in vitro* and *in vivo* in pilot experiments while creating the best ligand concurrently.

8 Attachment of Ligands

Generally, investigators attach ligands to PEG for incorporation into liposomes and other conjugates or for coating onto the surface of complexes after mixing. After extensive work with PEG in our lab, we have chosen or created alternative methods to use for the attachment of ligands, and have avoided the use of PEG due to its numerous disadvantages discussed above and below in this chapter.

Alternative strategies to the use of PEG include attachment of ligands through ionic interactions or by covalent attachment to "linker lipids". The ligands listed in the section above are generally negatively charged. Therefore, the ligands can simply be adsorbed onto the surface of complexes with encapsulated nucleic acids after mixing (Fig. 8b). Additional moieties can be added to the ligand to increase the amount of negative charge, and yet do not interfere with the ability of the modified ligand to efficiently bind to the appropriate cell surface receptor. For example, we used succinylated asialofetuin to target delivery of DNA–liposome complexes to the asialoglycoprotein receptor on hepatocytes in the liver. The succinic acid amides provided greater negative charge to asialofetuin, and, therefore, bound to the surface of complexes more efficiently than asialofetuin alone. The amount used for adsorption onto the surface of complexes is ligand dependent. Titration studies must be performed to determine the optimal amount of ligand to coat onto the surface of complexes. Ultimately, *in vivo* transfection experiments must be performed to verify the optimal amount of ligand to use to provide delivery to the target cells and the highest levels of gene expression in these cells with no generation of an immune response.

Ligands or modified ligands containing reactive groups can be covalently attached to linker lipids (Fig. 8a). These ligand–lipid conjugates must be checked for optimal activity of the ligand to bind to its receptor. Furthermore, the covalent linkage must not be immunogenic in animals or people after repeated administration. Ligand–lipid conjugates can be spontaneously inserted into the outside membrane of complexes in which the nucleic acids are encapsulated within liposomes (Figure 8b). The amount of ligand–lipid used for insertion into the surface of complexes is also ligand dependent. Titration studies must be performed to determine the optimal amount of ligand–lipid to insert into the surface of complexes. Again, *in vivo* transfection experiments must be performed to verify the optimal amount of ligand–lipid to use to provide delivery to the target cells and the highest levels of gene expression in these cells with no generation of an immune response.

Using the alternative approaches described above, we have produced complexes that provide delivery of nucleic acids to target cells. Furthermore, the gene expression in the target cells using the targeted complexes is higher than that using the "generic" extruded DOTAP:Chol complexes.

9 Serum Stability of Optimized Nucleic Acid–Liposome Complexes for Use *in vivo*

Serum stability of cationic complexes is complicated and cannot be assessed by simply performing studies at a random concentration of serum. Figure 9 shows results from serum stability studies of DNA–liposome complexes that have been optimized in our laboratory for systemic delivery. Serum stability of these complexes was studied at 37 °C out to 24 h at concentrations of serum ranging from 0 to 100%. Two different serum stability assays were performed. The first assay measured the OD400 of BIV DOTAP:Chol–DNA–liposome complexes added into tubes containing a different concentration of serum in each tube, ranging from 0 to 100%. The tubes were incubated at 37 °C, and small aliquots from each tube were removed at various time points out to 24 h. The OD400 of each aliquot was measured on a spectrophotometer calibrated to accurately measure turbidity. Previous work in our laboratory demonstrated that the OD400 predicted both the stability of the complexes and the transfection efficiency results obtained for multiple organs after intravenous injections. Percent stability for this assay is defined as the transfection efficiency that is obtained at a particular OD400 of the complexes used for intravenous injections. Therefore, this assay is rigorous because slight declines in OD400 of these complexes result in obtaining

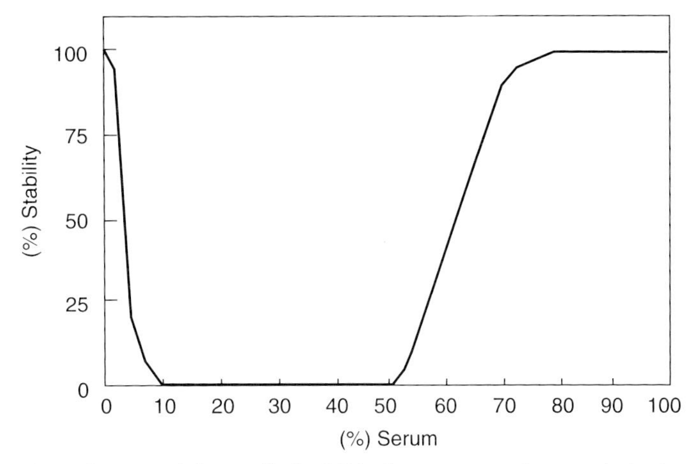

Fig. 9 Serum stability profile for DNA–liposome complexes optimized for systemic delivery. Serum stability of these complexes was studied at 37 °C out to 24 h at concentrations of serum ranging from 0 to 100%. Serum stability at the highest concentrations of serum, about 70 to 100%, that are physiological concentrations of serum found in the bloodstream is required.

no transfection *in vivo*. Declines in the OD400 also measure precipitation of the complexes.

A second assay was performed to support the results obtained from the OD400 measurements described above. A different concentration of serum, ranging from 0 to 100%, was placed into each well of a 96-well micro titer dish. BIV DOTAP:Chol–DNA–liposome complexes were added to the serum in the wells, and the plate was incubated at 37 °C. The plate was removed at various time points out to 24 h and complexes in the wells were observed under the microscope. Precipitation of complexes in the wells was assessed. 100% stability was set at no precipitation observed. Results from this assay were compared with those obtained in the first assay. 100% stability of complexes was set at no decline of OD400 in assay #1 and no observed precipitation in assay #2 at each % serum concentration, and the results were plotted (Fig. 9).

The results showed serum stability at the highest concentrations of serum, about 70 to 100%, that are physiological concentrations of serum found in the bloodstream. In addition, these complexes were also stable in no or low concentrations of serum. The complexes were unstable at 10 to 50% serum, perhaps due to salt bridging. Therefore, *in vitro* optimization of serum stability for formulations of cationic complexes must be performed over a broad range of serum concentration to be useful for applications *in vivo*.

10 Optimized Half-life in the Circulation

As stated above, the extruded BIV DOTAP:Chol–nucleic acid complexes are large enough so that they are not cleared rapidly by Kupffer cells in the liver and yet extravasate across tight barriers and diffuse through the target organ efficiently. Further addition of ligands to the surface of extruded BIV DOTAP:Chol–nucleic acid complexes does not significantly increase the mean particle size. Extravasation and penetration through the target organ and gene expression produced after transfection are not diminished. These modified formulations are positively charged and deliver nucleic acids efficiently into cells *in vitro* and *in vivo*. Because extruded BIV DOTAP:Chol–nucleic acid complexes with or without ligands have a 5-h half-life in the circulation, these complexes do not accumulate in the skin, hands, or feet. Extended half-life in the circulation is provided primarily by the formulation, preparation method, injection of optimal colloidal suspensions, and optimal nucleic acid:lipid ratio used for mixing complexes, serum stability, and size (200 to 450 nm). Therefore, these BIVs are ideal for use in the development of effective, targeted nonviral delivery systems that clearly require encapsulation of nucleic acids.

11
Broad Biodistribution of Optimized Liposome Formulations

Our "generic" BIV nucleic acid–liposome formulation transfects many organs and tissues efficiently after intravenous injection, and has demonstrated efficacy in animal models for lung cancer, pancreatic cancer, breast cancer, Hepatitis B and C (Clawson and Templeton, unpublished data), and cardiovascular diseases. Therefore, optimization of the morphology of the complexes, the lipids used, flexibility of the liposomes and complexes, colloidal suspension, overall charge, serum stability, and half-life in circulation allows for efficient delivery and gene expression in many organs and tissues other than the lung. Apparently, these extruded DOTAP:Chol BIV nucleic acid–liposome complexes can overcome the tendency to be adsorbed only by the endothelial cells lining the circulation surrounding the lungs described by other investigators. However, as discussed above and below, we can further direct delivery to specific target tissues or cells by our targeted delivery strategies in combination with reversible masking used to bypass nonspecific transfection.

12
Optimization of Targeted Delivery

Much effort has been made to specifically deliver nucleic acid–liposome complexes to target organs, tissues, and/or cells. Ligands that bind to cell surface receptors are usually attached to PEG and then attached to the cationic or anionic delivery vehicle. Owing to the shielding of the positive charge of cationic complexes by PEG, delivery to the specific cell surface receptor can be accomplished by only a small fraction of complexes injected systemically. Furthermore, delivery of PEGylated complexes into the cell occurs predominantly through the endocytic pathway, and subsequent degradation of the bulk of the nucleic acid occurs in the lysosomes. Thus, gene expression is generally lower in the target cell than when using the nonspecific delivery of highly efficient cationic complexes.

As discussed above, the vast majority of the injected PEGylated complexes bypasses the target cell. Apparently, the PEGylated complexes cannot utilize critical charge interactions for optimal transfection into cells by direct fusion. Inability to expose positive charge on the surface of optimized complexes results in the transfection of fewer cells. PEGylation was first used to increase the half-life of complexes in the circulation and to avoid uptake in the lung. However, this technology also destroys the ability to efficiently transfect cells. We were able to increase the half-life in circulation of BIVs to 5 h without the use of PEG. Because the extended half-life of BIVs is not too long, this delivery system does not result in the accumulation of complexes in nontarget tissues that circulate for one to three days. Some investigators have now reported targeted delivery that produces increased gene expression in the target cell over their nontargeted complexes. However, these nontargeted and targeted delivery systems are inefficient compared to efficient delivery systems such as the BIVs.

In using the extruded BIV DOTAP: Chol–nucleic acid:liposome complexes, we produced an optimal half-life in the circulation without the use of PEG. Extended half-life was produced primarily by the formulation, preparation method, injection of optimal colloidal suspensions, serum stability, and optimal nucleic acid:lipid ratio used for mixing complexes, and size (200 to 450 nm). Furthermore, we avoided uptake in the lungs using the negative charge of the ligands and "shielding/deshielding compounds" that can be added to the complexes used for targeting just prior to injection or administration *in vivo*. Our strategy to bypass nonspecific transfection is called *reversible masking*. Addition of ligands using the novel approaches that we developed, adequate overall positive charge on the surface of complexes is preserved. In summary, we achieve optimal circulation time of the complexes, reach and deliver to the target organ, avoid uptake in nontarget tissues, and efficiently interact with the cell surface to produce optimal transfection.

13 Efficient Dissemination Throughout Target Tissues and Migration Across Tight Barriers

A primary goal for efficient *in vivo* delivery is to achieve extravasation into and penetration throughout the target organ/tissue ideally by noninvasive systemic administration. Without these events, therapeutic efficacy is highly compromised for any treatment, including gene and drug therapies. Achieving this goal is difficult because of the many tight barriers that exist in animals and people. Furthermore, many of these barriers become tighter in the transition from neonates to becoming adults. Penetration throughout an entire tumor is further hindered because of the increased interstitial pressure within most tumors. We believe that nonviral systems can play a pivotal role in achieving target organ extravasation and penetration needed to treat or cure certain diseases. Our preliminary studies have shown that extruded BIV DOTAP:Chol–nucleic acid:liposome complexes can extravasate across tight barriers and penetrate evenly throughout entire target organs, and viral vectors cannot cross identical barriers. These barriers include the endothelial cell barrier in a normal mouse, the posterior blood retinal barrier in adult mouse eyes, complete and even diffusion throughout large tumors, and penetration through several tight layers of smooth muscle cells in the arteries of pigs. Diffusion throughout large tumors was measured by expression of ß-galactosidase or the proapoptotic gene *p53* in about half of the p53-null tumor cells after a single injection of BIV DOTAP:Chol – DNA–liposome complexes into the center of a tumor. Transfected cells were evenly spread throughout the tumors. Tumors injected with complexes encapsulating plasmid DNA encoding p53 showed apoptosis in almost all of the tumor cells by TUNEL staining. Tumor cells expressing p53 mediate a bystander effect on neighboring cells perhaps due to upregulation by Fas ligand that causes nontransfected tumor cells to undergo apoptosis. Currently, we are investigating the mechanisms used by extruded DOTAP:Chol–nucleic acid:liposome complexes to cross barriers and penetrate throughout target organs. By knowing more about these mechanisms, we hope to develop more robust nonviral gene therapeutics.

14 Optimization of Plasmids for *in vivo* Gene Expression

Delivery of DNA and subsequent gene expression may be poorly correlated. Investigators may focus solely on the delivery formulation as the source of poor gene expression. In many cases, however, the delivery of DNA into the nucleus of a particular cell type may be efficient, although little or no gene expression is achieved. The causes of poor gene expression can be numerous. The following issues should be considered independent of the delivery formulation, including suboptimal promoter-enhancers in the plasmid, poor preparation of plasmid DNA, and insensitive detection of gene expression.

Plasmid expression cassettes typically have not been optimized for animal studies. For example, many plasmids lack a full-length CMV promoter-enhancer. Over one hundred variations of the CMV promoter-enhancer exist, and some variations produce greatly reduced or no gene expression in certain cell types. Even commercially available plasmids contain suboptimal CMV promoters-enhancers, although these plasmids are advertised for use in animals. Furthermore, upon checking the company data for these plasmids, one would discover that these plasmids have never been tested in animals and have been tested in only one or two cultured cell lines. Conversely, plasmids that have been optimized for overall efficiency in animals may not be best for transfection of certain cell types *in vitro* or *in vivo*. For example, many investigators have shown that optimal CMV promoters-enhancers produce gene expression at levels several orders of magnitude less in certain cell types. In addition, one cannot assume that a CMV promoter that expresses well within the context of a viral vector, such as adenovirus, will function as well in a plasmid-based transfection system for the same cell context. Virus proteins produced by the viral vector are required for producing high levels of mRNA by the CMV promoter in specific cell nuclei.

Ideally, investigators design custom promoter-enhancer chimeras that produce the highest levels of gene expression in their target cells of interest. Recently, we designed a systematic approach for customizing plasmids used for breast cancer gene therapy using expression profiling. Gene therapy clinical trials for cancer frequently produce inconsistent results. We believe that some of this variability could result from differences in transcriptional regulation that limit expression of therapeutic genes in specific cancers. Our systemic liposomal delivery of a nonviral plasmid DNA showed efficacy in animal models for several cancers. However, we observed large differences in the levels of gene expression from a CMV promoter-enhancer between lung and breast cancers. To optimize gene expression in breast cancer cells *in vitro* and *in vivo*, we created a new promoter-enhancer chimera to regulate gene expression. Serial analyses of gene expression (SAGE) data from a panel of breast carcinomas and normal breast cells predicted promoters that are highly active in breast cancers, for example, the glyceraldehyde 3-phosphate dehydrogenase (GAPDH) promoter. Furthermore, GAPDH is upregulated by hypoxia, which is common in tumors. We added the GAPDH promoter, including the hypoxia enhancer sequences, to our *in vivo* gene expression plasmid. The novel CMV-GAPDH promoter-enhancer showed up to 70-fold increased gene expression in breast tumors compared to the optimized CMV promoter-enhancer alone. No significant increase in gene expression was observed in other tissues. These data demonstrate tissue-specific effects on gene expression after nonviral delivery, and suggest that gene delivery systems may require plasmid modifications for the treatment of different tumor types. Furthermore, expression profiling can facilitate the design of optimal expression plasmids for use in specific cancers.

Several reviews have stated that nonviral systems are intrinsically inefficient compared to viral systems. However, as discussed above, one must separate issues of the delivery vehicle versus the plasmid that is delivered. Case in point, we have shown that our extruded liposomes optimized for systemic delivery could outcompete delivery using a lentivirus. For example, we have compared SIVmac239, a highly noninfectious virus, with nonviral delivery of SIVmac239 DNA complexed to BIVs in adult rhesus macaques after injection into the saphenous vein of the leg. Our data showed that the monkeys injected with SIV DNA encapsulated in DOTAP:Chol BIVs were infected four days postinjection, and high levels of infection were produced in these monkeys at 14 days postinjection. Furthermore, higher levels of SIV RNA in the blood were produced using our BIV liposomes for delivery versus that using the SIV virus. CD4 counts were measured before and after injections. CD4 levels dropped in all monkeys to the lowest levels ever detected in the macaques in any experiment by 28 days postinjection, the first time point at which these counts were measured postinjection. All monkeys had clinical SIV infections and lost significant weight by day 28. These results were surprising because SIVmac239 is not highly infectious, and monkeys become sick with SIV infection only after several months or years postinjection with SIVmac239 virus. Therefore, we were able to induce SIV infection faster using our nonviral delivery of SIV plasmid DNA. In this case, we delivered a replication-competent plasmid so that gene expression increased over time posttransfection. Our delivery system was highly efficient and exceeded that of the lentivirus. The critical feature in this nonviral experiment was the plasmid DNA that was delivered.

Plasmids can be engineered to provide for specific or long-term gene expression, replication, or integration. Persistence elements, such as the inverted terminal repeats from adenovirus or adeno-associated virus, have been added to plasmids to prolong gene expression *in vitro* and *in vivo*. Apparently, these elements bind to the nuclear matrix thereby retaining the plasmid in cell nuclei. For regulated gene expression, many different inducible promoters are used that promote expression only in the presence of a positive regulator or in the absence of a negative regulator. Tissue specific promoters have been used for the production of gene expression exclusively in the target cells. As discussed in the previous paragraph, replication-competent plasmids or plasmids containing sequences for autonomous replication can be included that provide prolonged gene expression. Other plasmid-based strategies produce site-specific integration or homologous recombination within the host cell genome. Integration of a cDNA into a specific "silent site" in the genome could provide long-term gene expression without disruption of normal cellular functions. Homologous recombination could correct genetic mutations upon integration of wild-type sequences that replace mutations in the genome. Plasmids that contain fewer bacterial sequences and that produce high yield upon growth in *Escherichia coli* are also desirable.

15
Optimization of Plasmid DNA Preparations

The transfection quality of plasmid DNA is dependent on the preparation protocol and training of the person preparing the DNA. For example, we performed a blinded study asking three people to make DNA preparations of the same plasmid from the same box of a Qiagen Endo-Free Plasmid Preparation kit. One person then mixed all of the DNA–liposome complexes on the same morning using a single vial of liposomes. One person performed all tail vein injections, harvesting of tissues, preparation of extracts from tissues, and reporter gene assays on the tissue extracts. *In vivo* gene expression differed 30-fold among these three plasmid DNA preparations.

One source for this variability is that optimized methods to detect and remove contaminants from plasmid DNA preparations have not been available. We have identified large amounts of contaminants that exist in laboratory and clinical grade preparations of plasmid DNA. These contaminants copurify with DNA by anion exchange chromatography and by cesium chloride density gradient centrifugation. Endotoxin removal does not remove these contaminants. HPLC cannot detect these contaminants. Therefore, we developed three proprietary methods for the detection of these contaminants in plasmid DNA preparations. We can now make clinical grade good manufacturing practices (GMP) DNA that does not contain these contaminants. To provide the greatest efficacy and levels of safety, these contaminants must be assessed and removed from plasmid DNA preparations. These contaminants belong to a class of molecules known to inhibit both DNA and RNA polymerase activities. Therefore, gene expression posttransfection can be increased by orders of magnitude if these contaminants are removed from DNA preparations. The presence of these contaminants in DNA also precludes high dose delivery of DNA–liposome complexes intravenously. Our group and other investigators have shown that intravenous injections of high doses of improved liposomes alone cause no adverse effects in small and large animals.

Some investigators have removed the majority of CpG sequences from their plasmids and report reduced toxicity after intravenous injections of cationic liposomes complexed to these plasmids. However, only low doses containing up to 16.5 μg of DNA per injection into each mouse were shown to reduce toxicity. Therefore, no significant dose response to CpG motifs in plasmid DNA was demonstrated. To achieve efficacy for cancer metastases, particularly in mice bearing aggressive tumors, most investigators are interested in injecting higher doses in the range of 50 to 150 μg of DNA per mouse. Therefore, removal of CpG sequences from plasmid-based gene therapy vectors will not be useful for these applications because no difference in toxicity was shown after intravenous injections of these higher doses of plasmids, with or without reduced CpG sequences, complexed to liposomes. Therefore, we believe that removal of the other contaminants in current DNA preparations, discussed above, is the major block to the safe intravenous injection of high doses of DNA–liposome complexes.

16
Detection of Gene Expression

Thought should also be given to choosing the most sensitive detection method for every application of nonviral delivery rather than using the method that seems most simple. For example, detection of ß-galactosidase expression is far more sensitive than that for the green fluorescent protein (GFP). Specifically, 500 molecules of ß-galactosidase (ß-gal) per cell are required for detection using X-gal staining. Whereas, about one million molecules of GFP per cell are required for direct detection. Furthermore, detection of GFP may be impossible if the fluorescence background of the target cell or tissue is too high. Detection of chloramphenicol acetyltransferase (CAT) is extremely sensitive with little or no background detected in untransfected cells. Often, assays for CAT expression can provide more useful information than using ß-gal or GFP as reporter genes.

Few molecules of luciferase in a cell can be detected by luminescence assays of cell or tissue extracts posttransfection. The sensitivity of these assays is highly dependent on the type of instrument used to measure luminescence. However, luciferase results may not predict the therapeutic potential of a nonviral delivery system. For example, if several hundred or thousand molecules per cell of a therapeutic gene are required to produce efficacy for a certain disease, then production of only few molecules will not be adequate. If only few molecules of luciferase are produced in the target cell using a specific nonviral delivery system, then the investigator may be misled in using this system for therapeutic applications.

Furthermore, noninvasive detection of luciferase expression *in vivo* is not as sensitive as luminescence assays of cell or tissue extracts posttransfection. Recently, some colleagues of mine tried cooled charge coupled device (CCD) camera imaging on live mice after intravenous injection of other cationic liposomes complexed to plasmid DNA encoding luciferase, and they were not able to detect any transfection. However, these liposomal delivery systems had been used to detect luciferase by luminescence assays of organ extracts. My colleagues detected luciferase expression by CCD imaging after intravenous injections of BIV DOTAP:Chol-luciferase DNA–liposome complexes. Using the same CCD imaging system following intravenous injections of PEI-DNA complexes, extremely poor transfection efficiency was observed in all tissues. Because the luciferase protein is short-lived, maximal expression was detected at 5 h posttransfection. Whereas, detection of HSV-TK gene expression using microPET imaging in the same mice was highest at 24 h posttransfection. In contrast to luciferase, the CAT protein accumulates over time, and, therefore, the investigator is not restricted to a narrow time frame for assaying gene expression. Furthermore, detection of CAT seems to be more sensitive than CCD imaging of luciferase following intravenous injections of DNA–liposome complexes. However, the animals must be sacrificed in order to perform CAT assays on tissue or organ extracts. In summary, further work is still needed to develop *in vivo* detection systems that have high sensitivity and low background.

17 Optimization of Dose and Frequency of Administration

To establish the maximal efficacy for the treatment of certain diseases or for the creation of robust vaccines, injections, or administrations of the nonviral gene therapeutic, and so on via different routes may be required. For particular treatments, one should not assume that one delivery route is superior to others without performing the appropriate animal experiments. In addition, people with the appropriate expertise should perform the injections and administrations. In our experience, only a minority of people who claim expertise in performing tail vein injections can actually perform optimal injections.

The optimal dose should be determined for each therapeutic gene or other nucleic acid that is administered. The investigator should not assume that the highest tolerable dose is optimal for producing maximal efficacy. The optimal administration schedule should also be determined for each therapeutic gene or other nucleic acid. To progress faster, some investigators have simply used the same administration schedule that they used for chemotherapeutics, for example. The investigator should perform *in vivo* experiments to determine when gene expression and/or efficacy drops significantly. Most likely, readministration of the nonviral gene therapeutic is not necessary until this drop occurs. Loss of the therapeutic gene product will vary with the half-life of the protein produced. Therefore, if a therapeutic protein has a longer half-life, then the gene therapy could be administered less frequently.

18 Summary

Overcoming some hurdles remains in the broad application of nonviral delivery; however, we are confident that we will successfully accomplish the remaining challenges soon. Furthermore, we predict that eventually the majority of gene therapies will utilize artificial reagents that can be standardized and regulated as drugs rather than biologics. We will continue to incorporate the molecular mechanisms of viral delivery that produce efficient delivery to cells into artificial systems. Therefore, the artificial systems, including liposomal delivery vehicles, will be further engineered to mimic the most beneficial parts of the viral delivery systems while circumventing their limitations. We will also maintain the numerous benefits of the liposomal delivery systems discussed in this chapter.

Acknowledgment

I thank Dr. David D. Roberts at the National Cancer Institute, National Institutes of Health, Bethesda, MD for preparation of the figures.

See also Phospholipids; Plasmids.

Bibliography

Books and Reviews

Templeton, N.S. *Gene Targeting Protocols*, (2000) Vol. 133, Humana Press, Totowa, NJ, pp. 1–244.

Jain, R.K. (1994) Barriers to drug delivery in solid tumors, *Sci. Am.* **271**, 58–65.

Jain, R.K. (1999) Transport of molecules, particles, and cells in solid tumors, *Annu. Rev. Biomed. Eng.* **1**, 241–263.

Li, S., Ma, Z., Tan, Y., Liu, F., Dileo, J., Huang, L. (2002) Targeted Delivery Via Lipidic Vectors, in: Curiel, D.T., Douglas, J.T. (Eds.) *Vector Targeting for Therapeutic Gene Delivery*, Wiley-Liss Inc., Hoboken, NJ, pp. 17–32.

Marshall, E. (2003) Second child in French trial is found to have leukemia, *Science* **299**, 320.

Pirollo, K.F., Xu, L., Chang, E.H. (2002) Immunoliposomes: A Targeted Delivery Tool for Cancer Treatment, in: Curiel, D.T., Douglas, J.T. (Eds.) *Vector Targeting for Therapeutic Gene Delivery*, Wiley-Liss Inc., Hoboken, NJ, pp. 33–62.

Templeton, N.S., Lasic, D.D. (1999) New directions in liposome gene delivery, *Mol. Biotechnol.* **11**, 175–180.

Primary Literature

Aksentijevich, I., Pastan, I., Lunardi-Iskandar, Y., Gallo, R.C., Gottesman, M.M., Thierry, A.R. (1996) In vitro and in vivo liposome-mediated gene transfer leads to human MDR1 expression in mouse bone marrow progenitor cells, *Hum. Gene Ther.* **7**, 1111–1122.

Behr, J.-P., Demeneix, B., Loeffler, J.P., Perez-Mutul, J. (1989) Efficient gene transfer into mammalian primary endocrine cells with lipopolyamine-coated DNA, *Proc. Natl. Acad. Sci. U.S.A.* **86**, 6982–6986.

Felgner, P.L., Ringold, G.M. (1989) Cationic liposome-mediated transfection, *Nature* **337**, 387–388.

Felgner, J.H., Kumar, R., Sridhar, C.N., Wheeler, C.J., Tsai, Y.J., Border, R., Ramsey, P., Martin, M., Felgner, P.L. (1994) Enhanced gene delivery and mechanism studies with a novel series of cationic lipid formulations, *J. Biol. Chem.* **269**, 2550–2561.

Felgner, P.L., Gadek, T.R., Holm, M., Roman, R., Chan, H.W., Wenz, M., Northrop, J.P., Ringold, G.M., Danielson, H. (1987) Lipofection: a highly efficient lipid-mediated DNA transfection procedure, *Proc. Natl. Acad. Sci. U.S.A.* **84**, 7413–7417.

Gabizon, A., Catane, R., Uziely, B., Kaufman, B., Safra, T., Cohen, R., Martin, F., Huang, A., Barenholz, Y. (1994) Prolonged circulation time and enhanced accumulation in malignant exudates of doxorubicin encapsulated in polyethylene-glycol coated liposomes, *Cancer Res.* **54**, 987–992.

Gordon, K.B., Tajuddin, A., Guitart, J., Kuzel, T.M., Eramo, L.R., VonRoenn, J. (1995) Hand-foot syndrome associated with liposome-encapsulated doxorubicin therapy, *Cancer* **75**, 2169–2173.

Gustafsson, J., Arvidson, G., Karlsson, G., Almgren, M. (1995) Complexes between cationic liposomes and DNA visualized by cryo-TEM, *Biochim. Biophys. Acta* **1235**, 305–312.

Handumrongkul, C., Zhong, W., Debs, R.J. (2002) Distinct sets of cellular genes control the expression of transfected, nuclear-localized genes, *Mol. Ther.* **5**, 186–194.

Hildebrandt, I.J., Iyer, M., Wagner, E., Gambhir, S.S. (2003) Optical imaging of transferrin targeted PEI/DNA complexes in living subjects, *Gene Ther.* **10**, 758–764.

Hood, J.D., Bednarski, M., Frausto, R., Guccione, S., Reisfeld, R.A., Xiang, R., Cheresh, D.A. (2002) Tumor regression by targeted gene delivery to the neovasculature, *Science* **296**, 2404–2407.

Iyer, M., Berenji, M., Templeton, N.S., Gambhir, S.S. (2002) Noninvasive imaging of cationic lipid-mediated delivery of optical and PET reporter genes in living mice, *Mol. Ther.* **6**, 555–562.

Jain, R.K. (1991) Haemodynamic and transport barriers to the treatment of solid tumours, *Int. J. Radiat. Biol.* **60**, 85–100.

Leventis, R., Silvius, J.R. (1990) Interactions of mammalian cells with lipid dispersions containing novel metabolizable cationic amphiphiles, *Biochim. Biophys. Acta* **1023**, 124–132.

Liu, F., Qi, H., Huang, L., Liu, D. (1997) Factors controlling the efficiency of cationic lipid-mediated transfection in vivo via intravenous administration, *Gene Ther.* **4**, 517–523.

Liu, Y., Liggitt, D., Zhong, W., Tu, G., Gaensler, K., Debs, R. (1995) Cationic liposome-mediated intravenous gene delivery, *J. Biol. Chem.* **270**, 24864–24870.

Liu, Y., Mounkes, L.C., Liggitt, H.D., Brown, C.S., Solodin, I., Heath, T.D., Debs, R.J. (1997) Factors influencing the efficiency of cationic-liposome mediated intravenous gene delivery, *Nat. Biotechnol.* **15**, 167–173.

Loeffler, J.P., Behr, J.-P. (1993) Gene transfer into primary and established mammalian cell lines with lipopolyamine-coated DNA, *Methods Enzymol.* **217**, 599–618.

Lu, H., Zhang, Y., Roberts, D.D., Osborne, C.K., Templeton, N.S. (2002) Enhanced gene expression in breast cancer cells in vitro and tumors in vivo, *Mol. Ther.* **6**, 783–792.

Mislick, K.A., Baldeschwieler, J.D. (1996) Evidence for the role of proteoglycans in cation-mediated gene transfer, *Proc. Natl. Acad. Sci. U.S.A.* **93**, 12349–12354.

Papahadjopoulos, D., Allen, T.M., Gabizon, A., Mayhew, E., Matthay, K., Huang, S.K., Lee, K., Woodle, M.C., Lasic, D.D., Redemann, C., Martin, F.J. (1991) Sterically stabilized liposomes: improvements in pharmacokinetics and antitumor therapeutic efficacy, *Proc. Natl. Acad. Sci. U.S.A.* **88**, 11460–11464.

Philip, R., Liggitt, D., Philip, M., Dazin, P., Debs, R. (1993) In vivo gene delivery: efficient transfection of T lymphocytes in adult mice, *J. Biol. Chem.* **268**, 16087–16090.

Pinnaduwage, P., Huang, L. (1989) The role of protein-linked oligosaccharide in the bilayer stabilization activity of glycophorin A for dioleoylphosphatidylethanolamine liposomes, *Biochim. Biophys. Acta* **986**, 106–114.

Ramesh, R., Saeki, T., Templeton, N.S., Ji, L., Stephens, L.C., Ito, I., Wilson, D.R., Wu, Z., Branch, C.D., Minna, J.D., Roth, J.A. (2001) Successful treatment of primary and disseminated human lung cancers by systemic delivery of tumor suppressor genes using an improved liposome vector, *Mol. Ther.* **3**, 337–350.

Rose, J.K., Buonocore, L., Whitt, M.A. (1991) A new cationic liposome reagent mediating nearly quantitative transfection of animal cells, *Biotechniques* **10**, 520–525.

Senior, J., Delgado, C., Fisher, D., Tilcock, C., Gregoriadis, G. (1991) Influence of surface hydrophilicity of liposomes on their interaction with plasma protein and clearance from the circulation: studies with poly(ethylene glycol)-coated vesicles, *Biochim. Biophys. Acta* **1062**, 77–82.

Shi, H.Y., Liang, R., Templeton, N.S., Zhang, M. (2002) Inhibition of breast tumor progression by systemic delivery of the maspin gene in a syngeneic tumor model, *Mol. Ther.* **5**, 755–761.

Solodin, I., Brown, C.S., Bruno, M.S., Ching-Yi, C., Eun-Hyun, J., Debs, R., Heath, T.D. (1995) A novel series of amphiphilic imidazolinium compounds for in vitro and in vivo gene delivery, *Biochemistry* **34**, 13537–13544.

Sternberg, B. (1996) Morphology of cationic liposome/DNA complexes in relation to their chemical composition, *J. Liposome Res.* **6**, 515–533.

Templeton, N.S., Lasic, D.D., Frederik, P.M., Strey, H.H., Roberts, D.D., Pavlakis, G.N. (1997) Improved DNA: liposome complexes for increased systemic delivery and gene expression, *Nat. Biotechnol.* **15**, 647–652.

Templeton, N.S., Alspaugh, E., Antelman, D., Barber, J., Csaky, K.G., Fang, B., Frederik, P., Honda, H., Johnson, D., Litvak, F., Machemer, T., Ramesh, R., Robbins, J., Roth, J.A., Sebastian, M., Tritz, R., Wen, S.F., Wu, Z. (1999) Non-viral Vectors for the Treatment of Disease, *Keystone Symposia on Molecular and Cellular Biology of Gene Therapy*, Salt Lake City, UT.

Thierry, A.R., Lunardi-Iskandar, Y., Bryant, J.L., Rabinovich, P., Gallo, R.C., Mahan, L.C. (1995) Systemic gene therapy: biodistribution and long-term expression of a transgene in mice, *Proc. Natl. Acad. Sci. U.S.A.* **92**, 9742–9746.

Tirone, T.A., Fagan, S.P., Templeton, N.S., Wang, X.P., Brunicardi, F.C. (2001) Insulinoma induced hypoglycemic death in mice is prevented with beta cell specific gene therapy, *Ann. Surg.* **233**, 603–611.

Tsukamoto, M., Ochiya, T., Yoshida, S., Sugimura, T., Terada, M. (1995) Gene transfer and expression in progeny after intravenous DNA injection into pregnant mice, *Nat. Genet.* **9**, 243–248.

Uziely, B., Jeffers, S., Isacson, R., Kutsch, K., Wei-Tsao, D., Yehoshua, Z., Libson, E., Muggia, F.M., Gabizon, A. (1995) Liposomal doxorubicin: antitumor activity and unique toxicities during two complementary phase I studies, *J. Clin. Oncol.* **13**, 1777–1785.

Xu, Y., Szoka, F.C. (1996) Mechanism of DNA release from cationic liposome/DNA complexes used in cell transfection, *Biochemistry* **35**, 5616–5623.

Yew, N.S., Zhao, H., Przybylska, M., Wu, I.-H., Tousignant, J.D., Scheule, R.K., Cheng, S.H. (2002) CpG depleted plasmid DNA vectors with enhanced safety and long-term gene expression in vivo, *Mol. Ther.* **5**, 731–738.

Yotnda, P., Chen, D.-H., Chiu, W., Piedra, P.A., Davis, A., Templeton, N.S., Brenner, M.K. (2002) Bilamellar cationic liposomes protect adenovectors from preexisting humoral immune responses, *Mol. Ther.* **5**, 233–241.

Zhu, N., Liggitt, D., Liu, Y., Debs, R. (1993) Systemic gene expression after intravenous DNA delivery in adult mice, *Science* **261**, 209–211.

Liver Cancer, Molecular Biology of

Mehmet Ozturk and Rengul Cetin-Atalay
Bilkent University, Bilkent, Ankara, Turkey

Keywords

Aflatoxins
A family of natural toxins produced by *Aspergillus* species, inducing mutations in animals and involved in liver cancer development.

Axin1
The gene encoding a protein that facilitates the degradation of oncogenic β-catenin protein that is inactivated by mutation in liver cancers.

Encyclopedia of Molecular Cell Biology and Molecular Medicine, 2nd Edition. Volume 7
Edited by Robert A. Meyers.

ISBN: 3-527-30549-1

β-catenin
A proto-oncogene when mutated contributes to the development of liver cancer.

Cyclin D1
A gene-encoding cyclin partner of cyclin-dependent kinase 4 that is amplified in liver cancers.

Hepatitis B Virus
A small DNA virus that infects hepatocytes in the liver, causing acute or chronic hepatitis.

Hepatitis C Virus
A small RNA virus infecting hepatocytes in the liver, causing acute or chronic hepatitis.

$P16^{INK4a}$
The gene encoding a protein that inhibits the activity of cyclin-dependent kinase 4 that is inactivated by *de novo* methylation in different cancers, including liver cancers.

p53
A tumor suppressor gene that is inactivated by mutation in different cancers including liver cancers.

Primary liver cancer is the fifth most common cancer worldwide. Hepatocellular carcinoma (HCC), which constitutes 80% of liver cancers, is linked etiologically to hepatitis B virus (HBV), hepatitis C virus (HCV), and aflatoxins. Hepatocellular carcinoma results from malignant transformation of hepatocytes and/or its progenitor cells. Viral factors appear to contribute to hepatocellular carcinogenesis nonspecifically by inducing chronic liver regeneration leading to cirrhosis, followed by HCC. Major cellular genes that are mutated in HCC are *β-catenin* and *Cyclin D1* oncogene, as well as *p53* and *axin1* tumor suppressor genes. *$P16^{INK4a}$*, another tumor suppressor gene is inactivated by promoter hypermethylation. These mutations affect p53, retinoblastoma, and wnt pathways in HCC, leading to a loss of proliferation control, escape from apoptosis and senescence, and probably, aberrant differentiation process.

1 Epidemiology of Liver Cancers and Hepatocellular Carcinoma

Primary liver cancer is a major public health problem, worldwide. It is the fifth most common cancer in the world, and third most common cause of cancer-related death. Hepatocellular carcinoma, cholangiocarcinoma, and hepatoblastoma are the main forms of cancer of the liver. However, the great majority (more than

80%) of primary liver tumors are HCCs. More than 500 000 new cases of HCC are diagnosed each year. Long considered as a disease of people living in Africa and Asia, HCC is now becoming a serious health threat in Europe and North America also because of its rising incidence.

Chronic infection with HBV, HCV, as well as naturally occurring aflatoxins were classified as carcinogenic to humans by International Agency for Cancer Research (IARC) working groups. These three etiologic factors constitute the majority of liver cancer risk factors in the world. Other risk factors for liver carcinomas are alcohol, nonviral cirrhosis, autoimmune chronic active hepatitis, and metabolic diseases, such as α-1-antitrypsin deficiency and hemochromatosis. Hepatitis B virus is the major etiology in Asia and Africa, whereas HCV appears to be more involved in HCCs occurring in Japan, Europe, and North America.

All major etiological factors of HCC have in common the ability to cause chronic liver disease leading to viral cirrhosis, which represents the single most common cause of these tumors with a contribution to over 80% of these cancers. The chronic process leading to cirrhosis with further development into HCC may take many years. Therefore, when analyzing the molecular etiology of HCC, it is often to dissociate viral from host-related contributions. Although involved as major causes, HBV and HCV differ from *bona fide* oncogenic viruses such as human papillomavirus 16 in their contributions to cancer. Their contribution to HCC appears to be rather nonspecific, mainly through the chronic liver regeneration process that they are known to be more directly involved in.

2
Molecular Biology of Hepatocellular Carcinoma

The liver is the largest internal organ that exhibits endocrine, exocrine, metabolic, and detoxification activities. Hepatocytes carry out the main functions of the liver and account for approximately 80% of all liver cells. Hepatocellular carcinoma is believed to originate from these mature cells, and/or from their progenitors. Therefore, a short overview of hepatogenesis and homeostasis of normal and diseased liver may be helpful for a better understanding of HCC biology.

2.1
Hepatogenesis and Homeostasis in the Adult Liver

Hepatogenesis is a process that consists of distinct developmental stages: competence, commitment, differentiation, and morphogenesis of hepatocytes. The liver develops from the ventral gut endoderm juxtaposing the developing heart. The entire embryonic endoderm is competent to adopt a hepatic fate and expresses the transcription factor HNF-3. Signals, including FGFs, released by cardiac mesoderm act on the underlying ventral endoderm of the foregut, inducing it to commit to a hepatic cell fate. In response to this, the prehepatic endoderm begins to express genes associated with a differentiated hepatocyte phenotype. At this stage of early development, these cells are called *hepatoblasts* or embryonic liver stem cells as they are at least bipotential, giving rise to both hepatocyte and bile epithelial cell lineages. The mechanisms controlling liver stem cell phenotype and the bipotential cell lineage commitment are mostly unknown. As far as the hepatocyte lineage is concerned, expression of the full array of genes necessary for hepatocyte function requires the transcription factor HNF-4α. This differentiation step is followed by a period of rapid cell growth and morphogenesis that results in a functioning liver. Growth of the liver requires several transcription factors expressed in hepatocytes including NF-κB, c-Jun, and XBP-1.

The adult liver is estimated to be replaced by normal tissue renewal approximately once a year. This very slow process is maintained mainly by proliferation of differentiated hepatocytes. These hepatocytes are also able to restore the replacement of large volumes of liver tissue losses, following partial hepatectomy, for example, and during chronic liver diseases. However, under experimental conditions during which the proliferation of mature hepatocytes is suppressed, the liver mass can also be restored by adult liver cells, also called *oval* cells believed to reside in the liver. Furthermore, it has been shown that liver hepatocytes can also be derived from bone marrow cell populations in humans. Thus, the adult liver appears to be equipped with a powerful cell renewal system using mature hepatocytes, liver stem cells, or even bone marrow–derived stem cells.

2.2 Hepatocyte Regeneration During Chronic Liver Disease

The major etiological factors of HCC, namely, HBV and HCV may induce a chronic disease characterized by repeated inflammation, degeneration, and regeneration cycles in the liver. This chronic process is maintained by repetitive cycles of "exposure, inflammation, necrosis/apoptosis and regenerative response". After tens of years, the chronic hepatitis will lead to a cirrhosis stage, which is manifested by metabolic and replicative failure accompanied with the formation of fibrosis in the diseased liver. The HCC may arise at this stage from dysplastic nodules of hepatocytes without an overt stage of adenoma. The major limiting factors against the occurrence of malignant transformation are cell death and senescence. Therefore, HCC cells must acquire new abilities to escape from these limiting factors, namely, resistance to both apoptosis and senescence.

The process of inflammation in chronic liver disease creates a sustained state of oxidative stress. Reactive oxygen species can be generated during the inflammatory response to viral hepatitis or internally by the metabolic stress of hepatocytes. This oxidative stress is usually taken care of, at least initially, by antioxidant processes. Selenium is considered to play an indirect antioxidant role as an essential constituent of a selenoprotein, namely, glutathione peroxidase (Gpx), in elimination of cellular reactive oxygen species. Cytosolic or classical Gpx (Gpx-1) is considered as the most critical enzyme for the protective effects of selenium. It has recently been shown that selenium deficiency leads to apoptosis in hepatocyte-derived cells, due to oxidative stress. On the other hand, most HCC cells are tolerant to selenium deficiency and oxidative stress-inducing conditions. Thus, selenium deficiency in liver may confer growth advantage to malignant tumor cells, since they appear to adapt themselves to a selenium-deficient cellular environment.

Resistance to oxidative stress is probably one of the many ways in which malignant hepatocytes become resistant to apoptosis. Additional mechanisms involving the apoptotic machinery itself exist. Among others, resistance to *Fas*-mediated apoptosis, changes in the expression of *bcl-2* family of genes, as well as *p53* gene mutations appear to confer apoptosis resistance to HCC cells.

2.3 Genetic and Epigenetic Changes in Hepatocellular Carcinoma

Hepatocellular cancers display many chromosomal changes (aneuploidies) such as polyploidy, loss of heterozygosity, allelic imbalance, amplifications as well as translocations. Aneuploidy is a general feature of solid tumors, so its occurrence in HCC is not surprising. In addition, it is well established that HBV DNA integrates into the host genome causing many chromosomal rearrangements. It is expected that the chromosomal regions that undergo tumor-specific changes harbor critical genes involved in carcinogenesis.

Historically, the oncogenes were first discovered as transforming viral genes. It was later shown that most viral oncogenes are activated forms or homologs of cellular proto-oncogenes involved in positive regulation of cell proliferation. Cellular proto-oncogenes can also be activated by somatic or even germ-line mutations, in the absence of viral participation. There are several cellular oncogenes (β-catenin, cyclins, etc.), which are considered as oncogenic in human hepatocellular carcinogenesis. There are also suspected viral oncogenes of HBV (mainly *X* gene) and hepatitis C virus (mainly core gene), which may contribute to hepatocellular cancer development. However, despite endless efforts, the true oncogenic role of these viral genes in human hepatocellular carcinogenesis is still a matter of debate (Table 1).

Tab. 1 The status of oncogenes and tumor suppressor genes in hepatocellular cancers.

Gene	*Mutation (%)*	*Other alterations*
p53	28	HBx interaction
β-catenin	13–89[c]	–
APC[a]	0–62[c]	LOH
Axin	5	LOH
E-cadherin	–	LOH, methylation
Cyclin D[b]	11–13	–
Cyclin A[b]	19	HBV integration
RB1	15	LOH, HD
p15^{INK4B}	0	HD
p16^{INK4A}	0–55[a]	HD, methylation
p14^{ARF}	0	–
p21$^{KIP1/CIP1}$	5	–
M6P/IGF2R	0–33	HD
TGFRB2	0–44	–
Smad2	0–2	–
Smad4	0–6	–
IGF-2	–	LOI
c-met	30[d]	–
PTEN	0–5	LOH, HBx interaction
K-ras	0–17	–
N-ras	0–16	Amplification
H-ras	0–10	Methylation
c-myc[b]	0–50	–
N-myc[b]	0	–
BRCA2[a]	5	LOH
hMLH1	–	LOH
hMSH2	–	LOH
DLC-1	–	HD, hyper methylation

[a]Somatic & germ-line mutations.
[b]Amplification.
[c]In HCC and hepatoblastoma.
[d]In childhood hepatocellular carcinoma only.
Notes: LOH: Loss of heterozygosity; HD: homozygous deletion; LOI: Loss of imprinting.
Source: This table was constructed by the modification of a table presented in Ozturk M., Cetin-Atalay R. (2003) Biology of Hepatocellular Cancer, in: Rustgi A. (Ed.) *Gastrointestinal Cancers: Biology, Diagnosis and Therapy*, 2nd edition, Lippincott-Raven Press, New York. Ozturk & Cetin-Atalay, 2003, to include recent data for additional alterations.

Tumor suppressor genes (or anti-oncogenes) are another group of cellular genes whose inactivation contributes to carcinogenesis. These genes, discovered much later than oncogenes, appear to be the most frequently mutated genes in different cancers, including hepatocellular cancers. To date, many tumor suppressor genes including *p53*, *mannose-6-phosphate/insulin-like factor 2 receptor* (*M6P/IGF2R*), *retinoblastoma* (*RB1*), *p16*^{INK4A}, *adenomatosis polyposis coli* (*APC*), *axin* (*axin1*), and *BRCA2* are known to be mutated in hepatocellular cancers (Table 1).

The list of hepatocellular cancer-associated oncogenes and tumor suppressor genes will probably grow over the coming years to include many more genes. There are at least two reasons to explain the high number of altered genes in HCC. Firstly, solid tumors in the adult may need the accumulation of many genetic alterations before they become clinically detectable. Indeed, the well-known "latent period" between the first exposure to an etiological agent (i.e. infection with HBV) and the development of HCC is in favor of such a hypothesis. Secondly, the multiplicity of genetic alterations in HCC may indicate that different etiological factors affect different sets of target genes in hepatocytes. This etiologically defined genetic heterogeneity of HCC results in a phenotypic heterogeneity of these tumors.

Here, we will review major oncogenes and tumor suppressor genes involved in HCC development.

2.3.1 Major Oncogenes

***β-catenin* (*CTNNB1*) gene** This gene encodes a pivotal component of the *Wnt* signaling pathway, which controls cell fate decisions during animal development. Inappropriate deregulation of this pathway leads to cancer in a number of tissues. The stability of β-catenin protein in the cytoplasm is low in unstimulated cells because of proteasome-mediated degradation by a protein complex, which consists of Axin, Adenomatosis polyposis coli (APC), and glycogen synthase kinase 3β (GSK3β). Wnt signaling inhibits the kinase activity of the quaternary complex. As a consequence, β-catenin accumulates in the cytoplasm, translocates to the nucleus, and becomes a transcriptional coactivator of T cell factor (TCF), the ultimate nuclear target of Wnt signaling. The *β-catenin* gene has recently been identified as one of the most frequently mutated genes in both hepatocellular carcinomas and hepatoblastomas. Both missense mutations and in-frame deletions mostly affect only exon 3 of *β-catenin* gene. These mutations affect, as a rule, the N-terminal region of β-catenin protein, which carry serine/threonine phosphorylation motifs involved in "marking" of the protein for ubiquitin-mediated degradation. Reported frequencies of β-catenin mutations in human HCC vary between 13 and 41%. Nuclear accumulation of β-catenin was reported to occur in 35% of HCCs. This could be due to mutations on other genes involved in β-catenin degradation. *Axin* gene mutations were found in 5% of HCCs. We have recently reported that, p53 mutations cause *in vitro* accumulation of β-catenin. It appears that β-catenin mutations in HCCs are as frequent as p53 mutation rates. Unlike p53, β-catenin mutations are also frequent (about 30%) in chemically induced rodent tumors. Childhood hepatoblastomas carry an even higher rate of β-catenin mutations. The reported frequencies vary between 48 and 89%, the overall rate being 58%.

Cyclin genes Cyclin genes comprise a gene family involved in the stepwise activation of cyclin-dependent kinases (CDK) during the cell cycle. Historically, the first cyclin gene known to be mutated in HCC is *CYCLIN A2* (*CCNA2*), which was identified in the integration site of HBV in a single tumor. Although this type of cyclin gene alteration has not been found in other tumors, this observation provided new clues about the implications of cyclin genes in HCC. Another cyclin known to be altered in HCC is the Cyclin D1 (*CCND1*; also called *PRAD1*) gene whose product forms active complexes with CDK4 enzyme during early G1 phase. The amplification of *Cyclin D* gene was observed in 10 to 13% of HCCs. *Cyclin D* gene amplification in HCC appears to be a late event, associated with more aggressive tumors.

2.3.2 Major Tumor Suppressor Genes

***P53* (*TP53*) gene** Many reports now indicate that the *p53* gene, which is located at chromosome 17p, is mutated in nearly one third of HCCs, worldwide. All of the reported p53 mutations in HCC are somatic. Therefore, germ-line mutations of p53 appear not to predispose to HCC. Both the frequency and the type of p53 mutations are different depending on geographical location and suspected etiology of these tumors. An HCC-specific codon 249 mutation (AGG->AGT leading to R249S), suspected to be induced by aflatoxins, was found in most HCCs from geographical areas with high incidence of HCC and with a high risk of exposure to aflatoxins. The codon 249 mutation is present in 36% of tumors from Africa and 32% of tumors from China. These two regions of the world are known for high incidence of HCC, where both HBV and aflatoxins are known to be the main etiological factors. In contrast, the codon 249 mutation is seen in less than 4% of HCCs from Japan, Europe, and North America, where HBV and HCV, but not aflatoxins are the main etiological factors. The overall frequency of codon 249 mutations in the world is 11%. Mutations affecting other codons of the *p53* gene are detected in HCC and their worldwide frequency is 18%. The frequency of all p53 mutations in HCC varies between 15% in Europe and 42% in China with a worldwide frequency of 27%. Thus, *p53* gene is mutated in about a third of HCCs, but only a third of these can be etiologically linked to a high risk of aflatoxin exposure. Therefore, p53 mutations can occur in HCC independent of aflatoxin risk, as well as HBV or HCV infection.

Cyclin-dependent kinase inhibitor genes Two closely located genes at chromosome 9p, namely, Cyclin-dependent Kinase (CDK) inhibitor 2a (*CDKN2A*, also called CDK4 inhibitor p16^{INK4a} or MTS1) and cyclin-dependent kinase inhibitor 2b (*CDKN2B*, also called CDK4 inhibitor p15^{INK4b} or *MTS2*) display genomic alterations in different human cancers. A number of studies demonstrated that the *CDKN2A* gene is strongly implicated in HCC. This gene codes for two alternatively spliced transcripts. One of the transcripts is for p16^{CIP1KIP1} protein, an inhibitor of CDK4 and CDK6. The other transcript encodes for p14^{ARF} protein, mouse homolog of which was shown to regulate wild-type p53 activity. More than 50% of HCCs display *de novo* methylation of *CDKN2A* gene, whereas its mutations are rare. It is known that *de novo* methylation is a mechanism involved in gene silencing. Therefore, HCC cells with methylated *CDKN2A* gene are unable to express the gene, leading to the loss of a CDK inhibitor protein.

***Axin* (*AXIN1*) gene** *Axin* gene product is involved in the negative regulation of the *wnt* signaling. Like APC protein, the axin protein is required for active proteolytic degradation of β-catenin protein. *Axin* gene is mutated in about 5% of HCCs. The loss of the remaining *axin* gene allele was also demonstrated, suggesting that this gene acts as a tumor suppressor gene for HCC.

2.4 Molecular Mechanisms of Hepatocellular Carcinogenesis

The major genetic and epigenetic changes affecting mostly *p53*, *Retinoblastoma*, *p16*^{INK4a}, *β-catenin*, and *axin* genes allow us to link three major signaling pathways to malignant transformation of hepatocytes, namely, p53, retinoblastoma, and Wnt-β-catenin pathways. As shown in Fig. 1, these three pathways are involved in different processes that play key roles in malignant phenotype of not only HCC but also many other tumor types.

The p53 pathway, when it is active, controls three major cellular processes; namely, induction of apoptotic cell death, cell cycle arrest, and induction of senescence. Therefore, it is expected that p53 mutations that are observed at high frequency in HCC cells will lead to the inactivation of p53 pathway, together with a loss of p53-mediated apoptotic, antiproliferative, and antisenescence responses. Similarly, retinoblastoma pathway is involved in the control of cell cycle and permanent cell cycle arrest that is equivalent to senescence. The loss of p16^{INKa} expression in HCC may inactivate retinoblastoma pathway in these cells allowing them to escape from cell cycle control as well as from senescence.

The aberrant activation of Wnt-β-catenin pathway by mutations affecting β-catenin and Axin1 may confer new abilities to these cells such as autonomous proliferation ability, as well as cellular plasticity. Although hepatocellular effects of Wnt-β-catenin pathway activation are poorly

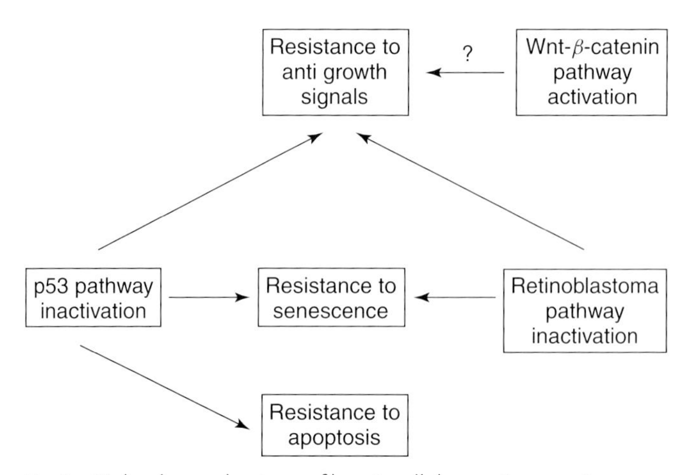

Fig. 1 Molecular mechanisms of hepatocellular carcinogenesis.

known, experimental data from colorectal cancers suggest that this activation may upregulate *Cyclin D1* and *c-myc* gene expression in these cells leading to an ability to proliferate even in the absence of mitogenic stimuli. In addition, activated Wnt-β-catenin pathway may confer HCC cells to gain a stem cell–like phenotype with the ability of unlimited self-renewal and generate progeny of differentiated cells that may count for an increased cell plasticity. With this regard, the origin of HCC is still under debate. According to the conventional theory, HCC results from the "dedifferentiation" of mature hepatocytes into a less differentiated state. A "stem cell origin" has been proposed as an alternative mechanism. According to this theory, HCC originates from "oval" cells displaying stem cell–like properties that are detected in the liver prior to the chemical induction of HCC in rats. Although "oval" cell-like structures have been described in association with different liver diseases, there is no convincing evidence for a direct contribution of liver stem cells to HCC in humans.

The molecular events leading to the development of HCC, whether it derives from a mature hepatocyte or a progenitor cell, are poorly understood. Future studies will probably focus on the role of Wnt-β-catenin pathway activation in HCC cells that will provide experimental evidence about whether such predictions are warranted.

Acknowledgments

The authors' work is supported by grants from TUBITAK, TUBA, and Bilkent University (Turkey).

See also Cancer Chemotherapy, Theoretical Foundations of; Epigenetic Mechanisms in Tumorigenesis; Intracellular Signaling in Cancer; Oncogenes; Oncology, Molecular.

Bibliography

Books and Reviews

Andrisani, O.M., Barnabas, S. (1999) The transcriptional function of the hepatitis B virus X protein and its role in hepatocarcinogenesis, *Int. J. Oncol.* **15**, 373–379.

Buendia, M.A., Hepatitis, B. (1998) viruses and cancerogenesis, *Biomed. Pharmacother.* **52**(1), 34–43.

Cadenas, E., Davies, K.J. (2000) Mitochondrial free radical generation, oxidative stress, and aging, *Free Radical Biol. Med.* **29**, 222–230.

Colombo, M. (1999) Natural history and pathogenesis of hepatitis C virus related hepatocellular carcinoma, *J. Hepatol.* **31**, 25–30.

Darlington, G.J. (1999) Molecular mechanisms of liver development and differentiation, *Curr. Opin. Cell Biol.* **11**, 678–682.

Fausto, N. (2000) Liver regeneration, *J. Hepatol.* **32**, 19–31.

Galle, P.R., Krammer, P.H. (1998) CD95-induced apoptosis in human liver disease, *Semin. Liver Dis.* **18**, 141–151.

Green, D.R., Reed, J.C. (1998) Mitochondria and apoptosis, *Science* **281**, 1309–1312.

Hanahan, D., Weinberg, R.A. (2000) The hallmarks of cancer, *Cell* **100**, 57–70.

Hurlstone, A., Clevers, H. (2002) T-cell factors: turn-ons and turn-offs, *EMBO J.* **21**, 2303–2313.

IARC. (1993) Some naturally occurring substances: food items and constituents, heterocyclic aromatic amines and mycotoxins, *IARC Monogr. Eval. Carcinog. Risks Hum.* **56**, 245–395.

IARC. (1994) Hepatitis viruses, *IARC Monogr. Eval. Carcinog. Risks Hum.* **59**, 39–221.

Llovet, J.M., Burroughs, A., Bruix, J. (2003) Hepatocellular carcinoma, *Lancet* **6**, **362**, 1907–1917.
Matsubara, K. (1991) Chromosomal Changes Associated with Hepatitis B virus DNA Integration and Hepatocarcinogenesis, in: McLacchlan, A. (Ed.) *Molecular Biology of Hepatitis B Viruses*, CRC Press, Boca Raton, FL, pp. 245–261.
Michalopoulos, G.K., DeFrances, M.C. (1997) Liver regeneration, *Science* **276**, 60–66.
Ozturk, M. (1999) Genetic aspects of hepatocellular carcinogenesis, *Semin. Liver Dis.* **19**, 235–242.
Ozturk, M., Cetin-Atalay, R. (2003) Biology of Hepatocellular Cancer, in: Rustgi, A. (Ed.) *Gastrointestinal Cancers: Biology, Diagnosis and Therapy*, 2nd edition, Lippincott-Raven Press, New York, pp. 575–596.
Sell, S., Dunsford, H.A. (1989) Evidence for the stem cell origin of hepatocellular carcinoma and cholangiocarcinoma, *Am. J. Pathol.* **134**, 1347–1363.
Shafritz, D.A., Dabeva, M.D. (2002) Liver stem cells and model systems for liver repopulation, *J. Hepatol.* **36**, 552–564.
Sherr, C.J., McCormick, F. (2002) The RB and p53 pathways in cancer, *Cancer Cell* **2**, 103–112.

Primary Literature

Bressac, B., Galvin, K.M., Liang, T.J., Isselbacher, K.J., Wands, J.R., Ozturk, M. (1990) Abnormal structure and expression of p53 gene in human hepatocellular carcinoma, *Proc. Natl. Acad. Sci. U.S.A.* **87**, 1973–1997.
Bressac, B., Kew, M., Wands, J., et al. (1991) Selective G to T mutations of p53 gene in hepatocellular carcinoma from southern Africa, *Nature* **350**, 429–431.
Cagatay, T., Ozturk, M. (2002) p53 mutation as a source of aberrant β-catenin accumulation in cancer cells, *Oncogene* **21**, 7971–7980.
Chaubert, P., Gayer, R., Zimmermann, A., Fontolliet, C., Stamm, B., Bosman, F., Shaw, P. (1997) Germ-line mutations of the p16INK4(MTS1) gene occur in a subset of patients with hepatocellular carcinoma, *Hepatology* **25**, 1376–1381.
de La Coste, A., Romagnolo, B., Billuart, P., Renard, C.A., Buendia, M.A., Subrane, O., Fabre, M., Chelly, J., Belford, C., Kahn, A., Perret, C. (1998) Somatic mutations of the beta-catenin gene are frequent in mouse and humanhepatocellular carcinomas, *Proc. Natl. Acad. Sci. U.S.A.* **95**, 8847–8851.
Hsu, I.C., Metcalf, R.A., Sun, T., Welsh, J.A., Wang, N.J., Harris, C.C. (1991) Mutational hotspot in the p53 gene in human hepatocellular carcinomas, *Nature* **350**, 427–428.
Hui, A.M., Sakamoto, M., Kanai, Y., Ino, Y., Gotoh, M., Yokota, J., Hirohashi, S. (1996) Inactivation of p16INK4 in hepatocellular carcinoma, *Hepatology* **24**, 575–579.
Irmak, M.B., Ince, G., Ozturk, M., Cetin-Atalay, R. (2003) Acquired tolerance of hepatocellular carcinoma cells to selenium deficiency: a selective survival mechanism? *Cancer Res.* **63**, 6707–6715.
Jung, J., Zheng, M., Goldfarb, M., Zaret, K.S. (1999) Initiation of mammalian liver development from endoderm by fibroblast growth factors, *Science* **284**, 1998–2003.
Miyoshi, Y., Iwao, K., Nagasawa, Y., Aihara, T., Sasaki, Y., Imaoka, S., Murata, M., Shimano, T., Nakamura, Y. (1998) Activation of the beta-catenin gene in primary hepatocellular carcinomas by somatic alterations involving exon 3, *Cancer Res.* **58**, 2524–2547.
Nishida, N., Fukuda, Y., Komeda, T., Kita, R., Sando, T., Furukawa, M., Amenomori, M., Shibagaki, I., Nakao, K., Ikenaga, M. (1994) Amplification and overexpression of the cyclin D1 gene in aggressive human hepatocellular carcinoma, *Cancer Res.* **54**, 3107–3110.
Ozturk, M. (1991) p53 mutation in hepatocellular carcinoma after aflatoxin exposure, *Lancet* **338**, 1356–1359.
Parkin, D.M., Bray, F., Ferlay, J., Pisani, P. (2001) Estimating the world cancer burden: GLOBOCAN 2000, *Int. J. Cancer* **94**, 153–56.
Satoh, S., Daigo, Y., Furukawa, Y., Kato, T., Miwa, N., Nishiwaki, T., Kawasoe, T., Ishiguro, H., Fujita, M., Tokino, T., Sasaki, Y., Imuoka, S., Murata, M., Shimano, T., Yamaoka, Y., Nakamura, Y. (2000) AXIN1 mutations in hepatocellular carcinomas, and growth suppression in cancer cells by virus-mediated transfer of AXIN1, *Nat. Genet.* **24**, 245–250.
Shimada, M., Hasegawa, H., Gion, T., Utsunomiya, T., Shitabe, K., Takenaka, K., Otsuka, T., Maehara, Y., Sugimachi, K. (2000)

The role of telomerase activity in hepatocellular carcinoma, *Am. J. Gastroenterol.* **95**, 748–752.

Ueda, H., Ullrich, S.J., Gangemi, J.D., Kappel, C.A., Ngo, L., Feitelson, M.A., Jay, G. (1995) Functional inactivation but not structural mutation of p53 causes liver cancer, *Nat. Genet.* **9**, 41–47.

Unsal, H., Yakicier, C., Marcais, C., Kew, M., Volkmann, M., Zentgraf, H., Isselbacher, K.J., Ozturk, M. (1994) Genetic heterogeneity of hepatocellular carcinoma, *Proc. Natl. Acad. Sci. U.S.A.* **91**, 822–826.

Wang, J., Chenivesse, X., Henglein, B., Brechot, C. (1990) Hepatitis B virus integration in a cyclin A gene in a hepatocellular carcinoma, *Nature* **343**, 555–557.

Yu, S.Y., Zhu, Y.J., Li, W.G. (1997) Protective role of selenium against hepatitis B virus and primary liver cancer in Qidong, *Biol. Trace Elem. Res.* **56**, 117–124.

Livestock Genomes (Bovine Genome)

John Lewis Williams
Roslin Institute, Roslin, Midlothian, Scotland, UK

Encyclopedia of Molecular Cell Biology and Molecular Medicine, 2nd Edition. Volume 7
Edited by Robert A. Meyers.

ISBN: 3-527-30549-1

Keywords

Genome
The complete nuclear DNA complement of an organism.

Genotype
The genetic make up of an individual, either at a specific locus, or considered across the whole genome.

Marker-assisted Selection (MAS)
Using DNA markers to aid the selection of animals of given genotype to achieve particular phenotypes.

Microsatellite Loci
A position in the genome where a short sequence (typically 2 or 3 bases long) is repeated several time in tandem.

Phenotype
The physical appearance or a measurable trait.

Quantitative trait loci (QTL)
A genetic locus that accounts for a proportion of the variation in a continuously distributed phenotype.

Selection
Choosing individuals with particular characteristics in order to improve the population.

Trait
A particular variable characteristic of individuals in a population.

Cattle have adapted to survive in a wide variety of environments, which has given rise to extensive diversity at the phenotypic and genetic level. Today, this diversity provides the opportunity for selective breeding and improvement of cattle for commercially desirable traits. Up to now, selection has been based on phenotype, focusing on the traits that are most easily measured. This approach has been successful, with spectacular improvement in some traits, although there have also been associated penalties for other traits. However, intensive selection based on a limited number of phenotypes could reduce the diversity present in the population, which may have important implications for the success of future breeding objectives. In addition, selective improvement has been focused on a limited number of breeds, which are now used internationally, with the consequence that many of the genetically different breeds of cattle are being lost worldwide. To make the selection process more efficient, it is important to gain knowledge of the genes controlling particular traits and to understand the way that variation within these genes affects the traits. The first step toward identifying the trait genes has been to develop genetic and physical maps of the bovine genome. This has been achieved at an international level using several genome-mapping methods. The ultimate genome map, the bovine genome sequence, is now being determined. This genomic information is being used in specific cattle populations to identify, first the genetic location of genes controlling particular traits, then, starting from the genomic locations, to identify the genes themselves. Knowledge of the genes controlling complex traits, such as feed-conversion efficiency, health, fertility, and product quality, would allow these traits to be included in breeding objectives, with the potential for improved commercial viability, while safeguarding welfare and genetic diversity.

1 Introduction

There are more than 750 genetically different breeds of cattle worldwide, in addition to innumerable crossbred and undefined populations. This wealth of genetic diversity has enabled cattle to be raised in most environments inhabited by man; there are cattle populations adapted to survive in extremes of temperature, in arid areas and in the face of disease and parasite challenge. This environmental adaptation has, for the most part, been achieved by natural selection, insofar as those individuals that were best suited to their environment thrived and were therefore used to breed subsequent generations. The uses to which cattle have been put are equally as diverse as the environments in which they are kept. Cattle provide meat and milk for feed, are used for traction, and as a source of building materials, clothing, tallow, and glue, with numerous other uses found for the by-products of cattle production. The wealth of genetic diversity in cattle populations has provided the opportunity for selective breeding and improvement of cattle for particular uses. Today, selection has moved beyond the need for adaptation to environmental conditions to selecting on commercially desirable traits, at least in the developed world. Traditionally, breeding was carried out at the local level and those cattle that were selected for breeding were the ones that best fulfilled the local uses. More recently, cattle and semen have been moved around the world and selection has been carried out to achieve peak production under ideal, rather than local, conditions of management.

In modern breeding, the selection criteria adopted differ depending on the goals; for example, whether the end market is beef or dairy production. However, in general, the traits used in the selection programs are those that are most readily accessible to measurement in commercial systems. Although, for example, feed efficiency is a very important factor in profitability for cattle production, this trait is often not taken into account because the trait is not readily recorded. Instead, selection is based on easily measured traits, in the case of dairy production on milk yield, protein, and fat content, and in the case of beef animals on growth rates and conformation. Over the past 40 years methods for breeding selection have improved, so that today selection of breeding sires is achieved by testing their progeny and applying sophisticated statistical methods, usually based on a "best linear unbiased prediction" (BLUP) model to calculate their relative genetic value for the desired traits. This approach is of greatest value in traits that are restricted by sex and age, for example, the prediction of the genetic merit of dairy-breed sires for milk traits. Application of this approach has given impressive results, for example, milk production from Holstein cows has doubled over the past 40 or so years. The progeny-testing approach has also resulted in significant improvements in beef-associated traits in the specialized beef breeds. Nevertheless, the gains achieved in a limited number of production traits come at the price of losses in other traits. The spectacular increases in milk yields achieved in the Holstein breed have been accompanied by decreased fertility and increased lameness. These losses are in part the result of an increased level of inbreeding, which is the inevitable penalty of a progeny-testing approach, as only a limited number of individuals can be tested in each generation. Along with losses in traits that are not under selection, the major problem associated with this approach, and associated with the consequent high level of inbreeding, is the potential concentration of deleterious recessive defects. As the level of inbreeding, defined by the inheritance of the same ancestral genes from both parents increases, the likelihood that an individual will inherit the same deleterious mutation from both parents also increases. This has been manifest in the elite Holstein dairy population by the appearance of genetic diseases such as DUMPS (deficiency of uridine mono-phosphate synthase), BLAD (bovine leukocyte adhesion deficiency), and most recently a bulldog calf syndrome.

In order to improve the process of genetic selection, it is important to develop a better knowledge of the genetic control of the various traits. The long-term goal is knowledge of the genes involved in particular traits, and an understanding of the variations within these genes, and how their products interact. This will ultimately allow the phenotype to be predicted from the genotype. In the shorter term, genetic maps and physical maps of the bovine genome are allowing markers for genes involved in particular traits and diseases to be localized, and in some cases, the genes themselves to be identified. The progress in unraveling the organization of the bovine genome is discussed below.

2 Domestication of Cattle

2.1 Origins of Domesticated Cattle

Analysis of both mitochondrial and nuclear DNA allows the relationships between cattle populations and their origins to be investigated. Domesticated cattle can be classified into two groups on the basis of phenotype: *Bos indicus*, or Zebu cattle, are characterized by a hump at the shoulder, and *Bos taurus*, or Taurine cattle, which are humpless cattle. These two groups are sometimes referred to as subspecies, but they freely interbreed. *Bos indicus* cattle are better adapted to arid regions and crossbreeding between the two subspecies is often used to produce cattle with improved productivity in unfavorable dry regions. Comparison of mitochondrial "D-loop" DNA sequence confirms that the phenotypic distinction is reflected at the molecular level. The D-loop sequences fall into two groups, or clades, that coincide broadly with their geographic distribution. The exceptions are the African Zebu cattle, which have humps and phenotypically appear to be *B. indicus*, but have mitochondrial DNA that is typically taurine.

The distinction between *B. indicus* and *B. taurus* is also found by examining sequence diversity of nuclear DNA. Comparison of nuclear DNA is most commonly achieved by examining allelic variation at microsatellite loci (see Sect. 3.3). Construction of phylogenetic relationships between cattle using microsatellite data also produces two major clusters, which correspond with the indicus and taurine cattle. In the case of nuclear DNA the African Zebu cattle cluster with the taurine breeds, suggesting that they are hybrids. Examination of sequence variation on the Y chromosome and comparison with the maternally derived mitochondrial DNA allows the migration and dynamics of breeding to be examined, and confirms the introgression of male indicus genes into the original African taurine breeds.

The history of cattle domestication can be traced using both the mitochondrial and microsatellite data, and suggests that the indicus and taurine lineages diverged at least 22 000 years ago. The data also suggest two sites for domestication of cattle, both in the Middle East, from where domesticated cattle radiated out: the eastern domestication spreading to India, founding the indicus breeds, which were subsequently imported into east Africa, and the western site of domestication which spread via north Africa into Europe, founding the taurine breeds as shown in Fig. 1.

2.2
Cattle Breeds

Since domestication, cattle have been raised in a large variety of environmental conditions. It is therefore not surprising that extensive diversity began to appear both at the phenotypic and at the genetic level. Traditionally, breeding was carried out at a local level, often using a limited number of shared bulls. Originally cattle were selected primarily for survival, and fulfilled a range of production requirements, meat, milk, traction and so on, with equal emphasis. For the most part this is still the case, with the 70% of cattle raised in the developing world. The selection of individuals with particular characteristics suited to local environments, needs, and preferences, led to the emergence of distinctive groups of cattle with characteristic phenotypes. This phenotypic distinctiveness of groups of cattle was pursued by further selection, and later these groups were formalized into breeds, which were characterized by fixing particular features, typically colors or markings. With the establishment of breeds and the formation of societies to promote them, particularly in Europe and North America, there came the tendency for selection aimed at breeding cattle to suit particular production niches, such as breeds specializing in dairy or meat production. Some breeds that started

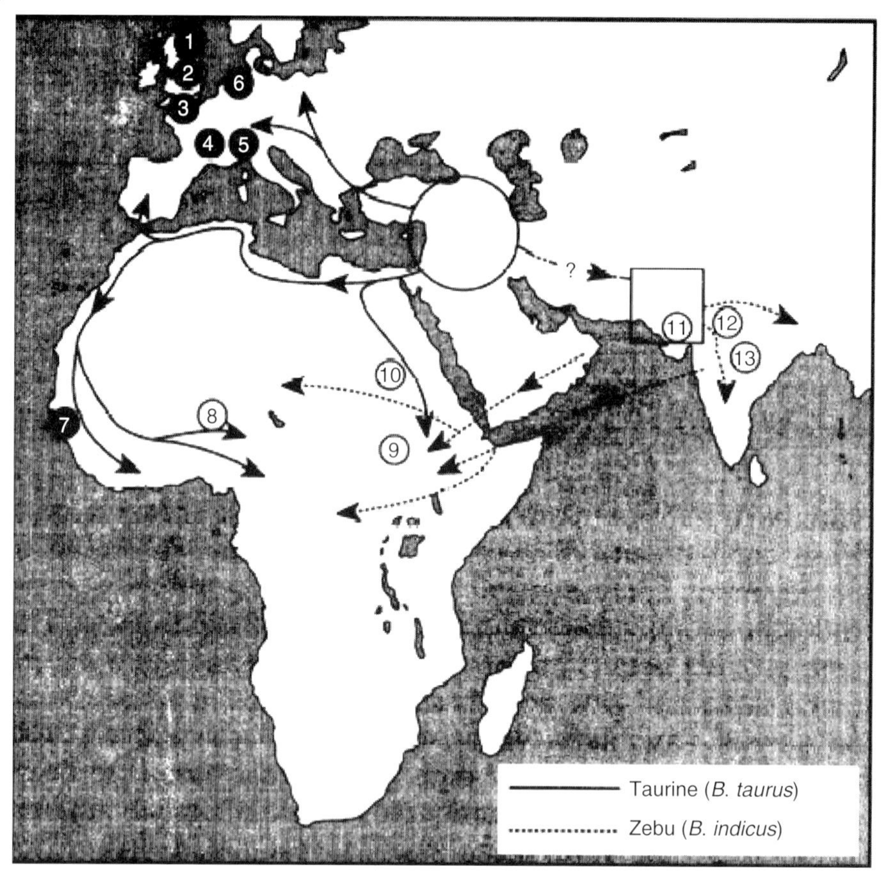

Fig. 1 Postulated migratory routes of cattle across western Asia, Africa, and Europe. Geographical origins of breed determined from mtDNA analysis of the following breeds (1) Aberdeen Angus, (2) Hereford, (3) Jersey, (4) Charolais, (5) Simmental (6) Friesian (7) N'Dama, (8) White Fulani, (9) Kenana, (10) Butana, (11) Tharparkar, (12) Sahiwal and 13 Hariana. *Bos taurus* breeds are illustrated by black symbols and *Bos indicus* breeds by open symbols (from Loftus, R.T., MacHugh, D.E., Bradley, D.G., Sharp, P.M., Cunningham, E.P. (1994) Evidence for two independent domestications of cattle, *Proc. Natl. Acad. Sci. U.S.A.* **91**, 2757–2761). Figure 1 (from Bradley et al. showing domestication origins).

with a dual function have more recently undergone strong selection for particular traits. The Friesian, which was used initially for both dairy and meat production, has become the leading dairy breed in the guise of the Holstein, in terms of quantity of milk produced. Modern dairy breeds produce far more milk than would be necessary to feed a calf, and often have trouble maintaining body weight during lactation. Conversely, selection of the meat-producing breeds has resulted in animals with little milk, in some cases insufficient milk for their calves, but with impressive muscular development, and in extreme cases are so muscle-bound that they have problems with locomotion. Thus, artificial selection has produced individuals that would not survive under natural selection.

The phenotypic distinctiveness of cattle breeds is also reflected at the genetic level. From the early 1960s through to the present day, biochemical markers, such as enzyme polymorphisms and later blood groups on the surface of red cells, detected using allo-antisera, has been used extensively by breed societies to verify the pedigrees of individuals and maintain the purity of their breeds. This biochemical information can also be used to track the genetic changes of breeds over this 40-year period, which coincides with the time during which the intensity of selection for increased productivity has also increased dramatically. Analysis of the blood-group data has shown that, in general, between-breed diversity is much larger than within-breed diversity. So individuals from the same breed cluster together. The phenotypic appearance of cattle breeds has changed in response to the intense selection that has been applied over the past 60 years. In addition, the livestock market has become international with semen, and to a lesser extent embryos, from many breeds being shipped all over the world. This has provided the opportunity for genetic divergence within the breeds as different selection criteria are applied in different regions; nevertheless, the genetic composition of breeds has remained relatively constant. Comparison of blood typing data over a 40-year period shows that although there has been some genetic drift, individuals remain more like their ancestors from the same breed, than individuals from other closely related or phenotypically similar breeds (see Fig. 2). Examination

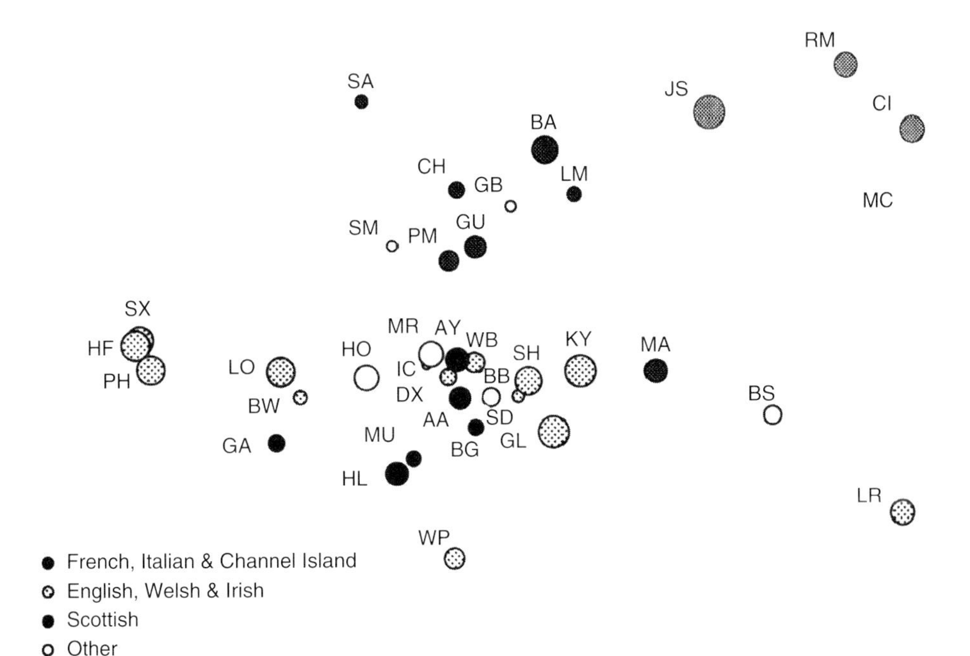

Fig. 2 Relationships between breeds based on a principal components analysis of differences between frequencies of red cell antigens (from Blott, S.C., Williams, J.L., Haley, C.S. (1998) Genetic relationships among European cattle breeds, *Anim. Genet.* **29**, 273–282). Figure 1 (from Bradley et al. showing domestication origins).

of individuals from different geographical locations reveals variations between the populations in different areas, most likely arising from a founder effect; however, the variation among individuals from a breed still remains smaller than the variation between breeds.

Other studies have used microsatellite loci to examine genetic diversity of cattle at the DNA level. A panel of 30 microstatellite loci has been selected, through a European Union network and have been adopted by the Food and Agriculture Organization of the United Nations for studies of diversity in cattle. This panel of markers is recommended for use in all studies of genetic diversity in cattle (see Table 1). The use of a common panel of markers should allow data to be integrated and compared between studies; this will allow changes in the genetic composition of breeds to be followed over time and genetic erosion to be monitored. The results from the microsatellite-based studies reflect the

Tab. 1 List of markers recommended for the study of genetic diversity in cattle.

Marker number	*Marker*	*Chromosome*	*Primer sequences (5′-3′)*
1	INRA063 (D18S5)	18	ATTTGCACAAGCTAAATCTAACC AAACCACAGAAATGCTTGGAAG
2	INRA005[2] (D12S4)	12	CAATCTGCATGAAGTATAAATAT CTTCAGGCATACCCTACACC
3	ETH225 (D9S1)	9	GATCACCTTGCCACTATTTCCT ACATGACAGCCAGCTGCTACT
4	ILSTS005 (D10S25)	10	GGAAGCAATGAAATCTATAGCC TGTTCTGTGAGTTTGTAAGC
5	HEL5[1] (D21S15)	21	GCAGGATCACTTGTTAGGGA AGACGTTAGTGTACATTAAC
6	HEL1 (D15S10)	15	CAACAGCTATTTAACAAGGA AGGCTACAGTCCATGGGATT
7	INRA035 (D16S11)	16	ATCCTTTGCAGCCTCCACATTG TTGTGCTTTATGACACTATCCG
8	ETH152 (D5S1)	5	TACTCGTAGGGCAGGCTGCCTG GAGACCTCAGGGTTGGTGATCAG
9	INRA023 (D3S10)	3	GAGTAGAGCTACAAGATAAACTTC TAACTACAGGGTGTTAGATGAACTC
10	ETH10[4] (D5S3)	5	GTTCAGGACTGGCCCTGCTAACA CCTCCAGCCCACTTTCTCTTCTC
11	HEL9 (D8S4)	8	CCCATTCAGTCTTCAGAGGT CACATCCATGTTCTCACCAC
12	CSSM66[3] (D14S31)	14	ACACAAATCCTTTCTGCCAGCTGA AATTTAATGCACTGAGGAGCTTGG
13	INRA032[5] (D11S9)	11	AAACTGTATTCTCTAATAGCTAC GCAAGACATATCTCCATTCCTTT
14	ETH3 (D19S2)	19	GAACCTGCCTCTCCTGCATTGG ACTCTGCCTGTGGCCAAGTAGG
15	BM2113 (D2S26)	2	GCTGCCTTCTACCAAATACCC CTTCCTGAGAGAAGCAACACC

Tab. 1 *(continued)*

Marker number	*Marker*	*Chromosome*	*Primer sequences (5′-3′)*
16	BM1824 (D1S34)	1	GAGCAAGGTGTTTTTCCAATC CATTCTCCAACTGCTTCCTTG
17	HEL13[5] (D11S15)	11	TAAGGACTTGAGATAAGGAG CCATCTACCTCCATCTTAAC
18	INRA037 (D10S12)	11	GATCCTGCTTATATTTAACCAC AAAATTCCATGGAGAGAGAAAC
19	BM1818 (D23S21)	23	AGCTGGGAATATAACCAAAGG AGTGCTTTCAAGGTCCATGC
20	ILSTS006 (D7S8)	7	TGTCTGTATTTCTGCTGTGG ACACGGAAGCGATCTAAACG
21	MM12 (D9S20)	9	CAAGACAGGTGTTTCAATCT ATCGACTCTGGGGATGATGT
22	CSRM60 (D10S5)	10	AAGATGTGATCCAAGAGAGAGGCA AGGACCAGATCGTGAAAGGCATAG
23	ETH185 (D17S1)	17	TGCATGGACAGAGCAGCCTGGC GCACCCCAACGAAAGCTCCCAG
24	HAUT24 (D22S26)	22	CTCTCTGCCTTTGTCCCTGT AATACACTTTAGGAGAAAAATA
25	HAUT27 (D26S21)	26	TTTTATGTTCATTTTTTGACTGG AACTGCTGAAATCTCCATCTTA
26	TGLA227 (D18S1)	18	CGAATTCCAAATCTGTTAATTTGCT ACAGACAGAAACTCAATGAAAGCA
27	TGLA126 (D20S1)	20	CTAATTTAGAATGAGAGAGGCTTCT TTGGTCTCTATTCTCTGAATATTCC
28	TGLA122 (D21S6)	21	CCCTCCTCCAGGTAAATCAGC AATCACATGGCAAATAAGTACATAC
29	TGLA53 (D16S3)	16	GCTTTCAGAAATAGTTTGCATTCA ATCTTCACATGATATTACAGCAGA
30	SPS115 (D15)	15	AAAGTGACACAACAGCTTCTCCAG AACGAGTGTCCTAGTTTGGCTGTG

Note: These markers were initially selected from those used across several projects and subsequently were adopted as the standard panel for work in cattle. More information can be obtained from http://www.projects.roslin.ac.uk/cdiv/markers.html.

same results as studies with biochemical markers, that is, for anonymous markers that are not subjected to selection: (1) breeds of cattle can be distinguished at the genetic level, and (2) within-breed variation is lower than between-breed variation. DNA markers are now replacing biochemical polymorphisms for pedigree verification, so a body of data is building up that could be used in the future to explore variations in the genetic diversity of cattle breeds.

The use of DNA markers has also been mooted for identifying the breed of an animal, or more appropriately in commercial terms the breed of origin of a meat sample. Unfortunately, although the analysis of DNA polymorphisms allows animals to be clustered by breed, there is overlap between the clusters. Therefore, although it is statistically possible to assign individuals to breeds, there is likely to be considerable error in the result, at least with the number of markers currently used.

2.3 Conservation of Diversity

The introduction of artificial insemination (AI) in the more developed countries during the 1950s, coupled with improvement in management, has resulted in rapid progress in the improvement of cattle for simple production traits. This increasing use of AI meant that particular bulls with desirable characteristics became widely used in preference to local bulls. Consequently, where the economic environment supports high input agriculture, there has been a dramatic increase in milk yield and meat produced from the improved stock. The unfortunate consequence has been the reduction of genetic diversity, both within the selected breeds, as superior individuals have been used preferentially as breeding stock, and also through the replacement of traditional, less productive, breeds. In less developed and environmentally less favored areas, the use of the improved breeds presents a great cause for concern. Local breeds are usually adapted to survive in their local environments, for example, with increased tolerance of extremes in temperature or with the ability to survive and remain productive in the face of particular disease or parasite challenge. The inappropriate and/or unmanaged attempts to introduce improved breeds into some areas have met with disastrous consequences. In 1993, 112 of the 783 cattle breeds worldwide were at risk of extinction. One of the greatest risks is the replacement of local stock that are adapted for survival in the face of disease challenge with disease sensitive stock, in areas where standards and resources to provide extensive veterinary care are not available. It is important that genetic diversity is maintained, as this is the source of variation that will allow further selection. The first essential task is to record and monitor diversity across cattle populations. Screening genetic diversity using a limited number of DNA markers, such as the FAO panel discussed above, provides a crude measure of genetic diversity; however, this approach does not identify genes that are involved in specific adaptation or that control unique phenotypes. Thus, to preserve genetic diversity, much more extensive studies that consider the whole genome in more detail and that take into account variations in phenotypes are necessary. To this end, genome-mapping programs have the goal of identifying the genetic control of phenotypic variation.

3 Developing Maps of the Bovine Genome

The bovine genome is contained within 29 autosomes and the X and Y sex chromosomes, and consists of about 3 000 Mb of DNA. The autosomes are all acrocentric, that is, the centromere is at the end, while the sex chromosomes are sub-metacentric. Like most other mammalian genomes, the bovine genome is composed of less than 10% sequence that codes for proteins and about 50% of repetitive sequences. One major repetitive component is the bovine A-2 repeat, which has a core 115-bp element and comprises around 1.6% of the bovine genome. The second most common repeat element is the bovine SINE or bovine – B repeat, which is characterized by a 560-bp band seen by electrophoresis following digestion with the restriction enzyme *Pst*1, the whole of this element is 3.1 kbp long and represents about 0.5% of the genome (see Fig. 3).

There has been considerable effort internationally to produce maps of the bovine genome driven for the most part by the objective of finding those genes that are involved in commercially important traits. Genomic maps can be compared between species to study the evolution of genomes and to reveal the organization and architecture of the DNA. This comparison will contribute to an understanding of the way the genome functions and the identification of those features that are important and that should be subjected to further study. There are several types of genomic map, which are either physical, or genetic. Physical maps can either be based on the localization of

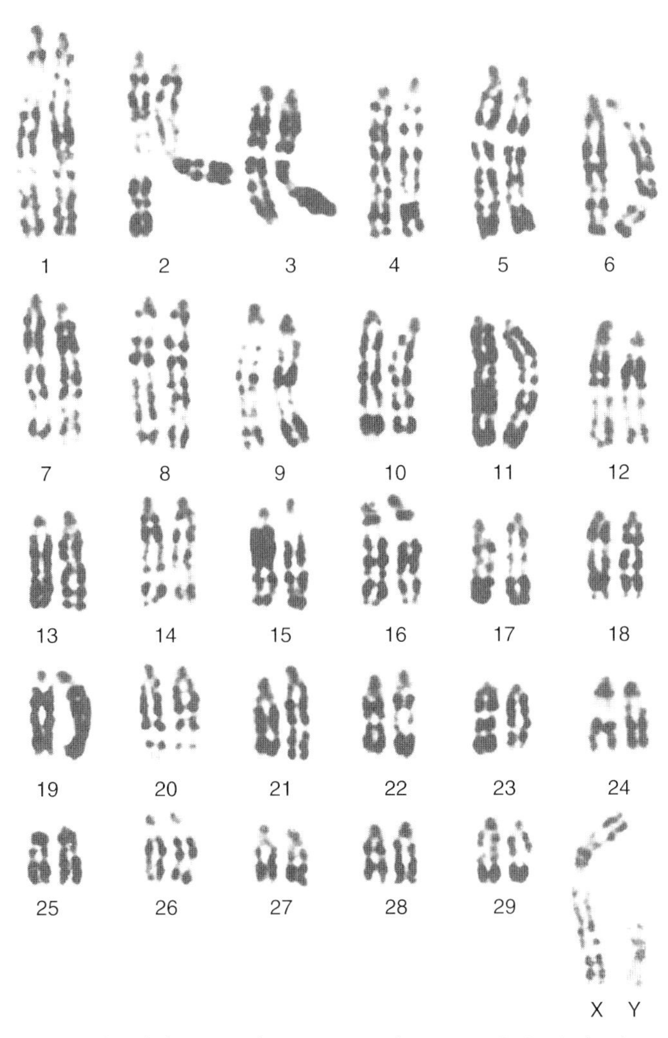

Fig. 3 R banded Bovine Chromosomes (from Fries, R., Ruvinsky, A. (1999) *The Genetics of Cattle*, CABI Publishing, Oxford, UK).

markers on the chromosomes by *in situ* hybridization, or they can be clone-based maps where the genome is fragmented and cloned and the clones physically reassembled so that the location and orientation of the pieces in relation to each other is known. A third type of physical map uses somatic cell hybrids, where fragments of the target genome are transferred into cells of another species; statistical methods are then used to calculate the proximity of markers from their coretention in the hybrid cells. In contrast, linkage maps are assembled on the basis of relationship between markers deduced from patterns of recombination in families. The ordering of loci should correspond between physical and linkage maps, but the distances between loci will vary due to the differences in the way the relationship between them is worked out. The ultimate genomic map of a species will be its DNA sequence.

3.1 Somatic Cell Hybrids

The first maps of the bovine genome were created using somatic cell hybrids made by fusing bovine cells with immortalized rodent cells (usually mouse or hamster). The immortalized cells are deficient in a selectable marker, usually thymidine kinase; thus, to survive in selection medium they need to retain the bovine Tk gene. In practice, the rodent cells retain large fragments of bovine chromosomes. Using a panel of somatic cell hybrids it is possible, using statistical methods, to identify which markers occur on the same chromosome, and to a certain extent the order of markers. In 1993, the first bovine genome map was published with 30 synteny groups, which were identified using a somatic cell hybrid panel. With the exception of the X chromosome, the synteny groups were not assigned to chromosomes and designated U1 to U29.

3.2 *In situ* Hybridization

In order to assign markers to bovine chromosomes, the method of *in situ* hybridization was initially used. Originally, radioactively labeled gene probes were hybridized to metaphase chromosomes spread out on microscope slides. The slides were then coated with photographic emulsion and following development the distribution of silver grains was analyzed. The distribution of silver grains was then compared with the banding pattern of the stained chromosomes using statistical analysis to assign the physical location of gene to cytogenetic bands identified on the standard karyotype. All the unassigned synteny groups identified by somatic cell hybrids were assigned to chromosomes using radioactive *in situ* hybridization. However, this approach is not very precise, as the silver grains are widely dispersed and a large number of hybridizations were required to assign a probe to a particular band with confidence, and even then the localization was at low resolution. A refinement of this methodology was to use fluorescently labeled probes, in a method called *fluorescent in situ hybridization* (FISH). With FISH, the hybridized probe is visualized directly; this increases the precision of the method and as a result reduces the number of chromosome hybridizations necessary to be confident of the localization. Currently there are around 500 genes localized on chromosomes by *in situ* mapping. In addition to mapping genes, it is possible to localize the anonymous markers used in linkage mapping to chromosomes using

FISH by labeling the cosmid clones from which the anonymous marker was derived, and using it as the probe. This approach allowed the physical bovine map and the linkage maps to be aligned.

The FISH approach allows the chromosomal band harboring the target sequence of the probe to be identified, and has a resolution at the DNA level of 2 Mbp or more. A higher resolution can be obtained if less condensed chromosomes, or indeed cloned DNA strands are used in the Fiber-FISH technique, which allows a localization of probes at a resolution of about a kilobase. The physical location of a probe to a chromosomal band obtained using the fiber-FISH approach also provides the information required for cloning of targeted regions by chromosomal microdissection.

Information obtained from somatic cell panels and *in situ* hybridization of genes demonstrated that in many cases the genes that came from the same chromosomal region in man or mouse also mapped to a single chromosome in cattle; indeed, genes from one region in one species was generally found close together on a chromosome of another, as shown in Fig. 4. This suggested the organization of the genomes of divergent species has been conserved during evolution. The extent to which this organization is conserved has been elegantly demonstrated using chromosomal probes derived from one species to "paint" the chromosomes of another using a ZOO-FISH approach. ZOO-FISH suggested that there are about 50 conserved segments of genome between the human and bovine genomes.

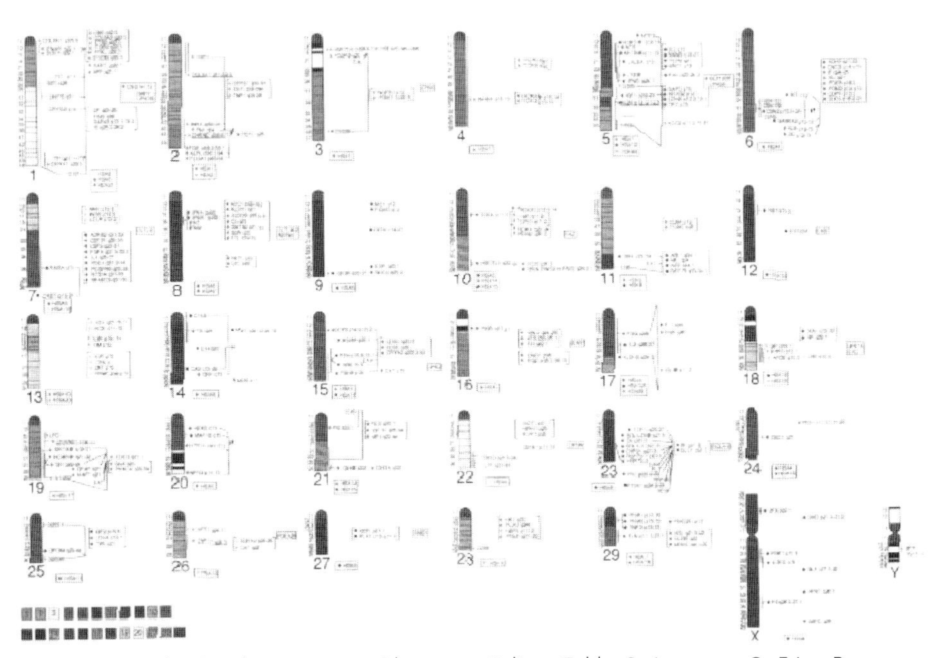

Fig. 4 Painting bovine chromosomes with human chromosome paints shows that large regions of chromosomes are conserved across species, at least at the gross level. (From Solinas-Toldo, S., Lengauer, C., Fries, R. (1995) Comparative Genome map of human and cattle, *Genomics* **27**, 489–596.)

3.3 Linkage Mapping

Linkage mapping relies on recombination occurring less frequently between markers that are close together than those that are more distant on chromosomes. Therefore, by analyzing the inheritance of markers in families it is possible to build a genetic map based on the frequency of recombination, or linkage, between the markers. To build a genetic map, there are two requirements: families of animals in which the inheritance of chromosomes can be tracked, and markers that are highly variable (polymorphic) so that in the majority of cases the parental origin of the marker can be determined. By combining the information on the inheritance of a large number of markers within families, it is possible to construct a genetic map by analyzing the recombination frequency between pairs of markers.

Creating genetic maps for cattle has two immediate problems: firstly, cows generally have singleton calves, and secondly, the generation for period for cattle is relatively long; therefore, cattle families are usually small. Producing a reasonable number of full siblings by natural breeding is both a long-term and expensive exercise. If conventional cattle families were used, in general, two parents and their calf would have to be typed for each marker. The situation in commercial herds is generally better as artificial insemination is used for breeding cattle and hence large half-sib sire families are readily available. Half-sib commercial populations have been extensively used for mapping the loci involved in the control of commercial traits, but in practice the contribution of the dam to the mapping information is limited, as she will only contribute one or two calves. For the construction of the genetic maps, large full-sib families are much more efficient. Fortunately, multiple ovulation embryo transfer (MOET) technology has been developed for cattle, which means a large number of progeny can be produced from two parents. MOET has allowed relatively large full-sib populations to be assembled as "reference populations" for genetic mapping. The advantage of full-sib families is that the effort required to type the markers is significantly reduced, while the resolution of the maps produced is increased. To obtain the highest density of markers on the genetic map, it is of benefit for all workers in the field to focus on a limited number of mapping populations. The international efforts to create bovine linkage maps focused on two reference populations: one, an international collaboration led by Australia with full-sib families provided by Australia, Kenya, Europe, and the United States to create the International Bovine Reference Pedigree (IBRP), and the second, focused on a reference population produced by the USDA Meat Animal Research Center in Nebraska. The distinction has been drawn between *reference* families, which have been specifically created and used to make genetic maps, and *resource* families, which have been assembled to localize the genes involved in particular traits.

The second requirement for constructing a genetic map is the markers. Whereas *in situ* hybridization and somatic cell hybrid mapping simply requires probes for the markers that can be hybridized to the DNA, the resolution of the linkage map is governed by the number of informative meiosis, that is, the more frequently the parental origin of markers can be determined the higher the resolution. Thus, it is important that the markers used for linkage mapping are highly polymorphic in the reference families. In general, genes

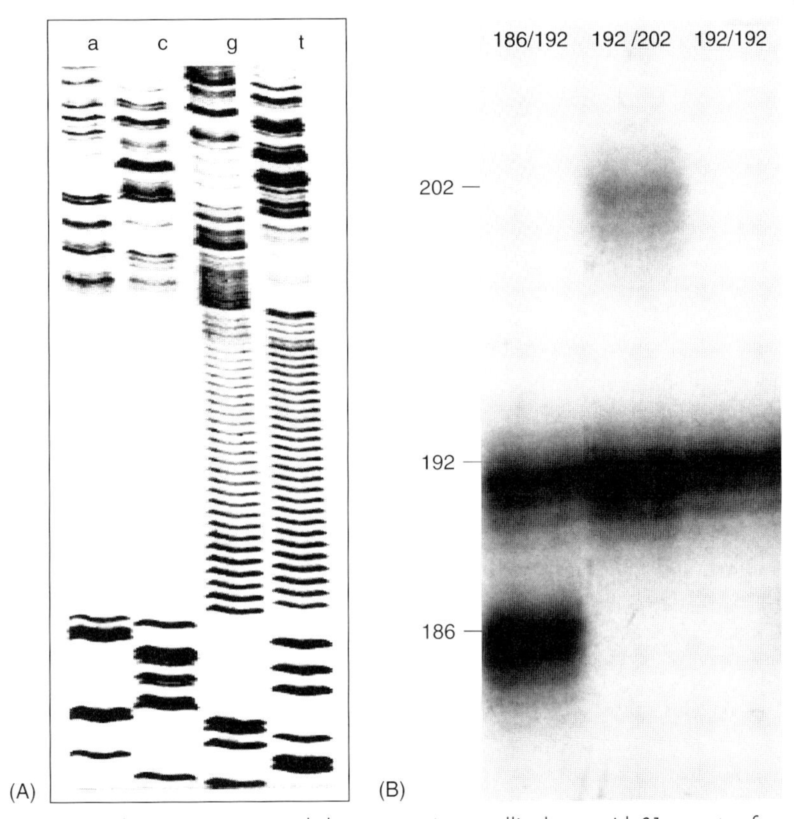

Fig. 5 Panel A. A sequencing gel showing a microsatellite locus with 31 repeats of the GT dinucleotide motif. Panel B. A microsatellite marker labeled with a radioactive isotope and visualized by exposure to photographic film. The marker has three alleles with sizes shown in base pairs along the side. The genotypes of each individual is shown at the top.

have less variation than the intervening DNA sequences; therefore, markers selected from anonymous stretches of DNA between genes have been mostly used to construct the linkage maps. The preferred type of "anonymous" marker is based on microsatellite loci. These loci are tandem repeats of simple sequences 1 to 4 bases in length, the most common being dinucleotide repeats of alternating C and T. Microsatellite loci seem to have a high mutation rate, which is thought to arise from DNA strand–slippage during replication, giving rise to variable numbers of the repeat (see Fig. 5). Therefore, each locus has a relatively large number of alleles, typically between 3 and 10. The number of alleles at a microsatellite locus is approximately proportional to the number of copies of the repeat unit. These microsatellite loci occur at about 40-kbp intervals and cover the whole genome, and therefore provide excellent markers for linkage mapping. By constructing primers flanking the microsatellite locus, the polymerase chain reaction (PRC) allows rapid genotyping of the marker.

The international mapping effort produced sequence information for a large number of these microsatellite loci during the early 1990s, which were then genotyped in the IBRP and MARC reference families to produce genetic maps. Each of these maps had around 1000 markers, mostly the microsatellite loci. Some of the markers were specific to one or the other map, while a large number of markers were common to both. In addition to the genetic maps produced using the reference families, resource families have been typed for a large number of microsatellite loci in order to localize the genes involved in various traits; some mapping studies have published genetic maps based on the information generated from the resource herd they used. As a significant number of the markers used to localize the trait loci are common between the studies, and are also in common with the genetic maps, the information can be combined to improve the resolution and confidence of ordering of the loci included in the maps. A series of international chromosome workshops, held under the auspices of the International Society for Animal Genetics, have assembled the various data into consensus maps. There are now well over 2400 microsatellite markers mapped on the bovine genome (see http://locus.jouy.inra.fr/cgi-bin/lgbc/mapping/common/intro2.pl?BASE=cattle or http://sol.marc.usda.gov/genome/cattle/cattl.html).

3.4 High-resolution Maps

Both *in situ* hybridization and linkage-mapping approaches have limitations with respect to creating detailed genome maps: the resolution of both is limited: the *in situ* approach, by the physical limits of observation, and linkage mapping, by the size of resource families used, and hence the number of informative meiosis available to order markers. A further drawback of the linkage-mapping approach is that the markers must be polymorphic, so that alleles on different parental chromosomes can be readily identified to track the inheritance of the chromosomal regions. The gross conservation of synteny between genomes discussed above may be disrupted at the level of gene order, with blocks of genes being rearranged within a broadly conserved framework. It is, therefore, important that mapping studies are able to include genes within the maps, which can then be used to compare genomic organization across species. Genes show much more limited levels of polymorphism than the microsatellite loci, and hence are more difficult to place on linkage maps accurately.

3.5 RH Mapping

Like linkage mapping, the radiation hybrid (RH) mapping approach relies on markers that are close together in the genome being separated less frequently when chromosomes are broken than more distant markers. In the case of RH mapping, the separation of markers is achieved by random fragmentation of the genome by radiation, rather than by recombination. An RH mapping-panel is constructed by irradiating cells from the target species, which are then fused with immortalized recipient hamster cells as shown in Fig. 6. The hamster cells used are deficient for a selectable marker (TK or HPRT); thus, growth in selection medium requires the retention of the region of the donor genome that carries the selectable marker; however, in addition many other fragments, typically between

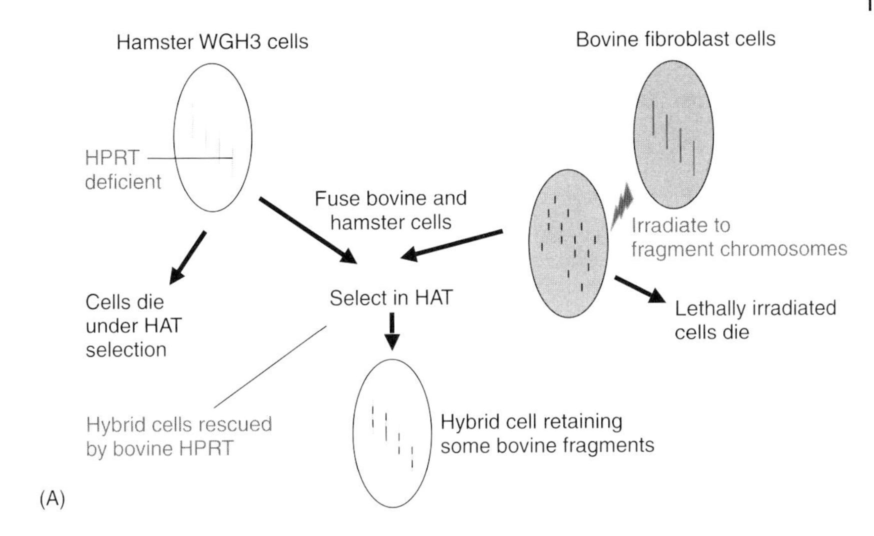

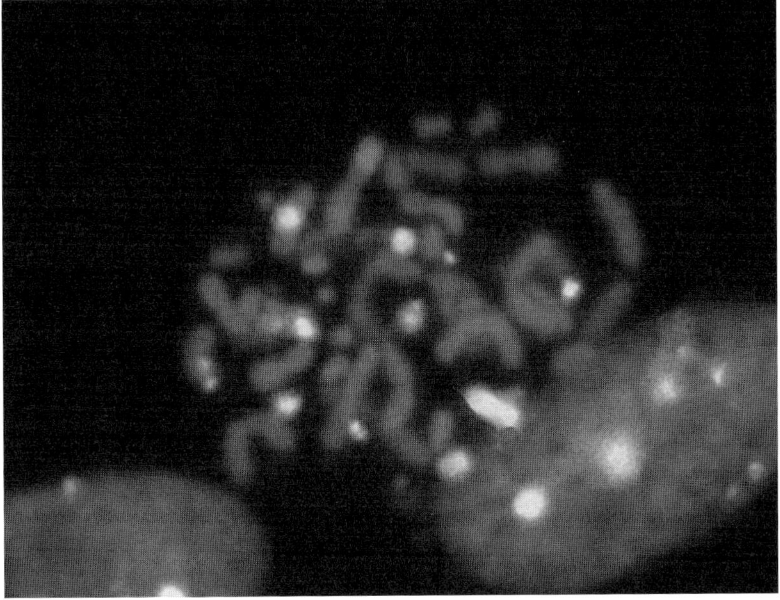

Fig. 8 Panel A. Construction of Radiation Hybrid cells. Fragments of the donor (bovine) chromosomes are retained in the immortalized rodent (hamster) cells that are rescued in selection medium by retaining the selectable marker from the donor genome. Panel B illustrates the presence of bovine genome fragments in the RH cells of the Bovine 3000 rad RH panel Williams, J.L., Eggen, A., Ferretti, L., Farr, C., Gautier, G., Amati, G., Ball, G., Caramori, T., Critcher, R., Costa, S., Hextall, P., Hills, D., Jeulin, A., Kiguwa, S.L., Ross, O., Smith, A.L., Saunier, K.L., Urquhart, B.G.D., Waddington, D. (2002) A Bovine Whole Genome Radiation Hybrid Panel and Outline Map, *Mamm. Genome* **13**, 469–474. (Photograph kindly provided by Dr C Farr Department of Genetics University of Cambridge).

15 and 40% of the donor genome is retained in each hybrid. A panel of RH cell lines is assembled in which each line contains different combinations of fragments from the donor genome. The presence or absence of markers in the cells of the hybrid panel is then used to calculate the relationship between the markers and build them into the RH map.

The RH approach has two major advantages over physical mapping by *in situ* hybridization and genetic-linkage mapping. Firstly, changing the radiation dose used in the construction of the panel allows panels with different resolutions to be created and secondly, the markers do not need to be polymorphic to be included in the map. Three RH panel have been described for cattle, two of which were constructed using a relatively low radiation dose (3–5000 rads) and one higher resolution panel (12 000 rads). There are now over 2000 markers mapped on the low-resolution panels and framework maps published. The inclusion of a proportion of microsatellite in the RH maps allows them to be aligned with the genetic maps (see Fig. 7). However, the main advantage of the RH panels is to allow the inclusion of genes or ESTs in the bovine map and so detailed comparative maps between the bovine and other genomes can be made.

3.6 Sequence Ready BAC Contigs

Identifying the genes involved in particular traits in cattle is currently carried out by linkage mapping followed by a positional candidate gene approach. This approach identifies genes that may be expected to be involved in the trait because of the function of their products and which map within the appropriate region. The candidate genes are then tests for involvement with the trait. This approach relies heavily on comparative mapping information to make use of data available for the human genome to provide the positional candidate genes. In the absence of positional candidate genes, or if the candidate genes identified do not prove to be the genes involved in the trait, the next approach is to construct a set of overlapping contiguous clones (called "Contigs") spanning the region of the chromosome that harbors the target gene. Contigs are usually constructed using clones containing large fragments of inserted genomic DNA, either cosmids (~40 kbp inserts), YACs (up to 1 Mbp inserts) or BACs (100+ kbp inserts). Construction of contigs is usually achieved by screening the large fragment libraries with markers that are available for the region of interest, then joining the clones together by "walking," which is achieved by using a probe from the end of one clone to find the next, and so on until clones covering the region of interest are found.

As the chromosomal locations for the genes controlling more traits are discovered, it becomes sensible to assemble a whole genome contig in which the clones from a large fragment library are placed in order along the each of the chromosomes. The whole genome contig then provides easy access to clones for particular regions for local sequencing and gene hunting projects. Selection of a "minimum tiling path" which contains the minimum number of overlapping clones to span all the chromosomes also provides the ideal underpinning resource from which to sequence the genome.

At the time of writing, the assembly of a whole genome BAC contig for the bovine genome is under way. The approach adopted is to use a BAC fingerprinting technique for the initial ordering

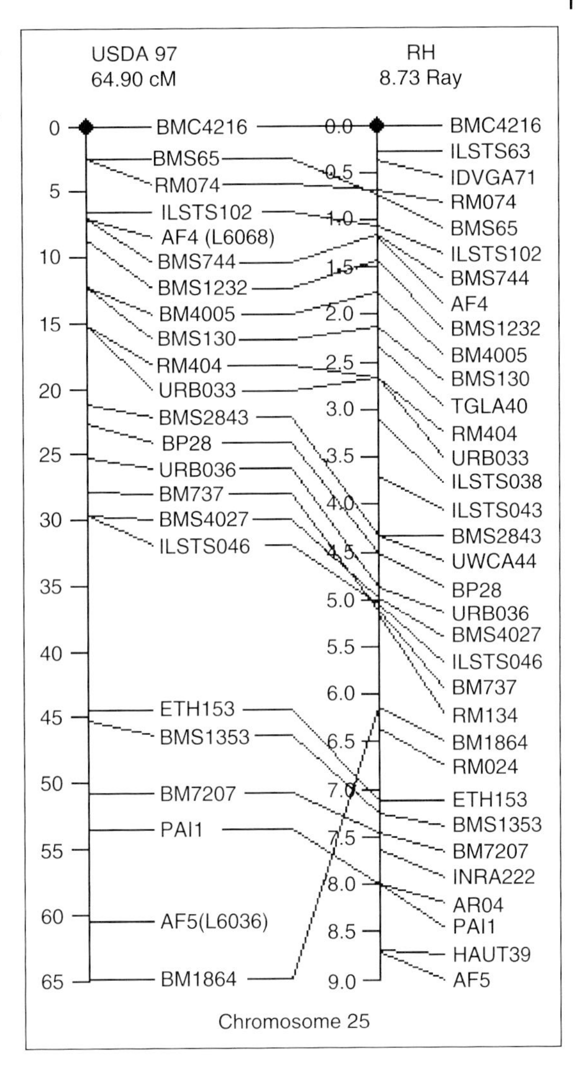

Fig. 7 Examples of genetic and RH maps for chromosome 25. The order of markers is broadly the same between methods; however, the relative distance between markers varies (from Williams et al. 2003).

of the clones. With this approach, BAC clones are digested with two restriction enzymes and sophisticated software is used to match fragment sizes and identify clones that contain regions of DNA that are common by identifying shared restriction fragments. The contig assembled from the fingerprinting is then confirmed by end sequencing the BACs and checking for overlapping sequence. With the availability of the human sequence, the BAC sequences can be used to align the bovine contigs with the human genome. However, interruptions in the conservation of synteny between human and bovine genomes may introduce errors into the ordering of the contigs; thus, it is unlikely that this approach will generate a continuous contig of all chromosomes. The order and orientation of contigs in the bovine genome will therefore be confirmed by aligning the BAC contig with the bovine

RH maps by typing markers in common between the BACs and the RH maps.

Two BAC libraries are currently being fingerprinted and the results merged. At the British Columbia Cancer Agency, 280 000 cattle BAC clones have been fingerprinted from libraries created at the BACPAC resources (Children's Hospital, Oakland Research Institute, Oakland, California), while at INRA in France, 90 000 clones have been fingerprinted. These data are currently being merged to produce the around 15-fold BAC clone coverage of the bovine genome. End sequencing of the BAC clones has just started and is being undertaken by the international bovine research community. The database of results will be maintained at The Genome Sciences Center, Vancouver, British Columbia, Canada. In parallel, the Texas and Roslin RH panels are being used to assist with the ordering of the clones.

3.7
Sequencing the Bovine Genome

The Human Genome Project announced the completion of the first draft of the entire human sequence in 2001 with the essential completion of the human genome sequence in 2003. This provides a vast resource for understanding the organization of the human genome and provides the template against which other genomes can be compared. The availability of the human sequence will also allow scientists to identify new genes and explore the evolutionary history of genomes. However, on its own the information contained within the human sequence is limited compared with the potential information that can be obtained by comparing and contrasting genomes between closely and distantly related species. Comparing the differences as well as the similarities between genomes is the key to understanding the functionally relevant features. Therefore, in addition to decoding the genetic makeup of humans, the sequence of a number of model organisms are also being analyzed, including baker's yeast, the roundworm, the laboratory mouse, and a rough draft of the sequence of the rat genome was produced in November 2002. The construction of the complete physical map of the bovine genome, by linking contiguous BAC clones to span all chromosomes, provides important resources that are the foundation for a bovine genome-sequencing project. The shotgun sequencing approach has inherent problems associated with assembling sequence data, whereas the identification of ordered BAC contigs and the clones constituting the minimum tiling path, provide the framework and the means to assemble the genome sequence efficiently. At the time of writing, a bovine genome-sequencing project is underway at Baylor College of Medicine Sequencing Center (see http://hgsc.bcm.tmc.edu/projects/bovine/)

4
Identification of Genes Controlling Traits

Genetic improvement of cattle has focused on traits that impact on profitability of agricultural enterprises; traditionally this meant increased output. Simple production traits are easiest to measure and so have been used in conventional selection programs with great success. However, using modern methods selection on a broader spectrum of traits can be considered and other traits that impact on profitability and welfare, such as feed-conversion efficiency, health, fertility, and product quality can be included in the selection objectives. If selection is phenotype-based, improvement in one trait is compromised by simultaneously including other traits in the breeding goals. However, using information on the genes controlling the traits potentially allows for selection on genotype and hence on several traits simultaneously. In theory, once sufficient knowledge is available, individuals carrying beneficial genes for several traits could be identified using DNA markers and mated to produce progeny with the desired characteristics. While we are a long way from this level of knowledge, work is under way to identify the genes involved in a wide range of traits. The ultimate goal of molecular genetics is to understand how allelic variations in the genes controlling various traits interact and result in the observed variations in the phenotype. Progress toward an understanding of the genetic control of traits can be achieved if the location in the genome of the genes involved is known, and markers linked to the genes are found. Once information has been accumulated to predict the alleles present in an individual, these linked markers can be used to aid breeding by marker-assisted selection (MAS).

Identifying the genes involved in the control of traits can be approached in a number of ways. Information on the physiology of the trait can be used to identify the biochemical pathways involved and hence postulate the controlling genes. This information can be coupled with patterns of expression among tissues to facilitate cloning of the gene(s) likely to affect the trait. These "candidate" genes are then studied in the context of the trait to identify if they play a role in controlling the observed variation. This approach clearly requires a good *a priori* knowledge of the trait and its physiology, but even so it could result in important genes being missed if they are not obviously involved in the known physiology. The second approach is to start with no prior assumptions regarding the physiology or genes controlling the trait, and to use a genome-mapping approach. The mapping approach is often necessary, particularly with complex traits where the various gene interactions cannot be predicted. In the same way as building a genetic map, mapping the genes that control particular traits requires families in which to track the inheritance of chromosomal regions. These families have to be segregating for the trait of interest and information quantifying the trait has to be collected. The mapping process involves tracking inheritance of chromosomal regions using DNA markers and then correlating inheritance of the markers with inheritance of characters associated with the trait. In practice, identification of the "trait genes" is achieved using a combination of the genetic mapping and candidate gene approaches.

4.1 Mapping Quantitative Trait Loci (QTL)

Production associated traits in cattle usually display continuous variation, such as growth or milk yield, and are called "Quantitative Traits." Variation in these traits is controlled by several loci, called *quantitative trait loci* (QTL). Two approaches have been adopted for mapping QTL, one is to make use of commercial cattle herds and the second is to use specifically bred cattle populations to examine particular traits. The advantage of using commercial populations is that they already exist and that simple measurements are recorded at a national level. In contrast, the resource population has to be specially bred;

however, this provides the opportunity of starting from diverse founders that allows the genetic and phenotypic variation in the trait(s) of interest to be maximized. Access to experimental resource herds also extends the possibility of recording the "difficult to measure" traits such as feed-conversion efficiency or immune response to pathogens, which would be impossible in a commercial context. The choice of using commercial or resource populations is to a large extent dependent on the trait that is being considered.

4.2 Dairy Traits

Both of the approaches described above, using candidate gene or genome-mapping approaches have been used to look for genes controlling traits that are important for milk production. The candidate gene approach has focused on the genes coding for the constituent proteins in milk. Milk is mainly water, typically around 88%, containing about 5% lactose, 3.5% fat, and 3.5% protein together with a number of minor components. Of the protein fraction, 80% are the caseins, which precipitate from milk at low pH and form the curd. The remaining whey proteins include β-lactoglobulin, α-lactalbumin and lactoferrin. There are 4 casein genes in cow, α_{S1}, α_{S2}, β and κ, which are clustered within a 25-kb locus on chromosome 6 (6q31), shown in Fig. 8.

Given the high proportion of casein forming the milk protein fraction, and that they are the chief constituents of the curd, which is used in cheese making, the casein genes have been extensively studied. The role of κ is in stabilization of casein micelles in liquid milk and cleavage of κ-casein (CSN3) leads to precipitation

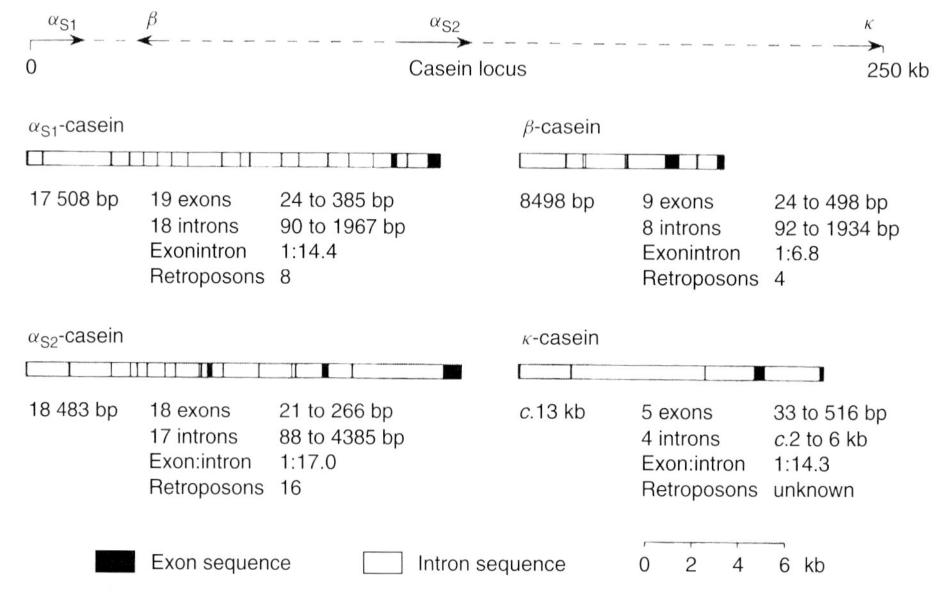

Fig. 8 Schematic representation of the bovine casein gene depicting the exon structure as well as their organization within the casein locus (from Fries, R., Ruvinsky, A. (1999) *The Genetics of Cattle*, CABI Publishing, Oxford, UK).

of the caseins in cheese production. Two alleles of the *CSN3* gene have been described, of which the B allele has been associated with improved manufacturing properties of milk. The B allele has also been associated with increased milk production, but only in Holstein cattle and not in other breeds. Determining the various alleles at each of the 4 casein loci allows casein haplotypes to be constructed, some of which have been associated with increased milk production in other breeds. This may suggest that genes linked to the casein locus, rather than the casein genes themselves are responsible for the variation. Additional interest has been focused on polymorphisms in the β-casein gene (*CSN2*), where there has been the suggestion that the A2 allele may be associated with protection from diabetes.

Selection for improved milk production has been achieved through progeny testing, which has resulted in milk yields in the Holstein breed nearly doubling in the past 40 years to 10 000 liters per 305-day lactation today. In developed countries, the Holstein represents the most numerous cattle breed and accounts for the majority of milk produced. The consequence of intensive selection has been the narrowing of the genetic base of the population and the wide-scale use of a relatively small number of "elite" sires, many of which are closely related. As a result, the dairy industry has very large half-sib families in commercial herds sired by a limited number of sires. While selection for milk yield has produced, in the Holstein breed, cattle with dramatically increased milk yield, there have been an associated decrease in protein content of the milk, decreased fertility, and increased incidence of lameness.

While resource herds have been used to address dairy associated traits, so far few of these results have been published. Most of the data available on QTL involved in dairy associated traits has come from studies on the commercial dairy population, the structure of which, as discussed above, is well suited to genetic mapping studies. However, the range of traits that is routinely recorded in commercial herds is limited, and the focus is on simple traits such as milk yield plus protein and fat composition, and somatic cell scores, the latter being related to the incidence of mastitis. In some cases, "linier type" (conformation) traits are also recorded. The major issue to be considered when addressing improvement in dairy traits is that they are often sex specific, and that the production animals are female, whereas selection is applied to the males. Thus, the industry requires females that produce large quantities of milk, have healthy udders and high fertility, while selection is on the bulls that display none of these traits. Genetic mapping studies could be carried out by tracking the inheritance of chromosomal regions from sires into their daughters, on which the traits are recorded, in a so-called daughter design study. However, this approach is inefficient for several reasons: firstly, a considerable number of daughters would need to be genotyped with DNA markers to track inheritance of the chromosomal regions from a single sire, and secondly, the sire has to be segregating for polymorphisms both in relevant genes and also for phenotypic variations in the trait, for many sires this will not be the case. An alternative approach, proposed by Weller et al., is the "Granddaughter design study" where a small number of males (bulls) are used to produce a large number of sons. DNA markers are used to track the chromosomal inheritance between the bull and his sons. The sons in their turn are used to sire a large number of daughters, on which the production traits are measured. The phenotypic measurements on the daughters are then used to calculate the genetic merit of their sire (the son of the elite sire). This is the basis of the progeny test used to select breeding sires. By correlating DNA marker information with the predicted breeding values, trait genes can be mapped.

The use of data collected on commercial herds to map QTL was pioneered by Georges et al., who used the US Holstein population to map QTL involved in milk yield and quality. The study used 1518 progeny tested bulls with production data from over 150 000 cows to derive their performance information. The sires were genotyped with 159 microsatellite loci, and analysis of the phenotypic and DNA marker information allowed 5 QTL to be identified, on chromosomes 1, 6, 9, 10, and 20. Each of the QTL was involved in different combinations of the traits examined, namely: milk yield, fat, and protein percentage, and fat and protein yield. As each of the QTL was identified in a subset of the sire families, it is unclear if these results indicate that the different QTL affect different traits, or whether the differences are because each of the sires and grand sires was segregating for different combinations of the QTL. Subsequent studies in other populations have independently confirmed one or more of the QTLs from the Georges study. An additional QTL for milk yield and composition has been reported on chromosome 14, which has been fine mapped and the likely trait gene identified (see DGAT1 below).

4.3 Beef Traits

The structure of the beef industry is different from the dairy industry in a number of ways. Beef is produced from a large number of breeds, which have not been under as intensive selection as seen with the dairy breeds. The conditions under which beef cattle are raised also vary considerably, maintaining the need for diversity of production animals. Breeding in beef units is often by natural service, and where AI, is used the number of progeny produced per bull is considerably fewer than seen with dairy breeds. There is some systematic recording of growth, fat, and conformation traits in a number of countries providing the opportunity for QTL mapping for beef production traits. However, the majority of published information on QTL for beef-associated traits comes from specifically bred resource herds. Several resource herds have been created worldwide, including herds at Texas A&M University (USA) and CSIRO (Australia) that have been produced by crossbreeding between *B. indicus* versus *B. taurus* to produce an F2 population to address meat quality traits. *Bos indicus* cattle, while more tolerant of drought, produce meat that is typically tougher than that of *B. taurus* breeds. Work by the Meat Animal Research Center in Nebraska using crosses between several breeds, has mapped QTL for several beef-associated traits, including growth rate, muscle mass, fat, and so on. A dairy x beef (Jersey vs Limousin) F2 resource herd has been bred in Adelaide (Australia) to specifically address traits associated with beef production. Progeny from this herd have also been raised in New Zealand, which will enable QTL affecting production in *intensive* systems (Australia) to be compared with those that are relevant in *extensive*, grass-based systems (New Zealand). Results of this study are due to be published. A conceptually similar herd, produced as F2 and reciprocal cross between dairy and beef breeds

(Holstein and Charolais) at the Roslin Institute, is also addressing the identification of QTL involved in beef production; however, in addition, an extensive portfolio of traits including health and fertility traits is being recorded on this population. This approach will allow pleiotropic interaction between traits and QTL to be addressed.

4.4 Health Traits

Whereas one of the primary drivers in natural selection will be health and resistance to disease challenge, the intervention of man in the selection of livestock species has, up to now, ignored health traits as selection criteria in favor of productivity traits. In some cases, productivity is dependent on health; however, with the increasing dependency on veterinary interventions, the natural defense mechanisms of the individual have been largely ignored. With the increasing demand from consumers for naturally produced products, together with the break down in effective antibiotic treatment caused by drug resistance in pathogens, there is an increased need for selection of animals with improved natural resistance to disease.

Much attention has been focused on the bovine major histocompatibility complex (MHC) as a potential locus that regulates immune response. The MHC region in cattle, in common with other mammals, contains genes coding for the class I and class II MHC molecules. T cells, that are responsible for eliciting an immune response do not recognize foreign antigens in their native form, but require specialist cells (B-cells, macrophages etc.) to process the antigen and present it bound to MHC molecules on their surface before a specific immune response is triggered. Variations in binding affinity of the antigens to the MHC molecules results in variations in the active immune response directed toward that antigen: failure of the antigen to bind to the MHC molecule means that it is not presented to T cells in the correct context and thus no immune response ensues. The bovine MHC (BoLA – bovine lymphocyte antigen locus) is located on chromosome 23 and in addition to the MHC antigens the region codes for 30 to 40 other genes including several immunologically active proteins, such as tumor necrosis factor and proteins of the complement system. There are two distinct classes of MHC antigen. The class I antigens are composed of two chains, a class I heavy chain that is encoded within the MHC region and β_2-microglobulin whose coding gene is located elsewhere in the genome. The class I molecules present antigen to cytotoxic T cells whose function is to kill infected cells, for example, with virus. There are around 20 MHC class I genes spanning about 1.5 Kb within the BoLA region. The class II molecules are composed of α- and β-chains, both encoded by genes within the MHC region. Several class II genes exist in cattle that resemble their counterparts in other mammalian species, namely, *DQ* α and β, and *DR* α and β genes, also DYA, and the class II-like chaperone molecule DMA. In the cattle, the *DR*β gene has been duplicated in some haplotypes, while the equivalent of the human, the DP locus appears to be absent in the BoLA region. The class II genes are involved in presentation of antigen to T-helper cells that promote both cellular and humoral immune responses. Given the central role of the MHC molecules in promoting immune response, it is not surprising that associations between MHC alleles and immune response to various infectious diseases and parasites have been sought.

The most common disease in dairy cattle is mastitis, which is an infection of the udder, and can be caused by several pathogens. Mastitis is both a welfare and an economic problem, resulting in decreased milk yield, increased veterinary treatments and is the most common cause of death in dairy cattle. The incidence of mastitis is not generally recorded, except in Scandinavian countries where a national health-recording scheme keeps individual records of disease for every cow. Studies using the Scandinavian data have identified putative QTL on chromosomes 3, 4, 14 and 27 associated with mastitis incidence. More commonly, somatic cell scores (SCS) are used as an indicator of mastitis incidence, although SCS are a measure of inflammatory response, and do not necessarily indicate clinical or subclinical infection. Consequently, the correlation between clinical mastitis and SCS range from 0.37 to 0.97. However, this is a measure that is routinely recorded in commercial dairy herds. Studies to examine associations between genetic loci and SCS suggest that some MRC DRβ alleles may be associated with increased SCS and hence increased risk of mastitis.

Breeds of cattle are known that show resistance to particular parasites, for example, the N'Dama cattle of West Africa are resistant to *Trypanosoma congolense*, which causes sleeping sickness, whereas other breeds of cattle are highly susceptible to this disease. The introduction of more productive breeds into endemic areas is currently not a feasible option, as the animals would either die or be too expensive to maintain through requirement for extensive veterinary treatment. However, the disease resistant breed is small and not very productive. One approach to allow the introduction of more productive breeds in to infected areas is to identify the genes controlling the disease resistance and introduce these resistance genes into more productive animals. In order to identify the genetic control of *T. congolense* infection, an F2 cattle resource population was bred at ILRAD (now ILRI) in Kenya by crossing N'Dama with Boran cattle, the latter being susceptible to trypanosomiasis. This study has identified QTL to be involved in resistance; and fine mapping of these loci is being pursued in a mouse model.

A new cattle disease was identified in the United Kingdom in 1986, called *bovine spongiform encephalopathy* (BSE). The BSE epidemic developed rapidly to peak in 1992. Today, more than 180,000 cases have been confirmed in the United Kingdom. In sheep incidence of the related transmissible spongiform encephalopathy, scrapie, is associated with polymorphisms in the *PrP* gene, at codons 136, 154, and 171. Sheep that are homozygous for glutamine and position 171 are most susceptible to scrapie, while individuals with arginine are most resistant. In cattle, two polymorphisms have been reported in the coding region of the *PrP* gene, a silent change that affects a *Hind*II restriction site, and a difference in the number of G-C rich octa-repeat elements, with alleles that have either 5, 6, or 7 copies of the repeat. Two case-control studies found no association between variations in the octa-peptide-repeat and incidence of BSE. Complete sequence information for a 25-kb region containing the bovine *PrP* gene revealed 9 polymorphisms in the PrP coding region, including the two mentioned above, and more than 40 SNPs and insertions in noncoding regions of the gene. However, there is no evidence to date for polymorphisms within the bovine *Prnp* coding region that affect susceptibility to BSE. A whole genome scan with microsatellite markers that examined transmission disequilibrium of alleles between BSE affected and control progeny bred from a small number of bulls revealed three loci, on chromosomes 5, 10, and 20 that were associated with incidence of BSE. The interval on bovine chromosome 10 coincides with the corresponding interval on mouse chromosome 9 that has been associated with resistance of mice to experimental scrapie and is now being investigated further.

4.5
Identifying the Trait Genes

The ultimate objective of the genome-mapping work is to identify the genes and variations within the genes that control the traits of interest. Up to now a reasonable number of genes controlling a diverse range of traits have been localized to fairly broad regions on many bovine chromosomes. Typically, linkage-mapping studies in cattle have identified, at best, a 20-cM interval (the whole bovine genome is 3000 cM) within which the gene affecting the trait of interest is located. This size of the interval represents 20 Mbp of DNA and, assuming a conservative estimate of around 30 000 genes in the bovine genome, is likely to contain at least 200 and possibly as many as 1000 genes. There are several ways of identifying the gene involved in controlling a trait from its map location; these include fine mapping using meta-analysis of data from different populations to refine the map location, examination of haplotypes to identify the limits of linkage disequilibrium between markers and the trait, positional cloning using large fragment clones to build contigs across the QTL region, and a positional candidate gene approach, where genes likely to be involved in the trait are identified within the QTL region. In conjunction with these approaches, comparative information can be used, taking information about the position and role of candidate genes from other species and using knowledge of conservation of synteny to identify candidate genes likely to be located within the QTL region in cattle.

Two studies have been published that are paradigms for identification of genes involved in meat and dairy traits in cattle. Both these studies used a mixture of linkage mapping, positional cloning, and comparative genomics to identify the gene controlling double muscling, and a gene for milk production in cattle.

Double muscling is an extreme form of muscle development that has been described in several breeds of cattle; the most extreme form of double muscling being seen in the Belgian Blue breed. Segregation patterns for the trait suggested that the phenotype was controlled by a single gene, which was subsequently mapped to chromosome 2 using a resource herd bred from double-muscled Belgian Blue cattle crossed to nondouble muscled cattle. The first approach for identifying the gene controlling the phenotype, starting from the map position, was to test candidate genes located close to the region identified from the mapping study. However, this approach failed to identify the trait gene. At the same time, unconnected work on mice identified a new member of the growth and differentiation factor β superfamily of genes (GDF-8) that was expressed in muscle and that was required for normal muscle development. Knocking-out the expression of GDF-8 resulted in hyper-muscularity in the mice and therefore was named myostatin. Using RH mapping, myostatin was shown to map to bovine chromosome 2 within the interval where the double-muscling gene had been mapped. Sequencing the myostatin gene from double-muscled individuals revealed an 11 bp deletion in Belgian Blue cattle that truncates the protein in the bioactive c-terminal domain, and a single nonconservative base change in the double-muscled Piedmontase breed. These mutations are strongly associated with the double-muscled phenotype and make myostatin the likely causative gene.

It has been suggested that double muscling originated in the British Beef Shorthorn breed. However, the myostatin gene in cattle is highly polymorphic, with 19 polymorphisms currently described within the bovine gene. Analysis of the distribution of these polymorphisms has allowed the possible origins of the alleles found in different breeds to be investigated. There is currently no molecular evidence to support the Shorthorn as the origin of the myostatin alleles responsible for double muscling in European cattle. The 19 polymorphisms can be assembled into 10 haplotypes, which are related in a broad phylogenetic tree with a shallow timescale, suggesting that most of the mutations have accumulated independently, and fairly recently. The 11-bp deletion seen in Belgian Blue cattle is the most common mutation responsible for double muscling and it is likely that this allele was introduced into several other breeds from a single source at about the same time. Independent, rather than a stepwise accumulation of mutations then followed.

The effects of the disruptive mutations in the myostatin gene on the phenotype vary between the different breeds. The same 11-bp deletion that is found in the Belgian Blue and which is associated with an extreme form of double muscling, is associated with a milder phenotype in other breed, such as the Spanish Asturiana and British South Devon cattle. This has lead to speculation that modifiers must interact with myostatin in the regulation of muscle development and are partly responsible for the double-muscled phenotype.

A QTL with effect on milk production, particularly fat content was localized to chromosome 14 in cattle using a linkage-mapping approach in several commercial populations. The fact that the same QTL location was identified independently gives confidence both in the existence of a locus with significant effects on milk synthesis, and also on the position of the QTL. By examining the extent of linkage disequilibrium on individual chromosomes, the QTL region was narrowed. Individual recombination events between marker and the phenotype allow most of the mapped QTL region to be discounted as the location of the gene, so that the trait gene could be localized to within 3 cM, representing 3 Mb of DNA. A BAC contig spanning this region was assembled in preparation for local sequencing to identify genes within the region.

Work carried out in parallel in mice identified diacyl-glycerol *O*-acyltransferase (DGAT1) as an enzyme involved in the formation of triglyceride, which is a major component of the fat found in milk. Studies in mice also showed that females lacking both copies of DGAT1 had impaired lactation. DGAT1 was mapped within the QTL region on chromosome 14 in cattle by RH mapping and also mapped with the refined 3 cM interval on the BAC contig, making it a very strong positional candidate gene. Sequencing the DGAT1 gene revealed a polymorphism at position 232, which substitutes alanine for lysine (K232A). Individuals with lysine have a higher milk production than those with alanine.

5
Functional Genomics

Having a detailed genome map, even the ultimate map in the form of the sequence, of the genome, is only the starting point for identifying the genes controlling traits in cattle and understanding their function. The next step will be to characterize the expression of the genes, and the functions of their protein products, in order to understand their role in controlling development and survival of the individual and the variations seen among individuals.

Advances in technology mean that we can now examine the expression of large numbers of genes simultaneously, using a macro- or microarray approaches. These arrays are created using either cDNA probes, or oligonucleotide probes designed from knowledge of the gene sequence. The probes are anchored to a solid matrix, a nylon membrane in the case of the macroarray, or a glass slide for the microarray. Complex probes are then created by reverse transcribing and labeling the RNA from the cells or tissues whose gene expression is being studied. The labeled probe is then hybridized to the arrays and expression patterns deduced, either by comparison with a reference, or in the case of microarrays by simultaneous hybridization of the samples being compared, each of which is labeled with a different fluorescent dye. The prerequisite for these arrays are the gene probes that are used. Although specialist oligonucleotide probes have been made for particular genes, for example, those involved in immune responses, currently most array probes are cDNAs that have been identified by sequencing clones from cDNA libraries produced from the target tissue or tissues. The data generated from the sequencing cDNA clones produces expressed sequence tags (ESTs). At the time of writing, there are over 122 000 cattle ESTs sequences recorded on the EMBL sequence database, mainly generated by the USDA Meat Animal Research Center in Nebraska. These arrays will enable gene expression to be explored in a range of tissues in different physiological states and from animals with variable phenotypes. This information will provide an extension to the genome-mapping studies to help identify the genes involved in the control of traits, and to understand their function

6
In Conclusion

The construction of maps for the bovine genome has allowed the chromosomal regions harboring genes controlling specific traits to be identified, and in some cases the genes themselves have been found using a variety of techniques, starting from their map position. Comparative mapping information is allowing the alignment of bovine chromosomes with high-density maps in other species, and with the human genome sequence. This will provide considerable information on the range of genes present in the genome and their location in cattle. In the foreseeable future, the sequence of the bovine genome will be available. Analysis and annotation of the bovine sequence along with the sequence from other species will, in many cases, confirm the predictions made from the comparative mapping. Differences observed between genomes will also lead to new findings with the identification of new genes and a greater understanding of the regulation and expression of the genes and their role in controlling phenotypic variation.

See also Genetics, Molecular Basis of; Immunoassays.

Bibliography

Books and Reviews

Fries, R., Ruvinsky, A. (1999) *The Genetics of Cattle*, CABI Publishing, Oxford, UK.

McCarthy, L.C. (1996) Whole genome radiation hybrid mapping, *Trends Genet.* **12**, 491–493.

Ng-Kwai-Hang, K.F. Grosclaude, F. (1992) Genetic polymorphism of milk proteins, in: P.F. Fox. (Ed.) *Advanced Dairy Chemistry* Vol. 1 Elsevier, Amsterdam pp. 405–455.

Primary Literature

Band, M.R., Larson, J.H., Redeiz, M., Green, C.A., Heyen, D.W., Donovan, J., Windish, R., Steining, C., Mahyuddin, P., Womack, J.E., Lewin, H.A. (2000) An ordered map of the cattle and human genomes, *Genome Res.* **10**, 1359–1368.

Barendse, W., Vaiman, D., Kemp, S., Sugimoto, Y., Armitage, S., Williams, J.L., Sun, H., Eggen, A., Agaba, M., Aleyasin, A., Band, M., Bishop, M., Buitkamp, J., Byrne, K., Collins, F., Cooper, L., Coppettiers, W., Denys, B., Drinkwater, R., Easterday, K., Elduque, C., Ennis, S., Erhardt, G., Ferretti, L., Flavin, N., Gao, Q., Georges, M., Gurung, R., Harlizius, B., Hawkins, G., Hetzel, J., Hirano, T., Hulme, D., Joergensen, C., Kessler, M., Kirkpatrick, B., Konfortov, B., Kostia, S., Kuhn, C., Lenstra, J., Leveziel, H., Lewin, H., Leyhe, B., Li, L., Martin Burriel, I., McGraw, R., Miller, R., Moody, D., Moore, S., Nakane, S., Nijman, I., Olsaker, I., Pomp, D., Rando, A., Ron, M., Shalom, A., Soller, M., Teale, A., Thieven, I., Urquhart, B., Vage, D-I., Van de Weghe, A., Varvio, S., Velmalla, R., Vilkki, J., Weikard, R., Woodside, C., Womack, J., Zanotti, M., Zaragoza, P. (1997) A medium density genetic linkage map of the bovine genome, *Mamm. Genome* **8**, 21–28.

Blott, S.C., Williams, J.L., Haley, C.S. (1998) Genetic relationships among European cattle breeds, *Anim. Genet.* **29**, 273–282.

Bradley, D.G., Loftus, R.T., Cunningham, P., McHugh, D.E. (1998) Genetics and domestic cattle origins, *Evol. Anthropol.* **6**, 79–86.

Dunner, S., Miranda, M.E., Amigues, Y., Cañon, J., Georges, M., Hanset, R., Williams, J.L., Ménissier, F. (2003) Haplotype diversity of the myostatin gene among beef cattle breeds, *Genetic Selection and Evolution* **35**, 103–118.

Ferretti, L., Urquhart, B.G.D., Eggen, A., Olsaker, I., Harlizius, B., Castiglioni, B., Mezzelani, A., Solinas-Toldo, S., Thieven, U., Zhang, Y., Morgan, A.L.G., Teres, V.M., Schwerin, M., Martin-Burriel, I., Chowdary, B., Erhardt, G., Nijman, I.J., Cribiu, E.P., Barendse, W., Leveziel, H., Fries, R., Williams, J.L. (1997) Cosmid-derived markers anchoring the bovine genetic map to the physical map, *Mamm. Genome* **8**, 29–36.

Fries, R., Eggen, A., Womack, J.E. (1993) The bovine genome map, *Mamm. Genome* **4**, 405–428.

Georges, M., Nielsen, D., Mackinnon, M., Mishra, A., Okimoto, R., Pasquino, A.T., Sargeant, L.S., Sorensen, A., Steele, M.R., Zhao, X., Womack, J.E., Hoeschele, I. (1995) Mapping quantitative trait loci controlling milk production in dairy cattle by exploiting progeny testing, *Genetics* **139**, 907–920.

Grisart, B., Coppieters, W., Farnir, F., Karim, L., Ford, C., Berzi, P., Cambisano, N., Mni, M., Ried, S., Simon, P., Spelman, R., Georges, M., Snell, R. (2002) Positional candidate cloning of a QTL in dairy cattle: identification of a missense mutation in the bovine DGAT1 gene with major effect on milk yield and composition, *Genome Res.* **12**, 222–231.

Hernandez-Sanchez, J., Waddington, D., Weiner, P., Haley, C.S., Williams, J.L. (2002) Genome-wide search for markers associated with bovine spongiform encephalopathy, *Mamm. Genome* **13**, 164–168.

Kappes, S.M., Keele, J.W., Stone, R.T., Sonstegard, T.S., Smith, T.P.L., McGraw, R.A., Lopez-Corrales, N.L., Beattie, C.W. (1997) A second-generation linkage map of the bovine genome, *Genome Res.* **7**, 235–249.

Klungland, H., Sabry, A., Heringstad, B., Olsen, H.G., Gomez-Raya, L., Vage, D.I., Olsaker, I., Odegard, J., Klemetsdal, G., Schulman, N., Vilkki, J., Ruane, J., Aasland, M., Ronningen, K., Lien, S. (2001) *Mamm. Genome* **12**, 837–842.

Lander, E.S., Linton, L.M., Birren, B., Nusbaum, C. et al (2001) Initial sequencing and analysis of the human genome, *Nature* **409**(6822), 860–921.

Lien, S., Gomez-Raya, L., Steine, T., Fimland, E., Rogne, S. (1995) Associations between casein haplotypes and milk yield traits, *J. Dairy Sci.* **78**, 2047–2056.

Loftus, R.T., MacHugh, D.E., Bradley, D.G., Sharp, P.M., Cunningham, E.P. (1994) Evidence for two independent domestications of cattle, *Proc. Natl. Acad. Sci. U.S.A.* **91**, 2757–2761.

McPheron, A.C., Lee, S-J. (1997) Double muscling in cattle due to mutations in the myostatin gene, *Proc. Natl. Acad. Sci. U.S.A.* **94**, 12457–12461.

Popescu, C.P., Long., S., Riggs, P., Womack, J., Schmutz, S., Fries, R., Gallagher, D.S. (1996) Standardization of cattle karyotype nomenclature: Report of the committee for the standardization of the cattle karyotype, *Cytogenet. Cell Genet.* **74**, 259–261.

Sharif, S., Mallard, B.A., Wilkie, B.N., Sargeant, J.M., Scott, H.M., Dekkers, J.C.M., Lesley, K.E. (1998) Associations of the major histocompatibility complex DRB3 (BoLA-DRB3) alleles with occurrence of disease and milk somatic cell score in Canadian dairy cattle, *Anim. Genet.* **29**, 185–193.

Solinas-Toldo, S., Lengauer, C., Fries, R. (1995) Comparative Genome map of human and cattle, *Genomics* **27**, 489–596.

Stone, R.T., Keele, J.W., Shackelford, S.D., Kappes, S.M., Koohmaraie, M. (1999) A primary screen of the bovine genome for quantitative trait loci affecting carcass and growth traits, *J. Anim. Sci.* **77**, 1379–1384.

Weller, J.I., Kashi, Y., Soller, M. (1990) Power of daughter and grand-daughter designs for determining linkage between markers and quantitative trait loci in dairy cattle, *J. Dairy Sci.* **73**, 2525–2537.

Williams, J.L., Eggen, A., Ferretti, L., Farr, C., Gautier, G., Amati, G., Ball, G., Caramori, T., Critcher, R., Costa, S., Hextall, P., Hills, D., Jeulin, A., Kiguwa, S.L., Ross, O., Smith, A.L., Saunier, K.L., Urquhart, B.G.D., Waddington, D. (2002) A Bovine Whole Genome Radiation Hybrid Panel and Outline Map, *Mamm. Genome* **13**, 469–474.

Winter, A., Kramer, W., Werner, F.A.O., Lokkers, S., Kata, S., Durstewitz, G., Buitkamp, J., Womack, J.E., Thaller, G., Fries, R. (2002) Association of a lysine-232/alanine polymorphism in a bovine gene encoding acyl-CoA:diacylglycerol acyltransferase (DGAT1) with variation at a quantitative trait locus for milk fat content, *Proc. Natl. Acad. Sci. U.S.A.* **99**, 9300–9305.

Living Organism (Animal) Patents

William Lesser
Department of Applied Economics and Management, Cornell University, Ithaca, NY, USA

Encyclopedia of Molecular Cell Biology and Molecular Medicine, 2nd Edition. Volume 7
Edited by Robert A. Meyers.

ISBN: 3-527-30549-1

Keywords

Animal Patent
The application of patent law to the protection of higher animals.

Intellectual Property Rights (IPR)
Legal systems for providing property protection for abstract ideas and knowledge, the systems generally including patents, trademark, and copyright, among others.

Patent
Instrument covering the grant of limited exclusive property rights to inventors; also utility patent.

Transgenic Animal
An animal to whose DNA (hereditary material) has been added DNA material from another source other than parental germplasm. Patented animals are typically transgenics.

Animal Patents refers to the granting of utility patents to higher animals. The application of patents to animals is a recent development in patent law with the first formal decision (in the United States) dating to only 1987. This extension of patent law raises for some legal, practical, and moral issues, yet others see patent protection as an essential incentive for the development of specialized animals for uses in medical research, medicine production, agricultural/aquaculture uses, and even pets. By 2004, some 600 animal patents had been granted worldwide, 80% in the United States, and most relating to "animal models" for medical research. Indeed, the first animal patent in 1987, for the "Harvard" mouse, was for cancer research. The World Trade Organization agreement, while specifying minimal Intellectual Property Rights protection allowable in member states, does specifically allow countries to exclude patents for animals while offering no alternative protection systems. Hence, besides the United States and European Union (EPO members), only three countries presently allow patents for animals. Animal patents are likely to remain geographically limited for some time, but scientific advances and commercial realities will eventually lead to pressures on governments to grant protection. Public opinion surveys show the best acceptance for patented animals providing clear benefits to humans.

1 Introduction

Animal Patents refers to the application of patents to higher animals. Patents are a means of granting property rights and, by providing limited monopoly control, an incentive for investing in inventive research and development (R&D). Patents, a form of Intellectual Property Rights (IPR), have been found to provide an especially significant incentive for easily copied inventions, which includes self-reproducible animals. As a background, the general functioning of patent systems (known more formally as utility patents) is described in Sect. 2. Also included is the description of a specialized patentlike system for plants only, which, while not directly applicable to animals, does allow some insights into the special considerations and approaches for applying IPR to self-reproducing life forms.

While, at the most basic level, there is nothing conceptually different about applying patents to animals, their use does raise two fundamental issues:

- Are animals discoveries or products of nature and hence excludable from patents?
- Should ethical/moral considerations for patenting animals exclude patentability?

The focus here is on issues associated with patented higher animals, as contrasted with other economic and social matters applied to animals in general. As a practical matter, this means the animals that have been genetically transformed. Methods are important, but only to the extent they affect the production of transgenic animals. Cloning, for example, is a method that can be applied to transgenic animals but is not associated with the animals themselves. Also, while there is no specific bar to the patenting of traditionally bred animals and, indeed, plants produced through conventional breeding approaches are routinely granted patents, as yet grants for animals have been limited to transgenics.

2 Concept and Operation of Patent and Related Systems

2.1 Origins of Intellectual Property Rights

The earliest form of Intellectual Property Rights has been traced by many scholars to Venice in the Middle Ages. Master craftsmen, it seems, were beset with competition from former apprentices. To control that competition, a law was passed that prohibited former apprentices from going into competition with licensed craftsmen for a period equivalent to the period of the apprenticeship. That period, of about 20 years, is said to be the origin of the current 20-year patent duration from first application. Certainly, craftsmen benefited from reduced competition. Apprentices benefited less so, but without the protection from competition, many may not have found masters willing to take on apprentices. Finally, the consuming public lost in the short run through higher prices, but with the long-term assurance of trained master craftsmen available to produce the products, eventually they would benefit as well.

2.2
Purposes of Intellectual Property Rights

2.2.1 Economic Justification

Over subsequent centuries, patent and related IPR laws have undergone multiple changes, but the essence of the public/private trade-off seen in long-ago Venice is still evident. Patents are temporary (typically 20 years). The inventor benefits from that period of reduced competition and thereby has an incentive to invest in inventive activities. The public benefits from the existence of products and processes that otherwise may not have existed, or whose appearance would have been delayed. The justification recognizes the copier, who does not have to recover the R&D expenses of the inventive process, can always undersell the inventor. Hence, no one outside the public sector has an incentive to invest in easily copied new products, and technological advances stagnate. Patent laws restore that incentive through the legal prohibition on direct copying.

Contemporary patent law adds a further dimension, the requirement that the invention be described. The resultant file of descriptions creates a public storehouse of technical information while helping assure competitive products can be produced soon after the patent expires. For these benefits, the public pays a higher price than would prevail in a competitive market. A higher price can come about only if supply is reduced so that the public has less of the product available than under competitive conditions. Stated differently, product use diffuses more slowly.

In addition to the mere existence of the novel product or process, the public benefits from the so-called spillover effect. Spillover recognizes that a private owner cannot capture all the benefits from a product. A large piece of property kept undeveloped by an owner not only benefits the owner but also the public, which enjoys the views and ecological benefits provided. Similarly, an invention (gasoline engine) can lead to derivative inventions (automobiles), which further benefit the public. Or the inventor, to meet unit sales and profit goals, may need to charge a lower price than many users would be willing and able to pay. Those users gain a direct benefit at the expense of the inventor.

The investment incentive is the principal justification for patents and related IPR. There is, however, another justification for patent-type laws: the provision of an incentive for firms to distribute products to other countries. Key to appreciating this aspect is a realization that intellectual property law is strictly national: protection must be held in each country to be effective. For example, a product protected in the United States has no protection in France or Japan unless specifically sought and received there. Inventors would be concerned about unauthorized copying in those markets as well, and are typically hesitant to supply them without the corresponding intellectual property protection. Hence, intellectual property protection can assist in the transfer of products and technologies.

Patents belong to a group of related rights known as *industrial property rights*, which also include Plant Breeders' Rights, trademarks, and trade secrets. Trademarks function somewhat differently, but with a general similarity of intent. Trademarks allow owners to develop goodwill in a product with the knowledge that it will not be dissipated by unauthorized imitators. For the consumer, a trademark serves as a signal of product quality – a Coca Cola or McDonald's hamburger has a consistency and standard taste consumers expect. True Rolex watches are generally of far higher quality than the copies. Thus, with trademarks too, both the owner of the trademark and the public benefit at least to some degree.

Copyrights fall into a separate class of protection mechanisms. Copyrights are applied to creative works like books and music, although important current uses include software. Copyrights are more of a protection by right and involve fewer formal procedures.

Trade secrets impose a penalty if secrets are acquired in an inappropriate way. The secret itself may be anything of value, such as customer lists or performance records. No formal process is required to achieve trade secret status, beyond a serious effort to keep the secret that way. In that way, a trade secret can conceivably be perpetual. The formula for Coca Cola is a well-known trade secret. In many cases, firms will achieve stronger (broader) protection by combining several forms of protection within a single product. For example, a patent might protect the product, but the least cost process for producing it is protected by a trade secret. Or a product is patented while the name is trademarked.

2.2.2 Moral Justification

A second justification for IPR is a moral, or perhaps more correctly, personal rights justification. This concept, often couched in the individualism philosophy of the Enlightenment, stresses that creation (whether artistic or technical) is from the self and hence should be granted the same rights as any personal property. IPR and personal rights are aspects of the self and hence should be granted as a right. Here, the economic justification will be stressed, as it is generally. The moral right concept, though, does continue in the idea of benefit sharing as well as requirements in most patent laws that the inventor(s) be identified by name. Perhaps, the inventor does not own the invention (employment contracts, for example, typically make any invention in the course of work the property of the employer). However, he/she/they must be at least identified on the patent as the inventor.

2.3 Forms of Intellectual Property Rights Applicable to Animals

The preceding subsection laid out the general justification for IPR and identified in broad terms the major forms. This section contains more detailed descriptions of the operation of the several systems potentially applicable to animals. Greatest attention is of course given to patents.

2.3.1 Patents

Patents are intended as an incentive and reward for developing something new and useful, but only if the development is a nontrivial extension. The system has developed specific terms and interpretations for these concepts.

Nonobviousness or inventive step: The invention should not be obvious to someone "skilled in the art." This is the nontrivial requirement.

Novelty: The invention must be new in the sense of not having been previously known through publication or public use. The United States uses a system with a one-year "grace period," which means simply that the invention could have been revealed up to one year prior to applying for a patent. Most countries do not allow any prior revealing of the invention – in what is known as absolute novelty. Absolute novelty, in particular, is difficult for public sector researchers whose work often requires publishing of results.

Utility: Utility requires that the invention must serve some identified purpose. This does not say the invention must be practical – patent offices do not judge practicality. Rather, the requirement is that a use be identified. This is an important component of the system for it led to the rejection of applications for human genes early in the Human Genome Project. Those applications read to a gene only, with no specific knowledge of a use for it. Hence, the rejection.

Issued patents are all of the same general type, although this masks some large differences. One is the *scope* of an individual patent – a wider scope means a related product must be more different so as not to be infringing. The scope of the first animal patent, the "Harvard" mouse, is quite broad, reading in the first claim, "all non-human mammals" that exhibit the identified trait. Often, patents in new technical fields are broad, narrowing as the techniques become more standardized. Typically, a person knowledgeable in the field is needed to interpret the scope of a patent.

A second distinction is made on whether the patent may be referred to as *per se*, process, product by process, or dependent. Those are not terms that will appear on the patent, but can be inferred from a close reading. For example, a claim which reads "a means of" is usually a patent for a process. A product by process patent covers the product only if it is produced in the described way. A different production technique lies outside the patent. The concept of *per se* and dependent patents can be explained by a factitious reference to aspirin. Imagine the initial aspirin patent was for headache relief (in fact, it was a natural product with a patent for a chemical synthesis). Because (in this example) it was the first, the patent owner had rights to any subsequently developed uses. One was identified – the use to help prevent and subsequently recover from a heart attack due to the blood thinning property – and patented. The later patent though is dependent on the first one, so it cannot be used without the permission of the initial patent owner. The owner of the initial patent, for his or her part, also cannot use the improvement without permission. Often, what could be an impasse is resolved with a cross-license between the two owners.

2.3.2 Plant Breeders' Rights

Plant variety protection (PVP) is a special purpose system only for plants. Technically, it is a form of *sui generis* or special purpose system owing to the exclusive focus on plants. Plant breeder's rights (PBR) are quite recent, with origins in the 1930s and 1940s. UPOV (the French acronym for the International Convention for the Protection of New Varieties of Plants), the international convention, was established only in 1961.

PBR replaces utility, nonobviousness, and novelty for patent systems (see above) with uniformity, stability, distinctness, and novelty (abbreviated as DUS). Novelty is the same in concept to patents, although the specific terms and use conditions differ. Uniformity and stability are technical requirements, which assure that the protected variety is identifiable after repeated multiplications. Clearly, if the protected product (variety) cannot be definitively identified, the system breaks down. For the future, genetic markers or a related technology may be used, but that is not the case at present.

Distinctness is akin to nonobviousness for patents. Distinctness may be claimed in one of a number of traits, such as disease resistance or flower color. Most countries actually grow out varieties and measure the difference from a base or reference variety. In the United States, the claims of the applicant are generally accepted, akin to a registration system. Moreover, the US Plant Variety Protection Office allows distinctness for traits of no practical merit. Hence, PBR in the United States are generally considered to be weaker than in other UPOV-member countries.

PVP have other components, which mean that they provide weaker protection than patents. These are the so-called Breeders' Rights and Farmers Privilege. Breeders' Rights specifically allow breeders to use protected varieties in a breeding program without the permission of the variety owner. For patented products, this is known as the *research exemption*. Because the research exemption is not statutory (written in the law), it is a matter of court interpretation, the focus of which is still under debate by scholars and practitioners. The Farmers Privilege expressly allows farmers to retain the crop for use as a seed source for a subsequent crop. The farmer, however, may not sell the seed or share with other farmers. The most recent UPOV Act (1991) makes the Farmers Privilege a national option, but most countries (including the United States) are choosing to allow it. In Europe, with the exception of "small farmers," a payment for seed savings set at 50% of the "normal" license fee must be paid.

PVP are intended to protect the entire plant as well as its propagating parts (seeds, etc.). This is in line with the specific focus of preventing direct competition for that variety. A weakness became apparent when overseas growers (where PVP was unavailable) shipped plant parts such as flower blossoms into a protected market. That was not illegal until the most recent version of PVP legislation was adopted, extending protection to plant parts, and potentially to products of those parts (say oil from a protected soybean variety).

For biotechnological applications though, the current scope of protection is insufficient. Breeders can simply use crossbreeding or other means to transfer a genetically engineered gene construct from one variety (protected) to another (unprotected). The breeders' rights exemption makes such a transfer perfectly legal. A patent plant would provide protection for it would read to the genetically engineered trait, so protection would not be voided through the breeding process. Moreover, the breeding process itself would be a likely infringement. Conversely, PVP covering the variety and a patent for the gene construct provides protection similar to a patented plant.

PVP-like systems have two advantages for animals. They are simpler and less costly to secure, and hence appropriate for sequential improvements such as livestock breeding. Second, many countries have specifically prohibited patents for plants or animals but are allowing PBR systems. However, a workable system requires identifying the target population – in short, a workable definition for a "variety." This exists for plants, but the relatively small number of progeny and inherent heterogeneity of an animal line (with the possible exception of inbred experimental lines) make a direct extension of PVP to animals inappropriate. One proposal for an "Animal Breeders' Rights" system is described in Sect. 3.4.

2.3.3 Trade Secrets

Since the 1930s, corn breeders have used trade secrets to protect first generation (F-1) hybrids. Since hybrids do not reproduce true-to-type, the sale of seed does not allow the duplication of the seed. Breeders need only protect the pure lines and crossings used to gain *de facto* protection of the resultant hybrid. Hence, breeders typically fence their pure line fields and of course do not reveal the crossing sequence.

For agricultural livestock, a similar approach is used by poultry and hog breeders. There, multiple crosses of pure lines produce synthetics with characteristics similar to plant hybrids. Poultry breeders keep tight control of their pure lines and release grandparent and even parent stock only under careful supervision. Of course, this strategy is not applicable for traditionally crossbred animals.

3 Issues with Protecting Living Organisms

The preceding text describes the general considerations for protecting inventions. Living organisms though bring to bear several additional considerations. This can be inferred from the general recentness of extensions of protection to living matter. An early example is a patent granted to Pasteur in the 1880s for a yeast strain, but the Patent Office seemingly considered the product to be inanimate rather than animate. The initial specialized law was the (US) Plant Patent Act of 1930 (subsequently incorporated into the Patent Act). The Plant Patent Act applies only to asexually propagated plants. Asexually propagated materials are stable, thus ending congressional uncertainty about the identification of protected materials over repeated regeneration.

Subsequently, European interests took the lead in protecting plant varieties, beginning in the 1940s. The first international convention, UPOV, was adopted in 1961 with the most recent version dating to 1991. The United States did not adopt a version, the Plant Variety Protection Act, until 1970, which was amended in 1980 and 1994.

3.1 Developments in the United States

Of major significance was the 1980 Supreme Court decision in *Chakrabarty*. While the particular decision related to a microorganism (for more rapid decomposition of spilled crude oil), the decision was made in sweeping terms. The Court declared, "everything under the sun that is made by man" is patentable subject matter. Shortly thereafter, in 1985, the patent and trademark office (PTO) in an internal decision, *Ex parte Hibberd*, declared (higher) plants to be patentable subject matter. "Patentable subject matter" simply means that the application cannot be rejected solely because of the area of application.

Interestingly, in 2001, there was a challenge heard by the Supreme Court to the provision of patents for plants (*J.E.M. AG Supply v. Pioneer Hi-Bred*). The challenge began (in typical fashion) as a defense for an infringement charge, declaring there was no infringement for the patent was invalid. The reason? The Congress passed the PVPA to protect plants and hence did not intend to allow for the extension of patents. The Supreme Court strongly endorsed *Hibberd,* declaring that patents and PBR could "mutually coexist."

Two years after *Hibberd,* in another internal decision in *Ex Parte Allen* (1987), the PTO declared higher animals to be patentable subject matter. Interestingly in that case, the underlying application for a "polypoidal" oyster was rejected for insufficient nonobviousness. The first animal patent was not issued until April 1988 for the "Harvard mouse," a laboratory research specimen.

3.2 Developments in Europe

Patent protection for animals (as well as other forms of living organisms) in Europe has been shaped first by the wording in the European Patent Convention (1973) (EPC), which, in Article 53(b), prohibited patents for "plant or animal varieties or essentially biological processes for the production of plants or animals." With an origin dating to an earlier and simpler era, it is now, with the advent of the new biotechnology, no longer easily determinable what is an "essentially biological process." Within Europe, there are two separate routes to patents: through national patent offices or through the European Patent Office (which grants "bundled" national patents). While there are small technical and practical differences between the patents received from the routes, the legislation harmonizing the requirement of the Convention means that the two systems operate in parallel for purposes here.

To clarify the interpretation of the phrase "essentially biological," the European Directive on the Legal protection of Biotechnological Inventions (Directive 98/44/EC, Rule 23(b)(5)) contains the definition:

"A process for the production of plants or animals is essentially biological if it consists entirely of natural phenomena such as crossing or selection."

As a consequence of that Directive, animals (and plants) that have been transformed genetically are patentable subject matter. Relevant Directive articles are as follows:

- Biological materials isolated from its natural environment or produced by means of a technical process may be the subject of an invention even if it previously occurred in nature (Article 6.3).
- The protection conferred by a patent on a biological material shall extend to any biological material derived from that biological material through propagation or multiplication in an identical or divergent form and possessing those same characteristics (Article 8).
- The protection conferred by a patent on a product containing or consisting of genetic information shall extend to all material (except the human body) in which the product is incorporated and in which the genetic information is contained and performs its function (Article 9).
- The protection shall not extend to biological material [] where the multiplication

or propagation necessarily results from the application for which the biological material was marketed, provided that material is not subsequently used for other propagation or multiplication (Article 10).

Note that, as a consequence of the genetic transformation, the products are no longer potentially construed as "discoveries," and hence are specifically unpatentable (EPC Article 52(2)(a)). Stated differently, an animal (or plant) is patentable in Europe (provided other patenting requirements are met) if the invention applies to more than the single variety.

The "Harvard mouse" invention, first to receive a US patent, also received the first animal patent in Europe, but the path was less than direct. The initial application (1985) was dismissed (1989) on the interpretation that it was a prohibited animal variety and then granted on appeal (1990, issued 1992). A subsequent opposition (1992) did not lead to a cancellation, but the revised patent was limited to rodents only (2001). Further appeals are likely.

The EPC (Article 53(a)) further allows exclusions from patentability for inventions that would be contrary to "ordre public" or "morality." Some attempts, generally unsuccessful, have been made to identify animal patenting as immoral. The Patent Office applies a test contrasting the public benefit from a successful invention with the potential suffering of the animals involved. Specifically, the Examining Division declared:

"In the case at hand three different interests are involved and require balancing: there is a basic interest of mankind to remedy widespread and dangerous diseases, on the other hand the environment has to be protected against the uncontrolled dissemination of unwanted genes and, moreover, cruelty to animals has to be avoided. The latter two aspects may well justify regarding an invention as immoral and therefore unacceptable unless the advantages, i.e. the benefit to mankind, outweigh the negative aspects."

3.3 Legislation in Other Leading Countries

Many patent statutes are silent on the patentability of higher life forms, making the patentability of animals a matter of legal interpretation. This point aside, 17 countries have granted the "Harvard mouse" patent (Austria, Belgium, Denmark, Finland, France, Germany, Greece, Ireland, Italy, Luxembourg, The Netherlands, Portugal, Spain, Sweden, the United Kingdom, and the United States; plus Japan has a related patent and New Zealand has issued a patent for another form of transgenic mouse). The issues in notable countries outside Europe and the United States are described below:

Australia: The Patent Office does not treat higher life forms (including animals) distinctly from lower life forms. And since lower life forms that are novel and inventive are patentable, it may be (and has been) higher animals can be patented too.

Canada: The "Harvard mouse" patent was initially rejected by the examiner on the grounds that higher life forms were not patentable subject matter in Canada. Following were a series of appeals:

- 1995 appeal to Commissioner of Patents, which upheld examiners rejection on the grounds that the creation of the mouse proper was controlled by the laws of nature.
- 1998 appeal to Trial Division of the Federal Court, which upheld the Commissioner's interpretation that higher

life forms were not patentable subject matter in Canada.
- 2000 appeal to the Federal Court of Appeals, which overturned the Trial Division decision, drawing on the reasoning of the US *Chakrabarty* decision.
- In 2002 (May), Commissioner of Patents appealed to the Supreme Court of Canada.
- In 2002 (December), the Supreme Court of Canada decided 5 to 4 that higher life forms are not patentable in Canada. The details are given below.

The sole issue before the Supreme Court was that the patentability of life forms constitute a "manufacture" or "composition of matter" within the context of Article 2 of the Canadian Patent Act. Interpreting the terms "manufacture" and "composition of matter," the former was understood to be a "nonliving mechanistic product or process," while the latter was construed, in the context of the proceeding terms of allowable subject matter (invention), "any new and useful art, process, machine, manufacture or composition of matter" (Patent Act, Section 2). The majority concluded that, while "composition of matter" can be interpreted broadly, the context of Section 2 means it is "best read as not including higher life forms."

"Higher life forms cannot be conceptualized as mere 'compositions of matter' within the context of the *Patent Act*. [] It is possible the Parliament did not intend to include higher life forms in the definition of 'invention.' It is also possible the Parliament did not regard cross-bred plants and animals as patentable because they are better regarded as 'discoveries.'"

The decision proceeds to note that patents for human life cannot be excluded judicially by the courts and that the existence of Canadian PVP legislation can be interpreted as possibly indicating the need for separate legislation to apply to higher life forms generally. Overall, "clear and unequivocal legislation [by the Parliament] is required for higher life forms to be patentable."

New Zealand: New Zealand uses the same term "manner of manufacture" to describe patentable subject matter as does Australia, so that applications in the two countries have progressed in parallel. Specifically, New Zealand has granted patents for higher animals.

Japan: Part VII of the Examination Guidelines for Patent and Utility Model in Japan (2000), Chap. 2, Part 4 describes when patents may be awarded for animals (Available at (verified 4/5/04) http://www.deux.jpo.go.jp/cgi/search.cgi?query=examination+guidelines&lang=en&root=short). Patents may be granted *per se*, for parts of animals or methods of creating animals. Humans are excluded. Also excluded are discoveries and those inventions not capable of industrial application (utility).

3.4 International Agreements

This European Patent Convention wording is also significant because it has carried through in modified form into the TRIPs (Trade-Related Aspects of Intellectual Property Rights) of the 1994 World Trade Organization. Signatories (all 147 of them currently) have agreed to adopt certain minimum forms of intellectual property protection. Some are straightforward – the provision of trade secret legislation is an example. The area of plants and animals though remains ambiguous, for in Article 27.3(b), countries have the option of excluding from patentability "plants and animals other than microorganisms, and essentially biological processes for the production of plants or animals []. However, Members shall provide for the protection of plant varieties either by patents or by an effective *sui generis* system or by any combination thereof." While the situation in some developed countries is ambiguous regarding the patenting of animals, as is noted above, virtually all developing countries are choosing to prohibit patents for plants and animals. In this case, the bar to the patenting of animals is more absolute than under the European Patent Convention for the wording reads as "animals," not "animal varieties." In consequence, it will likely be some time at best before animals are patentable in developing countries.

It is worth noting that while there is extant a form of *sui generis* (which means "special purpose") protection for plants in the form of PBR (see Sect. 2.3.2), nothing of the type exists for animals. Its absence means that there is no protection for animals not qualifying for a patent, which is to say nontransgenic animals. Yet the development of superior breeds is a costly and risky endeavor, just like other inventive activities, so animal breeding can be presumed to be under-funded as a consequence of the absence of a *sui geners* protection system. Lesser has attempted to conceptualize such a system, noting that the principal complexity is any workable definition of an "animal variety." Some animal lines such as those typically used in experimentation are sufficiently inbred to be stable and describable. Typically though, livestock, insects, fish, and pets are sufficiently heterogeneous as to defeat the description requirement of an IPR system. Further compounding the issue is the simple operation of heredity, which shows that only half the offspring of a simple trait will inherit the trait if just a single parent is a carrier, while more complex traits are inherited at far lower levels of probability. As a result, without testing individuals there is no assurance as to which will and which will not inherit a trait.

Lesser's approach in bypassing these characteristics of many higher animals functions more like a trademark than patent or PBR law. That is, protection is applied to the name of a parent and royalties are owing if the name is evoked, but not otherwise. Hence, one of *Man O'War's* foals would be charged royalties if he were specifically identified in a blood line, but not otherwise. Many beef cattle and purebred dogs use similar systems of identifying parent(s), while chickens and some fish and hogs (known as *synthetics*) are biologically more like hybrids than traditionally bred animals. They can be effectively protected through trade secrecy (see Sect. 2.3.3) by keeping the patent pure line stock and crossing sequence as confidential.

4 Issued Patents

On-line search engines of patent files at the US Patent and Trademark Office and the European Patent Office (EPO) make it possible to determine the numbers and types of patented animals. There is some uncertainty in the counts for it is not always entirely clear if the patent reads to an animal or a method for producing one. Also, when considering the species of the animals, some are definitive while other patents read to several species or all "nonhuman mammals." Nonetheless, from a review of patent files, it is possible to gain a clear understanding of the kinds of patents and uses thereof.

4.1 Animal Patents in the United States

US animal patents are classified in US class 800/13 – 20 (available at www.uspto.gov under patents, search). As of mid-2004, animal patents comprised:

Mammal (800/14)	88
Bovine (800/15)	22
Sheep (800/16)	21
Swine (800/17)	24
Mouse (800/18)	350
Bird (800/19)	1
Fish (800/20)	9

It is clear that mice (and rats) are the most commonly patented. As for the uses, animal disease models (as with the initial "Harvard mouse") dominate. An example is patent # 6,156,279, *HIV transgenic animals and uses therefore*, Dec. 5, 2000, which first claim reads (claims identify the invention):

1. A transgenic rat, whose genome contains at least one copy of a human immunodeficiency virus type 1 (HIV-1) proviral DNA [] and wherein said rat develops at least one symptom of acquired immune deficiency syndrome (AIDS).

Second in number are patents for the use of transgenic animals to produce useful compounds, most of which are excreted into milk. An example is patent # 6,727,405, *Transgenic animals secreting desired proteins into milk* (April 27, 2004):

1. A non-human mammal whose genome comprises a DNA construct [] wherein said mammal is selected from the group consisting of mouse, sheep, pig, goat and cow, and wherein said heterologous protein is expressed in the milk of the mammal.

Further in the medical area is a patent (# 6,331,658, *Genetically engineered mammals for use as organ donors*, Dec. 18, 2001) for producing a transgenic pig as a source of human transplant organs with a reduced likelihood of rejection:

10. A non-human transgenic mammal, wherein the genome of the mammal stably includes a nucleotide sequence [] such that the organ exhibits a decrease in antibody-mediated rejection...
11. The mammal of claim 10 wherein the mammal is a pig.

Examples of nonmedically related inventions are less common. One is (H1,065, Apr. 30, 2002, *Transgenic avian line resistant to avian leukosis virus*):

1. A chicken line, derived from a strain of chickens which is susceptible to avian leukosis virus infection, which chicken line has integrated in its genome a proviral sequence [] and is thus genetically resistant to avian leukosis virus, subgroup A.

This invention appears to anticipate commercial use, viruses being a cause of significant loss in poultry production.

Also included are – interestingly – transgenic pets. Led by patent # 6,380,458 (Apr. 30, 2002) is *Cell-lineage specific expression in transgenic zebrafish*. The claims leading to the commercial product, "GloFish," are:

1. A transgenic zebrafish that expresses a heterologous expression product, comprising a zebrafish cell lineage-specific expression sequence...

4. The transgenic zebrafish of claim 3 wherein the reporter protein is green fluorescent protein.

4.2 Animal Patents at the EPO

Animal patents (see Sect. 3.2) are indeed being issued in the EU. The UK Patent Office issued patent number GB2382579, "Assays and Disease Models Using Transgenic Zebrafish," in October 2003. The animal model screens suppressors that lessen the activity of a disease gene.

Current applications number approximately 70 (although not all can be expected to mature to a grant). Animal model patents represent 55 of applications, and expressions of proteins the remainder. The pattern is similar to the United States, and indeed many applications are based on inventions for which the first application was made in the United States. Examples from European inventors include:

"Transgenic Animal Model for Obesity Expressing FOXC2." Publication # EP1255774 11/02.

"C1 Inhibitor Produced in the Milk of Transgenic Mammals." Publication # 1252184, 9/01.

5 Issues with Patented Animals

This section addresses the numerous issues that arise or are associated with patented animals. The list of items includes other technical developments that are potentially patentable but have yet to be the subject of applications to what is known of effects, environmental, economic, and social, to attitudes.

5.1 Status of Technology

The current granted patents and applications (see Sect. 4) clearly indicate technologies and applications at a high level of development. It is possible that developments that are generally either (1) not sufficiently refined or novel to be the subject of an application, or (2) are of insufficient economic potential to justify the costs associated with a patent are absent. Yet, those technologies may in future years mature to become the subject of applications and so are worthy of notice. A technology of particular note is the replacement of microinjection (first applied in 1981) and retroviral vector systems (1985) as the transformation methods of choice to the use of sperm-mediated transgenics, or techniques involving nuclear transfer. Microinjection, while functional, was marked by a low level of success (less than 1%). That level means successful transformations are costly, and potentially contributes to unintended animal suffering.

Transgenic animals are produced for five underlying reasons:

- Improve animal health
- Increase productivity and improve product quality
- Mitigate environmental effects of animal production
- Produce therapeutics
- Provide models for medical research.

As is readily seen in Sect. 4, the final two reasons dominate current patents, while the first is also in evidence. The first three listed are nonetheless underrepresented in current patent documentation, and so warrant further attention here.

Sample developments are reported in Table 1.

Tab. 1 Examples of developments with transgenic animals.

Species	*Modification*	*Purpose*	*Main beneficiaries*
Tilapia	Growth hormone	Faster growth	Fish farmers
Salmon	Growth hormone	Faster growth	Fish farmers
Salmon	"Antifreeze" gene	Faster growth	Fish farmers
Sheep	Antibody in milk	Enriched milk	Consumers
Cow, ewe	Human protein in milk	Produce molecules	Humans
Cow	Higher casein	Enriched milk, heat resistance	Food industry
Pig	Phytase gene	Reduce phosphorus in excrement	Environment
Insects	Male sterility	Reduce pest populations	Farmers, public
Mosquitoes	*Plasmodium* (malaria) resistance	Health	Humans, esp. in tropics

Source: Findinier 2003, pp. 20–21.

A few of these require further explanation. Food quality/composition is a highly desirable development, but technically it has been difficult to achieve as traits are frequently controlled by multiple genes. Plant biotechnology has experienced similar complexities. Fish biotechnology is largely directed to enhanced growth rates, using several distinct approaches. Transgenic salmon have exhibited growth rates three to five times the nontransformed controls. Those with the antifreeze attribute show growth throughout the year, not only during the warmer months.

The transgenic Enviropig© uses the phytase enzyme transferred from a bacteria to help with the digesting of phosphorous in cereal grains. There is an 80 to 100% percent improvement over the figures for nontransformed pigs in the sense that there is no need for phosphorous supplements; Further, there are reductions of 57 to 64% of phosphorous in manure. Owing to the effect of phosphorous on stimulating aquatic algae growth, its reduction in manure will have distinct environmental benefits. However, the transgenic must compete in the marketplace with feed-based phosphorous-reducing approaches, so the practicality remains to be proven.

Genetic male sterility to control pest populations is seen as a more cost-effective alternative to radiation sterilization. It has the conceptual benefit of not reducing the competitiveness of the sterile males, but practicality varies across species because of the ease of sexing and degree of monogamous mating. Mosquitoes resistant to *Plasmodium* provide an alternative approach to controlling one of the greatest causes of loss of life worldwide. The need is greatest as populations are developing resistance to available prophylactic medications. However, studies have shown that resistance has to be near 100% to be effective, and resistance-delaying strategies may mean the engineering of multiple gene resistance, a complex and time consuming task.

5.2
Benefits

The preceding sections describe – or, at minimum, hint at – the benefits generated by patented animals. In this section, we have included further documentation of the benefits that are being generated and to whom they accrue.

5.2.1 Animal Models

Numerous animal models are being developed through the mechanism of biotechnology. Firms are presently designing new models on request, as with Charles River Laboratories, one of the large supplier firms:

> Charles River and Xenogen Biosciences formed a marketing collaboration []. Xenogen Biosciences provides custom programs for the development of transgenic animals []. (www.criver.com; verified 5/04)

Products are in considerable demand. By 1989, orders for the path breaking "Harvard mouse" amounted to 10 000 at $50 each, versus $3.50 for standard lines.

5.2.2 Production of Compounds

The actual production costs of compounds using transgenics versus traditional approaches are not publicized, making direct comparisons impossible. However, complex compounds are very costly to produce using bioreactors, such as blood clotting Factor IX, which costs $100 000 to 200 000 annually per patient for hemophiliacs. Pigs can secrete Factor IX into milk at a concentration 250 to 1000 times the level of bioreactors, making purification simpler. Indeed, production is so elevated that only several hundred transgenic pigs are needed to supply all hemophiliacs worldwide. Animal numbers, however, will be so small compared to standing herd sizes as to have no effect on the farm and agribusiness sectors.

5.2.3 Other Applications

Other transgenic technologies, and particularly those used for milk production, are insufficiently commercialized at this stage to have any reliable indication of effects on producers or consumers. One can however note that animal diseases are a major cost component for prevention, treatment, and associated losses so that enhanced resistance would have marked benefits for both producers and consumers of animal products. For example, US vet and medicine costs for hogs per hundredweight of gain in 2002 was $1.10, or 2% of total costs, but 14% of residual returns minus operating costs. Worldwide, small farmers are the major owners of livestock – in India, they own 70% of the livestock but just 30% of the land area – and could be a major beneficiary.

For fish, a genetic mutation (nonbiotech) leading to genetically male Nike tilapia increased production by 30%. For Philippine farmers utilizing them, the net income doubled.

5.3 Areas of Concern

Animal patents are, to understate the current situation, very controversial, at least with certain groups. Concerns are focused on

- ethics,
- food safety,
- health,
- environmental safety, and
- access to patented animals.

Here the underlying positions will be summarized in reverse order. No attempt is made to represent a consensus position as that may not exist, or is not discernible. Rather, in the following section, attitude surveys from several countries are reported as an indication of the acceptability of patented animals in particular, and the products of those creations.

5.3.1 Access to Patented Animals

Effective access involves a complex of cost, availability, and use provisions. Here, attention is on agriculture/aquaculture and other dispersed uses rather than medical applications, where commercial channels among limited number of users are of long-standing nature.

Cost cannot be reliably projected at this stage, other than to note that most food production have low margins, meaning high process would render the innovations uncompetitive in many cases. For example, with rBST, an injected biotech product used for stimulating milk production, analysis suggests that the supplier, Monsanto, prices the product near the breakeven point for users. That keeps adoption rates low – indeed, over time, use has been declining. Distribution of livestock-based technologies depends in part on the production rates by species. Poultry are high producers, hogs medium, while cattle naturally produce but one calf annually. For single gene traits, only half will inherit; for multiple gene ones, less. Embryo splitting and implantation speed up the rate, but at considerable cost. For cattle, then, dissemination through herds will be a slow process.

A second consideration relates to the collection of royalties, assuming that producer cash flow considerations will dictate royalties be paid when returns are realized at sale time. Lesser's analysis proposes that packing plants (livestock) or breed associations (pets) will be the royalty collectors of necessity. Until rapid, cheap tests are developed to determine the presence of traits, fees must be based on herd averages.

5.3.2 Environmental Safety

Environmental safety applies principally to concerns over the escape and competition by transgenic animals in the natural environment. Competition can affect the same species through crossbreeding, or other species by colonizing an environmental niche. The degree of the threat then depends in large measure on a species' ability to become feral, escape from captivity, and its mobility. Insects then are predicted to have substantial potential for environmental damage while domesticated cattle are ranked low (Table 2).

Some experience exists with nontransformed salmon raised in cages in the sea. Escapes are estimated at 15%, a number confirmed by the percent of farm-raised salmon caught by Norwegian fishermen. However, farm-raised salmon, like most nonnative species, have been found to be ineffective breeders compared to the wild stock. Nonetheless, if the alien species is a large portion of a local population, then a notable impact can be anticipated.

Procedures for assessing and, when needed, mitigating the environmental

Tab. 2 Environmental risk factors levels for transgenic animals.

Animal	*Ability to become feral*	*Likelihood of escape*	*Mobility*
Insects[a]	High	High	High
Fish[b]	High	High	High
Rodents	High	High	High
Cats	High	High	Moderate
Pigs	High	Moderate	Low
Rabbits	High	Moderate	Moderate
Chickens	Low	Low	Low
Cattle	Low	Low	Low

[a]Gypsy moth and African bee only.
[b]Excludes shellfish and crustaceans.
Source: FDA, reported in Findinier 2003, p. 17.

affects of transgenics are known as *biosafety*. (The term sometimes encompasses food safety, but as used here refers to environmental aspects only). National legislation specifies the review procedures involved, although they are better developed and more practiced with plants than animals. Internationally, there is an umbrella agreement, the Cartagena Protocol under the Biodiversity Convention (available at http://www.biodiv.org/biosafety/default.aspx; verified 5/04). The Protocol, established in 2000, has to date (5/04) been ratified by 98 countries. It establishes general procedures and requirements but does not contain specific provisions.

5.3.3 **Health**

Human health issues are associated with the development of new viruses, particularly if animal organs are used for human transplants. A retrovirus used as a transformation vector could combine with a latent virus in the new host, spreading through the food chain and into humans. This threat can be ended by discontinuing viral vectors.

Transplanted organs present a special risk for the same genetic modification that reduces the rejection of the foreign organ could also weaken the patient's immunity to the new virus. Thus situated, the new virus could adapt itself to the novel human host environment and become infectious. This possibility is not trivial for pigs – already the subject of a xenotransplantation patent (see Sect. 4.1) – have at least six known endogenous retroviruses. Overall, concerns about the generation of new diseases "require a chain of events of very low cumulative probability but for which there are some experimental precedents."

5.3.4 **Food Safety**

Internationally, food safety standards and procedures are established under FAO/WHO by the Codex Alimentarius Commission. The main purposes of this program are protecting health of the consumers and ensuring fair trade practices in the food trade, and promoting coordination of all food standards work undertaken by international governmental and nongovernmental organizations.

In 2003, the Commission adopted the "Principals for the Risk Assessment of Foods Derived from Modern Biotechnology" as an overarching agreement for understanding risk assessment and nutritional status of foods derived from biotechnology, including transgenic animals (available at http://www.fao.org/es/ESN/food/risk_biotech_taskforce_en.stm; verified 5/04). The principals are based on the following elements:

- Premarket assessment on a case-by-case basis
- Assessment based on a comparison with traditional foods (known as substantial equivalents)
- Risk management should be proportional to identified risks.

Food concerns of transgenic animals are focused largely on the existence of novel proteins. Those proteins can be added intentionally (as with a growth hormone), result from unanticipated compositional changes, or may be a consequence of the unpredictable trans gene expression in animals. For example, there may be a transfer from mammary glands to the blood and hence into the food supply. When testing for food safety, the traditional focus is on

- toxicity,
- allergenicity, and
- whole food value.

Toxicity is relatively straightforward to examine, although there is always a policy decision over which populations (elderly, children) are to be protected and at what presumed consumption levels. Allergenicity is problematic to screen for because very small portions of a population may be very allergic to a wide range of proteins. Whole food value – an assessment if the total nutritional balance has been altered – presents test complications because the subject product often constitutes a small portion of the diet so that feeding large-enough amounts to detect nutritional effects would be unfeasible or unhealthy. In any case, similar issues arise with transgenic plant-derived foods, and to date there is no documented example of any deleterious human health effects.

5.3.5 Ethics

Individuals may object to patenting animals either directly on the basis of a rejection of ownership and "playing God" by interfering with a natural creation or indirectly because patents can foster actions or behavior considered deleterious to animals. The first point is a difficult one for multicultural societies to resolve because opinions can be intense, but may vary widely. What is morally repugnant to some may be fully acceptable to others. The best indication of the prevalence of these objections is public opinion surveys (see Sect. 6 following).

The indirect effects of patents through transgenics can be discussed in the terms of animal rights and animal welfare. Animal rights supporters may be concerned with, for example, not changing what is essential about a pig, or allowing animals to live in a way consistent with their nature. Pigs, for example, are both social and programmed to rooting and, by one perspective, should not be prevented from those expressions. What is innate to a species is often difficult to conceptualize after millennia of being subject to human control. And rights have broad consequences that are likely not fully comprehended by many.

Animal welfare, however, has wide and growing support, largely directed to the avoidance of unnecessary suffering. Austria, for example, recently adopted legislation specifying that chickens must be reared on a floor (not in cages) and Dobermans may not have tails and ears trimmed for appearance reasons. Animal suffering is a possible issue with animal patents if, for example, territorial instincts can be reduced, allowing for more dense confinement, or if experimentation leads to suffering. How is that matter addressed?

A partial offset to overall suffering is the possibility of experimentation that will lessen future suffering. Developing disease treatment or resistance is an example. There are examples of disease-resistant animals becoming available (see Sect. 4). Yet, many experiments will benefit humans far more than the test species. In those cases, a common approach is to consider the balance of (animal) suffering compared to potential (human) benefit. In Europe, the EPO has devised a process for determining if the advantages outweigh the negative aspects (see Sect. 3.2). Outside patent offices, calls have been made for weighing "costs and benefits." For the foreseeable future, the utilitarian view of placing human needs in the primacy while minimizing the costs to animals seems likely to prevail, but views of animal rights are expanding with the general public and could affect fundamental research with transgenics in the longer term.

6 Public Attitudes

As a background, it is important to recognize that the general public is not particularly well informed about biotechnology. As examples:

- 1999 (Gallup) 50% heard little/nothing
- 1987 (Gallup) 63% heard little/nothing
- 2001 "heard a great deal about genetically modified foods" – 44%
- 2003 "heard a great deal about genetically modified foods" – 34%

If anything, awareness in the United States is declining in an era where there are no notable headlines regarding transgenics.

Indeed, the public indicates a low level of knowledge on the subject – in 2001, respondents self-selected a median rating of only 3 (1–10 scale of "knowledgeable") – and demonstrates a limited knowledge with such replies as:

United States	33% believe tomatoes do not contain genes, 58% say they have not eaten transgenic foods
Europe	65% believe tomatoes do not contain genes

Of course, knowledge does not necessarily influence opinions, which are based on emotions. If anything, a lack of knowledge makes would-be consumers less accepting as risk is associated with the new. For that reason, more educated consumers, who presumably are better able to educate themselves, typically are more accepting. The lack of knowledge does make respondents especially vulnerable to differences in terminology. Surveys have indicated that the label biotechnology is better accepted than using "genetically modified." Hence, traditional supporters of biotechnology like food companies use the former in surveys, while opponents such as some environmental groups favor genetically modified. Overall, as always, surveys should be interpreted with some care.

6.1 Attitudes in the United States

Whatever the comfort level with eating transgenics, consumers are accepting the transforming of life forms for other reasons. While transgenic animals for food are ranked at 2.8 out of 10 (compared with 6.8 for plants), other uses of transgenic animals is less than half that level, or 2.3 (Fig. 1). Seemingly, the closer the life form is to direct human experiences, the lower the acceptance.

Support for transgenic animals is thrown into sharper contrast when evaluated against transgenic plants. For plants, there is an 81% positive position, and a 67% point spread between favorable and unfavorable views (Fig. 2). For the same question applied to animals, the figures are 49% approval and only an eight-point spread. When the issue is less cost and more specific health and safety benefits, the divergence is smaller, but still notable. For example, 57% of respondents favor the use of animals for transplant organs, while 58% support the idea of producing spiderweb compounds in goat milk (which can make strong and light bulletproof vests). Another survey reports the approval for plant transgenics of 49% in 2003 versus 27% for animals.

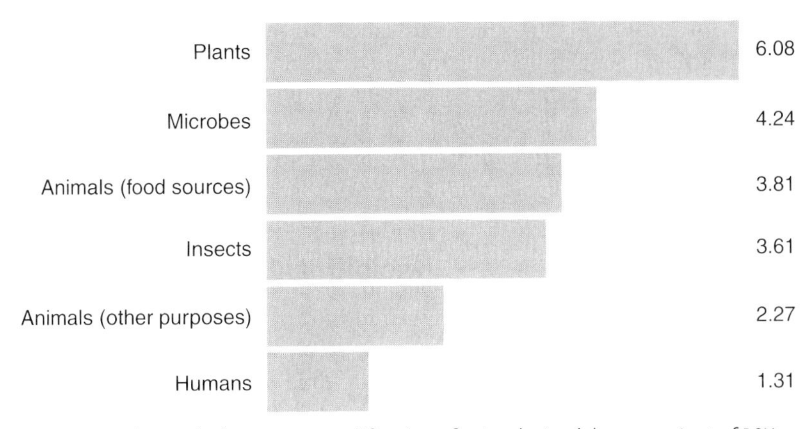

Fig. 1 Comfort with the genetic modification of animals (rank by mean (out of 10)). *Source*: Pew Initiative on Biotechnology (2003b) *Americans Are Far More Comfortable With Genetic Modifications Of Plants Than Animals*. Available at http://pewagbiotech.org/research/2003update/4.php; verified May 2004.

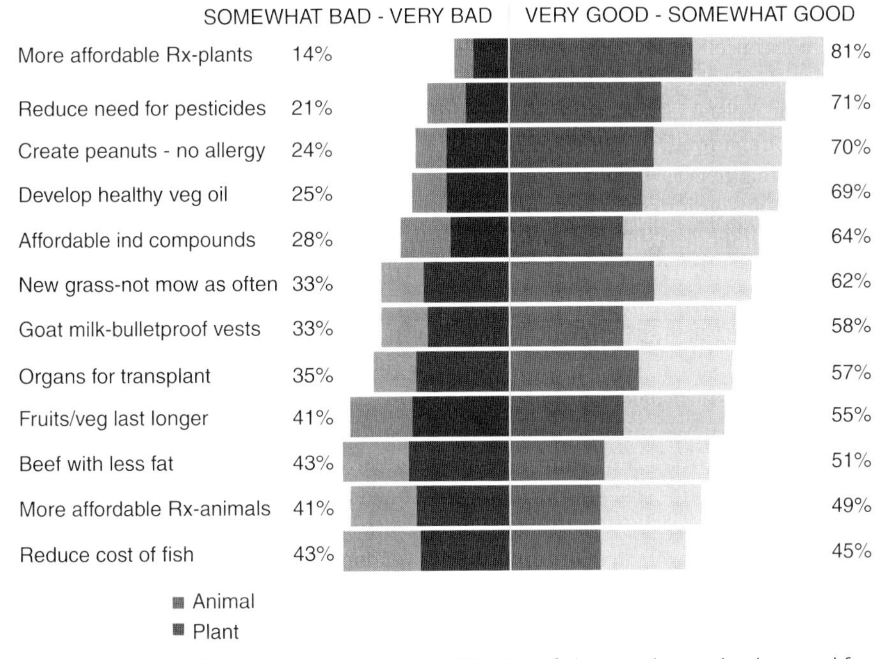

Fig. 2 Likelihood of supporting the genetic modification of plants and animals when used for (darker shading : stronger intensity). *Source*: Pew Initiative on Biotechnology (2003b) *Americans Are Far More Comfortable With Genetic Modifications Of Plants Than Animals*. Available at http://pewagbiotech.org/research/2003update/4.php; verified May 2004.

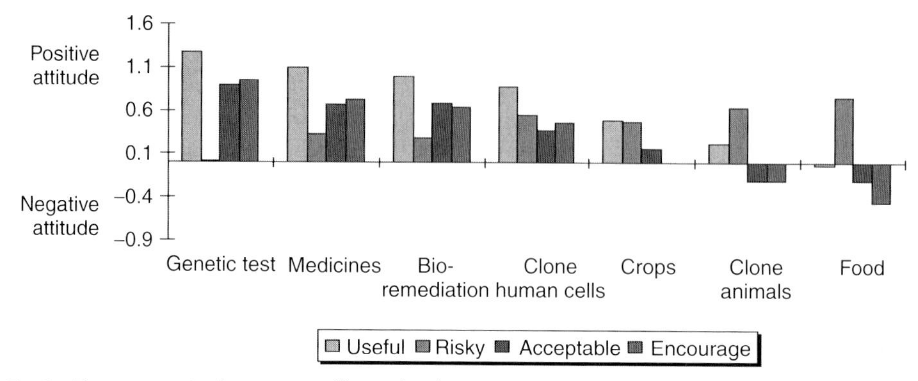

Fig. 3 European attitudes to uses of biotechnology, 2002. *Source*: Nature America, 2000, from Eurobarometer.

These opinions should be viewed in the context of public knowledge of animal biotechnology. In fact, 32% of Americans state that they do not know enough to form an opinion on the subject and another 11% are neutral (28% have a favorable impression, 29% unfavorable). Consumers distinguish among three areas of animal biotechnology. They are favorably disposed (55%) toward "genomics," the use of genetic information to improve animal care and food quality from animals. However, fewer consumers are favorably disposed toward genetic engineering in animals (36%), and even fewer toward cloning (18%).

6.2 Attitudes in Europe

While Europeans are more opposed to biotechnology than are North Americans, attitudes are nuanced by the use of the technology, as in North America. Support is far higher for medicines than food in general (Fig. 3).

By 2002, the periodic Eurobarometer series on attitudes to biotechnology had added a question about xenotransplantation, the only question with direct relevance to patented animals. The absolute rating was similar to crops, with useful and risky at nearly identical levels (Fig. 3). Weak support exists in 13 of the 15 countries polled, with Greece, Finland, and Austria in opposition (Table 2).

The question if there is a general change in attitude over time cannot be answered directly from the opinion surveys because of changes in the wording of the question. However, partitioning the respondents into supporters and risk acceptors, a general pattern of increases in support for "GM foods" is evident in all countries but France, Italy, and the Netherlands. Indeed, support in Italy actually declined – the only country where this happened.

See also Nuclear Transfer for Cloning Animals.

Bibliography

Books and Reviews

Beardmore, J.A., Porte, J.S. (2003) *Genetically Modified Organisms and Aquaculture*, FAO, Rome, Fisherise Circular No. 989 (FIRI/C989C).

Blattman, A., McCann, J., Bodkin, C., Naumoska, J. (2002) *Global Intellectual Property: International Developments in Animal Patents*, Ch. 5 63–84 in M.F.

European Patent Office (1992) *Official J.* **588**, 591–592.

Findinier, I. (2003) *Animal Biotechnologies: State of the Art, Risks and Perspectives*, FAO, Rome Available at www.fao.org/es/ESN/food/GenBiotech_en.

Gaskell, G., Allum, N., Stares, S. (2003) *Europeans and Biotechnology in 2002*. Eurobarometer 58.0. Available at http://europa.eu.int; verified 5/04.

Grubb, P.W. (1999) *Patents for Chemicals, Pharmaceuticals and Biotechnology*, Clarendon Press, Oxford.

Hallman, W.K., Hebdan, W.C., Aquino, H.L., Cuite, C.L., Lang, J.T. (2003) *Public Perceptions of Genetically Modified Foods: A National Study of American Knowledge and Opinion*, Food Policy Institute, Cook College, Rutgers, New Brunswick, NJ Available at www.foodpolicyinstitute.org; verified May 2004.

Houdebine, L.M. (2002) Transgenics to improve animal production, *Livest. Prod. Sci.* **74**, 255–268.

International Food Information Council (2004) *Support for Food Biotechnology Stable Despite News on Unrelated Food Safety Issues*. Available at http://ific.org/research/biotechres03.cfm; verified May 2004.

Ithaca J. (2004) Genetically modified pets making retailers uneasy, 8A.

Johnston, J.A., *Rochester Times-Union* (1989) B&L will make mice for cancer research, 1A, 8D.

Kesavan, P.C., Swaminathan, M.S. (2003) Ethical, Social, Environmental and Economic Issues in Animal Agriculture, *Book of Extended Synopses, FAO/IAEA International Symposium on Applications of Gene-Based Technologies for Improving Animal Production and Health in Developing Countries*, Vienna, 6–10 Oct. 2003, IAEA-CN-110/KN85.

Lavitrano, M., Bacci, M.L., Forni, M., Lazzereschi, D., Stefano, C.D., Foretti, D., Giancotti, P., Marfe., G., Pucci, L., Renzi, L., Wang, H., Stoppacciaro, A., Casa, G.D., Seren, E., Rossi, G. (2002) Efficient production by sperm-mediated gene transfer of human decay accelerating factor (hDAF) transgenic pigs for xenotransplantation, *Proc. Natl. Acad. Sci. U.S.A.* **99**, 14230–14235.

Lesser, W. (1987) The impacts of seed patents, *North Cent. J. Agr. Econ.* **9**, 37–48.

Lesser, W. (1993) Animal variety protection: a proposal for a US model law, *J. Pat. Trademark Off. Soc.* **76**, 697–715.

Lesser, W., Bernard, J., Billah, K. (1999) Methodologies for *ex ante* projections of adoption rates for agbiotech products: Lessons learned from rBST, *Agribusiness* **15**(2), 149–162.

Maskus, K.E. (2000) *Intellectual Property Rights in the Global Economy*, Institute for International Economic, Washington, DC.

National Research Council (2002) *Animal Biotechnology: Science-Based Concerns*, National Academies Press, Washington, DC.

Nature America (2000) *Biotechnology and the European Public*. Available at http://biotech.nature.com; verified May 2004.

Pew Initiative on Biotechnology (2002) *Biotechnology in the Barnyard: Implications of Genetically Engineered Animals*. Available at www.pew.org; verified 5/04.

Pew Initiative on Biotechnology (2003a) *Awareness of Genetically Modified Foods Remains Low*. Available at http://pewagbiotech.org/research/2003update/4.php; verified May 2004.

Pew Initiative on Biotechnology (2003b) *Americans Are Far More Comfortable With Genetic Modifications Of Plants Than Animals*. Available at http://pewagbiotech.org/research/2003update/4.php; verified May 2004.

Rothschild, M.F., Newman, S. (Eds.) (2002) *Intellectual Property Rights in Animal Breeding and Genetics*, CABI Publishing, Oxford.

Thompson, P.B. (1997) *Food Biotechnology in Ethical Perspective*, Blackie, London.

U.S. Department Agriculture (2004) *Hog Production Costs and Returns*. Available at http://usda.ers.gov/Data/CostsandReturns/data/current/C-hogs.xlts (verified 5/04).

Walterscheid, E.C. (1994) The early evolution of the united states patent law: antecedents (*Part I*), *J. Patent and Trademark Office Soc.* **75**, 398–426.

Primary Literature

Committee on the Judiciary, House of Representatives (1988) *Patents and the Constitution: Transgenic Animals*, U.S. Government Printing Office, Washington, DC.

Grubb, P.W. (1999) *Patents for Chemicals, Pharmaceuticals and Biotechnology*, Clarendon Press, Oxford.

Lesser, W. (Ed.) (1989) *Animal Patents: The Legal, Economic and Social Issues,* Stockton Press, New York.
National Research Council (2002) *Animal Biotechnology: Science-Based Concerns,* National Academies Press, Washington, DC.
Office Technology Assessment, Congress of the US (1989) *New Developments in Biotechnology: Patenting Life – Special Report,* OTA-BA-370. U.S. Goverment Printing Office, Washington, DC.
Rothschild, M., Newman, S. (Eds.) (2002) *Intellectual Property Rights in Animal Breeding and Genetics,* CABI Publishing, Oxford.

Lou Gehrig's Disease (Amyotrophic Lateral Sclerosis): *see* Motor Neuron Diseases: Cellular and Animal Models

Lymphokines: *see* Cytokines: Interleukins

Machines, Biological: *see* Motor Proteins

Macromolecules, X-Ray Diffraction of Biological

Albrecht Messerschmidt and Robert Huber
Max-Planck-Institut für Biochemie, Martinsried, Germany

Encyclopedia of Molecular Cell Biology and Molecular Medicine, 2nd Edition. Volume 7
Edited by Robert A. Meyers.

ISBN: 3-527-30549-1

Keywords

Anomalous scattering
If the wavelength of the incident X-ray beam is around the absorption edge of a given atom, its atomic scattering factor becomes complex with a normal, a dispersive, and an absorption component.

Atomic scattering factor
Mathematical expression for the X-ray scattering of an individual atom.

Crystallographic R-factor
Reliability factor for the X-ray structure determination, sum over the absolute values of the differences between the observed and calculated structure factors divided by the sum of the absolute values of the observed structure factors.

Electron density function
The spatial distribution of electrons in the crystal, which can be obtained by the Fourier transformation of the structure factors, represents the atomic arrangement in the crystal.

Ewald construction
Construction to visualize the diffraction of X-rays by a crystal; a sphere with a radius of the reciprocal of the incident X-ray wavelength is drawn with the origin of the reciprocal lattice of the crystal at the intersection of the incident beam with the surface of the sphere; diffraction occurs if a reciprocal lattice point intersects the surface of the so-called Ewald sphere; the direction of the diffracted beam is the connection between the center of the Ewald sphere and the intersection point of the relevant reciprocal lattice point.

MAD
Multiple anomalous diffraction, a method to exploit the anomalous scattering effect of a set of natural or artificially introduced anomalous scatterers for the solution of the phase problem.

MIR
Multiple isomorphous replacement, a method to solve the phase problem by preparing isomorphous heavy metal derivatives and measuring X-ray data sets of the native and the derivative crystals.

Phase problem
The phase of the diffracted waves is lost during the measurement as the X-ray intensities deliver the amplitude of the diffracted waves only.

Reciprocal lattice
A lattice in reciprocal space with lattice points, which are perpendicular to the corresponding set of lattice planes in the lattice in direct space with a length of the reciprocal of the lattice plane distance or its multiples.

Resolution
Minimal distance between two objects that can be resolved.

Structure factor
Mathematical expression for the X-ray wave diffracted by a crystal, which consists of amplitude and phase factor.

X-ray diffraction of biomolecules covers diffraction experiments on single crystals of small biological molecules such as oligopeptides, cofactors, steroid hormones, and so on, or biological macromolecules. But it also includes small angle scattering experiments on biological macromolecules in solution. This chapter is exclusively dedicated to the X-ray crystallography of biological macromolecules. It comprises the determination of the three-dimensional structures of proteins, nucleic acids, and other biological macromolecules at atomic resolution by diffraction of X-rays on crystals of such macromolecules. The atomic structure is obtained from the electron density distribution of the macromolecular crystal, which is the Fourier transform of the waves diffracted by the crystal. The amplitudes of the diffracted waves are determined directly from the diffraction experiment. The diffraction phases are revealed (1) from additional diffraction data of isomorphous heavy atom derivatives (multiple isomorphous replacement (MIR) technique), (2) from multiple anomalous diffraction (MAD), if suitable anomalous scatterers are in the crystal and using tunable synchrotron radiation, and (3) by Patterson search techniques (molecular replacement), if structural information on the biological macromolecule under investigation is available. X-ray crystallography of biological macromolecules is the unique method for the elucidation of the spatial structures at atomic resolution of complex biomolecules with molecular masses greater than 30 kDa. The structures represent a time and spatial average of molecules packed in a crystal. The preparation of suitable crystals may be a limiting factor.

1 Introduction

1.1 Crystals and Symmetry

In a crystal, atoms or molecules are arranged in a three-dimensionally periodic manner by translational symmetry. The crystal is formed by a three-dimensional stack of unit cells, which is called the *crystal lattice* (Fig. 1a and b). The unit cell is built up by three noncollinear vectors **a**, **b**, and **c**. In the general case, these vectors have unequal magnitudes and their mutual angles deviate from 90°. The arrangement of the molecule(s) in the unit cell may be asymmetrical, but very often it is symmetrical. This is illustrated in Fig. 2(a–e) in two-dimensional lattices for rotational symmetries.

In crystals, only 1-, 2-, 3-, 4-, and 6-fold rotations are allowed. This follows from the combination of the lattice properties with rotational operations. Other possible symmetry elements are mirror plane *m*, inversion center, and combination of rotation axis with inversion center (inversion axis). These are the point group symmetries. They can only occur among each other in a few certain combinations of angles. Other angle orientations would violate the lattice properties. The number

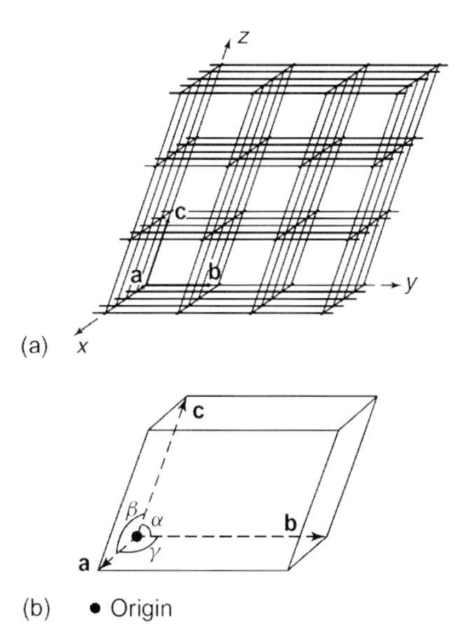

Fig. 1 (a) Crystal lattice and (b) unit cell.

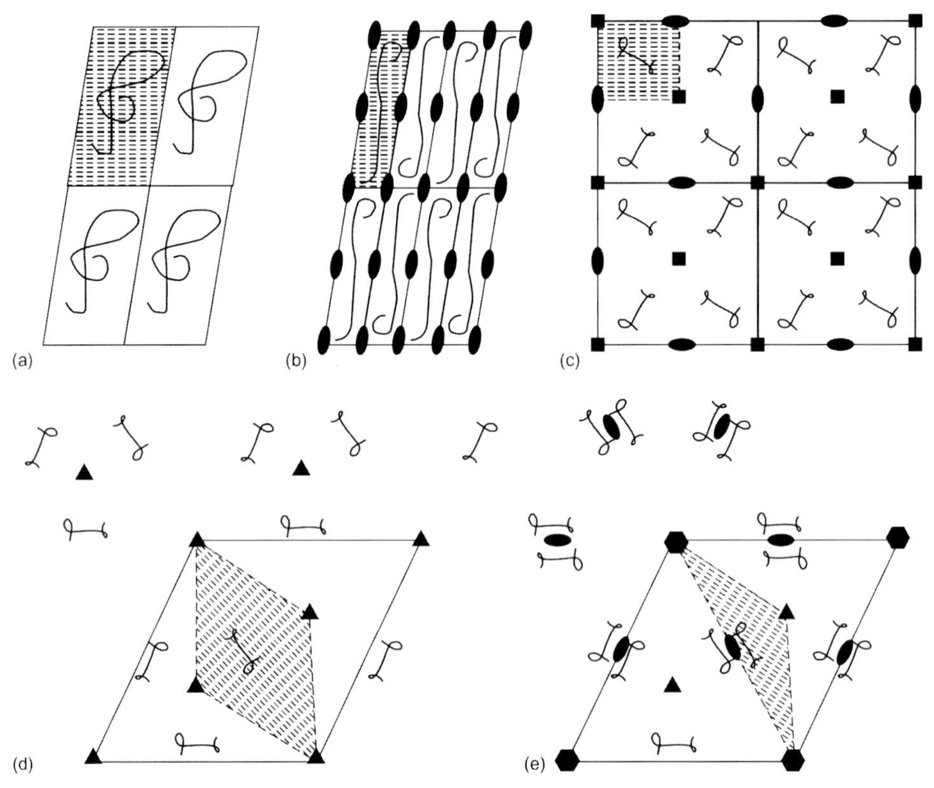

Fig. 2 Rotational symmetry elements in two-dimensional lattices: (a) 1, (b) 2, (c) 4, (d) 3, and (e) 62. The asymmetric unit is hatched.

of all possible combinations reveals the 32 point groups. The crystal morphology obeys the point group symmetries.

Adding an inversion center to the point group symmetry leads to the 11 Laue groups. These are of importance for the symmetry of X-ray diffraction patterns. Their symbols are 1, $2/m$, $2/mmm$, 3, 3 m, $4/m$, $4/mmm$, $6/m$, $6/mmm$, $m3$, and $m3m$. Proteins and nucleic acids are chiral molecules. Therefore, they can crystallize only in the 11 enantiomorphic point groups: 1, 2, 3, 4, 6, 23, 222, 32, 422, 622, and 432.

The combination of point group symmetries with lattices leads to seven crystal systems, triclinic, monoclinic, orthorhombic, trigonal, tetragonal, hexagonal, and cubic, with 14 different Bravais-lattice types, which can be primitive, face-centered, all-face-centered, and body-centered. Furthermore, additional symmetry elements are generated, having translational components such as screw axes or glide mirror planes. There exist 230 space groups of which 65 are enantiomorphic (for chiral molecules such as proteins). Figure 3 shows the graphical representation for the space group $P2_12_12_1$, as listed in the *International Tables for Crystallography*. The asymmetric unit is one-fourth of the unit cell and can contain one or several molecules. Multimeric molecules may have their own

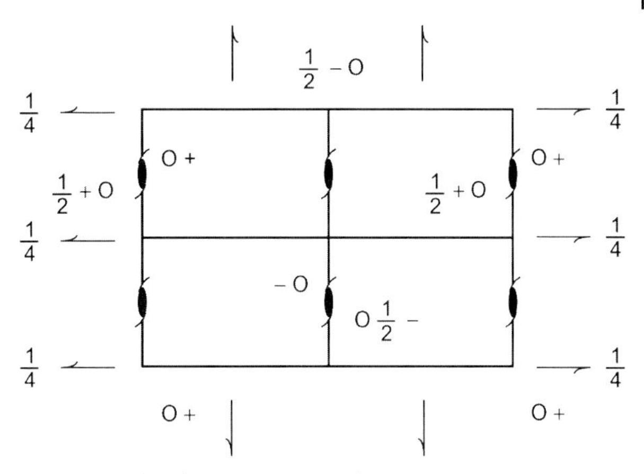

Fig. 3 Graphical representation of space group $P2_12_12_1$.

symmetries, which are called *noncrystallographic symmetries*. Here, axes that are 5-fold, 7-fold, and so on, are also allowed.

1.2 Protein Solubility

Figure 4 shows a typical phase diagram illustrating the solubility properties of a macromolecule. In the labile phase, crystal nucleation and growth compete, whereas in the metastable region, only crystal growth appears. In the unsaturated region, crystals dissolve. The solubility of proteins is influenced by several factors, as follows.

1.2.1 Ionic Strength

A protein can be considered as a polyvalent ion and, therefore, its solubility can be discussed on the basis of the Debye–Hückel theory. In aqueous solution, each ion is surrounded by an "atmosphere" of counter ions. This ionic atmosphere influences the interactions of the ion with water molecules and hence the solubility.

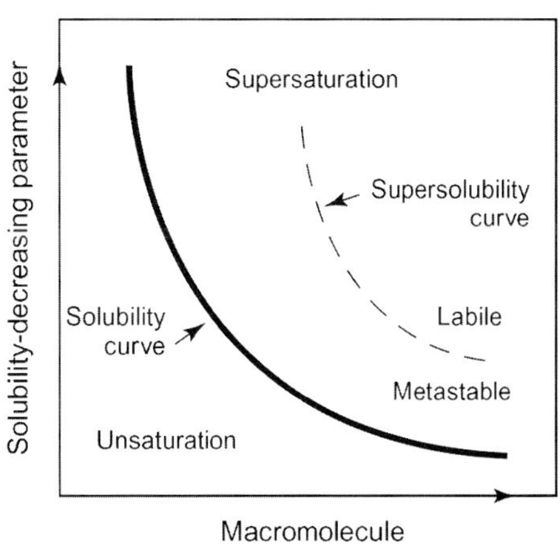

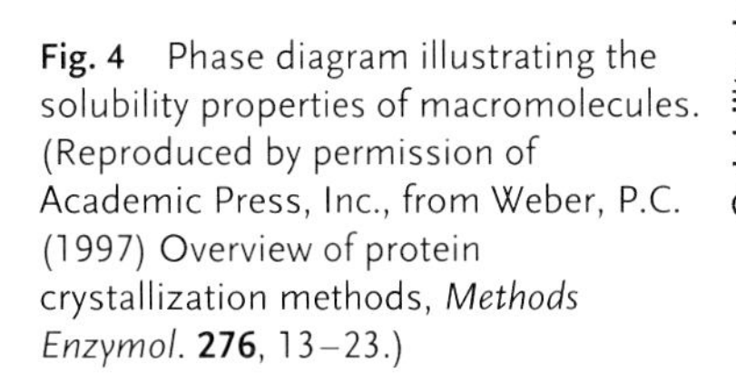

Fig. 4 Phase diagram illustrating the solubility properties of macromolecules. (Reproduced by permission of Academic Press, Inc., from Weber, P.C. (1997) Overview of protein crystallization methods, *Methods Enzymol.* **276**, 13–23.)

1.2.1.1 **"Salting-in"** At low ionic concentration, the "ionic atmosphere" increases the solubility as it increases the possibilities for favorable interactions with water molecules. We obtain (Eq. 1 and 2)

$$\log S - \log S_0 = \frac{AZ_+Z_-\sqrt{\mu}}{1 + aB\sqrt{\mu}} \quad (1)$$

$$\mu = \frac{1}{2}\sum c_j Z_j^2 \quad (2)$$

where μ = ionic strength, S = solubility of the salt at a given ionic strength μ, S_0 = solubility of the salt in absence of the electrolyte, Z_+, Z_- ionic charge of salt ions, A, B = constants depending on the temperature and dielectric constant, a = average diameter of ions, and c_j = concentration of the jth chemical component. Ions with higher charge are more effective for changes in solubility. Most salts and proteins are more soluble in low ionic strength than in pure water. This is termed as *salting-in* (Fig. 5).

1.2.1.2 **"Salting-out"** At higher ionic strength, the ions compete for the surrounding water. Therefore, water molecules are taken away from the dissolved agent and the solubility decreases according to Eq. (3):

$$\log S - \log S_0 = \frac{AZ_+Z_-\sqrt{\mu}}{1 + aB\sqrt{\mu}} - K_s\mu \quad (3)$$

The term $K_s\mu$ predominates at high ionic strengths, which means that "salting-out" is then proportional to the ionic strength (Fig. 5). In a medium with low ionic strength, the solubility of a protein can be decreased by increasing or decreasing the salt concentration. Salts with small, highly charged ions are more effective than those with large, lowly charged ions.

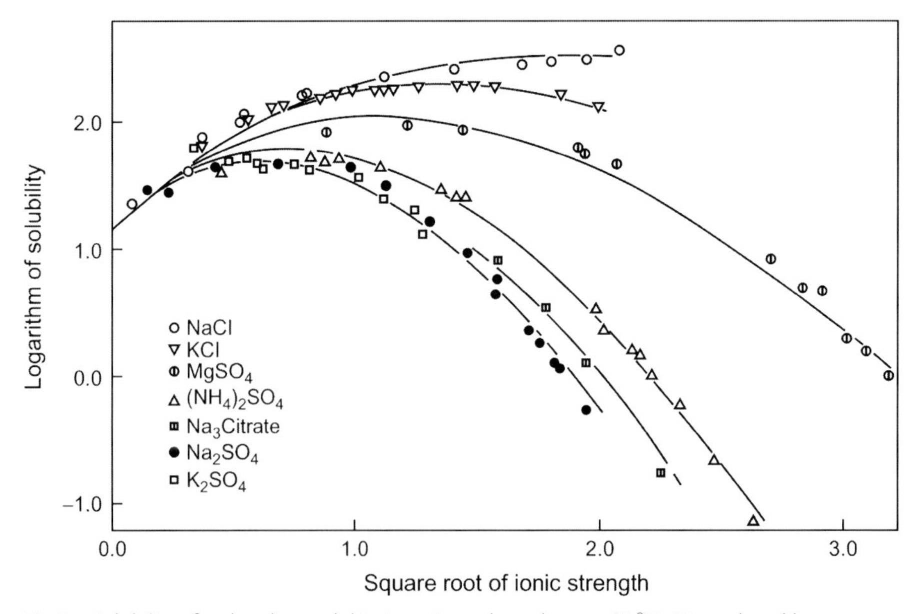

Fig. 5 Solubility of carboxyhemoglobin in various electrolytes at 25 °C. (Reproduced by permission of the American Society for Biochemistry and Molecular Biology, from Green, A.A. (1932) *J. Biol. Chem.* **95**, 47–66.)

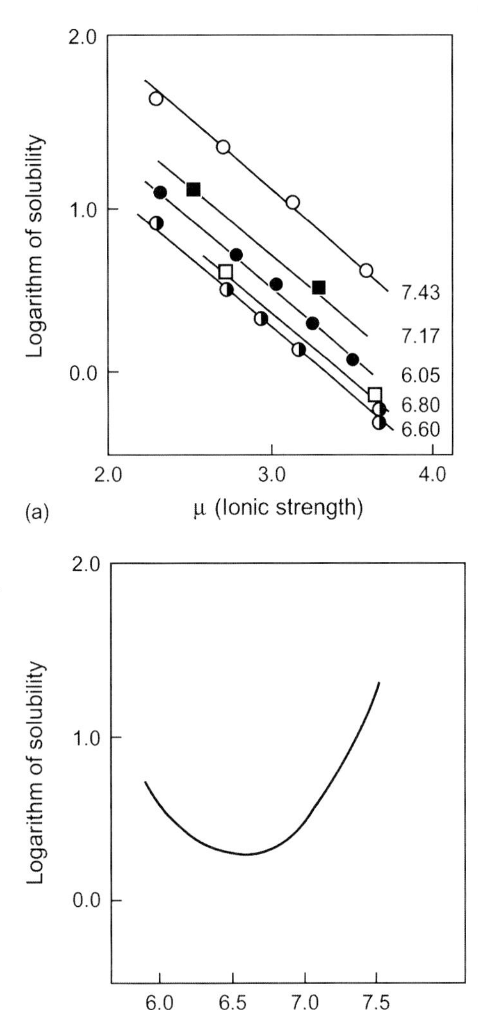

Fig. 6 (a) Solubility of hemoglobin at different pH values in concentrated phosphate buffers; (b) extracted from (a). (Reproduced by permission of the American Society for Biochemistry and Molecular Biology, from Green, A.A. (1931) *J. Biol. Chem.* **93**, 495–516.)

Ammonium sulfate is often used because of its high solubility.

1.2.2 pH and Counterions

A protein is more soluble when its net charge is larger. The minimum solubility is found at the isoelectric point. The net charge is zero and hence the packing in the solid state (in the crystal) is possible owing to electrostatic interactions without the accumulation of a net charge of high energy. All "salting-out" curves are parallel, K_s remains constant, and S_0 varies with pH (Fig. 6a and b). In some cases, the isoelectric point is different at low and high ionic strength owing to the interactions of the protein with counterions, which can cause a net charge at the pH of the isoelectric point.

1.2.3 Temperature

Many factors governing protein solubility are temperature dependent. The dielectric constant decreases with increasing temperature. In the solution energy, $\Delta G = \Delta H - T\Delta S$, the entropy term has an increasing influence with increasing temperature. The temperature coefficient of the solubility depends on other conditions (ionic strength, presence of organic solvents, etc.).

At high ionic strength, most proteins are less soluble at 25 °C than at 4 °C, for example the temperature coefficient is negative. The opposite is valid for low ionic strength.

1.2.4 Organic Solvents

The presence of organic solvents leads to a decrease in the dielectric constant. This causes an augmentation of the electric attraction between opposite charges on the surface of the protein molecule and hence to a reduction in solubility. In general, the solubility of a protein is reduced in the presence of an organic solvent if the temperature decreases. Often, organic solvents denature proteins. Therefore, one should work at low temperatures.

1.3 Experimental Techniques

The whole field of macromolecular crystallography has been excellently reviewed in Volumes 114 and 115 and Volumes 276 and 277 of *Methods in Enzymology*. A collection of review articles concerning the theory and practice of crystallization of biomacromolecules is given in Part A of Carter and Sweet.

A protein preparation to be used in crystallization should be "pure" or "homogeneous" at a level that established chromatographic methods are providing (protein content $\geq$95%). Furthermore, it should meet the requirements of "structural homogeneity." These requirements can be enumerated as follows. It is first necessary to prepare the protein in an isotypically pure state free from other cellular proteins. It may then be necessary to maintain the homogeneity of the protein preparation against covalent modification during crystallization by adding inhibitors of sulfhydryl group oxidation, by proteolysis, and by the action of reactive metals. It may be necessary to suppress the slow denaturation/aggregation of the protein and to restrict its conformational flexibility to reduce the entropic barrier to crystallization presented by extensive conformational flexibility.

For the crystallization of biomacromolecules, a broad spectrum of crystallization techniques exists. The most common techniques are described here. The oldest and simplest method is batch crystallization (Fig. 7a). In batch experiments, vials containing supersaturated protein solutions are sealed and left undisturbed. In microbatch methods, a small (2–10 μL) droplet containing both protein and precipitant is immersed in an inert oil, which prevents droplet evaporation. In the case that ideal conditions for nucleation and growth are different, it is useful to undertake the separate optimization of these processes. This can be done by seeding, a technique where crystals are transferred from nucleation conditions to those that will support only growth (Fig. 7b). For macroseeding, a single crystal is transferred to an etching solution, then to a solution of optimal growth. In microseeding experiments, a solution containing many small seed crystals, occasionally obtained by grinding a larger crystal, is transferred to a crystal growth solution.

The method of crystallization by vapor diffusion is depicted in Fig. 8(a). In this method, unsaturated precipitant-containing protein solutions are suspended over a reservoir. Vapor equilibration of the droplet and reservoir causes the protein solution to reach a supersaturation level where nucleation and initial crystal growth occur. Changes in soluble protein concentration in the droplet are likely to decrease supersaturation over the time course of the experiment. The vapor diffusion technique can be carried out as hanging drop or sitting drop method.

In crystallization by dialysis, the macromolecular concentration remains constant as in batch methods (Fig. 8b) because the molecules are forced to stay in a fixed volume. The solution composition is changed by diffusion of low-molecular-weight components through a semipermeable membrane. The advantage of dialysis is that the precipitating solution can be easily changed. Dialysis is also uniquely suited to crystallizations at low ionic strength and in the presence of volatile reagents such as alcohols.

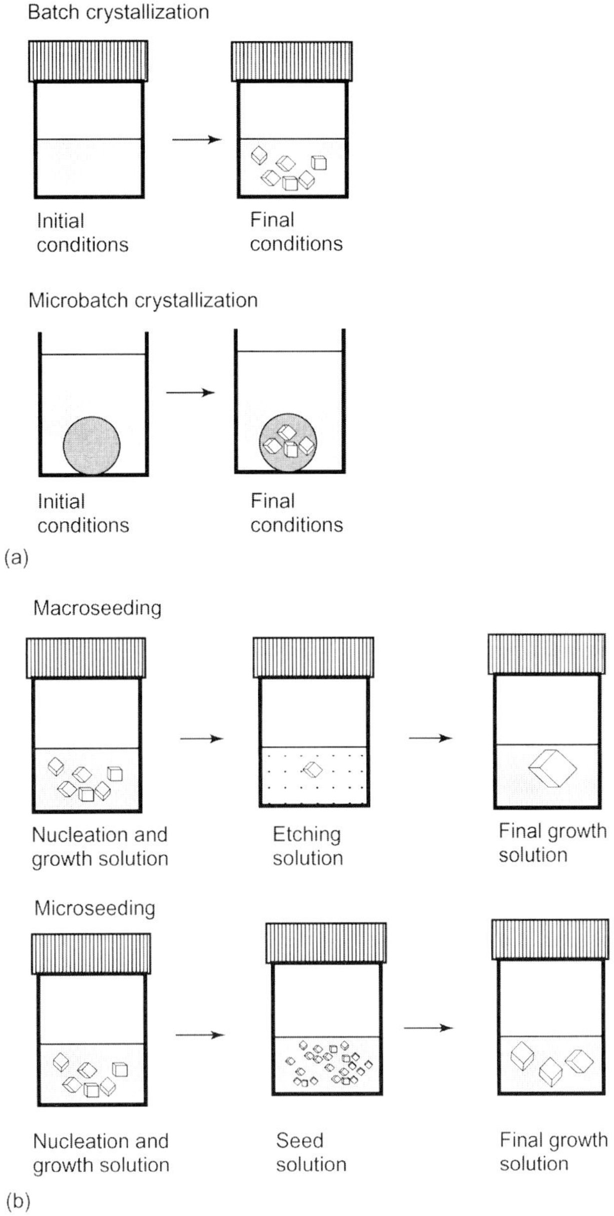

Fig. 7 Schematic presentation of (a) batch crystallization and (b) seeding techniques. (Reproduced by permission of Academic Press, Inc., from Weber, P.C. (1997) *Methods Enzymol.* **276**, 13–23.)

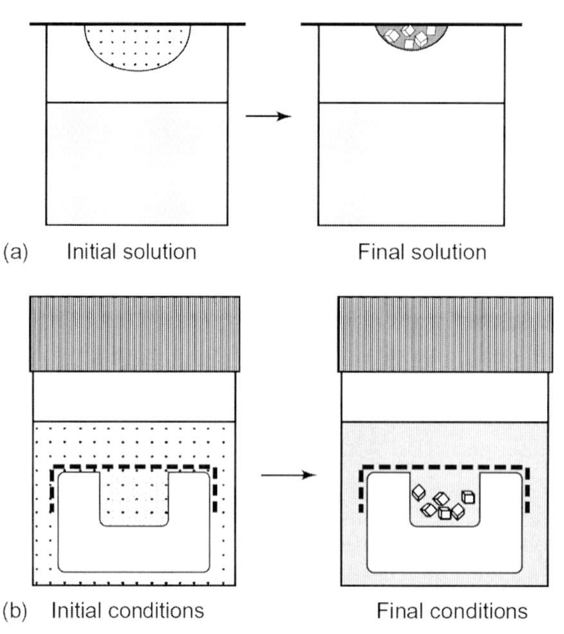

Fig. 8 Schematic representation of (a) vapor diffusion and (b) dialysis. (Reproduced by permission of Academic Press, Inc., from Weber, P.C. (1997) *Methods Enzymol.* **276**, 13–23.)

1.4 Crystallization Screenings

Screening schemes have been developed that change the most common parameters of this multiparameter problem such as protein concentration, the nature and concentration of the precipitant, pH, and temperature. Each screening can be extended by adding specific additives in low concentrations that affect the crystallization. Sparse matrix crystallization screens are widely applied. The sparse matrix formulation allows one to screen efficiently a broad range of the most popular and effective salts (e.g. ammonium sulfate, sodium and potassium phosphate, sodium citrate, sodium acetate, lithium sulfate), polymers (e.g. poly(ethyleneglycol) (PEG) of different molecular masses (from 400 to 8000)), and organic solvents (e.g. 2,4-methylpentanediol (MPD), 2-propanol, ethanol) versus a wide range of pH. Another approach is the systematic screening of the statistically most successful precipitants. A single precipitant is screened at four unique concentrations versus seven precise levels of pH between 4 and 10. Such grid screens can be done with ammonium sulfate, PEG 6000, MPD, and PEG 6000 in the presence of 1.0 M lithium chloride or sodium chloride. For the crystallization of membrane proteins, for each detergent that is necessary to make the membrane protein soluble, a whole grid screen or sparse matrix screen must be constructed. In principle, all three techniques can be applied for the different screening schemes, but mostly the vapor diffusion technique is applied because it is easy to use and the protein consumption is low. For a typical broad screening, about 2 mg of protein is sufficient. Chryschem plates (sitting drop) or Linbro plates (hanging drops) may be used for the vapor diffusion crystallization screening experiments. Once crystals have been obtained, their size and quality can be optimized by additional fine screens around the observed crystallization conditions. General rules do not exist that indicate which method for crystallization one has to use for which type of protein. Suggestions for crystallization conditions to be tested can be obtained from the Biological Macromolecule Crystallization Database.

2 Experimental Techniques

2.1 X-ray Sources

2.1.1 Conventional X-ray Generators

X-rays are produced when a beam of high-energy electrons, which have been accelerated through a voltage V in a vacuum, hit a target. An X-ray tube run at voltage V will emit a continuous X-ray spectrum with a minimum wavelength given by Eq. (4):

$$\lambda_{\min} = \frac{hc}{eV} = \frac{12398}{V} \tag{4}$$

with λ in angströms ($1\,\text{Å} = 10^{-10}$ m) and V in volts. The critical voltage, V_0, which is required to excite the characteristic line of a particular element, can be calculated from the corresponding wavelength for the appropriate absorption edge. For the copper absorption edge, $\lambda_{ae} = 1.380$ Å. Hence, we have (Eq. 5)

$$V_0 = \frac{12398}{\lambda_{ae}} = 8.98\ \text{kV} \tag{5}$$

Provided that $V > V_0$, the characteristic line spectra will be produced (Fig. 9). The oldest and cheapest X-ray sources are sealed X-ray tubes. The cathode and anode are situated under vacuum in a sealed glass tube, and the heat generated at the anode is removed by a water-cooling system. For the generation of higher intensities, as needed in protein crystallography, one has to use a rotating anode (Fig. 10). Here the anode is rotated, which allows a higher power loading at the focal spot. In protein crystallography, copper targets are usually taken. The used take-off angle is near 4°, which results in apparent focal spot sizes of about 0.3×0.3 mm.

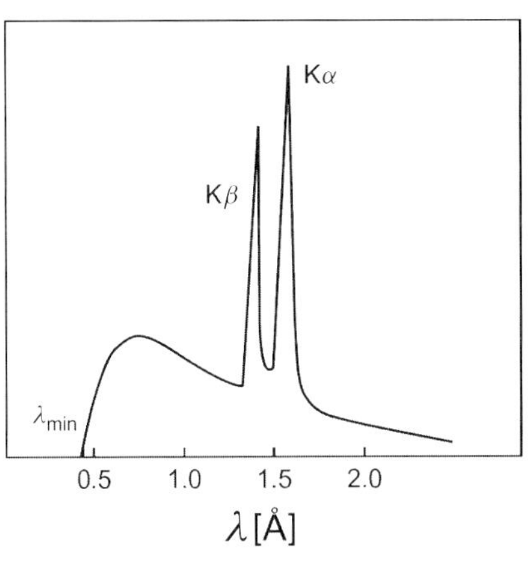

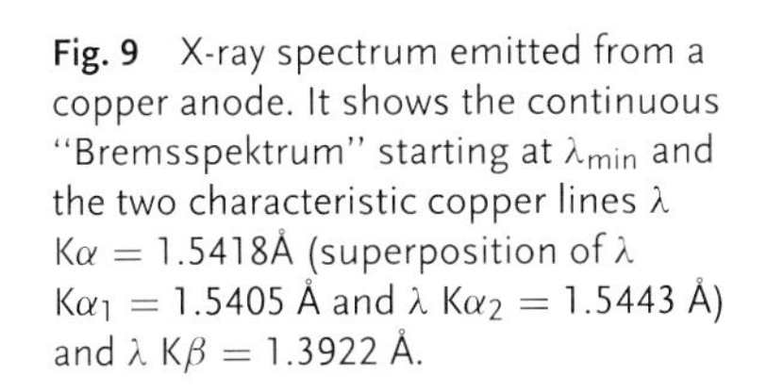
Fig. 9 X-ray spectrum emitted from a copper anode. It shows the continuous "Bremsspektrum" starting at $\lambda_{\min}$ and the two characteristic copper lines λ Kα = 1.5418Å (superposition of λ Kα_1 = 1.5405 Å and λ Kα_2 = 1.5443 Å) and λ Kβ = 1.3922 Å.

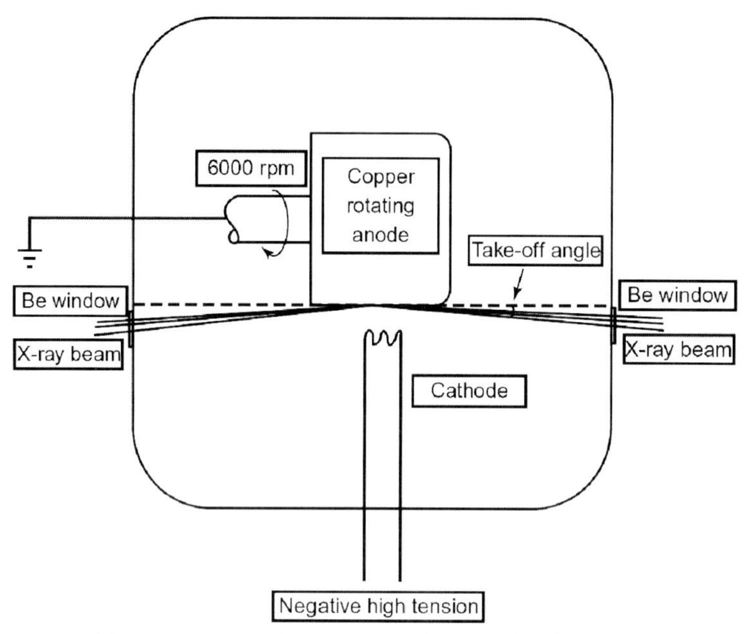

Fig. 10 Schematic drawing of a rotating anode tube. Take-off angle is near 4°. For copper, the tube is normally operated at 50-kV high tension and 100-mA cathode current.

2.1.2 Synchrotron Radiation

As electrically charged particles such as electrons or positrons of high energy are kept under the influence of magnetic fields and travel in a pseudocircular trajectory, synchrotron radiation is emitted and can be used in many different types of experiments. The particles are injected into the storage ring directly from a linear accelerator or through a booster ring. They circulate in a high vacuum for several hours at a relative constant energy. To keep the bunched particles traveling in a nearly circular path, a lattice of bending magnets is set up around the storage ring. As the particle beam traverses each magnet, the path of the beam is altered, and synchrotron radiation is emitted. The loss of energy of the particle beam is compensated by a radiofrequency input at each cycle. The synchrotron radiation can be channeled through different beamlines for use in research.

Other types of magnets – insertion devices called *wigglers* and *undulators* – can be assembled in the storage ring. Unlike the bending magnets, the primary purpose of which is to maintain the circular trajectory, wigglers and undulators are used to increase the intensity of the emitted radiation. Bending magnets and wigglers cause a continuous spectrum of radiation.

In contrast, the radiation produced by an undulator has a discontinuous spectrum, and can be tuned to various wavelengths. The importance of synchrotron radiation for macromolecular crystallography lies in the high brilliance (photons s^{-1} $mrad^{-2}$ mm^{-2} per $\Delta\lambda/\lambda$; that is, how small is the source and how well collimated are the X-rays?) of the beam, the high intensity, and the tunability of the wavelength in the relevant range from 0.5 to 3.0 Å. The time structure of the beam is of interest for time-resolved crystallography. The particles circulate in bunches with widths of 50 to 150 ps and repeat every few microseconds.

About 15 synchrotron radiation facilities equipped with beamlines for macromolecular crystallography are available throughout the world operated at energies from about 1.5 to 6 to 8 GeV for third-generation machines. An aerial view of the European Synchrotron Radiation Facility (ESRF) in Grenoble, a third-generation machine, is shown in Fig. 11. The ESRF storage ring is operated at 6 GeV and has a circumference of 844.39 m. Its critical wavelength, λ_c, is 0.6 Å.

2.1.3 Monochromators

In the majority of applied diffraction techniques, monochromatic X-rays are used. Therefore, the emitted white radiation of X-rays must be further monochromatized. With copper Kα radiation generated by a sealed or rotating anode tube, the Kβ radiation can be removed with a nickel filter. Much better results can be achieved with a monochromator. The simplest monochromator is a piece of a graphite crystal that reflects the copper Kα radiation at a Bragg angle of 13.1° and a glancing angle of 26.2°. Improved beam focusing is obtained by a double mirror system. The mirror assembly is composed of two perpendicular bent nickel-coated glass optical flats, each with translation, rotation, and slit components housed in a helium gas-flashed chamber, which is commercially available (Molecular Structure Corporation, The Woodlands, TX, USA). The prototype and basic theory in the use of this system were discussed in detail by Phillips and Rayment.

For synchrotron radiation with its much higher intensity, germanium or silicon single crystals can be applied as monochromators, which filter out a bandwidth of $\delta\lambda/\lambda$ from 10^{-4} to 10^{-5}, two orders of magnitude smaller than that with graphite. Single or double monochromators can be used, which are either flat or bent. The bent monochromators have the advantage that

Fig. 11 Aerial view of ESRF in Grenoble. Courtesy of ESRF.

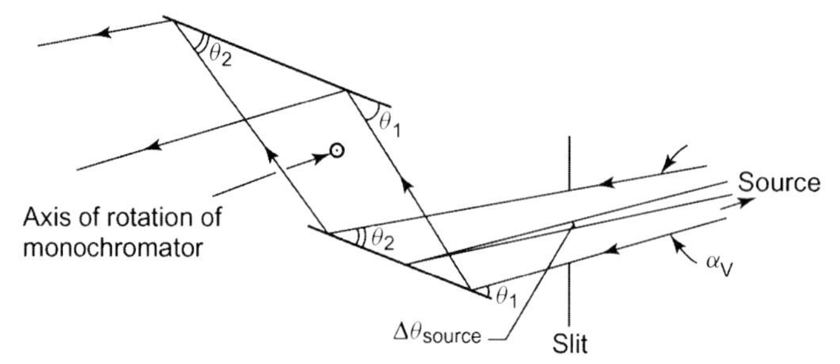

Fig. 12 Schematic drawing of a double monochromator system. (Reproduced by permission of Cambridge University Press, from Helliwell, J.R. (1992) *Macromolecular Crystallography with Synchrotron Radiation*, Cambridge University Press, Cambridge.)

they simultaneously focus the beam. The double monochromator (Fig. 12) has the advantage that the emergent monochromatic beam is parallel to and only slightly displaced from the incident synchrotron radiation beam. This makes necessary only small adjustments of the X-ray optics and detector arrangement when it is tuned to another wavelength compared to a single monochromator where the whole X-ray diffraction assembly must be moved.

2.2 Detectors

2.2.1 General Components of an X-ray Diffraction Experiment

A principal arrangement for a macromolecular X-ray diffraction experiment is depicted in Fig. 13. The primary beam leaves the X-ray source and passes the X-ray optics, which may be a simple collimator or the various types of monochromators or mirror systems described above,

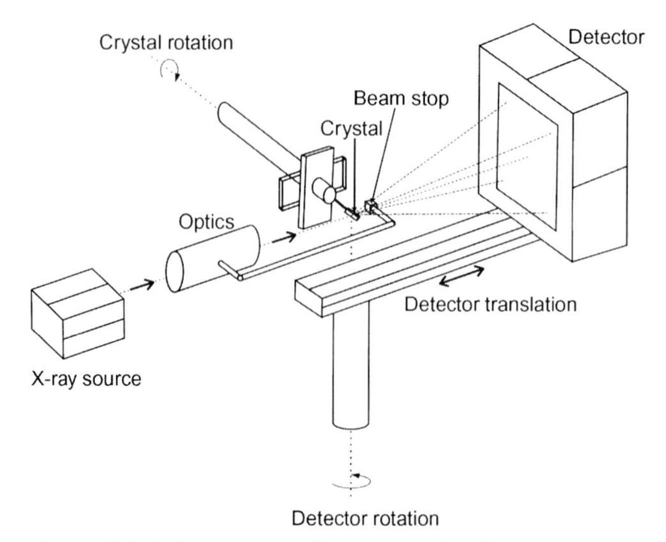

Fig. 13 Principal arrangement for a macromolecular X-ray diffraction experiment.

and terminated with a collimator. The crystal is mounted on a goniometer head either in a quartz capillary or in a cryo-loop shock-frozen at low temperature. The goniometer head is attached to a device that can perform spatial movements of the crystal around the center of the crystal. The simplest kind of such a movement is the rotation of the crystal about a spindle axis as indicated in Fig. 13. This device can be a multiple axis goniostat (2–4 axes), which allows the crystal to be brought into any spatial orientation around its center. The X-ray detector, which registers the diffracted intensities, is mounted on a device that permits the translation and rotation of the detector. If the active area of the detector is large enough to collect all generated diffracted beams at a given wavelength, detector rotation is not necessary and the detector is arranged normal to the primary beam. A small piece of lead is placed in the path of the primary beam just behind the crystal to prevent damage to the detector and superfluous gas scattering.

The classical detectors in macromolecular crystallography have been photographic films and single-photon counters. The photographic films were used on specially designed X-ray cameras and the single-photon counters on four-circle diffractometers. The main disadvantage of these detectors was their low sensitivity and with films the limited dynamic range (1 : 200). Over the last 15 years, powerful detectors have been developed, which will be discussed briefly. These new detectors have almost completely replaced photographic films and single-photon counters.

2.2.2 Image Plates

An image plate (IP) consists of a support (either a flexible plastic plate or a metal base) coated with a photostimulable phosphor (150 μm) and a protective layer (10 μm). The photostimulable phosphor is a mixture of very thin crystals of $BaF(Br,I):Eu^{2+}$ and an organic binder. This phosphor can store a fraction of the absorbed X-ray energy by electrons trapped in color centers. It emits photostimulated luminescence, whose intensity is proportional to the absorbed X-ray intensity, when later stimulated by visible light. The wavelength of the photostimulated luminescence ($\lambda \approx 390$ nm) is reasonably separated from that of the stimulating light ($\lambda \approx 633$ nm, in practice a red laser), allowing it to be collected by a conventional high quantum efficiency photomultiplier tube. The output of the photomultiplier is amplified and converted to a digital image, which can be processed by a computer. The residual image on the IP can be erased completely by irradiation with visible light, to allow repeated use.

IPs have several excellent performance characteristics as integrating X-ray area detectors that make them well suited in X-ray diffraction. The sensitivity is at least 10 times higher than for X-ray films and the dynamic range is much broader ($1:10^4$–10^5). Important for synchrotron radiation is their high sensitivity at shorter wavelengths (e.g. 0.65 Å). A disadvantage is the relatively long readout times for each exposure (from 45 s to several min). IP diffractometer systems are commercially available from several companies. All systems work reliably and deliver good-quality data. A photograph of the newest IP system produced by Mar Research (Norderstedt, Germany) is shown in Fig. 14.

2.2.3 Gas Proportional Detectors

As X-ray counters, gas proportional detectors provide unrivaled dynamic range and sensitivity for photons in the range important for macromolecular crystallography.

Fig. 14 Mar 345 IP diffractometer system from Mar Research, Norderstedt, Germany. The circular plate rotates for scanning and the laser is moved along a radial line. Courtesy of Mar Research.

The classical gas proportional detector is a multiwire proportional chamber (MWPC), widely used as an in-house detector with conventional X-ray sources. Two MWPC diffractometer systems are commercially available. Gas proportional detectors use as a first step the absorption of an X-ray photon in a gas mixture high in xenon or argon. This photoabsorption produces one electron–ion pair whose total energy is just the energy of the initial X-ray photon. The ion returns to its neutral state either by emission of Auger electrons or by fluorescence. Since the kinetic energy of these first electrons is far greater than the energy of the first ionization level of the xenon or argon atoms, fast collisions with atoms (or molecules) in the gas very quickly produce a cascade of new electron–ion pairs in a small region extending over a few hundred micrometers around the conversion point. The total number of primary electrons that are produced during this process is proportional to the energy of the absorbed X-ray photon and is thus a few hundred for ~10-keV photons. These primary electrons then drift to the nearest anode wire where an ionization avalanche of as many as 10 000 to 1 000 000 ion pairs results. The motion of the charged particles in this avalanche (chiefly the motion of the heavy positive ions away from the anode wire) causes a negative-going pulse on the anode wire and positive-going pulses on a few

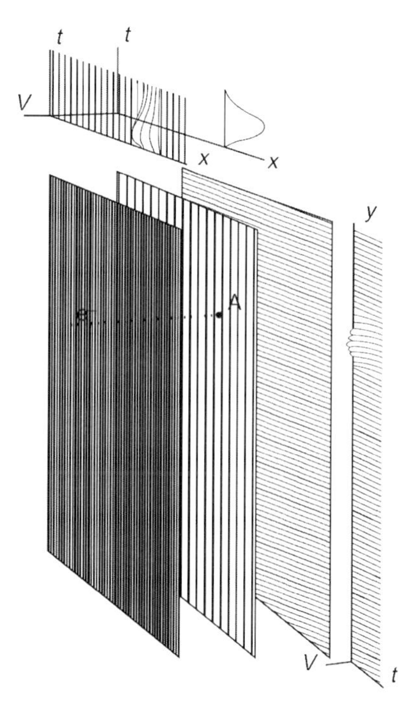

Fig. 15 Expanded view of an MWPC, showing the anode plane sandwiched between the two cathode planes. A is the position of the avalanche. The centers of the induced charge distributions are used to determine the coordinates, x and y, of the avalanche. (Reproduced by permission of Academic Press, Inc., from Kahn, R., Fourme, R. (1997) *Methods Enzymol.* **276**, 244–268.)

of the nearest wires in the back (cathode) wire plane (see Fig. 15).

Disadvantages of the MWPC detector are the limited counting rate due to the buildup of charges in the chamber and limitations in the readout electronics and the lower sensitivity at shorter wavelengths. This makes the application of MWPCs with synchrotron radiation poorly effective.

2.2.4 Charge-coupled Device-based Detectors

A remarkable development for the use with synchrotron radiation is the design and construction of charge-coupled device (CCD) detectors. CCDs were developed originally as memory devices, but the observation of localized light-induced charge accumulation in CCDs quickly led to their development as imaging sensors. These CCD detectors are integrating detectors like the conventional X-ray sensitive film, IPs, and analog electronic detectors using either silicon intensified target (SIT) or CCD sensors. Integrating detectors have virtually no upper rate limits because they measure the total energy deposited during the integration period (although individual pixels may become saturated if the signal exceeds its storage capacity).

The first commercially available analog electronic detector was the fast area television detector (FAST) detector produced by Enraf-Nonius (Delft, The Netherlands). This detector contained a SIT vidicon camera as an electronically readable sensor. The SIT vidicon exhibits higher noise than CCDs, which have therefore replaced SIT sensors during the past few years. Because of their high intrinsic noise, detectors with SIT vidicon sensors need an analog image-amplification stage and this limits the overall performance of such detectors. Several CCD detector systems have also been developed that incorporate image intensification. The most important development in detector design for macromolecular crystallography has been the incorporation of scientific-grade CCD sensors into instruments with no image intensifier. These detector designs are based on direct contact between the CCD and a fiber-optic taper. There are several commercial systems available based on this construction (Mar Research, Norderstedt, Germany; Hamlin Detector).

A schematic representation of such a detector is shown in Fig. 16. An X-ray phosphor (commonly $Gd_2O_2S:Tb$) is attached to a fiber-optic faceplate, which is tightly connected to a fiber-optic taper. The X-ray sensitive phosphor surfaces at the front convert the incident X-rays into a burst of visible-light photons. Although it is possible to permit the X-rays to strike the CCD directly, this method has several drawbacks, such as radiation damage to the CCD, signal saturation, and poor efficiency. The use of a larger phosphor as active detector area and the demagnifying fiber-optic taper is also necessary because the size of the scientific-grade CCD sensors is not as large as needed for the demands of the X-ray diffraction experiment. The fiber-optic taper is then bonded to the CCD, which is connected to the electronic readout system. The CCD must be cooled to temperatures ranging from -40 to $-90\,^{\circ}$C, depending on the various systems. The great advantage of CCD detectors is their short readout time, which lies in the range from 1 to a few seconds.

2.3 Crystal Mounting and Cooling

2.3.1 Conventional Crystal Mounting

The purpose of crystal mounting is to isolate a single crystal from its growth

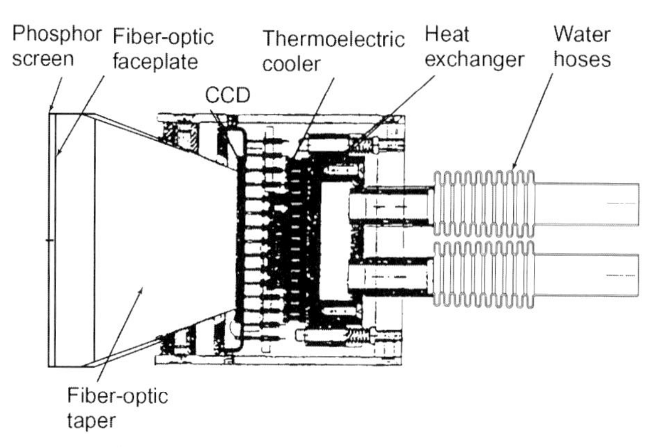

Fig. 16 Schematic representation of a CCD/taper detector. (Reproduced by permission of Academic Press, Inc., from Westbrook, E.M., Naday, I. (1997) *Methods Enzymol.* **276**, 268–288.)

medium so that it can be used in the X-ray diffraction experiment to study its diffraction properties. It is important that the manipulation of the crystal introduces as little damage as possible to its three-dimensional structure. The most important aspect of crystal mounting is to preserve the crystal in its state of hydration. This is accomplished by sealing the crystal in a thin-walled (0.001-mm thick) glass or quartz capillary tube. The important steps in conventional crystal mounting are illustrated in Fig. 17(a–c). The crystal must be dislodged from the surface on which it grew, and then it may be drawn into the capillary using suction from a small-volume (0.25 mL) syringe, micropipet, or mouth aspirator, which are connected to the funnel of the capillary by a flexible plastic hose of appropriate diameter. Next, the capillary should be inverted to allow the crystal to fall to the inner meniscus. Then, the surrounding solution may be removed using thin strips of filter paper or by a small glass pipet. The extent to which the crystal should be dried must be determined by experience. The final step is to place a small volume of mother liquor in the capillary and seal both ends. The capillary is then glued to a metal base, which can be attached to a goniometer head.

2.3.2 Cryocrystallography

Many macromolecular crystals suffer from radiation damage when exposed to X-rays with energies and intensities as used in macromolecular X-ray diffraction experiments with both conventional sources and synchrotron radiation. A possibility for reducing radiation damage of the crystal during the measurement is to cool the crystal to low temperatures, usually to 100 K. For this purpose, the crystal is flash-frozen to prevent ice formation or damage to the crystal. One method of crystal treatment is the removal of external solution by transferring the crystal in a small drop to a hydrocarbon oil and either teasing the liquid away or drawing it off with filter paper or a small pipet. The oil-coated crystal is then mounted on a glass fiber or small glass "spatula." Oil protects the crystal from drying and acts as an adhesive that hardens on cooling to hold the sample rigidly. Much more frequently used

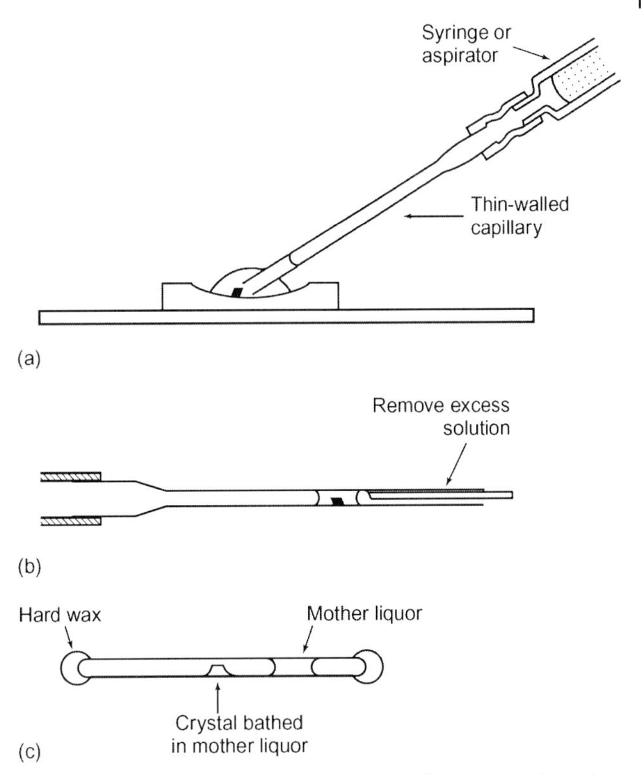

Fig. 17 Mounting of a crystal in a glass capillary. (Reproduced by permission of Academic Press, Inc., from Rayment, I. (1985) *Methods Enzymol.* **114**, 136–140.)

is a technique in which the crystal is suspended in a film of mother liquor in a small loop. This method avoids problems with damage by the oil or mechanical damage when removing the external liquid, and it has proven successful for most samples. It does, however, require the use of a cryoprotectant to prevent ice formation. The most commonly used cryoprotectants are glycerol, PEG of different molecular weights, glucose, and MPD.

The loop is produced from fine fibers that permit unobstructed data collection in nearly all sample orientations. The crystal is held within the loop, suspended in a thin film of cryoprotectant-containing harvest buffer. The loop is supported by a fine wire or pin, which itself is attached to a steel base used for placing the assembly on a goniometer head and in storing mounted crystals. Once in the loop, the crystal is cooled to a temperature at which the increasing viscosity of the liquid prevents molecular rearrangement. The rate of cooling must be rapid enough to reach this point before ice-crystal nucleation occurs. Two methods are used: cooling directly in the gas stream of a cryostat or plunging the crystal into a cryogenic liquid. The first method is explained in Fig. 18(a) and (b). The loop assembly (with crystal) is attached to the goniometer head, with the cold stream deflected. Then, the cold stream is unblocked to flash-freeze

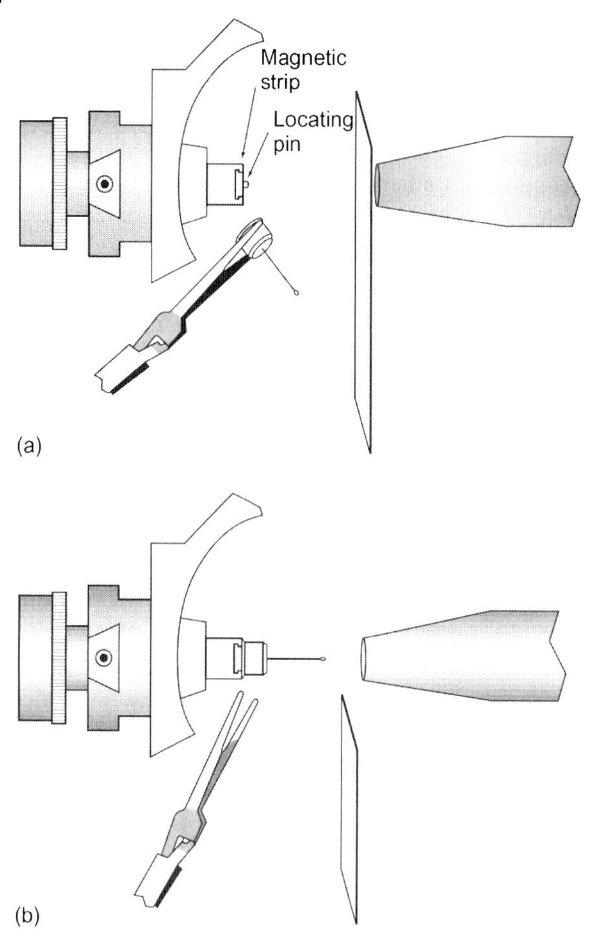

Fig. 18 Flash cooling in the direct cold gas stream of a cryostat. (Reproduced by permission of Academic Press, Inc., from Rodgers, D.W. (1997) *Methods Enzymol.* **276**, 244–268.)

the crystal. In this gas-stream position, the goniometer head must be heated to prevent ice formation on the goniometer head. Cryostats and cryocrystallographic tools are commercially available.

Cryocrystallography has had a great impact on macromolecular crystallography by dramatically increasing the lifetime of a crystal during the X-ray experiment allowing, for example, the collection of several data sets from one crystal at different wavelengths using synchrotron radiation.

2.3.3 Crystal Quality Improvement by Humidity Control

The crystal quality is of decisive significance for a successful X-ray crystal structure determination. Two principal cases must be distinguished: (1) the crystals diffract to a resolution only (>4.5 Å), which does not allow a structure determination at all and (2) the crystal quality is good enough to elucidate its 3-D structure (resolution <3.5 Å but not better than 2.5 Å) but a structure determination at higher resolution and accuracy therewith would be necessary, for example, for the characterization of a metal center in a metalloprotein. The control of the crystal packing via the solvent content of a crystal is a useful approach for the improvement of the crystal quality of biological macromolecules. For this purpose, a so-called Free Mounting System (FMS) has been developed and successfully applied. The principal construction of the FMS is shown in Fig. 19 (a). A main part of the FMS is the humidifier unit, which consists of the humidifier, the control, and power electronics. A stream of humid air with defined moisture is produced. A flexible teflon tube transports the humid air to the crystal holder, which is depicted in Fig. 19 (b). A very compact construction allows the sample to be mounted in a controlled environment with minimal restriction for the X-ray measurement. The head part is freely rotatable relative to the insert, without axial movement. A heating element and a temperature sensor are integrated into the head part. The humid air stream through the head part is adjusted to the temperature of the head part, independent of the ambient temperature. The crystal may either be mounted in a patch-clamp pipette or conventional cryo-loop (Fig 19 (c)).

In a typical experiment to increase the crystal quality, X-ray crystal diffraction is monitored at various defined crystal humidities. Usually, a positive effect is observed at lower humidity, which causes shrinkage of the unit cell volume and allows different and possibly more favorable crystal contacts. After having found the optimal condition, the crystal, mounted in a loop, can be shock-frozen for a subsequent data collection. A remarkable improvement of crystal quality by using the FMS could be observed with about 30% of different projects under investigation.

2.4 Data-collection Techniques

2.4.1 Rotation Method

Most macromolecular X-ray diffraction systems use the rotation method for data collection. For each crystal, a reciprocal lattice can be constructed, which is very useful in the interpretation of crystallographic crystal diffraction experiments. Diffraction theory (discussed later) tells us that an X-ray reflection is generated when a point of this reciprocal lattice lies on a sphere of radius $1/\lambda$ whose origin is $1/\lambda$ away from the origin of the reciprocal lattice in the direction of the primary beam (Fig. 20). The direction of such a diffracted beam is along the connection of the center of the so-called Ewald sphere (radius $1/\lambda$) and the intersection of the reciprocal lattice point on the Ewald sphere. Owing to certain factors, which will be discussed later, the apparent reciprocal lattice extends to a given radius only, which defines the resolution sphere. To bring all the reciprocal lattice points within the resolution sphere into the reflection position, the crystal must be rotated around its center. Nearly all macromolecular X-ray diffraction systems apply the rotation technique in the normal beam case where the rotation axis is normal to the incident X-ray beam. Rotating the crystal around 360° brings all reciprocal lattice points within the resolution sphere in the reflection position except for the region between the rotation axis and the Ewald sphere, which is therefore called the *blind region*. This region can be collected when the crystal has been brought into another orientation. The diffracted beams are usually registered with a flat detector at distance D from the crystal, which is also normal to the primary beam. To avoid overlapping of reflection

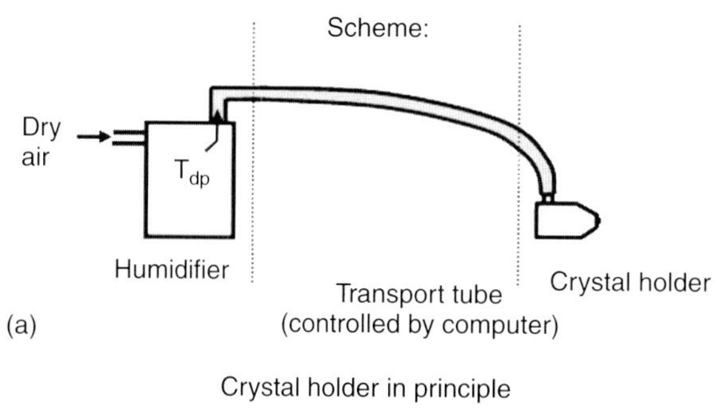

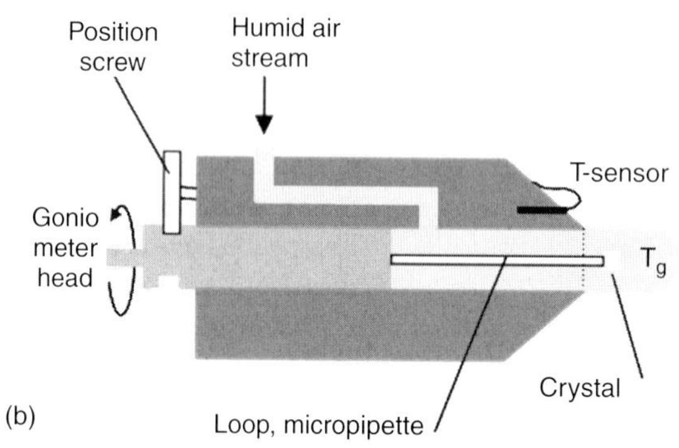

Fig. 19 The Free Mounting System: (a) principal construction; (b) schematic view of the crystal holder; (c) opened crystal holder with magnetic base and mounting loop. T_{dp} – temperature of dew point, T_g – temperature of goniometer head.

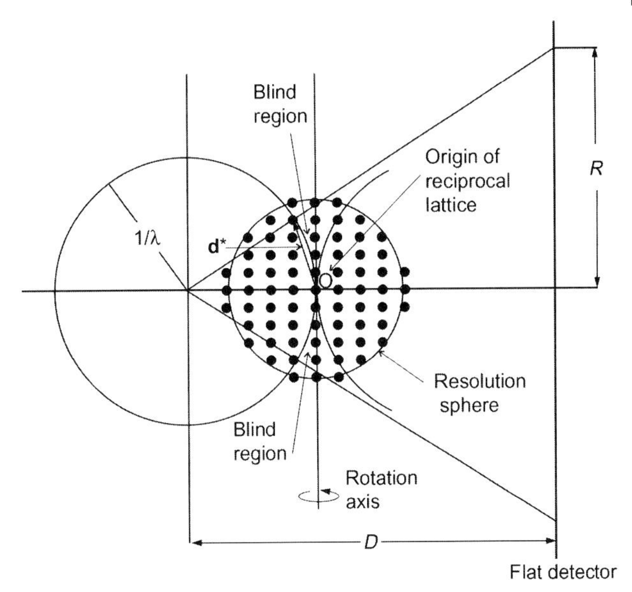

Fig. 20 Diffraction geometry in the rotation method usually applied in macromolecular X-ray diffraction systems.

spots on the detector, the crystal is rotated by rotation-angle increments, which can vary from tenths of degrees to 1 or to a few degrees, depending on the size of the crystal unit cell, crystal mosaicity, beam collimation, and other factors. Each individual exposure is processed and electronically stored in a computer. These raw data images are evaluated subsequently with relevant computer programs (discussed in some detail later) to give the intensities and geometric reference values (indices) for each collected intensity.

2.4.2 Precession Method

The rotation method delivers a distorted image of the reciprocal lattice for each geometry of the detector (flat or curved) and orientation with respect to the rotation axis. An undistorted image of the reciprocal lattice can be obtained by using the precession method. The principle of this technique is shown in Fig. 21. The detector is a flat film. During the motion of a given reciprocal lattice plane (in Fig. 21, a so-called zeroth plane going through the origin O of the reciprocal lattice), the flat detector must always be parallel to this reciprocal lattice plane to obtain an undistorted image of this plane. The normal of the reciprocal lattice plane and, consequently, also the detector are inclined with respect to the primary X-ray beam by an angle μ. When the normals of the reciprocal lattice plane and the detector carry out a concerted precession motion of angle μ around the primary X-ray beam, a circular region of the reciprocal lattice plane is registered on the detector (these regions are shown as dashed circles in Fig. 21). In a precession camera construction, the crystal and the film cassette are both held in a universal joint, which are linked so that the film and crystal move together in phase with precession angle μ. In Fig. 21, the joints are symbolized as forks and their linkage by a line. Parallel to

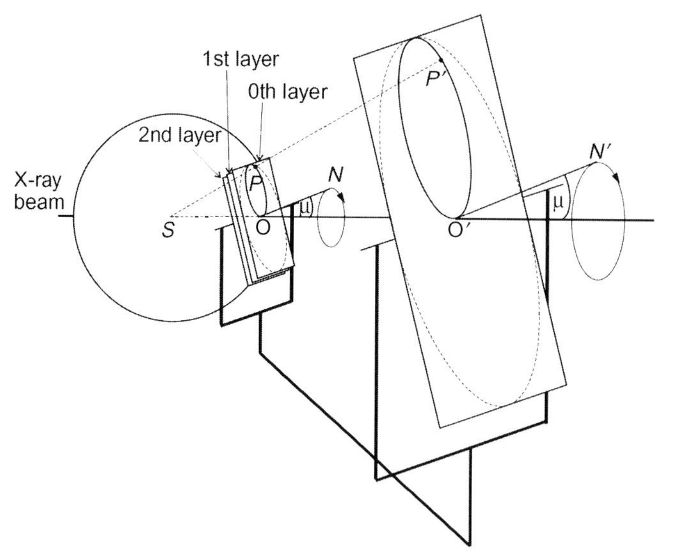

Fig. 21 Principle of the precession method (modified from Buerger, M.J. (1964) *The Precession Method in X-ray Crystallography*, John Wiley & Sons, New York).

the zeroth reciprocal lattice plane is a set of lattice planes that also carry out this precession movement. The parts of them that are swung through the Ewald sphere also give rise to an image on the flat detector (first and second reciprocal lattice layers are also indicated in Fig. 21). The images of these layers would superimpose on the detector, and the use of the technique in this way is denoted by the screenless precession method. The insertion of a screen with a suitable annular aperture between the crystal and the detector at an appropriate distance can be used to screen out the desired reciprocal layer. This screen is also inclined by the precession angle μ and is coupled to the concerted precession movement of crystal and film cassette. The strength of the precession technique is that in addition to the undistorted imaging of the reciprocal lattice planes, the indexing of the diffraction spots is straightforward and the symmetry of the diffraction pattern is readily obtained by inspection.

For this reason, the precession method has been broadly applied in macromolecular crystallography for a long time. The replacement of film by the new generation of detectors has almost completely stopped the use of the precession method. Nevertheless, the use of a precession camera has great teaching benefits in becoming familiar with the reciprocal lattice concept.

3
Principles of X-ray Diffraction by a Crystal

3.1
Scattering of X-rays by an Atom

A component of the electrical field of the incident wave has the following form (Eq. 6) in free space referred to an origin at $\mathbf{x} = 0$:

$$E(\mathbf{x}) = E_0 \exp[2\pi i(\nu t - \mathbf{x}\mathbf{s}_0)] \qquad (6)$$

where $\mathbf{s}_0$ = wave vector of incident wave, $\nu = c/\lambda$ = frequency, λ = wavelength, and $s_0 = 1/\lambda$ = absolute value of incident wave vector. This wave interacts with a scattering center at position $\mathbf{r}$ (Fig. 22). The electric field component of the incident wave causes the electron at position $\mathbf{r}$ to oscillate. Together with the positively charged nucleus of the atom, which does not oscillate, this can be considered as a classical dipole oscillator. This dipole oscillator emits a spherical wave, which is denoted as a scattered wave and can be given in the following form (Eq. 7 and 8):

$$E_{SC} = CE(\mathbf{r})\frac{\exp(-2\pi i s_0 r_1)}{r_1} \quad (7)$$

$$E_{SC} = CE_0 \exp(-2\pi i \mathbf{s}_0 \mathbf{r}) \times \frac{\exp(2\pi i(\nu t - s_0 r_1)}{r_1} \quad (8)$$

The phase-angle-dependent factor has been omitted. The amplitude of the scattered wave is proportional to the amplitude $E(\mathbf{x})$ of the wave incident at $\mathbf{r}$. This gives the factor $E(\mathbf{r})$. C is a proportionality factor, taking into account the peculiarities of the scattering center. The factor $1/r_1$ considers the conservation of the scattered energy flux. We add all wavelets scattered from different volume elements of an atom to get the total amplitude of the scattered wave at point $\mathbf{R}$ relative to the origin O in the atom.

Equations 9 to 12 follow from Fig. 22:

$$\mathbf{r} + \mathbf{r}_1 = \mathbf{R} \quad (9)$$

$$r_1^2 = (\mathbf{R} - \mathbf{r})^2 = R^2 + r^2 - 2rR\cos(\mathbf{r}, \mathbf{R}) \quad (10)$$

$$r_1 = R\left[1 - \frac{2r}{R}\cos(\mathbf{r}, \mathbf{R}) + \left(\frac{r}{R}\right)^2\right]^{1/2} \quad (11)$$

$$r_1 \approx R\left(1 - \frac{r}{R}\cos(\mathbf{r}, \mathbf{R}) + \ldots\right) \quad (12)$$

where $(r/R)^2$ and higher terms were neglected in the expansion of the square root. It follows that (Eq. 13)

$$r_1 \approx R - r\cos(\mathbf{r}, \mathbf{R}) \quad (13)$$

Now we can combine the spatial phase factors in the equation for E_{SC}. With the approximation for r_1, we obtain Eq. (14):

$$\exp(-2\pi i(\mathbf{s}_0\mathbf{r} + s_0 r_1) = \exp(-2\pi i s_0 \mathrm{R}) \times \exp(-2\pi i(\mathbf{s}_0\mathbf{r} - s_0 r\cos(\mathbf{r}, \mathbf{R})) \quad (14)$$

As $s_0 = s$ and $\mathbf{s}$ is parallel to $\mathbf{R}$, we obtain Eq. (15).

$$s_0 r\cos(\mathbf{r}, \mathbf{R}) = sr\cos(\mathbf{r}, \mathbf{s}) = \mathbf{sr} \quad (15)$$

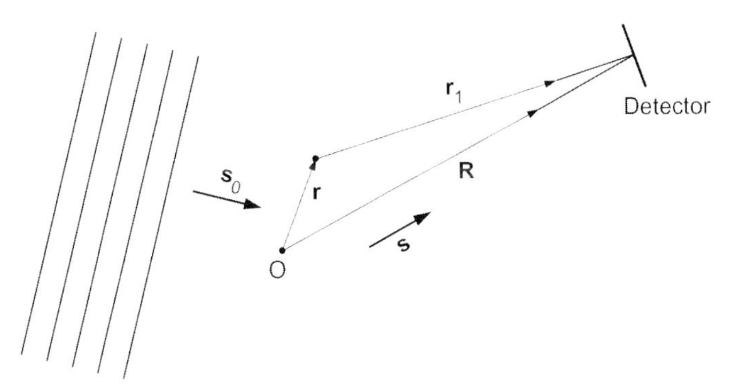

Fig. 22 Scattering of a planar X-ray wave by an electron at position **r** with respect to origin O.

It follows for the phase factor (Eq. 16):

$$\exp(-2\pi i(\mathbf{s}_0\mathbf{r} + s_0 r_1) = \exp(-2\pi i s_0 \mathrm{R})$$
$$\times \exp(-2\pi i(\mathbf{s}_0\mathbf{r} - \mathbf{sr}))$$
$$= \exp(-2\pi i \mathbf{S}_0 \mathrm{R})$$
$$\times \exp(2\pi i(\vec{\mathbf{r}}(\vec{\mathbf{S}} - \vec{\mathbf{S}}_0)) \quad (16)$$

Now we can write for the wave scattered by a center at **r** (Eq. 17):

$$E_{SC} = \left(\frac{CE_0 \exp(-2\pi i s_0 R) \exp(2\pi i \nu t)}{R}\right)$$
$$\times \exp(2\pi i \mathbf{r}(\mathbf{s} - \mathbf{s}_0)) \quad (17)$$

In a useful approximation, r_1 has been replaced by R in the denominator. In general, the scattering from an atom comes from the distribution of electrons in the atom. If the scattering of a volume element $\mathrm{d}\nu$ of the atom is proportional to the local electron density ρ (**r**), then the scattering amplitude will be proportional to the integral (Eq. 18):

$$f(\mathbf{S}) = \int_{\text{vol. of atom}} \rho(\mathbf{r}) \exp(2\pi i \mathbf{rS}) \mathrm{d}\nu \quad (18)$$

3.2 Scattering of X-rays by a Unit Cell

A unit cell may contain N atoms at positions of their internal origins at $\mathbf{r}_j$ ($j = 1, 2, 3, \ldots, N$) with respect to the origin of the unit cell (Fig. 23). For atom 1, we obtain Eq. (19):

$$\mathbf{f}_1 = \int_{\text{vol. of atom}} \rho(\mathbf{r}) \exp[2\pi i(\mathbf{r}_1 + \mathbf{r})\mathbf{S}] \mathrm{d}\nu$$
$$= f_1 \exp(2\pi i \mathbf{rS}) \quad (19)$$

with (Eq. 20)

$$f_1 = \int_{\text{vol. of atom}} \rho(\mathbf{r}) \exp 2\pi i \mathbf{r}_1 \mathbf{S} \mathrm{d}\nu \quad (20)$$

where f_1 is the atomic form factor for atom 1. It reflects the characteristics of the scattering of the individual atoms and is real if the wavelength of the incident X-ray is not close to an absorption edge of the atom. The atomic form factor f is equal to Z, the ordinary number of the scattering atom, at a diffraction angle of 0° and decreases with increasing diffraction angle. For N atoms, this adds up to the total scattered wave of a unit cell **G(S)** (Fig. 24) according to Eq. (21):

$$\mathbf{G}(\mathbf{S}) = \sum_{j=1}^{N} f_j \exp(2\pi i \mathbf{r}_j \mathbf{S}) \quad (21)$$

3.3 Scattering of X-rays by a Crystal

3.3.1 One-dimensional Crystal

In a one-dimensional crystal, the unit cells are separated by the unit cell vector **a**. The contribution of the scattered wave

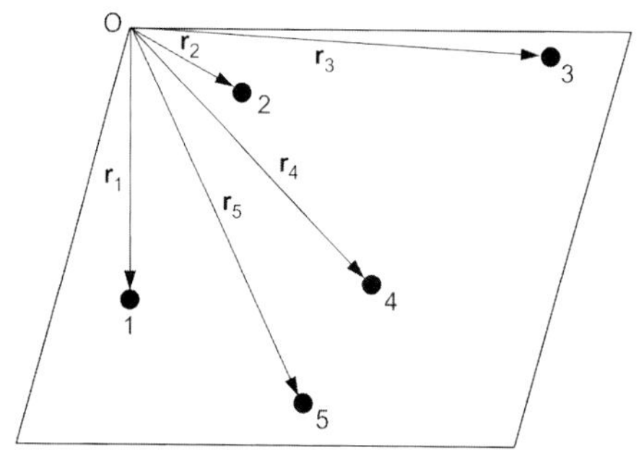

Fig. 23 Atomic positions in a unit cell.

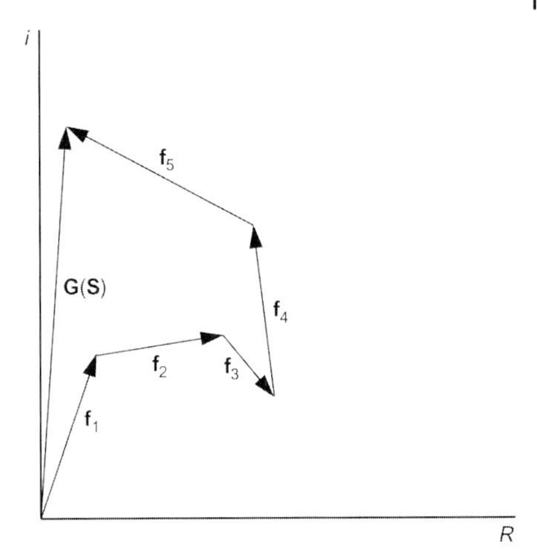

Fig. 24 Vector diagram for the total scattered wave in direction **S** added up for five atoms.

from the unit cell at the origin of the crystal is $\mathbf{G}(\mathbf{S})$. All scatterers in the second unit cell are displaced by the vector $\mathbf{a}$ relative to the origin, which introduces a corresponding phase factor and reveals for the second unit cell relative to the origin, $\mathbf{G}(\mathbf{S}) \exp 2\pi i\mathbf{aS}$. For the nth unit cell relative to the origin, we obtain $\mathbf{G}(\mathbf{S})$ $\exp 2\pi i(n-1)\mathbf{aS}$. This sums up for the total wave to Eq. (22):

$$\mathbf{F}(\mathbf{S}) = \sum_{n=1}^{T} \mathbf{G}(\mathbf{S}) \exp 2\pi i(n-1)\mathbf{aS} \quad (22)$$

Generally, $\mathbf{F}(\mathbf{S})$ is of the same order of magnitude as $\mathbf{G}(\mathbf{S})$, and no strong scattering effect is observed (Fig. 25a). However, when $2\pi\ \mathbf{aS} = 2\pi h$ or an integral multiple of 2π or $\mathbf{aS} = h$ (h is an integer), the waves add up constructively to a scattered wave proportional to $T|\ \mathbf{G}(\mathbf{S})|$ (Fig. 25b). $T = 10^5$ for a 1-mm long crystal with a 100-Å lattice constant. The intensity distribution of the scattered waves is concentrated around the values where $\mathbf{aS}$ is equal to an integer and depends on the number of contributing unit cells. The more unit cells are contributing, the sharper is the concentration of the intensity around these values.

3.3.2 Three-dimensional Crystal

In this case, the unit cell is spanned by the unit cell vectors $\mathbf{a}$, $\mathbf{b}$, and $\mathbf{c}$, and is repeated periodically by the corresponding vector shifts in the respective spatial directions. This means that we will obtain scattered waves of measurable intensities when the three subsequent conditions are fulfilled (Eq. 23):

$$\mathbf{aS} = h;\ \mathbf{bS} = k;\ \mathbf{cS} = l \quad (23)$$

These conditions are known as *Laue equations.*

If we neglect the proportionality constant T, we obtain Eq. (24) for the total scattered wave for a three-dimensional crystal with a unit cell containing N atoms:

$$\mathbf{F}(\mathbf{S}) = \sum_{j=1}^{N} f_j \exp 2\pi i\mathbf{r}_j\mathbf{S} \quad (24)$$

with (Eq. 25)

$$\mathbf{r}_j = \mathbf{a}x_j + \mathbf{b}y_j + \mathbf{c}z_j \quad (25)$$

Hence, we have (Eq. 26 and 27)

$$\mathbf{r}_j\mathbf{S} = x_j\mathbf{aS} + y_j\mathbf{bS} + z_j\mathbf{cS} = hx_j + hy_j + lz_j \quad (26)$$

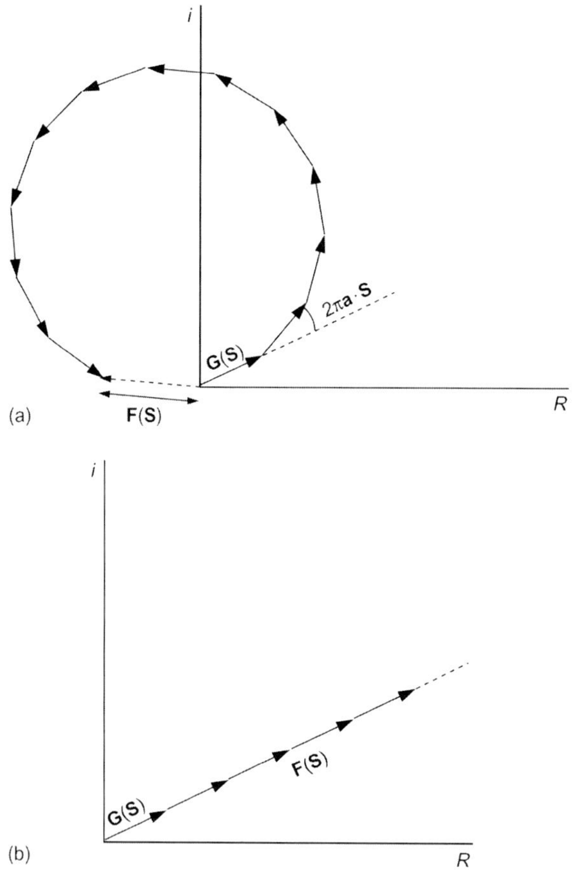

Fig. 25 Vector diagrams displaying the total wave scattered by a molecule in a crystal. (a) The phase differences between waves scattered by adjacent unit cells is $2\pi\,\mathbf{aS}$ and (b) the phase difference is an integral multiple of 2π (adapted from Blundell, T.L., Johnson, L.N. (1976) *Protein Crystallography*, Academic Press, New York).

(from Laue's equation) and

$$\mathbf{F}(hkl) = \sum_{j=1}^{N} f_j \exp 2\pi i(hx_j + hy_j + lz_j) = |\mathbf{F}(hkl)| \exp i\alpha(hkl) \qquad (27)$$

with $|\mathbf{F}(hkl)|$ – amplitude and α – phase angle. We obtain the intensity of the scattered wave as the structure factor $\mathbf{F}(hkl)$ multiplied by its complex conjugate value according to (Eq. 28):

$$I(hkl) = \mathbf{F}(hkl)\mathbf{F}^*(hkl) = |\mathbf{F}(hkl)|^2 \qquad (28)$$

3.4 The Reciprocal Lattice and Ewald Construction

In section 2.4, we mentioned the usefulness of the concept of the reciprocal lattice in understanding the diffraction of X-rays from a crystal. Now we have the necessary relations for the derivation of the reciprocal lattice. One can write the scattering vector **S** as Eq.(29):

$$\mathbf{S} = h_x\mathbf{a}^* + k_y\mathbf{b}^* + l_z\mathbf{c}^* \quad (29)$$

where **S** is a vector in reciprocal space with the metric $\mathbf{a}^*$, $\mathbf{b}^*$, $\mathbf{c}^*$. The relation to the direct space with metric **a**, **b**, **c** is still unknown. The vector **S** must obey the Laue equations (Eq. 30):

$$\begin{aligned} \mathrm{a}\mathbf{S} &= \mathbf{a}(h_x\mathbf{a}^* + k_y\mathbf{b}^* + l_z\mathbf{c}^*) = h \\ &= h_x\mathbf{aa}^* + k_y\mathbf{ab}^* + l_z\mathbf{ac}^* = h \quad (30) \end{aligned}$$

This is fulfilled only when $\mathbf{aa}^* = 1$, $h_x = h$, and $\mathbf{ab}^*$ and $\mathbf{ac}^* = 0$. Similar equations can be derived for the other two Laue conditions. Thus, vector **S** is a vector of a lattice in reciprocal space. The relation between direct and reciprocal lattice is given by the following set of nine equations (Eq. 31–39):

$$\mathbf{aa}^* = 1 \quad (31)$$

$$\mathbf{ba}^* = 0 \quad (32)$$

$$\mathbf{ca}^* = 0 \quad (33)$$

$$\mathbf{ab}^* = 0 \quad (34)$$

$$\mathbf{bb}^* = 1 \quad (35)$$

$$\mathbf{cb}^* = 0 \quad (36)$$

$$\mathbf{ac}^* = 0 \quad (37)$$

$$\mathbf{bc}^* = 0 \quad (38)$$

$$\mathbf{cc}^* = 1 \quad (39)$$

It follows from these that $\mathbf{a}^* \perp \mathbf{b}$; $\mathbf{c}$; $\mathbf{b}^* \perp \mathbf{a}$; $\mathbf{c}$; $\mathbf{c}^* \perp \mathbf{a}$; $\mathbf{b}$; and vice versa. The metric relations can also be derived from these relations. It means that the inverse lattice vectors are perpendicular to the plane that is spanned by the two other noninverse lattice vectors. Bragg's law can be derived now by inspection of Fig. 26. The wave vectors for the incident wave $\mathbf{s}_0$ and the scattered wave **S** have the same absolute value of $1/\lambda$. Vector **S** must be a vector of the reciprocal lattice and its absolute value is equal to d^*. From Fig. 26, we obtain Eq. 40 to 42:

$$\sin\theta = \frac{d^*}{2}\lambda \quad (40)$$

$$\lambda = \frac{2\sin\theta}{d^*} \quad (41)$$

$$\lambda = 2d\sin\theta \quad \text{for } n = 1 \quad (42)$$

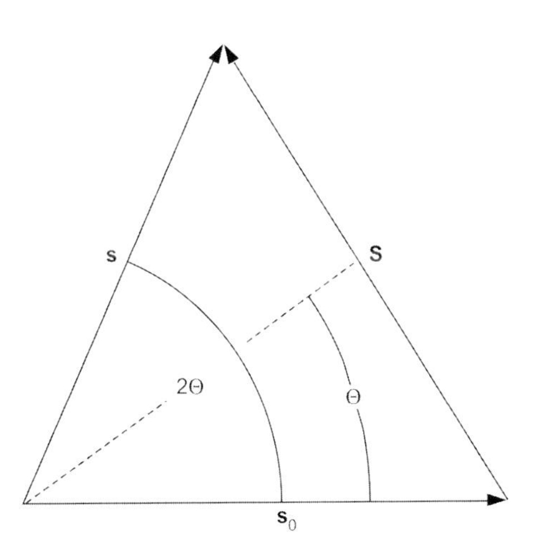

Fig. 26 Geometric representation of diffraction geometry, 2Θ, glance angle; Θ, Bragg angle.

The general equation for Bragg's law is Eq. (43):

$$2d \sin\theta = n\lambda \qquad (43)$$

where n is the order of reflection and d is the interplanar distance in the direct lattice.

The Ewald construction is contained in Fig. 20. A sphere of radius $1/\lambda$ is drawn. The origin of the reciprocal lattice is located where the wave vector $\mathbf{s}_0$ ends on the Ewald sphere. A diffracted beam is generated if a reciprocal lattice vector $\mathbf{d}^*_{hkl}$ with an absolute value of $1/d_{hkl}$ cuts the Ewald sphere. The beam is diffracted in the direction of the connection of the origin of the Ewald sphere and the intersection point of the reciprocal lattice point on the Ewald sphere. The diffraction pattern of a lattice is itself a lattice with reciprocal lattice dimensions.

3.5 The Temperature Factor

The thermal motion of the atoms causes a decrease of the scattering power by a factor of $\exp[-B(\sin^2\theta/\lambda^2)]$ with (Eq. 44)

$$B = 8\pi^2\overline{u}^2 \qquad (44)$$

where $\overline{u}$ is the mean displacement of the atoms due to the thermal motion. The atomic scattering factor f must be multiplied with this factor. In this model, the thermal motion has been assumed to be isotropic. Therefore, B is denoted as the isotropic temperature factor. In molecules, this is usually not the case and the thermal motion is described by a tensor ellipsoid. Here we obtain a set of six independent anisotropic temperature factors. In protein crystallography, isotropic B values for each atom of the molecules are used normally. The thermal motion of the atoms is one main reason for the falloff of the diffraction intensity especially at higher diffraction angles. This limits the possible recordable number of diffraction spots and, as will be seen later, the resolution of the diffraction experiment.

3.6 Symmetry in Diffraction Patterns

An X-ray diffraction data set from a crystal represents its reciprocal lattice with the corresponding diffraction intensities at the reciprocal lattice points (hkl). As the reciprocal lattice is closely related to its direct partner, it reveals symmetries, lattice properties, and other peculiarities (e.g. systematic extinctions) that are connected to the direct crystal symmetry such as unit cell dimensions and space group. A detailed discussion of this problem is given in Buerger.

In the case of real atomic scattering factors f, the diffraction intensities are centrosymmetric according to Friedel's law (Eq. 45):

$$I(hkl) = I(\overline{hkl}) \qquad (45)$$

This is illustrated in Fig. 27(a,b). The square of a complex number is the product of this number by its complex conjugate. This is shown for $\mathbf{F}$ (hkl) in Fig. 27(a) and for $\mathbf{F}(\overline{hkl})$ in Fig. 27(b). The resulting intensities are equal in both cases.

3.7 Electron Density Equation and Phase Problem

Inspection of the equation for the structure factor (Eq. 46)

$$\mathbf{F}(\mathbf{S}) = \sum_{j=1}^{N} f_j \exp 2\pi i \mathbf{r}_j \mathbf{S} = \int_{\substack{\text{vol. of} \\ \text{unit cell}}} \rho(\mathbf{r}) \exp 2\pi i \mathbf{r}\mathbf{S} \, dv \qquad (46)$$

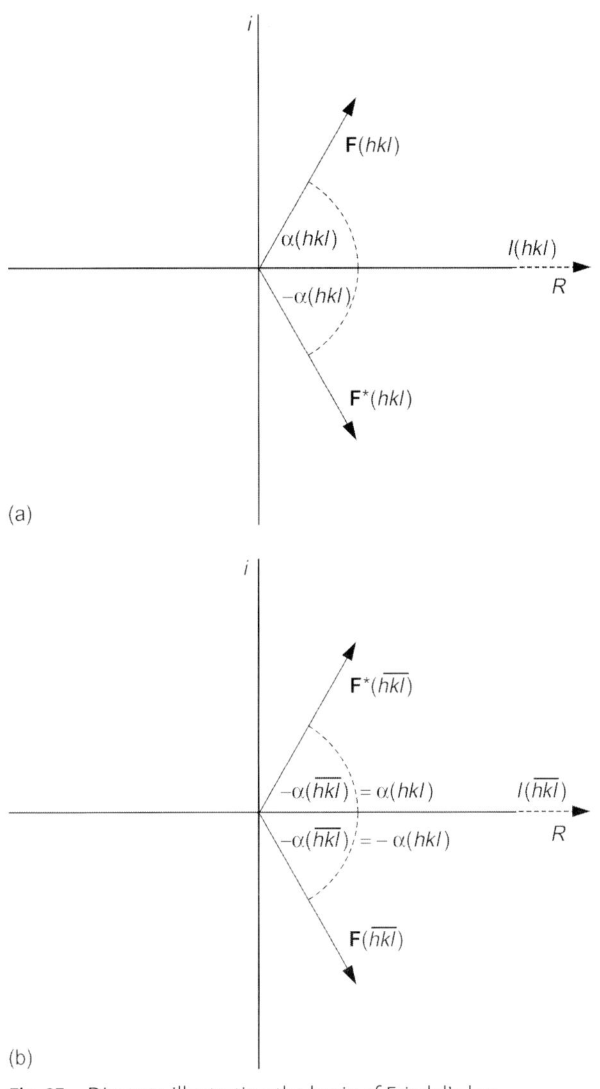

Fig. 27 Diagram illustrating the basis of Friedel's law.

shows that it is the Fourier transform of the electron density ρ (**r**).

The electron density ρ (**r**) is then the inverse Fourier transform of the structure factor **F**(**S**) according to Eq. (47):

$$\rho(\mathbf{r}) = \int_{\substack{\text{vol. of}\\ \text{diffraction space}}} \mathbf{F}(\mathbf{S}) \exp -2\pi i\mathbf{r}\mathbf{S} dv_s \tag{47}$$

The integration is replaced by summation since **F**(**S**) is not continuous and is nonzero only at the reciprocal lattice points. Hence, we have (Eq. 48)

$$\rho(xyz) = \frac{1}{V} \sum_{h=-\infty}^{\infty} \sum_{k=-\infty}^{\infty} \sum_{l=-\infty}^{\infty} \mathbf{F}(hkl) \times \exp[-2\pi i(hx + ky + lz) \tag{48}$$

Knowing the structure factors (Eq. 49)

$$\mathbf{F}(hkl) = |\mathbf{F}(hkl)| \exp i\alpha(hkl) \qquad (49)$$

one can calculate the electron density distribution in the unit cell and thus determine the atomic positions of the scattering molecule(s). Unfortunately, the measured quantities are only the absolute values $|\mathbf{F}(hkl)|$ of the structure factor. Information on the phase angles $\alpha(hkl)$ is lost during the diffraction experiment. The determination of these phases is the basic problem in any crystal structure determination, and we discuss the methods for solving the phase problem later.

3.8 The Patterson Function

The measured X-ray intensities are proportional to the square of the absolute value of the structure factor according to Eq. (28). Would it be possible to use the intensities directly to calculate from these a function that contains structural information? The answer is "yes." If one calculates a convolution of the electron density with itself, Patterson showed that this is just the Fourier transform of the intensities (Eq. 50):

$$P(uvw) = \int_{\substack{\text{vol. of}\\ \text{unit cell}}} \rho(xyz) \times \rho(x+u, y+v, z+w)\mathrm{d}v \qquad (50)$$

From this, it follows that (Eq. 51)

$$P(\mathbf{u}) = \frac{1}{V}\sum_{\mathbf{h}} F_{\mathbf{h}}^2 \exp -2\pi i\mathbf{hu} \qquad (51)$$

The function $P(\mathbf{u})$ will have maxima if the positions $\mathbf{x}$ and $\mathbf{x}+\mathbf{u}$ correspond to atoms. Thus, we obtain a function that contains the interatomic vectors as maxima. We expect N^2 peaks for N atoms. The maxima are proportional to $Z_i Z_j$. The Patterson function is a very useful tool to locate atoms when the number of atoms in the asymmetric unit of the unit cell is not too high (e.g. <20) or it contains a subset of heavy atoms among not too many (e.g. <100) light atoms such as C, N, O, or S. Here, the heavy atom–heavy atom vectors are clearly prominent. If a protein with 1000 or a multiple of that light atoms holds one or several heavy atoms per molecule, the signal resulting from the heavy atoms can no longer be resolved. However, when using the method of isomorphous replacement (discussed later), a Patterson function of the heavy atom structure can be calculated, from which it is possible to locate the heavy atoms.

3.9 Integrated Intensity Diffracted by a Crystal

Real crystals are not perfect. They can be regarded as consisting of small blocks of perfect crystals (sizes in the range of 0.1 μm), which have an average tilt angle among each other of 0.1 to 0.5° for protein crystals and which diffract independently of each other. Such a real crystal is denoted a mosaic crystal. The total energy, $\mathrm{E}(\mathbf{h})$, in a diffracted beam for a mosaic crystal rotating with uniform angular velocity w through the reflecting position in a beam of X-rays of incident intensity I_0 is given by Darwin's equation (Eq. 52):

$$\mathrm{E}(\mathbf{h}) = \frac{I_0}{\omega}\lambda^3 \frac{e^4}{m^2c^4} p \frac{LAV_x}{V^2} |\mathbf{F}(\mathbf{h})|^2 \qquad (52)$$

where λ = wavelength of X-rays, e = electronic charge, m = mass of electron, c = velocity of light, p = polarization factor, L = Lorentz factor (geometrical factor taking into account the relative time each reflection spends in the reflection position), A = absorption factor, V = volume of the unit cell, and V_x = irradiated crystal volume.

Owing to the mosaicity (0.1–0.5°), each reflection has a corresponding reflection width. The integrated intensity equation is valid under the assumption that apart from ordinary absorption, the incident intensity, I_0, is constant within the crystal (kinematic theory of X-ray diffraction) and the mosaic blocks are so small that no multiple scattering occurs within an individual mosaic block. The integrated intensity depends on λ to the third power. Increasing the wavelength causes appreciably stronger diffraction intensities but is accompanied by larger absorption. Copper Kα radiation with a wavelength of 1.5418 Å is an optimal choice for protein crystallography when using X-ray generator sources. Also important is the dependence of the integrated intensity on the unit cell volume V by its negative second power. Doubling of the unit cell volume with twice as many molecules, taking into account the increase in $|\mathbf{F}(\mathbf{h})|^2$ by having now $2n$ molecules per unit cell, reduces the average intensity for the reflected beams by a factor of two. In Eq. (52), $(\lambda^3/\omega V^2) \times (e^4/m^2c^4) \times V_x \times I_0$ is a constant for a given experiment. The corrected intensity on a relative scale $I(\mathbf{h})$ is obtained from Eq. (53):

$$I(\mathbf{h}) = \frac{E(\mathbf{h})}{p \times L \times A} \quad (53)$$

3.10
Intensities on an Absolute Scale

The corrected intensity on a relative scale $I(\mathbf{h})$ can be converted to an intensity given by Eq. (54)

$$I(abs, \mathbf{h}) = \mathbf{F}(\mathbf{h})\mathbf{F}(\mathbf{h})^* = |\mathbf{F}(\mathbf{h})|^2 \quad (54)$$

on an absolute scale by applying a so-called Wilson plot. The basis for this plot is an equation that connects the average intensity on an absolute scale with the average intensity on a relative scale by a scale factor C and considers the isotropic thermal motion of the scattering atoms by the temperature factor given in Eq. (44). This is written in the form of Eq. (55):

$$\ln \frac{\overline{I(\mathbf{h})}}{\sum_j (f_j)^2} = \ln C - 2B\frac{\sin^2\theta}{\lambda^2} \quad (55)$$

This is the equation of a straight line. B, the overall temperature factor, and C, the scale factor, can be obtained by plotting $\ln \overline{I(\mathbf{h})}/\sum_j (f_j)^2$ against $(\sin^2\theta)/\lambda^2$.

3.11
Resolution of the Structure Determination

The concept of resolution in X-ray diffraction has the same meaning as the concept in image formation in the optical microscope. After the Abbe theory, we obtain Eq. (56):

$$d_m = \frac{\lambda}{2NA} \quad (56)$$

where NA is the numerical aperture of the objective lens. In protein crystallography, the nominal resolution of an electron density map is expressed in d_m, the minimum interplanar spacing for which Fs are included in the Fourier series. The maximum attainable resolution at a given wavelength is $\lambda/2$. For Cu Kα radiation, it is 0.7709 Å and would suffice to determine protein structures at atomic resolution (the distance of a carbon–carbon single bond is about 1.5 Å). However, usually the thermal vibrations of the atoms in a protein crystal are so high that the diffraction data cannot be observed to the full theoretical resolution limit. The polypeptide chain fold can be determined at a resolution of better than 3.5 Å. A medium-resolution structure is in the resolution range of 3.0 to 2.2 Å, and makes the amino acid side chains clearly visible. A high-resolution structure has a nominal resolution better than 2.2 Å and can be as good as 1.2 Å. In such structures, the main-chain carbonyl oxygens become visible as prominent bumps and at a resolution better than 2.0 Å, aromatic side chains acquire a hole in the middle of their ring systems. For some very well diffracting crystals from small proteins, diffraction data extending to resolutions below 1.2 Å could be collected with synchrotron radiation. Such structures reveal real atomic resolution where each atom is visible as an isolated maximum in the electron density map.

3.12
Diffraction Data Evaluation

The analysis and reduction of diffraction data from a single crystal consists of seven main steps: (1) visualization and preliminary analysis of the raw, unprocessed data; (2) indexing of the diffraction patterns; (3) refinement of the crystal and detector parameters; (4) integration of the diffraction spots; (5) finding the relative scale factors between measurements; (6) precise refinement of crystal parameters using the whole data set; and (7) merging and statistical analysis of the measurements related by space-group symmetry. When using electronic area detectors with short readout times such as CCD or MWPC detectors, it is possible to collect diffraction images with small rotational increments (0.05–0.2°). In this case, the reflection profile over the crystal rotation angle can be registered, giving a three-dimensional picture of the spot. The evaluation of such diffraction data can be done with computer programs MADNES, XDS, the San Diego programs and related programs XENGEN, and X-GEN. IP systems with their longer readout times are operated in a film-like mode with rotational increments of 0.5 to 2.0°. Here, mainly the program systems MOSFLM and Denzo are applied. The most important developments in the data evaluation of macromolecular diffraction measurements are autoindexing, profile fitting, transformation of data to a reciprocal-space coordinate system, and demonstration that a single rotation image contains all of the information necessary to derive the diffraction intensities from that image.

Scaling, merging, and statistical analysis of the intensity data are either done with corresponding programs of the CCP4 program suite or with Scalepack. The principles of these operations are given in the manuals for these programs. With modern data-collection methods, the completeness should approach 100% (including the low-resolution data, which are very important for molecular replacement), the ratio $I/\sigma(I)$ should be significant even for the highest resolution shell, and undue emphasis should not be given to the reliability factor for merging the data (R-merge) unless factors such as multiplicity are taken into account. Nowadays, it is customary using synchrotron radiation techniques and fast CCD detectors to collect as much data as possible (before radiation damage becomes significant) in order to produce good statistics.

3.13 Solvent Content of Protein Crystals

The Matthews parameter V_M, which is defined according to Eq. (57):

$$V_M = \frac{V_{\text{unit cell}}}{M_{\text{Prot}}} \quad (57)$$

where $V_{\text{unit cell}}$ is the volume of the unit cell and M_{Prot} is the molecular mass of the protein in the unit cell, has values that are in the range 1.6 to 3.5 Å^3 Da^{-1} for proteins. This allows a rough estimation of the number of molecules in the unit cell. Furthermore, V_M can be used for the assessment of the solvent content of a protein crystal. Calling V_{Prot} the crystal volume occupied by the protein, V'_p its fraction with respect to the total crystal volume V, and M_{Prot} the mass of protein in the cell, we obtain Eq. (58):

$$V'_P = \frac{V_{\text{Prot}}}{V} = \frac{V_{\text{Prot}}/M_{\text{Prot}}}{M_{\text{Prot}}/V} \quad (58)$$

The first term is the specific volume of the protein, the second the reciprocal of V_M and, remembering that the molecular weight is expressed in daltons, we have Eq. (59):

$$V'_P = \frac{1.6604}{d_{\text{Prot}} V_M} \quad (59)$$

Taking 1.35 g cm^{-3} as the protein density, we obtain as a first approximation Eq. (60) and (61):

$$V'_P \approx \frac{1.23}{V_M} \quad (60)$$

$$V'_{\text{Solv}} \approx 1 - V'_P \quad (61)$$

The solvent content in a protein crystal may vary from 75 to 40%.

4 Methods for Solving the Phase Problem

4.1 Isomorphous Replacement

4.1.1 Preparation of Heavy Metal Derivatives

If one can attach one or several heavy metal atoms at defined binding site(s) to the protein molecules without disturbing the crystalline order, one can use such isomorphous heavy atom derivatives for the phase determination. The lack of isomorphism can be monitored by a change in the unit cell parameters compared with the native crystal and a deterioration of the quality of the diffraction pattern. The preparation of heavy atom derivatives is undertaken by soaking the crystals in mother liquor containing the dissolved heavy metal compound. Soaking times may be in the range from several minutes to months. Concentrations of the heavy metal compound may vary from tenths of millimolar to 50 mM. Favorite heavy atoms are Hg, Pt, U, Pb, Au, rare earth metals, and so on. Potential ligands can be classified as hard and soft ligands according to Pearson. Hard ligands are electronegative and undergo electrostatic interactions. In proteins, such ligands are glutamate, aspartate, terminal carboxylates, hydroxyls of serines and threonines, and in the buffer acetate, citrate, and phosphate. Soft ligands are polarizable and form covalent bonds such as cysteine, cystine, methionine, and histidine in proteins, and Cl^-, Br^-, I^-, S-ligands, CN^-, and imidazole in the buffer solution.

Metals are classified according to their preference for hard or soft ligands. Class (a) metals bind preferentially to hard ligands. They comprise the cations of A-metals such as alkali and alkaline earth metals, the lanthanides, some actinides, and groups IIIA, IVA, and VA of the transition metals. Class (b) metals are rather soft and polarizable and can form covalent bonds to soft ligands. They include heavy metals at the end of the transition metal groups such as Hg, Pt, and Au. Thus, in the protein, the class (b) metals Hg, Pt, and Au and complex compounds of them bind to soft ligands such as cysteine, histidine, or methionine and the class (a) metals U and Pb to hard ligands such as the carboxylate groups of glutamate or aspartate.

4.1.2 Single Isomorphous Replacement

The structure factor $\mathbf{F}_{\mathrm{PH}}$ for the heavy atom derivative structure (Fig. 28) becomes (Eq. 62)

$$\mathbf{F}_{\mathrm{PH}} = \mathbf{F}_{\mathrm{P}} + \mathbf{F}_{\mathrm{H}} \qquad (62)$$

where $\mathbf{F}_{\mathrm{P}}$ = structure factor of the native protein and $\mathbf{F}_{\mathrm{H}}$ = contribution of the heavy atoms to the structure factor of the derivative. The isomorphous differences, $F_{\mathrm{PH}} - F_{\mathrm{P}}$, which can be calculated from experimental intensity data sets of the native and derivative protein, correspond to the distance CB in Fig. 28, and are given by Eq. (63):

$$F_{\mathrm{PH}} - F_{\mathrm{P}} = F_{\mathrm{H}} \cos(\alpha_{\mathrm{PH}} - \alpha_{\mathrm{H}}) - 2F_{\mathrm{P}} \sin^2\left(\frac{\alpha_{\mathrm{P}} - \alpha_{\mathrm{PH}}}{2}\right) \qquad (63)$$

If F_{H} is small compared with F_{P} and F_{PH}, the sine term will be very small and we have (Eq. 64)

$$F_{\mathrm{PH}} - F_{\mathrm{P}} \approx F_{\mathrm{H}} \cos(\alpha_{\mathrm{PH}} - \alpha_{\mathrm{H}}) \qquad (64)$$

When vectors $\mathbf{F}_{\mathrm{P}}$ and $\mathbf{F}_{\mathrm{H}}$ are collinear, then (Eq. 65)

$$|\mathbf{F}_{\mathrm{PH}} - \mathbf{F}_{\mathrm{P}}| = F_{\mathrm{H}} \qquad (65)$$

The square of the isomorphous differences, $F_{\mathrm{PH}} - F_{\mathrm{P}}$, can be used as coefficients in a Patterson synthesis. We get

$$\begin{aligned}(F_{\mathrm{PH}} - F_{\mathrm{P}})^2 &= 4F_{\mathrm{P}}^2 \sin^4\left(\frac{\alpha_{\mathrm{P}} - \alpha_{\mathrm{PH}}}{2}\right) && \text{(i)} \\ &+ F_{\mathrm{H}}^2 \cos^2(\alpha_{\mathrm{PH}} - \alpha_{\mathrm{H}}) && \text{(ii)} \qquad (66) \\ &- 4F_{\mathrm{P}}F_{\mathrm{H}} \sin^2\left(\frac{\alpha_{\mathrm{P}} - \alpha_{\mathrm{PH}}}{2}\right) \times \cos(\alpha_{\mathrm{PH}} - \alpha_{\mathrm{H}}) && \text{(iii)}\end{aligned}$$

It is a theorem of Fourier theory that the Fourier transform of the sum

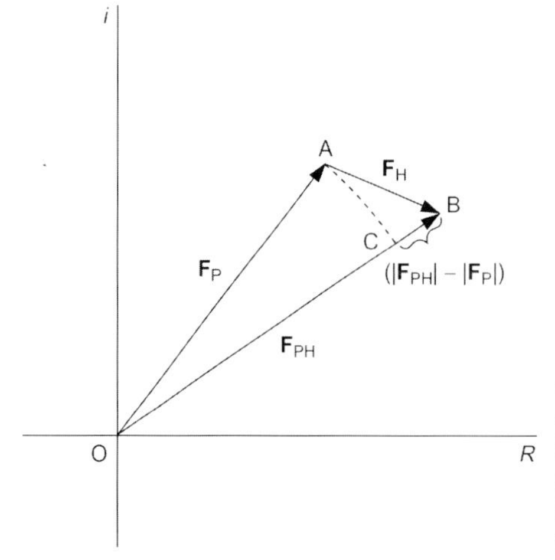

Fig. 28 Vector diagram for the vector addition of the structure factor of the native protein $\mathbf{F}_{\mathrm{P}}$ and the heavy atom contribution $\mathbf{F}_{\mathrm{H}}$ to the heavy atom derivative structure factor $\mathbf{F}_{\mathrm{PH}}$.

of Fourier coefficients is equal to the sum of the Fourier transforms of the individual Fourier coefficients. Here, there are three different terms. $(\alpha_P - \alpha_{PH})$ is small if F_H is small, and term (i), which gives the protein–protein interaction will be of low weight. The transform of term (iii) is zero if sufficient terms are included. However, if $F_H \ll F_P$, $(\alpha_P - \alpha_{PH})$ is effectively random, and term (ii) will give heavy atom vectors with half the expected peak heights (Eq. 67)

$$F_H^2 \cos^2(\alpha_{PH} - \alpha_H) = \frac{1}{2} F_H^2 + \frac{1}{2} F_H^2 \cos 2(\alpha_{PH} - \alpha_H) \quad (67)$$

with the second term on the right contributing only noise to the Patterson map because the angles α_{PH} and α_H are not correlated. Such an isomorphous heavy atom difference Patterson map allows the determination of the positions of the heavy metals on the condition of isomorphism and a not-too-large heavy atom partial structure. The interpretation of these difference Patterson maps is undertaken by vector verification routines, which are part of the CCP4 program suite. In these routines, the asymmetric unit of the unit cell is systematically scanned by calculating on each scan point the corresponding heavy atom–heavy atom vectors, determining their peak height in the Patterson map and evaluating a meaningful correlation value (e.g. the sum of the correlated maxima). Prominent heavy atom sites should show up with high correlation values.

It is important to know what intensity changes are generated by the attachment of heavy atoms to the macromolecule. According to Crick and Magdoff, the relative root-mean-square intensity change is given by Eq. (68) for centric reflections:

$$\frac{\sqrt{\overline{(\Delta I)^2}}}{\overline{I_P}} = 2 \times \sqrt{\frac{\overline{I_H}}{\overline{I_P}}} \quad (68)$$

and by Eq. (69) for acentric reflections:

$$\frac{\sqrt{\overline{(\Delta I)^2}}}{\overline{I_P}} = \sqrt{2} \times \sqrt{\frac{\overline{I_H}}{\overline{I_P}}} \quad (69)$$

where $\overline{I_H}$ is the average intensity of the reflections if the unit cell were to contain the heavy atoms only and $\overline{I_P}$ is the average intensity of the reflections of the native protein. Attaching one mercury atom ($Z = 80$) to a macromolecule with varying molecular mass and assuming 100% occupancy gives the following average relative changes in intensity: 0.51 for 14 000 Da, 0.25 for 56 000 Da, 0.18 for 112 000 Da, 0.13 for 224 000 Da, and 0.09 for 448 000 Da. From this estimation it is evident that with increasing molecular mass more heavy atoms or for large molecular masses heavy metal clusters such as $Ta_6Br_{12}^{2+}$ must be introduced to generate intensity changes that can be statistically measured (precision for intensity measurements between 5 and 10%) and which are sufficient for the phasing.

The phase calculation for single isomorphous replacement can be seen from the so-called Harker construction for this case (Fig. 29). $\mathbf{F}_H$, which can be calculated from the known heavy atom positions, is drawn in its negative direction from the origin O ending at point A. Circles are drawn with radii F_P and F_{PH} from points O and A respectively. The connections of the intersection points of both circles B and C with origin O determine two possible phases for $\mathbf{F}_P$. This means that the single isomorphous replacement leaves an ambiguity in the phase determination for the acentric reflections.

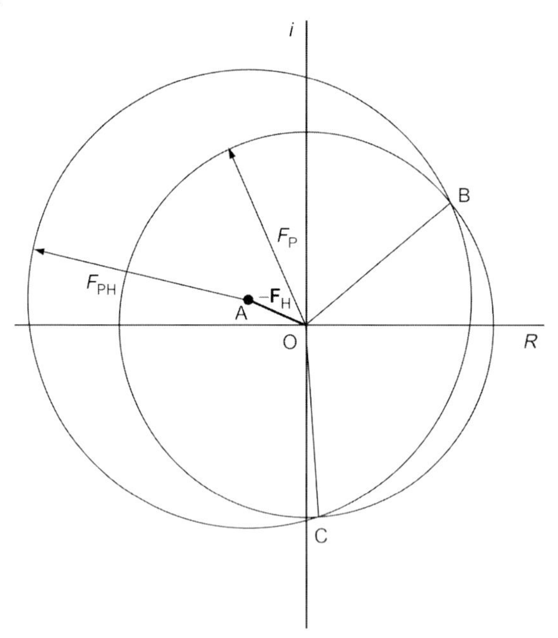

Fig. 29 Harker construction for the phase calculation by the method of single isomorphous replacement.

4.1.3 Multiple Isomorphous Replacement

The phase ambiguity can be overcome if two or more isomorphous heavy atom derivatives are used, which exhibit different heavy atom partial structures. In Fig. 30, the Harker construction for two different heavy atom derivatives is shown. In addition to Fig. 29, $-\mathbf{F}_{H2}$ is drawn from the origin O and a third circle with radius F_{PH2} is inserted around its endpoint B. The

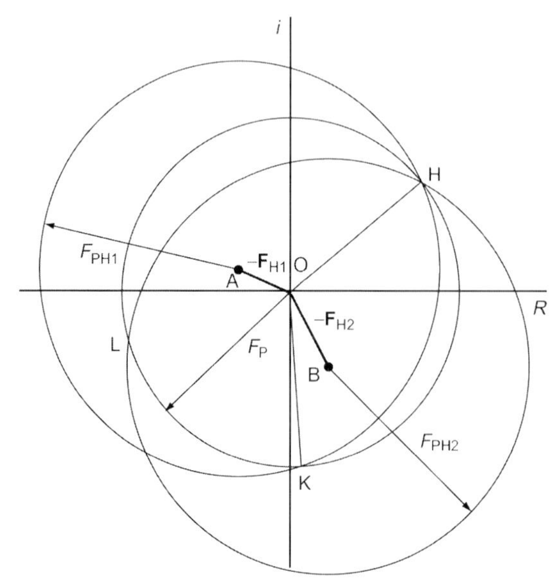

Fig. 30 Harker construction for the phase calculation by the method of MIR for two different heavy atom derivatives PH1 and PH2.

intersection point, H, of all three circles determines the protein phase, α_P. In the case of n isomorphous derivatives, there are $n+1$ circles, which have one common intersection point whose connection to origin O determines the protein phase, α_P.

4.2 Anomalous Scattering

4.2.1 Theoretical Background

So far, in the normal Thomson scattering of X-rays, the electrons in the atom have been treated as free electrons that vibrate as a dipole-oscillator in response to the incident electromagnetic radiation and generate elastic scattering of the X-rays. However, the electrons are bound to atomic orbitals in atoms, and this treatment is valid only if the frequency ω of the incident radiation is large compared to any natural absorption frequency ω_{kn} of the scattering atom. For the light atoms in biological macromolecules (H, C, N, O, S, P) with frequency ω of the used radiation (in the range of 0.4 to 3.5 Å), this condition is fulfilled and these atoms scatter normally. For heavier elements, the assumption $\omega \gg \omega_{kn}$ is no longer valid, and the frequency ω may be higher for some and lower for other absorption frequencies. If ω is equal to an absorption frequency ω_{kn}, absorption of radiation will occur, which is manifested by the ejection of a photoelectron with an energy corresponding to the ionization energy for this electron. This transition goes to a state in the continuous region because the discrete energy states are all occupied in the atom. The absorption frequencies for the K, L, or M shells are connected with the corresponding absorption edges, which are characterized by a sharp drop in the absorption curve (absorption vs λ) at the edge position. It is evident that the scattering from the electrons with their resonance frequencies close or equal to the frequency of the incident radiation will deliver a special contribution, which is called *anomalous scattering*.

The classical treatment is briefly outlined. It is assumed that the atoms scatter as if they contain electric dipole-oscillators having certain definite natural frequencies. The classical differential equation of the motion of a particle of mass m and charge e in an alternating electric field $\mathbf{E} = \mathbf{E}_0 e^{i\omega t}$ is Eq. (70):

$$\ddot{\mathbf{x}} + k\dot{\mathbf{x}} + \omega_s^2 \mathbf{x} = \frac{e\mathbf{E}_0}{m} e^{i\omega t} \tag{70}$$

where the damping factor, k, is proportional to the velocity of the displayed charge and ω_s is the natural circular frequency of the dipole if the charge is displaced. The steady state solution for this equation for the moment of the dipole that executes forced oscillations of frequency under the action of the incident wave is Eq. (71):

$$\mathbf{M} = e\mathbf{x} = \frac{e^2}{m} \frac{\mathbf{E}_0 e^{i\omega t}}{\omega_s^2 - \omega^2 + ik\omega} \tag{71}$$

The amplitude A of the scattered wave at unit distance in the equatorial plane is given by Eq. (72):

$$A = \frac{e^2}{mc^2} \frac{\omega^2 E_0}{\omega_s^2 - \omega^2 + ik\omega} \tag{72}$$

The scattering factor of the dipole, f, is now defined as the ratio of the amplitude scattered by the oscillator to that scattered by a free classical electron under the same conditions. This amplitude at unit distance and in the equatorial plane is given by Eq. (73):

$$A' = -\frac{e^2}{mc^2} E_0 \tag{73}$$

Hence, we obtain Eq. (74) for f:

$$f = \frac{\omega^2}{\omega^2 - \omega_s^2 - ik\omega} \quad (74)$$

If f is positive, the scattered wave has a phase difference of π with respect to the primary beam (introduced by the negative sign in the equation for A'). If $\omega \gg \omega_S$, f is unity. In the case of $\omega \ll \omega_S$, f is negative, and the dipole then scatters a wave in phase with the primary beam.

Equation (74) can be split into real and imaginary parts so that we obtain Eq. (75):

$$f = f' + if'' \quad (75)$$

with (Eqs. 76 and 77)

$$f' = \frac{\omega^2(\omega^2 - \omega_s^2)}{(\omega^2 - \omega_s^2)^2 + k^2\omega^2} \quad (76)$$

$$f'' = \frac{k\omega^3}{(\omega^2 - \omega_s^2)^2 + k^2\omega^2} \quad (77)$$

We now extend this for an atom consisting of s electrons, each acting as a dipole-oscillator with oscillator strength $g(s)$ and resonance frequency ω_S. We have to multiply the contribution for each electron by $g(s)$ and form the sum over all electrons. For the total real part of the atomic scattering factor, we obtain Eq. (78):

$$f' = \sum_s \frac{g(s)\omega^2}{\omega^2 - \omega_s^2} \quad (78)$$

which assumes that ω is not very nearly equal to ω_S, and a small damping. f' can be written as Eq. (79):

$$f' = f_0 + \Delta f' = \sum_s g(s) + \sum_s \frac{g(s)\omega_s^2}{\omega^2 - \omega_s^2} \quad (79)$$

For free electrons, we have $\omega_S = 0$ and $f' = f_0 = \sum_s g(s)$. The real part of the increment of the scattering factor is due to the binding of electrons. $\Delta f'$ is the dispersion component of the anomalous scattering.

If ω is comparable to ω_S but slightly greater, $ik\omega$ must not be neglected. f becomes complex (Eq. 80):

$$f = f' + if'' = f_0 + \Delta f' + i\Delta f'' \quad (80)$$

The imaginary part lags $\pi/2$ behind the primary wave, that is it is always $\pi/2$ in front of the scattered wave. $\Delta f''$ is known as the absorption component of the anomalous scattering. In the quantum mechanical treatment of the problem, the oscillator strengths are calculated from the atomic wave functions. Hönl, in theoretical work, used hydrogen-like atomic wave functions. In the frame of this approach, to each natural dipole frequency ω_S in the classical expression, there corresponds in the quantum expression a frequency ω_{kn}, which is the Bohr frequency associated with the transition of the atom from the energy state k to the state n in which it is supposed to remain during the scattering. Modern quantum mechanical calculations of anomalous scattering factors on isolated atoms, based on relativistic Dirac–Slater wave functions, have been carried out by Cromer and Liberman. It follows from the theory of the anomalous scattering of X-rays that f_0 is real, independent of the wavelength of the incident X-rays but dependent on the scattering angle. $\Delta f'$ and $\Delta f''$ depend on the wavelength, λ, of the incident radiation but are virtually independent of the scattering angle.

4.2.2 Experimental Determination

$\Delta f''$ is related to the atomic absorption coefficient μ_0 by Eq. (81):

$$\Delta f''(\omega) = \frac{mc\omega}{4\pi}\mu(\omega_0) \quad (81)$$

$\Delta f'$ can now be calculated by the Kramers–Kronig transformation (Eq. 82):

$$\Delta f''(\omega) = \frac{2}{\pi} \int_0^{\infty} \frac{\omega' \Delta f''(\omega')}{\omega^2 - \omega'^2} \mathrm{d}\omega' \quad (82)$$

As fluorescence is closely related to absorption, fluorescence measurements varying the X-ray radiation frequency are used to determine the frequency dependence of the dispersive components of the different chemical elements. Instead of the radiation frequency ω, the radiation is often characterized by its wavelength, λ, or photon energy, E. The dispersion correction terms $\Delta f'$ and $\Delta f''$ are often simply denoted f' and f''. Figure 31 shows the anomalous scattering factors near the absorption K edge of selenium from a crystal of *E. coli* selenomethionyl thioredoxin. The spectrum was measured with tunable synchrotron radiation. Apart from the "white line" feature at the absorption edge, f'' drops by about 4 electrons, approaching the edge from the short wavelength side; $\Delta f'$ exhibits a symmetrical drop of -8 electrons around the edge. Similar values can be observed at the K edges for Fe, Cu, Zn, and Br, whose wavelengths all lie in the range 0.9 to 1.8 Å, which is well suited for biological macromolecular X-ray diffraction experiments. For other interesting heavy atoms such as Sm, Ho, Yb, W, Os, Pt, and Hg, the LII (Sm) or LIII edges are in this range. Here, the effects are even greater. Considerably larger changes are found for several lanthanides, such as Yb, where the minimum f' is -33 electrons and the maximum f'' is 35 electrons.

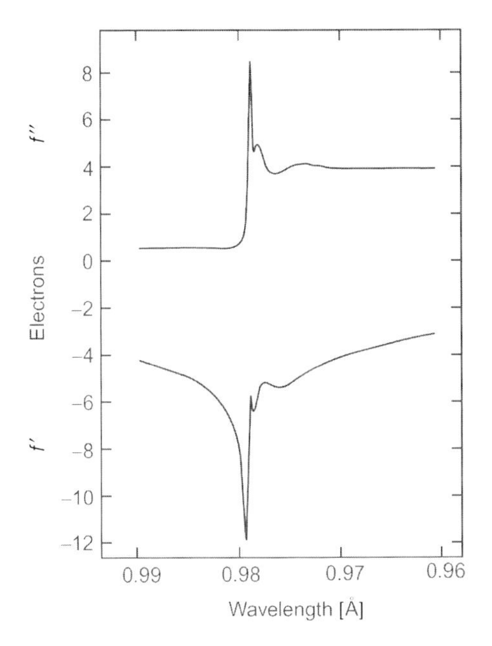

Fig. 31 Anomalous scattering factors near the absorption K edge of selenium from a crystal of *E. coli* selenomethionyl thioredoxin. (Reproduced by permission of Academic Press, Inc., from Hendrickson, W.A., Ogata, C.M. (1997) *Methods Enzymol.* **276**, 494–523.)

4.2.3 Breakdown of Friedel's Law

Under the assumption that the crystal contains a group of anomalous scatterers, one can separate the contributions from the distinctive components of the scattering factor according to Hendrickson and Ogata to obtain Eq. (83):

$$^{\lambda}\mathbf{F}(\mathbf{h}) = {}^{\circ}\mathbf{F}_{\mathrm{N}}(\mathbf{h}) + {}^{\circ}\mathbf{F}_{\mathrm{A}}(\mathbf{h}) + {}^{\lambda}\mathbf{F}'_{\mathrm{A}}(\mathbf{h}) + i^{\lambda}\mathbf{F}''_{\mathrm{A}}(\mathbf{h}) \quad (83)$$

where ${}^{\circ}\mathbf{F}_{\mathrm{N}}$ is the contribution of the normal scatterers and ${}^{\circ}\mathbf{F}_{\mathrm{A}}$, ${}^{\lambda}\mathbf{F}'_{\mathrm{A}}$, and ${}^{\lambda}\mathbf{F}''_{\mathrm{A}}$ are the contributions for the corresponding components of the complex atomic form factor. For the centrosymmetric reflection, we obtain Eq. (84):

$$\mathbf{F}(-\mathbf{h}) = {}^{\circ}\mathbf{F}_{\mathrm{N}}(-\mathbf{h}) + {}^{\circ}\mathbf{F}_{\mathrm{A}}(-\mathbf{h}) + {}^{\lambda}\mathbf{F}'_{\mathrm{A}}(-\mathbf{h}) + i^{\lambda}\mathbf{F}''_{\mathrm{A}}(-\mathbf{h}) \quad (84)$$

The geometric presentation for both structure factors is given in Fig. 32. The

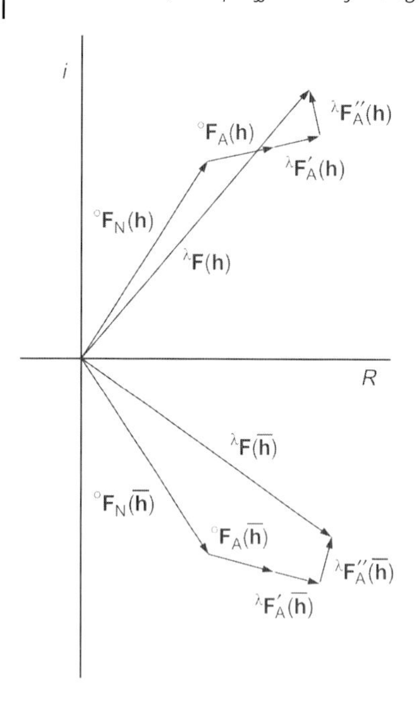

Fig. 32 Vector diagram explaining the breakdown of Friedel's law.

inversion of the sign of $\mathbf{h}$ causes a negative phase angle for all contributions where the components of the scattering factor are real. For the f''-dependent part, this is also valid, but owing to the imaginary factor i, this vector has to be constructed with a phase angle $+\pi/2$ with respect to ${}^{\circ}\mathrm{F}_{\mathrm{A}}(-\mathbf{h})$ and ${}^{\lambda}\mathbf{F}'_{\mathrm{A}}(-\mathbf{h})$. The resultant absolute values for ${}^{\lambda}\mathbf{F}(\mathbf{h})$ and ${}^{\lambda}\mathbf{F}(-\mathbf{h})$ are no longer equal, which means that their intensities (square of the amplitude) are different (breakdown of Friedel's law).

4.2.4 Anomalous Difference Patterson Map

One can show that (Eq. 85)

$$ {}^{\lambda}F(\mathbf{h}) - {}^{\lambda}F(-\mathbf{h}) \approx \frac{2}{k}[{}^{\circ}\mathrm{F}_{\mathrm{A}}(\mathbf{h}) + {}^{\lambda}F'_{\mathrm{A}}(\mathbf{h})]\sin(\alpha_{\mathbf{h}} - \alpha_{\mathrm{A}}) \qquad (85) $$

where $\alpha_{\mathbf{h}}$ is the phase angle of ${}^{\lambda}\mathbf{F}(\mathbf{h})$, α_{A} the phase angle of the anomalous scatterers, and (Eq. 86)

$$ k = \frac{{}^{\circ}\mathrm{F}_{\mathrm{A}}(\mathbf{h}) + {}^{\lambda}F'_{\mathrm{A}}(\mathbf{h})}{{}^{\lambda}F''_{\mathrm{A}}(\mathbf{h})} \qquad (86) $$

As coefficients for an anomalous difference Patterson map, we obtain Eq. (87):

$$ \Delta F^2_{\mathrm{ano}} = [{}^{\lambda}F(\mathbf{h}) - {}^{\lambda}F(-\mathbf{h})]^2 \sim \frac{4}{k^2}[{}^{\circ}\mathrm{F}_{\mathrm{A}}(\mathbf{h}) + {}^{\lambda}F'_{\mathrm{A}}(\mathbf{h})]^2\sin^2(\alpha_{\mathbf{h}} - \alpha_{\mathrm{A}}) \qquad (87) $$

The ΔF_{ano}s will be maximal if the phase angle α_{A} is perpendicular to the phase angle $\alpha_{\mathbf{h}}$ and zero if both vectors are collinear, which is opposite to the MIR case. The anomalous Patterson map contains peaks of the anomalous scatterers with heights proportional to half of $(4/k^2)[{}^{\circ}\mathrm{F}_{\mathrm{A}}(\mathbf{h}) + {}^{\lambda}F'_{\mathrm{A}}(\mathbf{h})]^2$ owing to the $\sin^2$ term and is therefore suited to determine the structure of the anomalous scatterers.

4.2.5 Phasing Including Anomalous Scattering Information

The combination of anomalous scattering information with isomorphous replacement permits the unequivocal determination of the protein phases, as shown in Fig. 33. Using the anomalous scattering information alone gives two possible solutions for the protein phase characterized by the intersection points H and L in Fig. 33. Combining it with the corresponding intensities from the native protein without the anomalous scatterers leaves only one solution for the protein phase (vector O-H in Fig. 33). The case in Fig. 33 is called *single isomorphous replacement anomalous scattering* (*SIRAS*). Having n isomorphous heavy atom derivatives, each with anomalous scattering contributions, the Harker

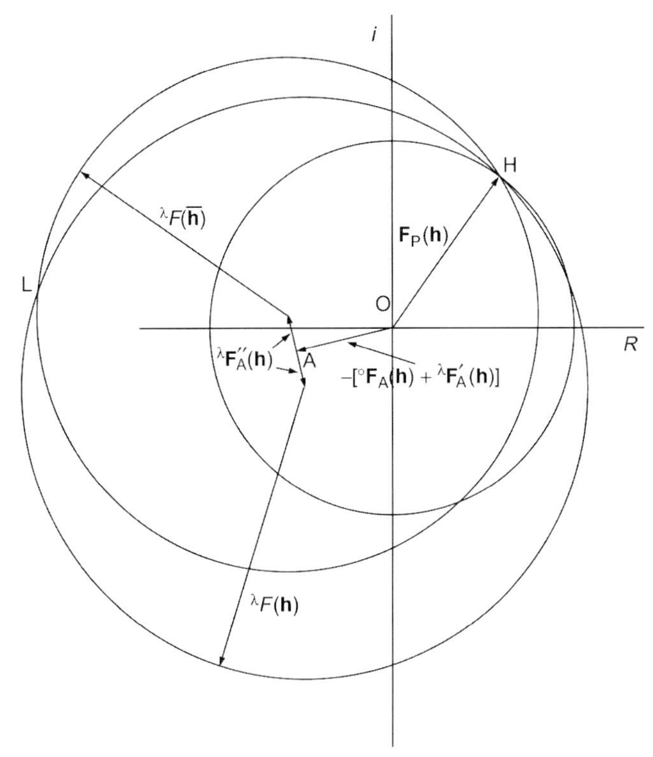

Fig. 33 Harker construction illustrating the phase determination combining information from anomalous scattering and isomorphous replacement.

construction can be extended for this situation, and the phasing method is then designated as multiple isomorphous replacement anomalous scattering (MIRAS).

4.2.6 Multiwavelength Anomalous Diffraction Technique

During the last few years, the MAD technique has matured to be a routine method and has led to a revolution in biological macromolecular crystallography. If there are one or a few anomalous scatterers in the biological macromolecule, it is possible to determine the whole spatial structure from one crystal (exact isomorphism) by the MAD technique. The anomalous scatterers may be intrinsic as in metalloproteins (e.g. Fe, Zn, Cu, Mo, Mn) or exogenous (e.g. Hg in a heavy atom derivative or Se in selenomethionyl proteins). A prerequisite for the MAD technique is well-diffracting crystals (resolution better than 2.8 Å) because the anomalous components of the atomic form factor are virtually independent of the diffraction angle and acquire increasing weight with increasing scattering angle. This advantageous property together with exact isomorphism serves for the determination of good phases down to the full resolution, and leads to the production of excellent experimental MAD-phased electron density maps. A typical MAD experiment is carried out at three different wavelengths (tunable synchrotron radiation), at minimum f' and maximum f'' at the absorption edge of the anomalous scatterer(s), and at a remote wavelength where anomalous scattering effects are small.

The basic equations for the MAD technique as formulated by Hendrickson and Ogata are as follows. Equation (83) can be written as Eq. (88):

$$^{\lambda}\mathbf{F}(\mathbf{h}) = {}^{\circ}\mathbf{F}_{\mathrm{T}}(\mathbf{h}) + {}^{\lambda}\mathbf{F}_{\mathrm{A}}(\mathbf{h}) + i^{\lambda}\mathbf{F}''_{\mathrm{A}}(\mathbf{h}) \quad (88)$$

where (Eq. 89)

$$^{\circ}\mathbf{F}_{\mathrm{T}} = {}^{\circ}\mathbf{F}_{\mathrm{N}} + {}^{\circ}\mathbf{F}_{\mathrm{A}} \quad (89)$$

with subscript T for the totality of atoms in the structure.

Furthermore, we have Eqs. (90) to (93):

$$^{\circ}\mathbf{F}_{\mathrm{T}}(f^{\circ}) = {}^{\circ}\mathrm{F}_{\mathrm{T}}\exp(i^{\circ}\phi_{\mathrm{T}}) \quad (90)$$

$$^{\circ}\mathbf{F}_{\mathrm{A}}(f^{\circ}) = {}^{\circ}\mathrm{F}_{\mathrm{A}}\exp(i^{\circ}\phi_{\mathrm{A}}) \quad (91)$$

$$^{\lambda}\mathbf{F}'_{\mathrm{A}} = f(f') \quad (92)$$

$$^{\lambda}\mathbf{F}''_{\mathrm{A}} = f(f'') \quad (93)$$

In the common case of a single kind of anomalous scatterer, we obtain Eqs. (94) and (95):

$$^{\lambda}\mathbf{F}'_{\mathrm{A}} = \frac{f'(\lambda)}{f^{\circ}}{}^{\circ}\mathbf{F}_{\mathrm{A}} \quad (94)$$

$$^{\lambda}\mathbf{F}''_{\mathrm{A}} = \frac{f''(\lambda)}{f^{\circ}}{}^{\circ}\mathbf{F}_{\mathrm{A}} \quad (95)$$

Separating the experimentally observable squared amplitude into wavelength-dependent and wavelength-independent terms gives Eq. (96):

$$\begin{aligned}^{\lambda}F(\pm\mathbf{h})^2 &= {}^{\circ}\mathrm{F}_{\mathrm{T}}^2 + a(\lambda)\,{}^{\circ}\mathrm{F}_{\mathrm{A}}^2 \\ &+ b(\lambda)\,{}^{\circ}\mathrm{F}_{\mathrm{T}}\,{}^{\circ}\mathrm{F}_{\mathrm{A}}\cos({}^{\circ}\phi_{\mathrm{T}} - {}^{\circ}\phi_{\mathrm{A}}) \\ &\pm c(\lambda)\,{}^{\circ}\mathrm{F}_{\mathrm{T}}\,{}^{\circ}\mathrm{F}_{\mathrm{A}}\sin({}^{\circ}\phi_{\mathrm{T}} - {}^{\circ}\phi_{\mathrm{A}}) \quad (96)\end{aligned}$$

with (Eqs. 97–99)

$$a(\lambda) = \frac{f'^2 + f''^2}{f^{\circ 2}} \quad (97)$$

$$b(\lambda) = 2\frac{f'}{f^{\circ}} \quad (98)$$

$$c(\lambda) = 2\frac{f''}{f^{\circ}} \quad (99)$$

The derivation of the formula for $^{\lambda}F(\mathbf{h})^2$ is illustrated in detail. $^{\lambda}F(\mathbf{h})^2$ is obtained from the triangle formed by the vectors $^{\lambda}\mathbf{F}(\mathbf{h})$, $^{\circ}\mathbf{F}_{\mathrm{T}}(\mathbf{h})$, and $\mathbf{a}$ (Fig. 34) by use of the cosine rule. The absolute values of the vectors are represented in italics, and the relevant angle is $(180^{\circ} - {}^{\circ}\phi_{\mathrm{A}} - \delta)$. We get

$$\begin{aligned}^{\lambda}F(\mathbf{h})^2 &= {}^{\circ}\mathrm{F}_{\mathrm{T}}^2 + \left(\frac{f'^2 + f''^2}{f^{\circ 2}}\right){}^{\circ}\mathrm{F}_{\mathrm{A}}^2 - 2 \\ &\times {}^{\circ}\mathrm{F}_{\mathrm{T}} \times \left(\frac{f'^2 + f''^2}{f^{\circ 2}}\right)^{1/2} \times {}^{\circ}\mathrm{F}_{\mathrm{A}} \\ &\times \cos(\pi - \delta - {}^{\circ}\phi_{\mathrm{A}} + {}^{\circ}\phi_{\mathrm{T}}) \quad (100)\end{aligned}$$

The cosine term in Eq. (100) can be obtained using some basic trigonometry:

$$\begin{aligned}&\cos(\pi + ({}^{\circ}\phi_{\mathrm{T}} - {}^{\circ}\phi_{\mathrm{A}} - \delta)) \\ &\quad = -\cos({}^{\circ}\phi_{\mathrm{T}} - {}^{\circ}\phi_{\mathrm{A}} - \delta) \quad (101)\end{aligned}$$

$$\begin{aligned}&\cos(({}^{\circ}\phi_{\mathrm{T}} - {}^{\circ}\phi_{\mathrm{A}}) - \delta) = \cos({}^{\circ}\phi_{\mathrm{T}} - {}^{\circ}\phi_{\mathrm{A}}) \\ &\quad \times \cos\delta + \sin({}^{\circ}\phi_{\mathrm{T}} - {}^{\circ}\phi_{\mathrm{A}}) \times \sin\delta \quad (102)\end{aligned}$$

with

$$\cos\delta = \frac{\left(\frac{f'}{f^{\circ}}\right) \times {}^{\circ}\mathrm{F}_{\mathrm{A}}}{\left(\frac{f'^2 + f''^2}{f^{\circ 2}}\right)^{1/2}}$$

$$\times\,{}^{\circ}\mathrm{F}_{\mathrm{A}} = \frac{\left(\frac{f'}{f^{\circ}}\right)}{\left(\frac{f'^2 + f''^2}{f^{\circ 2}}\right)^{1/2}} \quad (103)$$

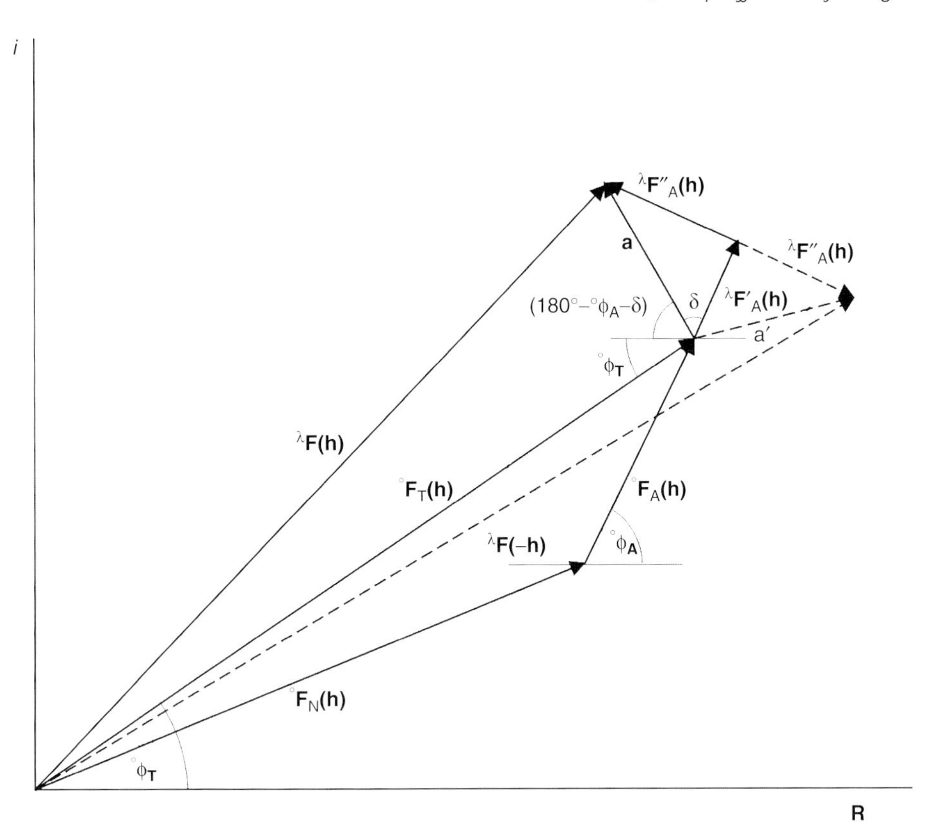

Fig. 34 Schematic drawing of structure factors of a biological macromolecule that contains one kind of anomalous scatterer.

$$\sin\delta = \frac{\left(\frac{f''}{f^{\circ}}\right) \times {}^{\circ}\mathrm{F}_{\mathrm{A}}}{\left(\frac{f'^2 + f''^2}{f^{\circ}2}\right)^{1/2}} \times {}^{\circ}\mathrm{F}_{\mathrm{A}} = \frac{\left(\frac{f''}{f^{\circ}}\right)}{\left(\frac{f'^2 + f''^2}{f^{\circ 2}}\right)^{1/2}} \quad (104)$$

Substituting this in Eq. 99, we obtain the expression for ${}^{\lambda}F(\mathbf{h})^2$. ${}^{\lambda}F(-\mathbf{h})^2$ is determined from the triangle formed by the vectors ${}^{\lambda}\mathbf{F}(-\mathbf{h})$, ${}^{\circ}\mathbf{F}_{\mathrm{T}}(\mathbf{h})$, and $\mathbf{a}'$ (Fig. 34) using a similar approach. Maximum anomalous scattering effects can be expected in intensity differences of reflections that would be equal for exclusively normal scattering. This is the case for Friedel pairs, $\mathbf{h}$ and $-\mathbf{h}$, or their rotational symmetry partners, and the relation for such differences is given in Eq. (85). Of further interest are dispersive differences between structure amplitudes at different wavelengths, Eq. (105):

$$\Delta F_{\Delta\lambda} \equiv {}^{\lambda i}F(\mathbf{h}) - {}^{\lambda j}F(\mathbf{h}) \quad (105)$$

The anomalous or dispersive intensity differences can be used to determine the structure of the anomalous scatterers. The methods are the same as for isomorphous replacement. They include vector verification procedures of difference Patterson maps or direct methods programs such as Shake and Bake and SHELXD.

4.2.7 Determination of the Absolute Configuration

As anomalous scattering destroys the centrosymmetry of the diffraction data, this effect can be used to determine the absolute configuration of chiral biological macromolecules. The most common method is to calculate protein phases on the basis of both hands of the heavy atom or anomalous scatterer structures and check the quality of the relevant electron density map that should be better for the correct hand. Furthermore, secondary structural elements in proteins (consisting of L-amino acids) such as α-helices should be right handed.

4.3 Patterson Search Methods (Molecular Replacement)

If the structures of molecules are similar (virtually identical) or contain a major similar part, this can be used to determine the crystal structure of the related molecule if the structure of the other molecule is known. This is done by systematically exploring the Patterson function of the crystal structure to be determined with the Patterson function of the search model. Let us first consider some important features of the Patterson function. The relation between two identical molecules in the search crystal structure (Fig. 35a) can generally be formulated as Eq. (106):

$$\mathbf{X}_2 = [C]\mathbf{X}_1 + \mathbf{d} \tag{106}$$

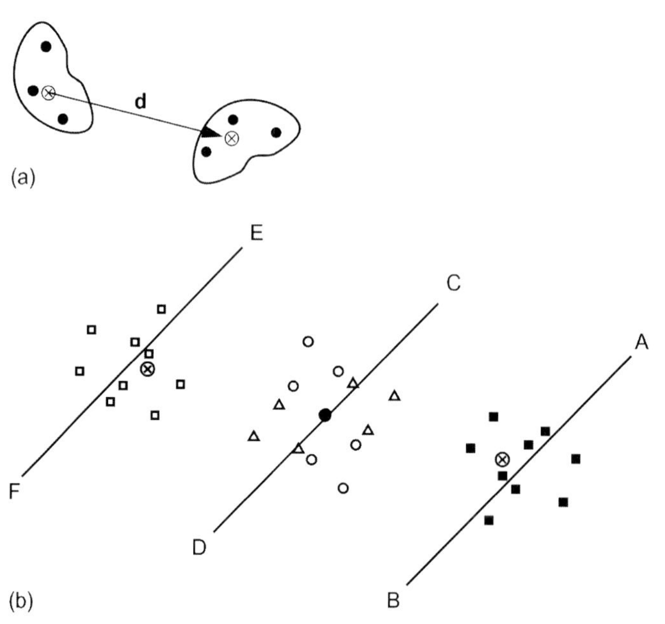

Fig. 35 Patterson function of two identical molecules separated by the spatial movement given in Eq. (106). (a) Positions of the two molecules; (b) interatomic vectors of structure in (a). △, ○, intramolecular vectors of the left and right molecule; □, ■, intermolecular vectors.

Equivalent positions $\mathbf{X}_1$ in molecule 1 are at positions $\mathbf{X}_2$ of molecule 2. $[C]$ is the rotation matrix and $\mathbf{d}$ is the translation vector of the movement of the molecule. Fig. 35(b) shows the Patterson function belonging to the molecular arrangement in Fig. 35(a). It is evident that around the origin, vectors are assembled that are intramolecular, whereas the vectors around lines AB and EF are intermolecular. The intramolecular vectors depend on the molecule orientation only and, therefore, can be used for its determination. Once the orientation of the molecule(s) has been elucidated, this can be used to reveal the translation of the molecule(s) by analyzing the intermolecular vector part of the Patterson function. The distinction between intra- and intermolecular vector sets and exploiting them for orientation and translation determination was given by Hoppe. The extension to protein crystallography and the first mathematical formulation of the rotation and translation functions were given by Rossmann and Blow.

4.3.1 Rotation Function

The intramolecular vectors are arranged in a volume around the origin of the Patterson function with a radius equal to the dimension of the molecule. The rotational search is then carried out in this volume u. The search Patterson (deduced either from the search model or from the crystal Patterson itself) is rotated to any possible rotational orientation $\mathbf{X}_2$, characterized by three rotational angles α, β, γ, which may be defined in different ways (polar angles, Eulerian angles, etc.). At each angular position, the actual functional values are correlated with those of the crystal Patterson all through the volume u and integrated over this volume. The correlation function may be the sum or the product of each corresponding pair of values. Rossmann and Blow proposed a product function, and the rotation function for this case is given by Eq. (107):

$$R(\alpha, \beta, \gamma) = \int_u P_2(\mathbf{X}_2 P_1(\mathbf{X}_1)\mathrm{d}\mathbf{X}_1 \quad (107)$$

The function has maxima if the intramolecular vector sets are coincident. The calculation can be carried out in both direct and reciprocal space.

The self-rotation function is a special form of the rotation function. If an asymmetric unit contains more than one copy of a molecule, the rotation matrix between the molecules can be determined by a self-rotation function. Here, the crystal Patterson is rotated against itself, and the integration is taken over the volume u around the origin in the same manner. The identical molecules may have an arbitrary orientation to each other or they may be related by local or the so-called noncrystallographic symmetries. Searching for local rotation axes is done best in a polar angle system. The search Patterson is brought into each polar orientation and then rotated around the angle value for the local axis being sought, for example 120° for a threefold local axis.

4.3.2 Translation Function

Once the orientation of the molecule(s) has been determined, the translation of the molecule(s) can be obtained from a translation function. The model Patterson $P_2(\mathbf{u})$ revealed from the model in the correct orientation is calculated for different translations $\mathbf{t}$ and correlated with the crystal Patterson $P_1(\mathbf{u})$. The translation function proposed by Crowther and Blow has the form shown in Eq. (108):

$$T(\mathbf{t}) = \int P_1(\mathbf{u}) P_2(\mathbf{u}, \mathbf{t})\mathrm{d}\mathbf{u} \quad (108)$$

$T(\mathbf{t})$ reveals a maximum peak at the correct translation $\mathbf{t}$ if the center of gravity of the search model was at the origin for $\mathbf{t} = 0$.

4.3.3 Computer Programs for Molecular Replacement

An early program for molecular replacement, working in direct space, was written by Huber. Nowadays, several program packages are available, either being exclusively dedicated to the molecular replacement technique or having integrated relevant modules. Pure molecular replacement programs are, for example, AMORE and GLRF. The rotational and translational search starting from the search model is fully automated in AMORE and includes a final rigid body refinement of each proposed solution. GLRF offers different types of rotation and translation functions, all operating in reciprocal space, and a Patterson correlation refinement. A peculiarity of the GLRF program is the locked rotation function. This function takes into account the possible noncrystallographic symmetries and is an average of n independent rotation functions with an improved peak-to-noise ratio. Frequently used program packages including molecular replacement modules are the CCP4 program suite, CNS, and PROTEIN.

4.4 Phase Calculation

4.4.1 Refinement of Heavy Atom Parameters

Before the protein phases can be calculated, it is necessary to refine the heavy atom parameters. These are the coordinates x, y, z, the temperature factor (either isotropic or anisotropic), and the occupancy. The refinement modifies the parameters in such a way that $|\mathbf{F}_{\mathrm{PH}}(\mathrm{obs})|$ becomes as close as possible to $|\mathbf{F}_{\mathrm{PH}}(\mathrm{calc})|$. Using the method of least squares, the refinement according to Rossmann minimizes Eq. (109):

$$\varepsilon = \sum_{\mathbf{h}} w(\mathbf{h})[(F_{\mathrm{PH}} - F_{\mathrm{P}})^2 - kF^2_{\mathrm{H\ calc}}]^2 \tag{109}$$

where k is a scaling factor to correct F^2_{Hcalc} to a theoretically more acceptable value because according to Eq. (64), $\mathbf{F}_{\mathrm{PH}} - \mathbf{F}_{\mathrm{P}}$ and $\mathbf{F}_{\mathrm{H}}$ have approximately the same length when $\mathbf{F}_{\mathrm{PH}}$, $\mathbf{F}_{\mathrm{P}}$, and $\mathbf{F}_{\mathrm{H}}$ point in the same direction. The probability for this case will be high if the difference between F_{PH} and F_{P} is large. An improvement can be obtained if the contribution from the anomalous scattering is included.

For the parameter refinement of anomalous scattering sites, the differences between the observed and calculated structure factor amplitudes for $^{\circ}\mathrm{F}_{\mathrm{A}}$ are subjected to minimization. Another approach treats the anomalous or dispersive contributions as in MIR phasing.

From the refined heavy atom parameters, preliminary protein phase angles α_{P} can be obtained as shown in the corresponding Harker construction. A further refinement of the heavy atom parameters can be achieved by the "lack of closure" method, incorporating this knowledge. The definition of this "lack of closure" ε is illustrated in Fig. 36(a) and (b). In the case of perfect isomorphism, the vector triangle $\mathbf{F}_{\mathrm{P}} + \mathbf{F}_{\mathrm{H}} = \mathbf{F}_{\mathrm{PH}}$ closes exactly (Fig. 36a). In practice, this condition will not be fulfilled, and a difference ε between the observed F_{PH} and the calculated F_{PH} will remain (Fig. 36b). $F_{\mathrm{PH}}(\mathrm{calc})$ can be obtained from the triangle OAB (Fig. 36b)

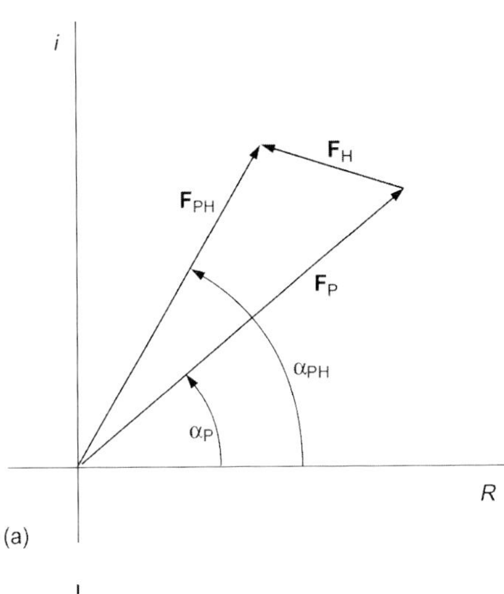

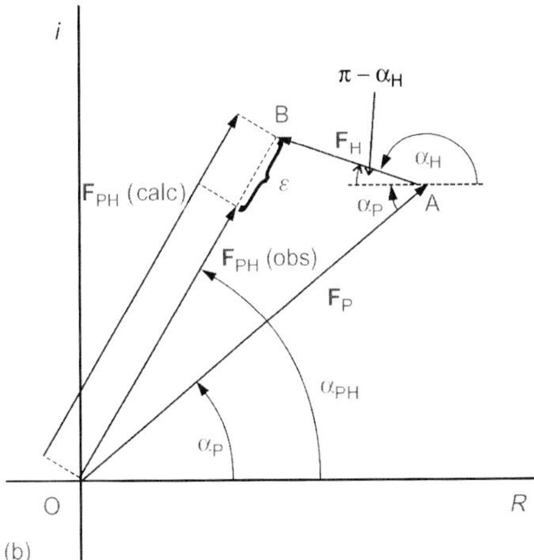

Fig. 36 Definition of "lack of closure." (a) Perfect isomorphism; (b) usually, the observed and calculated values for $\mathbf{F}_{PH}$ differ by the "lack of closure" ε.

with the cosine rule (Eq. 110):

$$F_{PH} = [F_P^2 + F_H^2 + 2F_P \times F_H \cos(\alpha_H - \alpha_P)]^{1/2} \quad (110)$$

The function that is minimized by the least-squares method is Eq. (111):

$$E_j = \sum_{\mathbf{h}} m_{\mathbf{h}} \varepsilon_j(\mathbf{h})^2 \quad (111)$$

where (Eq. (112))

$$\varepsilon_j = k_j F_{PHj}(\text{obs}) - F_{PHj}(\text{calc}) \quad (112)$$

is the "lack of closure" for the heavy atom derivative j, k_j is a scaling factor, and $m_{\mathbf{h}}$ is a weighting factor.

4.4.2 Protein Phases

As the structure factor amplitudes F_P, F_{PH}, F_H, and α_H are known, the protein phase angle α_P can be calculated. For the single isomorphous replacement situation (Fig. 28), ε is zero only for the two protein phase angles α_P where the two circles for F_P and F_{PH} intersect. In practice, all these observed quantities exhibit errors. For the treatment of these errors, it is assumed that all errors are in $\mathbf{F}_{PH}$ and that both $\mathbf{F}_H$ and $\mathbf{F}_P$ are error free. For each protein phase angle α, $\varepsilon(\alpha)$ is calculated. The smaller $\varepsilon(\alpha)$ is, the higher is the probability of a correct phase angle α. For each reflection of a derivative j, a Gaussian probability distribution is assumed for ε according to Eq. (113):

$$P(\alpha) = P(\varepsilon) = N \exp\left[-\frac{\varepsilon^2(\alpha)}{2E^2}\right] \quad (113)$$

where N is a normalization factor and E^2 is the mean square value of ε. Small values of E are related to probability curves with sharp peaks and well-determined phase angles, and the opposite is true for large E values. Such phase-angle probability curves can be calculated for each individual reflection and derivative. For single isomorphous replacement, this curve is symmetric with two high peaks corresponding to the two possible solutions for α_P. We obtain the total probability for each reflection with contributions from n heavy atom derivatives by multiplying the individual probabilities (Eq. 114):

$$P(\alpha) = \prod_{j=1}^{n} P_j(\alpha) = N' \exp\left[-\sum_j \frac{\varepsilon_j^2(\alpha)}{2E_j^2}\right] \tag{114}$$

These curves will be nonsymmetric with one or several maxima (see Fig. 37a and b).

The question arises of which phases should be taken in the electron density equation to calculate the best electron density function. An immediate guess would be to use the phases where $P(\alpha)$ has the highest value. This approach would be appropriate for unimodal distributions but not for bimodal distributions. Blow and Crick derived the phase value that must be applied under the assumption that the mean square error in electron density over the unit cell is minimal. For one reflection, this is given by Eq. (115):

$$\langle \Delta\rho^2 \rangle = \frac{1}{V^2}(\mathbf{F}_\mathrm{s} - \mathbf{F}_\mathrm{t})^2 \tag{115}$$

where $\mathbf{F}_\mathrm{t}$ is the true factor and $\mathbf{F}_\mathrm{s}$ is the structure factor applied in the Fourier synthesis. The mean square error is then obtained as Eq. (116):

$$\langle \Delta\rho^2 \rangle = \frac{1}{V^2} \frac{\int_{\alpha=0}^{2\pi} (\mathbf{F}_\mathrm{s} - F \exp i\alpha)^2 P(\alpha) \mathrm{d}\alpha}{\int_{\alpha=0}^{2\pi} P(\alpha) \mathrm{d}\alpha} \tag{116}$$

$\mathbf{F}_\mathrm{t}$ has a phase probability of $P(\alpha)$ and has been given as $\mathbf{F}_\mathrm{t} = F \exp i\alpha$. It can be shown that the numerator integral in Eq. (116) is minimal if (Eq. 117)

$$\mathbf{F}_{\mathrm{s(best)}} = F \frac{\int_{\alpha=0}^{2\pi} \exp(i\alpha) P(\alpha) \mathrm{d}(\alpha)}{\int_{\alpha=0}^{2\pi} P(\alpha) \mathrm{d}\alpha} = mF \exp(i\alpha_{\mathrm{best}}) \tag{117}$$

Equation (117) corresponds to the center of gravity of the probability distribution with polar coordinates (mF, α_{best}), where m is defined as magnitude of $\mathbf{m}$ given by Eq. (118):

$$\mathbf{m} = \frac{\int_{\alpha=0}^{2\pi} P(\alpha) \exp(i\alpha) \mathrm{d}\alpha}{\int_{\alpha=0}^{2\pi} P(\alpha) \mathrm{d}\alpha} \tag{118}$$

This magnitude of $\mathbf{m}$ is equivalent to a weighting function and is designated the "figure of merit." The electron density map calculated with mF and α_{best} is known

P(α)

(a) α

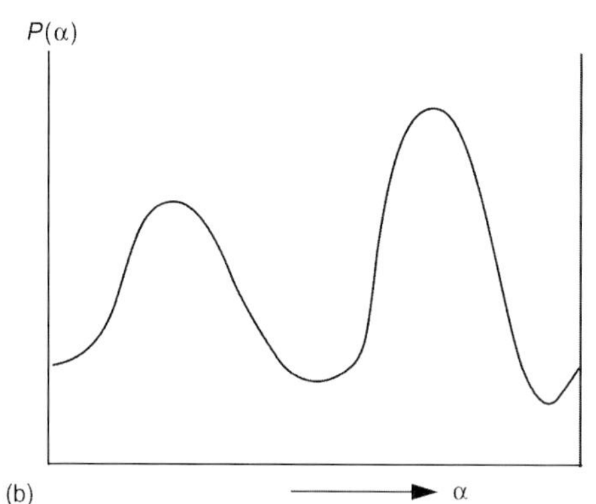

Fig. 37 Total probability curves $P(\alpha)$ for two different reflections: (a) for one derivative; (b) for more than one derivative.

as *best Fourier* and should represent a Fourier map with minimum least-squares error from the true Fourier map.

For the total error of the "best Fourier," Eq. (119) has been derived:

$$\langle \Delta\rho^2 \rangle = \frac{1}{V^2} \sum_{\mathbf{h}} F^2(\mathbf{h})(1 - m^2) \qquad (119)$$

The order of magnitude of this error may be illustrated by the example of the structure determination of lysozyme. The root-mean-square error in the Fourier synthesis was 0.35 e Å^{-3}, with values of 2.0-Å resolution for the diffraction data and a mean "figure of merit" of 0.6.

The program systems CCP4 and PROTEIN contain all routines necessary to calculate protein phases according to the MIRAS technique and a number of different kinds of Fourier maps. Alternative probabilistic approaches for the phase calculation are used in programs MLPHARE and SHARP. Both programs can also carry out MAD phasing. The MADSYS program is based on the algebraic approach outlined in the MAD section of this contribution and executes all tasks of a MAD analysis from scaling to phase-angle calculation.

4.5 Phase Improvement

With the methods so far described, an experimental electron density map can be calculated, and if its quality is high enough, the atomic model can be constructed. However, there are methods for further phase improvement available, which may be applied in general or depending on the given prerequisites. Such phase improvement routines have been used routinely over the last 10 years and have had a large impact on the advancement of biological macromolecular crystallography.

4.5.1 Solvent Flattening

Protein crystals have a solvent content of 75 to 40%. In a highly refined protein crystal structure, the solvent space between the molecules is rather flat owing to the dynamic nature of this region. Usually, the initial experimental starting phases are of lower quality than the final ones and as a result, the solvent region (if the molecular boundaries can be identified) contains noise peaks. It is now obvious to set the noisy solvent space to a low constant value and calculate new improved phases by Fourier back-transforming this corrected electron density map. However, it is evident that the definition of the molecular boundaries will be tedious and depend on the quality of the electron density map. Wang has proposed an automatic procedure that smooths the electron density to define the protein region. This smoothed electron density map is traced against a threshold value that separates this map into molecule and solvent space according to their ratio of volumes in the unit cell. The space inside the molecular envelope is polished to avoid internal voids. Now, a new electron density map is calculated using the observed structure factor amplitudes and the phases revealed from the solvent corrected map. The solvent corrected map is obtained by setting all electron density values inside the molecular envelope to those of the initial map and all values outside the envelope to a low constant value. These phases from the solvent flattening procedure can be combined with the MIR or MAD phases. This procedure can be repeated in several iterative cycles because after each cycle of solvent flattening, the quality of the electron density map is improved. There are no prerequisites for the application of the method of solvent flattening. It is evident that solvent flattening is most effective for crystals with a high solvent content.

4.5.2 Histogram Matching

Histogram matching is a technique emanating from image processing. In the application to electron density maps, it is assumed that a high-quality protein crystal structure has a characteristic frequency distribution of electron density, which serves as a standard reference distribution for other electron density maps. Such maps of lower quality exhibit a frequency distribution of electron density, which deviates from the standard distribution. The electron density map of low quality is then scaled in such a way that its frequency distribution of electron density now corresponds to the standard distribution. Histogram matching is normally used together with solvent flattening and is incorporated into the density modification programs SQUASH and DM from the CCP4 program package.

4.5.3 Molecular Averaging

If there is more than one identical subunit in the asymmetric unit of the crystal, molecular averaging can be used to improve the protein phases. The spatial relations between the single identical subunits in the asymmetric unit may be determined by Patterson search methods or from the arrangement of the heavy atoms or anomalous scatterers. The spatial relation between the identical subunits can be improper (the relevant spatial movement consists of a rotation about an unsymmetrical angle value and a translation component) or proper (the spatial movements form a symmetry group that is composed of rotational symmetry elements only). Such additional symmetries are called *noncrystallographic* or *local*, and there are no limitations concerning the Zähligkeit of the symmetry axes (e.g. five-, seven-, and higherfold axes are allowed). It is evident that averaging about the different related subunits, whose electron density should be equal in each subunit, must result in an improved electron density map and therefore in improved protein phases. Molecular averaging is best done in direct space and several programs (e.g. RAVE or MAIN) are available.

The procedure of molecular averaging is composed of several steps. First, the molecular envelope must be determined from the initial electron density map or from a molecular model that, for example, has been obtained from molecular replacement. Next, the particular electron density averaging between the related subunits is performed. This is followed by the reconstitution of the complete crystal unit cell with the averaged electron density. The space outside the molecular envelope is flattened. This map is then Fourier back-transformed. The obtained phase angles can either be taken directly or combined with known phase information to calculate a new and improved electron density map. This cycle can be repeated several times until convergence of the electron density map improvement has been reached. It is very useful to refine the local symmetry operations after every macrocycle of molecular averaging. Furthermore, molecular averaging can be applied if proteins crystallize in more than one crystal form.

Molecular averaging is especially efficient if a high noncrystallographic symmetry is present as in virus structures, but the averaging over two related subunits alone (the lowest case of local symmetry) can give a considerable improvement.

In special cases where high noncrystallographic symmetry exists and the phase information extends to low resolution only, cyclic molecular averaging can be used to extend the phase angles to the full resolution of the native protein. This was first shown in the structure analysis of hemocyanin from *Panulirus interruptus*. It is extensively used in the analysis of icosahedral structures and for large molecular assemblies.

4.5.4 Phase Combination

In the course of a crystal structure analysis of a biological macromolecule, phase information from different sources may be available, such as information from isomorphous replacement, anomalous scattering, partial structures, solvent flattening, and molecular averaging. An overall phase improvement can be expected when these factors are combined, and a useful method to do this was proposed by Hendrickson and Lattman. The probability curve for each reflection is written in an exponential form as Eq. (120):

$$P_S(\alpha) = N_S \exp(K_S + A_S\cos\alpha + B_S\sin\alpha + C_S\cos 2\alpha + D_S\sin 2\alpha) \qquad (120)$$

Subscript s stands for the source from which the phase information has been derived. K_S and the coefficients A_S, B_S, C_S, and D_S depend on the structure factor amplitudes and other magnitudes, for example, the estimated standard deviation of the errors in the derivative intensity, but are independent of the protein phase angles α. The overall probability function $P(\alpha)$ is obtained by a multiplication of the individual phase probabilities, and this turns out to be a simple addition of all K_S and of the related coefficients in the exponential term. We obtain Eq. (121):

$$\begin{aligned} P(\alpha) &= \prod_S P_S(\alpha) \\ &= N'\exp\left[\sum_S K_S + \left(\sum_S A_S\right)\cos\alpha + \left(\sum_S B_S\right)\sin\alpha + \left(\sum_S C_S\right)\cos 2\alpha + \left(\sum_S D_S\right)\sin 2\alpha\right] \end{aligned} \qquad (121)$$

K_S and the coefficients A_S to D_S have special expressions for each source of phase information.

4.6 Difference Fourier Technique

Supposing that one has solved the crystal structure of a biological macromolecule and has isomorphous crystals of this macromolecule, which contain small structural changes caused by a substrate-analog or inhibitor binding, a metal removal or replacement or a local mutation of one or several amino acids. Then, these structural changes can be determined by the difference Fourier technique. The difference Fourier map is calculated with the differences between the observed structure factor amplitudes of the slightly altered molecule F_{DERI}(obs) and the native molecule F_{NATI}(obs) as Fourier coefficients and the phase angles of the native molecule α_{NATI} as phases according to Eq. (122):

$$\rho_{DERI} - \rho_{NATI} \cong \frac{1}{V}\sum_{\mathbf{h}} m[F_{DERI}(\text{obs}) - F_{NATII}(\text{obs}) \times \exp(i\alpha_{NATI}) \times \exp(-2\pi i\mathbf{hx})] \qquad (122)$$

where m may be the figure of merit or another weighting scheme. The difference Fourier map can alternatively be calculated with coefficients $F_{DERI}(\text{obs}) - F_{DERI}(\text{calc})$ and phases $\alpha_{DERI}(\text{calc})$. $F_{DERI}(\text{calc})$ and $\alpha_{DERI}(\text{calc})$ do not include the unknown contribution of the structural change.

Figure 38(a) and (b) illustrate the relation for the structure factors involved in the difference Fourier technique. We assume that the structural change is small. If F_{NATI} is large, the structure factor amplitude of the structural change F_{SC} will be small compared to F_{NATI}, and α_{DERI} will be close to α_{NATI}. This is no longer valid if F_{NATI} is small. Now, F_{SC} is comparable to F_{NATI}, and α_{DERI} may deviate considerably from α_{NATI}. This implies the necessity to introduce a weighting scheme that scales down the contributions where the probability is high that α_{NATI} differs appreciably from the correct phase angle. Various weighting schemes have been elaborated such as those of Sim and Read. The weighting scheme of Sim has the following form (Eq. 123):

$$w = \frac{I_1(X)}{I_0(X)} \tag{123}$$

for acentric reflections and (Eq. 124)

$$w = \tanh \frac{X}{2} \tag{124}$$

for centric reflection with (Eq. 125)

$$X = \frac{2F_{DERI} \times F_{NATI}}{\sum_{1}^{n} f_j^2} \tag{125}$$

$I_0(X)$ and $I_1(X)$ are modified Bessel functions of zeroth and first order respectively. These equations and weighting schemes can also be used for the calculation of OMIT maps (where parts of the model have been omitted from the structure factor evaluation) or when a complete structure must be developed from a known partial model. F_{DERI} must be replaced by the observed structure factor F, F_{NATI} by the structure factor of the known or included part of the model F_K, and α_{NATI} by the phase angle α_K of the known or included part of the model.

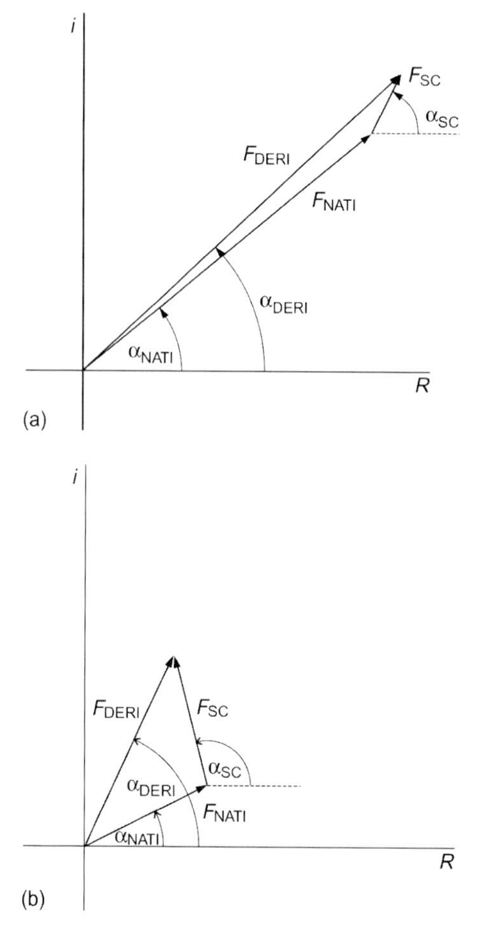

Fig. 38 Vector diagrams illustrating different situations (a) and (b) in the difference Fourier technique for the involved structure factors.

5 Model Building and Refinement

5.1 Model Building

Once the quality of the MIRAS or MAD maps is good enough, model building can be started. This is done on a computer graphics system, and the main modeling programs are O and TURBO-FRODO. An interesting alternative is the program MAIN, which additionally contains routines for molecular averaging, molecular docking, and other features. The visualization of the relevant electron density map on the computer graphics system is done as cagelike structures. For this purpose, the standard deviation from the mean value of the map is calculated, and the cagelike structure is built up for a given contour level (normally 1.0 σ). The first task in a *de novo* protein crystal structure analysis is to localize the trace of the polypeptide chain. This can be assisted by routines for automatic chain tracing such as the bones routine, an auxiliary program of O. Such automatic chain-tracing programs generate a skeleton of the electron density map. This representation was introduced by Greer. When the trace of the polypeptide chain has been identified, the atomic model can be built into the electron density. The atomic model is represented as sticks, connecting the atomic centers of bonded atoms. The individual building blocks (amino acids) of the protein molecule can be generated, interactively manipulated (e.g. linked with each other, moved, rotated, etc.), and fitted into the corresponding part of the electron density map. The geometry of the atomic model is regularized according to protein standard geometries.

The success of model building depends on the quality and resolution of the experimental electron density map. Usually, the quality of the electron density map is not so good that the complete model can be constructed in one cycle. In this case, the partial atomic model is refined crystallographically against the observed structure factor amplitudes. This phase information can be used directly to calculate a new electron density map commonly with $2F_{obs} - F_{calc}$ Fourier coefficient amplitudes. This kind of map is the sum of a normal F_{obs} Fourier and a difference Fourier synthesis. It displays the atomic model with normal weight and indicates errors in the model by its contribution of the difference Fourier map. The parallel determination and inspection of a difference Fourier map are also very helpful. As already mentioned, the model phases can be combined with phases present from other sources or incorporated in procedures of phase improvement. A further model-building cycle can be started with such new and improved electron density maps. After several cycles of model building and crystallographic refinement, the atomic model will be so well defined that the solvent structure of internally bound solvent molecules can be developed. The atomic model is complete now and the biochemical interpretation can be started.

5.2 Crystallographic Refinement

The structural model has to be subjected to a refinement procedure. In practical macromolecular crystallography, one does not always have atomic resolution. Therefore, the single atoms cannot be treated as moving independently. They must be refined using energy or stereochemistry restraints, taking care to maintain a reasonable stereochemistry of the macromolecule. There exist different approaches to structure refinement of macromolecules. The minimization of a potential energy function E together with a diffraction term D according to Eq. (126):

$$S = E + D \qquad (126)$$

where (Eqs. 127 and 128)

$$\begin{aligned} E = & \sum k_{\text{b}}[b_{j(\text{calc})} - b_j^0]^2 \\ & + \sum k_\tau [\tau_{j(\text{calc})} - \tau_j^0]^2 \\ & + \sum k_\Theta [1 + \cos(m\Theta_k + \delta] \\ & + \sum (Ar^{-12} + Br^{-6}) \qquad (127) \end{aligned}$$

$$D = \sum_i w_i [F_{i(\text{obs})} - kF_{i(\text{calc})}]^2 \qquad (128)$$

is applied in the programs EREF and CNS, which are now used frequently. The four terms of the right-hand side of E describe bond, valence angle, dihedral torsion angle, and nonbonded interactions, k_{b} is the bond stretching constant, k_τ is the bond angle bending force constant, k_Θ is the torsional barrier, m and δ are the periodicity and the phase of the barrier, A and B are the repulsive and long-range nonbonded parameters, D is the crystallographic contribution with w_i a weighting factor, F_{obs} is the observed structure factor, F_{calc} is the calculated structure factor, and k is a scaling factor. Programs TNT and PROLSQ use stereochemically restrained least-squares refinement. For both refinement schemes, parameters are employed that were derived from small molecule crystal structures of amino acids, small peptides, nucleic acids, sugars, fatty acids, cofactors, and so on. If noncrystallographic symmetry is present, a corresponding term may be introduced in the energy or stereochemistry part of the expression to be minimized. It is possible to divide the structural model into several individual parts and refine these parts as rigid bodies. This is especially useful with solutions from molecular replacement.

A measure of the quality of the crystallographic model is calculated from the crystallographic R-factor (Eq. 129):

$$R = \frac{\sum_i |F_{\text{obs}}| - |k||F_{\text{calc}}|}{\sum_i |F_{\text{obs}}|} \qquad (129)$$

Typical R-factors are below 0.2 for a well-refined macromolecular structure.

Beside the atomic coordinates x, y, z, the atomic temperature factor B may be refined at a resolution better than 3.5 Å. This is done in most programs in a separate step where, for example, in program CNS, the target function (Eq. 130)

$$T = E_{\text{XRAY}} + E_R \qquad (130)$$

is minimized, where (Eq. 131)

$$\begin{aligned} E_R = & W_B \sum_{(i,j)\text{-bonds}} \frac{(B_i - B_j)^2}{\sigma^2_{\text{bonds}}} \\ & + W_B \sum_{(i,j,k)\text{-angles}} \frac{(B_i - B_k)^2}{\sigma^2_{\text{angles}}} + W_B \\ & \times \sum_{k\text{-group}} \sum_{j\text{-equivalences}} \sum_{\substack{i\text{-unique} \\ \text{atoms}}} \frac{(B_{ijk} - \overline{B_{ijk}})^2}{\sigma^2_{\text{ncs}}} \end{aligned} \qquad (131)$$

The last term is used only if noncrystallographic symmetry restraints should be imposed on the molecules. Normally, isotropic B-factors are applied and refined in macromolecular crystallography only. Even for a high-resolution structure (1.7 Å), the ratio of observations (observed structure factors) to parameters to be refined (x, y, z, B for each atom) is only about 3. Therefore, as many additional "observations" (energy or stereochemistry restraints) as possible are incorporated. In some cases, it is useful to refine the individual occupancy of certain atoms such as bound metal ions or solvent atoms. This must be performed in a separate step.

All of the mentioned refinement procedures are based on the least-squares method. The radius of convergence for this method is not very high because it follows a downhill path to its minimum. If the model is too far away from the correct solution, the minimization may end in a local minimum corresponding to an incorrect structure. Brünger introduced the method of molecular dynamics (MD), which is able to overcome barriers in the S-function and find the correct global minimum. MD calculations simulate the dynamic behavior of a system of particles. The basic idea of the MD refinement technique is to increase the temperature sufficiently high for the atoms to overcome energy barriers and then to cool slowly to approach the energy minimum. This MD protocol is designated as simulated annealing (SA). The crystallographic application of the MD or SA technique includes a crystallographic term D, as given in Eq. (126), treated as a pseudoenergy term. A crystallographic MD or SA refinement is capable of overcoming a high-energy barrier occurring in the flipping of a peptide plane. It can be useful in removing model bias from the system.

A new approach is the refinement of macromolecular structures by the maximum likelihood method. Programs working on the basis of this method are REFMAC and CNS. The results derived using the maximum likelihood residual are consistently better than those from least-squares refinement.

If the resolution of a biological macromolecular crystal structure is equal to or better than 1.2 Å, it is in the range of real atomic resolution, and the ratio of observations to parameters is high enough to carry out, in principle, an unrestrained crystallographic refinement. Advances in cryogenic techniques, area detectors, and the use of synchrotron radiation enabled macromolecular data to be collected to atomic resolution for an increasing number of proteins. SHELXL is a program with all tools for the crystallographic refinement of biological macromolecules at real atomic resolution.

Since the advent of structural genomics, automation of all parts of structure analysis has become of paramount importance. A first step in this direction was the compilation of the Arp/Warp program system. This system allows the building and refinement of a protein model automatically and without user intervention, starting from diffraction data extending to a resolution higher than 2.3 Å and reasonable estimates of crystallographic phases. The method is based on an iterative procedure that describes the electron density map as a set of unconnected atoms and then searches for protein-like patterns. Automatic pattern recognition (model building) combined with refinement permits a structural model to be obtained reliably within few CPU hours.

5.3 Accuracy and Verification of Structure Determination

A measure of the quality of a structure determination is the crystallographic R-factor given in Eq. (129). For a high-resolution structure, for example 1.6 Å, it should not be much larger than 0.16. As this R-factor is an overall number, it does not indicate major local errors. This can be obtained by the evaluation of a real space R-factor, which is calculated on a grid for nonzero elements according to Eq. (132):

$$R_{\text{real space}} = \frac{\sum |\rho_{\text{obs}} - \rho_{\text{calc}}|}{\sum |\rho_{\text{obs}} + \rho_{\text{calc}}|} \quad (132)$$

where ρ_{obs} is the observed and ρ_{calc} is the calculated electron density.

It has been shown that the conventional R-factor may reach rather low values in a crystallographic refinement with structural models that turned out to be wrong later. To overcome this unsatisfactory situation, Brünger proposed the additional calculation of a so-called free R-factor. For this purpose, the reflections are divided into a working set (e.g. 90%) and a test set (e.g. 10%). The reflections in the working set are used in the crystallographic refinement. The free R-factor is calculated with reflections from the test set, which were not used for the crystallographic refinement and is thus unbiased by the refinement process. There exists a high correlation between the free R-factor and the accuracy of the atomic model phases.

The accuracy of the final model expressed by the mean coordinate error can be determined alternatively by a Luzzati or σ_A plot. The mean coordinate error for a macromolecular structure determined with a resolution of 2.0 Å and a crystallographic R-factor of 0.2 is in the region of ±0.2 Å.

The stereochemistry of the final model must also be checked. The root-mean-square deviation of bond lengths and bond angles from ideal geometry should not be greater than 0.015 Å and 3.0° respectively. The conformation of the main-chain folding is verified by a Ramachandran plot. The dihedral angles Φ and Ψ are plotted against each other for each residue. The data points should lie in the allowed regions of the plot, which correspond to energetically favorable secondary structures such as α-helices, ß-sheets, and defined turn structures. Exceptions are glycine residues, which may occur at any position in the Ramachandran plot. Further stereochemical parameters to be checked are bond lengths and angles, dihedral angles (e.g. determinating side-chain conformations), noncovalent interactions, geometry of H-bonds, and interactions in the solvent structure. This can be done with the programs PROCHECK or WHAT CHECK.

Nearly all spatial structures of biological macromolecules determined either by X-ray crystallography or nuclear magnetic resonance (NMR) techniques have been and will be deposited with the RCSB Protein Data Bank at Rutgers University. The information of the structural model is in a file that contains for each individual atom of the model a record with atom number, atom name, residue type, residue name, coordinates x, y, z, B-value(s), and occupancy. The header records hold useful information such as crystal parameters, amino acid sequence, secondary structure assignments, and references.

6 Applications

6.1 Enzyme Structure and Enzyme–Inhibitor Complex

6.1.1 X-ray Structure of Cystathionine ß-Lyase

Cystathionine ß-lyase (CBL) is a member of the γ-family of pyridoxal-5′-phosphate (PLP)-dependent enzymes that cleaves Cß-S bonds of a broad variety of substrates. The crystal structure of CBL from *Escherichia coli* has been solved using

MIR phases in combination with density modification. The enzyme has been crystallized by the hanging drop vapor diffusion method using either ammonium sulfate or PEG 400 as precipitating agent. The crystals belong to the orthorhombic space group $C222_1$ with unit cell parameters $a = 60.9$ Å, $b = 154.7$ Å, and $c = 152.7$ Å. There is one dimer per asymmetric unit. A native data set has been collected using synchrotron radiation (wavelength 1.1 Å) at the wiggler beamline BW6 at the storage ring DORIS at the Deutsches Elektronensynchrotron (DESY) in Hamburg, Germany. Data sets for three heavy atom derivatives (thiomersalate, 2-mercuri-4-diazobenzoic acid and platinum(II)-2,2′-6,6″-terpyridinium chloride) were registered on an imaging plate scanner (MAR Research, Norderstedt, Germany) using graphite monochromatized Cu Kα radiation from an RU200 rotating anode generator (Rigaku, Tokyo, Japan) operating at 5.4 kW. The reflection data were processed with the MOSFLM package and scaled with programs from the CCP4 program suite. All data sets reveal satisfactory symmetry consistency factors ($R_{merge} \leq 0.08$) and completenesses (>90%). The heavy atom positions were determined from isomorphous difference Patterson maps using the vector verification routines of program PROTEIN. All derivatives have one common heavy atom–binding site at Cys72, and one of the mercury derivatives shows a second binding site at Cys229. The dimer in the asymmetric unit is related by a local twofold axis that lies parallel to the x-axis. The translation of this local axis was determined from the distribution of the heavy atom sites in the unit cell. An initial MIR map was calculated, followed by solvent flattening, twofold averaging about the local symmetry, and density modification and phase extension to 2.5 Å resolution. Phase calculations, solvent flattening, density modification, and phase extension were done with programs of the CCP4 package. Program AVE was used for the averaging. The quality of the resulting electron density map was sufficiently high to build an almost complete atomic model. Model building was performed on an ESV-30 Graphic system workstation (Evans and Sutherland, Salt Lake City, UT, USA) using program O. The atomic model was refined by energy-restrained crystallographic refinement with XPLOR.

The final model of CBL is made up of two monomers with 391 amino acids each, one cofactor and one hydrogencarbonate molecule per monomer, and 581 solvent water molecules. The final crystallographic R-factor is 0.152 for data from 8.0- to 1.83-Å resolution, and the free R-factor is 0.221. The mean positional error of the atoms as estimated from a Luzzati plot is ±0.19 Å. A homotetramer with 222 symmetry is built up by crystallographic and noncrystallographic symmetry (Fig. 39). Each monomer of CBL (Fig. 40) can be described in terms of three spatially and functionally different domains. The N-terminal domain (residues 1–60) consists of three α-helices and one ß-strand. It contributes to tetramer formation and is part of the active site of the adjacent subunit. The second domain (residues 61–256) harbors PLP and has an α/ß structure with a seven-stranded ß-sheet as the central part. The remaining C-terminal domain (residues 257–395), connected by a long α-helix to the PLP-binding domain, consists of four helices packed on the solvent-accessible side of an antiparallel four-stranded ß-sheet. The fold of the C-terminal and the PLP-binding domain and the location of the active site are similar to aminotransferases. Most

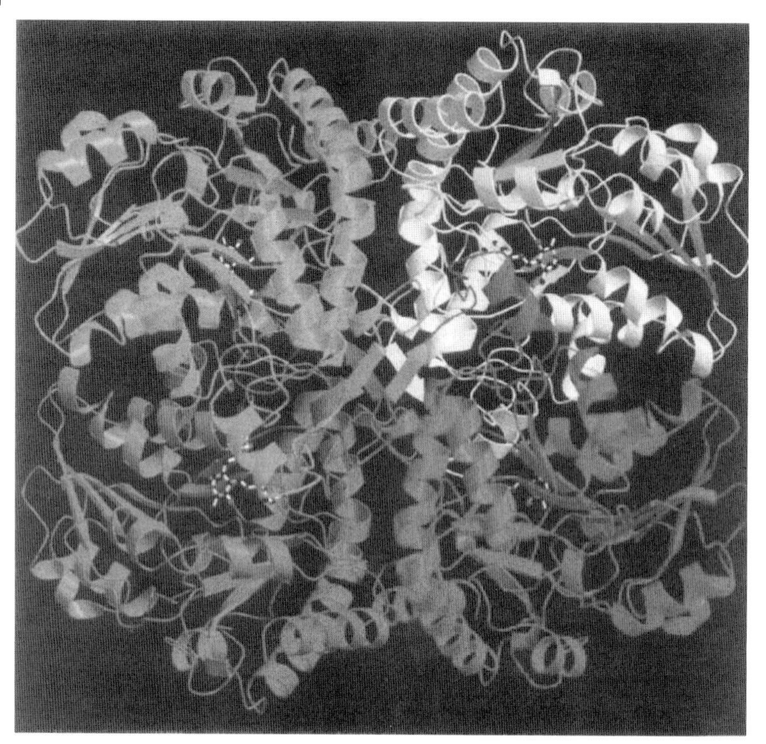

Fig. 39 Ribbon plot of the CBL tetramer viewed along the *x*-axis. The monomers are colored differently. The blue- and green-colored monomers, which are related by a crystallographic axis (horizontal, in the plane of the paper), build up one catalytic active dimer, and the yellow and red ones the other. The location of the PLP-binding site is shown in a ball-and-stick presentation; MOLSCRIPT and RASTER3D. (Reproduced by permission of Academic Press, Ltd., from Clausen, T. et al. (1996) *J. Mol. Biol.* **23**, 202–224.) (See color plate p. xxi).

of the residues in the active site are strongly conserved among the enzymes of the transulfuration pathway. Figure 41 shows the final $2F_{obs} - F_{calc}$ map superimposed with the refined atomic model around the active site of CBL. The cage-like structures for the representation of the electron density correspond to a contour level of 1.2σ. The protein part (main- and side-chain atoms), the PLP cofactor, and the hydrogencarbonate molecule are well defined in the electron density map.

The knowledge of the spatial structure of a given enzyme structure forms the basis for understanding its functional properties. It is now possible to design rational site-directed mutants or determine the structures of enzyme–substrate, enzyme–substrate analog, or enzyme–inhibitor complexes, which will deliver invaluable information in understanding the enzyme's functional properties.

6.1.2 Enzyme–Inhibitor Complex Structures of Cystathione ß-Lyase

The enzyme–inhibitor X-ray structures of ß,ß,ß-tri-fluoroalanine (TFA) and L-aminoethoxyvinylglycine (AVG) with CBL could be determined. In both cases,

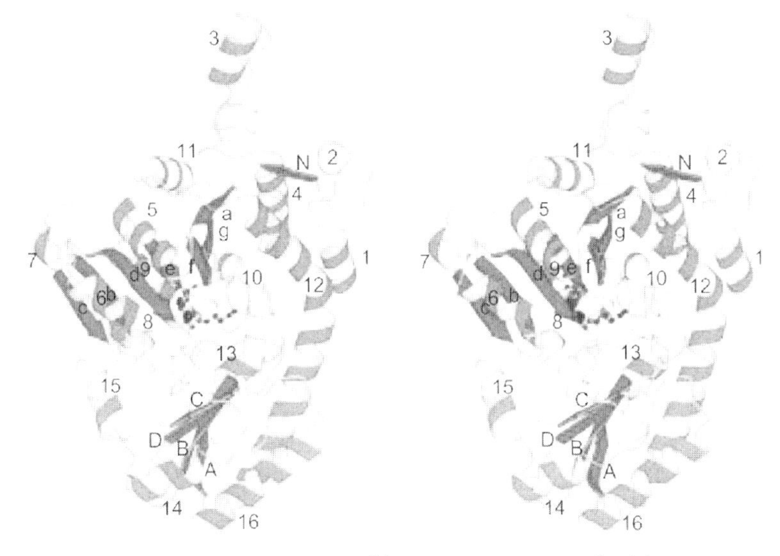

Fig. 40 Stereo ribbon presentation of the CBL monomer, emphasizing secondary structure elements. α-Helices are drawn as green spirals, ß-strands as magenta arrows. PLP and PLP-binding Lys210 are shown in a ball-and-stick representation; MOLSCRIPT and RASTER3D. (Reproduced by permission of Academic Press, Ltd., from Clausen, T. et al. (1996) *J. Mol. Biol.* **23**, 202–224.) (See color plate p. xxiii).

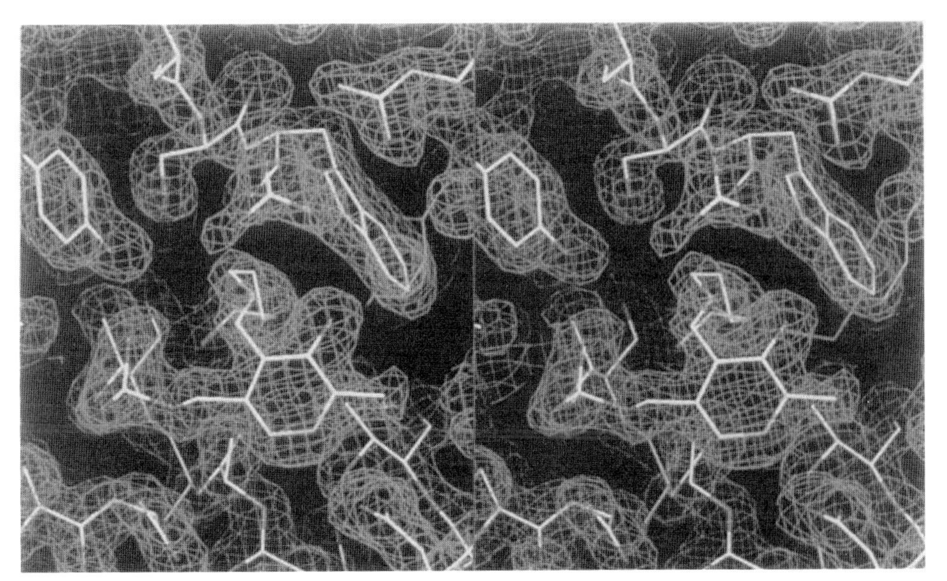

Fig. 41 Stereo plot of the electron density in the active site of CBL, superimposed with the refined model of the region around the cofactor. The $2F_{obs} - F_{calc}$ map is contoured at 1.2σ and calculated at 1.83-Å resolution. (Reproduced by permission of Academic Press, Ltd., from Clausen, T. et al. (1996) *J. Mol. Biol.* **23**, 202–224.)

crystals of the complexes were obtained by incubating the enzyme solution with inhibitor in the millimolar range and subsequent cocrystallization. The resultant crystals were isomorphous with the native enzyme, making it possible to apply the difference Fourier technique in the structure solution. The technical details for the structure analyses are given in the relevant references. The CBL/TFA complex structure was determined to substantiate that the ε-amino group of the active-site Lys210 can react with the nucleophile at the active site via Michael addition, which leads to covalent labeling and inactivation of the enzyme. The final $F_{obs} - F_{calc}$ and $2F_{obs} - F_{calc}$ electron density maps for the CBL/TFA complex around the active site are displayed in Figure 42. Clear, continuous electron density between the cofactor and Lys210 can be seen in this map, indicating a covalent lysine–inactivator–PLP product. In Fig. 42, the blue $F_{obs} - F_{calc}$ map reveals the well-defined electron density for the bound inhibitor. This binding mode of TFA to CBL corresponds to an intermediate in the reaction of TFA with CBL. The structure of the inactivation product proves that Lys210 is the active-site nucleophile reacting via Michael addition with the inactivator. It must also be the residue that transfers a proton from Cα to Sγ in the reaction with the substrates.

The CBL/AVG structure has been determined at 2.2-Å resolution and a crystallographic R-factor of 0.164. The X-ray structure shows that AVG binds to the PLP cofactor forming the external aldimine. Lys210 is no longer bound to the PLP cofactor. Figure 43 is an overlay of the atomic models of native CBL(magenta), CBL/TFA (yellow), and CBL/AVG (green). The main difference in inhibitor binding is the location of Cß and its substituents; in the

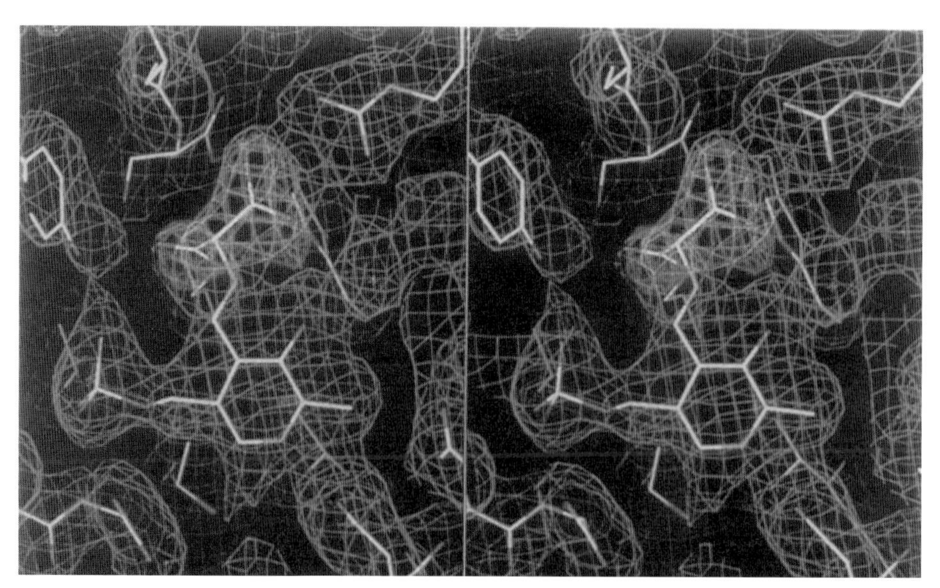

Fig. 42 $F_{obs} - F_{calc}$ (blue) and $2F_{obs} - F_{calc}$ (green) electron density map of the CBL/TFA complex around the active site contoured at 3.5σ and 1.0σ respectively, at 2.3-Å resolution. (Reproduced by permission of Academic Press, Ltd., from Clausen, T. et al. (1996) *J. Mol. Biol.* **23**, 202–224.) (See color plate p. xxiii).

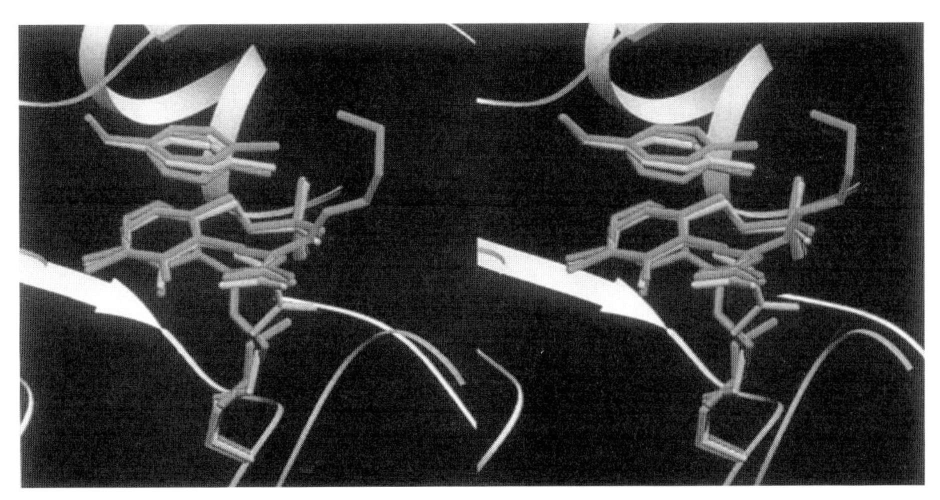

Fig. 43 Stereo view of a superposition of Tyr111 and the PLP derivative of the unliganded enzyme (magenta), the CBL/AVG adduct (green), and the TFA-inactivated enzyme (yellow); SETOR. (Reproduced by permission of the American Chemical Society, Clausen, T. et al. (1997) *Biochemistry* **36**, 12633–12643.) (See color plate p. xxiv).

TFA complex, the inactivator is directed toward the protein interior (the A face of the cofactor), whereas in CBL/AVG, Cß is located at the B side of the cofactor. The α-carboxylate group in CBL/AVG is located in the same position as the hydrogencarbonate in the native and the α-carboxylate group of TFA in the CBL/TFA complex. The terminal amino group of AVG is held in place mainly by interactions with the hydroxyl group of Tyr111.

The experimental determination of the external aldimine structure in the CBL/AVG complex is of high relevance because it can serve as a rational basis for modeling of substrate and inhibitor binding, leading to more effective herbicides.

6.1.3 **Crystal Structure of the Thrombin–Rhodniin Complex**

This complex structure is an example related to pharmaceutical research. The goal of this special application is the development of more efficient blood anticoagulants. The target enzyme is α-thrombin. This enzyme is a serine proteinase of trypsin-like specificity. α-Thrombin, the key enzyme in hemostasis and thrombosis, exhibits both enzymatic and hormonelike properties, and can be both pro- and anticoagulatory. Rhodniin is a highly specific inhibitor of thrombin isolated from the assassin bug *Rhodnius prolixus*. Such blood-sucking animals have developed various anticlotting mechanisms to prevent local clotting of the victim's blood. These natural thrombin inhibitors are polypeptides of 60 to 120 amino acid residues.

The crystal structure of the noncovalent complex between recombinant rhodniin and bovine a-thrombin has been determined at 2.6-Å resolution. Crystals were obtained by cocrystallization of thrombin with rhodniin in an approximately 1 : 1 molar ratio. The structure could be solved by molecular replacement because the spatial structure of the major constituent, bovine α-thrombin, was known. Only a diffraction data set of the complex crystal had to be collected. Rotational and translational searches for the orientation and position of the thrombin molecules in the unit cell were performed with the program AMORE. The rotational search showed two solutions with correlation values of 0.22 and 0.20 over 0.09 for the next highest peak. Translational search and rigid body fitting for these two solutions resulted in a correlation value of 0.54, with the two independent complex molecules in the asymmetric unit. The quality of the electron density map calculated from the thrombin phases was good enough in principle to build the model of the rhodniin molecule (noncrystallographic averaging was also applied). The structure was refined with XPLOR to an R-factor of 0.189 and a free R-factor of 0.262.

Figure 44 shows the structure of the complex between thrombin and rhodniin as a ribbon plot, with α-helices represented as ribbon spirals and ß-strands as arrows. The N-terminal domain binds in a substrate-like manner to the narrow active-site cleft of thrombin. The C-terminal domain, whose distorted reactive-site loop cannot adopt the canonical conformation, docks to the fibrinogen recognition exosite via extensive electrostatic interactions. The peculiarity of this complex structure is that the two KAZAL-type domains of rhodniin bind to two different sites of thrombin.

6.2 Metalloproteins

Metals bound as cofactors in proteins have a great variety of functions. They may be involved in the activation of small inorganic or organic molecules, in oxygen storage and transport, in electron transport, in regulation of biological processes, or in stabilizing a transition state during enzymatic catalysis. Their role may also be solely structural. The chemistry of metals in proteins has attracted the interest of inorganic chemists, and has led to the formation of

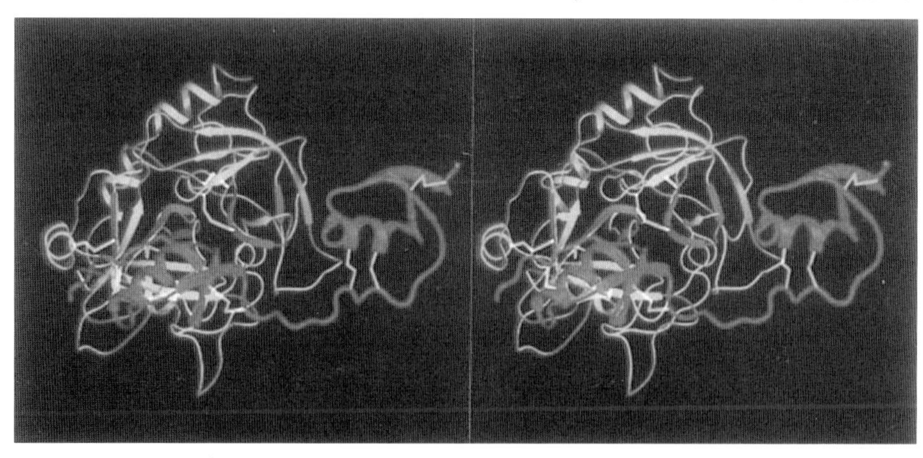

Fig. 44 Stereo view of the complex formed between thrombin (blue) and rhodniin (red) in the thrombin standard orientation, that is, with the active-site cleft facing the viewer and a bound inhibitor chain from left to right. Yellow connections indicate disulfide bridges. Rhodniin interacts through its N-terminal domain in a canonical manner with the active site and through its C-terminal domain with the fibrinogen recognition exosite of thrombin; SETOR. (Reproduced by permission of Oxford University Press, from van de Locht, A. et al. (1995) *EMBO J.* **14**, 5149–5157). (See color plate p. xxiv).

the field of biological inorganic chemistry. As an example of a complex metalloprotein, the multicopper oxidase ascorbate oxidase (AO) is presented.

6.2.1 Crystal Structure of the Multicopper Enzyme Ascorbate Oxidase

The blue protein AO belongs to the group of "blue" oxidases with laccase and ceruloplasmin. These are multicopper enzymes catalyzing the four-electron reduction of molecular oxygen with concomitant one-electron oxidation of the substrate. The crystal structure of AO has been solved by the MIR technique and refined to 1.9-Å resolution. The peculiarity of this structure determination is briefly described. It consists in the utilization of the information of two different crystal forms. In both crystal forms, the molecules arrange themselves as homotetramers with 222 symmetry, but in crystal form 1, one of these twofold axes is realized by a crystallographic twofold axis resulting in two subunits per asymmetric unit. In crystal form 2, one homotetramer is found per asymmetric unit. In crystal form 1, six isomorphous heavy atom derivatives could be found and interpreted. An initial MIR map was calculated, solvent flattened, and averaged about the local twofold axis. For crystal form 2, no phase information was available. From the averaged uninterpreted MIR map, a whole tetramer was selected and used for rotational and translational searches in crystal form 2. This was successful and provided the necessary phase information for crystal form 2 and, additionally, the local symmetry. Now, averaging could be performed both separately in the two crystal forms and subsequently between both crystal forms. The averaged electron density was transported into both unit cells and a new macrocycle of averaging could be started. These macrocycles were used to extend the phases from 3.5 Å to the full attainable resolution. This structure analysis was the first example where a molecular replacement was carried out with an uninterpreted MIR electron density–based model.

AO is a homodimeric enzyme with a molecular mass of 70 kDa and 552 amino acid residues per subunit (zucchini). The three-domain structure and the location of the mononuclear centers and trinuclear copper centers in the AO monomer as derived from the crystal structure are shown in Fig. 45. The folding of all three domains is of a similar ß-barrel type. The mononuclear copper site is located in domain 3 and the trinuclear copper species is bound between domains 1 and 3. The coordination of the mononuclear copper site is depicted in Fig. 46. It has the four canonical type-1 copper ligands (His, Cys, His, Met), also found in plastocyanin and azurin. The copper is coordinated to the ND1 atoms of His445 and His512, the SG atom of the Cys507, and the SD atom of Met517 in a distorted trigonal geometry. This unusual coordination geometry confers this copper site with its blue color. The trinuclear copper site (see Fig. 47) has eight histidine ligands symmetrically supplied by domains 1 and 3 and two oxygen ligands. The trinuclear copper site may be divided into a pair of copper (CU2, CU3), with six histidine ligands in a trigonal prismatic arrangement. The pair is bridged by an OH^-, which leads to a strong antiferromagnetic coupling and makes this copper pair electron paramagnetic resonance silent. The remaining copper has two histidine ligands and an OH^- or H_2O ligand. A binding pocket for the reducing substrate, which is complementary to an ascorbate molecule, is located near the mononuclear copper site and is accessible

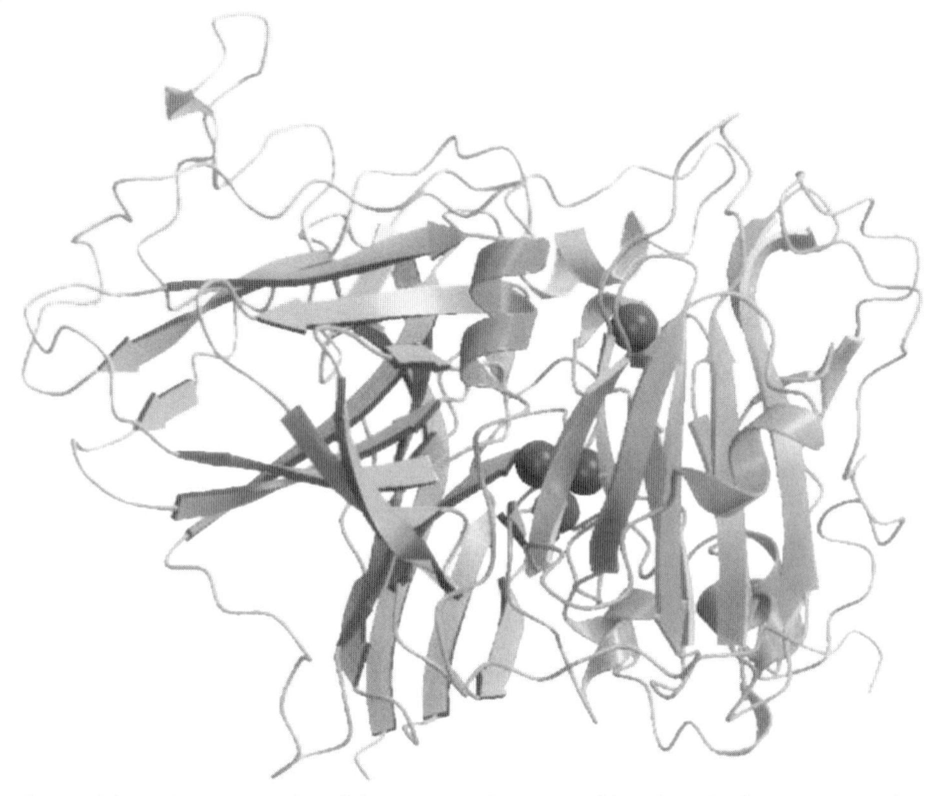

Fig. 45 Schematic representation of the monomer structure of AO. (Reproduced by permission of World Scientific Publishing Co., from Messerschmidt, A. (1997) Spatial Structures of Ascorbate Oxidase, Laccase and Related Proteins, in: Messerschmidt, A. (Ed.) *Multi Copper Oxidases*, World Scientific Publishing, Singapore.)

from solvent. A broad channel providing access from the solvent to the trinuclear copper species, which is the binding and reaction site for the dioxygen, is present in AO. During catalysis, an intramolecular electron transfer between the mononuclear copper site and the trinuclear copper cluster occurs. The distances between the mononuclear copper and the three coppers of the trinuclear center are 12.20, 12.69, and 14.87 Å, respectively. Furthermore, the crystal structures of functional derivatives of AO such as the reduced, azide, and peroxide form have been determined. They show considerable changes at the trinuclear copper site in the reduced form and the peroxide or azide binding to the trinuclear copper species in the relevant structures. The X-ray studies on AO delivered essential information in understanding the catalytic mechanism.

6.2.2 Crystal Structure of Cytochrome-*c* Nitrite Reductase Determined by Multiwavelength Anomalous Diffraction Phasing

The spatial structure of cytochrome-*c* nitrite reductase from *Sulfurospirillum deleyianum* has been solved by the MAD technique using synchrotron radiation and

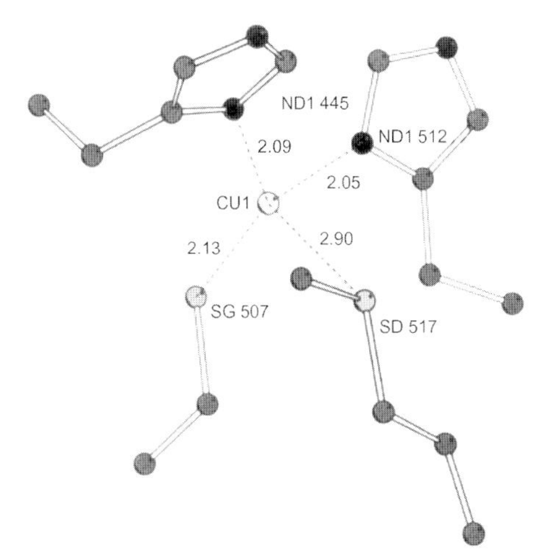

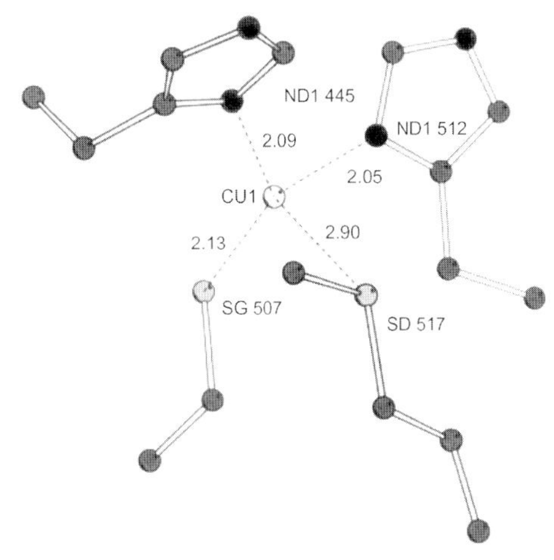

Fig. 46 Stereo drawing of the mononuclear copper site in domain 3 of AO. The displayed bond distances are for subunit A; MOLSCRIPT. (Reproduced by permission of World Scientific Publishing Co., from Messerschmidt, A. (1997) Spatial Structures of Ascorbate Oxidase, Laccase and Related Proteins, in: Messerschmidt, A. (Ed.) *Multi Copper Oxidases*, World Scientific Publishing, Singapore.)

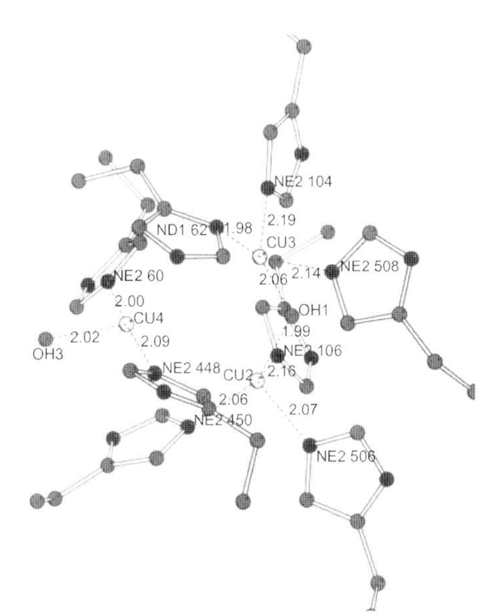

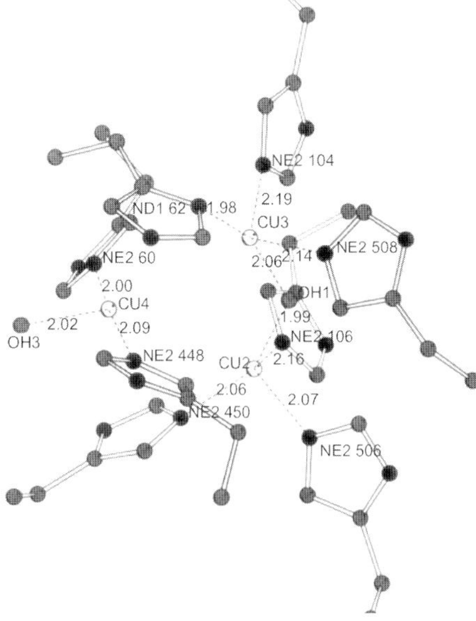

Fig. 47 Stereo drawing of the trinuclear copper site of AO. The displayed bond distances are for subunit A; MOLSCRIPT. (Reproduced by permission of World Scientific Publishing Co., from Messerschmidt, A. (1997) Spatial Structures of Ascorbate Oxidase, Laccase and Related Proteins, in: Messerschmidt, A. (Ed.) *Multi Copper Oxidases*, World Scientific Publishing, Singapore.)

refined at a resolution of 1.9 Å to a *R*-factor of 0.18. The data collection for the MAD phasing was performed at the wiggler beamline BW6 at the storage ring DORIS at DESY in Hamburg, Germany. At the beginning, an X-ray fluorescence spectrum around the Fe K absorption edge was registered using an NaI(Tl) scintillation counter. Evaluation of the spectrum with program DISCO gave the anomalous dispersion contributions, f' and f'', as a function of photon energy (Fig. 48). Subsequently, diffraction data were collected at three different photon energies. Two of them, 7141 and 7129 eV, correspond to maximum f'' and minimum f' respectively. A third high-resolution data set (1.90 Å resolution) was collected at a photon energy of 11 808 eV. All data sets were obtained from a single crystal that had been shock-frozen at 100 K. Anomalous difference Patterson maps were calculated from the data collected at f'' maximum of the Fe K absorption edge. Correlated peaks that appeared in all three Harker sections were chosen and used for phasing with MLPHARE. The highest peaks from the resulting Fourier map were analyzed for consistency with the anomalous difference Patterson map and, if correct, used for a new cycle of phase calculations. Thus, it was possible successively to find 15 iron positions, which divided into three groups of five. Phasing with program SHARP produced an interpretable electron density map, which was solvent flattened and threefold averaged with program AVE. The resulting electron density map at a resolution of 1.9 Å was of high quality and allowed the construction of the complete atomic model into that map.

Cytochrome-*c* nitrite reductase (58 kDa) catalyzes the six-electron reduction of nitrite to ammonia as one of the key steps in the biological nitrogen cycle, where it participates in the anaerobic energy metabolism of dissimilatory nitrate ammonification. The crystal structure shows that the enzyme is a homodimer

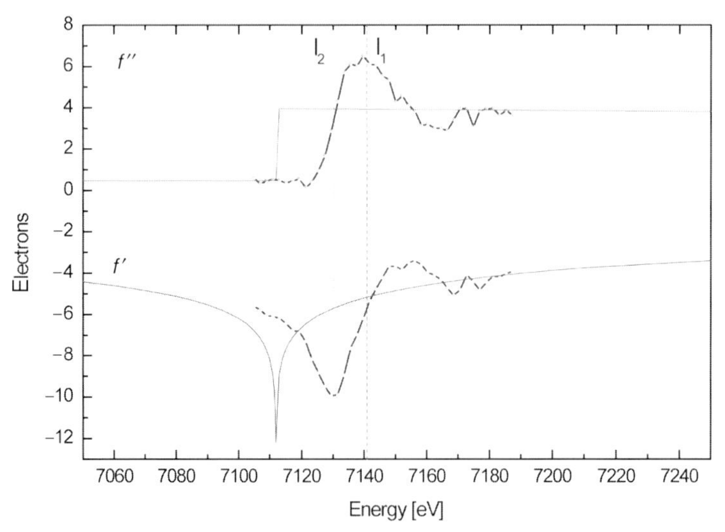

Fig. 48 Dispersion terms as a function of the photon energy as obtained from an X-ray fluorescence scan from cytochrome-*c* nitrite reductase sample crystal and evaluation with DISCO.

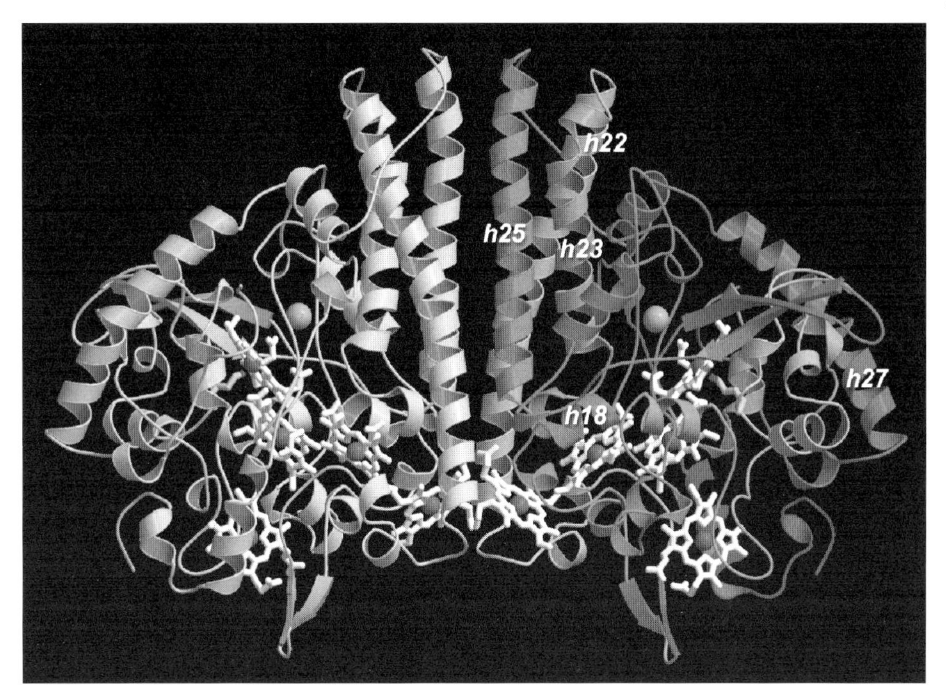

Fig. 49 Overall structure of the nitrite reductase dimer. A front view with the dimer axis oriented vertically, five hemes in each monomer (white), the Ca^{2+} ions (grey), and Lys133 that coordinates the active-site iron (yellow). The dimer interface is dominated by three long α-helices per monomer. All hemes in the dimer are covalently attached to the protein. (Reproduced by permission of Macmillan Magazines, Ltd., from Einsle, O. et al. (1999) *Nature (London)* **400**, 476–480.) (See color plate p. xxii).

of ~100 Å × 80 Å × 50 Å, with 10 hemes in a remarkably close packing (Fig. 49). The protein folds into one compact domain with α-helices as the predominant secondary structural motif, ranging from short helical turns to four long helices at the C-terminal end of the peptide chain (Fig. 49). There are eight 3_{10} helices three- or four-residues long, seven of which occur within the first 200 residues, ß-sheet structures are found only in two short, antiparallel strands, where one is part of a funnel-like cavity leading to the active site. Dimer formation is mediated by helices *h*22 and *h*25, with helix *h*25 as the key element, as it interacts with its counterpart in the other monomer over its full length of 28 residues, corresponding to 42 Å. The five hemes in the monomer of cytochrome-*c* nitrite reductase are in close contact, with Fe-Fe distances of between 9 and 12.8 Å. They are arranged as a group of three, almost coplanar, hemes, with heme 1 forming the active site. Hemes 2 and 5 are farther apart and are not coplanar with hemes 1, 3, and 4. All hemes except heme 1 are bis-histidynyl-coordinated, and are linked to the peptide backbone by thioether bonds to the cysteine residues of a classical heme-binding motif for periplasmic proteins, Cys-X_1-X_2-Cys-His. The propionate side chains of heme 1 form part of the active-site cavity. The site of nitrite reduction is clearly heme

1, with NZ atom of Lys133, replacing a histidine in the classical binding motif, and an oxygen atom of a sulfate ion, present in the crystallization buffer. An additional electron density maximum detected close to the active site was assigned to Ca^{2+}.

S. deleyianum cytochrome-*c* nitrite reductase reduces not only nitrite to ammonia but also the potential intermediates NO and hydroxylamine. No intermediates are released during nitrite turnover. Obviously, the active site accomodates anions and uncharged molecules and releases the ammonium cation only after full six-electron reduction. The preference for anions is reflected by a positive electrostatic potential around and inside the active-site cavity. The cationic product might make use of a second channel leading to the protein surface opposite to the entry channel. It branches before reaching the protein surface and ends with both arms in areas possessing a significant electrostatic surface potential. The existence of separate pathways for substrate and product with matched electrostatic potential could also contribute to the enzyme's high specific activity compared to the siroheme-containing reductases that catalyze the same reaction.

6.3 Large Molecular Assembly

6.3.1 Crystal Structure of 20S Proteasome from Yeast

The controlled degradation of proteins in the interior of cells is of central significance for many processes that range from cell cycle control and differentiation to cellular immune response. The target protein is labeled for destruction by the covalent linkage of a small protein called *ubiquitin* and is degraded after adenosine triphosphate (ATP)-driven unfolding in the closed internal chamber of a large protein complex, which is known as *26S proteasome* (molecular mass 2 000 000 Da). The core and the proteolytic chamber of the 26S proteasome are formed by the catalytic 20S particle (molecular mass 700 000 Da). This particle is flanked at each end by a so-called 19S cap, which seems to be responsible for the recognition and unfolding of the ubiquitin-labeled target proteins. The eukaryotic catalytic 20S machinery consists of 14 different but related protein subunits that assemble to the overall structure of 28 protein chains, a monster of the molecular world. The structure of the 20S proteasome from *Saccharomyces cerevisiae* has been elucidated at atomic resolution. This is one of the most complex structures that has been determined up to now, excluding symmetrical systems such as viral capsids. It shows that complexity itself is no limit to understanding a structure on the atomic level.

The structure analysis is briefly outlined. Crystals could be obtained by the hanging drop vapor diffusion method with a final precipitant concentration of 12% MPD in the drop (pH 6.5). The crystals are monoclinic, space group $P2_1$, cell parameters $a = 135.7$ Å, $b = 301.8$ Å, $c = 144.7$ Å, and ß $= 112.6°$, and have one 20S particle per asymmetric unit. Data sets with two different inhibitors were collected at the wiggler beamline BW6 at DESY with radiation of $\lambda = 1.1$ Å and at 90 K temperature. The mean attainable resolution was 2.4 Å. A self-rotation function calculated at 5-Å resolution revealed the orientation of the local twofold axis. The atomic model of the 20S proteasome from *Thermoplasma acidophilum* could be used for molecular replacement calculations (program AMORE) at 3.5-Å resolution, and a prominent solution could be found. The resultant electron density map was cyclically averaged about the local twofold axis. The individual subunits were identified according to their characteristic amino acid sequences and were built into the map. The final model with the lactacystin inhibitor was refined at a resolution of 2.4 Å to an *R*-factor of 0.26, consisting of 48 888 proteins, 30 inhibitors, 18 magnesium atoms, and 1800 solvent molecules.

The structure of the eukaryotic proteasome is important because it is much more complex than its archaebacterial relative, which has two types of subunits, α and ß, only. The α-subunits do not seem to be catalytic but they may self-assemble to form sevenfold rings. In contrast, the ß-subunits show catalytic activity but are not able to assemble themselves. The two components assemble themselves to a stack of four rings, two outer α-rings, which enclose an inner pair of ß-rings. The eukaryotic proteasome retains the important property, however, that the single subunits of the sevenfold rings are different, probably reflecting the increased biological functions. The exact sevenfold symmetry of the particle is lost and a twofold symmetry remains solely to give seven different α- and seven different ß-subunits (Fig. 50a and b). In the proteasome, the unfolded protein is cut

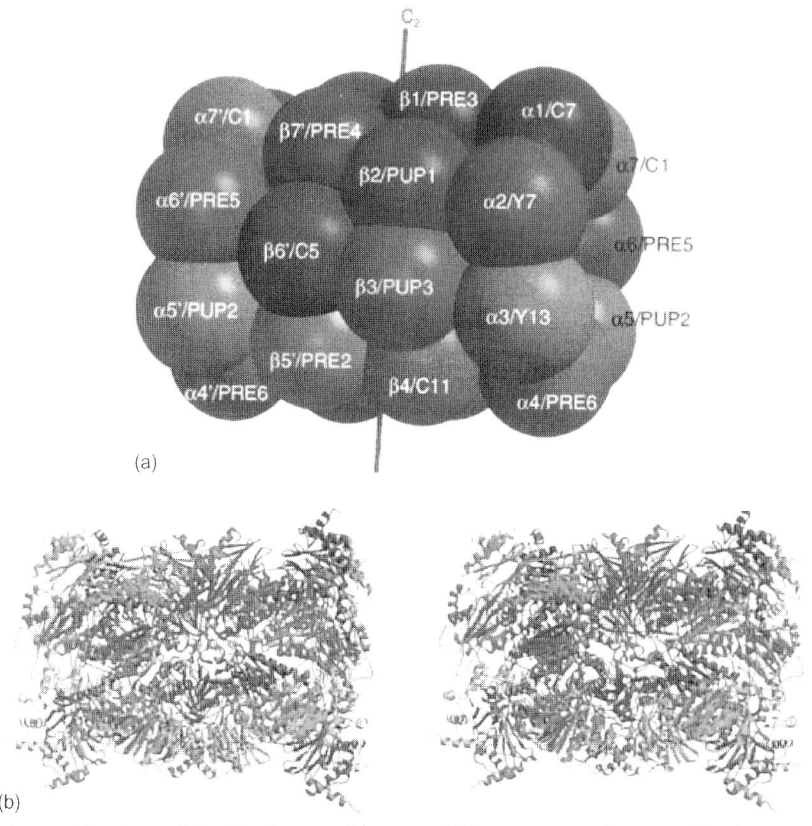

Fig. 50 Topology of the 28 subunits of the yeast 20S proteasome drawn as (a) spheres and (b) ribbon representation; MOLSCRIPT, RASTER3D. (Reproduced by permission of Macmillan Magazines Ltd., from Groll, M. et al. (1997) *Nature (London)*, **386**, 463–471.)

into peptide products in the inner chamber of the 20S particle, and these have a length distribution with a center around octa- or nonapeptides. These sizes are appropriate to bind to MHC-class-I molecules.

See also Chaperones, Molecular.

Bibliography

Books and Reviews

Arndt, U.W., Wonnacott, A.J. (Eds.) (1977) *The Rotation Method in Crystallography*, North Holland, Amsterdam, The Netherlands.

Blundell, T.L., Johnson, L.N. (1976) *Protein Crystallography*, Academic Press, New York.

Brünger, A.T. (1992) *XPLOR, Version 3.1, a System for Crystallography and NMR*, Yale University Press, New Haven, CT.

Buerger, M.J. (1961) *Crystal-structure Analysis*, John Wiley & Sons, New York.

Buerger, M.J. (1964) *The Precession Method in X-ray Crystallography*, John Wiley & Sons, New York, 1964.

Carter, C.W. Jr, Sweet, R.M. (Eds.) (1997) *Macromolecular crystallography*, Part A, *Methods Enzymol.* **276**, 1–700.

Carter, C.W. Jr, Sweet, R.M. (Eds.) (1997) *Macromolecular crystallography*, Part B, *Methods Enzymol.* **277**, 1–664.

Drenth, J. (1999) *Principles of Protein X-ray Crystallography*, Springer, New York.

Glusker, J.P., Lewis, M., Rossi, M. (1994) *Crystal Structure Analysis for Chemists and Biologists*, VCH Publishers, New York.

Hahn, T. (Ed.) (1997), *International Tables for Crystallography*, Vol. A. D. Reidel, Dordrecht, The Netherlands.

Helliwell, J.R. (1992) *Macromolecular Crystallography with Synchrotron Radiation*, Cambridge University Press, Cambridge, MA.

McRee, D. (1999) *Practical Protein Crystallography*, Academic Press, San Diego, CA.

Michel, H. (1991) *Crystallization of Membrane Proteins*, CRC Press, Boca Raton, FL.

Wyckoff, H.W., Hirs, C.H.W., Timasheff, S.N. (Eds.) (1985) *Diffraction Methods for Biological Macromolecules*, Part A, *Methods Enzymol.* **114**, 1–588.

Wyckoff, H.W., Hirs, C.H.W., Timasheff, S.N. (Eds.) (1985) *Diffraction Methods for Biological Macromolecules*, Part B, *Methods Enzymol.* **115**, 1–485.

Primary Literature

Berman, H.M., Westbrook, J., Feng, Z., Gilliland, G., Bhat, T.N., Weissig, H., Shindyalov, I.N., Bourne, P.E. (2000) The protein data bank, *Nucleic Acids Res.* **28**, 235–242.

Blow, D.M., Crick, F.H.C. (1959) The treatment of errors in the isomorphous replacement method, *Acta Crystallogr.* **12**, 794–802.

Brünger, A.T. (1990) Extension of molecular replacement: new search strategy based on Patterson correlation refinement, *Acta Crystallogr.* **A46**, 46–57.

Brünger, A.T. (1992) Free R value: A novel statistical quantity for assessing the accuracy of crystal structures, *Nature (London)*, **355**, 472–475.

Brünger, A.T. (1993) Assessment of phase accuracy by cross validation: the free R value. Methods and applications, *Acta Crystallogr.* **D49**, 24–36.

Brünger, A.T., Nilges, M. (1993) Computational challenges for macromolecular structure determination by X-ray crystallography and solution NMR spectroscopy, *Q. Rev. Biophys.* **26**, 49–125.

Brünger, A.T., Kuriyan, J., Karplus, M. (1987) Crystallographic *R*-factor refinement by molecular dynamics, *Science* **235**, 458–460.

Brünger, A.T., Adams, P.D., Clore, G.M., Delano, W.L., Gros, P., Grosse-Kunstleve, R.W., Jiang, J.S., Kuszewski, J., Nilges, M., Pannu, N.S., Read, R.J., Rice, L.M., Simonson, T., Warren, G.L. (1998) Crystallography and NMR system – a new software suite for macromolecular structure determination, *Acta Crystallogr.* **D54**, 905–921.

CCP4, (1994) Collaborative computational project No. 4, *Acta Crystallogr.* **D50**, 760–763.

Clausen, T., Huber, R., Laber, B., Pohlenz, H.-D., Messerschmidt, A. (1996) Crystal structure of the pyridoxal-5′-phosphate dependent Cystathionine ß-Lyase from *Escherichia coli* at 1.83 Å, *J. Mol. Biol.* **262**, 202–224.

Clausen, T., Huber, R., Messerschmidt, A., Pohlenz, H.-D., Laber, B. (1997) Slow-binding inhibition of *Escherichia coli* Cystathionine ß-Lyase by L-aminoethoxyvinylglycine: a kinetic and X-ray study, *Biochemistry* **36**, 12633–12643.

Crick, F.H.C., Magdoff, B.S. (1956) The theory of the method of isomorphous replacement, *Acta Crystallogr.* **9**, 901–908.

Cromer, D.T., Liberman, D. (1970) Relativistic calculation of anomalous scattering factors for X-rays, *J. Chem. Phys.* **53**, 1891–1898.

Crowther, R.A., Blow, D.M. (1967) A method of positioning a known molecule in an unknown crystal structure, *Acta Crystallogr.* **23**, 544–548.

Dauter, Z., Lamzin, V.S., Wilson, K.S. (1997) The benefits of atomic resolution, *Curr. Opin. Struct. Biol.* **7**, 681–688.

Diamond, R. (1969) Profile analysis in single crystal diffractometry, *Acta Crystallogr.* **A25**, 43–55.

Dickerson, R.E., Weinzierl, J.E., Palmer, R.A. (1968) A least-squares refinement method for isomorphous replacement, *Acta Crystallogr.* **B24**, 997–1003.

Dodson, E., Vijayan, M. (1971) The determination and refinement of heavy-atom parameters in protein heavy-atom derivatives: some model calculations using acentric reflections, *Acta Crystallogr.* **B27**, 2402–2411.

Eichhorn, K.D. (1985) DISCO: Calculation of the Anomalous Dispersion Corrections, f' and f'', to the Atomic X-ray Form Factor from Both EXAFS and Theory, Deutsches Elektronensynchrotron DESY, Hamburg.

Einsle, O., Messerschmidt, A., Stach, P., Bourenkov, G.P., Bartunik, H.D., Huber, R., Kroneck, P.M.H. (1999) Structure of cytochrome *c* nitrite reductase, *Nature (London)* **400**, 476–480.

Engh, R.A., Huber, R. (1991) Accurate bond and angle parameters for X-ray protein structure refinement, *Acta Crystallogr.* **A47**, 392–400.

Evans, S.V. (1993) SETOR: Hardware lighted three-dimensional solid model representation of macromolecules, *J. Mol. Graphics* **11**, 134–138.

Ford, G. (1974) Intensity determination by profile fitting applied to precession photographs, *J. Appl. Crystallogr.* **7**, 555–564.

Gaykema, W.P.J., Hol, W.G.J., Vereijken, J.M., Soeter, N.M., Bak, H.J., Beintema, J.J. (1984) 3.2 Å structure of the copper-containing, oxygen-carrying protein *Panulirus interruptus* haemocyanin, *Nature (London)* **309**, 23–29.

Gilliland, G.L., Tung, M., Blakeslee, D.M., Ladner, J.E. (1994) Biological macromolecule crystallization database, version 3.0: new features, data and the NASA archive for protein crystal growth data, *Acta Crystallogr.* **D50**, 408–413.

Green, A.A. (1931) Studies in the physical chemistry of the proteins. VIII. The solubility of hemoglobin in concentrated salt solutions. A study of the salting out of proteins, *J. Biol. Chem.* **93**, 495–516.

Green, A.A. (1932) Studies in the physical chemistry of the proteins. X. The solubility of hemoglobin in solutions of chlorides and sulfates of varying concentration, *J. Biol. Chem.* **95**, 47–66.

Greer, J. (1974) Three-dimensional pattern recognition: an approach to automated interpretation of electron density maps of proteins, *J. Mol. Biol.* **82**, 279–301.

Groll, M., Ditzel, L., Löwe, J., Stock, D., Bochtler, M., Bartunik, H.D., Huber, R. (1997) Structure of 20S proteasome from yeast at 2.4 Å resolution, *Nature (London)* **386**, 463–471.

Hendrickson, W.A. (1985) Stereochemically restrained refinement of macromolecular structures, *Methods Enzymol.* **115**, 252–270.

Hendrickson, W.A. (1991) Determination of macromolecular structures from anomalous diffraction of synchrotron radiation, *Science* **254**, 51–58.

Hendrickson, W.A., Lattman, E.E. (1970) Representation of phase probability distributions for simplified combination of independent phase information, *Acta Crystallogr.* **B26**, 136–143.

Hendrickson, W.A., Ogata, C.M. (1997) Phase determination from Multiwavelength Anomalous Diffraction measurements, *Methods Enzymol.* **276**, 494–523.

Hendrickson, W.A., Horton, J.R., LeMaster, D.M. (1990) Selenomethionyl proteins produced for analysis by Multiwavelength Anomalous Diffraction (MAD): a vehicle for direct determination of three-dimensional structure, *EMBO J.* **9**, 1665–1672.

Hönl, H. (1933) Zur Dispersionstheorie der Röntgenstrahlen, *Z. Phys.* **84**, 1–16.

Hooft, R., Vriend, G. (1996) WHAT IF Program Manual, EMBL, Heidelberg, Chapter 23.

Hoppe, W. (1957) Die Faltmolekülmethode: Eine neue Methode zur Bestimmung der Kristallstruktur bei ganz oder teilweise bekannten Molekülstrukturen, *Z. Elektrochem.* **61**, 1076–1083.

Howard, A.J., Nielsen, C., Xuong, Ng.H. (1985) Software for a diffractometer with multiwire area detector, *Methods Enzymol.* **114**, 452–472.
Howard, A.J., Gilliland, G.L., Finzel, B.C., Poulos, T.L., Ohlendorf, D.H., Salemne, F.R. (1987) The use of an imaging proportional counter in macromolecular crystallography, *J. Appl. Crystallogr.* **20**, 383–387.
Huber, R. (1965) Die Automatisierte Faltmolekülmethode, *Acta Crystallogr.* **19**, 353–356.
Jack, A., Levitt, M. (1978) Refinement of large structures by simultaneous minimization of energy and *R*-factor, *Acta Crystallogr.* **A34**, 931–935.
James, R.W. (1960) *The Optical Principles of the Diffraction of X-rays*, Vol. II The Crystalline State, Bell, London, 135–167.
Jancarik, J., Kim, S.H. (1991) Sparse matrix sampling: a screening method for crystallization of proteins, *J. Appl. Crystallogr.* **24**, 409–411.
Jones, T.A. (1978) A graphics model building and refinement system for macromolecules, *J. Appl. Crystallogr.* **15**, 24–31.
Jones, T.A., Zou, J.Y., Cowan, S.W., Kjelgaard, M. (1991) Improved methods for building protein models in electron density maps and location of errors in these models, *Acta Crystallogr.* **A47**, 110–119.
Kabsch, W. (1988) Automatic indexing of rotation diffraction patterns, *J. Appl. Crystallogr.* **21**, 67–81.
Kabsch, W. (1988) Evaluation of single-crystal X-ray diffraction data from position-sensitive detector, *J. Appl. Crystallogr.* **21**, 916–924.
Kabsch, W. (1993) *Data Collection and Evaluation with Program XDS*, in Data Collection and Processing, Proceedings of the CCP4 Study Weekend, 29–30 January 1993, Sawyer, L. Isaac, N. Bailey S., (Eds.) SERC Daresbury Laboratory, Warrington, UK, 63–70.
Kahn, R., Fourme, R. (1997) Gas proportional detectors, *Methods Enzymol.* **276**, 268–288.
Kiefersauer, R., Than, M.E., Dobbek, H., Gremer, L., Melero, M., Strobl, S., Dias, J.M., Soulimane, T., Huber, R. (2000) A novel free-mounting system for protein crystals: transformation and improvement of diffraction power by accurately controlled humidity changes, *J. Appl. Cryst.* **33**, 1223–1230.
Kleywegt, G.J., Jones, T.A. (1994) 'Halloween . . . Masks and Bones', in *From First Map to Final Model*, Bailey, S., Hubbard, R., Waller, D., (Eds.) Proceedings of the Study Weekend, SERC Daresbury Laboratory, Warrington, UK, 59–66.
Knäblein, J., Neuefeind, T., Schneider, F., Bergner, A., Messerschmidt, A., Löwe, J., Steipe, B., Huber, R. (1997) $Ta_6Br_{12}^{2+}$, a tool for phase determination of large biological assemblies by X-ray crystallography, *J. Mol. Biol.* **270**, 1–7.
Kraulis, P.J. (1991) MOLSCRIPT:A program to produce both detailed and schematic plots of protein structures, *J. Appl. Crystallogr.* **24**, 946–950.
Ladenstein, R., Schneider, M., Huber, R., Bartunik, H., Schott, K., Bacher, A. (1988) Heavy riboflavin synthase from bacillus subtilis: crystal structure analysis of the icosahedral Beta-60 capsid at 3.3 Å resolution, *J. Mol. Biol.* **203**, 1045–1070.
de La Fortelle, E., Bricogne, G. (1997) Maximum-likelihood heavy-atom parameter refinement for multiple isomorphous replacement and multiwavelength anomalous diffraction methods, *Methods Enzymol.* **276**, 472–494.
Laskowski, R.A., MacArthur, M.W., Moss, D.S., Thornton, J.M. (1993a) PROCHECK: A program to check the stereochemical quality of protein structures, *J. Appl. Crystallogr.* **26**, 283–291.
Laskowski, R.A., Moss, D.S., Thornton, J.M. (1993b) Main-chain bond lengths and bond angles in protein structures, *J. Mol. Biol.* **231**, 1049–1067.
Leslie, A. (1994) *Mosflm User Guide*, Mosflm Version 5.41', in Data Collection and Processing, Proceedings of the CCP4 Study Weekend, 29–30 January 1993, Sawyer, L., Isaac, N., Bailey, S., (Eds.) HRC Laboratory of Molecular Biology, Cambridge, UK, 44–51.
Löwe, J., Stock, D., Jap, B., Zwickl, P., Baumeister, W., Huber, R. (1995) Crystal structure of the 20S proteasome from the archaeon *Thermoplasma acidophilum* at 3.4 Å resolution, *Science* **268**, 533–539.
van de Locht, A., Lamba, D., Bauer, M., Huber, R., Friedrich, T., Kröger, B., Höffken, W., Bode, W. (1995) Two heads are better than one: crystal structure of the insect derived double domain Kazal Inhibitor rhodniin in complex with thrombin, *EMBO J.* **14**, 5149–5157.
Luzzati, V. (1952) Traitement Statistique des Erreurs dans la Determination des Structures Crystallines, *Acta Crystallogr.* **A5**, 802–810.

MacArthur, M.W., Laskowski, R.A., Thornton, J.M. (1994) Knowledge-based validation of protein-structure coordinates derived by X-ray crystallography and NMR spectroscopy, *Curr. Opin. Struct. Biol.* **4**, 731–737.

Merrit, E.A., Murphy, N.E.P. (1994) RASTER3D Version 2.0. A program for photorealistic molecular graphics, *Acta Crystallogr.* **D50**, 869–873.

Messerschmidt, A., Pflugrath, J.W. (1987) Crystal orientation and X-ray pattern prediction routines for area-detector diffractometer systems in macromolecular crystallography, *J. Appl. Crystallogr.* **20**, 306–315.

Messerschmidt, A., Luecke, H., Huber, R. (1993) X-ray structures and mechanistic implications of three functional derivatives of ascorbate oxidase from zucchini, *J. Mol. Biol.* **230**, 997–1014.

Messerschmidt, A., Ladenstein, R., Huber, R., Bolognesi, M., Avigliano, L., Petruzzelli, R., Rossi, A., Finazzi-Agro, A. Refined crystal structure of ascorbate oxidase at 1.9 Å resolution, *J. Mol. Biol.* **224**, 179–205.

Messerschmidt, A., Rossi, A., Ladenstein, R., Huber, R., Bolognesi, M., Gatti, G., Marchesini, A., Petruzzelli, R., Finazzi-Agro, A. (1989) X-ray crystal structure of the blue oxidase ascorbate oxidase from zucchini: analysis of the polypeptide fold and a model of the copper sites and ligands, *J. Mol. Biol.* **206**, 513–529.

Moffat, K. (1998) Ultrafast time-resolved crystallography, *Nature Struct. Biol., Synchrotron Suppl.* 641–643.

Murshudov, G.N., Vagin, A.A., Dodson, E.J. (1997) Refinement of macromolecular structures by the maximum likelihood method, *Acta Crystallogr.* **D53**, 240–255.

Navaza, J. (1994) AMoRe: An automated package for Molecular Replacement, *Acta Crystallogr.* **A50**, 157–163.

Otwinowski, Z. (1991) Maximum Likelihood Refinement of Heavy Atom Parameters, in *Isomorphous Replacement and Anomalous Scattering*, Wolf, W., Evans, P.R., Leslie, A.G.W., (Eds.) SERC Daresbury Laboratory, Warrington, UK, 80–86.

Otwinowski, Z. (1993) *The Denzo Program Package*, in Data Collection and Processing, Proceedings of the CCP4 Study Weekend, 29–30 January 1993, Isaac, N., Bailey, S., (Eds.) SERC Daresbury Laboratory, Warrington, UK, 56–62.

Otwinowski, Z., Minor, W. (1997) Processing of X-ray diffraction data collected in oscillation mode, *Methods Enzymol.* **276**, 307–326.

Patterson, A.L. (1934) A Fourier series method for the determination of the components of interatomic distances in crystals, *Phys. Rev.* **46**, 372–376.

Pearson, R.G. (1969) Hard and soft acids and bases, *Survey* **5**, 1–52.

Perrakis, A., Morris, R., Lamzin, V.S. (1999) Automated protein model building combined with iterative structure refinement, *Nat. Struct. Biol.* **6**, 458–463.

Phillips, W.C., Rayment, I. (1985) A systematic method for aligning double-focusing mirrors, *Methods Enzymol.* **114**, 316–329.

Ramachandran, G.N., Ramakrishnan, C., Sasisekharan, V.J. (1963) Stereochemistry of polypeptide chain configurations, *J. Mol. Biol.* **7**, 95–99.

Read, R.J. (1986) Improved Fourier coefficients for maps using phases from partial structures with errors, *Acta Crystallogr.* **A42**, 140–149.

Read, R.J. (1990) Structure-factor probabilities for selected structures, *Acta Crystallogr.* **A46**, 900–912.

Rodgers, D.W. (1997) Practical cryocrystallography, *Methods Enzymol.* **276**, 183–203.

Rossmann, M.G. (1960) The accurate determination of the position and shape of heavy-atom replacement groups in proteins, *Acta Crystallogr.* **13**, 221–226.

Rossmann, M.G. (1979) Processing oscillation diffraction data for very large unit cells with an automatic convolution technique and profile fitting, *J. Appl. Crystallogr.* **12**, 225–238.

Rossmann, M.G., Blow, D.M. (1962) The detection of subunits within the crystallographic asymmetric unit, *Acta Crystallogr.* **15**, 24–31.

Rossmann, M.G., Arnold, E., Erikson, J.W., Frankenberger, E.A., Griffith, J.P., Hecht, H.-J., Johnson, J.E., Kamer, G., Luo, M., Mosser, A.G., Rueckert, R.R., Sherry, B., Vriend, G. (1985) Structure of a human common cold virus and functional relationship to other picornaviruses, *Nature (London)* **317**, 145–153.

Roussel, A., Cambilleau, C. (1989) Turbo-Frodo in Silicon Graphics Geometry, Partners Directory, Silicon Graphics, Mountain View, CA.

Schneider, R.R., Sheldrick, G.M. (2002) Substructure solution with *SHELXD*, *Acta Crystallogr.* **D58**, 1772–1779.

Sheldrick, G.M. (1997) SHELX: High-resolution refinement, *Methods Enzymol.* **277**, 319–343.
Sim, G.A. (1959) The distribution of phase angles for structures containing heavy atoms. II. A modification of the normal heavy-atom method for noncentrosymmetrical structures, *Acta Crystallogr.* **12**, 813–815.
Steigemann, W. (1974) Die Entwicklung und Anwendung von Rechenverfahren und Rechenprogrammen zur Strukturanalyse von Proteinen am Beispiel des Trypsin-Trysininhibitorkomplexes, des freien Inhibitors und der L-Asparaginase, PhD Thesis, Technische Universität München.
Stubbs, M.T., Bode, W. (1993) A player of many parts: the spotlight falls on thrombin's structure, *Thromb. Res.* **69**, 1–58.
Teng, T.-Y. (1990) Mounting of crystals for macromolecular crystallography in a free-standing thin film, *J. Appl. Crystallogr.* **23**, 387–391.
Tong, L., Rossmann, M.G. (1990) The locked rotation function, *Acta Crystallogr.* **A46**, 783–792.
Tronrud, D.E., Ten Eyk, L.F., Matthews, B.W. (1987) An efficient general-purpose least-squares refinement program for macromolecular structures, *Acta Crystallogr.* **A43**, 489–501.
Turk, D. (1992) Weiterentwicklung eines Programmes für Molekülgrafik und Elektronendichte-Manipulation und seine Anwendung auf verschiedene Protein-Struktur-Aufklärungen, PhD Thesis, Technische Universität München.
Wang, B.-C. (1985) Resolution of phase ambiguity in macromolecular crystallography, *Methods Enzymol.* **115**, 90–112.
Westbrook, E.M., Naday, I. (1997) Charge-coupled device based area detectors, *Methods Enzymol.* **276**, 244–268.
Wilson, K.S. (1998) Illuminating crystallography, *Nat. Struct. Biol., Synchrotron Suppl.* 627–630.
Xu, H.L., Hauptmann, H.A., Weeks, C.M. (2002) Sine-enhanced *Shake-and-Bake*: the theoretical basis and applications to Se-atom substructures, *Acta Crystallogr.* **D58**, 90–96.
Yang, W., Hendrickson, W.A., Kalman, E.T., Crouch, R.J. (1990) Expression, purification, and crystallization of natural and selenomethionyl recombinant ribonuclease H from *Escherichia coli*, *J. Biol. Chem.* **265**, 13553–13559.
Zhang, K.Y.J., Main, P. (1990) The use of Sayre's equation with solvent flattening and histogram matching for phase extension and refinement of protein structures, *Acta Crystallogr.* **A46**, 377–381.

Malaria Mosquito Genome

Robert A. Holt[1] *and Frank H. Collins*[2]
[1]*Canada's Michael Smith Genome Science Centre, Vancouver, BC, Canada*
[2]*University of Notre Dame, Notre Dame, IN, USA*

Encyclopedia of Molecular Cell Biology and Molecular Medicine, 2nd Edition. Volume 7
Edited by Robert A. Meyers.

ISBN: 3-527-30549-1

Keywords

BAC
Bacterial artificial chromosome. A single copy cloning vector for large DNA fragments (on the order of 100 to 200 kbp).

Contig
A segment of contiguous sequence, uninterrupted by gaps, assembled from a number of overlapping sequence reads.

Cytogenetic Form
Subspecies of *Anopheles gambiae*, as classified by paracentric chromosomal inversions.

DDT
The pesticide dichlorodiphenyltrichloroethane.

Mate Pair
The two sequence reads derived from either end of a DNA fragment.

Orthologs
Genes in different species arising from a single ancestral gene (e.g. human α-globin and mouse α-globin).

Paralogs
Genes in the same species, arising from local gene duplication (e.g. human α-globin and human β-globin).

PEST
The pink eye standard laboratory strain of *Anopheles gambiae*.

Read
A DNA sequence, approximately 500 bp in length, that is derived from a template DNA molecule. The terms "read," "trace," "lane," and "electropherogram" are often used interchangeably.

SNP
Single nucleotide polymorphism.

Scaffold
A set of contigs separated by sequence gaps but of defined order and orientation.

Vector
An organism that carries disease-causing pathogens from one host to another.

WHO
World Health Organization.

Shotgun
A method of whole-genome sequencing where high molecular weight DNA from an organism is randomly fragmented and sequence reads are obtained from fragment ends. Sequencing is done at some level of redundancy (coverage) such that each nucleotide in the genome is represented in multiple reads. The consensus genome sequence is reconstructed computationally on the basis of overlaps among sequence reads.

Sporogeny
The stages of the *Plasmodium* development cycle that take place within the mosquito vector.

Mosquito-borne illnesses exact an enormous toll on human health; malaria parasites alone currently infect approximately 500 million people, more than 1 million of whom will die this year. *A. gambiae* is the principal malaria vector in sub-Saharan Africa, where 90% of the world's deaths due to malaria occur. Historically, the greatest successes in prevention of malaria, yellow fever, and other mosquito-borne disease have come from controlling the mosquito vector. Current work on vector control holds much promise and can be addressed more efficiently by genomic approaches. The reference genome sequence of *A. gambiae sensu stricto* (a representative member of the cryptic *A. gambiae* species complex) has yielded novel insecticide targets, has facilitated the identification of genes involved in insecticide resistance, and has led to the identification of genes that underlie the strong preference of this vector for human blood. Germ-line transformation of the mosquito has been achieved and the reference genome sequence will facilitate efforts to develop a tractable genetic control strategy for mosquitoes. The publication of the *A. gambiae* genome is a landmark in medical entomology, both because this was the first vector genome to be fully sequenced and because of the impact of this mosquito on public health.

1 The Mosquito and Infectious Disease

1.1 Scope and Impact of Mosquito-borne Disease

Mosquitoes, which are members of the insect family Culicidae of the order Diptera, the flies, are medically the most important group of arthropod vectors, or transmitters, of human pathogens. Mosquitoes transmit a large number of human viral pathogens, including the viruses that cause dengue, yellow fever, and West Nile encephalitis, as well as several nematodes species that cause lymphatic filariasis. But, by far, the most important mosquito-borne disease is malaria, which is caused by protozoan parasites of the genus *Plasmodium*. Malaria shares with HIV/AIDS and tuberculosis the distinction of being one of the three most important disease-specific causes of human mortality in the world today. Estimates of the public health impact of malaria are imprecise, but it is widely assumed that approximately half a billion people are currently infected with malaria parasites and that more than 1 million die each year, principally infants and very young children. Although four different *Plasmodium* species infect humans, nearly all of the mortality is caused by the parasite *Plasmodium falciparum*.

1.2
Strategies for Malaria Control

The etiologic agent of human malaria and the pathogen's mode of transmission by mosquitoes were both determined in the latter years of the nineteenth century by Charles Laveran (who discovered the parasite) and Ronald Ross and Battista Grassi (they independently identified the vector), but malaria has been recognized as an important human disease for centuries. By the fifth century, Hippocrates had clearly described the clinical symptoms associated with three different species of malaria parasites, and it was widely recognized in those times that the disease was caused by association with marshy environments with bad air (thus the name malaria).

Antimalarial drugs and interventions targeted at the vector remain the primary public health measures for dealing with malaria. Two very important discoveries emerged during World War II: the antimalarial drug chloroquine, and the insecticide DDT (dichlorodiphenyltrichloroethane). Both were extremely effective, very inexpensive to produce, unencumbered by patents, and had remarkably low levels of toxicity. (The adverse environmental impacts of DDT, which is a highly stable molecule that can accumulate to toxic levels in species at or near the top of the food chain, were entirely due to its widespread use for control of insects of agricultural importance.) In the decades of the 1950s and 1960s, chloroquine and DDT (and related types of antimalarials and insecticides) formed the basis for very effective, countrywide malaria-control programs that were implemented in the framework of a World Health Organization-directed worldwide Malaria Eradication Campaign. Malaria was permanently eradicated from most developed countries in the temperate-climate regions of the world and prevalence was significantly reduced in many other malaria-endemic countries. Unfortunately, worldwide eradication was an impossible goal, and by the 1970s, the responsibility for managing and funding malaria-control programs was returned to national health programs and budgets.

Several important lessons emerged from the WHO (World Health Organization) Malaria Eradication Campaign. Vector control using insecticides like DDT sprayed on the walls of houses was an extremely effective malaria-control strategy. However, it was costly and required well-managed teams of technically skilled people. Moreover, it was effective only in countries with sufficient infrastructures like roads and bridges for vector-control teams to reach the population that was at risk. Control programs based exclusively on antimalarial drugs like chloroquine were even more costly because of the requirement for active case detection. But as active control programs lapsed, effective and inexpensive antimalarial drugs like chloroquine found their way into malaria-endemic communities through either the national primary health systems or through the marketplace.

For most of sub-Saharan Africa, with the notable exceptions of some of the countries in southern Africa and around some of the major cities, the organized malaria-control programs developed during the WHO Malaria Eradication Campaign were not even attempted, particularly in rural areas. Nonetheless, inexpensive antimalarial drugs like chloroquine became widely available, even in relatively remote rural communities in most of Africa. Unfortunately, chloroquine-resistant *P. falciparum* appeared in coastal Kenya in 1978, and in the subsequent two decades, resistant parasites became the prevalent form almost throughout sub-Saharan Africa. Although the health impact of the publicly available and affordable antimalarials like chloroquine was never assessed, the emergence of parasite resistance to this drug has been coincident with an equally poorly documented perception that malaria-specific mortality in Africa has been increasing. Today, chloroquine is an effective antimalarial only in parts of west Africa, and unfortunately, replacement drugs like Fansidar (sulfadoxine/pyramethamine), mefloquine, and – more recently artemisinin based drugs (based on the ingredients in *Artemisia annua*) are either very costly or are themselves provoking the emergence of resistant parasite strains. As resistance to existing antimalarial drugs becomes more entrenched, there is increased need for novel and effective vector-control strategies.

2 *A. gambiae*, the Principal Vector of Malaria

2.1 The Mosquito and the Malaria Parasite

Malaria parasites are intracellular human parasites that undergo cycles of invasion, growth, and mitotic replication in erythrocytes. The parasites are transmitted among human hosts by *Anopheles* mosquitoes, which ingest parasites when they feed on the blood of an infected human host. The malaria parasites undergo a complex developmental cycle, referred to as the *sporogenic developmental cycle*, in the mosquito vector. A small fraction of the *Plasmodium*-infected erythrocytes are differentiated forms called *gametocytes* that are triggered by hypoxanthine in the mosquito's midgut to develop into male and female gametes. These fuse to form diploid zygotes (the only diploid stage in the *Plasmodium* life cycle), which traverse the mosquito's midgut cells and then undergo a meiotic division and differentiate further into growth stages called the *oocysts*. After approximately one week of growth and mitotic divisions, a single oocyst will produce several thousand specialized, haploid forms called *sporozoites*. On maturation of the oocysts, the sporozoites are released into the mosquito's open circulatory system through which they are transported passively to the salivary glands. These parasite stages invade the salivary gland cells and accumulate in the salivary ducts and acini of these cells, from where they are injected into a host capillary when the mosquito next takes a blood meal. In the human host, the sporozoites are transferred passively through the circulatory system to the liver, where they invade hepatocytes. In these cells, the parasite undergoes approximately a dozen rounds of mitotic replication that result in the production of several thousand *merozoites*. On completion of this cycle of replication, the infected hepatocytes rupture and release the merozoites into the circulatory system, where they invade erythrocytes and initiate the blood cell cycle of infection that causes disease.

Plasmodium falciparum malaria is widely distributed throughout the widely distributed throughout the world, especially in tropical regions bounded by approximately 30° north and south latitude. Over this broad distribution, parasite prevalence in human populations with endemic *P. falciparum* ranges from low levels of a few percent to peak endemicity levels, approaching 100% in rainy seasons when transmission is maximum. Of the approximately 500 or more *Anopheles* species found worldwide, several dozen are involved in the transmission of malaria parasites. The *Anopheles* vectors are the major determinant of local levels of disease endemicity. The majority of *Anopheles* species are probably physiologically competent to support sporogonic development of the parasite. In order for malaria parasites to remain endemic in a particular human community, the vector population must effectively propagate each infected human case to – on average – more than one new human host. With an effective case reproductive rate of less than 1, the parasite cannot be stably maintained in a human population.

Four important features of mosquito biology determine whether a particular *Anopheles* species will be effective as a malaria vector: (1) It must be a physiologically suitable host. (2) It must also be a moderately abundant species in order to maintain a density sufficient to achieve transmission. (3) The mosquito must also take blood meals from people. Because the vector must both acquire the infection in one blood meal and then take a second meal on another human host after the sporogonic cycle is complete, blood-meal host choice is an important determinant of vectorial effectiveness. (4) Finally, and most important, the age structure of the vector population must be such that a sufficient portion survives long enough to allow sporogonic development to be completed.

2.2 *A. gambiae* is a Polymorphic Species

Approximately 90% of the world's malaria-specific mortality occurs in sub-Saharan Africa, where the primary vector is *A. gambiae*. A second important African vector is the mosquito *Anopheles arabiensis*, which is a close relative of *A. gambiae* in a cluster of seven related African species known as the *A. gambiae sibling-species complex*. This extraordinary concentration of malaria in Africa is largely a consequence of the close ecological association of *A. gambiae* and *A. arabiensis* with people. *A. gambiae*, in particular, takes blood meals almost exclusively from people; its larval stages develop in temporary pools of water, like tire ruts, animal hoof-prints, and irrigated agricultural fields that are produced by human activity; and when not laying eggs or mating, the adult females spend most of their time in human dwellings either blood feeding or resting while eggs develop.

Perhaps, because of the close association this mosquito has with people, it is also a highly polymorphic species. Like a number of other insects, especially in the order Diptera, *A. gambiae* and its sibling species have giant polytene chromosomes in many of their tissues, and these have facilitated population analysis. During the past three decades, careful studies of the polytene chromosomes of field-collected *A. gambiae* from countries in west Africa have revealed patterns of distribution of a number of paracentric chromosome inversions that suggest the presence of at least five different reproductively isolated chromosome forms designated as Bissau, Bamako, Mopti, Forest, and Savanna. (All *Anopheles* genomes are based on three pairs of chromosomes, two autosomes, and an X and Y sex-determining pair, and in most *Anopheles* species, cells of either the larval salivary glands or nurse cells of developing ovaries contain polytene chromosomes that, when properly prepared, will show a reproducible, species-specific banding pattern. In addition to their value in species identification and population studies, polytene chromosomes are excellent for physical mapping of cloned DNA.) Most of the inversions that distinguish chromosomal forms are restricted to the right arm of chromosome 2. In geographic regions where two or more of the chromosome forms are present, inversion heterozygotes expected from interbreeding of forms are either totally absent or are present at levels significantly below Hardy–Weinberg expectations. Three of these forms, Bamako, Mopti, and Savanna, have been carefully studied in Mali, where they exhibit clear differences in ecology. In spite of considerable effort by several groups to identify molecular markers diagnostic of these chromosome forms, the only molecule found thus far that distinguishes different *A. gambiae* populations is the rDNA, where sequences in both the IGS and ITS regions can be used to identify what are referred to as *S and M ribosomal DNA* (rDNA) *types*. In Mali, the M rDNA type is associated exclusively with Mopti cytogenetic specimens and the S rDNA type is found only in Savanna and Bamako specimens. In other parts of west Africa, however, the association between karyotype and rDNA type is less predictable. Both S and M rDNA types are found in Forest, Bissau, and Savanna cytogenetic forms.

The frequencies of polymorphic paracentric inversions are clearly important characters in *A. gambiae* as well as in its sibling species, *A. arabiensis*. In both species, inversion frequencies have been shown to vary along clines associated with climatic and ecological parameters, and the data from Mali, supporting the presence of chromosomally distinct *A. gambiae* forms that are both ecologically and genetically differentiated is quite robust. Unfortunately, karyotyping of wild specimens is both difficult and possibly biased, and the lack of congruence between the chromosomal inversion and rDNA molecular markers means that the population structure of this important African vector remains unclear.

3
The *A. gambiae* Genome

3.1
The PEST Strain

The *A. gambiae* PEST (Pink Eye STandard) strain was chosen for genome sequencing because it had a fixed chromosomal arrangement, a sex-linked pink-eye mutation that can readily be used as an indicator of cross-colony contamination, and because clones from two different PEST-strain BAC (Bacterial Artificial Chromosome) libraries had already been end sequenced and physically mapped. The pink-eye mutation originated in a colony called *A. gambiae LPE*, established in 1951 at the London School of Hygiene and Tropical Medicine from mosquitoes collected in Lagos, Nigeria. In 1986, this mutation was introduced into a colony of *A. gambiae* from western Kenya by crossing males of the LPE strain with female offspring of wild caught Kenyan *A. gambiae* (the Savanna form), selecting males from the F2 of this cross and then crossing them again with additional female offspring of wild caught Kenyan *A. gambiae*. From the F2 offspring of this second outcross to Kenyan mosquitoes, a strain was selected that was fixed for pink eye. This outcrossing scheme was repeated once more in 1987, producing a pink-eye strain with a genetic composition largely constituted of the western Kenya Savanna cytogenetic form. In each of these crosses, several hundred female offspring of at least 20 wild caught mosquitoes were used in the cross. This strain, designated *A. gambiae PE*, was polymorphic for the inversions 2La (32%) and 2Rbc (19%). The 2Rbc inversion is characteristic of Mopti, indicating that the original LPE strain from Nigeria was the Mopti form. This inversion was apparently balanced by the uninverted form because no 2Rbc/bc individuals were detected in the colony. From this PE strain, Mukabayire and Besansky selected a set of nine families whose female parent and at least 20 female offspring were fixed for the standard chromosome karyotype. The progeny of these nine families were pooled to form the *A. gambiae* PEST strain. This strain may clearly have some Mopti-derived DNA as the standard karyotype is shared by Mopti and Savanna and the original PE strain did have the 2Rbc inversion rather than the 2Rb that is typical of Savanna. When tested, this colony was fully susceptible to *P. falciparum* from western Kenya.

3.2 The Draft Genome Sequence

Obtaining the draft genome sequence of *A. gambiae* was a collaborative effort, with sequence data provided by Celera, The Institute for Genomics Research, and the French National Sequencing Centre, Genoscope. The whole-genome shotgun method was adopted for this project, given its speed and efficiency and in consideration of its previously successful use for the *Drosophila* genome project. Paired end reads were derived from 2 254 546 plasmid clones with 2-, 10-, or 50-kb inserts, (50 kbp plasmid libraries had the bulk of their insert sequence excised during the initial steps of library construction, and only the insert ends were propagated in *Escherichia coli*.) and approximately 40 000 large-insert BAC clones, to give a total of 10.2-fold sequence coverage of the genome, estimated from CoT analysis to be about 260 Mbp. All plasmid and BAC libraries were constructed using DNA extracted from several hundred adult PEST-strain mosquitoes. Libraries from male and female mosquitoes were constructed separately and sequenced in equal proportion, giving approximately equal coverage of the autosomes and accordingly less coverage of the X and Y chromosomes. All electropherograms are downloadable from the NCBI trace archive (http://www.ncbi.nlm.nih.gov/Traces/trace.cgi?) or Ensembl trace server (http://trace.ensembl.org).

The WGS data set was assembled using Celera's assembly software, where algorithms are executed in a stepwise manner. An initial Blastn comparison of each sequence read to every other sequence read is used to detect overlaps (the overlap criteria being 40 or more base pairs with 6% or less mismatch). Reads within each cluster of overlapping sequences are then assembled into a contiguous consensus sequence (contig). Contigs that appear from read depth to be single copy in the genome are ordered and oriented into longer scaffolds, using mate pairs (two sequence reads derived from either end of a clone insert). Mate pairs are also used to guide computational intrascaffold gap filling, whereby one mate is located in the sequence flanking either side of a gap and its partner is placed within the gap. The resulting assembly is thus a set of typically very long scaffolds that are ordered and oriented sets of contigs whose intervening gaps are of known mean length and whose standard deviations are based on estimated library insert sizes.

The published draft of the *A. gambiae* whole-genome assembly was comprised of 8987 scaffolds, spanning 278 million base pairs (Mbp) of sequence. It is important to note that a small number of very large scaffolds cover most of the genome (91% of the sequence is represented by 303 scaffolds greater than 30 kb), and that many small scaffolds are expected to fit within intrascaffold gaps. In the entire assembly, there are 8986 gaps between scaffolds and 9964 gaps within scaffolds. Gaps within scaffolds are largely due to repetitive sequences that could not be uniquely placed on the basis of sequence overlap or mate-pair information and may also occasionally be due to a simple lack of coverage. Genbank accession numbers for the 8987 genome scaffolds are AAAB01000001 through AAAB01008987 and the scaffold sequences are downloadable as a set from ftp://ftp.ncbi.nih.gov/genbank/genomes/*Anopheles*_gambiae/Assembly_scaffolds.

While use of the whole-genome shotgun approach for this project was well justified by time and cost savings, it is important to understand that the present form of the assembly is that of a draft genome sequence and, as with all draft genomes, there are numerous gaps that will require focused finishing efforts. Further, because of the fact that the plasmid libraries were derived from whole adult mosquitoes, some contaminating sequence from commensal gut bacteria was expected and were in fact observed. Scaffolds with hits to bacteria were flagged in their Genbank record as containing putative contaminating sequence of possible bacterial origin. In total, 213 short scaffolds were flagged, which together comprised 36.5 kb or 0.013% of the genome. These scaffolds were not culled from the assembly because of the remote possibility that some may represent real horizontal transfer events.

3.3 Genome Assembly, Genome size, and Physical Mapping

The sum of all the assembled *Anopheles* genome scaffolds is 278 Mb, which is significantly larger than the genome size of 260 Mb predicted by CoT analysis. There are several reasons why the assembly is close to, but slightly larger than the genome size predicted by CoT analysis. First, many small scaffolds may fit into gaps within other larger scaffolds. Second, because the X chromosome accounts for close to 20% of the genome but is underrepresented in males relative to the autosomes, CoT analysis using a mixed-sex pool of mosquitoes is expected to underestimate genome size. The third likely cause of discrepancy is the unanticipated assembly of highly variant alleles into separate contigs, as discussed further below (Sect. 3.4 Genetic variation).

The final step in defining the reference genome for *A. gambiae* was to order and

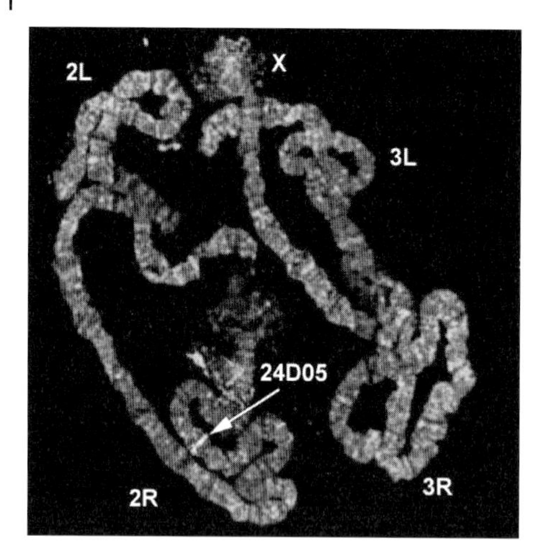

Fig. 1 Illustration of fluorescent *in situ* hybridization using a BAC clone probe. The *A. gambiae* polytene chromosome complement was isolated from the ovarian nurse cells of half-gravid mosquitoes. The probe, 24D08, is a BAC clone from the PEST library and is labeled with Cy3-dUTP red; the polytene chromosome is stained with Yoyo-1 iodide green (Molecular Probes).

orient the scaffolds on the three chromosome pairs, using physical mapping data. The chromosomal locations of assembled scaffolds were determined with sequence-tagged genomic DNA BAC clones that were mapped by *in situ* hybridization to ovarian nurse-cell polytene chromosomes (Fig. 1). For the initial assembly, approximately 2000 BAC clones were mapped and these data enabled the chromosomal location and orientation of scaffolds, constituting about 227 Mbp. Photo archive data showing the results of physical mapping are viewable at http://www.anobase.org. Since the initial assembly, the order and orientation of some of the smaller scaffolds have been confirmed or corrected, and ~7 Mbp more scaffolded sequence has been placed and oriented on the physical genome by *in situ* mapping of selected BAC and cDNA clones. Currently, 172 of the largest scaffolds have been assigned to chromosomal locations (a total of ~233 Mbp), and gaps between almost one-third of these scaffolds have been crossed by BAC clones or short BAC tiling paths (these BACs remain to be sequenced). Several dozen scaffolds (<10 Mbp) can be assigned only to nonspecific centromeric heterochromatic regions and about 1 Mbp of scaffolded sequence has been identified as possible Y-chromosome sequence; thus, about 243 Mbp of the 278 Mbp assembly has been assigned to locations with some level of confidence. Most of the remaining ~8700 unassigned scaffolds (~35 Mbp) are less than 20 kb. The majority of these appear to be repetitive heterochromatic sequences that, in *A. gambiae,* are clustered around the centromeres, the telomeres, and in a few islands in the euchromatic chromosome arms. Most of these scaffolds do not appear to contain genes and many are unlikely to be assigned to specific chromosomal locations by currently available approaches.

3.4 Genetic Variation

Analysis of high-quality mismatches within contig multiple sequence alignments revealed approximately half-a-million single nucleotide polymorphisms (SNPs) in the *A. gambiae* PEST-strain genome. The distribution of SNPs along the chromosomes is bimodal, with some regions showing roughly one SNP every 10 kb and others showing approximately one SNP every 200 bp. Given that the genome of the PEST strain appears to have resulted from a complex introgression of divergent Mopti and Savanna chromosomal forms, we expect that some genomic regions may be derived only from one or the other form, yielding a low density of SNPs, while other genomic regions may continue to segregate both divergent forms, yielding a high SNP density. Interestingly, the X chromosome has a markedly lower average level of polymorphism and does not show this bimodal pattern, perhaps due to male hemizygosity and hence, a lower rate of introgression is seen on this chromosome. In addition, heterozygosity of the X is expected to be depressed because of the selection for homozygosity of the X-linked pink-eye mutation.

As expected, the overwhelming majority of SNPs lie in intergenic regions, but there is still an abundance of SNPs within functional genes. Introns and intergenic regions have virtually identical heterozygosities, but the silent coding positions appear to have more than twofold enrichment of variability. Generally, silent coding sites are considered as having more stringent constraints than introns or intergenic regions because of biased codon usage, and this is reflected in a lower diversity of silent sites in most organisms. The reason for elevated silent variation in *A. gambiae* is at present unknown. Nucleotides with strong functional constraints, such as splice donors, splice acceptors, and stop codons have the lowest heterozygosity, and nonsynonymous (missense) positions are also evidently low in heterozygosity.

The bimodal SNP distribution in the PEST strain, and hence, the existence of two approximately equally frequent haplotypes in a substantial fraction of the genome, presented a unique challenge for assembly of the sequence data set, and it is important to note that some assembly artifacts have resulted. Where the sequence variation between the two haplotypes is small, they have been assembled together into a single contig, with the most frequent nucleotide at each position contributing to the consensus sequence. This is the manner in which all current whole-genome assembly algorithms handle genetic variation, even though the resulting consensus sequence does not precisely represent any native haploid genome. In the case of *A. gambiae*, where the haplotypes are more divergent, they have occasionally been split into two separate contigs. This is problematic because the assembler only allows a single consensus sequence to represent any given position in the genome. As a result, in the draft assembly, there are situations in which the second redundant contig (1) has been placed in the nearest gap, causing an artificial tandem duplication within the scaffold; (2) has been left out of the scaffold as singleton; or (3) where the problem is at a scaffold end, the duplicate has been placed at the beginning of the neighboring scaffold. The phenomenon described here was initially revealed as a larger separation of mate pairs in parts of the assembly than what the library insert sizes had predicted. Comprehensive screening of the 1 644 078 total mate pairs in the assembly revealed a total of 726 regions, with this diagnostic mate-pair signature covering a total of approximately 7.7% of the assembly. The amount of sequence that is doubly represented in the genome assembly because of this artifact is estimated at about 23.5 Mbp.

While the bimodal pattern of SNP variation is a unique feature of the PEST strain and is not representative of any wild mosquito population, individual variants represent real polymorphisms in *A. gambiae* and will provide valuable marker sets for genetic studies. The approximately 500 000 PEST-strain SNPs are downloadable from ftp://ftp.ncbi.nih.gov/genbank/genomes/Anopheles_gambiae or www.ensembl.org/anopheles_gambiae. A further 75 000 SNPs have been identified by comparing light sequence coverage (1.2×) from a lab strain of the Mopti cytogenetic form to the assembled PEST-strain sequence, and these SNP data are also available at www.ensembl.org/anopheles_gambiae.

Understanding the genetic diversity of anopheline mosquitoes and the molecular basis of cytogenetic form–associated traits will be accelerated by genomics and will likely point the way to more efficient strategies of vector-targeted malaria control. Efforts are presently underway to analyze large regions of the *A. gambiae* Y chromosome as population genetic markers and as indicators of the evolutionary relationships of the different *A. gambiae* forms and other species closely related to *A. gambiae*. Efforts to use other parts of the genome to define and identify such species and forms have been complicated by presumed interspecific and interform recombination over many regions of the different genomes. As a presumably nonrecombining part of the genome, the Y-chromosome sequence will be an invaluable source of material for teasing apart population structure and for developing rapid diagnostics for field studies. Studies are also underway to obtain wild isolates of *A. gambiae* from several different regions of Africa and sequence segments of DNA at regular intervals across the genomes of multiple individuals from each isolate. This survey of genetic diversity among different isolates of *A. gambiae* will facilitate the teasing apart of anopheline mosquito population structure and will play a crucial role in understanding mosquito biology, particularly those biological traits that influence vectorial capacity.

3.5 Gene Content

The initial automated annotation of the draft *A. gambiae* genome sequence was accomplished using established annotation pipelines at Celera and the Ensembl group at the European Bioinformatics Institute/Sanger Institute. Both pipelines use a combination of homology-based evidence and *ab initio* gene-finding algorithms to model gene structure. At the inception of the project, there were only approximately 6000 EST sequences available in public databases. However, given the evolutionary distance between *Drosophila* and *Anopheles* (~250 million years), obtaining wide representation from *Anopheles* EST libraries was essential for identifying the open reading frames of *A. gambiae* genes, and an additional approximately 80 000 EST sequences were derived from whole mosquitoes, as an integral part of the *Anopheles* genome project.

For proteome analysis, a nonredundant set of predicted protein sequences from the two pipelines (Celera and Ensembl) was constructed by selecting, for each locus, the annotation containing the largest number of exons. A total of 15 189 automated predictions were derived in this manner and, after removal of putative transposable elements and other contaminants, served as the set for initial analysis of the *A. gambiae* proteome. As is typical with automated genome annotations, it is likely that some false-positive predictions (pseudogenes, bacterial contaminants, and transposons) and false-negative predictions (*Anopheles* genes that were not computationally predicted) remain, and there are undoubtedly numerous errors in defining the precise boundaries of these putative gene structures. Automated annotation, while being a cost-effective method for an initial high-level view of the genome, does not substitute for the careful manual curation that will be undertaken by the community over a period of time. Since the initial publication of the *Anopheles* genome, the annotation has been under continuous development by the Ensembl group in coordination with the International *A. gambiae* Genome Consortium. The present version of the annotated genome is available at www.ensembl.org/*Anopheles*_gambiae. A tutorial on the Ensembl Anopheles site is available at www.ensembl.org/*Anopheles*-gambiae/documents/mosquito_doc.pdf.

3.6
Comparison of the *Anopheles* and *Drosophila* Proteomes

Analysis of the annotated gene set has revealed many interesting features of the proteome of *A. gambiae*, particularly in comparison with *Drosophila melanogaster*, the closest relative of this vector (*A. gambiae* and *D. melanogaster* shared a common Dipteran ancestor approximately 250 million years ago.) to have a completely sequenced genome. In total, 47.2% of the *Anopheles* protein set is composed of 1 : 1 orthologs (defined as *reciprocal best Blast matches*), and 13.8% is composed of "many-to-many" orthologs, where paralogous gene families have expanded/contracted unevenly. A further 17.9% of the *Anopheles* protein set is represented by proteins that have a homologous relationship to *Drosophila* proteins, but the match cannot be easily defined as orthologous or paralogous. For example, proteins in this group might share one or more protein domains or might be divergent members of large and diverse protein families. 10.0% of *Anopheles* proteins had their best match to a noninsect species and 11.1% of *Anopheles* proteins do not show homology to proteins from any other species. Of these 1437 *Anopheles*-specific genes, 575 are supported by EST evidence and 522 show homology to other *A. gambiae* proteins. While some of these proteins may represent *de novo Anopheles* genes or rapidly evolving ancestral genes that are now mutated beyond recognition, it is likely that homologous sequences for these proteins will be discovered as the genomes of organisms with closer evolutionary relationships to *A. gambiae* are sequenced. Interestingly, in comparing the *A. gambiae* and *D. melanogaster* proteomes, sequence similarity is highest among structural molecules, transporters, and enzymes and is lowest among genes involved in defense and immunity. This is consistent with the results from other recent whole-genome projects that suggest that the most plastic gene families are those that mediate the direct interaction of the organism with its environment, such as genes involved in olfaction, detoxification, and immune response. In comparison, developmental genes are highly conserved between *D. melanogaster* and *A. gambiae*, with 85% having single 1 : 1 orthologs (best reciprocal matches) between the two species.

Given the evolutionary distance between *A. gambiae* and *D. melanogaster* conserved gene order (synteny) is recognizable only over short distances. Approximately 34% of *A. gambiae* genes colocalize in small microsyntenic clusters, defined as chromosomal regions where at least two orthologs or orthologous groups are separated by no more than five unrelated genes in the intervening DNA. In total, there are 948 such microsyntenic blocks in the *A. gambiae* genome, when compared to *D. melanogaster*, the largest block containing 31 orthologous genes.

Most clusters contain substantially fewer genes and show evidence of local duplications, inversions, and translocations. In contrast, homology between *A. gambiae* and *D. melanogaster* on the level of whole-chromosome arms is easily recognizable upon mapping of 1 : 1 orthologs. The most conserved pair of chromosomal arms are *D. melanogaster*2L (Dm2L) and *A. gambiae*3R (Ag3R), with 67% of Ag3R orthologs and 83% of microsyntenic clusters being significantly similar to Dm2L. *A. gambiae* chromosomal arm 2L appears to be largely homologous to Dm3L and Dm2R, with approximately 42% of *A. gambiae* orthologs found on each of these two *D. melanogaster* arms. Similarly, Ag2R shares homology with Dm3R. Ortholog content of Ag3L indicates that this chromosomal arm has modest homology to both Dm2R (30%) and Dm3L (22%). Homology between the X chromosomes of the two species is limited, with significant shuffling of chromosomal segments to and from autosomes.

The predicted *Anopheles* protein set has been classified on the basis of protein domains, and their functional categories, using InterPro and the Gene Ontology (GO). The relative abundance of the majority of proteins containing InterPro domains is similar between the mosquito and fly. As expected, insect-specific cuticle and chitin-binding domains and the insect-specific olfactory receptors are overrepresented relative to noninsects. However, comparing *A. gambiae* and *D. melanogaster*, only 6 of the 200 most abundant protein domains differ significantly in abundance. However, some of these differences are dramatic, such as the striking expansion of fibrinogen domain–containing proteins observed in *A. gambiae* relative to *D. melanogaster*. There are 58 *A. gambiae* and 13 *D. melanogaster* fibrinogen domain–containing proteins and only a single pair is a 1 : 1 ortholog. It is tempting to speculate that fibrinogen domains, which were originally found in proteins that mediate blood coagulation in mammals, could be conferring some function to mosquitoes that enables blood feeding, such as inhibition of blood-meal coagulation. Further, the serine proteases (central effectors of proteolytic processes) are well represented in both insect genomes, but *Anopheles* has nearly 100 additional members, perhaps again reflecting the hematophagous adaptation of this Dipteran.

Major differences in gene number are seen in the peroxidase system, which mediates the oxidation of diverse substrates. A total of 18 peroxidases were observed in *A. gambiae* and only 10 were observed in *D. melanogaster*. The *Anopheles* peroxidases have high homology to the salivary peroxidases of the mosquito *A. albimanus*, suggesting the possibility that this gene family has expanded to facilitate the blood-feeding process.

In a separate comparative analysis using the LEK algorithm to identify protein families with different representation in *A. gambiae* and *D. melanogaster*, a prominent expansion of a 19-member family of odorant receptors (ORs) was detected in *A. gambiae*. The members of this family do not show substantial sequence similarity to ORs from *D. melanogaster* or any other organism. While between *A. gambiae* and *D. melanogaster*, there is significant conservation of the G protein-coupled receptor superfamily as a whole, this *A. gambiae* OR subfamily appears to be unique and may play an important role in mosquito-specific behavior that includes host seeking, as discussed further in Sect. 4.1. In addition to the conspicuous protein family expansions discussed above, many additional expansions were seen in diverse families. These include critical components of the visual system, structural components of the cell adhesion and contractile machinery, energy-generating glycolytic enzymes, anabolic and catabolic enzymes involved in protein and lipid metabolism, transporters, and detoxification enzymes.

Identification of gene family contractions have not been a focus of analysis of the *A. gambiae* genome, even though gene loss is considered to be an important mediator of evolution. It is ambiguous when doing a binary comparison whether there is expansion in one species or contraction in the other, relative to the state of the common ancestor. This can only be revealed by analysis of orthology across several species. Further, it is difficult to distinguish real gene losses from simple false-negative annotation results. Given that the overall gene count in *A. gambiae* and *D. melanogaster* is not significantly different, with each species having roughly 14 000 annotated gene loci, the overall number of expansions must be balanced by an equal number of reductions in each species, as they evolve to fill their ecological niches.

3.7
Genes Regulated by Blood Feeding

Newly emerged female mosquitoes are sustained by sugar meals as they seek the blood meal that is necessary to support egg development. Active host seeking is facilitated by olfactory, temperature, and visual cues. After locating a host and extracting several times their weight in blood, the female mosquito appears to undergo profound metabolic reprogramming, commensurate with the task of processing the massive blood bolus.

The *A. gambiae* genome project included an EST-sequencing program utilizing two source libraries. One library was constructed using DNA from whole adult female mosquitoes before blood feeding, when in a relatively quiescent metabolic state, and the second library was constructed 24 h after a blood meal, when the mosquitoes were under the metabolic burden of converting the blood meal into eggs. This approach facilitated identification of genes activated or inactivated by blood feeding in a hematophagous insect. Approximately 40 000 ESTs were sequenced from each library and assembled into approximately 7000 consensus sequences, each representing a distinct gene locus. 435 genes showed significantly altered transcription following blood-meal ingestion, as determined by Chi-squared analysis. Analysis of gene expression pre- and post blood feeding has revealed both expected and unexpected changes (Table 1). As expected, there is dramatic upregulation (up to 144 fold) of transcripts encoding serine proteases and other proteases, transcripts involved in transcriptional control and nuclear regulation, and transcripts representing the protein synthesis machinery that are involved in converting the digested blood proteins into new functional mosquito protein products. Further, transcripts involved in gluconeogenesis, lipid metabolism, and purine metabolism are upregulated, as are a small number of transcripts, such as ubiquitin-conjugating enzymes and cytosolic aminopeptidases, whose protein products mediate intracellular degradation of protein. Finally, there is upregulation of oogenesis-associated proteins and lipids 24 h post blood meal, including several vitellogenins (yolk constituents) and members of the phenoloxidase pathway that mediates eggshell melanization. In terms of decreased expression, genes involved in sugar metabolism (for example, salivary and midgut maltases, glycolysis, and oxidative phosphorylation) are downregulated, which is not unexpected, given the presumed decreased reliance on sugar metabolism with the high intake of a protein-rich blood meal. More intriguing is the marked downregulation post blood meal of striated muscle proteins, including actin, myosin, tafazzin, and troponin, as well as ATP-dependent cation transporters and Ca^{2+} binding proteins involved in motor signaling. Taken together with the observed decrease in transcription of opsins, photoreceptor associated proteins, cuticular proteins associated with the eye lens, and pheromone-binding proteins, the post blood-meal expression pattern suggests that after a blood meal, mosquitoes become detached from their environment and sacrifice mobility in order to devote metabolic attention to the monumental task of blood-meal processing and egg development. The set of genes upregulated by blood-meal digestion presents a rich source of novel insecticide targets and will allow identification of promoters that can activate transgenes in a blood-meal responsive manner.

4 Utility of the *A. gambiae* Genome

4.1 Odorant Receptors and Novel Repellents/Attractants

The distinct preference of *A. gambiae* for human blood meals contributes substantially to the menace of this vector. Very

Tab. 1 Representative messages significantly more transcribed or less transcribed in female *A. gambiae* mosquitoes 24 h after human blood-meal ingestion.

Category	↕	*Gene description*	*Gene symbol*	*EST count (no blood)*	*EST count (blood)*
Extracellular protein digestion	↑	Trypsin-2 precursor	AgCP10888	1	144
Intracellular protein degradation	↑	Ubiquitin	AgCP12328	0	47
Nuclear regulation	↑	Histone	AgCP10826	2	40
Protein synthesis	↑	Elongation factor 1a	AgCP3905	163	278
Oogenesis	↑	Vitellogenin	AgCP2518	0	878
Gluconeogenesis	↑	Aspartate amino transferase	AgCP8491	1	10
Glycolysis	↓	Succinate dehydrogenase	AgCP3730	24	7
Oxidative phosphorylation	↓	Glycerol-3-phosphate dehydrogenase	AgCP3306	51	11
Extracellular sugar digestion	↓	Salivary maltase	AgCP12790	10	0
Striated muscle	↓	Flightin (striatal muscle protein 27)	AgCP3724	71	2
Vision	↓	Opsin	AgCP12420	276	184
Chemosensation	↓	Pheromone-binding protein	AgCP11481	7	0

little is known about the molecular basis of this selective behavior, although it almost certainly involves specific human odors and mosquito pathways for detecting and responding to these odors. Continuing to explore the diversity of ORs and odorant binding proteins encoded in the *A. gambiae* genome will lead to a better understanding of the mechanisms of host preference and will uncover potential targets for a new generation of specific attractants/repellents. Preliminary analysis of the draft genome sequence identified a total of 276 G protein-coupled receptors, including 79 candidate ORs. The majority of these ORs (64 of 79) show expression only in olfactory tissues, as determined by RT-PCR experiments. Comparison to *D. melanogaster* revealed a similar total number of ORs, but only a single unequivocal orthologous pair of ORs between the two species. The extensive lineage-specific expansion of ORs likely reflects the ecological and physiological relevance of these receptors and the importance of each species being able to detect relevant chemicals – for example, rotting fruit odors for *D. melanogaster* and human host odors for *Anopheles*. AgOr1, a putative *A. gambiae* OR, has recently been identified as a possible mediator of human host finding. Initial clues that AgOr1 may be important in host finding include its expression exclusively in the olfactory tissue of female mosquitoes, and the downregulation of its expression in these tissues after a blood meal. Hallem EA et al. provided further evidence by engineering *D. melanogaster* neurons lacking endogenous ORs and by transfecting these with AgOr1. Electrophysiological measurements of the transfected cells showed that a component of human sweat, 4-methylphenol, elicited a strong response.

4.2 Insecticide Resistance Genes

During the 1950s and early 1960s, the WHO malaria eradication campaign succeeded in eradicating malaria from Europe and dramatically reduced its prevalence in many other parts of the world, primarily through programs that combined mosquito control agents like DDT with antimalarial drugs like chloroquine. However, malaria and anopheline mosquitoes remain entrenched in sub-Saharan Africa and the appearance of chloroquine-resistant parasites and insecticide-resistant mosquitoes is contributing to the current rise of malaria in Africa. Even control programs based on insecticide-impregnated bed nets, which are now advocated by WHO and are being widely implemented in Africa are threatened by the development of resistance to pyrethroids, the insecticide class of choice for this application. Insecticide resistance is often metabolic, where the insecticide will induce the expression of an enzyme that degrades it. Metabolic resistance is mediated largely by one of the three classes of enzymes, depending on the particular insecticidal agent in question. These are carboxylesterases (CEs), glutathione-*S*-transferases (GSTs) and cytochrome P450s. Ranson et al. cataloged members of these three major insecticide resistance–related families in the *A. gambiae* genome. Comparison to *D. melanogaster* revealed that a considerable expansion of these families has occurred in the mosquito. For example, there are 51 versus 31 CEs and 111 versus 90 cytochrome P450 enzymes in *A. gambiae* and *D. melanogaster* respectively. Secure orthologs between *D. melanogaster* and *A. gambiae*, identified by rigorous analysis of phylogenetic trees comprise less than

15% of these supergene families, indicating that gene families have radiated independently from common ancestral genes, presumably driven by the different metabolic requirements of the two species for natural compounds. Over the past decade, several resistance loci for different agents have been mapped in the mosquito genome, using classical genetic methods, and inspection of gene content within the boundaries of mapped resistance loci has been revealing. For example, it has been observed that several large clusters of cytochrome P450 genes colocalize with a known pyrethroid resistance locus involved in oxidative metabolism of the insecticide. Further, a major DDT resistance locus colocalizes with a cluster of eight epsilon (insect-specific) GST genes on chromosome 3R. Biochemical characterization of the members of this GST cluster has shown that only one of these eight enzymes, GSTE2-2, is able to metabolize DDT. Western blots using antibodies raised against this GST indicated that its expression is elevated in a DDT-resistant strain (ZAN/U) of *A. gambiae*, relative to an insecticide-sensitive strain (Kisumu). Subsequent and more comprehensive quantitative PCR studies in these two mosquito strains have indicated that not just GSTE2-2, but in total, five of the eight GSTs genes in this cluster are significantly overexpressed in the DDT-resistant strain. Gene dosage analysis showed no evidence for gene amplification, suggesting perhaps that coexpression of genes in the epsilon cluster is controlled by a common regulatory element.

In addition to metabolic resistance, insensitivity to an insecticide can arise through mutations that disrupt binding of the insecticide molecule to its target protein. A well-characterized example is the insensitivity to two major insecticide families, organophosphates and carbamates, that arises through mutation in their target protein Acetylcholinesterase (AChE). Only one AChE gene is present in *D. melanogaster* (called ace-2) and it is well established that resistance in *D. melanogaster* is caused by mutations at this locus. Two AChE genes, ace-1 and ace-2, have been identified in *A. gambiae*, through sequence searches of the reference genome, using human and *D. melanogaster* AChEs as queries. The *A. gambiae* ace-1 gene shows 52% amino acid identity and the *A. gambiae* ace-2 gene shows 83% amino acid identity to the single *D. melanogaster* AChE, and both *Anopheles* genes contain the conserved FGESAG active site motif. Phylogenetic analysis suggests that the presence of two AChE genes is the ancestral state, and one of these genes has been lost in *D. melanogaster*. It is well established that mutation of the target site of the single *D. melanogaster* AChE gene causes resistance to organophosphates and carbamates in this species, Interestingly, however, in organophosphate and carbamate resistant *Culex pipiens* and *A. gambiae* strains, insensitivity results from a substitution (G119S) near the catalytic site of the ace-1 gene, not a mutation of the ace-2 gene, which is presumably the direct *D. melanogaster* ortholog.

Understanding the common mechanisms of insecticide resistance in *A. gambiae* is of practical importance. The possibility now exists of screening new compounds for tolerance liability by determining whether they cause mutation in known hotspots in genes previously shown to confer resistance or induction of genes with a detoxifying function. Further, specific mutations in resistance-conferring genes can be used as markers to monitor the spread of resistance to insecticides in use in Africa, and can provide important information to help guide ongoing chemical-based vector-control programs.

4.3 Insecticide Targets

While most chemical agents used in the battle against mosquito-borne disease are broad-spectrum agents developed for the purpose of agricultural pest control, the possibility now exists of finding and exploiting physiological and biochemical systems that are unique to the mosquito, in order to produce new and highly selective agents for mosquito control. The reference mosquito genome sequence has helped reveal a variety of systems that may be good candidates for the development of more selective agents for mosquito control. For example, regulatory peptides act as neurochemicals and hormones that guide mosquito reproduction and development. A total of 35 genes encoding putative regulatory peptides have been annotated in the *A. gambiae* genome and agents that mimic or block the action of these peptides may be effective insecticides. Ecdysteroid peptide hormones, which govern gene expression during female reproduction and larval development are of particular interest. Ovary ecdysteroids released after a blood meal stimulate the fat body to begin secreting yolk proteins. The presence of an ortholog of *A. gambiae* ovary ecdysteroid hormone in the mosquito *Aedes aegypti* but the absence of an ortholog in *D. melanogaster* suggests that there is a unique role for this peptide in hematophagous insects and, thus, an opportunity for development of specific antimosquito agents. Similarly, diuresis is a process that has a critical function in the mosquito that may be exploited. After blood feeding, female mosquitoes engage in rapid diuresis to decrease the volume of the blood meal and allow flight. Genes for all four diuretic hormones known to regulate fluid secretion in insects have been identified in the *A. gambiae* genome and present additional targets for incapacitating the mosquito. Also of interest, the receptor for FRMF amide, a cardioexcitatory tetrapeptide first identified in clam, has been identified in the mosquito genome. In functional studies, the tetrapeptide significantly increased the frequency of spontaneous contractions of the heart in mosquito larvae, suggesting the possibility of developing agonists for this receptor as a larvicide.

Mining of reference genome sequences has revealed that enzyme glutathione reductase is absent from Diptera and is functionally substituted by the thioredoxin system. Thioredoxins are small ubiquitous thiol proteins that efficiently cleave disulfide bonds in a number of other proteins and take part in redox control of numerous cellular processes such as protein folding, signaling, and transcription. The key enzyme in this system is thioredoxin reductase, a single copy gene located on the *A. gambiae* X chromosome and shares 52% sequence identity with its human ortholog. The sequence of the catalytic redox center is different between the *A. gambiae* (Thr-Cys-Cys-SerOH) and human (Gly-Cys-selenocysteine-GlyOH) proteins and this difference provides an attractive avenue for the development of a selective inhibitor.

Differences between mammalian and insect metabotropic glutamate receptors have also been revealed by whole-genome comparison. The *D. melanogaster* metabotropic glutamate receptor (mGluR) shares a very similar pharmacological profile with its mammalian orthologs, being activated or inhibited by the same natural or synthetic ligands. However, a new receptor in mGluR subclass has been found in the mosquito, the honeybee *Apis mellifera*, and *D. melanogaster*, called AmXR, HBmXR, and *Dm*XR respectively. No direct orthologs of these novel insect receptors are detectable in *C. elegans* or in human genome. The native ligand for these receptors is unknown, although it has been shown that extract from *D. melanogaster* and *A. gambiae*, but not *C. elegans* or mouse brain, caused dose-dependent receptor activation. Sensitivity to formaldehyde (which masks amino groups) and insensitivity to HCl (which disrupts peptide bonds) suggests that the native ligand is an amino acid–like molecule, but the native ligand is clearly not glutamate, given that residues contacting γ-carboxyl group of glutamate are not conserved with the mGluR, and heterologous expression showed no stimulation by glutamate or mGluR agonists AMPA, kainate, and NMDA quisqualate. Because the XR receptors appear to be insect specific, compounds that disrupt their function may, if lethal, form a promising new class of specific insecticide.

4.4
Immune-response Genes and Genetic Vector Control Strategies

To successfully complete its complex life cycle, the malaria parasite must evade both the human and mosquito immune systems. Parasites ingested with the blood meal develop within the mosquito gut into ookinetes, which subsequently burrow into the gut epithelium and form oocysts. When the oocysts burst, sporozoites are released and these travel to the salivary glands to await transfer to a new host. However, large losses in parasite number occur during this invasion because of mosquito defenses such as melanotic parasite encapsulation. These defenses do not comprise an adaptive immune system, such as that found in mammals, but rather belong to the ancient innate immune system that most metazoans rely on for dealing with invading microorganisms. Facilitating these innate mosquito defenses offers a new and promising strategy for malaria control.

In the past decade, genetic selection in the laboratory has yielded several interesting *A. gambiae* strains, including some that are completely refractory to *Plasmodium*. These strains have been the subject of intensive study in a number of different laboratories, and the *A. gambiae* genome is facilitating the identification of candidate genes for *Plasmodium* resistance. More recently, Christophides et al. screened the *A. gambiae* genome for genes implicated in the innate immune response of the mosquito and identified 242 players from 18 different gene families including, for example, pattern-recognition receptors that bind pathogen-specific molecules, second messengers that can amplify or dampen immune signals, and effector molecules such as the prophenoloxidases that mediate melanization. There is marked diversification of these genes relative to their *Drosophila* homologs, likely reflecting the adaptation of *Anopheles and Drosophila* to different immune challenges represented by different ecological and physiological conditions. Recent functional screening (gene silencing using dsRNA) of a subset of these candidate immune-response genes in *A. gambiae* has yielded compelling findings for several pattern-recognition receptors. Functional knockout of the *A. gambiae* leucine-rich repeat protein (LRIM1) led to a substantial (3.6-fold) increase in oocyst numbers, indicating that this gene has a strong protective effect against the invading parasite. Similarly, functional knockout of the pattern-recognition receptor thioester-containing protein 1 (TEP1) in *A. gambiae* led to a fivefold increase in the number of oocysts developing on the midgut epithelium and a complete abolishment of parasite melanization. Conversely, functional knockout of two different *A. gambiae* C-type lectines (CTL4 and CTLMA2) resulted in a paradoxical enhancement of immune response, with massive melanization of invading ookinetes. It is not clear how these mosquito CTL genes act to protect the developing malaria parasite within the mosquito and no obvious homologs of these genes have been detected in any other organism. LRIM1, TEP1, CTL4, and CTLMA2 are the first mosquito genes to be identified as having an important role in the immune response of the mosquito to the invading malaria parasite. While it is unlikely that gene silencing using dsRNA will be practical as a malaria-control measure, these exciting observations are nonetheless the first steps along the road to the development of a genetic-based transmission-blocking strategy for malaria. Germ-line transformation of the mosquitoes *A. stephensi* and *A. gambiae* has recently been achieved, bringing the possibility of malaria control through the introduction of transgenic mosquitoes another step closer to reality. While genetic control strategies for malaria vectors are complicated by several factors, including the diversity of naturally occurring cytogenetic forms that are reproductively isolated, the selective pressure applied to the *Plasmodium* population, plus all of the usual risks associated with the release of transgenic insects, there is much promise that transgenics will play a key role in malaria control in the coming decades.

See also Genetics, Molecular Basis of; Immunoassays.

Bibliography

Books and Reviews

Adams, M.D., Celniker, S.E., Holt, R.A., Evans, C.A., Gocayne, J.D., Amanatides, P.G., Scherer, S.E., Li, P.W., Hoskins, R.A., Galle, R.F., George, R.A., Lewis, S.E., Richards, S., Ashburner, M., Henderson, S.N., Sutton, G.G., Wortman, J.R., Yandell, M.D., Zhang, Q., Chen, L.X., Brandon, R.C., Rogers, Y.H., Blazej, R.G., Champe, M., Pfeiffer, B.D., Wan, K.H., Doyle, C., Baxter, E.G., Helt, G., Nelson, C.R., Gabor, G.L., Abril, J.F., Agbayani, A., An, H.J., Andrews-Pfannkoch, C., Baldwin, D., Ballew, R.M., Basu, A., Baxendale, J., Bayraktaroglu, L., Beasley, E.M., Beeson, K.Y., Benos, P.V., Berman, B.P., Bhandari, D., Bolshakov, S., Borkova, D., Botchan, M.R., Bouck, J., Brokstein, P., Brottier, P., Burtis, K.C., Busam, D.A., Butler, H., Cadieu, E., Center, A., Chandra, I., Cherry, J.M., Cawley, S., Dahlke, C., Davenport, L.B., Davies, P., de Pablos, B., Delcher, A., Deng, Z., Mays, A.D., Dew, I., Dietz, S.M., Dodson, K., Doup, L.E., Downes, M., Dugan-Rocha, S., Dunkov, B.C., Dunn, P., Durbin, K.J., Evangelista, C.C., Ferraz, C., Ferriera, S., Fleischmann, W., Fosler, C., Gabrielian, A.E., Garg, N.S., Gelbart, W.M., Glasser, K., Glodek, A., Gong, F., Gorrell, J.H., Gu, Z., Guan, P., Harris, M., Harris, N.L., Harvey, D., Heiman, T.J., Hernandez, J.R., Houck, J., Hostin, D., Houston, K.A., Howland, T.J., Wei, M.H., Ibegwam, C., Jalali, M., Kalush, F., Karpen, G.H., Ke, Z., Kennison, J.A., Ketchum, K.A., Kimmel, B.E., Kodira, C.D., Kraft, C., Kravitz, S., Kulp, D., Lai, Z., Lasko, P., Lei, Y., Levitsky, A.A., Li, J., Li, Z., Liang, Y., Lin, X., Liu, X., Mattei, B., McIntosh, T.C., McLeod, M.P., McPherson, D., Merkulov, G., Milshina, N.V., Mobarry, C., Morris, J.,

Moshrefi, A., Mount, S.M., Moy, M., Murphy, B., Murphy, L., Muzny, D.M., Nelson, D.L., Nelson, D.R., Nelson, K.A., Nixon, K., Nusskern, D.R., Pacleb, J.M., Palazzolo, M., Pittman, G.S., Pan, S., Pollard, J., Puri, V., Reese, M.G., Reinert, K., Remington, K., Saunders, R.D., Scheeler, F., Shen, H., Shue, B.C., Siden-Kiamos, I., Simpson, M., Skupski, M.P., Smith, T., Spier, E., Spradling, A.C., Stapleton, M., Strong, R., Sun, E., Svirskas, R., Tector, C., Turner, R., Venter, E., Wang, A.H., Wang, X., Wang, Z.Y., Wassarman, D.A., Weinstock, G.M., Weissenbach, J., Williams, S.M., Woodage, T., Worley, K.C., Wu, D., Yang, S., Yao, Q.A., Ye, J., Yeh, R.F., Zaveri, J.S., Zhan, M., Zhang, G., Zhao, Q., Zheng, L., Zheng, X.H., Zhong, F.N., Zhong, W., Zhou, X., Zhu, S., Zhu, X., Smith, H.O., Gibbs, R.A., Myers, E.W., Rubin, G.M., Venter, J.C. (2000) The genome sequence of *D. melanogaster*, *Science* **287**, 2185–2195.

Besansky, N.J., Powell, J.R. (1992) Reassociation kinetics of *Anopheles gambiae* (Diptera: Culicidae) DNA, *J. Med. Entomol.* **29**, 125–128.

Catteruccia, F., Nolan, T., Loukeris, T.G., Blass, C., Savakis, C., Kafatos, F.C., Crisanti, A. (2000) Stable germline transformation of the malaria mosquito *Anopheles* stephensi, *Nature* **405**, 959–962.

Christophides, G.K., Zdobnov, E., Barillas-Mury, C., Birney, E., Blandin, S., Blass, C., Brey, P.T., Collins, F.H., Danielli, A., Dimopoulos, G., Hetru, C., Hoa, N.T., Hoffmann, J.A., Kanzok, S.M., Letunic, I., Levashina, E.A., Loukeris, T.G., Lycett, G., Meister, S., Michel, K., Moita, L.F., Muller, H.M., Osta, M.A., Paskewitz, S.M., Reichhart, J.M., Rzhetsky, A., Troxler, L., Vernick, K.D., Vlachou, D., Volz, J., von Mering, C., Xu, J., Zheng, L., Bork, P., Kafatos, F.C. (2002) Immunity-related genes and gene families in Anopheles gambiae, *Science* **298**, 159–165.

Coluzzi, M., Sabatini, A., Petrarca, V., Di Deco, M.A. (1979) Chromosomal differentiation and adaptation to human environments in the *Anopheles gambiae* complex, *Trans. R. Soc. Trop. Med. Hyg.* **73**, 483–497.

Hill, C.A., Fox, A.N., Pitts, R.J., Kent, L.B., Tan, P.L., Chrystal, M.A., Cravchik, A., Collins, F.H., Robertson, H.M., Zwiebel, L.J. (2002) G protein-coupled receptors in *Anopheles gambiae*, *Science* **298**, 176–178.

Holt, R.A., Subramanian, G.M., Halpern, A., Sutton, G.G., Charlab, R., Nusskern, D.R., Wincker, P., Clark, A.G., Ribeiro, J.M., Wides, R., Salzberg, S.L., Loftus, B., Yandell, M., Majoros, W.H., Rusch, D.B., Lai, Z., Kraft, C.L., Abril, J.F., Anthouard, V., Arensburger, P., Atkinson, P.W., Baden, H., de Berardinis, V., Baldwin, D., Benes, V., Biedler, J., Blass, C., Bolanos, R., Boscus, D., Barnstead, M., Cai, S., Center, A., Chaturverdi, K., Christophides, G.K., Chrystal, M.A., Clamp, M., Cravchik, A., Curwen, V., Dana, A., Delcher, A., Dew, I., Evans, C.A., Flanigan, M., Grundschober-Freimoser, A., Friedli, L., Gu, Z., Guan, P., Guigo, R., Hillenmeyer, M.E., Hladun, S.L., Hogan, J.R., Hong, Y.S., Hoover, J., Jaillon, O., Ke, Z., Kodira, C., Kokoza, E., Koutsos, A., Letunic, I., Levitsky, A., Liang, Y., Lin, J.J., Lobo, N.F., Lopez, J.R., Malek, J.A., McIntosh, T.C., Meister, S., Miller, J., Mobarry, C., Mongin, E., Murphy, S.D., O'Brochta, D.A., Pfannkoch, C., Qi, R., Regier, M.A., Remington, K., Shao, H., Sharakhova, M.V., Sitter, C.D., Shetty, J., Smith, T.J., Strong, R., Sun, J., Thomasova, D., Ton, L.Q., Topalis, P., Tu, Z., Unger, M.F., Walenz, B., Wang, A., Wang, J., Wang, M., Wang, X., Woodford, K.J., Wortman, J.R., Wu, M., Yao, A., Zdobnov, E.M., Zhang, H., Zhao, Q., Zhao, S., Zhu, S.C., Zhimulev, I., Coluzzi, M., della Torre, A., Roth, C.W., Louis, C., Kalush, F., Mural, R.J., Myers, E.W., Adams, M.D., Smith, H.O., Broder, S., Gardner, M.J., Fraser, C.M., Birney, E., Bork, P., Brey, P.T., Venter, J.C., Weissenbach, J., Kafatos, F.C., Collins, F.H., Hoffman, S.L. (2002) The genome sequence of the malaria mosquito *Anopheles gambiae*, *Science* **298**, 129–149.

Mongin, E., Louis, C., Holt, R.A., Birney, E., Collins, F.H. (2004) The *Anopheles gambiae* genome: an update, *Trends Parasitol.* **20**, 49–52.

Ranson, H., Claudianos, C., Ortelli, F., Abgrall, C., Hemingway, J., Sharakhova, M.V., Unger, M.F., Collins, F.H., Feyereisen, R. (2002) Evolution of supergene families associated with insecticide resistance, *Science* **298**, 179–181.

Zdobnov, E.M., von Mering, C., Letunic, I., Torrents, D., Suyama, M., Copley, R.R., Christophides, G.K., Thomasova, D., Holt, R.A.,

Subramanian, G.M., Mueller, H.M., Dimopoulos, G., Law, J.H., Wells, M.A., Birney, E., Charlab, R., Halpern, A.L., Kokoza, E., Kraft, C.L., Lai, Z., Lewis, S., Louis, C., Barillas-Mury, C., Nusskern, D., Rubin, G.M., Salzberg, S.L., Sutton, G.G., Topalis, P., Wides, R., Wincker, P., Yandell, M., Collins, F.H., Ribeiro, J., Gelbart, W.M., Kafatos, F.C., Bork, P. (2002) Comparative genome and proteome analysis of *Anopheles gambiae* and *Drosophila melanogaster*, *Science* **298**, 149–159.

Primary Literature

Altschul, S.F., Gish, W., Miller, W., Myers, E.W., Lipman, D.J. (1990) Basic local alignment search tool, *J. Mol. Biol.* **215**, 403–410.

Ashburner, M., Ball, C.A., Blake, J.A., Botstein, D., Butler, H., Cherry, J.M., Davis, A.P., Dolinski, K., Dwight, S.S., Eppig, J.T., Harris, M.A., Hill, D.P., Issel-Tarver, L., Kasarskis, A., Lewis, S., Matese, J.C., Richardson, J.E., Ringwald, M., Rubin, G.M., Sherlock, G. (2000) Gene ontology: tool for the unification of biology. The gene ontology consortium, *Nat. Genet.* **25**, 25–29.

Bauer, H., Gromer, S., Urbani, A., Schnolzer, M., Schirmer, R.H., Muller, H.M. (2003) Thioredoxin reductase from the malaria mosquito *Anopheles gambiae, Eur. J. Biochem.* **270**, 4272–4281.

Blandin, S., Shiao, S-H., Moita, L.F., Janse, C.J., Waters, A.P., Kafatos, F.C., Levashina, E.A. (2004) Complement-like protein TEP1 is a determinant of vectorial capacity in the malaria vector *Anopheles gambiae, Cell* **116**, 661–670.

Bryan, J.H., Di Deco, M.A., Petrarca, V., Coluzzi, M. (1982) Inversion polymorphism and incipient speciation in *Anopheles* gambiae s.str. in The Gambia, West Africa, *Genetica* **59**, 167–176.

Collins, F.H., Sakai, R.K., Vernick, K.D., Paskewitz, S., Seeley, D.C., Miller, L.H., Collins, W.E., Campbell, C.C., Gwadz, R.W. (1986) Genetic selection of a Plasmodium-refractory strain of the malaria vector *Anopheles gambiae, Science* **234**, 607–610.

Coluzzi, M., Petrarca, V., Di Deco, M.A. (1985) Chromosomal inversion intergradation and incipient speciation in *Anopheles gambiae, Boll. Zool.* **52**, 45–63.

Coluzzi, M., Sabatini, A., della Torre, A., Di Deco, M.A., Petrarca, V. (2002) A polytene chromosome analysis of the *Anopheles gambiae* species complex, *Science* **298**, 1415–1418.

Ding, Y., Ortelli, F., Rossiter, L.C., Hemingway, J., Ranson, H. (2003) The *Anopheles gambiae* glutathione transferase supergene family: annotation, phylogeny and expression profiles, *BMC Genomics* **4**, 35.

Duttlinger, A., Mispelon, M., Nichols, R. (2003) The structure of the FMRFamide receptor and activity of the cardioexcitatory neuropeptide are conserved in mosquito, *Neuropeptides* **37**, 120–126.

Favia, G., Lanfrancotti, A., Spanos, L., Siden-Kiamos, I., Louis, C. (2001) Molecular characterization of ribosomal DNA polymorphisms discriminating among chromosomal forms of Anopheles gambiae s.s, *Insect Mol. Biol.* **10**, 19–23.

Favia, G., Dimopoulos, G., della Torre, A., Toure, Y.T., Coluzzi, M., Louis, C. (1994) Polymorphisms detected by random PCR distinguish between different chromosomal forms of *Anopheles gambiae, Proc. Natl. Acad. Sci. U.S.A.* **91**, 10315–10319.

Favia, G., della Torre, A., Bagayoko, M., Lanfrancotti, A., Sagnon, N.F., Toure, Y.T., Coluzzi, M. (1997) Molecular identification of sympatric chromosomal forms of Anopheles gambiae and further evidence of their reproductive isolation, *Insect Mol. Biol.* **6**, 377–383.

Fournier, D., Mutero, A., Pralavorio, M., Bride, J.M. (1993) *D. melanogaster* acetylcholinesterase: mechanisms of resistance to organophosphates, *Chem. Biol. Interact.* **87**, 233–238.

Gentile, G., Slotman, M., Ketmaier, V., Powell, J.R., Caccone, A. (2001) Attempts to molecularly distinguish cryptic taxa in *Anopheles gambiae* s.s, *Insect Mol. Biol.* **10**, 25–32.

Gibbs, R.A., Weinstock, G.M., Metzker, M.L., Muzny, D.M., Sodergren, E.J., Scherer, S., Scott, G., Steffen, D., Worley, K.C., Burch, P.E., Okwuonu, G., Hines, S., Lewis, L., DeRamo, C., Delgado, O., Dugan-Rocha, S., Miner, G., Morgan, M., Hawes, A., Gill, R., Holt, R.A., Adams, M.D., Amanatides, P.G., Baden-Tillson, H., Barnstead, M., Chin, S., Evans, C.A., Ferriera, S., Fosler, C., Glodek, A., Gu, Z., Jennings, D., Kraft, C.L., Nguyen, T., Pfannkoch, C.M., Sitter, C., Sutton, G.G., Venter, J.C., Woodage, T., Smith, D., Lee, H.M., Gustafson, E.,

Cahill, P., Kana, A., Doucette-Stamm, L., Weinstock, K., Fechtel, K., Weiss, R.B., Dunn, D.M., Green, E.D., Blakesley, R.W., Bouffard, G.G., De Jong, P.J., Osoegawa, K., Zhu, B., Marra, M., Schein, J., Bosdet, I., Fjell, C., Jones, S., Krzywinski, M., Mathewson, C., Siddiqui, A., Wye, N., McPherson, J., Zhao, S., Fraser, C.M., Shetty, J., Shatsman, S., Geer, K., Chen, Y., Abramzon, S., Nierman, W.C., Havlak, P.H., Chen, R., Durbin, K.J., Egan, A., Ren, Y., Song, X.Z., Li, B., Liu, Y., Qin, X., Cawley, S., Cooney, A.J., D'Souza, L.M., Martin, K., Wu, J.Q., Gonzalez-Garay, M.L., Jackson, A.R., Kalafus, K.J., McLeod, M.P., Milosavljevic, A., Virk, D., Volkov, A., Wheeler, D.A., Zhang, Z., Bailey, J.A., Eichler, E.E., Tuzun, E., Birney, E., Mongin, E., Ureta-Vidal, A., Woodwark, C., Zdobnov, E., Bork, P., Suyama, M., Torrents, D., Alexandersson, M., Trask, B.J., Young, J.M., Huang, H., Wang, H., Xing, H., Daniels, S., Gietzen, D., Schmidt, J., Stevens, K., Vitt, U., Wingrove, J., Camara, F., Mar Alba, M., Abril, J.F., Guigo, R., Smit, A., Dubchak, I., Rubin, E.M., Couronne, O., Poliakov, A., Hubner, N., Ganten, D., Goesele, C., Hummel, O., Kreitler, T., Lee, Y.A., Monti, J., Schulz, H., Zimdahl, H., Himmelbauer, H., Lehrach, H., Jacob, H.J., Bromberg, S., Gullings-Handley, J., Jensen-Seaman, M.I., Kwitek, A.E., Lazar, J., Pasko, D., Tonellato, P.J., Twigger, S., Ponting, C.P., Duarte, J.M., Rice, S., Goodstadt, L., Beatson, S.A., Emes, R.D., Winter, E.E., Webber, C., Brandt, P., Nyakatura, G., Adetobi, M., Chiaromonte, F., Elnitski, L., Eswara, P., Hardison, R.C., Hou, M., Kolbe, D., Makova, K., Miller, W., Nekrutenko, A., Riemer, C., Schwartz, S., Taylor, J., Yang, S., Zhang, Y., Lindpaintner, K., Andrews, T.D., Caccamo, M., Clamp, M., Clarke, L., Curwen, V., Durbin, R., Eyras, E., Searle, S.M., Cooper, G.M., Batzoglou, S., Brudno, M., Sidow, A., Stone, E.A., Payseur, B.A., Bourque, G., Lopez-Otin, C., Puente, X.S., Chakrabarti, K., Chatterji, S., Dewey, C., Pachter, L., Bray, N., Yap, V.B., Caspi, A., Tesler, G., Pevzner, P.A., Haussler, D., Roskin, K.M., Baertsch, R., Clawson, H., Furey, T.S., Hinrichs, A.S., Karolchik, D., Kent, W.J., Rosenbloom, K.R., Trumbower, H., Weirauch, M., Cooper, D.N., Stenson, P.D., Ma, B., Brent, M., Arumugam, M., Shteynberg, D., Copley, R.R., Taylor, M.S., Riethman, H., Mudunuri, U., Peterson, J., Guyer, M., Felsenfeld, A., Old, S., Mockrin, S. and Collins, F. (2004) Genome sequence of the Brown Norway rat yields insights into mammalian evolution, *Nature* **428**, 493–521.

Githeko, A.K., Brandling-Bennett, A.D., Beier, M., Atieli, F., Owaga, M., Collins, F.H. (1992) The reservoir of Plasmodium falciparum malaria in a holoendemic area of western Kenya, *Trans. R. Soc. Trop. Med. Hyg.* **86**, 355–358.

Grossman, G.L., Rafferty, C.S., Clayton, J.R., Stevens, T.K., Mukabayire, O., Benedict, M.Q. (2001) Germline transformation of the malaria vector, Anopheles gambiae, with the piggyBac transposable element, *Insect Mol. Biol.* **10**, 597–604.

Hallem, E.A., Nicole Fox, A., Zwiebel, L.J., Carlson, J.R. (2004) Olfaction: mosquito receptor for human-sweat odorant, *Nature* **427**, 212–213.

Kanzok, S.M., Fechner, A., Bauer, H., Ulschmid, J.K., Muller, H.M., Botella-Munoz, J., Schneuwly, S., Schirmer, R., Becker, K. (2001) Substitution of the thioredoxin system for glutathione reductase in *D. melanogaster*, *Science* **291**, 643–646.

Knols, B.G., de Jong, R., Takken, W. (1995) Differential attractiveness of isolated humans to mosquitoes in Tanzania, *Trans. R. Soc. Trop. Med. Hyg.* **89**, 604–606.

Mason, G.F. (1967) Genetic studies on mutations in species A and B of the *Anopheles gambiae* complex, *Genet. Res.* **10**, 205–217.

Mitri, C., Parmentier, M.L., Pin, J.P., Bockaert, J., Grau, Y. (2004) Divergent evolution in metabotropic glutamate receptors. A new receptor activated by an endogenous ligand different from glutamate in insects, *J. Biol. Chem.* **279**, 9313–9320.

Mukabayire, O., Besansky, N.J. (1996) Distribution of T1, Q, Pegasus and mariner transposable elements on the polytene chromosomes of PEST, a standard strain of *Anopheles gambiae*, *Chromosoma* **104**, 585–595.

Mukabayire, O., Caridi, J., Wang, X., Toure, Y.T., Coluzzi, M., Besansky, N.J. (2001) Patterns of DNA sequence variation in chromosomally recognized taxa of *Anopheles gambiae*: evidence from rDNA and single-copy loci, *Insect Mol. Biol.* **10**, 33–46.

Myers, E.W., Sutton, G.G., Delcher, A.L., Dew, I.M., Fasulo, D.P., Flanigan, M.J., Kravitz, S.A., Mobarry, C.M., Reinert, K.H., Remington, K.A., Anson, E.L., Bolanos, R.A.,

Chou, H.H., Jordan, C.M., Halpern, A.L., Lonardi, S., Beasley, E.M., Brandon, R.C., Chen, L., Dunn, P.J., Lai, Z., Liang, Y., Nusskern, D.R., Zhan, M., Zhang, Q., Zheng, X., Rubin, G.M., Adams, M.D., Venter, J.C. (2000) A whole-genome assembly of *D. melanogaster*, *Science* **287**, 2196–2204.

Olson, M.V. (1999) When less is more: gene loss as an engine of evolutionary change, *Am. J. Hum. Genet.* **64**, 18–23.

Osta, M.A., Christophides, G.K., Kafatos, F.C. (2004) Effects of mosquito genes on plasmodium development, *Science* **303**, 2030–2032.

Ortelli, F., Rossiter, L.C., Vontas, J., Ranson, H., Hemingway, J. (2003) Heterologous expression of four glutathione transferase genes genetically linked to a major insecticide-resistance locus from the malaria vector *Anopheles gambiae*, *Biochem. J.* **373**, 957–963.

Price, D.A., Greenberg, M.J. (1977) Structure of a molluscan cardioexcitatory neuropeptide, *Science* **197**, 670–671.

Ribeiro, J.M. (2003) A catalogue of *Anopheles gambiae* transcripts significantly more or less expressed following a blood meal, *Insect Biochem. Mol. Biol.* **33**, 865–882.

Riehle, M.A., Garczynski, S.F., Crim, J.W., Hill, C.A., Brown, M.R. (2002) Neuropeptides and peptide hormones in *Anopheles gambiae*, *Science* **298**, 172–175.

Rubin, G.M., Yandell, M.D., Wortman, J.R., Gabor Miklos, G.L., Nelson, C.R., Hariharan, I.K., Fortini, M.E., Li, P.W., Apweiler, R., Fleischmann, W., Cherry, J.M., Henikoff, S., Skupski, M.P., Misra, S., Ashburner, M., Birney, E., Boguski, M.S., Brody, T., Brokstein, P., Celniker, S.E., Chervitz, S.A., Coates, D., Cravchik, A., Gabrielian, A., Galle, R.F., Gelbart, W.M., George, R.A., Goldstein, L.S., Gong, F., Guan, P., Harris, N.L., Hay, B.A., Hoskins, R.A., Li, J., Li, Z., Hynes, R.O., Jones, S.J., Kuehl, P.M., Lemaitre, B., Littleton, J.T., Morrison, D.K., Mungall, C., O'Farrell, P.H., Pickeral, O.K., Shue, C., Vosshall, L.B., Zhang, J., Zhao, Q., Zheng, X.H., Lewis, S. (2000) Comparative genomics of the eukaryotes, *Science* **287**, 2204–2215.

Shahabuddin, M., Pimenta, P.F. (1998) Plasmodium gallinaceum preferentially invades vesicular ATPase-expressing cells in Aedes aegypti midgut, *Proc. Natl. Acad. Sci. U.S.A.* **95**, 3385–3389.

Takken, W., Knols, B.G. (1999) Odor-mediated behavior of Afrotropical malaria mosquitoes, *Annu. Rev. Entomol.* **44**, 131–157.

Toure, Y.T., Petrarca, V., Traore, S.F., Coulibaly, A., Maiga, H.M., Sankare, O., Sow, M., Di Deco, M.A., Coluzzi, M. (1994) Ecological genetic studies in the chromosomal form Mopti of *Anopheles gambiae* s.str. in Mali, West Africa, *Genetica* **94**, 213–223.

Toure, Y.T., Petrarca, V., Traore, S.F., Coulibaly, A., Maiga, H.M., Sankare, O., Sow, M., Di Deco, M.A., Coluzzi, M. (1998) The distribution and inversion polymorphism of chromosomally recognized taxa of the *Anopheles gambiae* complex in Mali, West Africa, *Parassitologia* **40**, 477–511.

Venter, J.C., Adams, M.D., Myers, E.W., Li, P.W., Mural, R.J., Sutton, G.G., Smith, H.O., Yandell, M., Evans, C.A., Holt, R.A., Gocayne, J.D., Amanatides, P., Ballew, R.M., Huson, D.H., Wortman, J.R., Zhang, Q., Kodira, C.D., Zheng, X.H., Chen, L., Skupski, M., Subramanian, G., Thomas, P.D., Zhang, J., Gabor Miklos, G.L., Nelson, C., Broder, S., Clark, A.G., Nadeau, J., McKusick, V.A., Zinder, N., Levine, A.J., Roberts, R.J., Simon, M., Slayman, C., Hunkapiller, M., Bolanos, R., Delcher, A., Dew, I., Fasulo, D., Flanigan, M., Florea, L., Halpern, A., Hannenhalli, S., Kravitz, S., Levy, S., Mobarry, C., Reinert, K., Remington, K., Abu-Threideh, J., Beasley, E., Biddick, K., Bonazzi, V., Brandon, R., Cargill, M., Chandramouliswaran, I., Charlab, R., Chaturvedi, K., Deng, Z., Di Francesco, V., Dunn, P., Eilbeck, K., Evangelista, C., Gabrielian, A.E., Gan, W., Ge, W., Gong, F., Gu, Z., Guan, P., Heiman, T.J., Higgins, M.E., Ji, R.R., Ke, Z., Ketchum, K.A., Lai, Z., Lei, Y., Li, Z., Li, J., Liang, Y., Lin, X., Lu, F., Merkulov, G.V., Milshina, N., Moore, H.M., Naik, A.K., Narayan, V.A., Neelam, B., Nusskern, D., Rusch, D.B., Salzberg, S., Shao, W., Shue, B., Sun, J., Wang, Z., Wang, A., Wang, X., Wang, J., Wei, M., Wides, R., Xiao, C., Yan, C., Yao, A., Ye, J., Zhan, M., Zhang, W., Zhang, H., Zhao, Q., Zheng, L., Zhong, F., Zhong, W., Zhu, S., Zhao, S., Gilbert, D., Baumhueter, S., Spier, G., Carter, C., Cravchik, A., Woodage, T., Ali, F., An, H., Awe, A., Baldwin, D., Baden, H., Barnstead, M., Barrow, I., Beeson, K., Busam, D., Carver, A., Center, A., Cheng, M.L., Curry, L., Danaher, S., Davenport, L., Desilets, R., Dietz, S., Dodson, K.,

Doup, L., Ferriera, S., Garg, N., Gluecksmann, A., Hart, B., Haynes, J., Haynes, C., Heiner, C., Hladun, S., Hostin, D., Houck, J., Howland, T., Ibegwam, C., Johnson, J., Kalush, F., Kline, L., Koduru, S., Love, A., Mann, F., May, D., McCawley, S., McIntosh, T., McMullen, I., Moy, M., Moy, L., Murphy, B., Nelson, K., Pfannkoch, C., Pratts, E., Puri, V., Qureshi, H., Reardon, M., Rodriguez, R., Rogers, Y.H., Romblad, D., Ruhfel, B., Scott, R., Sitter, C., Smallwood, M., Stewart, E., Strong, R., Suh, E., Thomas, R., Tint, N.N., Tse, S., Vech, C., Wang, G., Wetter, J., Williams, S., Williams, M., Windsor, S., Winn-Deen, E., Wolfe, K., Zaveri, J., Zaveri, K., Abril, J.F., Guigo, R., Campbell, M.J., Sjolander, K.V., Karlak, B., Kejariwal, A., Mi, H., Lazareva, B., Hatton, T., Narechania, A., Diemer, K., Muruganujan, A., Guo, N., Sato, S., Bafna, V., Istrail, S., Lippert, R., Schwartz, R., Walenz, B., Yooseph, S., Allen, D., Basu, A., Baxendale, J., Blick, L., Caminha, M., Carnes-Stine, J., Caulk, P., Chiang, Y.H., Coyne, M., Dahlke, C., Mays, A., Dombroski, M., Donnelly, M., Ely, D., Esparham, S., Fosler, C., Gire, H., Glanowski, S., Glasser, K., Glodek, A., Gorokhov, M., Graham, K., Gropman, B., Harris, M., Heil, J., Henderson, S., Hoover, J., Jennings, D., Jordan, C., Jordan, J., Kasha, J., Kagan, L., Kraft, C., Levitsky, A., Lewis, M., Liu, X., Lopez, J., Ma, D., Majoros, W., McDaniel, J., Murphy, S., Newman, M., Nguyen, T., Nguyen, N., Nodell, M., Pan, S., Peck, J., Peterson, M., Rowe, W., Sanders, R., Scott, J., Simpson, M., Smith, T., Sprague, A., Stockwell, T., Turner, R., Venter, E., Wang, M., Wen, M., Wu, D., Wu, M., Xia, A., Zandieh, A., Zhu, X. (2001) The sequence of the human genome, *Science* **291**, 1304–1351.

Vernick, K.D., Fujioka, H., Seeley, D.C., Tandler, B., Aikawa, M., Miller, L.H. (1995) Plasmodium gallinaceum: a refractory mechanism of ookinete killing in the mosquito, *Anopheles gambiae*, *Exp. Parasitol.* **80**, 583–595.

Waterston, R.H., Lindblad-Toh, K., Birney, E., Rogers, J., Abril, J.F., Agarwal, P., Agarwala, R., Ainscough, R., Alexandersson, M., An, P., Antonarakis, S.E., Attwood, J., Baertsch, R., Bailey, J., Barlow, K., Beck, S., Berry, E., Birren, B., Bloom, T., Bork, P., Botcherby, M., Bray, N., Brent, M.R., Brown, D.G., Brown, S.D., Bult, C., Burton, J., Butler, J., Campbell, R.D., Carninci, P., Cawley, S., Chiaromonte, F., Chinwalla, A.T., Church, D.M., Clamp, M., Clee, C., Collins, F.S., Cook, L.L., Copley, R.R., Coulson, A., Couronne, O., Cuff, J., Curwen, V., Cutts, T., Daly, M., David, R., Davies, J., Delehaunty, K.D., Deri, J., Dermitzakis, E.T., Dewey, C., Dickens, N.J., Diekhans, M., Dodge, S., Dubchak, I., Dunn, D.M., Eddy, S.R., Elnitski, L., Emes, R.D., Eswara, P., Eyras, E., Felsenfeld, A., Fewell, G.A., Flicek, P., Foley, K., Frankel, W.N., Fulton, L.A., Fulton, R.S., Furey, T.S., Gage, D., Gibbs, R.A., Glusman, G., Gnerre, S., Goldman, N., Goodstadt, L., Grafham, D., Graves, T.A., Green, E.D., Gregory, S., Guigo, R., Guyer, M., Hardison, R.C., Haussler, D., Hayashizaki, Y., Hillier, L.W., Hinrichs, A., Hlavina, W., Holzer, T., Hsu, F., Hua, A., Hubbard, T., Hunt, A., Jackson, I., Jaffe, D.B., Johnson, L.S., Jones, M., Jones, T.A., Joy, A., Kamal, M., Karlsson, E.K., Karolchik, D., Kasprzyk, A., Kawai, J., Keibler, E., Kells, C., Kent, W.J., Kirby, A., Kolbe, D.L., Korf, I., Kucherlapati, R.S., Kulbokas, E.J., Kulp, D., Landers, T., Leger, J.P., Leonard, S., Letunic, I., Levine, R., Li, J., Li, M., Lloyd, C., Lucas, S., Ma, B., Maglott, D.R., Mardis, E.R., Matthews, L., Mauceli, E., Mayer, J.H., McCarthy, M., McCombie, W.R., McLaren, S., McLay, K., McPherson, J.D., Meldrim, J., Meredith, B., Mesirov, J.P., Miller, W., Miner, T.L., Mongin, E., Montgomery, K.T., Morgan, M., Mott, R., Mullikin, J.C., Muzny, D.M., Nash, W.E., Nelson, J.O., Nhan, M.N., Nicol, R., Ning, Z., Nusbaum, C., O'Connor, M.J., Okazaki, Y., Oliver, K., Overton-Larty, E., Pachter, L., Parra, G., Pepin, K.H., Peterson, J., Pevzner, P., Plumb, R., Pohl, C.S., Poliakov, A., Ponce, T.C., Ponting, C.P., Potter, S., Quail, M., Reymond, A., Roe, B.A., Roskin, K.M., Rubin, E.M., Rust, A.G., Santos, R., Sapojnikov, V., Schultz, B., Schultz, J., Schwartz, M.S., Schwartz, S., Scott, C., Seaman, S., Searle, S., Sharpe, T., Sheridan, A., Shownkeen, R., Sims, S., Singer, J.B., Slater, G., Smit, A., Smith, D.R., Spencer, B., Stabenau, A., Stange-Thomann, N., Sugnet, C., Suyama, M., Tesler, G., Thompson, J., Torrents, D., Trevaskis, E., Tromp, J., Ucla, C., Ureta-Vidal, A., Vinson, J.P., Von Niederhausern, A.C., Wade, C.M., Wall, M., Weber, R.J., Weiss, R.B., Wendl, M.C., West,

A.P., Wetterstrand, K., Wheeler, R., Whelan, S., Wierzbowski, J., Willey, D., Williams, S., Wilson, R.K., Winter, E., Worley, K.C., Wyman, D., Yang, S., Yang, S.P., Zdobnov, E.M., Zody, M.C., Lander, E.S. (2002) Initial sequencing and comparative analysis of the mouse genome, *Nature* **420**, 520–562.

Weill, M., Lutfalla, G., Mogensen, K., Chandre, F., Berthomieu, A., Berticat, C., Pasteur, N., Philips, A., Fort, P., Raymond, M. (2003) Comparative genomics: insecticide resistance in mosquito vectors, *Nature* **423**, 136–137.

Yeates, D.K., Wiegmann, B.M. (1999) Congruence and controversy: toward a higher-level phylogeny of Diptera, *Annu. Rev. Entomol.* **44**, 397–428.

Zdobnov, E.M., Apweiler, R. (2001) InterProScan: an integration platform for the signature-recognition methods in InterPro, *Bioinformatics* **17**, 847–848.

Zheng, L., Cornel, A.J., Wang, R., Erfle, H., Voss, H., Ansorge, W., Kafatos, F.C., Collins, F.H. (1997) Quantitative trait loci for refractoriness of *Anopheles gambiae* to Plasmodium cynomolgi B, *Science* **276**, 425–428.

Zieler, H., Nawrocki, J.P., Shahabuddin, M. (1999) Plasmodium gallinaceum ookinetes adhere specifically to the midgut epithelium of Aedes aegypti by interaction with a carbohydrate ligand, *J. Exp. Biol.* **202**(Pt 5), 485–495.

Male Reproductive System: Testis Development and Spermatogenesis

Kate A.L. Loveland and David M. de Kretser
Monash Institute of Reproduction and Development, Monash University, Clayton, Victoria, Australia and The Australian Research Council Centre of Excellence in Biotechnology and Development

Encyclopedia of Molecular Cell Biology and Molecular Medicine, 2nd Edition. Volume 7
Edited by Robert A. Meyers.

ISBN: 3-527-30549-1

Keywords

Gonocyte
A diploid male germ cell in fetal testis that undergoes mitosis and then becomes quiescent until the onset of spermatogenesis at puberty after birth.

Leydig Cell
A somatic cell producing testosterone, found in the interstitium between seminiferous tubules.

Peritubular Myoid Cell
A somatic cell type with contractile capacity; forms a layer around each seminiferous tubule and contributes to its basement membrane.

Primordial Germ Cell
Indifferent germ cells that exist before the onset of gonadal differentiation in a fetal testis or ovary.

Sertoli Cell
Somatic cells that surround and nourish the developing male germ cells that form the scaffolding of the seminiferous epithelium.

Spermatid
A haploid male germ cell that is transformed from a round shape to an elongated shape during spermiogenesis, as nuclear remodeling, acrosome formation, and tail elongation occur.

Spermatocyte
The male germ cell undergoing meiosis I and II; may be tetraploid (primary spermatocyte) or diploid (secondary spermatocyte).

Spermatogonium
A diploid male germ cell that can undergo mitosis.

Spermatozoon
A motile, elongated, haploid male gamete released from the seminiferous epithelium.

The determination of gonadal sex establishes the basic structure of the testis. This results in the formation of seminiferous cords containing immature Sertoli cells and gonocytes, with the testosterone-secreting Leydig cells lying within the intertubular region. Gonadotrophic stimulation of the testis and the resultant testosterone secretion stimulates spermatogenesis, which involves mitosis, meiosis, and the production of spermatozoa. Germ cell function is crucially dependent on the Sertoli cells that control the intratubular environment, partly by forming the blood–testis barrier. The biochemical nature of the signals that regulate somatic cell function and germ cell maturation is now partially understood, though the complexities of the regulatory pathways and their complete significance are yet to be clarified. This brief overview of the current state of knowledge regarding the controlling signals presents some aspects of their impact on the target cells, both somatic and germinal.

1 Testicular Architecture, Cellular Composition, and Principles

1.1 The Adult Testis

The synthesis of spermatozoa is initiated during puberty and is continuous throughout adult life. Male fertility is dependent on the maintenance of a self-renewing stem cell population and a complex set of signals derived from hormonal and local factors that drive spermatogenesis in a highly regulated manner.

1.1.1 Two Functional Compartments

The adult mammalian testis consists of two structurally distinct but functionally interrelated compartments: the *seminiferous* epithelium, wherein germ cells divide and differentiate surrounded by the Sertoli "nurse" cells, and the *interstitial compartment*. Spermatogenesis occurring within the seminiferous tubules can be divided into three processes: (1) replication of stem cells by mitosis (primordial germ cells in the fetus, gonocytes in the juvenile, and spermatogonia in the sexually mature individual); (2) reduction of chromosomal number by the process of meiosis (primary and secondary spermatocytes); and (3) a complex metamorphosis, called *spermiogenesis,* involving the transformation of round cells (spermatids) into elongated spermatozoa.

The developing germ cells, other than spermatogonia, are enveloped by Sertoli cells, which in turn have access to the interstitial space that contains the Leydig cells,

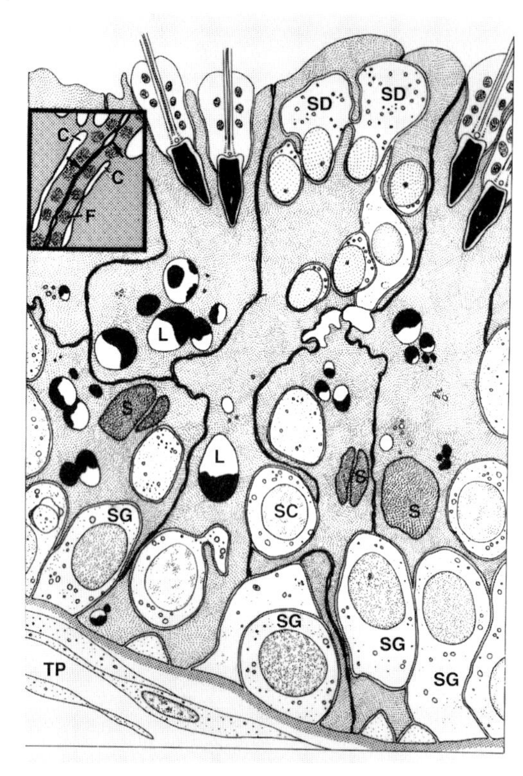

Fig. 1 Diagrammatic representation of the epithelium of the seminiferous tubule. Illustrated are spermatogonia (SG), primary spermatocytes (SC), spermatids (SD), mature spermatozoa, and Sertoli cells (S) with lipid inclusions (L). The peritubular myoid cells lie outside the tubule in the tunica propria (TP). The Leydig cells and vasculature lie external to the tunica propria. *Inset*: organization of the tight junctions between Sertoli cells that form the blood–testis barrier.

peritubular myoid cells, macrophages, and the vasculature fluids (Fig. 1).

The adjacent Sertoli cells are connected by a unique cluster of junctional specializations comprised of tight and gap junctions at their base. This junction divides the seminiferous epithelium into two functional compartments, apical and basal. Mitotic germ cells are in the basal compartment, and they cross the tight junction as they enter their meiotic phase of differentiation. Within the apical compartment of the seminiferous epithelium, the spermatocytes and spermatids are exclusively reliant on Sertoli cell products and are also invisible to the surveillance of the immune system, which might otherwise view them as "foreign" cells. Differentiating germ cells are connected to their clonal siblings by cytoplasmic bridges until the time of their release as spermatozoa into the tubule lumen as the result of incomplete cytokinesis. Thus they remain functionally diploid even when their nuclear compartments contain only one parental copy of each chromosome.

It is important to recognize that the seminiferous tubule is an avascular compartment. Intercellular transport into the epithelium cannot occur because there are tight junctions between adjacent Sertoli cells luminal to the spermatogonial layer. Consequently, germ cells other than spermatogonia are dependent on the Sertoli cells for transport and for the regulation of the internal milieu of the tubule.

1.1.2 Spermatogenesis is Cyclical

The process of spermatogenesis is highly organized and ordered, with precise

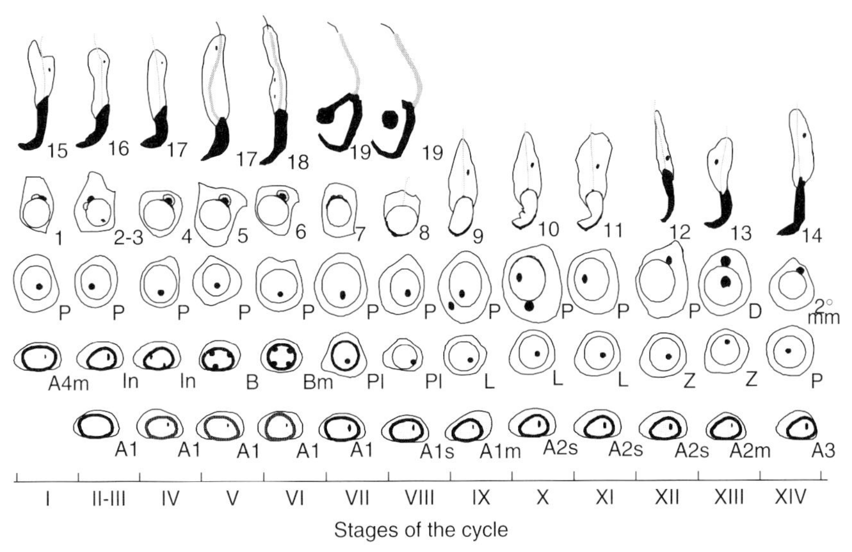

Fig. 2 Stages of the cycle of the seminiferous epithelium observed in the adult rat testis. Each column represents the invariant set of germ cell maturation types present simultaneously within a cross-section of a seminiferous tubule. The least mature germ cell types, present at the base of the seminiferous epithelium, (stem cells $\{A_s\}$, A_{pr}, A_{al}, A_8 and A_{16} spermatogonia) are infrequently observed and are not illustrated here. Type A_1, A_2, A_3 and A_4 spermatogonia (A_1–A_4) are morphologically indistinguishable, while Intermediate (In) and Type B spermatogonia have distinctive heterochromatin. Stages of spermatocytes : preleptotene (PL), leptotene (L), zygotene (Z), pachytene (P), diplotene (D) and metaphase (2°mm). Spermatid steps are indicated by numbers 1 through 19, with 1 to 8 being round spermatids and 9 to 19 elongating spermatids. Adapted from Russell et al. 1990.

associations of germ cells at specific stages of maturation (Fig. 2), creating invariant cycles within the epithelium and waves of stages along the tubule. Although disruption of germ cell development can be induced experimentally, leading to an arrest at discrete stages of spermatogenesis, the normal arrangement of cellular associations is rapidly reestablished upon recovery. The mechanisms that maintain these highly ordered patterns are poorly understood, though they are believed to be essential for attainment of the full spermatogenic potential of the testis. The underlying basis for the spermatogenic wave appears to be established in fetal life, as illustrated by the cyclic expression of the galectin gene in fetal Sertoli cells, prior to the onset of germ cell differentiation. That germ cells can in turn influence the cyclic function of the adult seminiferous epithelium has been demonstrated by interspecific transfer of rat sperm into the mouse testis; the rat sperm specifies the timing of the epithelial cycle when embedded in the mouse Sertoli cell.

The preceding description indicates that the testis contains germ cells and somatic cells involved in discrete, characteristic patterns of macromolecular synthesis. This arises from the steps in spermatogenesis that require replication of stem cells by mitosis, the reduction of chromosomal number and DNA content to the haploid state, and the complex processes constituting spermiogenesis. The latter requires specific protein synthesis by the haploid genome and the cessation of transcription as the chromatin condenses to form the sperm head. These events and their implications are emphasized in the following sections.

1.2 The Developing Testis

1.2.1 The Fetal Testis

The indifferent gonad The gonads arise from urogenital ridges that form on either side of the midline of the developing embryo. Formed from a thickening of the coelomic epithelium on the surface of the mesonephros, the gonad is initially devoid of germ cells and has the potential to develop as either a testis or an ovary, and hence it is termed *indifferent*. The migration of cells into this area contributes to both somatic (fetal Leydig cell and peritubular myoid cell lineages from the mesonephros; some Sertoli cells from the coelomic epithelium) and germ cell (primordial germ cells from the endoderm of the yolk sac near the allantoic invagination) lineages.

The indifferent gonad is present during approximately weeks 4 to 7 postcoitum in the human and 9 to 12 days postcoitum (dpc) in the mouse. Establishment of this structure requires expression of steroidogenic factor 1 (SF-1; a transcription factor), the Wilm's tumor 1 (WT1) protein isoforms, and fibroblast growth factor-9 (Fgf-9). For each of these, the absence of the gene results in initial formation of the indifferent gonad in the mouse embryo, but the tissue degenerates and does not support germ cell survival.

Genes controlling testis formation The primary molecular trigger for gonadal sex determination is SRY (sex-determining region on the Y chromosome) gene expression, observed in the mouse gonad at 10.5 to 12.5 dpc within the somatic cells that are precursors to Sertoli cells. Introduction of the mouse SRY gene onto a mouse somatic chromosome through transgenesis produced sex-reversed male mice, that is, XX males, having male external genitalia, male sexual behavior, and testes that lack germ cells. Thus, SRY expression directs testis formation but is not sufficient to provide the environment required for spermatogenesis. A highly conserved portion of this gene encodes an HMG (high mobility group) box, characteristic of a family of DNA-binding proteins, and mutations in this region of SRY can cause sex reversal in mouse and humans. The definitive identification of SRY target genes is yet to be achieved.

The first visible sign of testicular differentiation occurs at day 12.5 postcoitum in mouse, with the organization of Sertoli cells into cords surrounding the male primordial germ cells, now termed *gonocytes*. In the female, the initiation of germ cell meiosis at day 13.5 postcoitum marks ovarian development.

One likely downstream target of SRY is SOX9, and, similar to SRY, expression of SOX9 in pre-Sertoli cells is sufficient to drive development of an XX gonad to form a testis. Also containing an HMG box, SOX9 is present in both male and female gonads when indifferent, but it is upregulated in males and downregulated in females following SRY expression onset. As testis differentiation begins, SOX9 moves to the nucleus to effect changes in transcription of downstream genes. SOX9 may switch on the genes required to form testicular cords, and it is known to induce expression of Müllerian Inhibiting Substance

(MIS). The growth factor MIS is produced by fetal Sertoli cells during cord formation, and it mediates regression of the female duct components (see section on Mullerian inhibiting substance). The MIS-gene promoter includes a binding site for SF-1. Expression of fgf9 in the mouse male gonad begins immediately after SRY, and it is required for normal Sertoli cell differentiation, germ cell proliferation, and migration of mesonephric cells into the developing gonad. The expression of neurotropin-3 (NT3) in Sertoli cells appears to regulate mesonephric cell migration and can positively regulate SOX9 expression.

The germ cells are directed to form either gonocytes or oocytes by signals from their somatic cell environment. This commitment occurs in the testis between 11.5 and 12.5 dpc, and, in the ovary, with the commencement of meiosis at 13.5 dpc. Evidence of a positive feedback loop is derived from the production of prostaglandin D2 by germ cells, which will stimulate XX somatic cells to synthesize MIS *in vitro* and form cords, indicative of their differentiation into the Sertoli cell lineage.

This period of development is classically held to be gonadotrophin-independent. The subsequent emergence of a fetal-type Leydig cell population depends on expression of the type A platelet-derived growth factor (PDGF) receptor, and retinoic acid has been implicated in the formation of cords by affecting laminin deposition. These data highlight the concept that, during fetal life, a diversity of locally derived signals are required in order to orchestrate development of the full complement of somatic cell lineages required for the support of spermatogenic cells.

The testis continues to grow through fetal life while the gonocyte remains quiescent. The rate of Sertoli cell proliferation in the rodent peaks just before birth, and the interstitial cell populations continue to mature and multiply.

1.2.2 Neonatal and Juvenile Somatic and Germ Cell Differentiation

The onset of spermatogenesis occurs in a testis that contains quiescent germ cells engulfed by proliferating Sertoli cells in the testis cords (Fig. 3). The cords are surrounded by layers of immature myoid cells, which, along with the Sertoli cells, contribute to the expanding and changing basement membrane at the cord perimeter. Nests of fetal Leydig cells are evident. Within a few days after birth in rodents, and in the prepubertal period, the germ cells are stimulated to reenter the cell cycle and migrate to contact the cord basement membrane. The gonocytes that do not migrate will die by apoptosis; those that do migrate will form the stem cell population that is the progenitor of all adult spermatogenesis. The stimulus for this differentiation, termed *the first wave of spermatogenesis*, appears to be a combination of hormonal (e.g. follicle-stimulating hormone; FSH) and growth-factor (e.g. transforming growth factor β superfamily members) signaling to both germ cells and somatic cells. In the rat testis, germ cell migration is mediated by cell adhesion (N-CAM) and growth factor signaling, with Sertoli cell–derived stem cell factor (SCF) interacting with the c-kit tyrosine kinase receptor on the maturing germ cells (Sect. 3.3.2). The situation in the mouse may be slightly different, and perhaps independent of SCF signaling at this step, with the expansion of this stem cell population being profoundly affected by the presence of glial-derived neurotropic factor (GDNF). Other factors implicated in events at this time include NT3, nerve growth factor

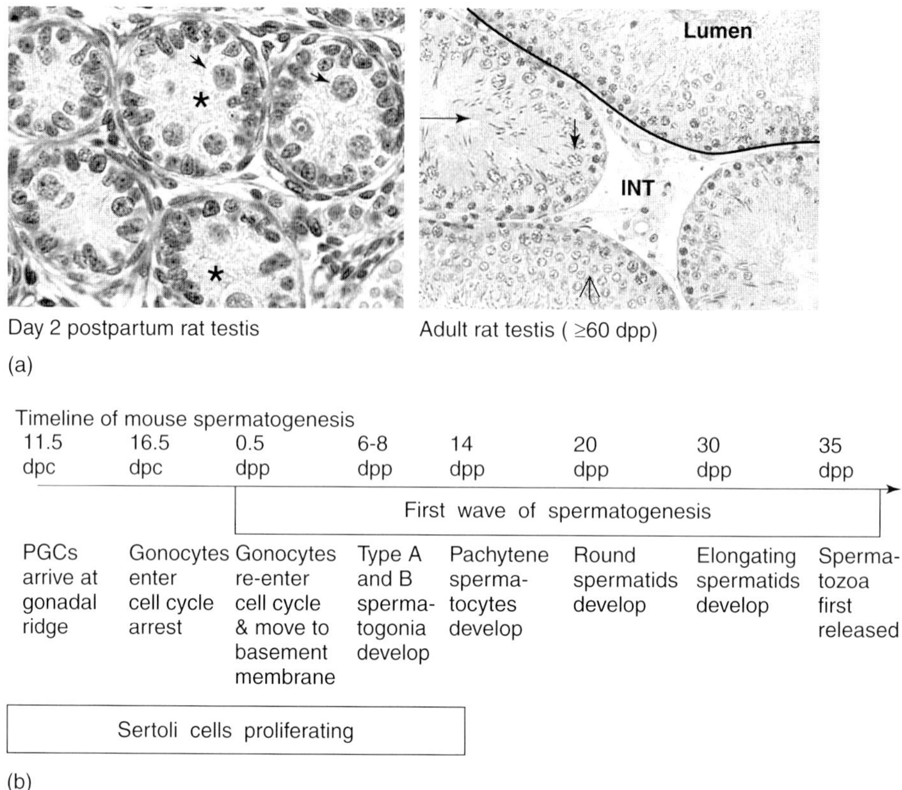

Fig. 3 Progressive changes in testis during fetal and postnatal development. Micrographs (a) and timeline (b) illustrating changes in the developing rodent testis. (a) newborn rat testis is comprised of cords (shown in cross-section) surrounded by maturing peritubular myoid cells. Gonocytes (arrows) are the only germ cell type present; these are surrounded by Sertoli cell cytoplasm (asterisk). Adult testis tubules (shown in cross-section) are surrounded by flattened layer of peritubular myoid cells; line indicates tubule boundary. Interstitium (INT) contains Leydig cells, macrophages, lymphatics, and blood vessels. Pachytene spermatocytes (arrowheads), round spermatids (open arrows) and elongating spermatids (dashed arrow) are progressively closer to the tubule lumen where mature spermatozoa are released. (b) timeline of murine male germ cell developmental events through to initial production of spermatozoa. Adapted from McCarrey (1993).

(NGF), epidermal growth factor (EGF), PDGF and follistatin (an activin and BMP antagonist).

Examining the rodent testis at discrete time points after birth allows one to observe the appearance of progressively maturing germ cell types. (Fig. 3). In contrast, these time points are highly variable in primates.

Somatic cell differentiation is an important feature of the first wave of spermatogenesis. The cessation of Sertoli cell division during this interval establishes the full complement of these cells that are present throughout the individual's life. This is of crucial importance for establishing the maximum capacity for adult sperm production (discussed further in Sect. 3.1.2). Sertoli cell proliferation is controlled by interactions between hormones and growth factors, with FSH, thyroid hormone, GDNF, and activin representing some key *in vivo* modulators. A new Leydig cell population arises and differentiates in the interstitium, and the cells surrounding the cords gradually flatten to form the thin layer of peritubular myoid cells observed in the adult testis. Once again, the differentiation and function of these cells appear to respond to a network of hormonal and local factor cues.

To date, a limited number of genes have been identified that show differential expression within a particular cell type between the immature (i.e. first wave) testis and the adult testis. This presumably reflects the changing nature of the seminiferous epithelium with the onset of germ cell, Sertoli cell and peritubular myoid cell differentiation. For example, mRNAs encoding the prosurvival members of the bcl-2 family, bcl-2 and bcl-$_{XL}$, are present in spermatogonia and Sertoli cells during the first two postnatal weeks of mouse testis development but are absent thereafter. The bcl-2 mRNA is switched off entirely, while bcl-$_{XL}$ shifts to the newly emerging spermatocyte population. Such differences are not surprising within the Sertoli cells that transform from mitotic to terminally differentiated cells. Less predictable was the observation that spermatogonia during the first wave of spermatogenesis are functionally different from those present in adulthood. These findings emphasize the notion that germ cell mRNA and protein synthesis are governed by changing environmental cues, including those from the adjacent clones of the progressively more and less mature germ cell clones within the seminiferous epithelium.

2 Regulation of Gene Expression During Spermatogenesis

2.1 DNA

During the preleptotene stage of Meiosis I, there is a peak of DNA synthesis coincident with the peak in DNA polymerase activity that transforms the diploid spermatogonium into a tetraploid spermatocyte. A low level of DNA synthesis documented in pachytene spermatocytes is associated with the DNA repair that follows recombination between homologous chromosomes. DNA methylation patterns vary, with some single-copy genes being methylated at normal levels, some regions of repetitive DNA sequences being under-methylated, and others, such as the genes encoding the chromatin-associated protamine and transition proteins, and the germ cell–specific phosphoglycerate kinase (PGK-2) undergoing changes in methylation during spermatogenesis.

Chromatin remodeling occurs in the haploid germ cells to enable DNA compaction in spermatozoa. The chromosomes become highly organized in hairpin structures of DNA arranged into looped domains along a track of nuclear matrix, with the centromeres found in the center and the telomeres at the perimeter of the sperm nucleus. During spermiogenesis, the histones that package genomic DNA in somatic cells, spermatogonia, and spermatocytes are replaced by transition proteins that are in turn replaced by protamines. The protamines are small, basic proteins represented by 1 or 2 family members. In mouse, both protamine-1 and protamine-2 are required for normal sperm formation. A relatively small proportion of sperm chromatin remains associated with a nucleohistone; in humans, this accounts for about 15% of sperm DNA. Association of DNA with histones has been proposed to reflect the involvement of genes in these regions with the early postfertilization events.

Inactivation of the X chromosome occurs during meiosis, but it is incomplete. In particular, the Xp terminus remains active, and includes genes encoding the GM-CSF receptor, mic2, and steroid sulfatase in humans. Coincident with X chromosome inactivation, there is activation of autosomal homologs of several genes, including those encoding PGK-2, the EI subunit of pyruvate dehydrogenase, lactate dehydrogenase (LDH-X), and Zfa. These appear to be the result of retroposon activities, as they are characterized by a lack of introns, unlike their counterparts on the X chromosome.

2.2 RNA

RNA synthesis occurs in all spermatogenic cells up through the round spermatid stage, with a constant ratio of DNA to RNA content per cell through meiosis. One study provided the estimate that one half of all mRNAs made in the adult testis increase in abundance after meiosis, and one quarter of adult testicular mRNAs are synthesized only in postmeiotic cells, indicating the requirement for specialized proteins to build a functional spermatozoon.

An important feature of postmeiotic gene expression is that mRNAs produced by the haploid male germ cell are shared between clonal siblings, rendering them functionally diploid while still in the seminiferous epithelium. This was illustrated by the observation of human growth hormone (hGH) transcripts in all spermatids of heterozygote mice bearing a transgene that encodes the hGH gene linked to the protamine promoter. The protamine-transcript sharing between spermatids has also been described, and proteins are also exchanged between clonal siblings at this stage.

As is apparent in every developmental system, unique interactions between many sets of molecules dictate the specific physiological response of the differentiating cell. Because spermatogenesis involves morphological and physiological events that do not happen in any other cell type, it is easy to anticipate that a unique set of genes and regulatory mechanisms will function in the testis. There are also many genes that are expressed in somatic cells that are important for sperm development. Control of the production of proteins required for spermatogenesis is exhibited during both transcription and translation. Many meiotic and postmeiotic gene transcripts are characterized by being structurally distinct from their premeiotic and somatic cell counterparts. The examples in Table 1 illustrate the variety of causes and consequences of this feature. The potential activity of unique RNA-splicing machinery has been postulated on the basis of this information.

2.2.1 Overview of Transcription

Regulation of RNA polymerase II activity in eukaryotic cells is effected by interaction with transcription factors that bind to specific sequences of DNA in each gene. In each cell, the expression level of a particular gene is determined by the nature and abundance of transcription factors at a specific time, and it is common that several factors are influential. Transcription factors are classified on the basis of the predicted structures of their DNA-binding

Tab. 1 Alternative splicing of mRNAs is a feature common to many meiotic and haploid transcripts. This table highlights the diversity of examples.

Protein name	*Known activity*	*Meiotic and/or haploid mRNA splicing features*	*Size [kb]*	
			Somatic or Premeiotic transcript	*Postmeiotic transcript*
c-abl	Cytoplasmic and nuclear tyrosine kinase	3′UTR uses alternative polyadenylation site; no change in protein detected	5.3, 6.5	4.7
c-kit	Receptor tyrosine kinase	Use of alternative splice site; truncated protein encoded lacking all but part of intracellular domain; protein expression not documented (see Sect. 3.4.2)	5.5	3.2, 2.3
CREM	Transcription factor	Protein encoded by spermatocyte mRNA (CREM-τ) has stimulatory rather than inhibitory effect on cAMP-response element (see Sect. 2.2.2)	2 (unique product size detected by PCR)	2
Proenkephalin	Precursor to opiods	Longer 5′UTR; no effect on protein product	1.5	1.9

domains. The common transcription factor motifs are all detected in spermatogenic cells, indicating that the mechanisms employed to effect transcription are conserved between somatic and germ cells. On the other hand, the presence of unique mRNA sizes in spermatids suggests that novel RNA-processing mechanisms exist in the haploid male germ cell.

2.2.2 Transcription Factors

The ubiquitous transcription factor SP1 contains the zinc-finger motif for DNA binding, and it binds to a GC-rich promoter region (GC box) to mediate transcription in genes lacking the TATA box motif. High levels of this protein are present in spermatids. Zinc finger–containing proteins expressed in the testis are encoded on the Y chromosome (*ZFY* in humans and *Zfy-1* and *Zfy-2* in mouse), on autosomes (*Zfa* in mouse) and on the X chromosome (*ZFX* in human and *Zfx* in mouse). *Zfy-1* expression is first detected in mouse primordial germ cells (PGCs), while *Zfy-2* becomes abundant first between 7 and 14 days of age. Messenger RNA levels in both these increase at the beginning of meiosis. A strain of mouse (*Sxr'*) that lacks the *Zfy-1* and *Zfy-2* genes shows no progression of spermatogenesis into meiosis.

Members of the hormone receptor superfamily also bind to DNA through a zinc-finger domain. Retinoic acid receptor mRNAs are found in Sertoli cells (*Rara*) and (*Rarb*) and in germ cells (*Rara*). Vitamin A deprivation results in blocked proliferation of type A spermatogonia and meiotic prophase entry, while subsequent administration of retinol has been used to produce testes in which all germ cells progress in synchrony through the stages of the seminiferous epithelium.

Homeobox domain and POU domain family members have been detected in the adult testis. Messenger RNAs encoding several OCT proteins have been found in male germ cells at all developmental stages. *Oct-4* gene expression is an important marker of undifferentiated germ cells, as it is present specifically in cells of the embryo epiblast and then in the germ cells that arise from them. In the adult, the *Hox 1.4* gene is expressed in meiotic and haploid germ cells, where its expression may be regulated by retinoic acid.

The leucine zipper family of transcription factors includes *fos* and *jun*, which dimerize to form the AP1 transcription factor that binds to the TPA-response element (TRE). The activity of AP1 varies in cells in response to the activation of protein kinase C, through stimulation by growth factors or phorbol esters. *Jun-D* mRNA is present in all spermatogenic cells, and is predominant in spermatocytes and round spermatids, while the relatively low levels of *c-jun* and *junB* mRNA are highest in type B spermatogonia. *Fra-1* and *c-fos*, and mRNAs are also found in spermatogenic cells. Reversible induction of *c-fos*, *c-jun* and *junB* transcripts occurs in response to the cell dissociation procedures used to produce purified germ cells for analysis, an important observation for those wishing to study gene expression in purified germ cell populations.

Cellular responses to cAMP levels are mediated through changes in protein kinase A (PKA) activity in order to regulate transcription factors that bind to the cAMP-response element (CRE) in the promoter regions of many genes. Active PKA typically phosphorylates one of the CREB family of bZIP transcription factors, CREB, CREM, or ATF-1 isoforms, enabling its association with a CREB binding protein to affect transcription.

A distinct mechanism of CRE activation occurs in meiotic and postmeiotic germ cells, wherein PKA activation promotes association of CREM with activator of CREM in Testis (ACT), a member of the LIM-only family of transcription factors; this does not involve the activation step *via* phosphorylation required in other cell types. CREs are present in many genes that are specifically or preferentially expressed in the testis or in meiotic germ cells, including those encoding pyruvate dehydrogenase a-2, phosphoglycerate kinase-2, testis angiotensin-converting enzyme, polyubiquitin, transition proteins 1 and 2, protamines 1 and 2, calspermin, and RT-7. In addition, the expression of PKA regulatory and catalytic subunit mRNAs is elevated in meiotic and postmeiotic germ cells, and PKA regulatory subunit isoforms have discrete localizations in the mature human spermatozoan.

The CREB protein exists in a wide range of tissues at low levels, which suggests that its intracellular level is not the primary regulator of its activity, and a modulator protein (CREM) has been characterized (see below). Multiple isoforms of CREB are found in the testis, including those that lack the leucine zipper domain and the nuclear translocation signal; these would presumably regulate the activity of the other isoforms of CREB. CREM also is present in the testis in multiple isoforms. In somatic and premeiotic cells, the presence of CREM α, β, and γ isoforms antagonizes the action of cAMP by downregulating expression of genes bearing the CRE binding motif. Alternative splicing of the *CREM* mRNA in primary spermatocytes produces germ cell–specific forms, designated CREMτ, τ1, and τ2, which act as agonists in round spermatids, where these proteins are found. It is thus apparent from these specific changes in patterns of gene expression in primary spermatocytes that a cascade of changes in intracellular signaling via cAMP occurs. Mice lacking the CREM gene are sterile, with defects evident in round spermatids and a complete absence of elongating spermatids. Expression of CREM and CREB is dependent upon both FSH and androgen, while the switch of CREM from antagonist to activator is effected by FSH. Transcription of CREB appears to undergo positive autoregulation, as its promoter contains three consensus CREs.

Other DNA-binding proteins that are unique to spermatogenesis have been identified using a combination of DNA footprinting, gel-retardation assays, *in vitro* transcription assays, and transgenic mouse production. Promoter regions active in regulating transcription specifically in postmeiotic germ cells were initially identified from regions of homology in the 5'UTR of mouse protamine genes, *Prm-1* and *Prm-2*, which are transcribed only in haploid male germ cells. One of these (region D) confers negative regulation of these genes in somatic cells, while regions B and E appear to confer positive regulatory affects. D element motifs have been identified in the *RT7* promoter, in close proximity to a CREB binding site, where at least one of them appears to positively regulate transcription in nuclear extracts of seminiferous tubules. *RT7* encodes a component of the sperm tail outer dense fiber, also named Odf1, which is required for motility. A 39-kDa protein that is present in nuclear extracts of seminiferous tubules, TTF-D, appears to bind the D_1 element of both *RT7* and *c-mos* promoters. Enhancer and repressor elements have been identified in the germ cell–specific *PGK-2* gene, which appear to coordinately determine the appropriate levels of expression for somatic and germ cells. Other DNA-binding proteins detected in association with developing germ cells include members of the HMG box (e.g. SRY; see section on Genes controlling testis formation), *ets* proto-oncogene, and *c-abl* families.

2.2.3 Sex Chromosome–encoded Genes

The absence of spermatogonia in the *SRY* transgenic *males* highlights the involvement of other gene products in germ cell development. In as many as 40% of cases of human infertility attributed to the male partner, no specific cause, such as endocrinological failure or presence of antisperm antibodies, can be identified. The concept that other genes required for the successful completion of spermatogenesis are located on the Y chromosome emerged from observations of patients with azoospermia who had small Y chromosomes. Several genes, termed azoospermia factors (AZFs), have been identified in humans by intensive mapping of the Yq11 region of such patients. Deletions in the Y chromosome are highly correlated with a low-to-absent sperm count. ($\leq$1 million sperm per ml).

One group of genes identified at multiple subintervals of Yq11encodes RNA-binding proteins, including DAZ and RBMY. Some of these (e.g. RBMY) appear required for storing and then facilitating translation of mRNAs required for spermiogenesis in elongating spermatids when the genomic DNA has already been packaged with protamine and compacted for long term storage in spermatozoa. New information derived from detailed analysis of the human Y chromosome sequence has indicated that a unique combination of mechanisms are likely to have mediated evolution of genes encoded on the male chromosome. In addition to derivation of genes from common ancestors of the X and Y chromosomes, such as RBMY, the transposition and amplification of autosomal genes (e.g. DAZL, a DAZ paralog) and retrotransposition (CDY) have been implicated.

The X chromosome also harbors genes that are specifically used in male germ cells. Following subtractive hybridization to identify mRNAs expressed in mouse spermatogonia, a disproportionate number of those uniquely expressed in male germ cells (i.e. not in ovary or in other somatic tissues) were subsequently identified as products of X-encoded genes.

2.2.4 Translational Regulation

A predominant feature of gene expression in the testis is the storage of mRNAs synthesized in the meiotic and postmeiotic germ cells for translation at later stages. Relatively high levels of certain proteins are essential at discrete stages of spermiogenesis, such as during tail formation and acrosome development. Accessibility of genes encoding the proteins required for transcription becomes limited during spermiogenesis as nuclear DNA condensation occurs. Transcription has been detected in round spermatids up to stage 8, though translation continues during acrosome and tail formation, finally ceasing with the cytoplasmic rearrangements that precede sloughing of residual cytoplasm prior to spermiation.

There are many mRNAs that are synthesized in meiotic cells but are not translated until later on, in spermiogenesis. For example, an autosomal form of the enzyme phosphoglycerate kinase, PGK-2, is synthesized in spermatids following inactivation of the X chromosome, when the X-linked *PGK-1* gene is no longer transcriptionally active. The *PGK-2* mRNA is synthesized in premeiotic and meiotic cells but is not associated with ribosomes, and hence not translated until after meiosis. The mRNAs encoding the proteins required for chromatin packaging (transition proteins and protamines), acrosome formation, and sperm-tail structures are also under translational regulation.

Sequences that regulate transcription are present in the 5′ and 3′ UTRs of mRNAs. At the 5′ end, interactions between initiation factors (eIF-4F) and the 5′ cap must occur for translation to begin, and the level of phosphorylation of initiation and elongation factors (which determines their activity) is regulated by extracellular signals. Sequences in the 3′ UTR also influence transcription rates, and the poly(A) tail length of some mRNAs fluctuates in response to signals that mediate changes in their translation rates. Poly(A)-binding proteins (PABPs) increase mRNA stability and hence increase the amount of protein produced, while 3′ AU-rich sequences contribute to the destabilization of mRNAs.

Increased poly(A) tail length correlates with higher levels of association with polyribosomes (and therefore increased translation rates) of some meiotic cell–specific mRNAs, including cytochrome c_T and LDH C, while others show no difference. In contrast, poly(A) tail shortening occurs in association with translation of mRNAs encoding the transition proteins and protamines. These mRNAs are synthesized in round spermatids and stored until spermatid elongation, when they sequentially replace histones during nuclear rearrangement. Premature synthesis of protamine 1 protein in a transgenic mouse caused arrest of spermatid development. Through analysis of *Tpap-/-* mice, a testis-specific cytoplasmic poly(A) polymerase, TPAP, has been identified as essential for progression of round to elongated spermatids. The absence of TPAP is associated with decreased expression of genes required for spermiogenesis, including transition protein 1, protamines 1 and 2, sperm fibrous sheath component 1 (Fsc1), sperm outer dense fiber protein RT7 (Odf1), and heat shock protein Hsc70t. The *Tpap-/-* male mouse phenotype is comparable to that in mice lacking the TATA-binding protein-related factor-2 (TRF2). Analysis of the mRNA-encoding TRF2 and other transcription factors involved in RNA polymerase II-mediated transcription revealed a reduction in the length of the poly(A) tails in the *Tpap-/-* mice, and impairment of transport into the nucleus of the TAF10 protein, all contributing to the reduction in TFIID complex activity required for synthesis of the gene products that are essential for spermiogenesis.

Postmeiotic mRNA translation is mediated by RNA-binding proteins, and several of these have been shown, by gene knockout studies in mice, to be essential for postmitotic germ cell differentiation including Prpb, TLS/FUS, TB-RBP, and SPNR. Testis brain RNA-binding protein (TB-RBP/ translin), present in round spermatids but absent from elongated spermatids, binds specifically to conserved sequences (Y and H elements) in many mRNAs including transition protein 1, protamines 1 and 2, AKAP4, glyceraldehyde 3-phosphate dehydrogenase-S, myelin basic protein, calmodulin kinase II, and tau protein. TB-RBP appears to have a role in storage and transport of these mRNAs, binding them to microtubules and potentially moving these mRNAs between clonal siblings through their intracytoplasmic bridges. It has also been characterized as a DNA-binding protein that may link consensus sequences during recombination and repair during meiosis in male germ cells and in lymphoid tumors. Spermatid perinuclear RNA-binding protein (SPNR) is also a microtubule-associated RNA-binding protein that is essential for spermatogenesis. The *SPNR* mRNA itself appears to be translationally regulated, as it is synthesized in spermatocytes, and the protein is present only after meiosis. *SPNR-/-* mice produce abnormally shaped sperm in reduced numbers, indicating that SPNR is required for normal completion of spermiogenesis. Phosphorylation of RNA-binding proteins, TLS/FUS, TB-RBP, and MSY2, mediates their dissociation from target mRNAs to enable their translation. The phosphatase that effects this translational activation step in male germ cells has yet to be identified. A potential role for a phosphatase antagonist, the Styx protein, which contains a point mutation in the catalytic domain to render it inactive, is suggested by the phenotype of abnormal and reduced spermatid differentiation and infertility.

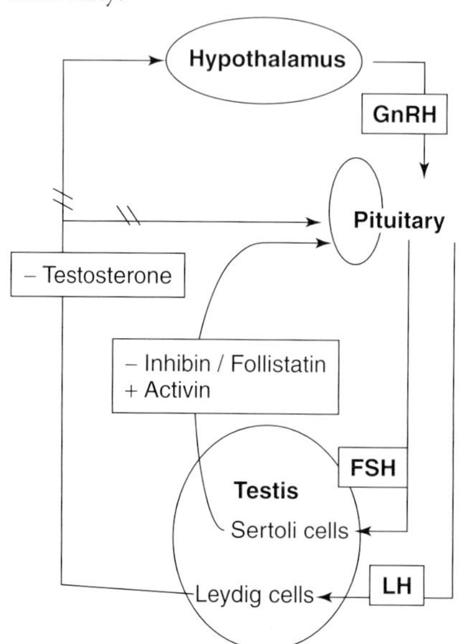

Fig. 4 Illustration of primary components of the feedback loops comprising the hypothalamal–pituitary axis.

3 Control Mechanisms

3.1 Extragonadal and Hormonal Control of Spermatogenesis

3.1.1 Overview

The production of normal sperm numbers in the adult is dependent on the actions of the pituitary gonadotrophins, follicle-stimulating hormone (FSH) and luteinizing hormone (LH). The secretion of these is regulated by gonadotrophin-releasing hormone (GnRH) from the hypothalamus and feedback from the testis *via* testosterone and inhibin (Fig. 4). In the absence of FSH or LH, spermatogenesis is compromised to an extent that varies between species and with the stage of sexual maturity at removal. The receptors for FSH and LH within the testis are present only in the somatic Sertoli cells and Leydig cells, respectively, and hence these hormones affect germ cell development indirectly. The hormone testosterone is also essential for normal spermatogenesis, and its secretion by the Leydig cell is under the influence of LH from the pituitary. Receptors for testosterone (T) are abundant on Sertoli cells, peritubular myoid cells, and Leydig cells. As is true for LH and FSH, the influence of testosterone on spermatogenesis is clearly through the local production of intermediary compounds. In addition, thyroxine has a profound influence on Sertoli cell maturation in immature animals. Recent evidence for the impact of estrogen on testis development has arisen from the analysis of infertile male mice that lack the components of the estrogen synthesis and signaling pathway; data that reinforce the findings from prior studies that described the impact of estrogen and xenoestrogens.

3.1.2 The Action of FSH and Thyroid Hormone

The action of FSH in the testis is through its impact on Sertoli cell physiology, since receptors for FSH are exclusively located on these cells, while thyroid hormone receptors have been identified in both somatic and germ cells. The impact of these hormones on adult testicular function has been studied using approaches that manipulate their levels and function at defined periods of development and in adulthood. Through recent investigations on humans and mice that are deficient in FSH production and signaling, we now understand that spermatogenesis can proceed in the complete absence of this hormone. These individuals achieve complete spermatogenesis in the lifelong absence of FSH, although the testis size is reduced and fertility is compromised due to suboptimal sperm production and quality.

FSH in the immature testis In immature rats, FSH stimulates Sertoli cell proliferation and the onset of maturation; suppression of FSH levels prior to the normal cessation of Sertoli cell division by day 16 in the rat reduces Sertoli cell numbers in the adult testis. The timing of Sertoli cell proliferation in the juvenile has been established using observation of mitotic figures in histochemical sections and quantitative autoradiography following administration of ^{3}H-thymidine to monitor DNA synthesis. Cell division is evident in the rat fetus at day 16, and the mitotic index of Sertoli cells is maximal between 18 and 21 dpc, corresponding to the period when fetal Leydig cells increase their output of testosterone. The rate of Sertoli cell proliferation decreases after birth and ceases by day 16 in the rat. Thus, the full complement of Sertoli cells in the adult testis is present by this time. It is not known whether Sertoli cell proliferation becomes slower for all cells after birth or whether there exists a discrete population of Sertoli cells that continues to divide until a specific signal to arrest division is received. The response of Sertoli cells to FSH changes during the postnatal period in rats. The initial postnatal period is characterized by being singularly FSH-dependent for proliferation, while between days 5 and 13 postpartum, Sertoli cell proliferation can be transiently stimulated by activin. This effect of activin is augmented by the presence of FSH. Other local factors have been observed to affect Sertoli cell proliferation, including GDNF.

With the termination of mitosis, the Sertoli cell enters a phase of differentiation, wherein continued germ cell differentiation (i.e. onset of meiosis) within the seminiferous epithelium becomes evident, and the lumen of the mature epithelium is formed as Sertoli cells develop tight junctions at the base of the epithelium (forming the blood–testis barrier) and establish a pattern of vectorial secretion. The changed expressions of several mRNAs and proteins are regarded as hallmarks of the now differentiated Sertoli cell, for example, androgen-binding protein.

FSH in the adult testis The impact of FSH in adult spermatogenesis appears to be greatest in regulating spermatogonial development, though its influence on spermatid adherence to the Sertoli cell ectoplasmic specialization has also been documented. However, as stated previously, FSH is not essential for completion of spermatogenesis, but it is required to achieve quantitatively normal sperm production. In the adult rat, expression of FSH receptor mRNA is greatest at stages IX–XII in the cycle of the seminiferous epithelium, while the protein is apparently most abundant between stages XIII–I.

The impact of FSH on testicular function in the adult varies dramatically between species, with neutralization of FSH having a limited effect on sperm production in rats and a much greater effect in nonhuman primates. Suppression of FSH action has been achieved using passive immunization, injection of high doses of testosterone, and by active immunization against GnRH. In rats, this results in a decrease in testis weight, germ cell numbers, and sperm output of about 10%; in monkeys, a 50% reduction in testis size results from active and passive immunization against FSH. Following hypophysectomy in rats, which removes LH and FSH, germ cell degeneration is most evident at stage VII and can be partially halted by administration of FSH. In this model, in the absence of exogenous testosterone, FSH supports spermatid development to step 8, while the addition of testosterone permits completion of spermiogenesis. This correlates with the finding that androgen receptors are most abundant on stage VII Sertoli cells.

Several *in vitro* studies have documented the effects of the administration or deprivation of FSH on adult rat Sertoli cell protein secretion (e.g. transferrin and inhibin), mRNA synthesis, and cytoskeletal architecture. An increase in spermatogonial and spermatocyte apoptosis, and a reduction in the number of type B spermatogonia entering meiosis is observed in response to FSH withdrawal in adult rats, while administration of FSH increases spermatogonial proliferation and DNA synthesis in spermatogonia and spermatocytes. *In vitro* studies of adult Sertoli cell function indicate that FSH stimulates inhibin secretion both in intact seminiferous tubules and in purified cell preparations. Inhibin will act as a hormone to reduce FSH secretion from the pituitary, but it clearly may have local effects through its impact on TGFβ and activin signaling (see Sect. 3.3.3), thereby affecting germ cell proliferation and differentiation. Some FSH-regulated products are known to act directly on spermatogonia to promote survival, most notably stem cell factor (See section on Interactions between SCF and c-kit affecting spermatogenesis).

Effects of FSH on adult sperm output and quality: *in vivo* and *in vitro* models In the adult rat testis, each Sertoli cell has the capacity to support approximately six to eight round spermatids. Therefore, the potential sperm output of the adult testis (i.e. number of sperm produced per day) is fixed during the postnatal period of Sertoli cell division. Intratesticular administration of an antimitotic drug to reduce neonatal Sertoli cell division resulted in a testis containing a smaller number of germ cells that were present in proportion to number of Sertoli cells at a ratio identical to that in untreated animals.

In addition to the apparent variation in impact of FSH between species, an assessment of the role of FSH on spermatogenesis is complicated by issues such as the purity and source of the FSH administered in older studies, the physiological status of the animal under treatment, and the complexity of intercellular events that occur within the normal testicular environment. With the availability of recombinant FSH for experimental work, and the use of genetically modified mouse models, a more convincing picture of the interactions between hormones in testicular physiology is developing.

Abrogation of FSH signaling in mice was achieved through the production of mice lacking the genes encoding the FSH β subunit or the FSH receptor. In both these models, spermatogenesis proceeded to completion, but the number of spermatozoa was reduced (to 45–65% of the number in the controls), sperm morphology was abnormal, and fertility was impaired. Similar observations are recorded for patients with homozygous inactivating mutations in the FSH receptor.

Effects of the thyroid hormone on adult sperm output: *in vivo* and *in vitro* models The thyroid hormone receptor is most abundant in rat Sertoli cells during the period when Sertoli cell division is decreasing, between the end of fetal development and the first two weeks after birth. The results of *in vitro* and *in vivo* studies indicate that thyroxine normally decreases Sertoli cell proliferation and promotes Sertoli cell maturation. Thyroid hormone levels affect the cessation of Sertoli cell proliferation. *In vivo* and *in vitro* administration of T3 in rat models promotes premature reduction in Sertoli cell mitosis and the synthesis of products associated with their terminal differentiation. Information from related studies on the influence of thyroid hormone on testicular physiology has reinforced this observation that the Sertoli cell population determines the number of sperm produced in the adult. The induction of neonatal hypothyroidism by administration of prophylthiouracil to the drinking water of pregnant rats and mice results in extension of the period of Sertoli cell proliferation to day 35, with a corresponding delay in their maturation. The extended period of Sertoli cell proliferation results in larger testes in adulthood, with a marked increase in Sertoli and germ cell numbers. Evidence from *in vitro* studies supports the concept that there is synergism between FSH and tri-iodothyronine (T3) in effecting some aspects of this change; thus, the balanced influence of thyroid hormone and FSH actions on Sertoli cell proliferation will have a profound effect on adult male fertility.

3.1.3 The Action of Testosterone

Leydig cell populations and their development The pattern of Leydig cell development varies between species, though the common observation is that Leydig cells are first observed in the interstitium after cord formation. Fetal Leydig cells have been shown to derive from mesenchymal cells originating in the mesonephros under the influence of local factors, including PDGF. Their rapid proliferation is stimulated by chorionic gonadotrophin and LH. This population appears to be static during perinatal life. At the time of puberty, a different population of interstitial cells is stimulated by LH to differentiate into adult-type Leydig cells. It appears that this effect is partly mediated through the actions of FSH. In immature rats, treatment with FSH, but not LH, promoted transformation of interstitial cells into adult Leydig cells, and FSH has also been shown to stimulate both LH receptor activity and T-production in response to LH stimulation. Fetal and adult Leydig cells share the characteristic of containing large amounts of smooth endoplasmic reticulum, and they can also be identified by the presence of steroidogenic enzymes and their capacity to produce androgens. Fetal Leydig cells are easily distinguished by the presence of lipid inclusions, which, in the rat, are mostly absent in the adult Leydig cell.

The abundance of intracellular lipid stores varies in adult Leydig cells with the degree of LH stimulation and availability of dietary cholesterol.

Receptors for LH are found only on Leydig cells, and the addition of LH stimulates production and secretion of testosterone (T). The amount of T produced varies *in vitro* with the addition of various growth factors present *in vivo*, including bFGF, PDGF, and TGFα. Thus, production of T by the Leydig cell is modulated by both local and hormonal signals.

Actions of testosterone The understanding from recent information is that testosterone is required for providing the basal support for spermatogenesis, while FSH affects the quantity and quality of sperm output through its actions in the immature as well as the adult animal.

The presence of testosterone is essential for complete spermatogenesis, though the likely importance of dihydroxytestosterone (DHT) as an alternative ligand for the androgen receptor has been strongly suggested. The binding affinity of the androgen receptor for DHT is 2 to 10 times greater than for T, but T is 5 times more abundant than DHT within the testis and is hence more likely to be available for binding to the receptor. The intratesticular level of T is 50 times greater than what is measured in serum, and the physiological basis for this requirement by the testis of such high levels of T is not known. Androgen-binding protein and albumin in interstitial fluid can both bind T, and may contribute to the two to threefold higher concentrations of T measured in interstitial fluid relative to the testicular venous blood in normal adult rats.

The most profound effects of T on spermatogenesis are exerted at stages VII–VIII of the seminiferous epithelium. Conversion of step 7 spermatids to step 8 spermatids is specifically dependent on T *in vivo*. The absence of T *in vivo* leads to a change in the interaction between Sertoli cells and elongating spermatids. The mature germ cells are not released into the tubule lumen (spermiation), and are instead phagocytosed by the Sertoli cell. Genes encoding CAMs and other components of the ectoplasmic specialization, which mediates attachment of elongating spermatids to Sertoli cells, are regulated by local concentrations of T and FSH.

Restoration of T levels in long-term suppression models leads to rapid restoration of round to elongated spermatid conversion ratios, though the numbers of spermatogonia and spermatocytes are not fully restored. In contrast, the presence of high testosterone levels can inhibit spermatogonial proliferation in models of damaged or recovering testis, such as the *jsd* mouse or the irradiated rat testis, thereby indicating a role for T in facilitating stem cell proliferation or recovery.

While Sertoli cells possess androgen receptors (see above), they may also respond to factors produced by interstitial cells in response to T, to produce the androgen dependency of spermatogenesis. Proteins produced by peritubular myoid cells, such as P-Mod-S, have been implicated in such a scenario, as they stimulate secretion of transferrin, inhibin, and retinol-binding protein by Sertoli cells.

The *hpg* mouse, which bears a naturally occurring deletion of the GnRH gene, has been exploited to elucidate the distinct roles of T and FSH. Administration of T can restore spermatogenic capacity to these animals that are chronically FSH-deficient, in accord with the findings from analyses of FSHβ and FSH receptor knockout animals, though the number of sperm produced is subnormal. A human FSHβ-subunit transgene was introduced to the *hpg* mouse, resulting in restoration of spermatogonia numbers. Restoration of pachytene spermatocytes to normal and spermatid numbers to near normal numbers was achieved with T administration beginning on day 21 postpartum for 6 weeks. These findings further reinforce the importance of T in the postmitotic differentiation of germ cells.

3.1.4 Estrogen in Spermatogenesis

There is long-standing recognition of the detrimental effects of exogenous estrogen exposure on various aspects of male reproductive tract development and function from the studies of animal models and humans. Fetal and neonatal exposure to estrogenic compounds has been implicated in an increased rate of testicular cancer, cryptorchidism, epididymal abnormalities, and decreased fertility.

In recent years, the influence of estrogens on spermatogenesis has gained a deeper appreciation. Animals lacking the genes encoding the estrogen receptor-α and -β genes (ERαKO and ERβKO mice) and the aromatase cytochrome P450 gene (*CYP19;* ARKO mice) have been generated and examined, and reagents have been developed that can discern the cellular localization of each receptor subunit. The ERαKO mice have impaired fertility due to impaired efferent duct structure and function, while the ERβKO mice are fertile. The ARKO mice suffer from a progressive loss in spermatogenic capacity; the animals are fertile and testicular histology is normal at 14 weeks postpartum, while abnormalities in both the interstitial and seminiferous tubule compartments are apparent after 18 weeks. The double ERKO mice (ER$\alpha\beta$KO) lacking both receptor subunits revealed a phenotype similar to that of the ERαKO, but with distinctively affected mounting behavior.

Because each receptor subunit bears distinct ligand binding and responsive properties, the collection of information about receptor distribution and ligand production is required in order to fully understand the etiology of these phenotypes. Much of this information remains to be sorted out, and there are conflicting results in the literature, some of which are likely to highlight species-specific differences. The ERα appears to be restricted to Leydig and accessory duct cells. ERβ is more widely distributed, found on Leydig and Sertoli cells, in accessory ducts, and in spermatogonia, pachytene spermatocytes, and round spermatids. The aromatase enzyme distribution is similarly broad, having been detected in Leydig and Sertoli cells, pachytene spermatocytes, and in round and elongating spermatids.

3.2 Molecules Involved in Cell Cycle Control

3.2.1 Cyclins

The identification of control points in spermatogenic progression has been achieved by determining cyclin family member expression patterns. Cyclins, proteins originally identified in invertebrate eggs, undergo tremendous changes in levels of expression immediately following fertilization and in subsequent rounds of cell division, with passage through the cell cycle. Individual cyclins have been associated with specific stages of the cell cycle and, to date, at least 8 familes have been identified. Family members are classified as cyclins A through H based on their amino acid homologies and/or their time of appearance during the cell cycle. Cyclin B associates with a serine/threonine kinase (a cyclin-dependent kinase (Cdk), Cdk1; named $p34^{cdc2}$ kinase in yeast) to form the maturation promoting factor (MPF) which allows the cell to progress through the M-phase of the cell cycle in mitosis and meiosis. The kinase activity of each Cdk is regulated by a complex series of phosphorylation and dephosphorylation events, dependent upon its association with cyclin. Following M-phase, cyclin B is degraded by proteolysis following cross-linking to ubiquitin. This scenario represents the paradigm of how cell cycle passage is regulated, through a sequence of cyclin activation and deactivation events.

Analysis of cyclin gene expression during murine spermatogenesis shows that members of the cyclin A, B, and D families are expressed at discrete sites within the seminiferous epithelium in association with distinct stages of spermatogenic differentiation. There are some differences between reports of their cellular localizations that most likely reflect the nature of the reagent used (i.e. the specificity of the antibody) and the detection method applied. Functional data are predominantly derived from analyses of mice lacking a specific cyclin gene.

The cyclin A family members are usually associated with the G1/S transition of mitotic cells. Cyclin A2 mRNA is present in spermatogonia and preleptotene spermatocytes, and it appears to regulate entry into the meiotic prophase. Cyclin A1 mRNA is synthesized in late pachytene spermatocytes and is essential for the pachytene-to-diplotene transition in mouse spermatocytes. Cyclin B family members participate in the G2/M transition. Cyclin B3 is present in preleptotene and zygotene spermatocytes, and cyclin B2 protein is most abundant in pachytene spermatocytes. Cyclin B1 mRNA is most abundant in round spermatids, although it too is detected in pachytene and more advanced spermatocytes. Cyclin D family members are most abundant in G1 stage cells and are involved with the G1/S transition in many cellular contexts. In the murine testis, cyclin D3 has been observed in gonocytes, round spermatids, Sertoli cells, and Leydig cells. Cyclin D2 expression has been reported as low to negligible, with protein being detected in limited numbers of adult mouse spermatogonia and in Sertoli cells. Its expression in spermatogonia appears to be associated with those cells that are entering the differentiation pathway. The cyclin D2 knockout male mouse testes show a severe loss of germ cells. Cyclin D1, the expression of which is restricted to Sertoli cells, has also been identified as an estrogen–receptor ligand, the significance of which remains to be established.

3.2.2 Cyclin-dependent Kinases and Regulatory Proteins

Cyclin activity is clearly linked to association with a specific Cdk. Cdk4, the binding partner to cyclins D2 and D3, is most highly expressed in the mouse testis during the first spermatogenic wave, predominantly in spermatogonia and Sertoli cells. The absence of this gene in mouse leads to a loss of spermatogonia and spermatocytes. The predominant expression of *Cdc2* and *Cdk2* mRNAs in spermatocytes and their presence in postmeiotic germ cells suggests that these molecules are involved in germ cell differentiation. Other *Cdk* family genes are expressed in adult Sertoli cells, which are nonproliferating, indicating that roles other than cell cycle regulation are played by these kinases.

The cdks that function during G1 of the cell cycle are regulated by specific cyclin-dependent inhibitors (CKIs) belonging to two families, and many of these functions facilitate rather than inhibit cyclin action. The cip/kip CKIs preferentially affect the action of cyclins A and E through Cdk2. The p27^{Kip1} protein is found in terminally differentiated Sertoli cells, while p21$^{Cip1/Waf1}$ is in pachytene spermatocytes and round spermatids. The INK4 CKIs bind to and inhibit Cdk4 and Cdk6, thereby blocking the activation of the D-type cyclins, the susbsequent phosphorylation of Rb, and the entry into the S-phase of the cell cycle. The INK4 proteins p18^{Ink4c} and p19^{Ink4d} are synthesized in spermatocytes. The absence of either of these genes in mice results in fertile adults with smaller-than-normal testes, increased germ cell apoptosis, and reduced sperm numbers. The combined absence of these two CKIs results in a more severely infertile phenotype in male mice that exhibit delayed entry into meiosis and elevated numbers of spermatocytes undergoing apoptosis. These mice are also characterized by Leydig cell hyperplasia (as seen in the p18^{Ink4c} KO mouse) and elevated FSH. The impact of deregulated cyclin D function, therefore, affects both somatic and germ cells in these animals. This finding may be related to how cyclin D/CDK4 can prevent the inhibitory action of p27^{Kip1} on cyclin E/CDK2. With the loss of CDK4, p27^{Kip1} activity is elevated, and cyclin E/CDK2 function is diminished.

Dephosphorylation of Cdk1 at specific tyrosine residues by the phosphatase *cdc 25* increases its serine/threonine kinase activity to allow progression through mitosis. The expression of *cdc25C* in the adult mouse is highest in the testis, with the germ cells being the predominant site of expression of its mRNA. It is suggested that interactions between Cdc25, Cdk1, and CycB2 in meiotic cells mediate progression through the G2-M transition of meiosis.

3.2.3 Other Regulators of Cell Cycle Progression

The association of a cyclin with a Cdk results in the formation of an active enzyme complex that phosphorylates the retinoblastoma protein (Rb). Rb generally acts to prevent progression through the cell cycle by binding to E2F and blocking its ability to drive transcription. Phosphorylation of Rb leads to its dissociation from E2F, and gene products required for DNA synthesis are then produced. Rb has been detected in Sertoli cells, spermatogonia, and late-stage elongated spermatids in the rat testis, and it undergoes phosphorylation in a cyclic manner through the cycle of the seminiferous epithelium, indicating that its function is regulated during spermatogenesis.

Progression through meiosis depends on the presence of Hsp70.2, a molecular chaperone required for cyclin B/Cdk1 function. Mice lacking Cks2, a mammalian homolog of a yeast Cdk-1 binding protein, enter prophase I of meiosis but fail to complete the first meiotic division (MI), indicating that a specific cyclin–cdk complex is crucial for spermatocyte progression into anaphase. Deletion of the gene encoding the DNA repair protein, Ercc1, affects murine germ cells at all stages of development, leading to the removal of damaged germ cells by apoptosis.

The c-mos protein was originally identified as a component of cytostatic factor with influence during meiosis in testes and ovaries. C-mos regulates progression through the cell cycle at least partly by limiting cyclin degradation. The testis-specific isoform is a 43-kDa protein detected predominantly in spermatocytes. Overexpression of c-mos protein in the pachytene spermatocytes of transgenic mice resulted in no overt changes in phenotype, including fertility. Overexpression in spermatids resulted in enlargement of testicular size, apparently due to increased numbers of germ cells.

Expression of the pin1 protein has been implicated in enhanced expression of cyclin D1 in several tissues. Its deletion in a mouse knockout model produced a phenotype that was strikingly similar to that of the cyclin D1 knockout mouse in mammary and retinal tissue. However, the testicular phenotype of hypoplasia was similar to that observed in the cyclin D2 knockout mouse, suggesting that pin1 may interact with D2 in the testis.

An understanding of the roles of these and other proteins that interact with the machinery that regulates mitosis and meiosis will reveal much about how hormones and growth factors combine to regulate spermatogenesis. One example is derived from an analysis of mice lacking both *p27*^{Kip1} and *inhibin* α (a member of the transforming growth factor β superfamily; see section on Activin, Inhibin, and Follistatin). The absence of inhibin α leads to development of Sertoli cell and granulosa cell tumors beginning 4 weeks after birth, while those lacking p27^{kip1} have enlarged organs, including testes. Inhibin acts as a *tumor suppressor* in normal tissue by promoting association of p27^{Kip1} with cyclinD2/E1 and Ck4/2, holding them in an inactive complex. In a normal testis, the balance of signals between FSH and inhibin drive Sertoli cell terminal differentiation and exit from the cell cycle, leading to the onset of p27^{Kip1} expression. In the absence of either one of these factors in *p27*^{Kip1}*-/-/inhibin* α*-/-* mice, this signaling pathway is lost, resulting in the early onset of tumorigenesis.

3.2.4 The Machinery of Meiosis

A plethora of gene products expressed in somatic cells also affect testis development and/or germ cell maturation. The genes that govern the events of meiosis, however, have a unique relevance to the biology of gametogenesis and hence to spermatogenesis. The meiotic prophase begins with DNA replication by the preleptotene spermatocytes. The chromosomes condense and axial elements of the synaptonemal complex are formed during the leptotene stage, followed by pairing of the homologous chromosomes during zygotene. The longest stage of meiosis in the male, pachytene, is characterized by fully aligned, synapsed chromosomes that have undergone the recombination that is required for progression through diplotene (when the chiasmata resolve) and into the successive meiotic divisions, M1 and M2.

Checkpoints to ensure the fidelity of replication and DNA rejoining are key features of meiosis. While some of the crucial genes are known from analysis of infertile mice, others represent mammalian homologs of highly conserved genes that mediate cell division in yeast. DNA mismatch repair proteins, DMC1, COR1, MSH4, MSH5, and MLH1 facilitate formation of the synaptonemal complex that links homologous chromosomes during Meiosis I. The murine Brca1 protein is required for crossing over and repair of DNA damage during spermatogenesis – a function known to be required for suppressing breast tumor formation in humans. The sister chromatid proteins, SCP1, SMC1, and SMC3, facilitate DNA recombination, and the heat shock protein, HSP70-2 mediates disassembly of the synaptonemal complex.

The unique passage of the male germ cell through the G2/MI transition of

Meiosis I has been examined by treating mouse spermatocytes with the protein phosphatase inhibitor, okadaic acid. The pachytene (but not leptotene or zygotene) spermatocyte is competent to proceed through MI all through the several days in which the homologous chromosomes remain fully paired *in vivo*, and the specific machinery required to proceed through the meiotic metaphase is currently being clarified.

3.2.5 Regulation of Apoptosis

It has been estimated that up to 75% of germ cells entering the differentiation pathway will die by apoptosis without completing spermatogenesis. As in the developing ovary, there are waves of germ cell apoptosis in the rodent testis that coincide with the commitment to spermatogenesis by gonocytes and entry of spermatocytes into meiosis during the first wave. This cell death appears related to the need to maintain the optimal numerical ratio of Sertoli to germ cells, to ensure maximal production of fully fertile sperm, while also serving as a quality control mechanism to delete damaged germ cells.

Two pathways for triggering apoptosis are well understood, and both have been identified and characterized within the mammalian testis. The death receptor–mediated pathway is triggered by a signal from one of several extracellular ligands in the tumor necrosis superfamily (e.g. FasL, TNFα) upon binding to a specific cell surface receptor (e.g. Fas, TNFR1). Ligand binding causes receptor trimerization and recruitment of an adaptor protein (e.g. FADD), which leads to association with, and activation of, the central executioner complex of caspase enzymes, initially through activation of caspase 8. The complementary expression of Fas in Leydig and germ cells (quiescent primordial germ cells in the fetus and spermatocytes in the adult) and FasL in Sertoli cells, and of TNFα in spermatocytes and TNFR1 in Sertoli cells, can govern survival of each of these cell types. Treatments causing upregulation of only Fas (radiation, hyperthermia, ischemia/reperfusion) induce germ cell apoptosis, while those increasing Sertoli cell apoptosis (treatment with specific toxicants) induce upregulation of both Fas and FasL.

A separate pathway is triggered in response to a variety of extracellular insults (e.g. irradiation, cytotoxic drugs, serum deprivation) and intracellular events (DNA damage, inappropriate expression of cell cycle regulatory molecules). In this pathway, the balance of prosurvival and pro-apoptotic Bcl-2 family proteins dictates the fate of each cell. In response to injury (e.g. local heating), the upregulation of pro-apoptotic family members leads to cell death. The pro-apoptotic proteins can drive release of cytochrome c from the mitochondrion that, then, binds to Apaf-1, enabling it to oligomerize and recruit procaspase-9. This *apoptosome* now functions to recruit and activate downstream caspases. In both pathways, the end result of caspase activation is degradation of proteins required for maintenance of cell integrity, including nuclear lamins and cytoskeletal elements.

Conflicting reports regarding the cellular localization of Bcl-2 family proteins stem from the limited availability of reliable antibodies; however, it is clear that the expression of these proteins is cell-specific and alters during testis development. For example, while prosurvival *Bcl-2* mRNA is readily detected in spermatogonia and Sertoli cells of the newborn and developing mouse testis, it is reportedly undetectable in adult testis. Prosurvival Bcl_{XL} and pro-apoptotic Bad are initially synthesized in spermatogonia and Sertoli cells but shift to being synthesized exclusively within spermatocytes as the testis progresses through the first wave of spermatogenesis. Prosurvival Bcl-w and its partner, pro-apoptotic Bax, are present in spermatogonia and Sertoli cells throughout development and adulthood, while the pro-apoptotic Bim is widely distributed.

The functional importance of the balanced expression of the Bcl-2 family proteins has been repeatedly demonstrated though production of transgenic animals that misexpress a prosurvival family member or knockout mice that lack one. In the case of Bcl-2 overexpression in spermatogonia or deletion of Bax, the enhanced survival of germ cells leads to disruption of the seminiferous epithelium about 3 weeks into development, apparently due to the inability of the seminiferous epithelium to function with the excess number of immature germ cells. In the case of Bcl-w deletion, there is a delay in the onset of germ cell death to 3 to 4 weeks postpartum, presumably due to the compensatory function of other prosurvival family members, Bcl-2 and Bcl_{XL}, during the first wave of spermatogenesis.

Expression levels of Bcl-2 family members is probably influenced by the normal milieu of signals that affect testis cells, though to date there are only limited data concerning this point. The levels of bcl-w and bcl-$_{XL}$ in stage VII tubule segments of adult rats are stimulated by *in vitro* exposure to stem cell factor, a known germ cell survival factor normally synthesized by the Sertoli cell under partial regulation by FSH (see section on Stem cell factor). Addition of FSH to these cultures supports germ cell survival through the SCF/c-kit signaling pathway. The rodent testis responds to a variety of insults with changes in apoptotic regulators. Exposure to mild hyperthermia (testicular heating to 43 °C for 15 min) causes redistribution of bax protein from the cytoplasmic to the perinuclear area of the cell within 30 min.

There are conflicting reports describing the effects of testosterone withdrawal by administration of a specific Leydig cell toxicant, ethane dimethylsulfonate, and changes in bcl-2 family members as well as in Fas, have been reported.

The p53 mRNA and protein are found in spermatocytes. A protein with a complex set of functions, inactive p53 is bound in the cell to the heat shock chaperone complex, and when activated, it can regulate expression of bcl-2 and bax to drive apoptosis. Its absence in mice has been reported to impede correct completion of meiotic division I and slightly affect sperm morphology and male fertility. Conflicting reports regarding its functional impact on spermatogenic progression are thought to arise from the differences in the mouse strains examined. Apoptosis of spermatocytes but not spermatogonia appears to be dependent on the presence of p53 in the mouse under normal conditions, whereas the converse has been reported in the case of irradiation-induced DNA damage.

Heat shock factor 1(HSF1) is also bound to the heat shock chaperone complex when inactive. Its release is triggered by various cellular stressors, and the translocation of HSF1 to the nucleus promotes synthesis of heat shock proteins. Similar to p53, HSF1 also appears to have different effects on spermatogonia and spermatocytes, protecting the former and driving apoptosis of the latter in response to heating. Unlike in other cell types, these responses appear to be independent of Hsp70 protein production.

3.3 Growth Factors in Testicular Function

3.3.1 Overview

The term growth factor is loosely applied to molecules that mediate cell proliferation, and yet it is clear that they may instead abrogate proliferation in some circumstances and facilitate differentiation in others. Given the diversity of cell types that are present simultaneously within the testis, it is difficult to distinguish the specific actions of a certain molecule. Highly purified cell preparations and the development of *in vitro* model systems continue to be important for studying the effects of growth factors on testicular cell types. Many examples of potential paracrine and autocrine interactions have been described, and the list of these continues to grow. Many have been detected in more than one cell type, a fact that should inspire caution in investigators as they assess the significance of experimental responses to modulating a growth factor signaling event. Localization and physiological data should also be addressed and interpreted with caution, because of the known artifacts induced by testis cell dissociation and culture. The availability of natural mutants or *genetically engineered* mice with altered expression of growth factors and receptors has provided some information about the roles of growth factors in the testis, but these avenues are limited by the normally diverse biological roles of these same substances outside the testis. Mice with tissue- and cell type-specific ablation of individual genes and the manipulation of immortalized somatic and germ cells *in vitro* are providing new information about local regulation of spermatogenesis.

It is not possible to review the roles of many of the growth factors that are progressively being identified in the testis. The sections that follow will provide an overview of the two growth factor systems in the testis that have been studied with relative intensity.

3.3.2 c-Kit and Stem Cell Factor

The c-kit receptor tyrosine kinase Mutations at the *White-spotting (W)* and *Steel (Sl)* loci in mice can result in animals that are anemic, sterile, and white in coat color. Naturally occurring mutations at these loci have provided a superb source of information concerning two genes that are essential for normal germ cell development. The *W* locus encodes c-kit, a receptor tyrosine kinase glycoprotein in the PDGF receptor family. In humans, the *C-kit* gene covers 20 kb and is comprised of 21 exons. Mutations involving this locus include large deletions, rearrangements and point mutations, and these have been documented in humans, mice and rats. These mutations can affect the amount of c-kit protein expressed and the level of kinase activity; all naturally occurring point mutations described to date are within the kinase domain of the protein. Phenotypic changes accompanying heterozygous mutations affect the melanocyte lineage in humans, but to date no association with human infertility has been documented, nor have homozygous mutations been documented. Dimerization of both ligand and receptor subunits is required for signal transduction, causing autophosphorylation of the c-kit receptor and subsequent recruitment of signaling molecules and activation of Akt by phosphatidylinositol-3 kinase (PI-3K), and of extracellular-signal-related kinases 1/2 (ERK1/2) through SH2 binding proteins including Grb2. This activation drives translocation of cyclin D3 to the nucleus where it effects cellular progression through G1 and into S phase.

Stem cell factor The ligand for c-kit is an integral membrane glycoprotein encoded at the *Sl* locus, and it has been named stem cell factor (SCF; also known as mast cell growth factor (MGF), kit ligand (KL) and steel factor). SCF is encoded by at least 8 exons in mouse, rat and humans. Exon 6 encodes an amino acid sequence that is readily cleaved by proteases on the surface of cells *in vitro*, and this cleavage produces a soluble form of SCF. Two SCF mRNA forms are common to every tissue examined, one including and one excluding exon 6. Differential responses by the c-kit-positive target cell to the soluble and membrane-anchored SCF isoforms have been recorded. The relative proportions of the two SCF mRNA forms changes during testicular maturation, with the membrane-anchored form predominant in early development and the soluble form prevalent in the adult. This alternative splicing is regulated *in vitro* by changing the pH of the culture medium, and it is thought to reflect *in vivo* changes in the local environment as the Sertoli cells mature. Synthesis of SCF is promoted by FSH, and this is a key mechanism by which FSH facilitates spermatogonial survival in the postnatal testis.

Interactions between SCF and c-kit affecting spermatogenesis The involvement of c-kit and SCF in spermatogenesis was originally observed during analyses of *W* and *Sl* mice. Phenotype characteristics shared by mice with mutations in either of these genes reflected a loss of stem cell function in hematopoietic, melanocyte, and germ cell lineages. These mice had varying degrees of anemia, white fur patches, and infertility.

In both *Sl*/*Sl* and *W*/*W* homozygotes, the PGCs form normally and are present in normal numbers until 8.5 dpc, when the number of PGCs declines to about 2% of normal by 12.5 dpc. By 14 dpc, the *Sl*/*Sl* mice are devoid of germ cells, and the PGCs have not migrated from the hindgut to the gonadal ridge, while some migration is seen in the *W*/*W* individuals. *SCF* mRNA is detected in mouse mesodermal cells along the migratory pathway of the PGCs from about 9 dpc and in the genital ridge at 12.5 dpc. Isolated primordial germ cells of mice have been grown in culture with SCF and other growth factors. In these systems, the addition of SCF enhances PGC survival; more limited effects on PGC proliferation and adhesion have been described. In addition, recombinant human leukemia inhibitory factor (LIF) stimulates PGC proliferation *in vitro*, and it is probable that other locally produced factors including BMPs (see sections 'The indifferent gonad' and 'Bone morphogenetic proteins'.) regulate migration, proliferation, and differentiation of these cells.

SCF/c-kit interactions are also essential in postnatal spermatogenesis. Messenger RNA encoding c-kit is found in Leydig cells, differentiating spermatogonia, primary spermatocytes, and round spermatids. C-kit receptor protein has been detected on Leydig cells and spermatogonia, and the postmitotic cell types contain mRNA encoding a truncated protein, tr-kit, consisting of only part of the intracellular domain. Tr-kit accumulates in the post-acrosomal region of the sperm head and in the mid-piece of the sperm tail. It is believed to have a role after fertilization in triggering embryo development, as its microinjection into mouse eggs leads to their activation. In contrast to the differentiating type A spermatogonia that have c-kit mRNA and a functional c-kit signaling receptor, stem cell spermatogonia are characterized by the absence c-kit mRNA and protein. The mechanisms that govern upregulation of c-kit, as these cells become committed to differentiate, are clearly important to identify, and Bmp4 has recently been identified as playing a role (see section on Bone morphogenetic proteins).

Membrane-anchored SCF produced by Sertoli cells can interact with c-kit on spermatogonia, while the cleaved form could reach c-kit on additional distal targets such as interstitial Leydig cells. Binding of murine spermatogonia to Sertoli cells, specifically *via* membrane-anchored SCF interaction with c-kit, has been demonstrated *in vitro*. An antiserum that blocks binding of SCF to c-kit disrupts proliferation of type A_{1-4} spermatogonia *in vivo* and *in vitro*, causing increased apoptosis of spermatogonia and spermatocytes. Genetically modified mice bearing a specific point mutation required for PI-3K-mediated c-kit signaling are infertile due to the specific loss of type A spermatogonia. Abrogation of the PI-3K signaling pathway does not cause the pleiotropic phenotypes of white patches and anemia observed in other *W* locus mutants. These data indicate that Sertoli cell signaling through c-kit is required for survival and may impact on mitotic progression of spermatogonia.

3.3.3 Transforming Growth Factorβ Superfamily

The transforming growth factor, βs (TGFβs), belongs to a family of at least 50 proteins sharing a common dimeric structure that is built upon monomers containing seven conserved cysteines. Two common receptor subtypes with serine/threonine kinase activity are involved in most TGFβ superfamily signaling, type I and type II. The ligands in this family have both distinct and overlapping pathways for signal transduction (Fig. 5). Signaling by each of these many ligands is affected by the local availability of a host of soluble and inhibitory ligands, competition for receptor subunits, and competition from inhibitory and stimulatory intracellular signaling components. Owing to the wide variety of effects of these proteins on a broad spectrum of biological systems, it is clear that their expression and action in the testis will be important in a multiplicity of ways that pose a significant challenge to experimenters. In this section, we will review some aspects of this family of growth factors and receptors that relate to our understanding of their roles in testicular function.

TGFβ Three TGFβ genes have been identified in mammals, encoding TGFβ1–3. These are synthesized as large, inactive glycoproteins that must be cleaved and/or released from binding proteins in order to become activated. TGFβ has an additional type III receptor subunit, betaglycan, a transmembrane proteoglycan with the capacity to bind all three TGFβs that is required to enable TGFβ binding to the type II receptor. Betaglycan is also a coreceptor for inhibin A (see section Activin, Inhibin, and Follistatin). Endoglin is another type III receptor for TGFβs 1 and 3 that can also bind activin A and BMPs 2 and 7.

TGFβ receptors on mouse primordial germ cells may mediate suppression of their proliferation, as has been shown *in vitro* on dissociated cells. Later in fetal development, TGFβ application to cultured rat testis fragments induced apoptosis of gonocytes at the ages when the germ cells were proliferating (13.5 dpc and day 3 postpartum), but not at an age when these cells are quiescent (17.5 dpc). Rat gonocytes possess both type I and type II receptors and synthesize TGFβ3. In the adult rat, TGFβ receptor types are

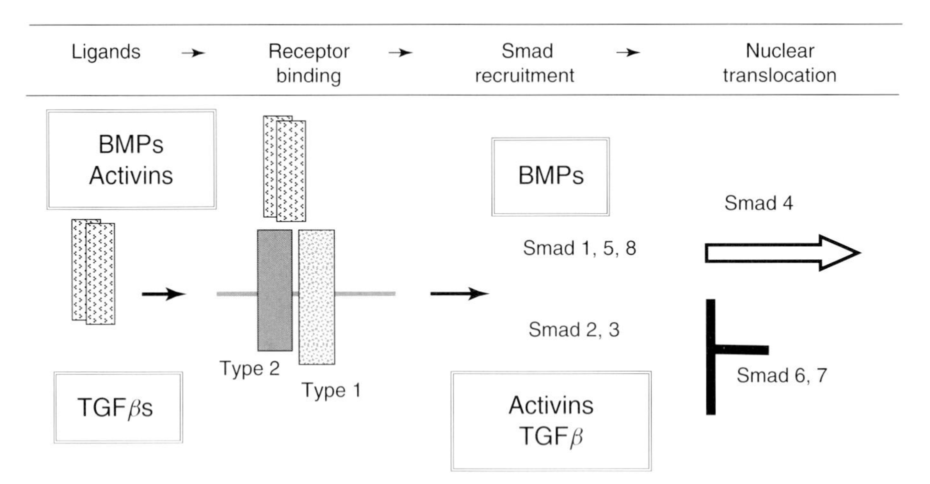

Fig. 5 The TGFβ superfamily members have overlapping and distinct signaling machinery. The local concentration of ligands, inhibitors, receptors, and signaling molecules will create a unique situation in each cell type, with the potential to generate a multiplicity of signals depending on the physiological situation. This diagram illustrates the shared use of receptor subunits by activin and BMPs, while TGFβs use their own set of type I and II receptors and rely on betaglycan for signaling. Inside the cell, activation by activin and TGFβ recruits Smads 2 and 3, while Bmp-signaling recruits Smads 1, 5, and 8. These pathways converge with their requirement for binding the shared Smad 4 to achieve nuclear translocation required for affecting gene transcription, and they are all inhibited by Smads 6 and 7.

coexpressed in pachytene spermatocytes and round spermatids.

Immature Sertoli cells do not appear to have TGFβ receptors, and only the TGFβII receptor has been detected on adult Sertoli cells. However, these cells do make TGFβ1 and 2. Sertoli cell tight junction formation and synthesis of junction-associated proteins *in vitro* is affected by TGFβ3, with *TGFβ*2 and *TGFβ*3 mRNA levels decreasing as these junctions are being established and increasing thereafter. Secretion of TGFβ2 by Sertoli cells and peritubular cells may relate to how they interact to form the testicular cords *in vivo*, since the addition of exogenous TGFβ2 to cocultures, but not to monocultures, promotes formation of cellular aggregates. TGFβ1 can reduce growth of the embryonic and newborn rat testis, and TGFβ3 has been observed in the cells surrounding the seminferous cords of the early postnatal testis.

TGFβ receptors and ligands are also synthesized in Leydig cells. TGFβ3 immunoreactivity is reported in these cells as early as 16.5 dpc, with functional inhibition of LH-stimulated testosterone synthesis by TGFβ described at day 20.5 postpartum. The *in vitro* regulation of Leydig cell testosterone production by TGFβ1 is at least partially achieved by modulating the numbers of cell surface–LH receptors.

Mullerian inhibiting substance (MIS) At the time of transformation of the indifferent gonad into a testis (see section on Genes controlling testis formation), SRY triggers MIS production in Sertoli cells. Acting through the Bmpr1α receptor located on mesenchymal cells of the Müllerian ducts, MIS is essential for regression of this duct that contains the precursors of the female uterus, fallopian tube, and upper vagina. At this time, the onset of testosterone secretion by Leydig cells induces differentiation of the Wolffian duct and male external genitalia.

MIS expression is confined to Sertoli cells, with much higher levels measured in fetal and neonatal testes than those observed after Sertoli cell division has ceased. An inactivating mutation in the murine *MIS* gene results in males with normal testis and male reproductive tract development. However, these animals are infertile owing to the presence of Müllerian duct derivatives that present a physical barrier to normal sperm efflux during coitus. This phenotype can be rescued by mating homozygous *MIS* null mice with those that overexpress the human *MIS* gene, thereby demonstrating the conservative structure of the encoded proteins. MIS-deficient females exhibit normal fertility. Foci of Leydig cell hyperplasia were observed in the testes of some (27%) adult MIS-deficient mice, but no other abnormalities were evident in testis weight and growth of seminal vesicles. From results such as these, a role for MIS in regulation of Leydig cell proliferation has been proposed.

Bone morphogenetic proteins Although there is currently limited information concerning the expression and cellular localization of the 15 BMP ligands and BMP receptor subunits, the impact of BMP signaling has been demonstrated at multiple stages in spermatogenesis. Deletion of murine *BMP4* results in the absence of primordial germ cells, as both Bmp4 and Bmp8 signals from the surrounding somatic environment are required for cells to be allocated to this lineage from the embryo epiblast. A potential role for Bmps2 and 4 in governing the fate of murine stem cells has been revealed through the use of *in vitro* treatment followed by transplantation into a recipient host. In contrast to glial-derived neurotropic factor, a distantly related TGFβ superfamily member, the addition of Bmp4 (or activin; see section on Activin, Inhibin, and Follistatin) reduces the stem cell potential of spermatogonia from cryptorchid animals when cultured on a feeder layer prior to transplantation. This suggests that Bmp signaling may provide a signal to differentiate at the onset of the spermatogenic pathway in the postnatal testis. Sertoli cell–derived Bmp4 has been identified as a candidate mediator of the developing mouse testis by upregulating c-kit at this time. Deletion of *Bmp8b* leads to male mouse infertility due to reduced germ cell proliferation at the onset of spermatogenesis and increased spermatocyte loss in the adult testis.

Activin, Inhibin, and Follistatin Inhibin and activin were first identified and purified on the basis of their role in feedback regulation of the testicular–hypothalamic–pituitary axis. The secretion of inhibin by Sertoli cells acts to reduce FSH secretion by the pituitary, while activin has the capacity to stimulate FSH production (Fig. 4). The signaling activity of activin is modified by the binding of other proteins, such as $\alpha_2$macroglobulin, a component of serum, and follistatin. Follistatin is produced within the testis by Sertoli cells under the influence of FSH, and the protein is present in spermatogonia, spermatocytes, and Leydig cells.

Documenting the pattern of activin and inhibin synthesis, the action within the testis is complicated by the fact that they are each dimers, which share a common subunit, with activin being a homo- or heterodimer of two β subunits (βA, βB, βC, βD, and βE subunits have all been identified), while inhibin is composed of one α and one β subunit (either βA or βB). Inhibin α subunit expression has been detected in Sertoli and Leydig cells, while β subunits have been observed in Sertoli cells, Leydig cells, peritubular myoid cells, gonocytes (βA only), spermatocytes (βB), and round spermatids (βB). The adult Sertoli cell produces predominantly activin B (dimers of βB), while peritubular myoid cells synthesize activin A (βA dimers). *In vivo*, the ratio of α and βB subunits changes during the cycle of the seminiferous epithelium, leading to the suggestion that the relative proportions of inhibin and activin also vary along the length of the tubule. During Sertoli cell maturation, the level of α subunit mRNA synthesis is relatively constant during the proliferative and nonproliferative (maturational) phases, while the α subunit mRNA level decreases as the cells cease to proliferate and begin to mature.

Preferential binding of ^{125}I-labelled inhibin to rat Leydig cells has been reported, and at least two genes have been identified that encode specific inhibin receptor subunits, betaglycan and InhBP/p120. Inhibin, binding to its coreceptor betaglycan (TGFβ type III receptor; see section TGFβ.) blocks activin access to its type II receptor subunit. Betaglycan and InhBP/p120 are both present in Leydig cells. The addition of inhibin to stage dissected–tubule segments in culture results in a decrease in DNA synthesis, specifically in intermediate spermatogonia and in preleptotene spermatocytes, while the addition of activin increases DNA synthesis in the same cell types in this system. This data suggests that either there are other as-yet-unidentified inhibin receptors on germ cells, or that inhibin blocks activin signaling by direct competition for activin receptor binding sites. The effect of activin βA addition on immature rat Sertoli cells is to stimulate their proliferation during a discrete window of postnatal life. Activin addition to Sertoli cell cultures also regulates androgen receptor expression and FSH-induced aromatase activity.

Messenger RNAs encoding activin receptor subunits have been detected in Sertoli and germ cells at all stages of maturation, with stage-dependent peaks during the cycle of the seminiferous epithelium in the adult. Activin, like TGFβ1, inhibits proliferation of mouse primordial germ cells in culture. In contrast, activin increases gonocyte numbers in cultures of testis fragments taken from the newborn rat testis, while blocking their differentiation into spermatogonia at the onset of spermatogenesis. This differentiation was stimulated by the combined addition of FSH and follistatin. Other studies with mouse tissues have shown that activin, like BMP, reduces the stem cell potential of germ cells in culture. The addition of activin to cocultures of Sertoli and germ cells leads to increased spermatogonial proliferation and cellular aggregation. Hence activin may play a role in regulating the proliferation or differentiation of germ cells at many stages of development, further demonstrating the intricacy of regulation, which must contribute to achievement and maintenance of normal spermatogenesis. Messenger RNA encoding a membrane-bound inhibitor of TGFβ superfamily signaling, Bambi, is upregulated as germ cells transform into spermatogonia at the onset of spermatogenesis. In the adult, Bambi mRNA appears highly expressed in pachytene spermatocytes and round spermatids, making it another potential player in the complex set of interactions that govern regulation of TGFβ signaling in the testis.

Another role of activin in the testis may be to maintain the immunosuppressed nature of the interstitium. Activin has been shown to suppress proliferation of peripheral T lymphocytes, an activity that would limit the influence of lymphocytes when presented with novel antigens such as haploid germ cell products, as might occur during testicular damage.

Our understanding of how inhibin and activin act are regulated within the testis is incomplete. In a genetically engineered mouse lacking the inhibin α subunit gene, no effect on testicular development or adult fertility was observed, suggesting that the physiological role of inhibin is either limited or at least partially met by another protein. However, the subsequent development of stromal cell tumors indicates that inhibin can function as a tumor suppressor (see Sect. 3.2.3). Disruption of the *βB subunit* gene has no affect on male fertility, but it has a profound influence on female fecundity manifested as perinatal lethality in offspring of homozygous females. The *βA-/-* mouse dies at birth, so the importance of the βA protein on postnatal development cannot be studied in these animals. Insertion of the *βB* gene into the *βA* locus has been performed to create mice with reduced activin bioactivity (activin A is reportedly 10-fold more bioactive than activin B in terms of signaling), and the animals survive to adulthood. The males are fertile but exhibit a delay in completion of the first wave of spermatogenesis. This further highlights the importance of activin A in the first wave of spermatogenesis, correlating with downregulation of this mRNA during the first week of postnatal life. *In vitro* studies and detailed mRNA analyses conducted on newborn rats has shown that the quiescent gonocytes synthesize activinA in the fetal testis, and the protein is stored until the cells transform into spermatogonia. At this time, the germ cells begin to synthesize the activin antagonists, follistatin and Bambi, indicating that the commencement of their differentiation requires a downregulation of activin signaling.

4 Conclusion

It is clear that a complicated set of hormones and locally produced factors determines the pathway of testis development and spermatogenesis. It is a challenge for the future to further develop model systems that will allow germ and somatic cell differentiation to proceed *in vitro* and to use the techniques that allow manipulation of gene expression in discrete cell types *in vivo*. Already the potential exists, with the production of immortalized somatic and germ cell lines which can support some degree of progression through meiosis. Advances in germ cell and organ transplantation have enabled sperm development in the testis of a host. We are just learning how to identify, store, culture, and stimulate the development of spermatogonial stem cells. This new information is being applied to better understand the source of the complex regulatory signals that guide spermatogenesis, to identify contraceptive targets, to understand the genetic basis of male infertility, and to restore fertility in genetically deficient individuals.

See also Endocrinology, Molecular; Female Reproduction System, Molecular Biology of.

Bibliography

Books and Reviews

de Rooij, D.G., Russell, L.D. (2000) All you wanted to know about spermatogonia but were afraid to ask, *J. Androl.* **21**, 776–798.

Eddy, E.M. (2002) Male germ cell gene expression, *Recent Prog. Horm. Res.* **57**, 103–128.

Handel, M.A. (Ed.) (1998) *Meiosis and Gametogenesis*, Academic Press, CA.

Russell, L.D., Ettlin, R.A., Sinha Hikim, A.P., Clegg, E.D. (1990) *Histological and Histopathological Evaluation of the Testis*. Cache River Press, Clearwater, FL.

Russell, L.D., Griswold, M.D. (Eds.) (1993) *The Sertoli Cell*, Cache River Press, Clearwater, FL.

Soder, O. (Ed.) (2003) *The Developing Testis. Physiology and Pathophysiology*, Karger, Basel.

Primary References and Specialist Reviews

Adams, I.R., McLaren, A. (2002) Sexually dimorphic development of mouse primordial germ cells: switching from oogenesis to spermatogenesis, *Development* **129**, 1155–1164.

Almiron, I., Chemes, H. (1988) Spermatogenic onset. FSH modulates mitotic activity of germ and Sertoli cells in immature rats, *J. Androl.* **11**, 235–246.

Beumer, T.L., Kiyokawa, H., Roepers-Gajadien, H.L., van den Bos, L.A., Lock, T.M., Gademan, I.S., Rutgers, D.H., Koff, A., de Rooij, D.G. (1999) Regulatory role of p27kip1 in the mouse and human testis, *Endocrinology* **140**, 1834–1840.

Beumer, T.L., Roepers-Gajadien, H.L., Gademan, I.S., Kal, H.B., de Rooij, D.G. (2000) Involvement of the D-type cyclins in germ cell proliferation and differentiation in the mouse, *Biol. Reprod.* **63**, 1893–1898.

Blanco-Rodriguez, J. (2001) Mitotic/meiotic Checkpoints and Germ Cell Apoptosis, *Proceedings of the VII International Congress of Andrology*, MEDIMOND, pp. 173–184.

Brennan, J., Tilmann, C., Capel, B. (2003) Pdgfr-alpha mediates testis cord organization and fetal Leydig cell development in the XY gonad, *Genes Dev.* **17**, 800–810.

Brinster, R.L., Avarbock, M.L. (1994) Germline transmission of donor haplotype following spermatogonial transplantation, *Proc. Natl. Acad. Sci.* **91**, 11303–11307.

Burns, K.H., Agno, J.E., Sicinski, P., Matzuk, M.M. (2003) Cyclin D2 and p27 are tissue-specific regulators of tumorigenesis in inhibin {alpha} knockout mice, *Mol. Endocrinol.* **17**, 2053–2069.

Clarke, T.R., Hoshiya, Y., Yi, S.E., Liu, X., Lyons, K.M., Donahoe, P.K. (2001) Mullerian inhibiting substance signaling uses a bone morphogenetic protein (BMP)-like pathway mediated by ALK2 and induces SMAD6 expression, *Mol. Endocrinol.* **15**, 946–959.

Clermont, Y., Perey, B. (1957) Quantitative study of the cell population of the seminiferous tubules in immature rats, *Am. J. Anat.* **100**, 241–267.

Colvin, J.S., Green, R.P., Schmahl, J., Capel, B., Ornitz, D.M. (2001) Male-to-female sex reversal in mice lacking fibroblast growth factor 9, *Cell* **104**, 875–889.

Cooke, H.J., Saunders, P.T. (2003) Mouse models of male infertility, *Nat. Rev. Genet.* **3**, 790–801.

Cooke, P.E., Zhao, Y.-D., Bunick, D. (1994) Triiodothyronine inhibits proliferation and stimulates differentiation of cultured neonatal Sertoli cells: possible mechanism for increased adult testis weight and sperm production induced by neonatal goitrogen treatment, *Biol. Reprod.* **51**, 1000–1005.

Crackower, M.A., Kolas, N.K., Noguchi, J., Sarao, R., Kikuchi, K., Kaneko, H., Kobayashi, E., Kawai, Y., Kozieradzki, I., Landers, R., Mo, R., Hui, C.C., Nieves, E., Cohen, P.E., Osborne, L.R., Wada, T., Kunieda, T., Moens, P.B., Penninger, J.M. (2003) Essential role of Fkbp6 in male fertility and homologous chromosome pairing in meiosis, *Science* **300**, 1291–1295.

Cupp, A.S., Kim, G.H., Skinner, M.K. (2000) Expression and action of neurotropin-3 and nerve growth factor in embryonic and early postnatal rat testis development, *Biol. Reprod.* **63**, 1617–1628.

Fragale, A., Puglisi, R., Morena, A.R., Stefanini, M., Boitani, C. (2001) Age-dependent activin receptor expression pinpoints activin A as a physiological regulator of rat Sertoli cell proliferation, *Mol. Hum. Reprod.* **7**, 1107–1114.

Gnessi, L., Fabbri, A., Spera, G. (1997) Gonadal peptides as mediators of development and functional control of the testis: an integrated system with hormones and local environment, *Endocr. Rev.* **18**, 541–609.

Habert, R., Lejeune, H., Saez, J.M. (2001) Origin, differentiation and regulation of fetal and adult Leydig cells, *Mol. Cell Endocrinol.* **179**, 47–74.

Handelsman, D.J., Spaliviero, J.A., Simpson, J.M., Allan, C.M., Singh, J. (1999) Spermatogenesis without gonadotropins: maintenance has a lower testosterone threshold than initiation, *Endocrinology* **140**, 3938–3946.

Haywood, M., Spaliviero, J., Jimemez, M., King, N.J., Handelsman, D.J., Allan, C.M. (2003) Sertoli and germ cell development in hypogonadal (hpg) mice expressing transgenic follicle-stimulating hormone alone or in combination with testosterone, *Endocrinology* **144**, 509–517.

Heckert, L.L., Griswold, M.D. (2002) The expression of the follicle-stimulating hormone receptor in spermatogenesis, *Recent Prog. Horm. Res.* **57**, 129–148.

Hsia, K.T., Millar, M.R., King, S., Selfridge, J., Redhead, N.J., Melton, D.W., Saunders, P.T. (2003) DNA repair gene Ercc1 is essential for normal spermatogenesis and oogenesis and for functional integrity of germ cell DNA in the mouse, *Development* **130**, 369–378.

Kashiwabara, S., Noguchi, J., Zhuang, T., Ohmura, K., Honda, A., Sugiura, S.A., Miyamoto, K., Takahasi, S., Inoue, K., Ogura, A., Baba, T. (2002) Regulation of spermatogenesis by testis-specific, cytoplasmic poly(A) polymerase TPAP, *Science* **298**, 1999–2002.

Kent, J., Wheatley, S.C., Andrews, J.E., Sinclair, A.H., Koopman, P. (1996) A male-specific role for SOX9 in vertebrate sex determination, *Development* **122**, 2813–2822.

Kumar, T.R., Varani, S., Wreford, N.G., Telfer, N.M., de Kretser, D.M., Matzuk, M.M. (2001) Male reproductive phenotypes in double mutant mice lacking both FSHbeta and activin receptor IIA, *Endocrinology* **142**, 3512–3518.

Kurihara, Y., Tokuriki, M., Myojin, R., Hori, T., Kuroiwa, A., Matsuda, Y., Sakurai, T., Kimura, M., Hecht, N.B., Uesugi, S. (2003) CPEB2, a novel putative translational regulator in mouse haploid germ cells, *Biol. Reprod.* **69**, 261–268.

Lee, J., Richburg, J., Shipp, E.B., Meistrich, M.L., Boekelheide, K. (1999) The Fas system, a regulator of testicular germ cell apoptosis, is differentially up-regulated in Sertoli cell versus germ cell injury of the testis, *Endocrinology* **140**, 852–858.

Lewis, K.A., Gray, P.C., Blount, A.L., MacConell, L.A., Wiater, E., Bilezikjian, L.M., Vale, W. (2000) Betaglycan binds inhibin and can mediate functional antagonism of activin signaling, *Nature* **404**, 411–414.

Liu, D., Liao, C., Wolgemuth, D.J. (2000) A role for cyclin A1 in the activation of MPF and G2-M transition during meiosis of male germ cells in mice, *Dev. Biol.* **224**, 388–400.

Livera, G., Rouiller-Fabre, V., Durand, P., Habert, R. (2000) Multiple effects of retinoids on the development of Sertoli, germ, and Leydig cells of fetal and neonatal rat testis in culture, *Biol. Reprod.* **62**, 1303–1314.

Martineau, J., Nordqvist, K., Tilmann, C., Lovell-Badge, R., Capel, B. (1997) Male-specific cell migration into the developing gonad, *Curr. Biol.* **7**, 958–968.

McCarrey, J.R. (1993) Development of the Germ Cell, in: Desjardins, C., Ewing, L.L. (Eds.), *Cell and Molecular Biology of the Testis*, New York Oxford University Press, pp 58–89.

McCoshen, J.A., McCallion, D.J. (1975) A study of the primordial germ cells during migratory phase in steel mutant mice, *Experientia* **31**, 589–590.

McGuinness, M.P., Orth, J.M. (1992) Reinitiation of gonocyte mitosis and movement of gonocytes to the basement membrane in testes of newborn rats in vivo and in vitro, *Anat. Rec.* **233**, 527–537.

McLachlan, R.I., O'Donnell, L., Meachem, S.J., Stanton, P.G., de Kretser, D.M., Pratis, K., Robertson, D.M. (2002) Identification of specific sites of hormonal regulation in spermatogenesis in rats, monkeys, and man, *Recent Prog. Horm. Res.* **57**, 149–179.

Meehan, T., Loveland, K.L., de Kretser, D., Cory, S., Print, C.G. (2001) Developmental regulation of the bcl-2 family during spermatogenesis: insights into the sterility of bcl-w-/- male mice, *Cell Death Differ.* **8**, 225–233.

Meehan, T., Schlatt, S., O'Bryan, M.K., de Kretser, D.M., Loveland, K.L. (2000) Regulation of germ cell and Sertoli cell development by activin, follistatin, and FSH, *Dev. Biol.* **220**, 225–237.

Meng, X., Lindahl, M., Hyvonon, M.E. Parvinen, M., de Rooij, D.G., Hess, M.W., Raatikainen-Ahokas, A. Sainio, K., Rauvala, H., Lakso, M.,

Pichel, J.G., Wstphal, H., Saarma, K., Sariola, H. (2000) Regulation of cell fate decision of undifferentiated spermatogonia by GDNF. *Science* **287**, 1489–1493.

Meachem, S.J., Mclachlan, R.I., Stanton, P.G., Robertson, D.M., Wreford, N.G. (1999) FSH immunoneutralization acutely impairs spermatogonial development in normal adult rats, *J. Androl.* **20**, 756–762; discussion 755.

Morrish, B.C., Sinclair, A.H. (2002) Vertebrate sex determination: many means to an end, *Reproduction* **124**, 447–457.

O'Donnell, L., McLachlan, R.I., Wreford, N.G., de Kretser, D.M., Robertson, D.M. (1996) Testosterone withdrawal promotes stage-specific detachment of round spermatids from the rat seminiferous epithelium, *Biol. Reprod.* **55**, 895–901.

Orth, J.M., Gunsalus, G.L., Lamperti, A.A. (1988) Evidence from Sertoli cell-depleted rats indicates that spermatid number in adults depends on numbers of Sertoli cells produced during perinatal development, *Endocrinology* **122**, 787–794.

Orth, J.M., Jester, W.F., Li, L.H., Laslett, A.L. (2000) Gonocyte-Sertoli cell interactions during development of the neonatal rodent testis, *Curr. Top. Dev. Biol.* **50**, 103–124.

Orth, J.M., Qiu, J., Jester, W.F. Jr., Pilder, S. (1997) Expression of the c-kit gene is critical for migration of neonatal rat gonocytes in vitro, *Biol. Reprod.* **57**, 676–683.

Osborne, L.R., Wada, T., Kunieda, T., Moens, P.B., Penninger, J.M. (2003) Essential role of Fkbp6 in male fertility and homologous chromosome pairing in meiosis, *Science* **300**, 1291–1295.

Pellegrini, M., Grimaldi, P., Rossi, P., Geremia, R., Dolci, S. (2003) Developmental expression of BMP4/ALK3/SMAD5 signaling pathway in the mouse testis: a potential role of BMP4 in spermatogonia differentiation, *J. Cell Sci.* **116**, 3363–3372.

Print, C.G., Loveland, K.L. (2000) Germ cell suicide: new insights into apoptosis during spermatogenesis, *BioEssays* **22**, 423–430.

Ravnik, S.E., Wolgemuth, D.J. (1999) Regulation of meiosis during mammalian spermatogenesis: the A-type cyclins and their associated cyclin-dependent kinases are differentially expressed in the germ-cell lineage, *Dev. Biol.* **207**, 408–418.

Raymond, C.S., Murphy, M.W., O'Sullivan, M.G., Bardwell, V.J., Zarkower, D. (2000) Dmrt1, a gene related to worm and fly sexual regulators, is required for mammalian testis differentiation, *Genes Dev.* **14**, 2587–2595.

Rossi, P., Dolci, S., Albanesi, C., Grimaldi, P., Ricca, R., Geremia, R. (1993) Follicle-stimulating hormone induction of steel factor (SLF) mRNA in mouse Sertoli cells and stimulation of DNA synthesis in spermatogonia by soluble SLF, *Dev. Biol.* **155**, 68–74.

Rossi, P., Dolci, S., Sette, C., Geremia, R. (2003) Molecular mechanisms utilized by alternative c-kit gene products in the control of spermatogonial proliferation and sperm-mediated egg activation, *Andrologia* **35**, 71–78.

Russell, L.D., Peterson, R.N. (1984) Determination of the elongate spermatid-Sertoli cell ratio in various mammals, *J. Reprod. Fertil.* **70**, 635–641.

Rozen, S., Skaletsky, H., Marszalek, J.D., Minx, P.J., Cordum, H.S., Waterston, R.H., Wilson, R.K., Page, D.C. (2003) Abundant gene conversion between arms of palindromes in human and ape Y chromosomes, *Nature* **423**, 873–876.

Schrans-Stassen, B.H.G.J., van de Kant, H.J.G., de Rooij, D.G., van Pelt, M.M. (1999) Differential expression of *c-kit* in mouse undifferentiated and differentiating Type A spermatogonia, *Endocrinology* **140**, 5894–5900.

Schrans-Stassen, B.H., Saunders, P.T., Cooke, H.J., de Rooij, D.G. (2001) Nature of the spermatogenic arrest in Dazl-/- mice, *Biol. Reprod.* **65**, 771–776.

Shuttlesworth, G.A., de Rooij, D.G., Huhtaniemi, I., Reissmann, T., Russell, L.D., Shetty, G., Wilson, G., Meistrich, M.L. (2000) Enhancement of A spermatogonial proliferation and differentiation in irradiated rats by gonadotropin-releasing hormone antagonist administration, *Endocrinology* **141**, 37–49.

Sicinski, P., Donaher, J.L., Geng, Y., Parker, S.B., Gardner, H., Park, M.Y., Robker, R.L., Richards, J.S., McGinnis, L.K., Biggers, J.D., Eppig, J.J., Bronson, R.T., Elledge, S.J., Weinberg, R.A. (1996) Cyclin D2 is an FSH-responsive gene involved in gonadal cell proliferation and oncogenesis, *Nature* **384**, 470–474.

Sinclair, A.H., Berta, P., Palmer, M.S., Hawkins, J.R., Griffiths, B.L., Smith, M.J., Foster, J.W., Frischauf, A.M., Lovell-Badge, R., Goodfellow, P.N. (1990) A gene from the human sex-determining region encodes a protein with homology to a conserved DNA-binding motif, *Nature* **346**, 240–244.

Spruck, C.H., de Miguel, M.P., Smith, A.P., Ryan, A., Stein, P., Schultz, R.M., Lincoln, A.J., Donovan, P.J., Reed, S.I. (2003) Requirement of Cks2 for the first metaphase/anaphase transition of mammalian meiosis, *Science* **300**, 647–650.

Tadokoro, Y., Yomogida, K., Hiroshi, O., Tohda, A., Nishimune, Y. (2002) Homeostatic regulation of germinal stem cell proliferation by the GDNF/FSH pathway, *Mech. Dev.* **113**, 29–39.

Tsutsui, T., Hesabi, B., Moons, D.S., Pandolfi, P.P., Hansel, K.S., Koff, A., Kiyokawa, H. (1999) Targeted disruption of CDK4 delays cell cycle entry with enhanced p27(Kip1) activity, *Mol. Cell Biol.* **19**, 7011–7019.

Venables, J.P., Elliott, D.J., Makarova, O.V., Makarov, E.M., Cooke, H.J., Eperon, I.C. (2000) RBMY, a probable human spermatogenesis factor, and other hnRNP G proteins interact with Tra2beta and affect splicing, *Hum. Mol. Genet.* **9**, 685–694.

Vergouwen, R.P., Jacobs, S.G., Huiskamp, R., Davids, J.A., de Rooij, D.G. (1991) Proliferative activity of gonocytes, Sertoli cells and interstitial cells during testicular development in mice, *J. Reprod. Fertil.* **93**, 233–243.

Vidal, V.P., Chaboissier, M.C., de Rooij, D.G., Schedl, A. (2001) Sox9 induces testis development in XX transgenic mice, *Nat. Genet.* **28**, 216–217.

Wang, P.J., McCarrey, J.R., Yang, F., Page, D.C. (2001) An abundance of X-linked genes expressed in spermatogonia, *Nat. Genet.* **27**, 422–426.

Wu, X.Q., Hecht, N.B. (2000) Mouse testis brain ribonucleic acid-binding protein/translin colocalizes with microtubules and is immunoprecipitated with messenger ribonucleic acids encoding myelin basic protein, alpha calmodulin kinase II, and protamines 1 and 2, *Biol. Reprod.* **62**, 720–725.

Wykes, S.M., Krawetz, S.A. (2003) The structural organization of sperm chromatin, *J. Biol. Chem.* **278**, 29471–29477.

Yan, W., Kero, J., Suominen, J., Toppari, J. (2001) Differential expression and regulation of the retinoblastoma family of proteins during testicular development and spermatogenesis: roles in the control of germ cell proliferation, differentiation and apoptosis, *Oncogene* **20**, 1343–1356.

Yan, W., Suominen, J., Samson, M., Jegou, B., Toppari, J. (2000) Involvement of Bcl-2 family proteins in germ cell apoptosis during testicular development in the rat and pro-survival effect of stem cell factor on germ cells in vitro, *Mol. Cell Endocrinol.* **165**, 115–129.

Yang, J., Chennathukuzhi, V., Miki, K., O'Brien, D.A., Hecht, N.B. (2003) Mouse testis brain RNA-binding protein/Translin selectively binds to the messenger RNA of the fibrous sheath protein glyceraldehyde 3-phosphate dehydrogenase-S and suppresses its translation in vitro, *Biol. Reprod.* **68**, 853–859.

Zhang, F.P., Poutanen, M. (2001) Normal prenatal but arrested postnatal sexual development of luteinizing hormone receptor knockout (LuRKO) mice, *Mol. Endocrinol.* **15**, 172–183.

Zhao, G.Q., Liaw, L., Hogan, B.L.M. (1996) Bone morphogenetic protein 8A plays a role in the maintenance of spermatogenesis and the integrity of the epididymis, *Development* **125**, 1103–1112.

Zhao, M., Shirley, C.R., Yu, Y.E., Mohapatra, B., Zhang, Y., Unni, E., Deng, J.M., Arango, N.A., Terry, N.H., Weil, M.M., Russell, L.D., Behringer, R.R., Meistrich, M.L. (2001) Targeted disruption of the transition protein 2 gene affects sperm chromatin structure and reduces fertility in mice, *Mol. Cell Biol.* **21**, 7243–7255.

Zhou, Q., Nie, R., Prins, G.S., Saunders, P.T., Katzenellenbogen, B.S., Hess, R.A. (2003) Localization of androgen and estrogen receptors in adult male mouse reproductive tract, *J. Androl.* **23**, 870–881.

Zsebo, K.M., Williams, D.A., Geissler, E.N., Broudy, V.C., Martin, F.H., Atkins, H.L., Hsu, R.Y., Birkett, N.C., Okino, K.H., Murdock, D.C. Jacobsen, F.W., Langley, K.E., Smith, K.A., Takeishi, T., Cattanach, B.M., Galli, S.J., Suggs, S.V. (1990) Stem cell factor is encoded at the *Sl* locus of the mouse and is the ligand for the *c-kit* tyrosine kinase receptor, *Cell* **63**, 213–224.

Mammalian Cell Culture Methods

Dieter F. Hülser
University of Stuttgart, Stuttgart, Germany

Encyclopedia of Molecular Cell Biology and Molecular Medicine, 2nd Edition. Volume 7
Edited by Robert A. Meyers.

ISBN: 3-527-30549-1

Keywords

Cell Cloning
Generation of a colony from a single cell. Subculturing results in a cell strain.

Cell Cycle
Ordered sequences (G_1, S, G_2, and M) of cellular syntheses between two cell divisions.

Cell Line
Subcultured primary cultures. A cell line can be finite or continuous.

Cell Strain
Cell line that has been purified by physical separation, selection, or cell cloning.

Continuous Cell Line
Indefinite proliferation. This immortalization of a cell line may be induced by a viral gene transfer or was already acquired by some cancer cells before cultivation.

Primary Culture
Freshly isolated cells in culture until the first passage into a subculture.

Serum
Blood fluid without cells and clotting factors.

Suspension Culture
Cells proliferate isolated from each other when suspended in growth medium.

Tissue Culture
Accustomed term for the cultivation of animal cells. Originally: Fragments of tissues maintained *in vitro*.

Mammalian cell cultures originate from tissue explants or cell suspensions as primary cell cultures that can be subcultured with a limited life span. By transformation, these cells might lose some of their original properties and establish permanent growth. Many of these continuous cell lines are aneuploid and genetically unstable, nevertheless, our knowledge of molecular, physiological, biochemical, and biophysical properties of cells is notably based on investigations with such cell lines. Since the synthesis of various bioproducts such as vaccines, monoclonal antibodies, enzymes, and hormones is accomplished with cell cultures, many efforts were made to develop and improve cell culture technology. Besides this impetus, cancer research also stimulated the progress of cell culture methods as can be seen with the development of three-dimensionally growing cultures. These cultures provided a better understanding of tumor invasion and revealed the importance of the extracellular matrix for physiological regulations of cell–cell interactions that cannot be observed with monolayer or suspension cultures. This knowledge helped to improve the cultivation of cells that are used in clinical treatments as is the case for wound healing with implantation of epidermis or for defect organs such as liver, which can be supported with bioartificial organs during a temporary extracorporal bypass.

1 Principles

1.1 Development of Cultured Cells

Animals are three-dimensionally organized complex multicellular creations. Their tissues and organs maintain specific internal milieus and are separated from each other and from their environment by specialized endothelial and epithelial cells. The adult human body consists of hundreds of cell types and altogether of approximately 50 to 100 × 10^{12} cells from which about 10^9 cells vanish within one hour. This sounds more alarming than it really is because most of these cells are replaced within the same time. The highest turnover is found for blood and epithelial cells; after wounding, organs and tissues might start proliferation albeit they are normally almost quiescent. Even when most cells have the potential for proliferation, it is not an easy task to cultivate them *in vitro*, that is, in Petri dishes, flasks, or bioreactors.

A stimulus to study isolated cells was given in 1858 when Rudolf Virchow postulated that pathological characteristics may be detected on a cellular level. First experiments to cultivate cells were performed with non-mammalian cells such as amphibians with their high capacity to regenerate lost limbs and with chicken embryos that are easily accessed and proliferate with a high activity. Successful attempts to maintain cells *in vitro* date back to 1885 when Wilhelm Roux kept isolated nerve fibers of chicken embryos alive in a warm saltbroth. Much effort was put in the establishment of an appropriate medium in which the cells could survive and even proliferate. In 1895, Paul Ehrlich succeeded to grow and propagate mouse tumor cells by intraperitoneal injection in mice where they grew as single-cell suspension in ascites. This *in vivo* cultivation revealed that the liquid of the peritoneum obviously contains all nutrients necessary for cell proliferation. Consequently, human ascites was used to keep pieces of human skin alive *in vitro*, as was demonstrated in 1898 by Ljunggren who grafted them back to the donors, including himself. A significant step forward was achieved in 1907 by Ross Harrison, who cultivated spinal cords of frogs in a hanging drop of coagulated frog's lymph and studied the outgrowth of nerve fibers during several weeks. In 1912, Carrel reported on the permanent life of tissues outside the organism when he cultivated embryonic chicken cells in an extract of chicken embryos. He introduced aseptic methods and cultivated the cells as monolayers in glass flasks, but that he has kept these cells for more than 30 years is certainly not true since *in vitro* proliferation of normal cells is limited to about 50 divisions, which became known later by the investigations of Hayflick and Moorhead. Carrel and coworkers had regularly fed their cultures with insufficiently filtered extracts of chicken embryos, which contained fresh cells.

Different media were adopted for specific cells, which grew as monolayers on different substrata such as glass, polypropylene, ceramics, and so on. Earle and coworkers treated mouse fibroblasts with methylcholanthrene and cultured these transformed cells in medium with a bicarbonate/CO_2 buffered salt solution. These L-cells have been growing as a continuous cell line since 1943, and at present their use might only be outnumbered by the first continuous human cell line HeLa, which was derived from a human cervical carcinoma by Gey and coworkers in 1952 (Fig. 1). Modified medium formulations were used by Dulbecco to grow animal viruses in cultured cells and Eagle described in 1955 the minimum requirements of nutrients for appropriate synthetic media, which, however, still required the addition of serum from young animals. Another important step was made by Levi-Montalcini and colleagues who isolated of a protein, which stimulates the growth of certain nerve cells. Similarly, to this nerve growth factor, other growth factors were later isolated from serum (see Sect. 2.2). With better-defined media, more complex cells could be cultivated as was the case with the successful development of a hybridoma cell line for the production of monoclonal antibodies by Köhler and Milstein.

Even from the very beginning of tissue culture, three-dimensional explants were always used, and most experiments are still performed with monolayer and suspension cultures. This might change in the near future since many regulatory processes are better studied in three-dimensionally growing cells that perfectly match the *in vivo* conditions.

1.2 Selection of Cells

Most animal cells are cultivated as monolayers since the majority of cells is anchorage dependent and not very selective with regard to a substratum. This type of cell cultivation offers advantages for microscopic inspections during growth or during experiments and it allows micromanipulation of the cells with electrodes. Normal cells in a primary culture usually stop proliferation by a so-called contact inhibition when they come in close contact on a flat surface. By serum deprivation, the proliferation of these cells can also be inhibited and they remain quiescent in the G_1- or G_0-phase

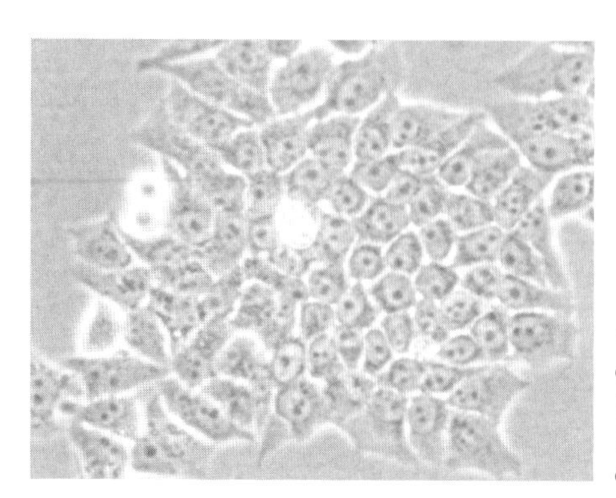

Fig. 1 Monolayer of HeLa cells during logarithmic growth. Diameter of a cell is about 15 μm. Light-microscope phase contrast picture.

of their cell cycle. After addition of serum to these cultures, the cells reenter the cell cycle and progress in a synchronized manner. Under these conditions, signals from individual cells are multiplied since effects occur synchronously in all cells as has been demonstrated for growth factor–induced channel openings in embryonic rat fibroblasts. In spite of the high variance of samples, cells of primary cultures are recommended when specialized functions should be investigated under controlled culture conditions.

Cultures with selected properties are available with transformed cells. They are not contact inhibited and continue to grow as long as the medium is not depleted of nutrients. Numerous investigations in cell biology as well as in cancer research were performed with these continuous cell lines and cell strains, and much of our knowledge on cellular regulation, synthesis, and proliferation is based on these cells. For the production of cellular material, cells were selected after transformation, hybridization, or transfection. In catalogs of the American Type Culture Collection (ATCC), of the European Collection of Cell Cultures (ECACC), or of other national culture collections, appropriate cells for many purposes can be found.

Since production is usually optimized for a high yield, cells were further selected for anchorage-independent growth, which qualifies them to proliferate and to synthesize the requested product in single-cell suspension in large volumes. However, as is evident from intact organisms, producer cells must not necessarily be in a proliferative status. This understanding finally led to cultivation methods that allow the cells to grow as three-dimensional cell aggregates. This might end in organ-like cultures where the cells stop to proliferate but still synthesize substances or function as bioartificial organs as is the case for extracorporal cultivated liver cells.

1.3
Cell Cycle and Growth Curves

Under optimal conditions in culture, cells may divide every 10 to 20 hours except when cells may be stressed by transferring from one culture vessel into the next or by exhausted medium. Between two cell divisions, a cell must synthesize all its materials such as DNA, RNA, proteins, lipids, and so on. These processes could occur continuously or in discrete phases. As has been demonstrated by Quastler and Sherman, DNA is replicated in a separate phase that starts a few hours after division and ends a few hours before the next division. This DNA synthesizing period is called *S-phase*, it is preceded by the G_1-and followed by the G_2-phase, whereas the M-phase covers the mitosis and the cell segregation between the two G-phases. Similarly, as in an adult organism, cells may stop proliferation and rest in the G_0-phase.

When cells are cultivated as suspension culture, they exist as isolated single cells and can easily be harvested without enzymatic treatment and transferred directly into a new culture vessel. Under these conditions, cells may immediately continue to proliferate at a high rate. During mitosis, one cell divides into two cells and, therefore, the number of cells increases to $N + 1$. Since one "old" cell disappears after its doubling time and two "young" cells are instead added to the culture, we must distinguish between growth rate $G = (\mathrm{d}N/\mathrm{d}t)$, which indicates how many cells are added to a culture during a certain time period, and the birth rate

$B = 2G$, which gives the number of cells that are "born" at a certain time point. A culture of continuously proliferating cells has, therefore, twice as much young cells as old cells and its age distribution declines exponentially.

Under these conditions, an exponential growth is observed, which is described by

$$N = N_0 \cdot e^{\mu \cdot t} \tag{1}$$

with N: number of cells; N_0: number of cells at the start of the culture; growth parameter $\mu = ln2/t_d$; t: time; t_d: doubling time of cells. Every cell has an individual cell age τ that starts after mitosis at $\tau = 0$ and ends at $\tau = t_d$ with the next division. Under optimal conditions, the individual doubling time of a cell could be identical with the population doubling time. However, during growth, the medium will gradually be exhausted and some cells may stop proliferating or are prolonged in their cell cycle because one or more restriction points cannot be passed. Furthermore, when monolayer cells are transferred from one vessel into the next by trypsinization and mechanical treatment, they might additionally be stressed by temperature shifts and their proliferation in the new vessel starts delayed. These delays result in characteristic growth curves as is shown in Fig. 2 for BICR/M1R_k cells. For fitting these points with a curve, the growth parameter μ must be modified to $\mu = \mu_1 - \mu_2 N$ and leads to the Verhulst–Pearl equation:

$$N = \frac{\mu_1 \cdot N_0 \cdot e^{\mu_1 \cdot t}}{\mu_1 + \mu_2 \cdot N_0(e^{\mu_1 \cdot t} - 1)} \tag{2}$$

This equation represents realistic growth conditions in which the proliferation also depends on the number of cells and the available amount of nutrients in a medium. A typical growth curve starts with a lag-phase in which the number of cells is rather constant before the cells start with exponential growth and enter the log-phase. For practical reasons, the ordinate of a growth curve is logarithmically divided and, therefore, the log-phase is characterized by a straight line. When the medium is deprived of nutrients, cells stop proliferating and for a while the number of cells is constant – the culture is in its stationary phase before cells start to die away.

1.4
Cell Profiling

Altogether, several ten thousands of cell lines may be kept worldwide; therefore, a reliable identification of the cells becomes mandatory. Besides microbial contamination and phenotypic drift, cross-contamination between cells occurs more

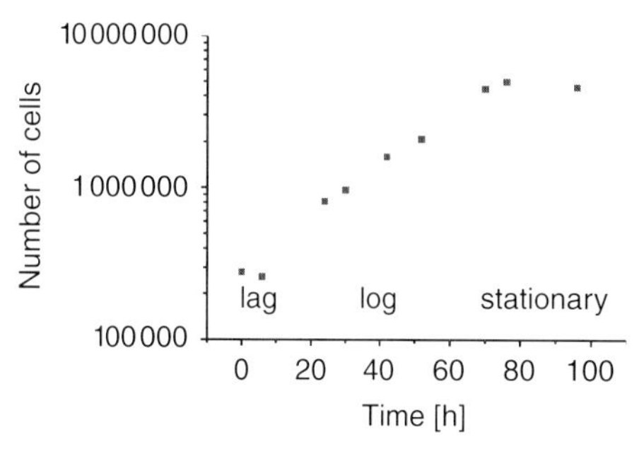

Fig. 2 Growth curve of BICR/M1R_k cells with a typical lag-, log-, and stationary-phase. This continuous cell line originates from a mammary carcinoma of the Marshall rat.

often than it is detected and leads to severe misidentifications of cell cultures. Within 10 years of the start in culture of the human cancer cells HeLa, other human and animal cell lines were contaminated with these fast-growing HeLa cells as was detected by genetic markers. These alarming reports did not stop this carelessness and in 1976, Nelson-Rees and Flandermeyer reported more than 90 cell lines that were contaminated with HeLa cells. It is still estimated that up to one-third of all cell lines is of different origin or species to that claimed. These cross-contaminations result from careless handling as might be the case when several cell lines are transferred simultaneously and medium is added to all cultures with the same pipette. Aerosols and small droplets with cells might adhere to the outside of the pipette and are thus transferred into the stock medium from where they can contaminate other cultures. Cross-contamination could be detected by experienced molecular biology laboratories with several time-consuming and expensive methods, but none was suitable for use as an international reference standard. With the development of new PCR-based forensic techniques, an inexpensive method, the short tandem repeat (STR) profiling, has been applied to DNA from human cell lines. STR generates standard numerical codes for lengths of polymorphic DNA loci, which qualify as universal reference standard for human cell lines. Masters and colleagues recommend DNA profiling of cell lines as a normal practice and suggest this analysis as a prerequisite for publication. This DNA fingerprint will not only help to identify cell lines and greatly enhance the confidence in these studies but should also allow comparisons of published results on a more solid basis as momentarily possible.

1.5 Ethical Rules

Working with continuous cell cultures does not seem to cause severe ethical problems; however, when primary cells are acquired from humans and animals, several critical implications arise. For clinical treatments as well as for research, the use of human stem cells from embryos, fetuses, or adults is strictly regulated by national and international legislation. However, research with other human cellular material must also be justified and can only be approved by local research ethical committees when the demands of national guidelines are enforced. Similarly, the use of animal tissue requires ethical and legal approval, often by local animal welfare committees. Animal welfare aspects include the minimizing of pain, suffering, distress, and lasting harm as well as the restriction to the absolutely necessary number of animals. Experiments with cultures of permanently growing cells do not exclude animal welfare considerations since most cell lines are still maintained with purchased fetal bovine or newborn calf serum. Even when little or no influence can be taken on the collecting process, there are also technical arguments for considering a replacement of serum (see Sect. 2.2).

2 Techniques

2.1 Preparation of Primary and Continuous Cultures

A primary culture starts with biopsies from solid tissues or organs, preferentially from embryos, but tumor cells are also often selected because of their high proliferation capacity. These firmly attached cells have to be disaggregated to single cells, a procedure that combines mechanical dissociation and enzymatic detachment. After chopping the tissue with fine surgical scalpels, proteolytic enzymes such as trypsin, collagenase, or pronase are applied to open proteinaceous bonds between cells, a process that might be accelerated by the appropriate temperature and thorough pipetting. This cell suspension will be diluted with medium plus serum; centrifuged and resuspended in medium for several times before the cells are transferred into Petri dishes or flasks. Most cells are anchorage dependent and will adhere to an appropriate substratum after a few hours. At that time point, the cultures must be washed free of all cellular debris and damaged cells, since only a small portion of cells may have survived the isolation procedure. Microscopic inspection will then reveal the yield of attached single cells, but also aggregates of cells will be found from which an outgrowth of cells can be observed after a couple of days. Usually, a mixture of cell types will be present in these primary cultures, the majority being fibroblasts that will proliferate better than other cell types. Since they are also motile, they can grow out from cell aggregates and spread over the substratum. Epithelial cells, on the other hand, are more or less immobile and tend to form patches on the substratum. Figure 3 shows a primary culture of embryonic mouse brain cells in which nerve cells have spread on top of other cells that are out of focus.

Adherent cells can be detached by trypsin, and enough cells might be harvested from these primary cultures to proceed with their propagation in multiple dishes or flasks. These subcultures can be used to multiply the selected cells and generate a new cell line that contains all the different cells of the primary culture. If these cells are selected from normal tissues, these cultures have a limited life span and must, therefore, be multiplied by several cultivation passages, collected, and stored in liquid nitrogen so that planned experiments can always rely on identical cell populations (see Sect. 2.5). Critical cells may not proliferate, especially when the density of transferred cells is too low. This problem can sometimes be solved either with conditioned medium or with feeder layers. The so-called conditioned medium is taken from a culture of logarithmically growing cells and added to freshly propagated cells. This medium contains cell specific factors that mimic the presence of numerous cells and stimulate the growth of critical cells. A similar effect can be obtained with irradiated feeder layer cells, which form a monolayer and serve as substratum for critical cells.

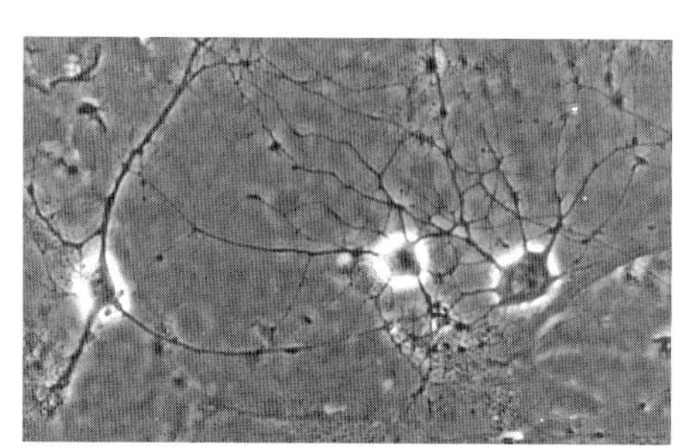

Fig. 3 Primary culture of embryonic mouse brain cells. Nerve cells have spread on top of other cells, which are out of focus. Light-microscope phase contrast picture.

By transforming and cloning cell lines, permanently growing cell strains with specific properties can be derived, and many of them are available from cell banks. A majority of experiments is, therefore, performed with these continuous cells.

2.1.1 Monolayer Cultures

Plasma membranes of animal cells carry fixed negative charges that allow cell–cell adhesion by divalent cations (Ca^{++}, Mg^{++}), but cells may also adhere to artificial surfaces that are either negatively or positively charged. Cells form a monolayer when they adhere not only to each other but also to the surface of culture vessels. Different cells form differently arranged monolayers: epithelial cells (Fig. 1) show a typical cobble stone appearance, whereas transformed fibroblasts (Fig. 4) may grow criss-cross and pile up when a high density is reached. The cells' morphology might indicate that the medium is exhausted and its replacement or a subculture is necessary. For propagating the cells, the medium is withdrawn and trypsin is added with a Ca^{++}/Mg^{++} free balanced salt solution. The time and temperature required for this treatment depend on the sensitivity of the cells, and other enzymes might also be required for a gentle disaggregation. In any case, the enzymatic treatment should be as short as possible to avoid unnecessary stress. Therefore, the cells must be washed with medium to block the activity of trypsin by its serum. The cells can then be detached from the surface by repetitive pipetting of medium over the monolayer, a procedure that also disperses floating cell aggregates. Portions of this resulting single-cell suspension are used for subcultures, and an aliquot might be used for determining the concentration of cells. The number of cells that are necessary for starting a new culture depends on the cell type, on the kind of planned experiments, and on the type of culture vessels. A few hours after starting the incubation of new cultures, they can be microscopically inspected to exclude any damage that might have occurred during the passage.

2.1.2 Suspension Cultures

Several monolayer cell lines have lost their anchorage dependence by transformation and are now also kept as suspension cultures. HeLa cells are amongst these cultures, which grow as monolayers and in suspension. Primary suspension cultures can be obtained from normal lymphocytes, which are anchorage independent. They have a limited life span, but lymphoblastoid cells with unlimited growth in culture are also available as are other ascites tumor cells that have lost their anchorage dependence already *in vivo*.

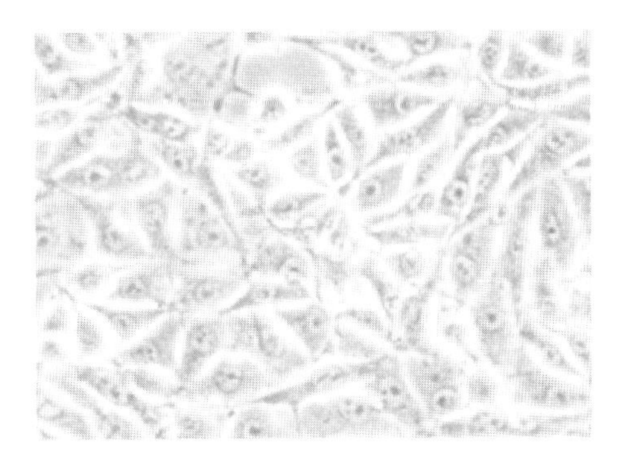

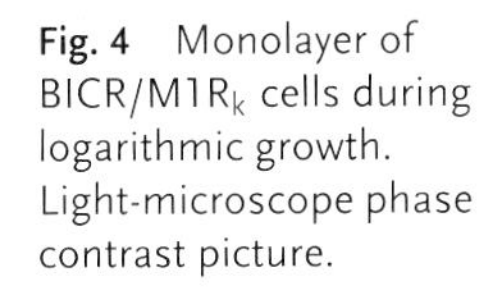

Fig. 4 Monolayer of BICR/M1R$_k$ cells during logarithmic growth. Light-microscope phase contrast picture.

2.1.3 Three-dimensionally Growing Cultures

Not only may primary explants grow three-dimensionally but also transformed cells can be cultured as so-called multicellular spheroids in suspension. For a few anchorage-dependent cell lines, it is sufficient to start a culture of these cells in a stirrer flask for suspension cultures. When cells meet each other in this suspension, they will stay together and thus form larger spheroidal aggregates within a couple of weeks. Most other cell lines must first be kept in dishes with a nonadhesive surface to induce this three-dimensional growth. These Petri dishes are commercially available and are mainly used for cultivating bacteria on agar. After a few days of cultivation in these dishes, the cells have formed irregular aggregates, which are transferred into stirrer flasks in which they form multicellular spheroids and may be kept and treated for several weeks (Fig. 5). Spheroids can be harvested directly with pipettes, and medium is easily replaced since the spheroids precipitate quickly when they are no longer stirred. When such a spheroid is placed on a culture dish, the cells quickly adhere to this charged surface and within a day many cells will have migrated from the spheroid and proliferate again as monolayer cells.

When biopsies of organs are cultured in stirrer flasks, they will maintain morphological and physiological properties, and many cells remain fully differentiated and proliferate sparsely. Even when fresh explants are required for new experiments, these cultures offer some advantages for complex experiments that might, for instance, include immunological reactions as has been shown with lymphoid tissue and HIV infection.

Epithelium, which separates an organism from its environment or an organ from its surrounding fluid, is formed by layers of different cells. When cultured *in vitro*, epithelial cells often form flat monolayers of different cells but epidermal cells may be kept as organotypic flat multilayers in a liquid–gas interphase. This *in vitro* skin is another example for the many types of three-dimensionally growing cultures that are not only used to study tissue regeneration but are also applied in biomedical treatments.

2.2 Cell Culture Media

The basic components of cell culture media are inorganic salts, glucose, and organic substances, which are also present in blood plasma of animals. Their concentration varies with different compositions of

Fig. 5 Multicell spheroids of BICR/M1R_k cells. The diameter of the larger spheroids is about 350 μm. Scanning electron microscope picture.

media that were designed for specific cells with special properties. Most of these media are commercially available as solutions and also as powder that has to be dissolved in water. The selection of water, however, is not trivial since it may contain organic and inorganic material, which will not be removed by simple distillation so that other purification processes must be included. Differences in cell activity may be due to different water qualities; therefore, the use of ultrapure water is recommended. In any case, powdered media require sterilization by filtration through a pore size of 0.22 μm, preferably during bottling and before storage in a refrigerator.

All basic salt broths of cell culture media contain NaCl, KCl, $CaCl_2$, and $MgCl_2$. The concentration of these inorganic salts is set to adjust the osmotic balance of the cells and maintain their membrane potential. When this solution is buffered to a pH value of 7.2 to 7.4 with the pH buffering phosphates NaH_2PO_4 and Na_2HPO_4, it can be used to keep the cells alive for a short while. A complete medium requires much more additives for the functioning of inorganic salts as mediators for cell attachment or as enzyme cofactors. Carbohydrates (glucose) are added as energy source, and amino acids serve as nitrogen sources, although glutamine can also provide a carbon source via transamination. Essential and nonessential amino acids and water-soluble vitamins are also necessary as nutrients and additional energy sources; their concentrations vary with different medium formulations.

For proliferating mammalian cells, the buffering capacity of the medium is increased by bicarbonate ($NaHCO_3$). In this case, the medium's CO_2/HCO_3 content must be balanced by gaseous CO_2, which requires gassed incubators in which an atmosphere of 5–10% CO_2 in air can be maintained. Another buffering system is given by the zwitterion HEPES, which is often used together with the "natural" bicarbonate buffering system. With this buffer, cells can be manipulated under atmospheric conditions without drastic pH changes. For a rapid and easy control of the medium's pH, phenol red is added as indicator.

Many laboratories supplement their medium with antibiotics such as streptomycin sulfate and penicillin G. This reduces the risk of bacterial contamination but increases the development of resistant organisms and encourages the use of inadequate aseptic techniques. Transfected cells, however, may have been selected as resistant to puromycin or geneticin, and the use of these antibiotics is recommended for maintaining a selection pressure.

A very important component of a cell culture medium is serum, and in most cases newborn calf or fetal bovine serum is added. Serum is a complex mixture that contains not only essential growth factors but also plasma proteins, hormones, metabolites, and growth inhibitors. Since serum components are affected by age, health, and nutrition of the donor animals, the quality of serum will change from batch to batch and, therefore, every new batch has to be tested for the required quality standards. When bovine spongiform encephalopathy (BSE) spread in cattle, guidelines became effective, which should minimize the risk of BSE transmission via medicinal products and the necessity for media without any material from animal origin has increased.

At the beginning, the so-called serum-free media contained factors that were isolated from serum and stimulated the growth of certain cell types, which can be seen from their name: fibroblast growth factor (FGF), nerve growth factor (NGF), epithelial growth factor (EGF), platelet-derived growth factor (PDGF), and insulin-like growth factor (IGF). Other serum-isolated additives include interferons, hormones, and attachment factors. With these components, a serum-free medium was still an undefined medium since it contained components that were subject to variation. Effects to develop defined media where the chemical structure and the concentration of every component are known were successful. These media are consistent from batch to batch and are optimized for growth of specific cell types and product synthesis. Almost every cell line requires its own expensive serum-free medium, its use is therefore still limited to productions with expensive purification processes and to some biomedical treatments.

2.3 Equipment

2.3.1 Vessels

Monolayer cells need a substratum to adhere, and this material must be of good optical quality since microscopic inspection of the cells is mandatory for most experiments, but other material might be advantageous when it comes to production. Cells preferentially adhere to negatively charged surfaces and, therefore, the plastic vessels for cell culturing are charged during the manufacturing process. Many laboratories use transparent disposable multiwell plates, Petri dishes, and T-flasks made of polystyrene. Their volume varies between 0.1 and 250 mL medium and with the corresponding surface areas. However, other mechanisms are also effective in cell adhesion. When cells are cultivated on glass surfaces, they might not adhere well when the glass flasks are first in use. After several passages of cells, the adhesion is enhanced, indicating a conditioning of these glass surfaces. This is due to specific receptors for cellular adhesion, which are part of the extracellular matrix of plasma membranes. They partly remain adherent to the glass surfaces when the cells move or are harvested. Even after cleaning and sterilizing, material is firmly connected with the glass and thus facilitates the attachment of cells of sequential cultures. This finally led to the use of special mixtures of extracellular matrix glycoproteins and proteoglycans by which any substratum could be made available for the cultivation of critical cells. Continuous cell lines such as HeLa adhere to different substrata, for example, carbon-coated glass (Fig. 6) and polished stainless steel (Fig. 7).

Fig. 6 Monolayer of HeLa cells during logarithmic growth on carbon- coated glass. Attached particles originate from serum and/or dead cells' debris. Scanning electron microscope picture, bar = 30 μm.

Fig. 7 HeLa cells grown on stainless steel. Note the amoeboid protrusions onto the nonpolished gaps. Scanning electron microscope picture, bar = 10 μm.

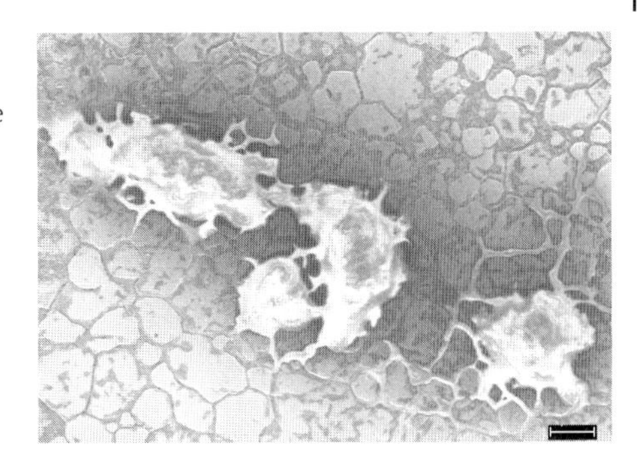

Many sophisticated systems have been developed to improve the yield of monolayer cells. Since the area of the substratum on which the cells can grow must be increased and the volume of the vessels must be expanded, risks of contamination and problems of regulation will also simultaneously rise. These tasks are solved with different strategies. A classical technique is the use of cylindrical bottles that are slowly rotated so that the total inner glass or plastic surface is available for cultivating monolayers with a small amount of medium. Many of these roller culture bottles can be rotated simultaneously on racks and thus increase the total amount of products. Other mass culture techniques make use of separated compartments in which the cells grow on permeable membranes and medium is provided in a compartment separate from the products, a strategy that reduces the cost of product purification. An example is the hollow fiber, which is not only used for adherent but also for suspended cells. The surface of substratum can also be increased when the cells are grown on microcarriers. These small beads with diameters of about 200 μm are fabricated from different materials, such as polystyrene, silica, glass, or dextran and may be coated with collagen or mixtures of cell adhesive substances. Cells can grow as monolayers on these beads, which are kept as suspension and thus mediate mass cultivation of monolayer cells.

Cells can also grow isolated in suspension cultures in which they must be kept floating with a minimum of shear stress and optimal conditions for growth and production. For small volumes, this is achieved with stirrer flasks or gyratory shakers, but cells may also be cultivated in air-lift bioreactors that may hold several ten thousand liters. In small stirrer flasks, the medium is kept in motion with magnetic stirrers and rotating paddles or pendulums, necessary controls are maintained by the incubator's devices. This is no longer possible when the volume of a stirrer flask exceeds one liter. Scaling-up in volume requires better controls of medium conditions; therefore, bioreactors are equipped with sensors for the regulation of gas, temperature, pH, and metabolites.

To prevent outgrowth of cells, multicellular spheroids and three-dimensionally growing biopsies are constantly kept in motion in stirrer flasks or gyratory shakers. For this cultivation in small volumes, an interesting alternative is given by rotating wall vessels that were designed for cell growth in low or zero gravity. Their rotation speed is so adjusted that the cells remain stationary, which not only facilitates the formation of three-dimensional aggregates but also explants from organs and tissues can be maintained under these conditions.

2.3.2 Appliances

Essential equipments that are used in tissue culture laboratories are shortly mentioned here. A laminar flow hood protects not only cells from microbial infections but also the operator from hazardous materials (see Sect. 2.4). Incubators must allow regulation of temperature, CO_2, and humidity and provide good internal convection. An autoclave is necessary for sterilizing solutions, but medium must be filtered since many ingredients will not tolerate the high temperature in an autoclave. Wasted dishes and other disposable material that came in contact with pathogens or with cultures of primate or transfected cells must also be autoclaved before they are trashed. Glassware and glass pipettes can be sterilized in a dry heat at temperatures higher than 160 °C for at least one h. Besides refrigerators and freezers for storing substances and solutions, liquid nitrogen canisters should also be available for deep-freezing and storing cell lines. Ultrapure water might not always be necessary, but a cell culture laboratory should have access to a supply; in any case, deionized and distilled water must be available. Well-adjusted phase contrast microscopes, both upright and inverted, are important tools in a cell culture laboratory since microscopical inspection of the cells during growth and before an experiment can save time and costs when inappropriate cells are to be used. A balance should not be missing as well as centrifuges and many other helpful instruments such as pH meter, cell counter, dispensers, pipettors, and a glassware washing machine.

2.4
Safety and Biohazards

For any workplace, national regulations are effective that prevent operators from safety hazards of physical or chemical origin. Working in a tissue culture laboratory adds further sources of potential safety hazards, which originate from cell cultures contaminated with latent viruses or occult pathological organisms. A known risk comes from transfected cells for which not only the origin of the cell line but also the transfection vector outlines the risk. A low individual and community risk is given with microorganisms that are unlikely to cause disease in healthy operators or animals and with cell lines that are not of human or primate origin. These risk group I agents can be handled in a containment level I laboratory that matches a functionally designed standard microbiology laboratory. Aseptic techniques must be used and all liquid waste should be treated with bleach or similar detergents. Since during pipetting aerosols are formed that might carry cells and unidentified pathogens, it is good laboratory practice to protect operators with a laminar flow hood with an air barrier at the front opening and an exhaust filter.

Agents that are of moderate individual and limited community risk are classified as risk group II. This includes cultures of primate and human origin, recombinants, transfectants, and animal tumor cells. A separate containment level II laboratory with separate equipments is obligatory, and cells must be handled in a class II laminar-flow hood. Human cell lines that are virus-producing or are infected with pathogens have a high individual and a low community risk and are classified as risk group III. These cells can also be handled in a class II laminar-flow hood, but any waste and material that came in contact with cells must be sterilized before leaving the separate containment level III laboratory. Only designated laboratory staff is allowed to work in this room.

The risk group IV is given when a high individual and a high community risk must be expected from the experiments. This includes human pathogens that produce very serious human or animal diseases that might be untreatable and readily transmitted. Biopsies and cell cultures carrying known human pathogens must be handled in a containment level IV laboratory. In addition to the requirements for a level III laboratory, this laboratory must be physically isolated and functionally independent of other areas and must have an air lock for entry and exit. Class III biological safety cabinets and/or positive pressure ventilated suits protect designed operators for whom a shower and change of clothes are obligatory when entering or leaving the laboratory. Authorization for handling transfected cells and human pathogens as well as the routine inspection of containment laboratories are regulated by national laws.

2.5 Storage

Primary cells have a limited life span and continuous cell lines may change properties by transformation, dedifferentiation, or because of genetic instability and infection. When production or research depend on certain properties, it is therefore essential to have a stock of cells with the original properties. Cell lines as well as small multicellular organisms and embryos can be preserved by freezing since much of the cellular volume is water that is essential for cellular metabolism. When water becomes to ice, cellular metabolism is stopped and cells can be kept for years under cryopreserved conditions when the temperature is deep enough.

Before freezing, cells should be maintained under routine conditions and must be inspected for contamination. Monolayer cells are harvested during exponential growth by trypsinization; suspension cultures are concentrated by centrifugation. Cells must be thoroughly washed and resuspended to a concentration of about 1 to 10 million cells per milliliter freezing medium. This freezing medium is composed of the appropriate growth medium plus a high concentration of serum (50% or more) plus dimethyl sulfoxide or glycerol. This cell suspension is transferred in plastic ampoules wherein cells will be frozen at a slow rate till they reach $-70\,^{\circ}C$ and will be kept overnight before they are transferred into liquid nitrogen and stored at $-196\,^{\circ}C$.

For thawing, an ampoule is placed in a $37\,^{\circ}C$ water bath and slowly agitated. Its content is transferred into a flask and diluted with medium. After 6 to 8 h, when viable cells are firmly attached, medium with dead cells and cellular debris should be aspirated and fresh medium is added.

3 Applications

Considerable knowledge of cellular structures and functions is the result of investigations with cells in culture. This includes intracellular synthesis processes for numerous macromolecules, such as nucleic acids and proteins, as well as for most other relevant cellular molecules. Intracellular energy and signal transfer, cytoskeletal architecture, transport through membranes, product formation, but also cell–cell interactions, malignant transformations, and microbial infections are representatively mentioned for the numerous issues for which answers were found in cell cultures. A few examples for applications in research and production are given that concentrate on the importance of three-dimensionally grown cell aggregates, a method that might become the standard technique in the future.

3.1 Cellular Regulations

Complex cell interactions and epithelial differentiation were studied with organotypic cocultures of flat epidermal multilayers and revealed the normal regulation and balance of these cells. For this *in vitro* epidermalization, a collagen matrix at the basal side is required, whereas the keratinocytes are exposed to air. The collagen layer also supports dermal fibroblasts that synthesize the right extracellular matrix proteins. Together with the metabolites of the medium, which is separated from the collagen layer by a filter, all necessary components are available for proliferation and differentiation of the epidermal cells.

Intracellular regulations are not only dependent on the combination of cells but also on the culture conditions. This is trivial as long as necessary metabolites or messenger molecules are missing or are too low in concentration; however, with the same concentrations available in medium, cells may behave in a completely different manner when grown three-dimensionally as cell aggregates instead of as two-dimensional monolayers. An example is the so-called contact effect, which was observed for multicellular spheroids in which the tumor cells are more sensitive to irradiation than in two-dimensional growth. Later, it was detected that cells that were coupled to their neighbors via gap junction channels showed this effect more pronounced. Gap junction channels regulate an intercellular exchange of ions and molecules of up to 900 dalton and are normally found to be open in monolayer cells. However, in multicellular spheroids of several tumor cell lines, the permeability of these channels will be regulated, they were found open in two-day old spheroids and were closed two days later. It is still unknown what controls this channel closing, but channels may open again when the cells were allowed to grow as monolayer. Another growth-dependent regulation was demonstrated with β-galactosidase synthesizing L-cells. For this production, the cells had been transfected with a lacZ gene, which was under control of a β-actin promoter. This constitutive transfection led to a constant production when the cells were cultured as monolayer but β-galactosidase activity was downregulated in cells that were cultivated as spheroids. Since monolayer L-cells are motile, a permanent synthesis of cytoskeletal elements such as β-actin is required, but in spheroids the cells are rather immobile and the synthesis of β-actin is downregulated, which also effects the β-galactosidase activity. In contrast to the layered growth of epidermal cells, multicellular spheroids contain only one type of continuously growing cells, and these are exposed to gradients of metabolites, protons, and oxygen. Furthermore, spheroids establish an extracellular matrix, which is not found when the same cells are cultivated as monolayers.

3.2 Cancer Research

Malignant growth of cells can only be understood correctly when the modes for normal growth are known. Both cases can be investigated under the same experimental conditions with cell cultures. A successful model for studying carcinogenesis *in vitro* is again the multilayered epidermis, which allowed examination of malignant transformation and tumor progression. A problem of metastatic behavior of tumor cells is invasion, it can be investigated *in vitro* with cocultures of three-dimensionally grown aggregates of normal and malignant cells. First, attempts were made with organ cultures from chick embryo or fragments of human endometrium, which were confronted with suspended tumor cells. When precultured embryonic chicken heart fragments were confronted with tumor cell spheroids, invasive behavior could be analyzed with immuno-histological techniques, and comparative studies revealed that this *in vitro* invasiveness matched the *in vivo* situation. Interestingly, it was shown that invasive tumor cells were coupled to the host via gap junction channels and that noncoupled HeLa cells became invasive after transfection with a connexin gene, which led to an expression of host-compatible gap junction channels. Several other prerequisites for invasion have been detected with similar cell culture models: Metalloproteases were found to be active during invasive processes of tumors from brain and the cell adhesion molecule e-cadherin plays a role when breast tumor cells invade the host. Weakening the cadherin-mediated cell–cell contacts facilitates detachment and migration of single tumor cells in a mesenchymal type of movement. Recent studies with new fluorescent probes allowed a time-resolved investigation of invasive processes in three-dimensional collagen matrices and revealed a supramolecular plasticity mechanism with a transition from a proteolytic mesenchymal toward a nonproteolytic amoeboid movement.

3.3
Production

Two modes of cellular production are possible: either the product remains within the cells or it is secreted into the medium. In any case, purification processes are necessary and, therefore, the culture conditions should be adjusted and optimized to increase the yield and avoid unnecessary costs. For biomedical treatments, the cells are often the requested product, which requires safe culture conditions. These should not only perfectly match the host's physiological conditions but must also be free of nonhuman additives that may be due to limited cleaning processes.

Monoclonal antibodies are produced by hybridoma cells in suspension culture. They are secreted into the medium and can be harvested with the culture supernatant. For mass production, hybridomas are often grown in hollow-fiber cultures that facilitates harvesting; the highest yield, however, might be achieved when hybridomas are grown as ascites culture in mice.

Amongst the first products of monolayer cultures were therapeutically important substances such as viral vaccines and tissue plasminogen activator, which were harvested from roller bottle cultures or from microcarriers in suspension cultures. With the availability of genetically manipulated cells, the production of therapeutically relevant proteins with animal cell cultures increased considerably. The list of these substances encloses insulin, interferon, interleukin, plasminogen activators, blood clotting factors, hemopoietic growth factors, hormones, vaccine, and so on. Many of these proteins have a complex tertiary structure and require posttranslational modifications that can only be synthesized in animal cells but not in bacteria. For mass production, producing cells are often additionally modified so that they tolerate high shear stress when cultivated as suspension in large bioreactors.

An often-applied method in biomedical treatment is transplantation of skin. For this purpose, small pieces of skin will be taken from the patient, cut, and expanded to a mesh. They will then be cultivated, and keratinocytes will fill the interstice. These pieces can then be retransplanted onto the wounds of the patient. Another approach, which is still under development, is the temporary support of defect organs such as liver by bioartificial organs via an extra-corporal bypass.

4 Perspectives

Biomedical treatment with primary or continuous cell cultures requires complex growth conditions, and some of the current approaches might lead to new therapeutic applications. Three-dimensionally growing cells do not completely represent the cells' behavior in human or animal, and monolayer cells are not necessarily representative for more complex culture systems. This does not argue against investigations with cell cultures as long as possible limitations are taken into account.

A great potential of research with monolayer cultures is given by certain cellular functions such as channel activities, intracellular transport, or protein synthesis. Examples are cells that were transfected with a fusion construct that leads to the synthesis of proteins that are labeled with a fluorescent protein. This labeling enables investigations of supramolecular dynamics by microscopic techniques such as internal reflection fluorescence or fluorescence correlation spectroscopy. Results obtained with these techniques offer insights into general cellular transport and diffusion processes.

Cellular production has been optimized with selected cell types that grow under adequate suspension- or monolayer-conditions and might still be expanded with other transfected cells and new recipes for synthetic media. A great potential for production can be expected from cell cultures of coelenterates and arthropods because pathogens are unlikely to transfer to human cells. Successful cultivation of three-dimensional aggregates of sponges has already been reported, and continuously growing insect cells are available for a longer time.

See also Antibody Molecules, Genetic Engineering of; Bacterial Cell Culture Methods; Immunology; Prokaryotic and Eukaryotic Cells in Biotech Production.

Bibliography

Books and Reviews

Abbot, A. (2003) Biology's new dimension (news feature), *Nature* **424**, 870–872.

Alberts, B., Johnson, A., Lewis, J., Raff, M., Roberts, K., Walter, P. (2002) *Molecular Biology of the Cell*, 4th edition, Garland Press, New York.

Bjerkvig, R. (Ed.) (1992) *Spheroid Culture in Cancer Research*, CRC, Boca Raton, FL.

Davis, J.M. (Ed.) (2002) *Basic Cell Culture: A practical Approach*, 2nd edition, Oxford University Press, London.

Freshney, R.I. (2000) *Culture of Animal Cells: A Manual of Basic Techniques*, 4th edition, Wiley-Liss, New York.

Freshney, R.I., Freshney, M.G. (Eds.) (2002) *Culture of Epithelial Cells*, 2nd edition, Wiley-Liss, New York.

Lodish, H., Baltimore, D., Berk, A., Darnell, J., Matsudaira, P., Zipursky, S.L. (1999) *Molecular Cell Biology*, 4th edition, W. H. Freeman, New York.

Masters, J.R.W. (Ed.) (2000) *Animal Cell Culture: A Practical approach*, 3rd edition, Oxford University Press, London.

O'Brien, S.J. (2001) Cell culture forensics, *Proc. Natl. Acad. Sci. U.S.A.* **98**, 7656–7658.

Pollack, R. (Ed.) (1981) *Readings in Mammalian Cell Culture*, 2nd edition, Cold Spring Harbor Laboratory, Cold Spring Harbor.

Primary Literature

Abercrombie, M. (1970) Contact inhibition in tissue culture, *In Vitro* **6**, 128–142.

Abercrombie, M., Heaysman, J.E.M. (1954) Observations on the social behaviour of cells in tissue culture. II. "Monolayering" of fibroblasts, *Exp. Cell Res.* **6**, 293–306.

Behrens, J., Mareel, M.M., Van Roy, F.M., Birchmeier, W. (1989) Dissecting tumor cell invasion: epithelial cells acquire invasive properties after the loss of uvomorulin-mediated cell-cell adhesion, *J. Cell Biol.* **108**, 2435–2447.

Bernstein, L.R., Liotta, L.A. (1994) Molecular mediators of interactions with extracellular matrix components in metastasis and angiogenesis, *Curr. Opin. Oncol.* **6**, 106–113.

Birchmeier, W. (1991) E-Cadherin-mediated cell-cell adhesion prevents invasiveness on human carcinoma cells, *J. Cell Biol.* **113**, 173–185.

Bjerkvig, R., Lærum, O.D., Mella, O. (1986) Glioma cell interactions with fetal rat brain aggregates *in vitro* and with brain tissue *in vivo*, *Cancer Res.* **46**, 4071–4079.

Bräuner, T., Hülser, D.F. (1990) Tumor cell invasion and gap junctional communication. II. Normal and malignant cells confronted in multicell spheroids, *Invasion Metastasis* **10**, 31–48.

Carrel, A.(1912) On the permanent life of tissues outside the organism, *J. Exp. Med.* **15**, 516–528.

Custodio, M.R., Prokic, I., Steffen, R., Koziol, C., Borojevic, R., Brümmer, F., Nickel, M., Müller, W.E. (1998) Primmorphs generated from dissociated cells of the sponge Suberites domuncula: a model system for studies of cell proliferation and cell death, *Mech. Ageing Dev.* **105**, 45–59.

de Ridder, L.J., Lærum, O.D. (1981) Invasion of rat neurogenic cell lines in embryonic chick heart fragments *in vitro*, *J. Natl. Cancer Inst.* **66**, 723–728.

Dertinger, H., Hülser, D. (1981) Increased radioresistance of cells in cultured multicell spheroids. I. Dependence on cellular interaction, *Rad. Envir. Biophys.* **19**, 101–107.

Dulbecco, R. (1952) Production of plaques in monolayer tissue culture by single particles of an animal virus, *Proc. Natl. Acad. Sci. U.S.A.* **38**, 747–752.

Durand, R.E., Sutherland, R.M. (1972) Effects of intercellular contact on repair of radiation damage, *Exp. Cell Res.* **71**, 75–80.

Eagle, H. (1955) Nutrition needs of mammalian cells in tissue culture, *Science* **122**, 501–504.

Earle, W.R., Schilling, E.L., Stark, T.H., Straus, N.P., Brown, M.F., Shelton, E. (1943) Production of malignancy *in vitro*; IV: The mouse fibroblast cultures and changes seen in the living cells, *J. Natl. Cancer Inst.* **4**, 165–212.

Eckert, R., Dunina-Barkovskaya, A., Hülser, D.F. (1993) Biophysical characterization of gap-junction channels in HeLa cells, *Pflügers Arch – Eur. J. Physiol.* **424**, 335–342.

Elfgang, C., Eckert, R., Lichtenberg-Fraté, H., Butterweck, A., Traub, O., Klein, R.A., Hülser, D.F., Willecke, K. (1995) Specific permeability and selective formation of gap junction channels in connexin-transfected HeLa cells, *J. Cell Biol.* **129**, 805–817.

Engebraaten, O., Bjerkvig, R., Lund-Johansen, M., Wester, K., Pedersen, P.-H., Mørk, S., Backlund, E.-O., Lærum, O.D. (1990) Interaction between human brain tumour biopsies and fetal rat brain tissue *in vitro*, *Acta Neuropathol.* **81**, 130–140.

Fusenig, N., Boukamp, P. (1998) Multiple stages and genetic alterations in immortalization, malignant transformation, and tumor progression of human skin keratinocytes, *Mol. Carcinog.* **23**, 144–158.

Gerlach, J.C., Mutig, K., Sauer, I.M., Schrade, P., Efimova, E., Mieder, T., Naumann, G., Grunwald, A., Pless, G., Mas, A., Bachmann, S., Neuhaus, P., Zeilinger, K. (2003) Use of primary human liver cells originating from discarded grafts in a bioreactor for liver support therapy and the prospects of culturing adult liver stem cells in bioreactors: a morphologic study, *Transplantation* **76**, 781–786.

Gey, G.O., Coffman, W.D., Kubicek, M.T. (1952) Tissue culture studies of the proliferative capacity of cervical carcinoma and normal epithelium, *Cancer Res.* **12**, 264–265.

Good, N.E., Winget, G.D., Winter, W., Conolly, T.N., Izawa, S., Singh, R.M.M. (1966) Hydrogen ion buffers and biological research, *Biochemistry* **5**, 467–477.

Graeber, S.H., Hülser, D.F. (1998) Connexin transfection induces invasive properties in HeLa cells, *Exp. Cell Res.* **243**, 142–149.

Grümmer, R., Hohn, H.P., Mareel, M.M., Denker, H.W. (1994) Adhesion and invasion of three human choriocarcinoma cell lines into human endometrium in a three-dimensional organ culture system, *Placenta* **15**, 411–429.

Hansson, A., Bloor, B.K., Sarang, Z., Haig, Y., Morgan, P.R., Stark, H.J., Fusenig, N.E., Ekstrand, J., Grafström, R.C. (2003) Analysis of proliferation, apoptosis and keratin expression in cultured normal and immortalized human buccal keratinocytes, *Eur. J. Oral. Sci.* **111**, 34–41.

Harrison, R.G. (1907) Observations on the living developing nerve fiber, *Proc. Soc. Exp. Biol. Med.* **4**, 140–143.

Hayflick, L., Moorhead, P.S. (1961) The serial cultivation of human diploid cell strains, *Exp. Cell Res.* **25**, 585–621.

Hülser, D.F., Brümmer, F. (1982) Closing and opening of gap junction pores between two and three-dimensionally cultured tumor cells, *Biophys. Struct. Mech.* **9**, 83–88.

Hülser, D.F., Frank, W. (1971) Stimulation of embryonic rat cells in culture by a protein fraction isolated from fetal calf serum. I. Electrophysiological measurements on cell surface membranes [in German], *Z. Naturforsch. B* **26**, 1045–1048.

Klünder, I., Hülser, D.F. (1993) Beta-galactosidase activity in transfected Ltk-cells is differentially regulated in monolayer and in spheroid cultures, *Exp. Cell Res.* **207**, 155–162.

Knedlitschek, G., Schneider, F., Gottwald, E., Schaller, T., Eschbach, E., Weibezahn, K.F. (1999) A tissue-like culture system using microstructures: influence of extracellular matrix material on cell adhesion and aggregation, *J. Biomech. Eng.* **121**, 35–39.

Köhler, G., Milstein, C. (1975) Continuous cultures of fused cells secreting antibodies of predefined specificity, *Nature* **256**, 495–497.

Laerum, O.D., Nygaar, S.J., Steine, S., Mork, S.J., Engebraaten, O., Peraud, A., Kleihues, P., Ohgaki, H. (2001) Invasiveness *in vitro* and biological markers in human primary glioblastomas, *J. Neuro-Oncol.* **54**, 1–8.

Levi-Montalcini, R., Angeletti, P.U. (1968) Nerve growth factor, *Physiol. Rev.* **48**, 534–569.

Lin, C.Q., Bissel, M.J. (1993) Multi-faceted regulation of cell differentiation by extracellular matrix, *FASEB J.* **7**, 737–743.

Maas-Szabowski, N., Szabowski, A., Stark, H.J., Andrecht, S., Kolbus, A., Schorpp-Kistner, M., Angel, P., Fusenig, N.E. (2001) Organotypic cocultures with genetically modified mouse fibroblasts as a tool to dissect molecular mechanisms regulating keratinocyte growth and differentiation, *J. Invest. Dermatol.* **116**, 816–820.

Mareel, M., Kint, J., Meyvisch, C. (1979) Methods of study of the invasion of malignant C3H-mouse fibroblasts into embryonic chick heart *in vitro, Virchows Arch., B, Cell Pathol.* **30**, 95–111.

Mareel, M.M., Bracke, M.E., Van Roy, F.M., de Baetselier, P. (1997) Molecular mechanisms of cancer invasion, *Encyclopedia of Cancer* **II**, 1072–1083.

Margolis, L.B., Fitzgerald, W., Glushakova, S., Hatfill, S., Amichay, N., Baibakov, B., Zimmerberg, J. (1997) Lymphocyte trafficking and HIV infection of human lymphoid tissue in a rotating wall vessel bioreactor, *AIDS Res. Hum. Retroviruses* **13**, 1411–1420.

Margolis, L., Hatfill, S., Chuaqui, R., Vocke, C., Emmert-Buck, M., Linehan, W.M., Duray, P.H. (1999) Long term organ culture of human prostate tissue in a NASA-designed rotating wall bioreactor, *J. Urol.* **161**, 290–297.

Masters, J.R., Thomson, J.A., Daly-Burns, B., Reid, Y.A., Dirks, W.G., Packer, P., Toji, L.H., Ohno, T., Tanabe, H., Arlet, C.F., Kelland, L.R., Harrison, M., Virmani, A., Ward, T.H., Ayres, K.L., Debenham, P.G. (2001) Short tandem repeat profiling provides an international reference standard for human cell lines, *Proc. Natl. Acad. Sci. U.S.A.* **98**, 8012–8017.

Minuth, W.W., Sittinger, M., Kloth, S. (1998) Tissue engineering – generation of differentiated artificial tissues for biomedical applications, *Cell Tissue Res.* **291**, 1–11.

Moolenaar, W.H., Jalink, K. (1992) Membrane potential changes in the action of growth factors, *Cell. Physiol. Biochem.* **2**, 189–195.

Moscona, A.A. (1968) Cell aggregation: properties of specific cell-ligands and their role in the formation of multicellular systems, *Dev. Biol.* **18**, 250–277.

Nelson-Rees, W.A., Flandermeyer, R.R. (1976) HeLa cultures defined, *Science* **191**, 96–98.

Quastler, H., Sherman, F. (1959) Cell population kinetics in the intestinal epithelium of the mouse, *Exp. Cell Res.* **17**, 420–438.

Schleich, A., Tchao, R., Frick, M., Mayer, A. (1981) Interaction of human carcinoma cells with an epithelial layer and the underlying basement membrane. A new model, *Arch. Geschwulstforsch.* **51**, 40–44.

Smith, J.A., Martin, L. (1973) Do Cells Cycle? *Proc. Natl. Acad. Sci. U.S.A.* **70**, 1263–1267.

Smola, H., Stark, H.J., Thiekotter, G., Mirancea, N., Krieg, T., Fusenig, N.E. (1998) Dynamics of basement membrane formation by keratinocyte-fibroblast interactions in organotypic skin culture, *Exp. Cell Res.* **239**, 399–410.

Sutherland, R.M., McCredie, J.A., Inch, W.R. (1971) Growth of multicell spheroids in tissue culture as a model of nodular carcinomas, *J. Natl. Cancer Inst.* **46**, 113–120.

Todaro, G.J., Green, H. (1963) Quantitative studies of the growth of mouse embryo cells in culture and their development into established lines, *J. Cell Biol.* **17**, 299–313.

Wolf, E. (1970) Organ Chimeras and Organ Culture of Malignant Tumors, in: Thomas, J.A. (Ed.) *Organ Culture*, Academic press, New York, pp. 459–505.

Wolf, K., Mazo, I., Leung, H., Engelke, K., von Adrian, U.H., Deryugina, E.I., Strongin, A.Y., Bröcker, E.-B., Friedl, P. (2003) Compensation mechanism in tumor cell migration: mesenchymal-amoeboid transition after blocking of pericellular proteolysis, *J. Cell Biol.* **160**, 267–277.

Mass Spectrometry of Proteins (Proteomics)

Hiroyuki Matsumoto[1], *Sadamu Kurono*[1], *Masaomi Matsumoto*[2], *and Naoka Komori*[1]
[1] *The University of Oklahoma Health Sciences Center, Oklahoma City, OK, USA*
[2] *Oklahoma Department of Agriculture, Oklahoma City, OK, USA*

Encyclopedia of Molecular Cell Biology and Molecular Medicine, 2nd Edition. Volume 7
Edited by Robert A. Meyers.

ISBN: 3-527-30549-1

Keywords

Abductive Inference or Abduction
Inference to the best explanation as proposed by Charles S. Peirce.

Collision Induced Dissociation or MS/MS
In CID, a selected precursor ion is activated by collisions with neutral gas molecules and fragmented into daughter ions.

Electrospray Ionization
A method for the ionization of nonvolatile molecules such as proteins, peptides, sugars, and nucleic acids by spraying the sample solution onto a small orifice in the mass spectrometer.

Endopeptidase
An enzyme that cleaves a polypeptide chain of proteins at a specific site.

Genome
The complete DNA sequence of an organism.

Mass Spectrometry/Mass Spectrometer
In mass spectrometry, a mass spectrometer measures the mass of almost any molecule that can be ionized in the gas phase.

Matrix-Assisted Laser Desorption/Ionization
A method for the ionization of nonvolatile molecules such as proteins, peptides, sugars, and nucleic acids by irradiating the mixed crystals between sample molecules and laser dye (matrix) molecules in the mass spectrometer.

Peptide Mass Fingerprinting
Identification of a protein through the measurement of the masses of endoprotease-cleaved fragments in reference to genome information.

Post Source Decay
The fragmentation of ions after the acceleration in the ion source prior to entering the reflectron that separates the daughter ions.

Posttranslational Modification
Covalent modification of amino acids in proteins taking place during and after the translational process.

Proteome
The full complement of proteins produced by a particular genome.

Proteomics
The study of the full complement of proteins encoded by a genome.

Two-dimensional Gel Electrophoresis
A gel electrophoresis system to separate a complex mixture of proteins by an isoelectric focusing in the first dimension followed by the sodium dodecylsulfate polyacrylamide gel electrophoresis in the second dimension.

U
The symbol for a mass unit representing one-twelfth the mass of carbon; equivalent to dalton (Da) used in biochemistry.

The completion of genome projects represents a pinnacle of human enterprise in the biosciences. The orchestration of technologies in mass spectrometry, genome science, and computer sciences has created a new branch of molecular bioscience called proteomics. A proteomics approach to the study of protein expression and posttranslational modification does not require foreknowledge of the identity of target proteins. A proteomic investigation begins with the discovery of unidentified proteins of interest under well-defined physiological conditions. The approach involves (1) protein display by 2-D gel electrophoresis or other separation technique; (2) determination of protein entities of interest; (3) peptide mass fingerprinting (PMF); and (4) genome/proteome database search. The methodology of the proteomics approach is characterized neither by *deduction* nor by *induction* in the traditional sense, but is a clear example of what C. S. Peirce described as *abductive inference* a century ago. *The investigation of molecular and cellular events is often intractable to deductive and inductive methods due to its extreme complexity and nonlinearity*. Proteomics is a powerful tool to study complex biological systems because of its characteristics: (1) no *a priori* knowledge of the protein's identity is required to initiate a project; and (2) a holistic approach is possible for investigation. It is expected that proteomics will substantially contribute to the future development of molecular medicine.

1 Proteomics: the Primary Application of Mass Spectrometry to Biomolecules

In the last decade, mass spectrometry, which has longer than a century of history in physical sciences, entered a new era: two new ionization methods that enable mass measurement of nonvolatile biomolecules such as proteins, nucleic acids, and sugars have been invented and perfected. The new types of mass spectrometry, together with the development of genome science, opened up a totally new type of inquiry into the molecular aspects of life. "Proteomics" represents such emerging genre of molecular biosciences that utilizes biomolecular mass spectrometry and genome information.

1.1
What is Proteomics?

Proteomics is a new discipline in biosciences, which aims to study the proteome. The new term "proteome" was first used in late 1994 at the Siena 2-D Electrophoresis Meeting and was first printed in a scientific journal in 1995 by Wasinger et al. There is an inherent relationship between genome and proteome and the creation of the former precedes the latter. Therefore it would be appropriate to briefly explain genome before defining proteome. The term genome was first used around 1965 to represent the whole set of information encoded by chromosomal DNA. Some of the early literature includes the following: Mauer (1965), Breeze and Cohen (1965), Winocour (1965), Diamandopoulos and Enders (1965), Easton and Hiatt (1965), and Heisenberg and Blessing (1965). Since each species of living organism possesses a particular set of chromosomal DNA, it can be said that each living species possesses one genome. Moreover, since the chromosomal DNA remains the same in every cell under any physiological conditions within a certain species, the genome also remains the same. The usage of the term genome implies that each species possesses only one genome, that is, human genome, rat genome, *Drosophila* genome, yeast genome, *Escherichia coli* genome, rice genome, and so on. Those terms are well defined scientifically although there could exist some variations in genome even among the same species. In contrast to the well-defined nature of "genome" by itself, "proteome" needs to be defined with caution. The first usage of the term proteome is as follows: "proteome" refers to the total complement of a genome or the complement able to be encoded by a given genome. Thus, the term proteome was coined about three decades after the term genome was introduced. The MEDLINE (OVID) entries of relevant manuscripts reflect this historical fact: a search with "genome" yields more than 7800 entries, whereas the search for "proteome" yields about 1700 entries as of late 2002. The early literature that used the word "proteome" include Wasinger et al. (1995), Kahn (1995), Wilkins et al. (1996a), Yan et al. (1996), Qi et al. (1996), O'Brien (1996), Wilkins et al. (1996b), Wilkins et al. (1996c), Shevchenko et al. (1996), Celis et al. (1996), Wimmer et al. (1996), and Wilkins et al. (1996d). In their 1995 paper, Wasinger et al. carefully added that it is unlikely that the totality of this potential for protein expression will be realized at any one given instant. According to the "central dogma" of gene expression, DNA encodes mRNAs, and the mRNAs encode proteins. In the past half a century, since the establishment of the double helix model of DNA and its replication, many modes of molecular mechanisms that affect and regulate the expression of proteins have been discovered. The entire process of protein expression consists of three distinguished steps: transcription, translation, and posttranslation. Every process that affects any of these steps modulates the ability of the genome to express itself, thereby modulating the proteome. Under a certain condition a certain type of cell expresses a group of proteins that is a subset of the total complement of a given genome. Because changes in physiological parameters even in the same type of cells most likely affect protein expression transcriptionally, translationally, and/or post-translationally, the kinds and amounts of proteins expressed under one set of physiological parameters will be different from those under another set of physiological parameters. Thus, in contrast to genome, which is a rather static entity under all physiological conditions, the proteome is a dynamic entity: there is only one genome for a given species, but there are many realizations of proteomes even for the same single species. This characteristic of the proteome makes it essential to define precisely the samples for proteome analysis because without defining the biological parameters such as cell types and physiological conditions the interpretation of the proteome analysis will be ambiguous. In a certain sense, all the traditional experiments performed on protein targets can be called "proteomics studies." However, there appear to be intrinsic differences between the traditional experimental inquiries in biochemistry and molecular biology and those of proteomics studies in two ways: (1) proteomics renders high throughput of inquiries because of the advance in technology of protein separation and analysis by mass spectrometry and genome database, and (2) the advent of technology to investigate proteins on a massive scale has enabled us to utilize a type of investigational logic, which is distinct from the traditional biochemical inquiries. A new logical inference appropriate for proteomics will be discussed in the following Sect. 3.1 Proteomics and Abductive Inference.

In contrast to the static aspect of genome, the dynamic aspect of genome will be studied by a new area of discipline called "functional genomics." In functional genomics, the expressed mRNAs will be studied under a well-defined condition compared to a control. The pool of mRNAs will be reverse-transcribed into cDNAs and the resulting cDNAs will be analyzed mainly by a DNA array technique. In the interpretation of DNA array experiments, the amount of a protein and that of the corresponding mRNA are assumed to be correlated, which is not always the case, at least in the case of monocellular eukaryote yeast. Further studies need to be performed to evaluate the studies of mRNA expression in reference to the protein expression, especially in higher eukaryotes. In order to achieve this goal, a sensitive and reliable methodology for quantitative proteomics of expressed proteins will be essential. (See Sect. 5 Quantitative Proteomics.)

2
Technical Basis of Proteomics

2.1
Protein Display for Proteomic Investigation

The early proteomic inquiries can be found in works in which a mixture of proteins were separated by 2-D gel electrophoresis. Although there had been several precedents for 2-D gel systems before O'Farrell (1975) invented the modern and powerful method of separating a mixture of proteins, the O'Farrell-type 2-D gel system has been proven to be a powerful method for the separation of a complex mixture of proteins. O'Farrell's protocol combines isoelectric focusing (IEF) gel under a nonionic detergent NP-40 and urea in the first dimension and the traditional Laemmli-type sodium dodecyl sulfate polyacrylamide gel electrophoresis (SDS-PAGE) in the second dimension. Almost at the same time, Ames and Nikaido reported a similar 2-D gel system that makes the use of sodium dodecyl sulfate (SDS) possible for solubilizing membrane proteins. Since O'Farrell's protocol does not solubilize many of the intrinsic membrane proteins in the sample, Ames and Nikaido's method complements that of O'Farrell's. Early work using 2-D gel electrophoresis clearly demonstrated that protein expression and posttranslational modification *in vivo* are dynamic processes (for example, see Lee et al., 1979; Giometti and Anderson, 1981; Willard and Anderson, 1981; Kosik et al., 1982; Neukirchen et al., 1982; Matsumoto et al., 1982; Matsumoto and Pak, 1984). Although these examples and others indicated the power of 2-D gel electrophoresis for studying the protein expression, its weakness is that the researchers at that time did not have the power to identify the proteins of interest. This identification of proteins of interest would have to wait another decade for the development of new technologies. In the 1980s, many new and powerful technologies in biochemistry and molecular biology were developed. Modern biotechniques relevant to early proteomic inquiries include monoclonal antibody production, membrane blotting techniques both of nucleic acids and proteins, gas phase Edman degradation performed on the membrane blot, and cDNA cloning techniques including λ*gt11*-based protein expression vectors that enable library screening using antibodies. It should be noted that in the 1980 to early 1990 genome databases did not exist. Therefore, cloning of each gene encoding the protein of interest was the only way to identify the gene. The premise of proteomics is that as long as a protein of interest is distinctively displayed by a 2-D gel or other methods, one can identify the protein and its gene without gene cloning. This capability of proteomics approach gives us an opportunity to catalog all the genes expressed in a particular tissue such as, for example, retina; see Fig. 1.

2.2
Mass Spectrometry: an Essential Tool for Proteomics

In late 1980s, two important ionization methods were invented and revolutionized the modern mass spectrometry: Koichi Tanaka invented matrix-assisted laser desorption/ionization (MALDI) and John Fenn invented electrospray ionization (ESI), for which both scientists received Nobel Prize in Chemistry in 2002 (http://www.nobel.se/chemistry/laureates/2002/). Before the invention of ESI and MALDI, the analyte molecules had to be volatile for any mass spectroscopic analysis. Biopolymers such as proteins, peptides, nucleic acids such as DNA and RNA, and polysaccharides cannot be analyzed by traditional mass spectrometers unless they are chemically modified to render them volatile. This is not always possible except for some shorter polymers.

A mass spectrometer consists of three important components: (1) the ion source that ionizes the analyte molecules (for example, MALDI and ESI are two types of ion sources that can ionize nonvolatile molecules such as proteins, peptides, and nucleic acids); (2) the ion analyzer that separates each ion from the mixture according to the mass-to-charge ratio (m/z); and (3) the detector that counts the number of ions characterized by a specific m/z value. Several types of ion analyzers are currently available: for example, magnetic sector–type ion analyzer, quadrupole (Q) ion analyzer, time-of-flight (TOF) ion analyzer, quadrupole ion-trap (QIT) ion analyzer, and Fourier transform ion cyclotron resonance (FTICR) ion analyzer. Theoretically, any combination of ion source

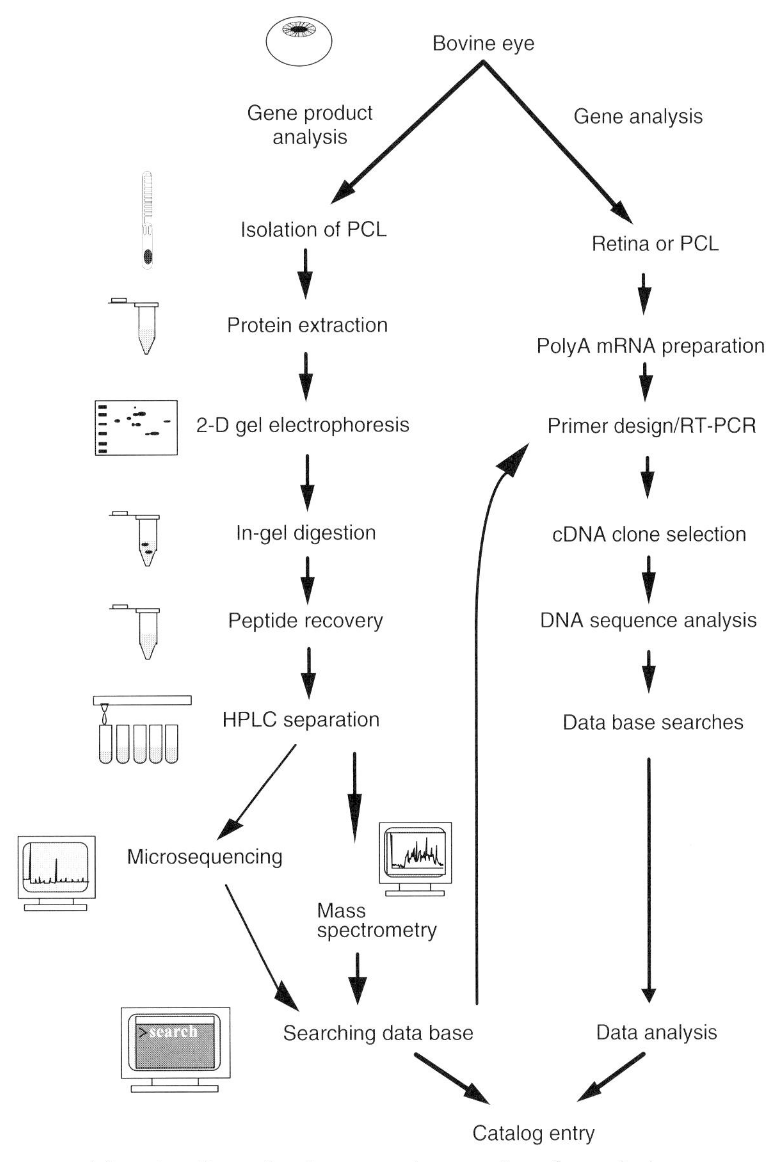

Fig. 1 A flow chart illustrating the proteomics procedures for cataloging photoreceptor proteins from the photoreceptor monolayer of bovine retina (see Fig. 9).

and ion analyzer is possible. However, the most common combinations include ESI-tandem-Q, ESI-Q-TOF, ESI-QIT, MALDI-TOF, and MALDI-QIT-TOF. FTICR is a rather expensive mass spectrometer and both ESI and MALDI ionization methods are common to FTICR. In any case, what a mass spectrometer does is to measure the mass of ionized molecules in the form of m/z.

2.3
Genome Information: a Prerequisite for Peptide Mass Fingerprinting

The ability of mass spectrometry to measure the m/z value of peptides with high sensitivity and accuracy even in a mixture constitutes the experimental basis for the identification of proteins in proteomics. The method is called "*peptide mass fingerprinting*," which is often abbreviated as PMF. In the early 1990s, several groups reported the possibility of identifying proteins based on their peptide mass fingerprints. All proteins have their unique amino acid sequences because they are all encoded by unique genes. Highly ambitious endeavors to sequence whole genome of a certain organism were planned in late 1980s. In the beginning, the genome projects progressed rather slowly. Through the early 1990s, however, the advent of automated DNA sequencing technologies together with the advancement in microcomputers and their software gradually accelerated the progress of genome projects. It is interesting to point out that since the late 1990s, the competition between the government-funded genome project led by National Institute of Health and the genome project operated by a private-sector Celera Genomics significantly accelerated the development of the human genome project. Now complete genome information is available for many bacteria, monocellular eukaryote yeast, *Caenorhabditis elegans* (worm), *Drosophila melanogaster* (fruit fly), and other model eukaryotes. The human genome has been sequenced to its rough draft in 2001, followed by its near completion in 2003. The genome projects of rodents such as rat and mice, both of which serve as excellent model systems for biomedical research, will also be completed in the near future.

In the case of *D. melanogaster*, the whole DNA sequence of its genome has been known and available through the Internet since 2000. Annotation of the DNA sequence combined with all the information assembled from studies of cloning individual genes predicts 14 431 open reading frames (ORFs). This indicates that *Drosophila* is likely to have about 14 000 different kinds of proteins expressed during its entire life cycle. Suppose one discovers a protein of interest and uncovering its identity appears to be important for the further development of the ongoing project. For the practice of PMF, it is necessary that the protein of interest is isolated from other proteins and displayed in a certain reproducible way. In the early phase of proteomics around 1995, 2-D gel electrophoresis has been the major technical tool for the display and discovery of a protein of interest. However, in recent years, the capabilities of 2-D gel electrophoresis, when applied to a total protein extract of biological samples, has been challenged. The first weakness is the limitation in the capacity of 2-D gel electrophoresis to display a large number of proteins at once. The human genome is estimated to encode more than 30 000 proteins. Currently, there is no 2-D gel electrophoresis system to display proteins in the order of tens of thousands. However, as discussed in the previous chapter, it is unlikely for one type of cell to express all the proteins. Therefore, in practice, the total number of proteins that one has to display on one 2-D gel is a fraction of 30 000. We have no reliable data regarding how many proteins are actually expressed in one type of cell under a given physiological condition. It is speculated that probably the number is 5000 $\sim$ 10 000. The maximum number of proteins that can be displayed depends on the gel system used. However, it would be safe to say that $3000 \sim 4000$ is the maximum displaying capacity of a 2-D gel at best. The second weakness, which is somehow related to the first problem, is the difficulty of displaying minor proteins on a 2-D gel when major proteins are present. It is speculated that the dynamic range of protein expression can be in the order of $10^5 \sim 10^6$. Therefore, if the protein sample contains several major proteins, then the minor protein components tend to be masked on a 2-D gel. This type of situation will be observed in the analysis of body fluids such as blood. The third weakness is the difficulty of displaying membrane-bound proteins on a 2-D gel, especially of the O'Farrell's type. As mentioned earlier, the 2-D gel system developed by Ames and Nikaido would ameliorate this difficulty. However, each membrane protein has its own characteristics; there is no guarantee for the complete solubilization and the subsequent display of membrane-bound proteins on a 2-D gel. Because of these disadvantages of 2-D gel electrophoresis, other techniques for protein display and analysis have been actively sought. One major advance in this direction, sometimes claimed as a technique "beyond 2-D gel," is the development of multidimensional high performance liquid chromatography (HPLC). In this technique, the whole protein mixture will be extracted and subjected to HPLC at least twice (two-dimensional or 2-D) or more times (multidimensional). The resins packed in the HPLC columns have "orthogonal" analytical properties so that the separation in each step will not be redundant, but rather complementary, making the separation of each protein thorough. It should be noted, however, that even a carefully designed and performed multidimensional HPLC protocol could suffer from technical imperfections such as incomplete solubilization of proteins and peptides or failure in effective elution of those molecules from the columns. The effectiveness of multidimensional HPLC for displaying the proteome must be tested empirically for each proteomics application.

2.4
Peptide Mass Fingerprinting

Suppose that, after an intensive investigation, a protein is isolated and speculated to be responsible for a certain physiological and/or pathological activity. For example, the protein of interest can be displayed as a spot on a 2-D gel or can be fractionated as a distinctive elution peak on an HPLC. It should be noted that a 2-D gel that has been run and archived many years ago can still serve as a sample source for PMF (Matsumoto and Komori, the procedure for PMF is summarized in Table 1). There are three steps in PMF:

1. *Endopeptidase digestion of the protein of interest.* In most of the PMF, trypsin, which digests the carboxyl termini of lysine (K) and arginine (R), is the endopeptidase of choice. Other endopeptidases can also be used. In Fig. 2, an example is shown when the protein of interest was isolated on a 2-D gel. In-gel digestion is preferentially used for the digestion of proteins isolated on either one-dimensional or 2-D gel electrophoresis; although other methods such as digestion after electroelution of the protein or digestion after blotting on a membrane filter can be used, the in-gel digestion is less labor-intensive and yet effective. In-gel digestion of a 2-D gel protein spot followed by mass spectrometry was reported in the mid-1990s by several groups.

Tab. 1 Peptide mass fingerprinting (PMF) consists of three steps of experimental operation; (1) in-gel digestion of protein by trypsin (1 ∼ 10), 2) MALDI-TOF MS (11 ∼ 12) and 3) PMF search through the Internet database (13 ∼ 15).

In-gel digestion

1. Excise the protein spot out. Destain the gel piece in 50% acetonitrile (MeCN)/100 mM NH_4HCO_3, pH 8.0
2. Dry the gel piece briefly.
3. Apply 2 µL (1 µL each for two times) of TPCK-treated trypsin (Promega, Madison, WI, USA) solution (1 µg/µL in 50 mM acetic acid) to the gel piece.
4. Cover the gel piece with 50 µl of 2.5 mM NH_4HCO_3.
5. Digest at 30 °C for ca. 12 h.
6. Suck up the solution and keep it.
7. Extract the digested peptides with 60% MeCN/0.1% trifluoroacetic acid (TFA) with shaking for 1 h.
8. Mix above 6 and 7 solutions and freeze-dry the peptide mixture.
9. Dissolve the residue in 5-µL water.
10. Submit 1 µL of the obtained sample to MALDI-TOF MS.

Mass spectrometry by MALDI-TOF MS

11. Mix 1 µL of the tryptic peptides with 1 µL of matrix solution consisting of 10 mg/ml α-cyano-4-hydroxycinnamic acid (α-CHCA) in 50% MeCN/0.1% TFA in a plastic tube and apply 1 mL on a sample plate.
12. Dry the sample and measure the MALDI-TOF spectrum.

Peptide mass fingerprinting (PMF)

13. Read the mass numbers of the major peaks.
14. Search PMF through MS-Fit, and so on. If the search hits a promising candidate for protein, retrieve the full sequence.
15. Reevaluate the PMF in reference to the full sequence of the candidate.

2. *Mass spectrometry.* Mass spectrometric measurement of the tryptic digest of a target protein can be made by either of the two types of mass spectrometers: electrospray-ionization (ESI) mass spectrometer or matrix-assisted laser desorption/ionization (MALDI) mass spectrometer. The ESI ion source is often interfaced with a quadrupole, quadrupole ion trap, or time-of-flight mass analyzer. The MALDI ion source is commonly interfaced with a TOF mass analyzer. The ESI ion source has a tendency to produce multiply charged peptide ions. In contrast, the MALDI ion source produces mainly singly charged peptide ions and thus the interpretation of the mass spectrum is straightforward. For the initial stage of PMF, a MALDI-TOF mass spectrometer (MALDI-TOF MS) is often used. In Fig. 3, a PMF procedure by MALDI-TOF MS is illustrated.
3. *PMF by database search.* After measuring the peptide masses created by an endoprotease such as trypsin, the mass numbers will be searched through a database. A conceptual scheme for fingerprint matching is illustrated in Fig. 4. The search is usually performed through the Internet, although the use of stand-alone databases for the search is also possible. In this particular case shown

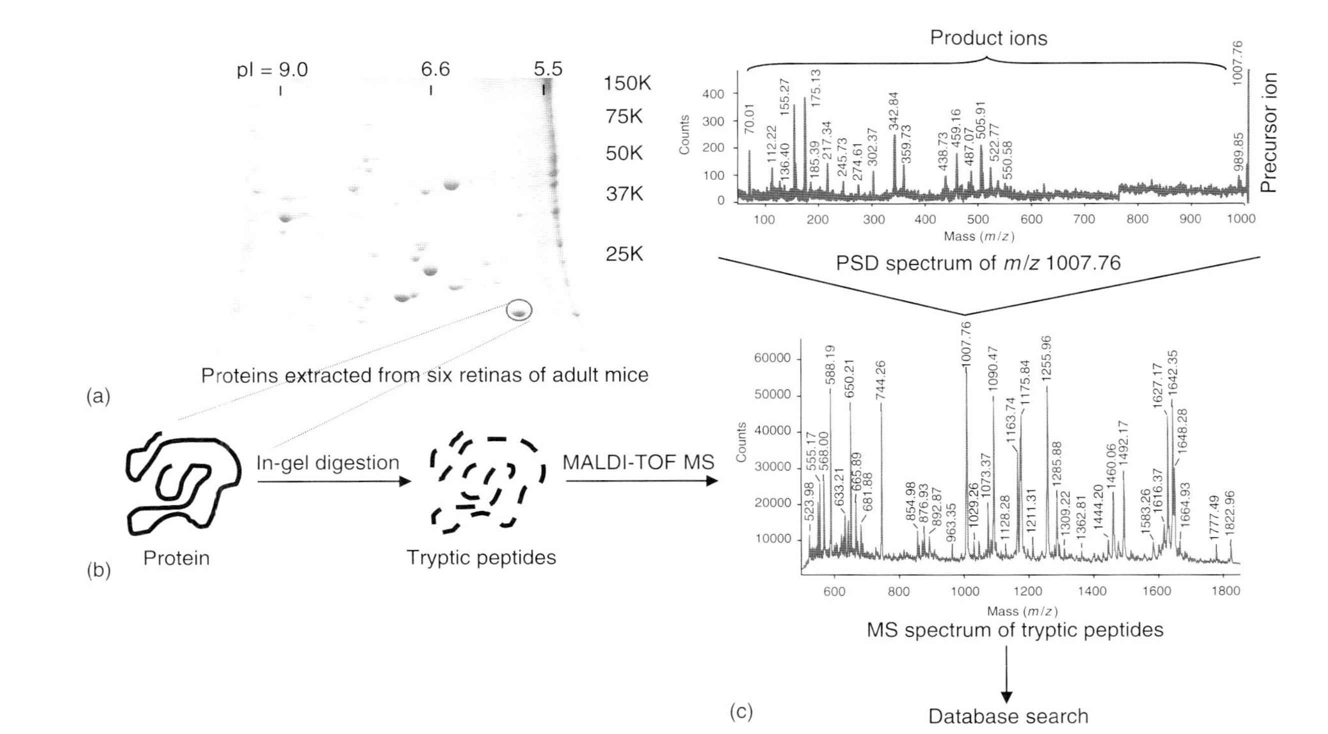

Fig. 2 (a)Outline of proteomics investigation starting from a protein displayed on a two-dimensional gel (a). The protein of interest is excised and digested by trypsin and the peptide mass fingerprint is taken by MALDI-TOF MS (b). Database search will identify the candidate protein, which will be confirmed by a PSD spectrum analysis (c).

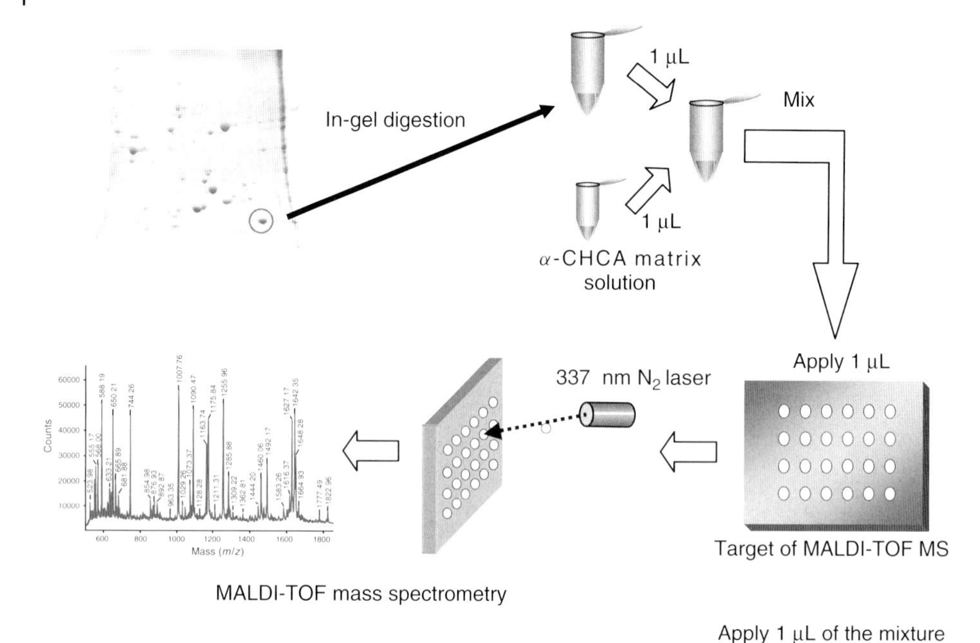

Fig. 3 Procedural outline of peptide mass fingerprinting by MALDI-TOF MS. After the in-gel digestion of the protein of interest, the mixture of tryptic digests is mixed with matrix solution containing α-cyano-4-hydroxycinnamic acid (α-CHCA) and spotted on a MALDI-TOF MS sample plate. The irradiation of the mixed crystals between the peptides and α-CHCA by 337 nm N_2 laser produces mostly singly charged ions. The m/z value of each peptide ion is measured at high accuracy and sensitivity.

Amino acid sequence of a target protein

MDVTIQHPWFKRALGPFYPSRLFDQFFGEGLFEYDLLPFLSSTISPYYRQSLFRTVLDSGISE
VRSDRDKFVIFLDVKHFSPEDLTVKVLEDFVEIHGKHNERQDDHGYISREFHRRYRLPSNVD
QSALSCSLSADGMLTFSGPKVQSGLDAGHSERAIPVSREEKPSSAPSS

(a)

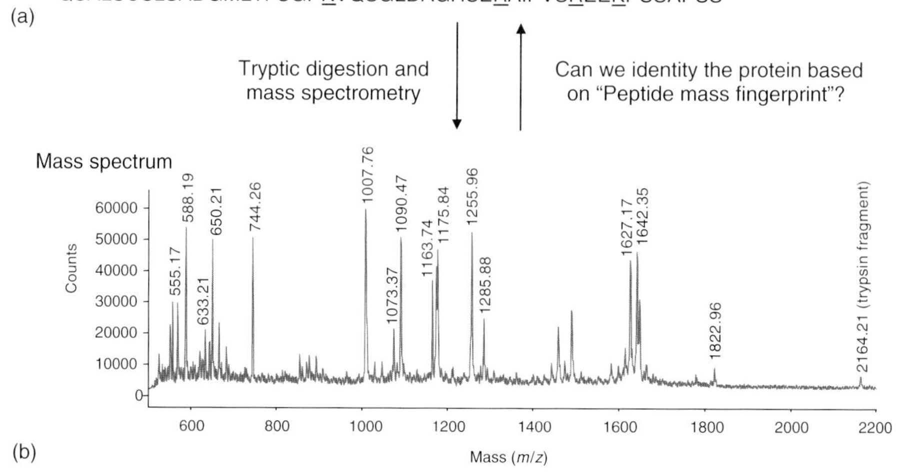

(b)

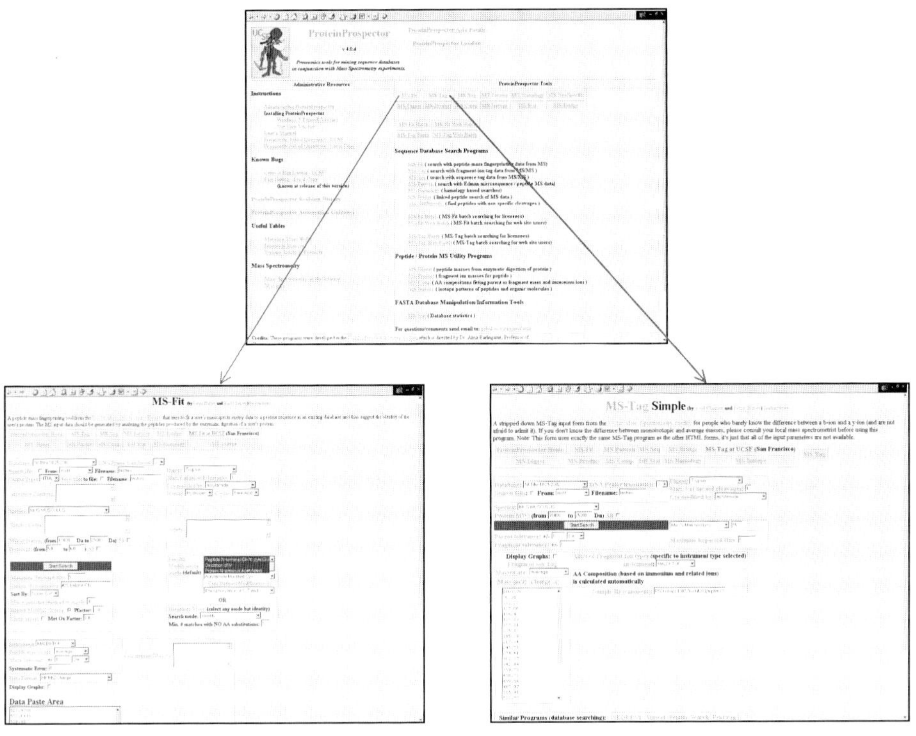

Fig. 5 An example of peptide mass fingerprinting database search: MS-Fit and MS-Tag. In MS-Fit (Lower left), the *m/z* values of the peptide ions are used for inquiries. In MS-Tag (Lower right), the *m/z* values of fragment ion tags are used for inquiries. In both cases of database search, other parameters such as "species," "the protein molecular weight range," and "the tolerance of mass measurement," and so on can be specified. All the print outs shown in Figs. 5, 6, and 8 are the results of peptide mass fingerprinting using Protein Prospector (http://prospector.ucsf.edu/).

in Fig. 5, MS-Fit in Protein Prospector at http://www.prospector.ucsf.edu/ was used. Other database sites usable for PMF include MASCOT at http://www.matrixscience.com/, ProFound at http://prowl.rockefeller.edu/, and PeptIdent at http://www.expasy.org/. The output of MS-Fit search

Fig. 4 The concept of peptide mass fingerprinting: The amino acid sequence of the target protein is initially unknown (a). The digestion of the protein by trypsin creates a mixture of peptides. The following mass spectrometry gives the *m/z* values of tryptic peptide ions (b). If the amino acid sequence of the protein is known, all the peptides generated by the tryptic digest can be predicted by a theoretical (*in silico*) digestion. In a real experiment, however, it is unlikely that the mass spectrometric measurement will reveal all the digested fragments. In this particular case, the matched peptides covered only 72% (126 out of 173 amino acids) of the protein. "Peptide mass fingerprinting" is a practice to predict the identity of the target protein based on the observed masses of the peptide fragments.

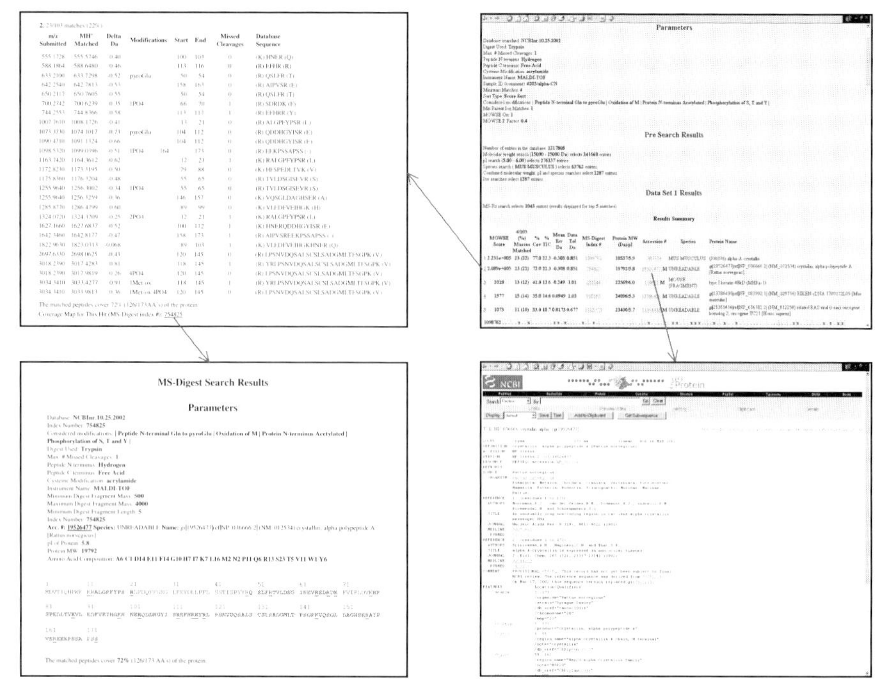

Fig. 6 The results of MS-Fit database search indicated in Fig. 5 (shown in the upper right corner). Clicking each hyperlink will output: (1) NCBI Protein Database Entry (by clicking Accession #; shown in the lower right), (2) the calculated and the observed fragments including posttranslational modifications (by clicking the candidate listing; shown in the upper left), and (3) the MS-Digest Search Result (by clicking MS-Digest Index #; shown in the lower left).

shown in Fig. 5 is shown in Fig. 6.

2.5 Confirmation of the Candidate Protein in Peptide Mass Fingerprinting

In an ideal case of PMF, all the predicted (*in silico*) fragments of the protein of interest are supposed to be confirmed in the mass spectrum. Practically, though, a total coverage of peptide mass fingerprints is unlikely to be achieved from many reasons, including incomplete digestion, low ionization efficiencies of some of the proteolytic peptides, loss of material during sample processing, and posttranslational modifications of amino acid side chains. Observation of peptides with unexpected masses, that is, "orphan masses," is rather common and is attributed to (1) posttranslational or artifactual modification, imperfect proteolysis, and/or contaminated proteins in the case of successful protein identification and (2) false positive identification of a protein.

2.5.1 Confirmation by "Orthogonal" Type of Data

In order to increase the confidence of the PMF, further experiments need to be performed. One way is to incorporate "orthogonal data types" in addition to the original PMF. The orthogonal data types include (1) chemical modification such as methyl esterification (increase of

14 mass units) on aspartic acid, Glutamic acid, and C-terminus carboxyl groups, (2) iodination on tyrosine and histidine residues (increase of 126 mass units), (3) hydrogen/deuterium exchange on all amino acids (increase of one mass unit per exchanged atom), and others. One may extend the orthogonal data types into other data types; given the amino acid sequence of the candidate protein, one may predict many properties and results that can be derived by further experiments. For example, if an antibody specific to the candidate protein is available, a western immuno blot will give further evidence. Alternatively, partial Edman sequencing of the N-terminus or the N-termini of internal peptides after appropriate endopeptidase digestion and peptide isolation will also give orthogonal type data for further confirmation. Most importantly, after this initial stage of discovery, the candidate protein will emerge and a hypothesis will be made. In all the scientific investigations, further experiments must be carried out to confirm the hypothesis. If the assignment of the protein identity is incorrect, the following experiments will produce contradictory results, resulting in the correction of the old hypothesis to a new hypothesis. This cyclic process was declared by Peirce as part of abductive inference.

2.5.2 Confirmation by Peptide Fragmentation by MS/MS

Mass spectrometry can create further evidence for the confirmation of the candidate protein if the machine is equipped with collision-induced dissociation (CID) capability. For this purpose, the mass spectrometer needs to have at least two ion mass analyzers. This can be achieved in two ways: (1) by connecting two ion mass analyzers in series, for example, such as a tandem quadrupole (tandem Q) mass spectrometer or a tandem time-of-flight mass spectrometer (TOF-TOF), and (2) by connecting two different types of ion mass analyzers, for example, such as quadrupole time-of-flight (Q-TOF) mass spectrometer or quadrupole ion trap time-of-flight (QIT) mass spectrometer. These mass spectrometers enable a real CID fragmentation to confirm the candidate proteins from the daughter ions and they are at the higher end in cost.

A MALDI-TOF mass spectrometer equipped with a reflectron can also obtain a fragmentation spectrum through "post-source decay (PSD)" mechanism. In the PSD process, the ion to be analyzed will be selected and separated through an electric mirror called "reflectron." An example is shown to confirm a peptide "ALGPFYPSR" in Fig. 7. All the expected ions in the fragmentation process can be predicted by computer algorithm, as illustrated in Fig. 8.

3 Logic in Proteomics

3.1 Proteomics and Abductive Inference

The advancement of medical biosciences in the past decades has accumulated massive amounts of information due to the rapid technological development in every relevant area. The increase in knowledge and technologies in molecular biology has played an especially significant role in directing the researchers toward molecular medicine. Molecular biology in the modern context can be said to have started when the double helical structure of DNA implying the mode of its replication was revealed by Watson and Crick in 1953. Half a century later, the draft of human

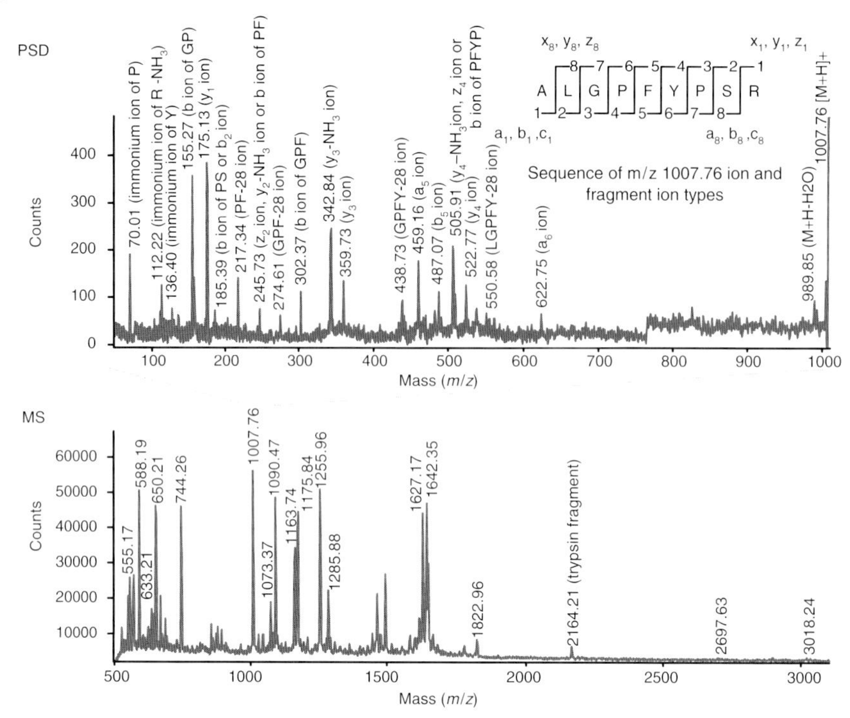

Fig. 7 An example of peptide fragmentation by PSD by MALDI-TOF MS. The ion at $m/z = 1007.76$ (shown in MS) was selected and analyzed by PSD (shown in PSD). The fragment ions shown support the assignment of the $m/z = 1007.76$ peak to be the peptide α-A-crystallin amino acid 13–21 with the sequence "ALGPFYPSR."

genome was completed in 2001. The draft of the human genome marks an epoch and summarizes all the efforts in the participating areas in biomedical sciences. During the past half century, the logical approaches that molecular biologists used are based mainly on "reductionism"; a scientific strategy and belief that a deeper understanding would be possible by investigating entities at more and more miniscule dimensions (see, for example, "Methodological reductionism" pp. 750, The Oxford Companion to Philosophy). The first and most triumphant reductionist approach in modern biochemistry took place when Eduard Büchner (1860–1917) showed that the cell-free systems of yeast are still capable of fermenting glucose in a test tube. The reductionist approach flourished since then toward the middle of the twentieth century, establishing the major metabolic pathways including glycolysis, gluconeogenesis, glycogen metabolism, citric acid cycle, pentose phosphate cycle, biosynthetic pathways of amino acids and nucleotides, catabolism of fatty acids, and others. Since the emergence of the Watson-Crick's double helix model of DNA, the trend of biochemical investigations shifted toward topics other than metabolism and a new area called molecular biology was gradually established in 1970s. The completion of the draft of human genome announced in 2001 is an epoch-making event, eloquently summarizing the past half a century of molecular

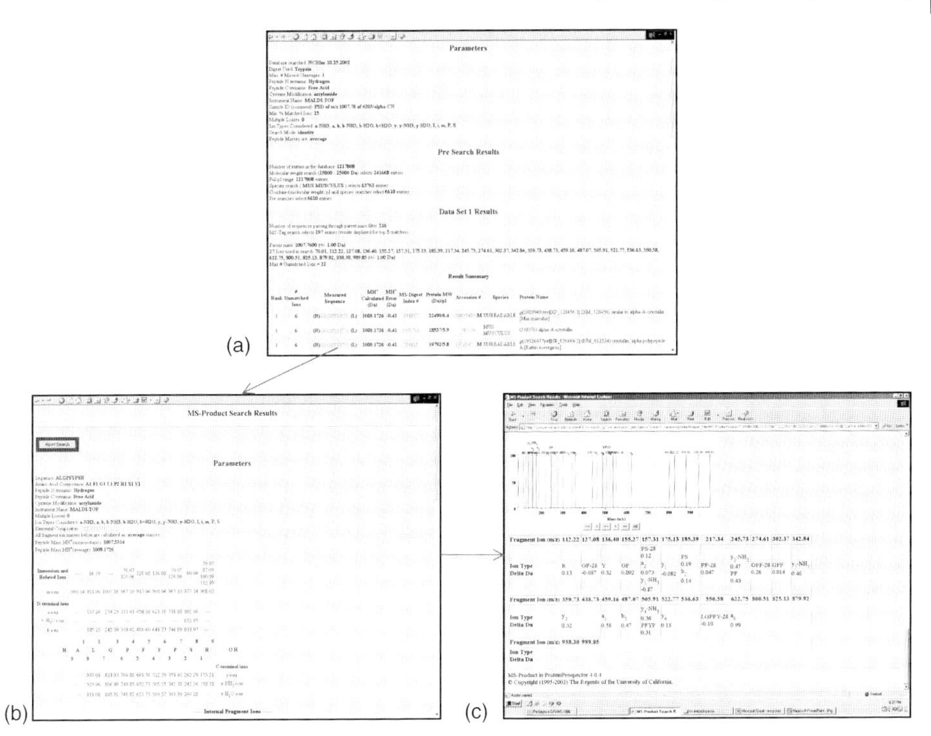

Fig. 8 The peptide fragments induced in the MS/MS process can be predicted by computer algorithm. Clicking the measured sequence (in the upper figure) outputs the MS-Product Search Results shown in (b) and (c).

biology. The initiation and substantial acceleration of progress in genome projects of bacteria, simple eukaryote models such as yeast, nematode, and fruit fly, and vertebrate models such as rodents took place in the late 1980s through the early 1990s. It is in this period that the proteomics was prepared and initiated.

The logic and strategies of proteomics investigation are novel and unique compared to those in conventional biochemistry and molecular biology. The reductionist approach in these areas has its origin in the use of cell-free systems and all subsequent efforts to fractionate the components. This approach constitutes the backbone of most of the disciplines in modern natural sciences including physics and chemistry. The arguments that have been employed in modern science are said to be classified in two categories; deduction and induction. However, it appears that the logical backbone of proteomics approach is "abductive inference or abduction," which is completely different from either deduction or induction.

Umberto Eco explains the difference between deduction, induction, and abduction as follows: "Suppose I enter a room and there find a number of bags, containing different kinds of beans. On the table there is a handful of white beans; and after some searching, I find one of the bags contains white beans only. I at once infer as a probability, or as a fair guess, that this handful was taken out of that bag. This sort of inference is called making a hypothesis". Eco continues as follows:

In the case of logical *deduction* there is rule from which, given a case, I deduce a result: *All the beans from this bag are white – These beans are from this bag – These beans are white.*

In the case of *induction,* given a case and a result, I infer a rule: *These beans are from this bag – These beans are white – All the beans from this bag are white* (probably).

In the case of hypothesis or *abduction,* there is the inference of a case from a rule and a result: *All the beans from this bag are white – These beans are white – These beans are from this bag* (probably).

– Umberto Eco, "A Theory of Semiotics"

If we follow Eco's format, the logical structure of PMF can be represented by the following three statements:

1. Gene X is translated into protein Y. Protein Y, if digested, will produce a peptide mass fingerprint represented by a set of mass (y1, y2, y3, . . .). (This is what Eco designates a "rule.")
2. The digestive products of protein A gave a peptide mass fingerprint represented by a set of mass (y1, y2, y3, . . .). (This is what Eco designates a "result.")
3. Protein A is the same entity as protein Y and is encoded by gene X. (This is what Eco designates a "case.")

In Table 2, the relationship between deduction, induction, and abduction is summarized. Thus, the argument involved in PMF is an example of abduction as explained in Table 2. Note that neither rule nor result requires precise knowledge of the identity of protein encoded by gene X.

Tab. 2 Deduction, induction, and abduction.

In the case of deduction, the argument follows as indicated below:
Rule: Gene X is translated into protein Y. Protein Y, if digested, will produce a peptide mass fingerprint represented by a set of mass (y1, y2, y3, . . .).
Case: Protein A is the same entity as protein Y and is encoded by gene X.
Result: The digestive products of protein A gave a peptide mass fingerprint represented by a set of mass (y1, y2, y3, . . .).
The conclusion derived by deduction is always true.

In the case of induction, the argument follows as indicated below: Case: Protein A is the same entity as protein Y and is encoded by gene X.
Result: The digestive products of protein A gave a peptide mass fingerprint represented by a set of mass (y1, y2, y3, . . .).
Rule: Gene X is translated into protein Y. Protein Y, if digested, will produce a peptide mass fingerprint represented by a set of numbers (y1, y2, y3, . . .).
The conclusion derived by induction could be either true or false.

In the case of abduction, the argument follows as indicated below:
Rule: Gene X is translated into protein Y. Protein Y, if digested, will produce a peptide mass fingerprint represented by a set of numbers (y1, y2, y3, . . .).
Result: The digestive products of protein A gave a peptide mass fingerprint represented by a set of numbers (y1, y2, y3, . . .).
Case: Protein A is the same entity as protein Y and is encoded by gene **X**.
The conclusion derived from abduction is actually a hypothesis that needs to be proven by further experiments.

3.2
Advantage of Proteomics Approach over Conventional Hypothesis-based Investigation

The logical structure of proteomics investigation as explained in 3.1 renders a unique advantage that does not exist in conventional hypothesis-based investigation. The distinguishing characteristic of abductive inference is the fact that it requires no initial hypothesis. This is because *abduction is a process of inferring to the best explanation. Abduction, instead of being based on a hypothesis, will create a hypothesis.* This characteristic in the logical structure of proteomics approach makes it suitable for the investigation of complex systems such as cellular signaling and gene expression. The wider the system to be investigated in biomedical sciences, the more complex and nonlinear the underlying molecular mechanisms will be. The most extreme cases include diagnosis of diseases. Abduction has been evaluated to be the right and practical logic of diagnosis. In a similar context, abductive inference in proteomics investigation will be a powerful tool to study the diagnostic and pharmacological aspects of human diseases. There is no surprise in observing the fact that many pharmaceutical/biotechnology companies have initiated and invested substantial resources along this line, resulting in new areas such as pharmaceutical proteomics or proteomics of diseases. A unifying theme in all these efforts is the holistic character of approach. Expected prospective areas using proteomics approach include the following: cancer diagnosis, disease diagnosis from body fluids, pharmacological screening, toxicology, hormone research, and diagnosis of infection. In fact, the applicability of proteomics investigation to medicine is unlimited.

4
Proteomics under a Particular Theme: Ocular Proteomics

Although proteomics techniques are high throughput in nature, it is more sensible to reduce the variables in the original samples as much as possible in order to make the interpretation of the results simpler. Practically speaking, each proteomics investigator is likely to have a specific target. "Ocular proteomics" is illustrated here as an example. All the visual information is initiated when photons reflected from the visible objects get absorbed by the retinal (i.e. vitamin A aldehyde) chromophore in the visual pigment molecules rhodopsin and cone visual pigments. These visual pigments constitute the photosensitive membrane structure of photoreceptor cells in the retina. The retina contains, in addition to photoreceptor cells, other types of neural and glial cells in order to maintain its function. The neural retinal cells process visual signals prior to the visual cortex of the CNS. There are numerous diseases reported in which malfunction and deterioration of retinal cells occur. Some retinal degenerative diseases are apparently caused by mutations by retinal specific genes, implying that protein malfunction causes the retinal degeneration. Therefore, cataloguing all the proteins expressed in the retina will be a useful resource for the understanding of retinal function and its malfunction at the protein level.

An effort has been made to initiate cataloguing proteins expressed in the photoreceptor cell layer of bovine retina compared to the rest of the retina from which the photoreceptor layer has been depleted. In order to dissect the retina, a tissue printing method was used, as illustrated in Fig. 9. The variation of this

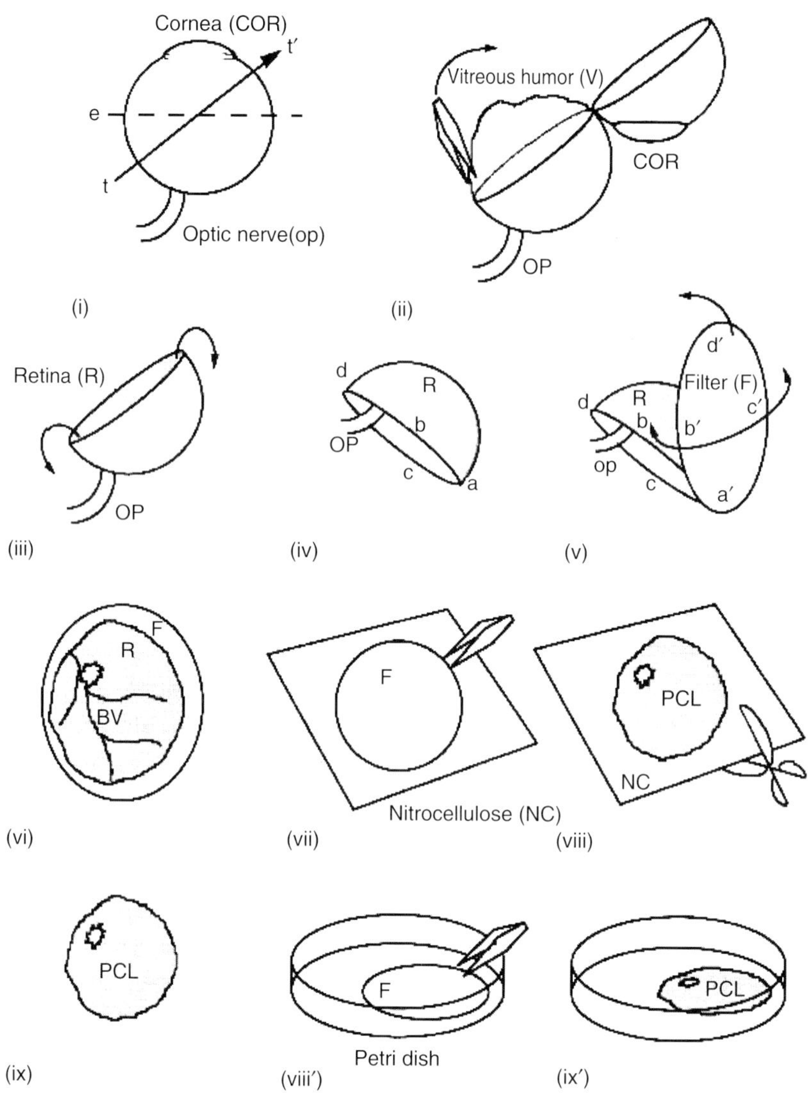

Fig. 9 A schematic diagram for preparation of the photoreceptor cell monolayer. (i) The eyeball is cut obliquely along the line t to t'. Cornea (COR), Optic nerve (OP). (ii) Vitreous humor (V) is removed from the eyecup. (iii) & (iv) The dissected eyecup is inverted. Retina (R). (v) & (vi) The retina is transferred to a Whatman filter paper (F) and lifted. (vii) The filter paper is overlaid face-side down onto a nitrocellulose membrane (NC). (viii) As the filter paper is lifted, the photoreceptor cell monolayer (PCL) remains on the nitrocellulose membrane. The excess margin of NC is removed with scissors. (ix) The PCL preparation is obtained. (viii′) As a modification, a plastic Petri dish is used instead of the nitrocellulose membrane. (ix′) The PCL transfers onto the Petri dish.

method can be used for other tissues if cells of significant similarity form a layer detachable from other types of cells. After tissue printing of the photoreceptor cells, the separated layers were examined by morphology (Fig. 10, left). Each layer of preparation was processed for 2-D gels (Fig. 10, right). The major 2-D gel spots on the photoreceptor layer (PCL) were excised and PMF was performed. The results of PMF are summarized in Table 3. By comparing the 2-D gels in Fig. 10 and Table 3, we conclude that four protein spots are expressed more abundantly or even exclusively in the PCL layer; Spot 1, Spot 8, Spot 9, and Spot 12. The protein represented by Spot 1 is IRBP, which is an extracellular protein assumed to be responsible for carrying retinoids between the pigment epithelial cells and the photoreceptor cells. The proteins represented by Spot 8 and Spot 9 are aspartate aminotransferase (AAT) and creatine kinase (CK) respectively, both of the mitochondrial type. Both of these enzymes are crucial components of the energy metabolism in the mitochondria. The protein represented by Spot 12 is the β subunit of transducin, which is involved in the G protein–coupled receptor signaling in the photoreceptor excitation and, therefore, is highly specific to the photoreceptor cells. The protein spots other than 1, 8, 9, and 12 also exist abundantly in the rest of the retina (Fig. 10). Those proteins expressed in high abundance in the entire retina are components of the glycolysis, as illustrated

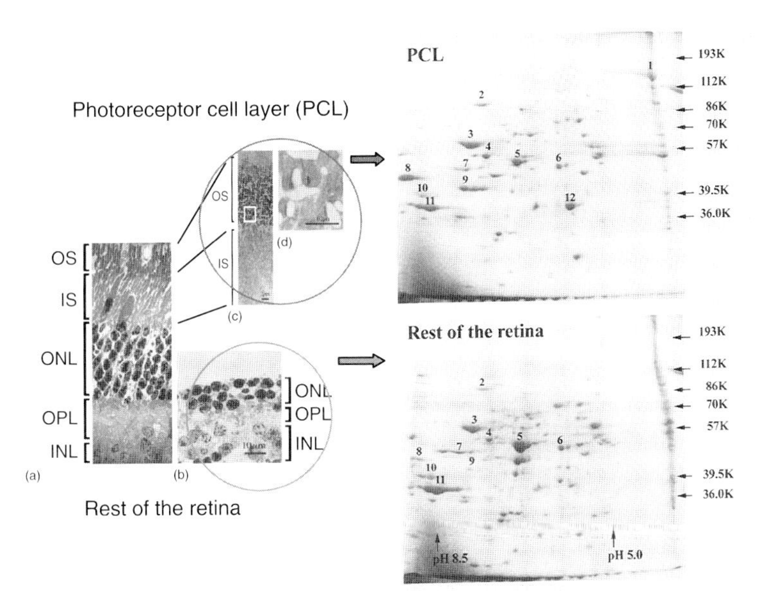

Fig. 10 Electron microscope images (Left) of the bovine retina (a), residual retina after removal of the photoreceptor cell monolayer (b), PCL adhered to nitrocellulose (c). Figure (d) illustrates the enlarged portion of OS as indicated by a square in Fig. (c). OS, outer segments; IS, inner segments; ONL, outer nuclear layer; OPL, outer plexiform layer; INL, inner nuclear layer. Each bar in the figures indicates dimension. The PCL sample and the rest of the retina were then subjected to 2-D gel electrophoresis (Upper and Lower Right).

Tab. 3 Major photoreceptor cell proteins assigned. MW, molecular weight in kilodaltons; pI, isoelectric point.

Spot #	*Assignment*	*MW/pI*	*Comments*
1	IRBP	138 kDa/5.0	Retinoid metabolism/extracellular
2	cis-Aconitase	82.7 kDa/7.6	TCA cycle/mitochondrial
3	Pyruvate kinase	55 kDa/7.7	Glycolysis/cytosolic
4	F1-ATPase	55.3 kDa/8.6	Respiratory/mitochondrial
5	Enolase	50 kDa/5.9	Glycolysis/cytosolic
6	Unknown		
7	Phosphoglycerate kinase	47 kDa/8.0	Glycolysis/cytosolic
8	(AAT)	45 kDa/9.2	Mitochondrial
9	Creatine kinase	43 kDa/7.2	Mitochondrial
10	Pyruvate dehydrogenase	41 kDa/8.6	Glycolysis/cytosolic
11	Glyceraldehyde 3-phosphate dehydrogenase	36 kDa/8.5	Glycolysis/cytosolic
12	Transducin-β	37 kDa/5.9	Phototransduction/membrane

in Fig. 11. From these results, one may conclude at least two things: (1) in bovine retina, the most abundant classes of proteins include the enzymes that are responsible for basic metabolism such as glycolysis, and (2) in the photoreceptor layer, two enzymes, that is, AAT and CK in the mitochondria responsible for the energy metabolism, are abundant compared to the rest of the retinal layer excluding the photoreceptor cells. Interesting conjectures emerge from these results. These conjectures can be evaluated by questions such as follows: (1) Do the mitochondria in the photoreceptor cells express larger quantities of AAT and CK compared to those in the rest of the retina? (2) Is the role of AAT in the photoreceptor cells to bypass the tricarboxylic acid cycle? These questions are being asked on the basis of a new set of hypotheses that did not exist before the experiments. Therefore, the whole process of proteomics approaches follows the method of *abductive inference* (or *making hypotheses*) that Peirce formulated. It should be noted that, based on these new hypotheses, another set of experiments will be performed and that the experiments should be conducted on the basis of hypotheses through conventional investigational logic, hence not by the "abductive inference" *per se* used in the very beginning of inquiries.

5 Quantitative Proteomics

The first stage of proteomics is to compare two samples, that is, a standard and the unknown sample. In order to obtain quantitative results in the comparison, there are at least two types of techniques available.

First is a rather straightforward method of obtaining quantitative data by scanning the 2-D gels by a densitometric scanner and quantifying the density of each individual spot by integration of pixels. In order to perform this type of quantitation, a scanning densitometer with an accurate calibration and software to perform the gel image analysis and pixel integration of each spot are required. Currently, there are several companies selling a scanning

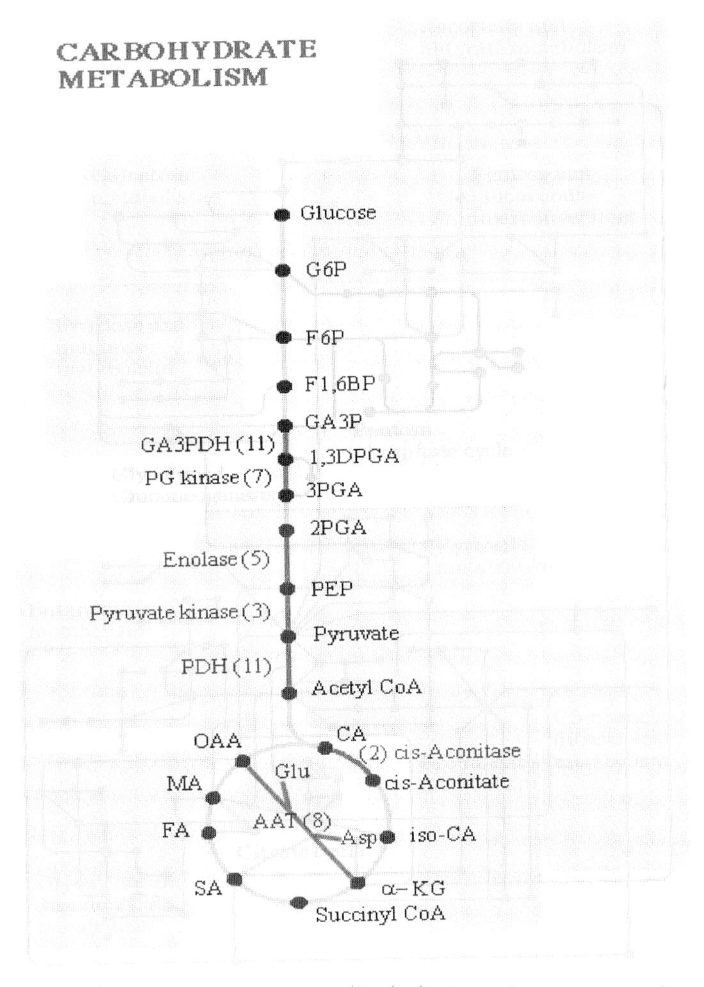

Fig. 11 The major proteins expressed in the bovine retina are enzymes in carbohydrate metabolism. The mitochondrial AAT is expressed in a larger quantity in the PCL compared to the rest of the retina. This result produces a hypothesis that the tricarboxylic cycle in the PCL operates differently from that in the rest of the retinal cells, that is, the interconversion of glutamate (Glu) and aspartate (Asp) via the bypass reaction between oxaloacetic acid (OAA) and α-ketoglutaric acid (α-KG) is more active in the PCL than the rest of the retina. The metabolic scheme is adapted from a website at Kyoto Encyclopedia of Genes and Genomes (KEGG).

densitometer and software that are suitable for this purpose.

Second group of quantitation methods utilizes modification of amino acid side chains with a couple of agents that have different physical properties such as different emission properties in a pair of fluorescent dyes or different masses caused by the incorporation of stable isotopes such as deuterium (D), which has 2 u compared to H (1 u). Carbon-13 (13 u) instead of carbon-12 (12 u) and nitrogen-15

(15 u) instead of nitrogen-14 (14 u) can also be used. In this type of quantitation, one sample (control) is labeled with a reagent with a certain property and the second sample (experiment) is labeled with a reagent with a property distinguishable from the control. Endopeptidase digestion needs to be done after or before the modification of amino acid side chains, depending on which system will be used.

For example, in a protocol called "2-D Fluorescence Differential Gel Electrophoresis (DIGE)", two protein samples will be modified by two different fluorescent dyes that emit at distinguishable wavelengths, and mixed. The mixture of fluorescent dye-labeled proteins will be separated on a 2-D gel. Since the pair of fluorescent dyes are designed so that the changes in the isoelectric point (pI) and the molecular weight are the same between the control and the experiment, each protein of the same kind, though originating from different samples, migrates and focuses at the same spot on a 2-D gel. With the use of a two-color scanning fluorescence densitometer, the amount of the modified proteins can be quantified.

Another example of quantitation in the second group is a mass spectrometric method in which a pair of light or heavy peptide modifier is used. "Isotope Coded Affinity Tag (ICAT)" is an especially successful technique that belongs to this category. In the ICAT technique, two different samples will be labeled by a derivative of a cysteine-modifier iodoacetoamide that carries an affinity tagged biotin and 8 atoms of deuterium (d_8-ICAT). A control ICAT reagent was synthesized without the 8 atoms of deuterium, making the reagent 8 u lighter than the d_8-ICAT. The two samples modified with d_8-ICAT and ICAT will be digested. The advantage of ICAT is that the affinity tag biotin can be used to purify the modified peptide fragment. After affinity purification by an avidin column, the eluate will be analyzed by an HPLC-interfaced ESI mass spectrometer. A deuterium-labeled *N*-alkylmaleimides also can be used in this type of analysis without the sophistication of affinity tag. Another method to quantify protein involves the digestion of one protein sample in the presence of ^{18}O-labeled water. The other sample to be compared will be digested in regular water (^{16}O). The former digests will make each peptide 4 u heavier than the latter. After mixing the digests and a high-resolution mass spectrometric analysis, the ratio between the two peaks separated by 4 u indicates the ratio of the protein existing in the two samples.

6
Proteomics and Molecular Medicine

The unique characteristic of proteomics approach that does not require presuppositions or hypotheses makes it an excellent tool for molecular medicine. Instead of starting from one particular protein that is supposed to be important for the understanding of molecular mechanisms underlying a disease, a proteomics approach can start from evaluating the protein expression and modification and their abnormal variance without knowing what they are in the beginning. Development of technologies in many areas in the past decades enabled us to apply a proteomics approach to virtually any area of medical biosciences. Each area of proteomics inquiry will be defined well by the set of parameters in its particular project. One can choose any input parameters that affect the system to be investigated, and, depending on the variable parameters, proteomics can be classified as, for example, "pharmaceutical proteomics," "physiological proteomics," "pathological proteomics," and so on. For example, with a proteomics approach, it is now feasible to initiate a project to evaluate the effect of drugs on the protein expression and posttranslational modification in a particular type of cells under a particular set of physiological parameters (pharmaceutical proteomics). Another example is that a comparison of proteins from cancerous tissues and those from normal tissues may give us a hint to initiate a new project based on a new hypothesis (cancer proteomics). Thus, the applicability and versatility of proteomics approach to the study of diseases appear to be unlimited.

Acknowledgment

This work is supported by grants from National Institutes of Health EY06595, EY13877, EY12190, and RR17703.

See also Gel Electrophoresis, 2D-difference; Mass Spectrometry, High Speed DNA Fragment Sizing; Nucleic Acids (DNA) Sequencing, Transcriptional.

Bibliography

Books and Reviews

Buchler, J. (Ed.) (1955) *Philosophical Writings of Peirce*, Dover Publications, New York.
Copi, I.M. (1972) *Introduction to Logic*, 4th edition, Collier-Macmillan Canada, Ltd., Toronto, Canada.
Dowling, J.E. (1987) *Retina: An Approachable Part of the Brain*, The Belknap Press of Harvard University Press, Cambridge, MA.
Eco, U. (1979) *A Theory of Semiotics*, Indiana University Press, Bloomington, IN.
Grayson, M.A. (Ed.) (2002) *Measuring Mass: From Positive Rays to Proteins*, Chemical Heritage Press, Philadelphia, PA.
Josephson, J.R., Josephson, S.G. (Eds.) (1996) *Abductive Inference. Computation, philosophy, technology*, Cambridge University Press, Cambridge, MA.
Liebler, D.C. (2002) *Introduction to Proteomics*, Humana Press, Totowa, NJ.
Matsumoto, H., Komori, N. (2000) Ocular proteomics: cataloging photoreceptorproteins by 2-D gel electrophoresis and mass spectrometry, *Methods Enzymol.* **316**, 492–511.
Matsumoto, H., Kahn, E.S., Komori, N. (1999) Emerging Role of Mass Spectrometry in Molecular Biosciences: Studies of Protein Phosphorylation in Fly Eyes as an Example, in: Yoshizawa., T. (Ed.) *Rhodopsins and Phototransduction*, Novartis Foundation Symposium No. 224., pp. 225–248.
Morange, M. (1998) *A History of Molecular Biology*, Translated by Cobb, M., Harvard University Press, Cambridge, MA.
Rabilloud, Th. (Ed.) (2000) *Proteome Research: Two-dimensional Gel Electrophoresis and Identification Methods*, Springer, Berlin, Germany.
Siuzdak, G. (1996) *Mass Spectrometry for Biotechnology*, Academic Press, San Diego, CA.
Sparkman, O.D. (2000) *Mass Spectrometry Desk Reference*, Global View Publishing, Pittsburgh, PA.
Wilkins, M.R., Williams, K.L., Appel, R.D., Hochstrasser, D.F. (Eds.) (1997) *Proteome Research: New Frontiers in Functional Genomics*, Springer, Berlin, Germany.

Primary Literature

Ames, G.F., Nikaido, K. (1976) Two-dimensional gel electrophoresis of membrane proteins, *Biochemistry* **15**, 616–623.
Breeze, D.C., Cohen, P.S. (1965) Evidence for the exponential reproduction of the genome of encephalomyocarditis virus, *Virology* **27**, 227–229.
Corthals. G.L., Gygi, S.P., Aebersold, R., Patterson, S.D. (2000) Identification of Proteins by Mass Spectrometry, in: Rabilloud, Th., (Ed.) *Proteome Research: Two-dimensional Gel Electrophoresis and Identification Methods*, Springer, Berlin, Germany pp. 197–231.

Celis, J.E., Gromov, P., Ostergaard, M., Madsen, P., Honore, B., Dejgaard, K., Olsen, E., Vorum, H., Kristensen, D.B., Gromova, I., Haunso, A., Van Damme, J., Puype, M., Vandekerckhove, J., Rasmussen, H.H. (1996) Human 2-D PAGE databases for proteome analysis in health and disease, *FEBS Lett.* **398**, 129–134.

Diamandopoulos, G.T., Enders, J.F. (1965) Studies on transformation of Syrian hamster cells by simian virus 40 (SV40): acquisition of oncogenicity by virus-exposed cells apparently unassociated with the viral genome, *Proc. Natl. Acad. Sci. U. S. A.* **54**, 1092–1099.

Easton, J.M., Hiatt, C.W. (1965) Possible incorporation of SV40 genome within capsid proteins of adenovirus 4, *Proc. Natl. Acad. Sci. U. S. A.* **54**, 1100–1104.

Fenn, J.B., Mann, M., Meng, C.K., Wong, S.F., Whitehouse, C.M. (1989) Electrospray ionization for mass spectrometry of large biomolecules, *Science* **246**, 64–71.

Fountoulakis, M., Langen, H. (1997) Identification of proteins by matrix-assisted laser desorption ionization-mass spectrometry following in-gel digestion in low-salt, nonvolatile buffer and simplified peptide recovery, *Anal. Biochem.* **250**, 153–156.

Gharbi, S., Gaffney, P., Yang, A., Zvelebil, M.J., Cramer, R., Waterfield, M.D., Timms, J.F. (2002) Evaluation of two-dimensional differential gel electrophoresis for proteomic expression analysis of a model breast cancer cell system, *Mol. Cell. Proteomics* **1**, 91–98.

Giometti, C.S., Anderson, N.G. (1981) Muscle protein analysis. III. Analysis of solubilized frozen-tissue sections by two-dimensional electrophoresis, *Clin. Chem.* **27**, 1918–1921.

Gygi, S.P., Rochon, Y., Franza, B.R., Aebersold, R. (1999b) Correlation between protein and mRNA abundance in yeast, *Mol. Cell. Biol.* **19**, 1720–1730.

Gygi, S.P., Rist, B., Gerber, S.A., Turecek, F., Gelb, M.H., Aebersold, R. (1999a) Quantitative analysis of complex protein mixtures using isotope-coded affinity tags, *Nat. Biotechnol.* **17**, 994–999.

Han, D.K., Eng, J., Zhou, H., Aebersold, R. (2001) Quantitative profiling of differentiation-induced microsomal proteins using isotope-coded affinity tags and mass spectrometry, *Nat. Biotechnol.* **19**, 946–951.

Heisenberg, M., Blessing, J. (1965) Unvollstandige Viruspartikel als Folge von Mutationen im Genom des Bacteriophagen fr., *Zeitschrift fur Naturforschung – Teil B – Anorganische Chemie, Organische Chemie, Biochemie, Biophysik, Biologie* **20**, 859–864.

Henzel, W.J., Billeci, T.M., Stults, J.T., Wong, S.C., Grimley, C., Watanabe, C. (1993) Identifying proteins from two-dimensional gels by molecular mass searching of peptide fragments in protein sequence databases, *Proc. Natl. Acad. Sci. USA* **90**, 5011–5015.

Hochstrasser, D.F. (1997) Clinical and Biomedical Applications of Proteomics, in: Wilkins, M.R., Williams, K.L., Appel, R.D., Hochstrasser, D.F. (Eds.) *Proteome Research: New Frontiers in Functional Genomics*, Springer, Berlin, Germany, pp. 187–219.

Hookway, C.J. (1995) Abduction, in: Honderich, T. (Ed.) *The Oxford Companion to Philosophy*, Oxford University Press, Oxford, UK, pp. 1.

Hurley, J.B., Yarfitz, S. (1994) Transduction mechanisms of vertebrate and invertebrate photoreceptors, *J. Biol. Chem.* **269**, 14329–14332.

International Human Genome Sequencing Consortium (2001) Initial sequencing and analysis of human genome, *Nature* **409**, 860–921.

James, P., Quadroni, M., Carafoli, E., Gonnet, G. (1993) Protein identification in DNA databases by peptide mass fingerprinting, *Biochem. Biophys. Res. Commun.* **195**, 58–64.

Ji, H., Whitehead, R.H., Reid, G.E., Moritz, R.L., Ward, L.D., Simpson, R.J. (1994) Two-dimensional electrophoretic analysis of proteins expressed by normal and cancerous human crypts: application of mass spectrometry to peptide-mass fingerprinting, *Electrophoresis* **15**, 391–405.

Kahn, P. (1995) From genome to proteome: looking at a cell's proteins, *Science* **270**, 369–370.

Kinumi, T., Jackson, K.W., Ohashi, M., Tobin, S.L., Matsumoto, H. (1997) The phosphorylation site and desmethionyl N-terminus of *Drosophila* phosrestin I *in vivo* determined by mass spectrometric analysis of proteins separated on two-dimensional gel electrophoresis, *Eur. Mass Spectrom.* **3**, 367–378.

Kosik, K.S., Gilbert, J.M., Selkoe, D.J., Strocchi, P. (1982) Characterization of postmortem human brain proteins by two-dimensional gel electrophoresis, *J. Neurochem.* **39**, 1529–1538.

Larsson, T., Norbeck, J., Karlsson, H., Karlsson, K.A., Blomberg, A. (1997) Identification of two-dimensional gel electrophoresis resolved yeast proteins by matrix-assisted laser desorption ionization mass spectrometry, *Electrophoresis* **18**, 418–423.

Lee, C.Y., Charles, D., Bronson, D., Griffin, M., Bennett, L. (1979) Analyses of mouse and Drosophila proteins by two-dimensional gel electrophoresis, *Mol. Gen. Genet.* **176**, 303–311.

Li, G., Waltham, M., Anderson, N.L., Unsworth, E., Treston, A., Weinstein, J.N. (1997) Rapid mass spectrometric identification of proteins from two-dimensional polyacrylamide gels after in gel proteolytic digestion, *Electrophoresis* **18**, 391–402.

Lindemann, T., Hintelmann, H. (2002) Identification of selenium-containing glutathione S-conjugates in a yeast extract by two-dimensional liquid chromatography with inductively coupled plasma MS and nanoelectrospray MS/MS detection, *Anal. Chem.* **74**, 4602–4610.

Lockhart, D.J., Winzeler, E.A. (2000) Genomics, gene expression and DNA arrays, *Nature* **405**, 827–836.

Macdonald, N., Chevalier, S., Tonge, R., Davison, M., Rowlinson, R., Young, J., Rayner, S., Roberts, R. (2001) Quantitative proteomic analysis of mouse liver response to the peroxisome proliferator diethylhexylphthalate (DEHP), *Arch. Toxicol.* **75**, 415–424.

Matsui, N.M., Smith, D.M., Clauser, K.R., Fichmann, J., Andrews, L.E., Sullivan, C.M., Burlingame, A.L., Epstein, L.B. (1997) Identification of two-dimensional gel electrophoresis resolved yeast proteins by matrix-assisted laser desorption ionization mass spectrometry, *Electrophoresis* **18**, 409–417.

Matsumoto, H., Komori, N. (1999) Protein identification on two-dimensional gels archived nearly two decades ago by in-gel digestion and matrix-assisted laser desorption ionization time-of-flight mass spectrometry, *Anal. Biochem.* **270**, 176–179.

Matsumoto, H., Pak, W.L. (1984) Light-induced phosphorylation of retina-specific polypeptides of *Drosophila in vivo*, *Science* **223**, 184–186.

Matsumoto, H., Kurono, S., Komori, N. (2002) Proteomics as a Tool for Studying Complex Systems and the Abductive Inference of C.S. Peirce, *J. Mass Spectrom. Soc. Jpn.* **50**, 116–125.

Matsumoto, H., O'Tousa, J., Pak, W.L. (1982) Light-induced modification of *Drosophila* retinal polypeptides *in vivo*, *Science* **217**, 839–841.

Matsumoto, H., Kurien, B., Takagi, Y., Kahn, E.S., Kinumi, T., Komori, N., Yamada, T., Hayashi, F., Isono, K., Pak, W.L., Jackson, K.W., Tobin, S.L. (1994) Phosrestin I undergoes the earliest light-induced phosphorylation by a calcium/calmodulin-dependent protein kinase in *Drosophila* photoreceptors, *Neuron* **12**, 997–1010.

Mann, M., Hojrup, P., Roepstorff, P. (1993) Use of mass spectrometric molecular weight information to identify proteins in sequence databases, *Biol. Mass Spectrom.* **22**, 338–345.

Mauer, I. (1965) Chromosome and genome as conceptual models for certain types of mental retardation, *Am. J. Mental Deficiency* **70**, 191–203.

Neukirchen, R.O., Schlosshauer, B., Baars, S., Jackle, H., Schwarz, U. (1982) Two-dimensional protein analysis at high resolution on a microscale, *J. Biol. Chem.* **257**, 15229–15234.

Nishizawa, Y., Komori, N., Usukura, J., Jackson, K.W., Tobin, S.L., Matsumoto, H. (1999) Initiation of ocular proteomics for cataloging bovine retinal proteins: microanalytical techniques permit the identification of proteins derived from a novel photoreceptor preparation, *Exp. Eye Res.* **69**, 195–212.

Niwayama, S., Kurono, S., Matsumoto, H. (2001) Synthesis of d-labelled N-alkylamides and application to quantitative peptide analysis by isotope differential mass spectrometry, *Bioorg. Med. Chem. Lett.* **11**, 2257–2261.

O'Brien, C. (1996) Protein fingerprints, proteome projects and implications for drug discovery, *Mol. Med. Today* **2**, 316.

O'Farrell, P.H. (1975) High resolution two dimensional electrophoresis of proteins, *J. Biol. Chem.* **250**, 4007–4021.

Pappin, D.J.C., Hojrup, P., Bleasby, A.J. (1993) Rapid identification of proteins by peptide-mass fingerprinting, *Curr. Biol.* **3**, 327–332.

Patterson, S.D., Aebersold, R. (1995) Mass spectrometric approaches for the identification of gel-separated proteins, *Electrophoresis* **16**, 1791–1814.

Patton, W.F. (2002) Detection technologies in proteome analysis, *J. Chromatogr., B* **771**, 3–31.

Qi, S.Y., Moir, A., O'Connor, C.D. (1996) Proteome of Salmonella typhimurium SL1344: identification of novel abundant cell envelope proteins and assignment to a two-dimensional reference map, *J. Bacteriol.* **178**, 5032–5038.

Rosenfeld, J., Capdevielle, J., Guillemot, J.C., Ferrara, P. (1992) In-gel digestion of proteins for internal sequence analysis after one- or two-dimensional gel electrophoresis, *Anal. Biochem.* **203**, 173–179.

Sechi, S. (2002) A method to identify and simultaneously determine the relative quantities of proteins isolated by gel electrophoresis, *Rapid Commun. Mass Spectrom.* **16**, 1416–1424.

Shevchenko, A., Jensen, O.N., Podtelejnikov, A.V., Sagliocco, F., Wilm, M., Vorm, O., Mortensen, P., Boucherie, H., Mann, M. (1996) Linking genome and proteome by mass spectrometry: large-scale identification of yeast proteins from two dimensional gels, *Proc. Natl. Acad. Sci. USA* **93**, 14440–14445.

Smolka, M.B., Zhou, H., Purkayastha, S., Aebersold, R. (2001) Optimization of the isotope-coded affinity tag-labeling procedure for quantitative proteome analysis, *Anal. Biochem.* **297**, 25–31.

Tanaka, K., Waki, H., Ido, Y., Akita, S., Yoshida, Y., Yoshida, T. (1988) Protein and polymer analyses up to m/z 100000 by laser ionization time-of-flight mass spectrometry, *Rapid Commun. Mass Spectrom.* **2**, 151–153.

Tonge, R., Shaw, J., Middleton, B., Rowlinson, R., Rayner, S., Young, J., Pognan, F., Hawkins, E., Currie, I., Davison, M. (2001) Validation and development of fluorescence two-dimensional differential gel electrophoresis proteomics technology, *Proteomics* **1**, 377–396.

Turecek, F. (2002) Mass spectrometry in coupling with affinity capture-release and isotope-coded affinity tags for quantitative protein analysis, *J. Mass Spectrom.* **37**, 1–14.

Unlu, M., Morgan, M.E., Minden, J.S. (1997) Difference gel electrophoresis: a single gel method for detecting changes in protein extracts, *Electrophoresis* **18**, 2071–2077.

Venter, J.C., Adams, M.D., Myers, E.W., et al. (2001) The sequence of the human genome, *Science* **291**, 1304–1351.

Von Eggeling, F., Gawriljuk, A., Fiedler, W., Ernst, G., Claussen, U., Klose, J., Romer, I. (2001) Fluorescent dual colour 2D-protein gel electrophoresis for rapid detection of differences in protein pattern with standard image analysis software, *Int. J. Mol. Med.* **8**, 373–377.

Wagner, K., Miliotis, T., Marko-Varga, G., Bischoff, R., Unger, K.K. (2002) An automated on-line multidimensional HPLC system for protein and peptide mapping with integrated sample preparation, *Anal. Chem.* **74**, 809–820.

Wasinger, V.C., Cordwell, S.J., Cerpa-Poljak, A., Yan, J.X., Gooley, A.A., Wilkins, M.R., Duncan, M.W., Harris, R., Williams, K.L., Humphery-Smith, I. (1995) Progress with gene-product mapping of the Mollicutes: Mycoplasma genitalium, *Electrophoresis* **16**, 1090–1094.

Wilkins, M.R., Sanchez, J.C., Williams, K.L., Hochstrasser, D.F. (1996b) Current challenges and future applications for protein maps and posttranslational vector maps in proteome projects, *Electrophoresis* **17**, 830–838.

Wilkins, M.R., Sanchez, J.C., Gooley, A.A., Appel, R.D., Humphery-Smith, I., Hochstrasser, D.F., Williams, K.L. (1996c) Progress with proteome projects: why all proteins expressed by a genome should be identified and how to do it, *Biotechnol. Genet. Eng. Rev.* **13**, 19–50.

Wilkins, M.R., Ou, K., Appel, R.D., Sanchez, J.C., Yan, J.X., Golaz, O., Farnsworth, V., Cartier, P., Hochstrasser, D.F., Williams, K.L., Gooley, A.A. (1996a) Rapid protein identification using N-terminal "sequence tag" and amino acid analysis, *Biochem. Biophys. Res. Commun.* **221**, 609–613.

Wilkins, M.R., Pasquali, C., Appel, R.D., Ou, K., Golaz, O., Sanchez, J.C., Yan, J.X., Gooley, A.A., Hughes, G., Humphery-Smith, I., Williams, K.L., Hochstrasser, D.F. (1996d) From proteins to proteomes: large scale protein identification by two-dimensional electrophoresis and amino acid analysis, *Bio-Technology* **14**, 61–65.

Willard, K.E., Anderson, N.G. (1981) Two-dimensional analysis of human lymphocyte proteins: I. An assay for lymphocyte effectors, *Clin. Chem.* **27**, 1327–1334.

Wimmer, K., Kuick, R., Thoraval, D., Hanash, S.M. (1996) Two-dimensional separations of the genome and proteome of neuroblastoma cells, *Electrophoresis* **17**, 1741–1751.

Winocour, E. (1965) Attempts to detect an integrated polyoma genome by nucleic acid hybridization II. Complementarity between polyoma virus DNA and normal mouse synthetic RNA, *Virology* **27**, 520–527.

Yan, J.X., Wilkins, M.R., Ou, K., Gooley, A.A., Williams, K.L., Sanchez, J.C., Golaz, O., Pasquali, C., Hochstrasser, D.F. (1996) Large-scale amino-acid analysis for proteome studies, *J. Chromatogr., A* **736**, 291–302.

Yao, X., Freas, A., Ramirez, J., Demirev, P.A., Fenselau, C. (2001) Proteolytic ^{18}O labeling for comparative proteomics: model studies with two serotypes of adenovirus, *Anal. Chem.* **73**, 2836–2842.

Yates, III, J.R., Speicher, S., Griffin, P.R., and Hunkapiller, T. (1993) Peptide mass maps: a highly informative approach to protein identification, *Anal. Biochem.* **214**, 397–408.

Zhou, G., Li, H., DeCamp, D., Chen, S., Shu, H., Gong, Y., Flaig, M., Gillespie, J.W., Hu, N., Taylor, P.R., Emmert-Buck, M.R., Liotta, L.A., Petricoin, E.F., III, Zhao, Y. (2002) 2D differential in-gel electrophoresis for the identification of esophageal scans cell cancer-specific protein markers, *Mol. Cell. Proteomics* **1**, 117–124.

Mass Spectrometry, High Speed DNA Fragment Sizing

Chung-Hsuan Chen
Oak Ridge National Laboratory, Oak Ridge, Tennessee, USA

Encyclopedia of Molecular Cell Biology and Molecular Medicine, 2nd Edition. Volume 7
Edited by Robert A. Meyers.

ISBN: 3-527-30549-1

Keywords

Matrix-assisted Laser Desorption and Ionization (MALDI)
Biomolecules that are cocrystallized with a much larger amount of organic compounds on a substrate are laser-ablated to accomplish desorption and ionization for mass spectrometry detection.

Time-of-Flight Mass Spectrometer (TOF)
A mass spectrometer that is used to determine the mass-to-charge ratio of gas-phase ions by measuring the flight time in a drift tube for ions.

Gel Electrophoresis
Migration of charged molecules in an electric field with gel as a medium, which is obtained for separation and/or purification of biomolecules.

DNA Hybridization
Two single-stranded DNAs with complementary sequences combine to produce a double-stranded, with hydrogen bonding between two single-stranded DNAs. Hybridization is often used for DNA sequence identification. Microarray hybridization is often used for high throughput–DNA analysis.

DNA Sequencing
A process that determines the sequence of bases in a DNA segment.

Abbreviations

ASPCR:	Allele-specific polymerase chain reaction
CF:	Cystic fibrosis
CFTR:	Cystic fibrosis transmembrane conductance regulator
ds-DNA:	Double-stranded deoxyribonucleic acid
DNA:	Deoxyribonucleic acid
DRPLA:	Dentatorubral-pallidoluysian atrophy
EDTA:	Ethylene diamine tetraacetic acid
IR-MALDI:	MALDI with an infrared laser beam for desorption
MALDI:	Matrix-assisted laser desorption/ionization

MS:	Mass spectrometry
PCR:	Polymerase chain reaction
RFLP:	Restriction fragmented length polymorphism
ss-DNA:	Single-stranded deoxyribonucleic acid
SNP:	Single nucleotide polymorphism
STR:	Short tandem repeat
TOF:	Time-of-flight
TOFMS:	Time-of-flight mass spectrometry
UV-MALDI:	MALDI with an ultraviolet laser beam for desorption
VNTR:	Variable number of tandem repeats

Recently, high-speed DNA fragment sizing has been in critical need due to its important application on genomic function, DNA sequencing, disease diagnosis, DNA typing for forensic use, and microbial analysis for bioagent identification. Mass spectrometry has played a key role in rapid DNA sizing. This article discusses the fundamental principles of mass spectrometry for DNA analysis and the advantages and disadvantages of mass spectrometry compared to other approaches. It also gives a brief historical review of the development of mass spectrometry for DNA analysis. Applications in DNA sequencing, genetic disease diagnosis, and DNA finger printing are also presented.

1 Introduction

Since the discovery of the double-helix structure of DNA by Watson and Crick, research on DNA has been a key component of biological and medical research. The major functions of DNA comprise the transmission of information through a genetic code and self-replication. Many important advances have been made in understanding the complex organization of genetic material in cells and applying genetic engineering to manipulate the replication capability. However, because of the extreme complexity of the human genome, a worldwide Human Genome Project on mapping and sequencing the entire human genome was initiated in 1988 and has been more or less completed recently. The interest in DNA sequencing has expanded to include microbial genome, mouse genome, rice genome, and others. It is also expected that the need to sequence and resequence known genes will be increased by many orders of magnitude in the coming decades. Nevertheless, nearly all the data relating to sequencing of DNA have been obtained by gel electrophoresis, which is still relatively slow and somewhat expensive. It is clear that faster and cheaper methods are in critical need. For most resequencing work, the sequencing speed and cost can be more critical than the size of DNA that can be sequenced. Mass spectrometry for DNA sizing and analysis can serve this purpose because of its short analysis time.

1.1 Time-of-flight Mass Spectrometry

The conventional method used for DNA measurement is gel electrophoresis. In this process, an electric field is applied to the gel medium to cause the drift of DNA fragments, which typically exist as charged particles. The drift velocity tends to be lower as DNA fragments become larger. However, this process is slow and needs hours to separate different sizes of DNA fragments. The DNA-sizing speed by capillary gel electrophoresis is significantly faster than slab-gel electrophoresis. Nevertheless, mass spectrometry DNA sizing is much faster than either of the methods of electrophoresis. Mass spectrometry has been used for several decades to measure the molecular weight of gas samples by mass-to-charge ratio. In order to use a mass spectrometer for molecular weight determination, the relevant molecules need to be produced in gas phase. The pressure inside the mass spectrometer needs to be less than 10^{-4} Torr in order to prevent collisions between the ions and other gaseous molecules. Owing to the practically zero-vapor pressure of DNA at room temperature, a method of producing DNA gas-phase molecules is required. Desorption of DNAs by a laser beam oran ion beam is often needed. Another approach is to introduce solvated DNA samples through electrospray or ionspray processes. Since mass spectrometry is based on the detection of gas-phase ions, a means to produce such ions is obviously needed. Ions can be produced by several different processes. These processes include charged particle ionization, photoionization, chemi-ionization, and electron-impact ionization. All these processes involve the removal of an electron from a molecule to form a positive ion. However, positive or negative ions can also be produced by attaching or detaching a charged particle such as a proton. This process is quite often achieved by a chemical process. Once the ions have been produced, various sizes of ions may be separated by electric or magnetic fields. The mass-resolved ions are then accelerated to impinge on a solid surface of low work function to generate secondary electrons, which are subsequently detected by an electron multiplier, or a channeltron or microchannel plates (MCP). Since these electron detection devices can have an amplification factor of 10^8, even a single secondary electron can be detected.

Owing to the similarity of principles between a time-of-flight mass spectrometer (TOFMS) and a gel-electrophoresis device, a time-of-flight mass spectrometer is most often used for DNA detection. A TOFMS is a device in which ions of different masses are given the same energy and are allowed to travel in a field-free space. Because of their different velocities, the ions of differing masses arrive at the end of the field-free space at different times, where an ion detector produces a time-varying electrical signal proportional to the number of ions impacting it. The amplified current signal is typically displayed on a fast digital oscilloscope, where the molecular weights can be determined by measuring the time interval between the creation of ions and the detection of ions. The ions need to be produced in a very short time interval and in a very small volume in order to achieve high resolution mass spectra. Thus, ionization by a short-pulsed laser beam is an ideal choice for a TOFMS. A schematic of a typical TOFMS is shown in Fig. 1. Ions are produced by a short-pulsed laser beam, which is typically focused to a small area in the ionization region. Ions

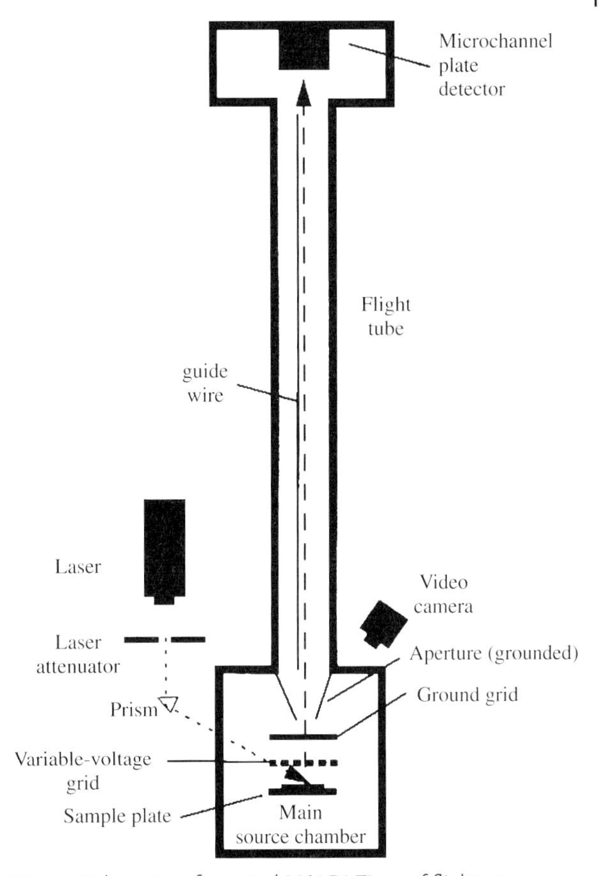

Fig. 1 Schematic of a typical MALDI Time-of-flight mass spectrometer. A laser beam is focused onto a sample for desorption and ionization through protonation/deprotonation process. UV, IR, and visible lasers have all been used for MALDI. However, most works have been done with UV-laser desorption. A video camera is typically used to observe the sample location so that the laser beam can be focused on the desired location. Pulsed-ion extraction is often used to enhance mass resolution. A guide wire is used to achieve the oscillation of desorbed ions to reduce the desorbed biomolecular ions hitting the wall of vacuum chamber. A charged particle detector such as microchannel plate is used for ion detection. Most commercial instruments have sample holder with array configuration so that multiple samples can be placed in the mass spectrometer for analysis.

are then accelerated to a uniform energy by an electric field before they go into the free drift zone. An electron multiplier or a microchannel plate is placed at the end of the drift tube to detect the ions. Since ions have different initial velocities when they are produced, the higher the acceleration energy, the lesser the effect the initial velocity spread will have on the flight time, and the higher the mass resolution will be. On the other hand, a higher voltage gradient in the area of the ionizing laser will result in ions produced in different parts of the laser beam being accelerated through different potentials, reducing the resolution of the flight times. Better resolution is achieved with a smaller ratio of ionization radius to acceleration length. There are several ways to improve the resolution of a TOFMS. A popular approach is the reflectron TOFMS. An electrostatic lens to bend ion trajectory is typically placed at the end of the drift tube to *reflect* the ions toward the detector. This design can be used to compensate for either the initial velocity spread or the spatial spread of the ions. However, it is difficult to compensate for both velocity spread and spatial spread. Resolution ($M/\Delta M$) of 10^5 for a reflectron TOFMS has been achieved by a few groups of researchers. One major advantage of a TOFMS over other types of mass spectrometers is its speed in measuring all the masses in a sample in a very short period of time, and another is its capability of measuring ions with very high molecular weights. In principle, there are no limitations on the size of molecules that can be measured by a TOFMS if the ions can be accelerated to a high enough velocity that secondary electrons can be ejected due to the ion's impact on the ion detector. This property is critically important for the detection of DNA fragments since most DNA fragments have very high molecular weights.

A mass spectrometer can be used for molecular weight determination only under the condition that the molecular ions are produced in space. Thus, a mass spectrometer, usually, can only be used for materials with vapor pressure higher than 10^{-14} Torr at room temperature. Although the minimum vapor pressure required is very low, the vapor pressures of many materials at room temperature are even lower (lower than 10^{-14} Torr). DNA fragments are in this category. In order to use mass spectroscopy for DNA measurement, it is necessary to develop a method that places DNA fragments in space without breaking them up. Recent developments in laser desorption have been significantly successful in achieving this. A frozen double-stranded DNA fragment on a copper plate was successfully desorbed by a dye-laser beam. Gel electrophoresis was used to prove the success of laser desorption of a large amount of DNA. However, the use of mass spectrometry for DNA analysis requires both desorption and ionization.

1.2
Matrix-assisted Laser Desorption/Ionization (MALDI) for DNA Analysis

Since a mass spectrometer can only be used to detect ions, some process of ionization is necessary if only neutral DNA segments are produced in the desorption process. Ionization can be achieved either by electron detachment to form a neat positive ion or by electron attachment to form a neat negative ion. Ions can also be produced by protonation to form $(M + H)^+$ or deprotonation to produce $(M - H)^-$. Since the size of a DNA segment is usually large and some of the DNA bonding (such as the glycosidic bond) is fragile, successful soft ionization to produce parent ions without breakup is a challenge. In addition to the *soft* desorption and *soft* ionization, the third major problem is the difficulty in detecting very heavy ions. In order to use an electron multiplier or an MCP in the detection of an ion efficiently, the ion velocity needs to be greater than 1.5×10^6 cm/s. For example, a 1000-bp DNA segment has a molecular weight of ~600 000 daltons. These ions need to be accelerated to 700 KeV in order to reach a velocity of 1.5×10^6 cm/s. Most mass spectrometer designs are not suitable for accelerating very heavy ions to the velocity required for detection. Recently, the secondary electron emission efficiency at various velocities for a few bio-organic molecules was measured and the conclusion reached was that the threshold velocity could be somewhat lower than 1.5×10^6 cm/s for larger molecules. The successful detection was attributed, at least partially, to the ejection of smaller ions when large ions hit the metal plates. Nevertheless, the low detection efficiency for very large molecular ions is a concern when using mass spectrometry for the detection of very large DNA segments.

It is clear that the use of mass spectrometers for DNA detection will require the solution of several challenging problems. However, there is a major advantage in using mass spectrometry because of the potential to achieve ultrafast speed. Using gel electrophoresis for DNA separation typically takes a few hours. If radioactive tagging is used, it can be done overnight. However, a mass spectrum can be obtained within one second. For example, a TOFMS usually needs less than one millisecond to separate different sizes of DNA fragments, thus speed is improved by many orders of magnitude. In addition, hundreds of samples can be placed in a mass spectrometer for fast analysis if an array sample holder is designed to hold multiple targets. For gel electrophoresis, tagging of radioactive materials or organic chromophore to DNA fragments is required for detection. Thus, radioactive or chemical wastes are produced. Mass spectrometry analysis eliminates the need for tagging; therefore, no waste will be produced. Although a mass spectrometer can be more expensive than a gel electrophoresis setup, the much faster speed of analysis by a mass spectrometer can drive the cost much lower on a per sample basis. For example, 1000 samples of DNA fragments can possibly be analyzed by a TOFMS within an hour. This can be particularly important for medical applications, since it may be necessary for large hospitals to analyze thousands of samples each day.

For the analysis of materials with no vapor pressure at room temperature, several desorption methods have been used to carry the analyte into space for mass analysis. Major approaches include desorption by fast-atom bombardment (FAB), secondary-ion bombardment, and laser desorption. However, none of these have been very successful for analysis of large biomolecules. Since most biomolecules are fragile, when subjected to heat or bombardment, a fragmentation process almost always occurs. However, in 1987, Hillenkamp and his coworkers discovered that large protein molecular ions can be produced without much fragmentation by laser desorption if these biomolecules are mixed with smaller organic compounds that serve as a matrix for strong absorption of a laser beam. This process is now called *matrix-assisted laser desorption* (MALD). The typical preparation technique for MALD is to dissolve biomolecular samples in solution, then prepare another solution that contains small organic compounds such as 3-hydroxypicolinic acid (3-HPA). These two solutions are subsequently mixed, and a small amount of the resulting solution is placed on a metal plate to dry. After the crystallization of the sample, the sample plate is placed in the mass spectrometer for analysis. The molar ratio of matrix-to-analyte is typically more than 100 to 1. During a MALD process, matrix materials strongly absorb the laser energy and become vaporized, and large biomolecules are carried outward during the fast-vaporization process. Large biomolecules can be delivered into space without breakup, probably due to minimal direct absorption of laser energy; thus, *soft* desorption can be achieved. However, it has been found that biomolecule parent ions are also produced during the MALD process in addition to the expected neutral molecules. Thus, these desorbed ions can be directly detected by a mass spectrometer. This process involving matrix-assisted laser desorption and ionization at the same time is referred to in an abbreviated form as MALDI (matrix-assisted laser desorption and ionization).The mechanism of ion production has been speculated by many researchers to involve proton transfer.

Up to now, one major disadvantage of MALDI for DNA analysis has been its relatively poor mass resolution especially for large DNA fragments. In order to study the cause of poor mass resolution, velocity distributions of desorbed oligomers were studied, and it was found that the distribution more or less follows the distribution of the matrix. In general, the mass resolution of MALDI spectrum of DNA is relatively poor compared to that of small organic compounds. This relatively poor resolution is due to (1) high initial velocity distribution, (2) the inhomogeneity of DNA samples on metal plates, (3) the adducts of matrix materials or metal ions to DNAs, and (4) fragmentation of DNA molecules due to the laser ablation process. Since the velocity distribution of the DNAs are passively carried out by matrix materials, the velocity spreads are expected to be similar to the velocity distribution of the matrix material. If the molecular weight of a DNA fragment is 200 000 daltons and the molecular weight of the matrix is 100, the desorbed energy of DNA molecules can reach 50 eV, even assuming that the average kinetic energy of desorbed matrix material is near thermal. Thus, the broad distribution of ion energies of desorbed biomolecules is one important factor that limits the resolution. When a biomolecule sample mixed with matrix materials begins to dry on the plate, the process of crystal growth makes the sample distribution inhomogeneous. The thickness of the sample can also cause lower resolution. However, a reflectron TOFMS can compensate for the ionization volume to achieve a resolution higher than 1000. Since most spectra by MALDI have resolutions of only a few hundred, the limited resolution is primarily a result of adduct and fragment formation. If a biomolecule can be attached to a matrix molecule, or part of a matrix molecule, the resolution then depends on the molecular weight of the attached particle. It is also not unusual to observe alkali metal ions attached to DNA molecules. The adduct formation is strongly dependent on the matrix used. Detection sensitivity for small DNA segments ($<$10mer) by MALDI can reach 1 femtomole. The bigger the size of the DNA segment, the lower is the detection sensitivity.

The chemical characteristic of matrix material used in MALDI for DNA segment analysis is probably the single most critical parameter. Only a few compounds have been found to be useful matrices for oligomers. A good matrix material needs to have good solubility in the solvents used for the analyte, strong absorption at the laser wavelength, and no chemical reactivity with the analytes. In general, the matrix needs to dissolve in the solvent in concentrations higher than 0.01 M solution. Otherwise, a high molar matrix-to-analyte ratio is difficult to achieve. The absorption coefficient at the laser wavelength needs to be higher than 3000 L $mole^{-1}cm^{-1}$; otherwise, the laser energy deposition cannot be concentrated in the top few hundred molecular layers to produce efficient desorption. It is clear that the matrix materials used cannot chemically react with the analytes. For example, strong acids tend to remove purine bases. Thus, they would not be good matrices. The glycosidic bonding of DNA is fragile and easily subject to chemical interaction. This can be a factor that makes MALDI for DNA analysis much more difficult than for protein analysis. Although a few criteria can be set up to rule out certain chemicals as good candidates for MALDI of DNA segments, the search for good matrices is still more or less trial and error.

In addition to the chemical properties of a matrix material, the amount of energy deposited by the laser is also critical for achieving good spectra. It has been shown that the relationship between biomolecule ion production and the intensity of the laser pulse is highly nonlinear. Below a threshold, no biomolecule ion signal is produced. Above a threshold, the ion production increases rapidly to a plateau. The threshold for ion production of DNA segments was found to be a few megawatts per square centimeter. However, ion signals tend to disappear when the laser intensity is so high that intense plasma is produced.

Detailed mechanisms for ion production of biomolecules by MALDI are still not well understood. However, a simple model of photoionization and photochemistry processes has been proposed to explain the production of biomolecular ions. The model suggests that a multiphoton absorption process leads to the ionization of matrix molecules, which have subsequent chemical reactions with biomolecule analytes to form ions.

The existence of metastable states of negative oligomer ions during the MALDI process has been experimentally proved. The lifetimes of small negative oligomer ions were measured to be a few hundred microseconds. The lifetimes become shorter for larger negative oligomer ions compared to that of smaller ones. These results indicate that the detection of large negative oligomer ions will be very difficult for Fourier transform, ion-cyclotron, reflectron TOF and ion-trap mass spectrometers with MALDI because of the long ion lifetimes inherently required by these instruments.

Although MALDI has been applied to detect large biomolecules with a good determination of molecular weights, it has been difficult to achieve quantitative measurements of biomolecules. The primary reasons for poor reproducibility are related to the inhomogeneous distribution of samples on substrates and the high sensitivity to changes in laser fluence. A small change in laser fluence can cause a very large fluctuation in analyte signals. When samples are prepared in a MALDI experiment, the crystallization process can make the distribution of matrix materials inhomogeneous. For example, nonhomogeneous needle-shaped crystals were observed when using 2,5-dihydroxybenzoic acid as a matrix. The signal level of analytes can be a strong function of the exact spot at which the laser desorbs. It is also known that an impurity in a MALDI process can reduce the signal levels of analytes significantly. In addition, the surface condition of the substrate can make a significant change in the threshold for the production of plasma. In brief, it is difficult to obtain reliable measurements of absolute quantities of analytes by MALDI. If MALDI is to become a routine analytical tool, it is obvious that its capability to make quantitative measurements of analytes must be improved. When the quantity of the matrix is fixed, signals from oligomers are not linearly proportional to the quantity of the analyte. This very nonlinear relation between ion signal and analyte is mostly due to the number of layers of analyte absorbing laser energy and the detailed interaction between DNA and matrix molecules during the crystallization process. Too much of the analyte on the samples may not produce bigger signals. However, if a known biomolecule with similar chemical properties is used for an internal calibration, the signal level can become more or less independent of the laser fluence and the inhomogeneous distribution of matrix materials.

2
DNA Sizing and Sequencing by Mass Spectrometry

2.1
DNA Sizing by MALDI

Since the discovery of MALDI, many research groups have succeeded in measuring various proteins and large organic compounds by MALDI. MALDI has also been applied to DNA segments. Initially, success was limited to the detection of small DNA fragments. Then, 3-hydroxypicolinic acid (3-HPA) was found to be a good matrix for oligonucleotide detection. Both positive and negative ions were observed using 3-HPA as matrix (Fig. 2). Restriction fragment length polymorphism (RFLP) and polymerase chain reactions (PCR) are two of the most important DNA analysis methodologies discovered in the past few decades. PCR is broadly used in

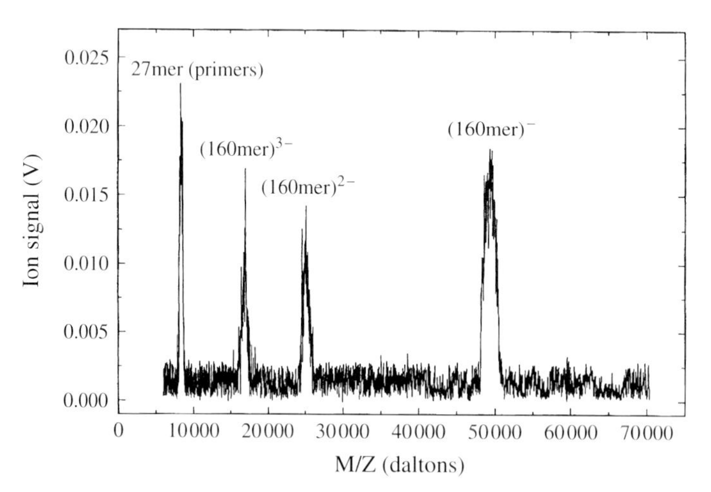

Fig. 2 Negative-ion mass spectrum of a 160-bp DNA amplified from pLB 129. The desorption laser was a fourth harmonic of a Nd-Yag laser with wavelength at 266 nm. In addition to single charged ions, doubly and triply – charged ion peaks were also observed. Although double-stranded DNA was used, only the peaks corresponding to single-stranded DNA ions were observed. It indicates that ds-DNA was probably dissociated into ss-DNA due to the interaction with acidic matrix molecule.

nearly every DNA laboratory to replicate a required segment of DNA from genomic DNAs. MALDI mass spectrometry has also been applied for RFLP and PCR products analysis. The success in using mass spectrometry for measurements of RFLP and PCR products paved the way for several important applications including disease diagnosis and DNA typing for forensic applications. With the development of an instrument to give high ion energy and the use of mixtures of matrices, MALDI was successfully used to detect double-stranded DNA (ds-DNA) of 500 bp, which was produced by a PCR amplification process (Fig. 3). However, only a single-stranded DNA was observed. This marked the first success of using MALDI for a large DNA fragment. Recently, the detection of DNA with the size of 3199 bp was reported (Fig. 4). It can be concluded that large DNA fragments can be measured by MALDI. However, the mass resolution for large DNA fragments (>300 nucleotides) is still very poor ($M/\Delta M < 50$). RFLP measurement by MALDI for S16 rDNA for a microbe has also been successfully demonstrated. (Fig. 5). It indicates that mass spectrometry is emerging as a valuable tool for DNA sizing for various applications.

2.2 DNA Sequencing

DNA sequencing is a method that not only determines the size of a DNA fragment, but also its structure. DNA contains four primary bases – adenine (A), guanine (G), cytosine (C), and thymine (T). Their chemical structures and molecular weights are shown in Fig. 6. Since there is a mechanism for translating the nucleotide sequence in DNA into the amino acid sequence of protein, DNA sequencing is critically important for biological and medical research. Current methods of DNA sequencing such as those of Sanger and Maxam-Gilbert have proven to be very useful for sequencing small DNA segments. These methods of sequencing usually

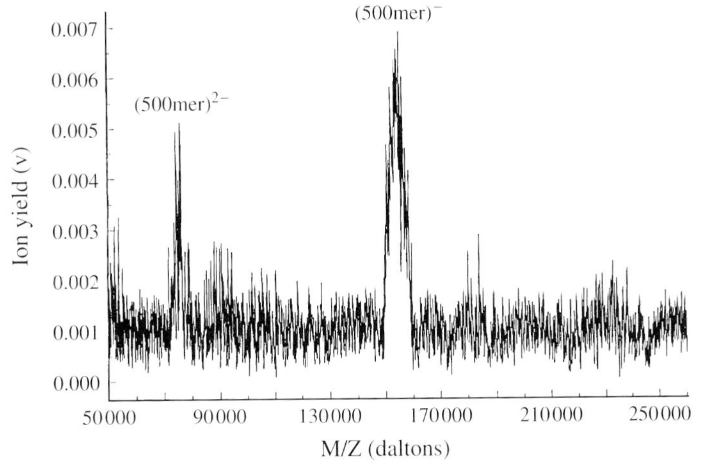

Fig. 3 Mass spectrum of a 500-bp DNA amplified by PCR from bacteriophage lamda genome. The laser fluence used was 200 mJ cm^{-2}. No peaks corresponding to ds-DNA were observed.

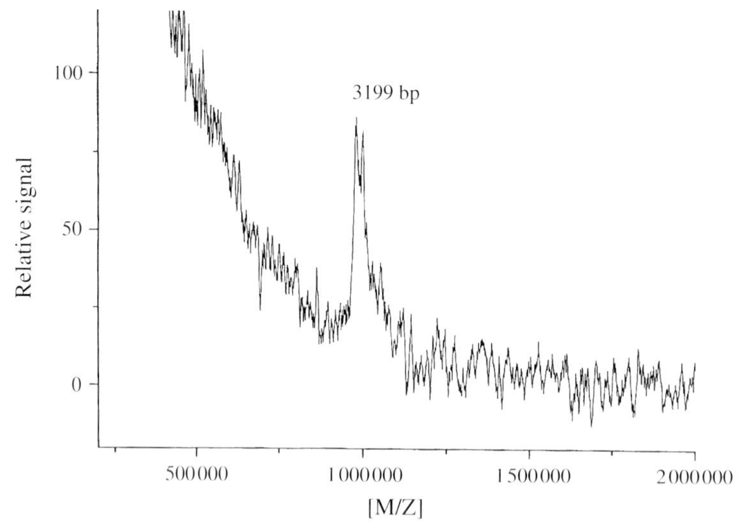

Fig. 4 Mass spectrum of a large DNA fragment of 3199 bp. It clearly indicates that large DNA segments can be detected by MALDI. However, the detection efficiency is very low due to the low yield of secondary electrons when the ion velocity is very low ($< 1 \times 10^5$cm s^{-1}).

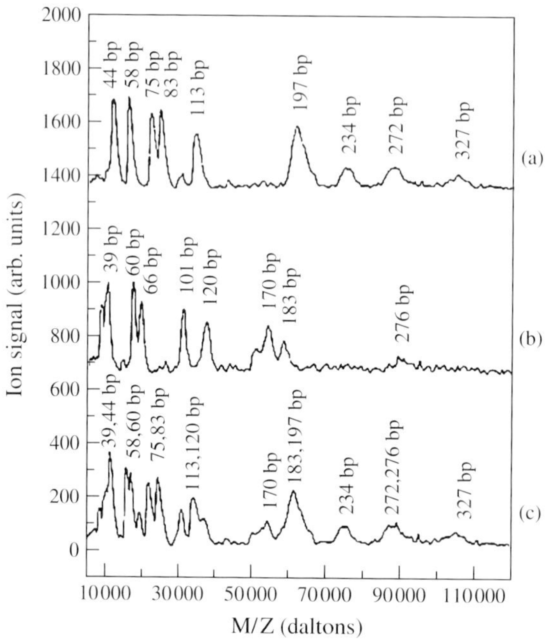

Fig. 5 MALDI spectra from enzyme digestion of PCR products of 16S rDNA from various microbial samples. (a) and (b) show the mass spectra of DNA fragments from RFLP with enzyme digestion by *Hha*I and *Has*III and *Msp*I and *Rsa*I, respectively. The pattern of fragmentation can be used to help in bacteria identification. (c) shows the results of DNA fragments from the mixtures of RFLP from samples contributing to (a) and (b).

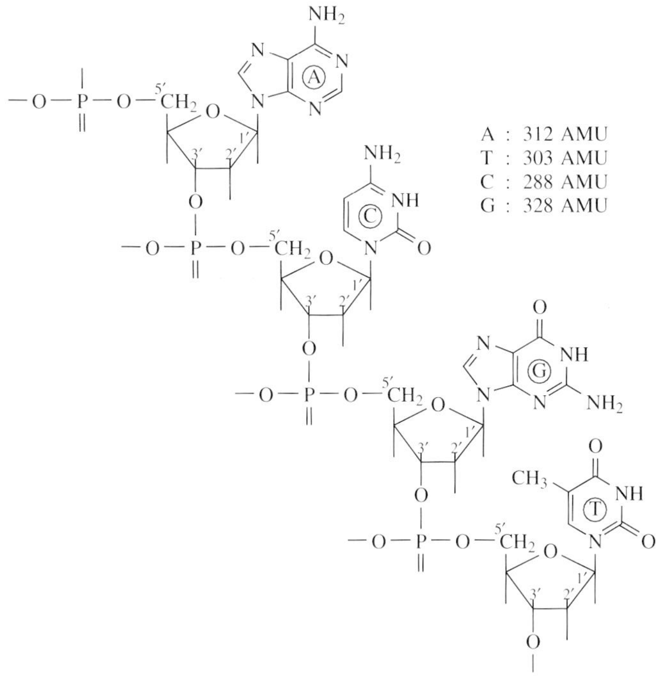

A : 312 AMU
T : 303 AMU
C : 288 AMU
G : 328 AMU

Fig. 6 Chemical structures and molecular weights of adenine (A), thymine (T), guanine (C) and cytosine (C).

involve labeling fragments of DNA for identification following gel electrophoresis. Either radioactive labels such as ^{32}P or ^{35}S, or chemical labels such as fluorescent dyes are currently used. At the present time, routine DNA sequencing in most laboratories relies on the use of radioactive isotopes. The fragments are generated either chemically or enzymatically to represent all possible positions of each of the four nucleotides (A, G, C, and T). Following electrophoresis, the radioactive labels are located by autoradiography. This method is very time-consuming. With radioactive labeling, four lanes of gel are usually required to separate the fragments from a given DNA segment that terminates in A, G, C, or T. Substitution of fluorescent labels for radioactive isotopes allows the DNA fragments to be detected continuously during electrophoresis. This method usually employs four different fluorescent labels. Using these four labels, all four types of DNA fragments (A, G, C, and T) can be run in one electrophoresis lane. The uncertainty in comparing the label positions in four adjacent lanes of the gel is eliminated. Automatic sequencing instruments based on fluorescence measurements are currently commercially available. Recently, capillary gel electrophoresis coupled with the fluorescence method has also been successfully used for improving sequencing speed. However, both radioactive methods and fluorescent dye labeling methods for DNA sequencing require the use of the time-consuming gel electrophoresis. Thus, it is natural to consider using TOFMS for DNA sequencing to reduce the sequencing time. Instead of taking several hours by gel electrophoresis, the TOFMS takes less than one millisecond to finish the separation of different sizes of DNA segments. However, the parent DNA ions must be produced with high efficiency and without serious fragmentation. Since the MALDI process has been successfully used in producing parent ions of small DNA segments without serious fragmentation, it can have an important role in the future of fast DNA sequencing especially for re-sequencing.

2.2.1 DNA Sequencing by Mass Spectrometry with DNA Ladders

DNA sequencing has been broadly used for biomedical research and clinical applications during the past two decades. Rapid and reliable DNA sequencing can also be very valuable in forensic applications. In the conventional sequencing approach, different sizes of DNA ladders, which are produced by either Sanger's enzymatic method or Maxam–Gilbert's chemical cleavage method, are separated by gel electrophoresis to achieve sequencing. With the use of MALDI for sequencing, the speed can be significantly greater than with gel electrophoresis. In addition, MALDI sequencing does not require labeling for identification, which saves both time and cost.

During the past few years, mass spectrometry for sequencing short DNA fragments has been pursued. In 1995, the first sequencing of a 45mer ss-DNA using MALDI-TOF with enzymatic preparation of DNA ladders was demonstrated. Recently, sequencing of ss-DNA higher than 100 nucleotides by using cycle sequencing to produce DNA ladders was achieved. Figure 7 shows the negative-ion mass spectra of DNA ladders from A, C, G, and T reactions using double-stranded DNA 130 bp as the template with reverse primer. Forward primers were also used for sequencing with results similar to the data shown in Fig. 7. Since sequencing by using both forward and backward primers

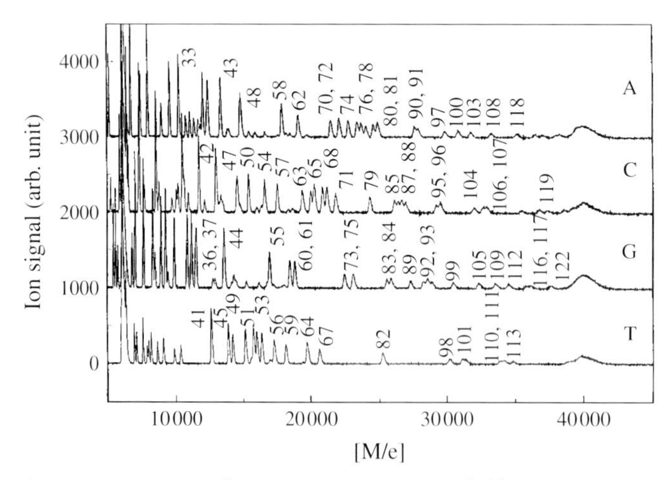

Fig. 7 Mass spectra of Sanger sequencing. DNA ladders were produced by dideoxynucleotide chain termination of the PCR product of a double-stranded template. Owing to the limitation of mass resolution (M/Δ M), the largest ss-DNA can be sequenced is limited to ~ 100 nt.

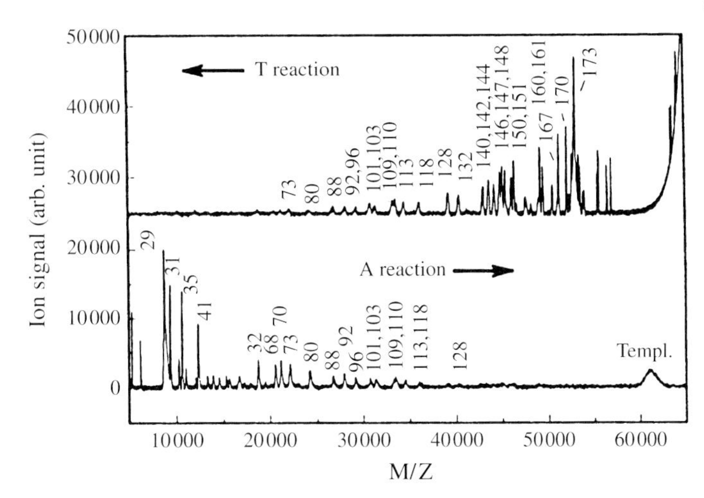

Fig. 8 Negative-ion mass spectra of Sanger sequencing. Ladders were produced by a dideoxynucleotide-chain termination of the PCR product of a double-stranded 200-bp template. The results were from the A- reaction with the reverse primer and T-reaction with forward primer.

can reach 120mer, complete sequencing of 200-bp DNA can be achieved. Figure 8 illustrates the idea of combining the spectra from forward and backward primers to sequence ds-DNA with a 200-bp template. Since mass is measured to identify the size of DNAs, there is no concern about a missing band, which can occur in the gel electrophoresis method. Thus, redundant sequencing can possibly be eliminated using MALDI DNA sequencing.

At present, automatic gel sequencers can read up to 1000 bp, but most routine sequencing ranges from 300 to 500 bases. Since the size of DNA to be sequenced by mass spectrometry is smaller than in the gel method, MALDI may not be an efficient method for *de novo* DNA sequencing for genome projects. However, with the sequencing of many different genomes completed, mass spectrometry is emerging as a useful tool for DNA resequencing due to its speed and the absence of a requirement for DNA labeling.

2.2.2 **Direct Sequencing**

When using mass spectrometry for sequencing DNA with ladders, the time needed for a TOFMS to separate and detect various ladders is only a few seconds. However, the time needed to prepare DNA ladders is many minutes to hours. Thus, the ideal sequencing method would be to use mass spectrometry to carry out direct DNA sequencing without the need to produce DNA ladders by chemical or enzymatic methods. This also implies that DNA ladders need to be produced during the laser ablation process.

Direct DNA sequencing can be easily understood from the illustrations in Figs. 9 and 10. Assuming the selective cleavage of P–O bond at 3′-linkage for an oligonucleotide of 5′-CTGTGA-3′, the primary fragmentation of such breaking of this bond will result in two series of fragments. These are labeled as 5′- and 3′-termini (Fig. 9). Each series has six members ranging from 1- to 6-mers. The vertical dotted line in Fig. 9 represents the position of the

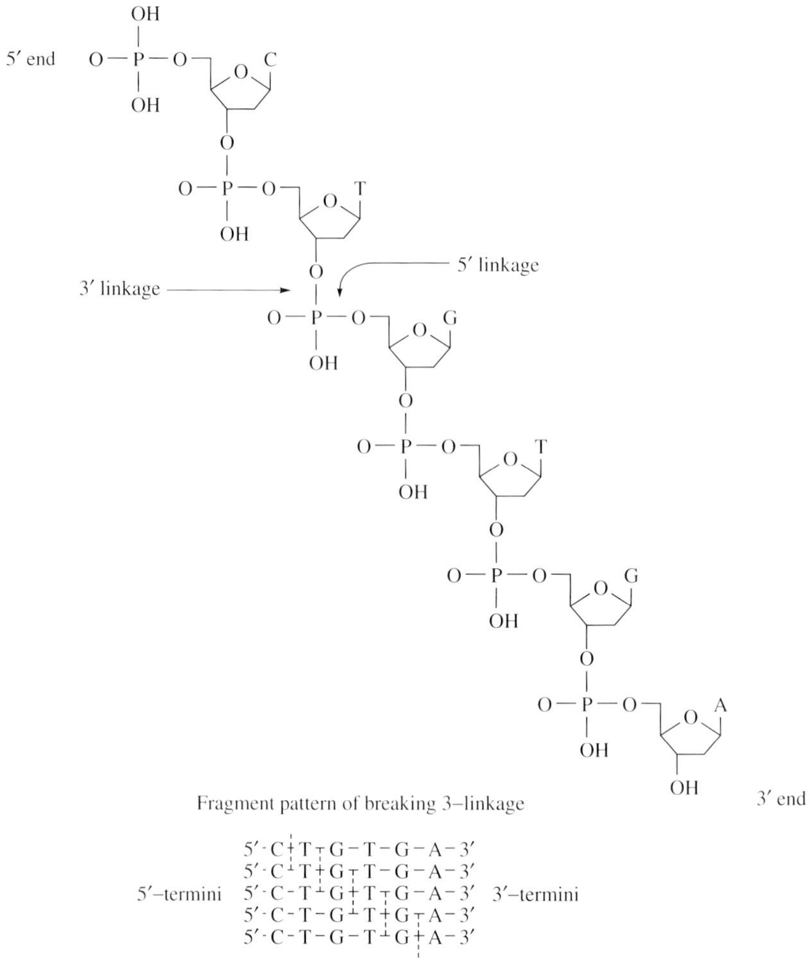

5′ end

3′ linkage

5′ linkage

3′ end

Fragment pattern of breaking 3–linkage

5′-termini
5′-C|T-G-T-G-A-3′
5′-C-T|G-T-G-A-3′
5′-C-T-G|T-G-A-3′
5′-C-T-G-T|G-A-3′
5′-C-T-G-T-G|A-3′
3′-termini

Fig. 9 Illustration of the selective cleavage P–O bond and the fragmentation pattern for a 5′-CTGTGA–3′ oligonucleotide.

cleavage. The fragments in both series can be ionized and resolved in MALDI spectra. Figure 10 shows the simulated mass spectra for these two series. Trace 'a' in Fig. 10 is the spectrum of the series with 3′-termini. The mass difference between two adjacent peaks in the series provides the information for each extra base. For example, the first member of this series is A, and the second member is GA. Owing to the selective cleavage of 3′-linkage, the mass difference between these two peaks will be exactly 329.2 daltons, which represents the mass of dGMP. The sequence information can be obtained by analyzing all values of mass difference in this series together with the total mass of 5′-CTGTGA-3′. The same information can also be obtained from the series with 5′-termini (Fig. 10, trace 'b'). If a complete series (either 5′- or 3′- termini) of such fragmentation for a DNA sample can be

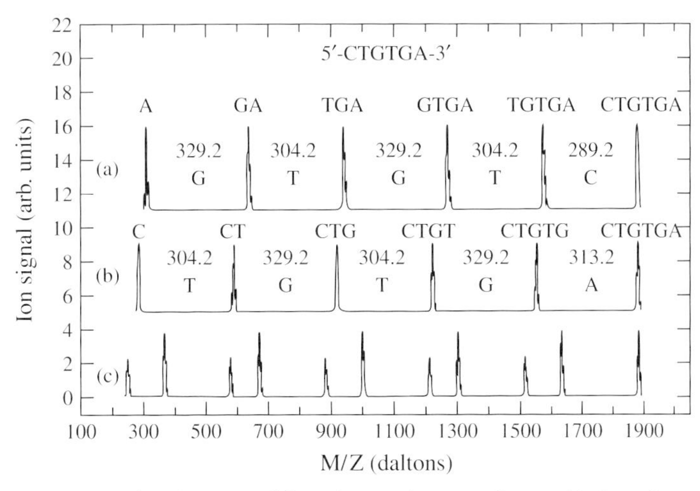

Fig. 10 Simulated spectra of the selective cleavage of P–O 3′-linkage for 5′-CTGTGA-3′ oligonucleotide: (a) simulated spectrum of fragment series with 3′-termini by the 3′-cleavage; (b) series with 5′-termini by 3′ cleavage; and (c) is the combination of (a) and (b).

experimentally resolved, the full sequence information will be obtained. Trace 'c' of Fig. 10 shows a mass spectrum of two series from the cleavage of 5'-linkage P–O bond. This sequencing method is very reliable since the sequence result from one series can be reconfirmed by the results from other series. Recently, UV-MALDI was attempted in order to achieve direct DNA sequencing by carefully adjusting the experimental conditions, which include the use of newly discovered two-component matrices and adequate laser fluence. Figure 11 shows the direct sequencing of DNA probes up to 35 nucleotides. Similar results by IR-MALDI and by electrospray high-resolution Fourier transform mass have also been reported. At present, direct sequencing is still limited to short oligonucleotides, owing to the limited mass resolution, since the mass difference between different bases needs to be identified. If a 60mer oligonucleotide is to be sequenced, the accuracy of the ion peaks needs to be better than 1 in 2000 since the mass difference between A (adenine) and T (thymine) is only 9 daltons. When a DNA segment is longer than 100mer, the effect of the abundance of ^{13}C isotope makes the distinction between A and T extremely difficult.

3 MALDI for Disease Diagnosis

DNA mutations that causes diseases can be simply classified into three major categories. These are (1) deletion or insertion, (2) point mutation, and (3) dynamic mutation. MALDI has been applied to all these different categories.

3.1 Base Deletion

In 1994, mutations in the cystic fibrosis (CF) gene in clinical samples were first demonstrated. Several measurements of CF were also carried out by mass spectrometry. In the North American

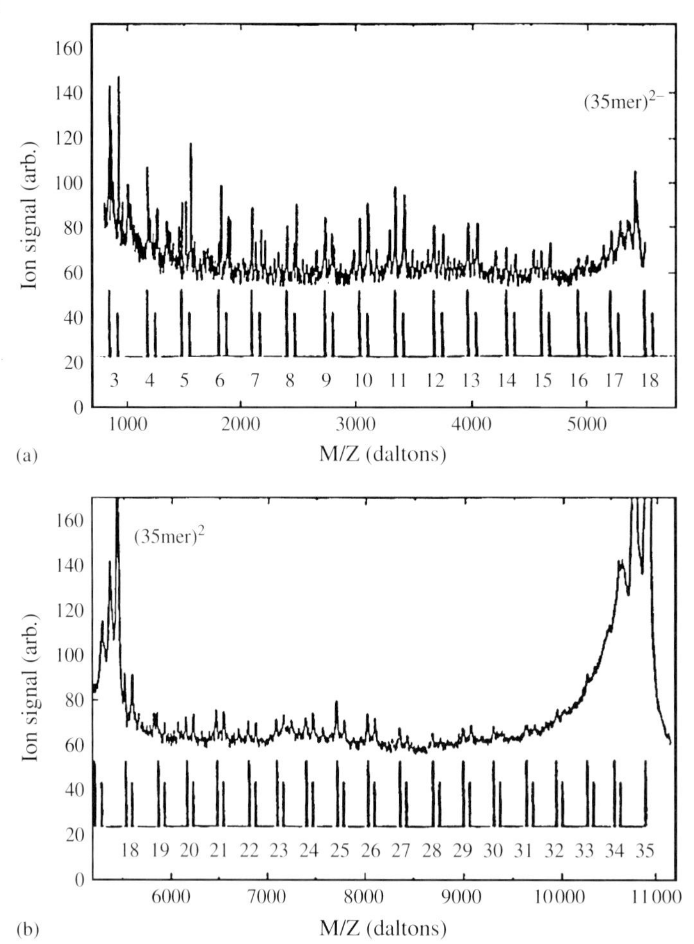

Fig. 11 MALDI mass spectrum with selective fragmentation for direct sequencing of a 35-nt oligonucleotide, 5′-GCGTGATGGAATCGATGACGTGCGATGTCGTGTTT-3′ The higher intensity series in the simulated spectra is 5′ termini and 3′ cleavage series and the lower intensity is 3′ termini and 3′ cleavage series. A mixture of 3-HPA and 2,4,6- trihydroxyacetophenone was used as a matrix. (a) and (b) are for low and high mass regions respectively.

population, about 70% of CF carriers have a 3-bp deletion in exon 10, resulting in the loss of phenylalanine residue at codon 508 (ΔF508). Two oligonucleotide primers, CF1 and CF2, were designed to amplify a DNA segment spanning the deletion, thus generating a 59-bp and 56-bp fragment of the normal CF gene and ΔF508 mutation, respectively (Fig. 12). Analysis of the PCR amplified products by MALDI resolved the 3-bp difference between the normal and the ΔF508 alleles in the spectra shown in Figure 12. To validate the efficiency of MALDI-TOFMS, a total of 30 genomic DNAs from patient samples were used as templates to amplify DNA segments

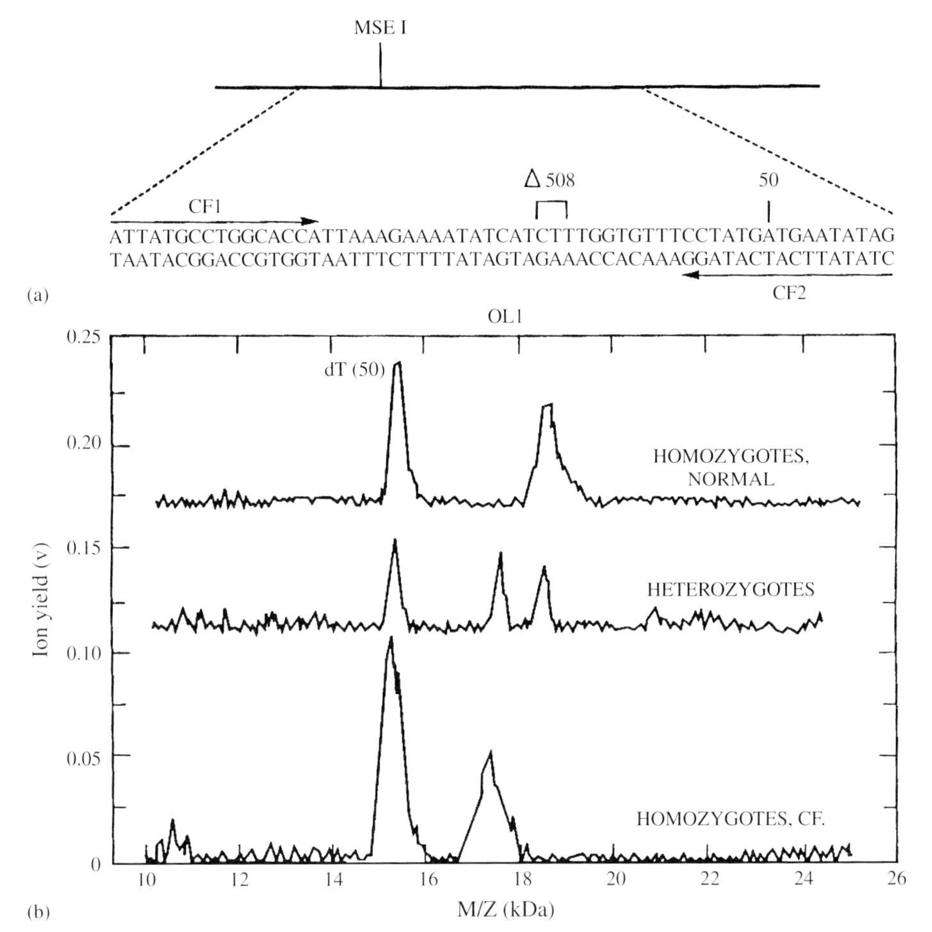

Fig. 12 Differential diagnosis of individuals with the normal and Δ F508 alleles by MALDI-TOF mass spectrometry. (a) Nucleotide sequence of 59 bp in exon 10 of the CFTR gene with the primers CF1 and CF2 indicated by arrows. A deletion of 3 bp in Δ F508 mutation is marked by the bracket. The DNA fragment amplified by PCR with forward and backward primers is schematically represented by the horizontal line on the top of the nucleotide sequence with a unique MseI site indicated. (b) MALDI-TOFMS of DNA amplified from a healthy volunteer, a Δ F508 carrier, and a CF patient. A synthetic dT_{50} was included as an internal standard for mass peak calibration.

from exon 10 or other exon/intron regions (as negative controls) of the CFTR gene. The samples were coded numerically and amplified by PCR. All the samples with ΔF508 mutations were verified earlier by size fractionation in a 10% acrylamide gel or by sequencing analysis. In a double-blind study (both PCR and mass analyses were performed without knowing whether the CFTR was normal or mutated), results from the mass spectrometry and conventional assays were in total agreement. These results indicate that MALDI mass spectrometry has the potential to become a valuable method of clinical CF diagnosis.

3.2 Point Mutation

Most DNA mutations due to contaminants are point mutations. Thus, the capability to detect point mutations by mass spectrometry is essential. MALDI has also been applied for point mutation detection. The approach is to amplify the desired region of a DNA template by PCR using two primers that have their 3'-ends extended to the site of expected mutation. To test this approach, a synthetic oligo with 60 nucleotides, whose sequence is derived from human p53 mRNA from codon 144 to 163, was used as a template with two normal allele-specific primers in a PCR (ASPCR). The lengths of the primers used were 16 and 23 nucleotides. By using a normal template and normal allele-specific primer pair, a normal PCR DNA product is expected to be produced. The size of the expected PCR product should be equal to the sum of both primers minus one. On the other hand, little PCR product will be made if the normal template is replaced with the mutant template, because the 3′-end base of normal primers results in a mismatch with the mutant template. Since Taq DNA polymerase lacks a 3′ exonucleolytic proofreading activity, the mismatch will block the amplification during PCR. The sequence of the template and the primers used were

Primer 1 (23mer)
5'-CAGCTGTGGGTTGATTCCACACC-3'
Template (60 bp)
5'-CAGCTGTGGGTTGATTCCACACCC
CCGCCCGGCACCCGCGTCCGCG
CCATGGCCATCTAC-3'
3'-GGGGCGGCCGTGGGC-5'
Primer 2 (16mer)

As shown in Fig. 13, the PCR products with the expected size of 38 nucleotides could be clearly observed. On the other hand, no PCR products were detected when the normal templates were replaced with mutant templates (Fig. 14). Since mutation of the p53 gene has been

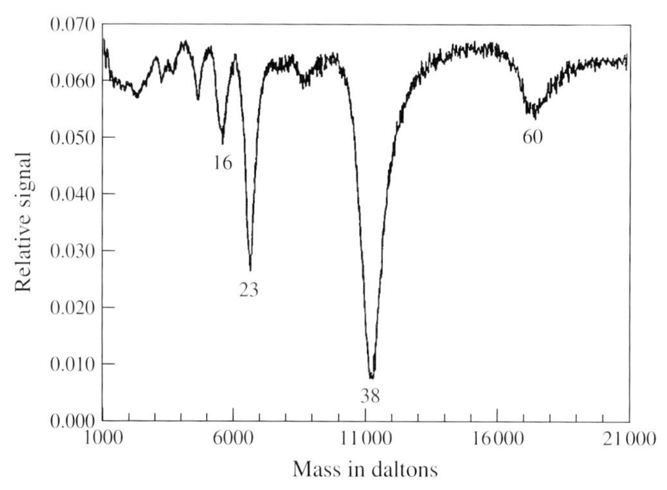

Fig. 13 MALDI mass spectrum of 38-bp PCR product with two primers overlapping at one base. The lengths of primers are 16 and 23 nucleotides.

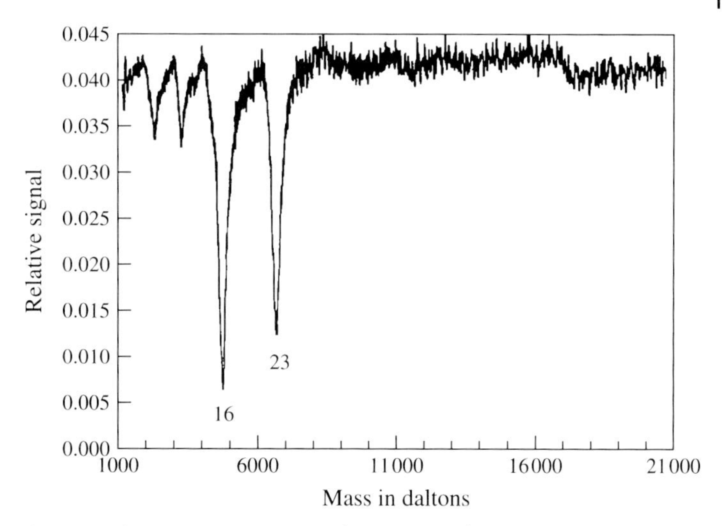

Fig. 14 MALDI mass spectrum with primers at the mutation site not complementary to the template. No PCR products, namely 38 bp were observed.

found to be responsible for more than 60% of cancers, the capability to detect p53 mutation is valuable for giving early warning of possible cancer development.

Measurements of G551D point mutation in cystic fibrosis transmembrane conductance regulator (CFTR) have also been attempted. The G551D mutation in the CFTR gene involves a G to A mutation, which results in a glycine to a spartic amino acid substitution. The approach is similar to the one adopted for point mutation in the synthetic template simulating a portion of the p53 gene. If the two primers based on the normal sequence match the target DNA sequence, a normal PCR product will be produced. However, if the alternate primers that match the mutant hybridize, a PCR product specific for mutant sequencing will be produced. Thus, the mass spectrometer can be used to identify people who are homozygous-normal, heterozygous, or homozygous-abnormal at a mutation site. A typical result is shown in Fig. 15. A heterozygous template was used. The peaks of 37 and 46 bp represent the PCR products from both normal and mutant plates. Four primers with different lengths were used in this work.

3.3 Dynamic Mutation

There are numerous genetic diseases such as Huntington's disease (HD), dentatorubral-pallidoluysian atrophy (DRPLA), Kennedy's disease, and a number of spinal–cerebellar ataxias, which are due to the abnormal trinucleotide expansion. Rapid measurements of the number of these trinucleotide repeats can give accurate identification of normal, heterozygous carrier, and homozygous patients. Examples of MALDI mass spectrometry for rapid diagnosis of Huntington's disease and DRPLA are given in Figs. 16 and 17.

From the above, it is clear that MALDI mass spectrometry can be used in the diagnosis of various diseases.

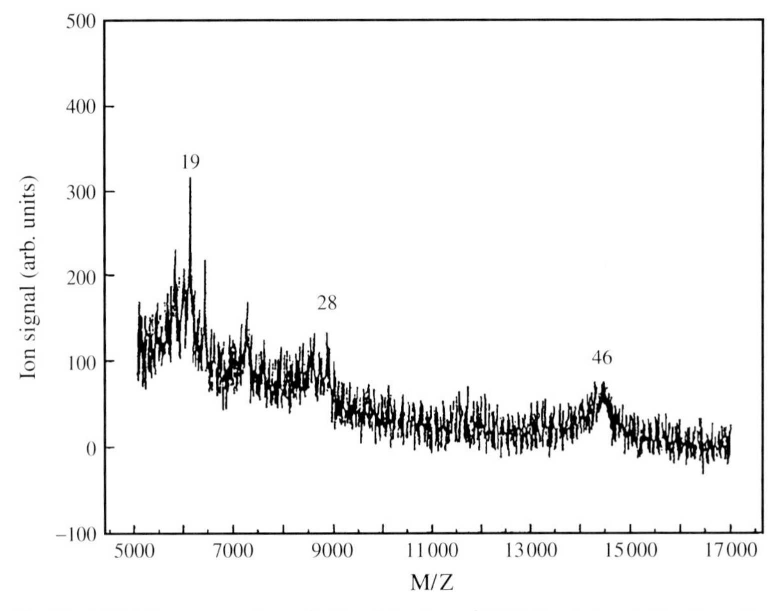

Fig. 15 MALDI mass spectrum for the detection of G551D point mutation in cystic fibrosis transmembrance conductance (CFTR). The G551D mutation in the CFTR gene involves a G to A mutation that results in a glycine to aspartic amino acid substitution. Mass spectrum of PCR products in 46 bp indicates the sample has a heterozygous template.

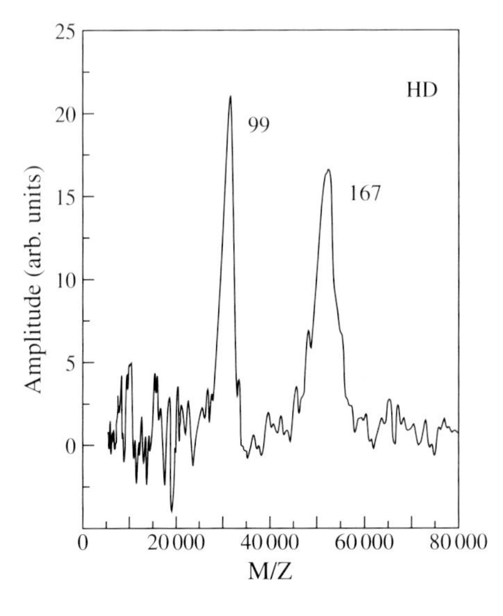

Fig. 16 Mass spectrum of a Huntington disease (HD) patient sample. Experimental results show the sample is heterozygous with CAG repeats of 17 and 40.

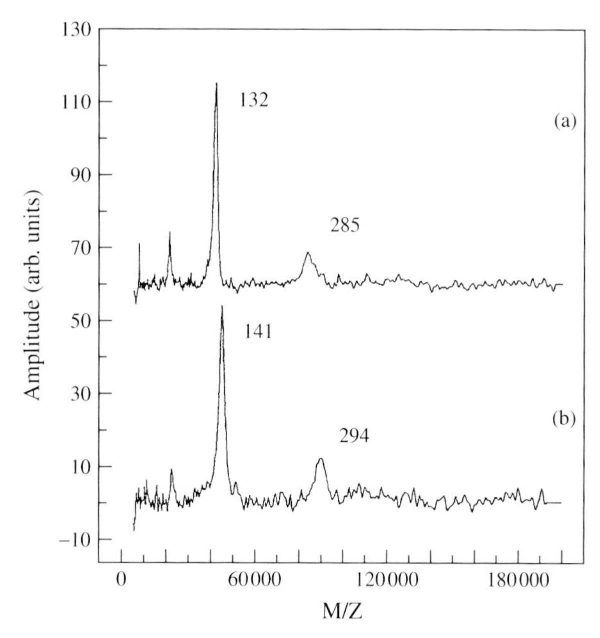

Fig. 17 Mass spectra for dentatorubral–pallidoluysian atrophy (DRPLA) patient samples. Both samples in (a) and (b) are heterozygous with the CAG repeats as 12/63 and 15/66 respectively.

4 DNA Typing for Forensic Applications

DNA fingerprinting is a technology that identifies individuals based on patterns of DNA markers detected in the genomic DNA. It can be used in forensic analysis to identify suspects or victims. It is particularly valuable at scenes of violent crimes where a body may not be available, or in instances where decomposition or dismemberment excludes the use of standard forensic techniques. There are several different approaches available to obtain DNA typing for forensic applications. These include restriction fragment length polymorphism (RFLP), short tandem repeats (STR), and single nucleotide polymorphism (SNP).

The foundation of DNA typing for person identification was established by the observation of restriction fragment length polymorphism (RFLP) in which polymorphic DNA loci are determined from restriction enzyme digestion. In 1985, Jefferys and his collaborators discovered that 33- and 16-bp repeated structures that vary in number between different individuals, referred to as *minisatellites*, could be used for fingerprinting. In 1987, the isolation and characterization of variable numbers of tandem repeats (VNTR) for single-locus DNA profiling was reported. More recently, the emphasis has shifted to the typing of STRs because of the high reliability arising from the large number of STRs in the human genome. During the past few years, it was discovered that SNPs) occur once every 300 to 1000 bp in DNA fragments. Thus, SNP typing is expected to become a very reliable tool for person identification in the future.

4.1 Forensic DNA Sample Preparation

In the process of using DNA typing for forensic applications, it is critically important to have very reliable and stringent protocols established. The major steps for DNA typing often include (1) Sample collection: blood, buccal cells, and hair follicles are the general sources for fresh DNA. For dried stains, samples should be collected and maintained in a dry, cool state. For autopsy tissue, it can be more reliable to have samples from more than one organ. All samples must be labeled as to source, tissue type, and the time and date of sample collection. (2) Storage and Transport: DNA samples can be stored and shipped as nonextracted tissue, as a lysate with tissue in an appropriate EDTA (ethylene diamine tetraacetic acid) solution, or as fully extracted pure DNA. For nonextracted tissues, sample refrigeration during shipping is often recommended. (3) DNA extraction: standard phenol–chloroform extraction and alcohol precipitation processes are often used to extract DNA from biological samples. An automated DNA extractor can achieve high quality extraction. It is desirable to complete DNA extraction as soon as possible. (4) DNA amplification: DNA amplification is often necessary to increase the quantity of sample available for analysis. PCR is the standard technique for the amplification of a selected DNA segment. Since an amplification of 10^6 or more can be achieved by PCR, any contamination by other DNAs can lead to a false conclusion. Thus, great care must be taken to prevent cross-contamination of sample material at the collection site and at any time during the analysis. (5) DNA size or sequence determination. This is the ultimate step leading to the conclusion of person identification. Until now, gel electrophoresis has been used for both DNA size determination and sequencing. However, analysis by mass spectrometry can be faster and more reliable.

The primary goal of DNA profiling for forensic application is to get the assurance of a match (inclusion) or mismatch (exclusion) of the evidence DNA sample with the suspect or victim's sample. In general, DNA profiling can be used to easily *rule out* the suspect when DNA typing between suspect and evidence shows clear differences. Thus, DNA typing is a very powerful tool for proving that an individual *does not* match the DNA collected at a crime scene. On the other hand, if the DNA profile from the evidence sample matches with the suspect, there is still a possibility that the evidence DNA could have come from another person. It might be a coincidental match between two persons having similar DNA profiles. Thus, population genetics plays an important role in suspect identification. The simplest approach is to assume that mating is random. The random mating of persons produces the same genotypes as random combinations of sperm and egg. Although the mating in local populations is not usually strictly random, the random match model provides a good approximation when no special genetic linkage is involved. For subpopulation groups such as people in an isolated village, the random mating approach is not generally a reliable assumption. However, it can probably be used as a guide for the minimum number of loci to be analyzed for confirming the matching of the suspect with the evidence.

Since the confirmation of matching the suspect with the evidence DNA often requires the analysis of several loci, the time needed for DNA profiling can be long and the cost, high. At present, all forensic DNA samples for court evidence are analyzed by gel electrophoresis. Visualization of DNA in this procedure requires the use of either radioactive material or dye tagging. VNTR electrophoretic analysis, especially with radioactive probes, can take several weeks to complete the analysis of different loci for forensic confirmation. The human factor introduced by lengthy handling increases the risk of error or contamination and reduces reliability.

4.2 Mass Spectrometry DNA Detection for Forensic Applications

Each DNA typing involves (1) measurements of DNA fragments associated with genetic markers; (2) match determination for various genetic markers for different samples; and (3) statistical analysis for the type match to determine the possibility of an accidental match. All sequencing or size determination of DNA fragments for forensic applications to date have been carried out using gel electrophoresis. The various approaches among the forensic community include the analysis of VNTR, STR, SNP, and DNA sequencing. Laboratory protocol and standard procedures for data analysis were set up for forensic sample analysis. With the recent progress in development of mass spectrometry for DNA analysis, there is a good potential for the use of mass spectrometry for all the above applications. Since the use of mass spectrometry for DNA analysis is still in the early stages of development, standard protocols for its use in forensic analysis have not yet been established. The potential applications of using mass spectrometry in DNA fingerprinting for forensic applications is presented and discussed below.

4.2.1 RFLP and VNTR Analysis

DNA typing relies on the use of DNA markers. Among the most widely used DNA markers are restriction fragment length polymorphisms. RFLP analysis is based on differences in the positions of restriction endonuclease recognition sites. Restriction endonucleases are enzymes, which recognize and cut double-stranded DNA at sequence-specific recognition sites. Homologous DNA fragments from different individuals will contain different cut sites based on random mutations within the stretch of nucleotides. Consequently, fragments differing in length and number will be generated upon restriction digestion. These constitute RFLPs, and because they differ in size, the RFLP fragments can be readily separated by routine agarose electrophoresis.

When an individual's entire genomic DNA is digested by a restriction endonuclease, hundreds to thousands of restriction fragments are produced. When digested genomic DNA is subjected to gel electrophoresis, individual bands are not discernible. In such a case, analysis is generally facilitated by the Southern blotting technique. In Southern blotting, digested DNA is separated by electrophoresis. The digested fragments are transferred to a nitrocelluose or nylon membrane through capillary action. The DNA binds to the membrane by electrostatic attraction. Labeled probes specific for the RLFP of interest are then hybridized to the DNA-containing membrane. A banding pattern can be visualized using autoradiography or chemical detection techniques. Each individual will exhibit a characteristic pattern. Several RFLPs must be used to statistically ensure that an individual's pattern is unique. The fact that RFLPs can occur anywhere in the genome and that there are potentially millions of RFLPs detectable with different probes makes RFLP analysis a tremendously powerful tool for forensic use.

A special class of RFLPs is based on DNA sequences that occur in tandem repeats. Tandem repeats arise when slippage mutation occurs during DNA replication. It has been observed that this type of mutation occurs frequently enough to generate significant variations over many generations. Polymorphism for these markers is so high in humans that only identical twins will have the same patterns. This factor makes the use of tandem repeats analysis extremely valuable to forensic scientists.

Variable number tandem repeat (VNTR) analysis has been used extensively in forensic applications. VNTRs consist of approximately 10 to 100 bp repeated several times in tandem. For example, a tandem repeat of the trinucleotide CTG would appear in the genome as CTGCTGCTGCTGCTGCTGCTG. In this particular example, this sequence has seven repeats. VNTRs are generally bordered by unique sequence DNA, a characteristic that can be exploited by current technology. In VNTR analysis, genomic DNA is digested with a restriction endonuclease that does not contain a cut site within the tandem repeat sequence. Since the restriction enzyme cuts outside of the tandem repeat region, the tandem repeat is left intact. Tandem repeat fragments are separated by gel electrophoresis and analyzed with Southern blotting. The length of the tandem repeat will vary between individuals. Individuals can be homozygous or heterozygous for a particular VNTR. The large number of VNTR-containing loci creates millions of possible pattern combinations. If there are n loci, there should be n homozygotes and $n(n-1)/2$ heterozygotes. If m loci are analyzed, the number of possible genotypes is $[n(n+1)/2]^m$. If $n = 20$ and $m = 4$, there are more than 1 billion genotypes. As a result, VNTR provides the forensic analyst a powerful tool for inclusion and/or exclusion.

The need for Southern blotting can be eliminated if the VNTR region is selectively amplified with the polymerase chain reaction (PCR). PCR primers complementary to sequences flanking the tandem repeat region can be designed and used to amplify the tandem repeat region. The length of an individual's tandem repeat can then be obtained by gel electrophoresis. It is PCR amplification of VNTRs that shows a lot of promise for MALDI detection. By using MALDI for RFLP detection, gel electrophoresis, Southern blotting and radioactive material or dye tagging can be eliminated. Figure 18 shows the mass spectra for restrictionenzyme–digested DNA samples. A difference of 12 bases with three 4-base tandem repeats can clearly be observed. Both primary and doubly charged ion peaks are observed. The spectrum indicates that one chromosome in this region is 12 bp longer than the same region in the other chromosomes. Thus, the result indicates the sample is from a person with heterozygous vWFII.

4.2.2 **Short Tandem Repeats (STR)**

For VNTR analysis by measuring RFLP without the use of PCR, one major disadvantage is that the quantity of the DNA sample is so small that a long analysis time is required. One promising alternative is to amplify and measure loci containing short tandem repeats (STR).

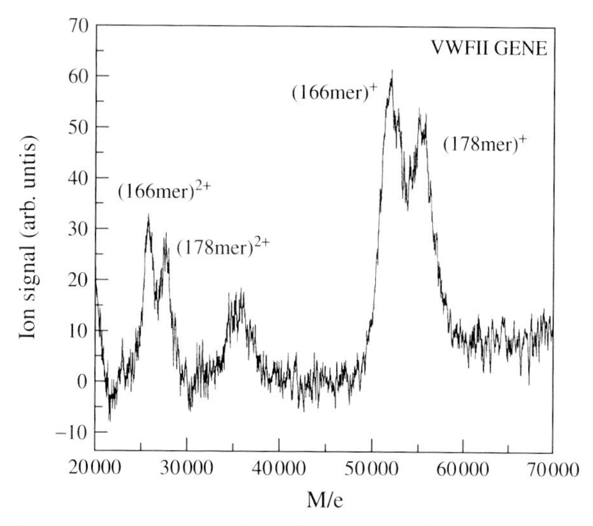

Fig. 18 Mass spectrum from the digestion of a part of *vWFII* gene. Peaks are observable for a difference of 12 bases. Both primary and doubly charged ion peaks are observed. The spectrum indicates that one chromosome in this region is 12 bp longer than the same region in the other chromosome.

There is a large number of STR distributed throughout the human genome. STR analysis can be used for as many loci as are needed for reliable identification. PCR is nearly always used in STR analysis. Since PCR can amplify the selected DNA segments by more than six orders of magnitude, only a very small quantity of DNA sample is required. However, the disadvantage is that any procedure using PCR amplification is highly susceptible to contamination. Even a trace of foreign DNA contamination can be amplified and detected to potentially lead to a false conclusion. However, careful handling of DNA samples or applying quantitative PCR techniques can more or less resolve the concern regarding DNA contamination. Since the analysis of STR is by measurement of the sizes of DNA products from the PCR process, the sizes of the DNA fragments are often less than 300 bp and the quantity is often in the range of a few picomoles, which can be readily detected by MALDI.

Typical mass spectra of a DNA sample amplified by PCR from parts of different genes are shown in Fig. 19. Since different people have different numbers of repeats in each of these genes, measurements of the number of repeats for several loci can be used for person identification. There are many tetranucleotide repeats, which can be used for forensic identification. HUMTH01 is one that is often used in forensic applications.

4.2.3 Single Nucleotide Polymorphism (SNP)

There are several types of DNA sequence variation, including deletion, insertion, difference in number of repeats such as STR and single base-pair differences. The single base difference is the most common type. This single base pair difference is often referred to as *single nucleotide*

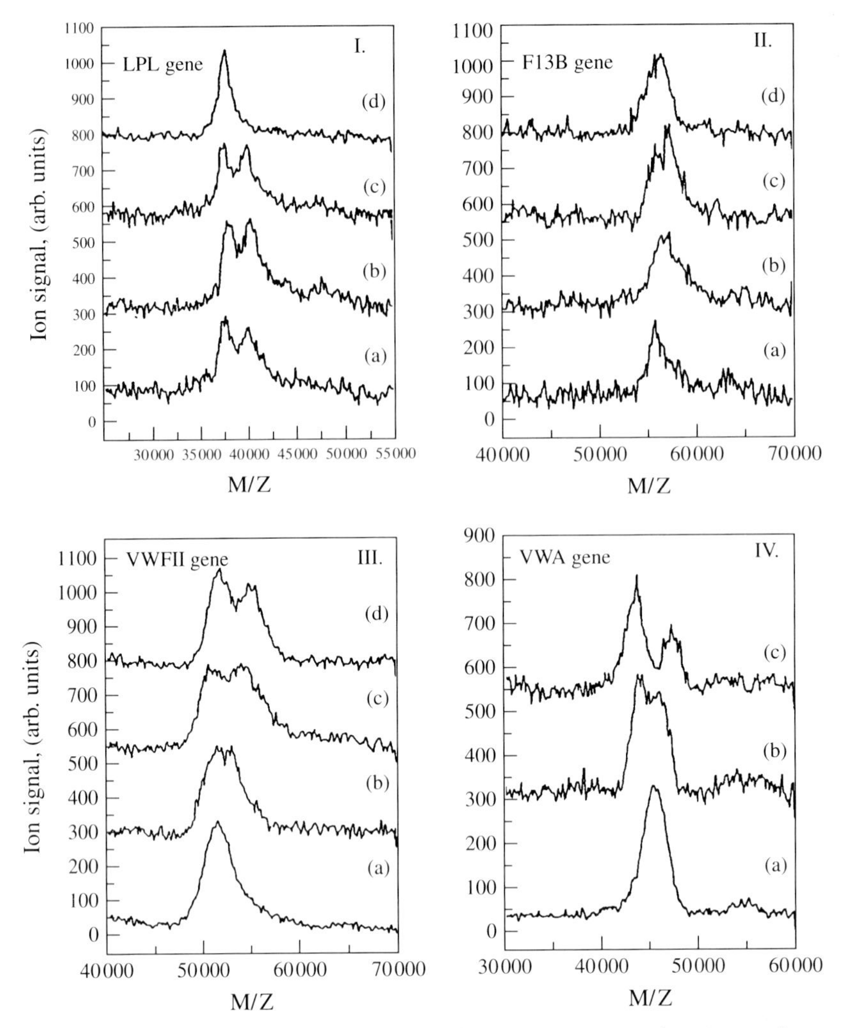

Fig. 19 MALDI mass spectra of PCR products of *LPL*, *F13B*, *vWFII*, and *vWA* genes from different persons. Homozygous and heterozygous as well as the number of the short tandem repeats can be determined. This indicates MALDI can be used for short tandem repeat determination for forensic applications.

polymorphism (SNP). There are several million SNPs in the human genome. The predominance of SNPs and their much lower mutation rate make SNP analysis extremely valuable for forensic applications. ASPCR, described above for point mutation detection, can also be applied to SNP detection.

Another approach to using MALDI for SNP detection is to hybridize a primer adjacent to the polymorphic locus and extend it by a single base by dideoxynucleotide addition. MS can be used to accurately measure the mass of the extension product to identify the base. For homozygotes, one peak will be observed.

However, two peaks should be observed for a heterozygote.

4.2.4 Gender Determination

Sex determination is often important for identifying suspects and victims. A reliable gender test is PCR amplification of a segment of the X–Y homologous gene amelogenin. A single pair of primers spanning part of the first intron generates 106-bp and 112-bp PCR products from the X and Y chromosomes respectively. MALDI can be subsequently used for the measurements of PCR products. Recently, it was found that the deletion of the Y-encoded gene occurs in a small percentage of Y chromosomes. Thus, it is more reliable to test both amelogenin and the male sex-determining gene SRY for gender determination. Male DNA generates three products of lengths 93 bp (SRY), 106 bp (Amelogenin in X chromosome) and 112 bp (amelogenin in Y chromosome). Coamplification of female DNA generates only the 106-bp product. As shown in Fig. 20, the distinction of genotyping between male and female is quite clear. Figure 20 shows examples of mass spectra of PCR products for (a) female gender, (b) male gender, and (c) mixture of male and female gender. Of special interest in forensic analysis, especially for sexual assault, is the detection of a trace amount of male-specific DNA in a female sample. Such detection can be performed by demonstrating the presence of male-derived products of 93 bp and 112 bp. When MALDI spectra of a composite sample containing mostly female species (10 ng) and only a trace amount of male sample (0.01 ng) was checked, the trace of the male sample could be detected. It

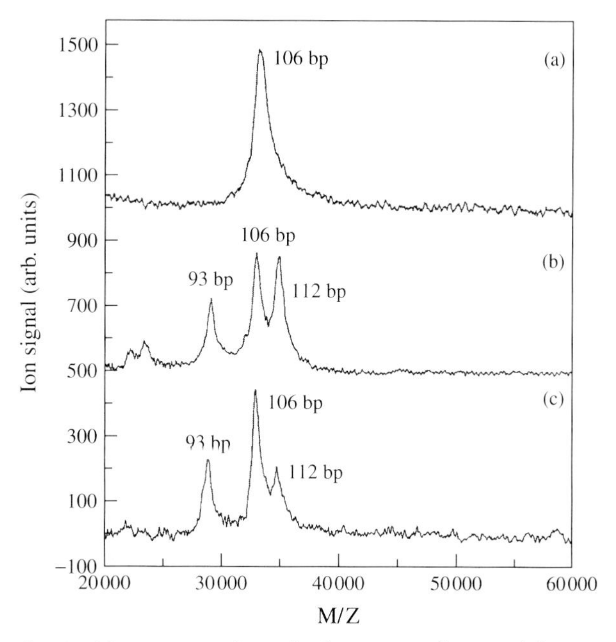

Fig. 20 Mass spectra of samples for sex-specific coamplification (Amel X/Y plus SRY) of loci for (a) female (b) male, and (c) female plus male (with sample ratio of 1 : 1) respectively.

indicates that as little as 0.1 percent of male DNA in a female DNA sample can be detected. It also indicates that the detection sensitivity can reach 0.01 ng or less which is about one order of magnitude better than by gel electrophoresis.

5 Conclusion

Mass spectrometry provides a unique method of nucleic acid detection and sequencing. Applications relating to disease diagnosis, and forensic applications in person identification have been clearly demonstrated. However, improvement in mass resolution and detection sensitivity of large DNA fragments is still required in order to enable MS to be routinely used for broad applications. A better understanding of the mechanisms of desorption and ionization will provide the basis for further improvement of MALDI.

Acknowledgment

This work is primarily supported by the National Institute of Justice, grant number 97-LB-VX-A047, and supported in part by the Office of Biological and Environmental Research, U.S. Department of Energy under contract DE-AC05-96OR22464 with UT Bartell LLC.

See also Genomic Sequencing (Core Article); Human Genetic Variation and Disease; Mass Spectrometry of Proteins (Proteomics); Nucleic Acid and Protein Sequence Analysis and Bioinformatics.

Bibliography

Books and Reviews

Chen, C.H. (1996) in: Meyers, R.A. (Ed.) *Mass Spectrometry High Speed DNA Fragment Sizing Encyclopedia of Molecular Biology and Molecular Medicine*, VCH Publication, VCH, Weinheim pp. 1–11.

Lay, J.O. Jr. (2001) MALDI-TOF mass spectrometry of bacteria, *Mass Spectrom. Rev.* **20**, 172–194.

Lechner, D., Lathrop, G.M., Gut, I.G. (2001) Large-scale genotyping by mass spectrometry: experience, advances and obstacles, *Proteomic Genomic* **6**, 31–38.

Limbach, P.A. (1996) Indirect mass spectrometric methods for characterizing and sequencing oligonucleotides, *Mass Spectrom. Rev.* **15**, 297–336.

Murray, K.K. (1996) DNA sequencing by mass spectrometry, *J. Mass Spectrom.* **31**, 1203–1215

Primary Literature

Amexis, G., Oeth, P., Abel, K., Ivshina, A., Pelloquin, F., Cantor, C.R., Braun, A., Chumakov, K. (2001) Quantitative mutant analysis of viral quasispecies by chip-based matrix-assisted laser desorption/ionization time-of-flight mass spectrometry, *Proc. Natl. Acad. Sci. U.S.A.* **98**, 12097–12102.

Berkenkamp, S., Kirpeka, F., Hillemkamp, F. (1998) Infrared MALDI mass spectrometry of large nucleic acid, *Science* **281**, 266–268.

Buetow, K.H., Edmonson, M., Macdonald, R., Clifford, R., Yip, P., Kelley, J., Little, D.P., Strausberg, R., Koster, H., Cantor, C.R., Braun, A. (2001) High throughput development and characterization of a genomewide collection of gene-based single nucleotide polymorphism markers by chip-based matrix-assisted laser desorption/ionization time-of-flight mass spectrometry based strategies and studies among those approaches, *Proc. Natl. Acad. Sci. U.S.A.* **98**, 581–584.

Chang, L.Y., Tang, K., Schell, M., Ringelberg, C., Matteson, K.J., Allman, S.L., Chen, C.H. (1995) Detection of ΔF508 mutation of the cystic fibrosis gene by laser mass spectrometry, *Rapid Commun. Mass Spectrom.* **9**, 772–774.

Chen, C.H., Chang, L.Y., Tang, K. (1994) Laser mass spectrometry for molecular medicine applications, *Opto News Lett.* **47**, 8–11.

Chen, C.H., Golovlev, V.V., Taranenko, N.I., Allman, S.L., Isola, N.R., Potter, N.T., Matteson, K.J., Chang, L.Y. (1999) MALDI for DNA sequencing, disease diagnosis, and forensic applications, *Am. Biotech. Lab.* **17**, 14–18.

Chen, C.H., Taranenko, N.I., Zhu, Y.F., Allman, S.L. (1995) MALDI for fast DNA analysis and sequencing, *Lab. Robotics Automat.* **8**, 87–99. Chiu, N.H.L., Tang, K., Yip, P., Braun, A., Koster, H., Cantor, C.R. (2000) Mass spectrometry of single-stranded restriction fragments captured by an undigested complementary sequence, *Nucleic Acids Res.* **28**, e31.

Ding, C., Cantor, C.R. (2003) A high throughput gene expression analysis technique using competitive PCR and matrix-assisted laser desorption ionization time-of-flight MS, *Proc. Natl. Acad. Sci. U.S.A.* **100**, 3059–3064.

Doktycz, M.J., Hurst, G.B., Habibi-Goudarzi, S., McLuckey, S.A., Tang, K., Chen, C.H., Uziel, M., Jacobson, K.B., Woychik, R.P., Buchanan, M.V. (1995) Analysis of PCR-amplified DNA products by mass spectrometry using matrix-assisted laser desorption and electrospray: current status, *J. Anal. Biochem.* **230**, 205–214.

Elso, C., Toohey, B., Reid, G.E., Potter, K., Simpson, R.J., Foote, S.J. (2002) Mutation detection using mass spectrometric separation of tiny oligonucleotide fragments, *Genomic Res.* **12**, 1428–1433.

Faulstich, K., Woemer, K., Brill, H., Engels, J.W. (1997) A sequencing method for RNA oligonucleotides based on mass spectrometry, *Anal. Chem.* **69**, 4349–4353.

Golovlev, V.V., Allman, S.L., Garrett, W.R., Taranenko, N.I., Chen, C.H. (1997) Laser induced acoustic desorption, *Int. J. Mass Spectrom. Ion Process.* **169/170**, 69–78.

Golovlev, V.V., Lee, S.-S.H., Allman, S.L., Taranenko, N.I., Chen, C.H. (2001) Non-resonance MALDI of oligonucleotides: mechanism of ion desorption, *Anal. Chem.* **73**, 809–812.

Haff, L.A., Smirnov, I.P. (1989) Single-nuvcleotide polymorphism identification assays using a thermostable DNA polymerase and delayed extraction MALDI/TOF mass spectrometry, *Genome Res.* **7**, 378–388.

Hartmer, R., Storm, N., Boecker, S., Rodi, C.P., Hillenkamp, F., Jurinke, C., Boom, D.V. (2003) Rnase T1 mediated base-specific cleavage and MALDI-TOF MS for high-throughput comparative sequence analysis, *Nucleic Acids Res.* **31**, e47.

Isbell, D.T., Gusev, A., Taranenko, N.I., Chen, C.H., Hercules, D.M. (1999) Analysis of nucleotides directly from TLC plates using MALDI-MS detection, *Fresenius' J. Anal. Chem.* **365**, 625–630

Isbell, D.T., Gusev, A., Taranenko, N.I., Chen, C.H., Hercules, D.M. (1999) Separation and detection of carcinogen-adducted oligonucleotides *J. Mass Spectrom.* **34**, 774–776.

Isola, N.R., Allman, S.L., Golovlev, V.V., Chen, C.H. (1999) Chemical cleavage sequencing of DNA using matrix-assisted laser desorption/ionization time-of-flight mass spectrometry, *Anal. Chem.* **71**, 2266–2269.

Isola, N.R., Taranenko, N.I., Allman, S.L., Golovlev, V.V., Chen, C.H. (2001) MALDI-TOF mass spectrometric method for detection of hybridized DNA oligomer, *Anal. Chem.* **73**, 2126–2131.

Jones, R.B., Allman, S.L., Tang, K., Garrett, W.R., Chen, C.H. (1993) Neutralization of negatively charged oligonucleotides, *Chem. Phys. Lett.* **212**, 451–456.

Jurinke, C., van den Boom, D., Cantor, C.R., Koster, H. (2001) The use of Mass ARRAY technology for high throughput genotyping, *Adv. Biochem. Eng. Biotechnol.* **77**, 57–74.

Kirpekar, F., Nordhoff, E., Kristiansen, K., Roepstorff, P., Lezius, A., Hahner, S., Karas, M., Hillenkamp, F. (1993) Matrix-assisted laser desorption/ionization mass spectrometry of enzymatically synthesized RNA up to 150 kDa, *Nucleic Acids Res.* **22**, 3866–3870.

Koster, H., Tang, K., Fu, D., Braun, A., van den Boom, D., Smith, C.L., Cotter, R.J., Cantor, C.R. (1996) A strategy for rapid and efficient DNA sequencing by mass spectrometry, *Nature/Biotechnology* **14**, 1123–1128.

Kwon, Y., Tang, K., Cantor, C.R., Koster, H., Kang, C. (2001) DNA sequencing and genotyping by transcriptional synthesis of chain-terminated RNA, *Nucleic Acids Res.* **29**, e11.

Lefmann, M., Honisch, C., Bocker, S., Storm, N., Wintzingerode, F., Schloteburg, C., Moter, A., Boom, D.V., Gobel, U.B. (2004) Novel mass spectrometry-based tool for genotypic identification of mycobacteria, *J. Clin. Microbiol.* **42**, 339–346.

Little, D.P., Braun, A., O'Donnell, M.J., Koster, H. (1997) Mass spectrometry from miniaturized arrays for full comparative DNA analysis, *Nat. Med.* **12**, 1413–1416.

Little, D.P., Chorush, R.A., Spier, J.P., Senko, M.W., Kelleher, N.L., McLafferty, F.W. (1994) Rapid sequencing of oligonucleotides by high-resolution mass spectrometry, *J. Am. Chem. Soc.* **116**, 4893–4897.

Maya, L., Chen, C.H., Stevenson, K.A., Kenik, E.A., Allman, S.L., Thundat, T.G. (2002) Mass spectrometric analysis of water-soluble gold nanoclusters, *J. Nanoparticle Res.* **4**, 417–422.

Muddiman, D.C., Anderson, G.A., Hofstadler, S.A., Smith, R.D. (1997) Length and base composition of PCR-amplified nucleic acids using mass measurements from electronspray ionization mass spectrometry, *Anal. Chem.* **69**, 1543–1549.

Nordhoff, E., Karas, M., Cramer, R., Hahner, S., Hillenkamp, F., Kirpekar, F., Lezins, A., Muh, J., Meier, C., Engela, J.W. (1995) Direct mass spectrometric sequencing of low picomole amount of oligonucleotides up to 21 bases by matrix-assisted laser desorption/ionoization mass spectrometry, *J. Mass Spectrom.* **30**, 99–112.

Null, A.P., Hannis, J.C., Muddiman, D.C. (2001) Genotyping of simple and compound short tandem repeat loci using electrospray ionization Fourier transform ion cyclotron resonance mass spectrometry, *Anal. Chem.* **73**, 4514–4521.

Tang, K., Allman, S.L., Chen, C.H. (1993) Matrix-assisted laser desorption ionization (maldi) of oligonucleotides with various matrices, *Rapid Commun. Mass Spectrom.* **7**, 943–948.

Tang, K., Allman, S.L., Chen, C.H., Chang, L.Y., Schell, M. (1994) Matrix-assisted laser desorption/ionization of restriction enzyme-digested DNA, *Rapid Commun. Mass Spectrom.* **8**, 183–186.

Tang, K., Allman, S.L., Jones, R.B., Chen, C.H. (1992) Comparison of rhodamine dyes as matrices for matrix-assisted laser desorption/ionization mass spectrometry, *Org. Mass Spectrom. Lett.* **27**, 1389–1392.

Tang, K., Allman, S.L., Jones, R.B., Chen, C.H. (1993) Laser mass spectrometry of polydeoxyribothymidylic acid mixtures, *Rapid Commun. Mass Spectrom.* **7**, 63–66.

Tang, K., Allman, S.L., Jones, R.B., Chen, C.H. (1993) Quantitative analysis of biopolymers by matrix-assisted laser desorption, *Anal. Chem.* **64**, 2164–2166.

Tang, K., Allman, S.L., Jones, R.B., Chen, C.H., Araghi, S. (1993) Laser mass spectrometry of oligonucleotides with isomer matrices, *Rapid Commun. Mass Spectrom.* **7**, 435–439.

Tang, K., Oeth, P., Kammerer, S., Denissenko, M.F., Ekblom, J., Jurinke, C., Boom, D.V., Cantor, C.R. Mining disease susceptibility genes through SNP analyses and expression profiling using MALDI-TOF mass spectrometry, *J. Proteomic Res.* (in Press).

Tang, K., Opalsky, D., Abel, K., van den Boom, D., Yip, P., DelMistro, G., Braun, A., Cantor, C.R. (2003) Single neucleotide polymorphism analyses by MALDI-TOF MS, *Int. J. Mass Spectrom.* **226**, 37–54.

Tang, K., Taranenko, N.I., Allman, S.L., Chang, L.Y., Chen, C.H. (1994) Detection of 500 – nucleotide DNA by laser desorption mass spectrometry, *Rapid Commun. Mass Spectrom.* **8**, 727–730.

Tang, K., Taranenko, N.I., Allman, S.L., Chen, C.H., Chang, L.Y., Jacobson, K.B. (1994) Picolinic acid as a matrix for laser mass spectrometry of nucleic acid and proteins, *Rapid Commun. Mass Spectrom.* **8**, 673–677.

Tang, W., Zhu, L., Smith, L.M. (1997) Controlling DNA Fragmentation in MALDI-MS by chemical modification, *Anal. Chem.* **69**, 302–312.

Taranenko, N.I., Allman, S.L., Golovlev, V.V., Taranenko, N.V., Isola, N.R., Chen, C.H. (1998) Sequencing DNA using mass spectrometry for ladder detection, *Nucleic Acid Res.* **26**, 2488–2490.

Taranenko, N.I., Chung, C.N., Zhu, Y.F., Allman, S.L., Golovlev, V.V., Isola, N.R., Martin, S.A., Haff, L.A., Chen, C.H. (1997) Matrix-assisted laser desorption/ionization for sequencing single-stranded and double-stranded DNA, *Rapid Commun. Mass Spectrom.* **11**, 386–392.

Taranenko, N.I., Golovlev, V.V., Allman, S.L., Taranenko, N.V., Chen, C.H., Hong, J., Chang, L.Y. (1998) Matrix-assisted laser desorption/ionization for short tandem repeat loci, *Rapid Commun. Mass Spectrom.* **12**, 413–418.

Taranenko, N.I., Hurt, R., Zhou, J.Z., Isola, N.R., Huang, H., Lee, S.H., Chen, C.H. (2002) Laser desorption mass spectrometry for microbial DNA analysis, *J. Microbiol. Methods* **48**, 101–106.

Taranenko, N.I., Matteson, K.J., Chung, C.N., Zhu, Y.F., Chang, L.Y., Allman, S.L., Haff, L.,

Martin, S.A., Chen, C.H. (1996) Laser desorption mass spectrometry for point mutation detection, *Genet. Anal.: Biomol. Eng.* **13**, 87–94.

Taranenko, N.I., Potter, N.T., Allman, S.L., Chen, C.H. (1999) DNA Analysis for Gender Identification by MALDI, *Anal. Chem.* **71**, 3974–3976.

Taranenko, N.I., Potter, N.T., Allman, S.L., Golovlev, V.V., Chen, C.H. (1999) Detection of trinucleotide expansion in neurodegenerative disease by matrix-assisted laser desorption/ionization time-of-flight mass spectrometry, *Genet. Anal.: Biomol. Eng.* **15**, 25–31.

Taranenko, N.I., Tang, K., Allman, S.L., Chang, L.Y., Chen, C.H. (1994) 3-Aminopicolinic acid as a matrix for MALDI of DNA, *Rapid Commun. Mass Spectrom.* **8**, 727–730.

Vallone, P.M., Devaney, J.M., Marino, M.A., Butler, J.M. (2002) A strategy for examining complex mixture of deoxyoligonucleotide using IP-RP HPLC, MALDI-TPF MS and Iinformatics, *Anal. Biochem.* **304**, 257–265.

Wada, Y., Yamamoto, M. (1997) Detection of single-nucleotide mutation including substitution and deletion by matrix-assisted laser desorption/ionization time-of-flight mass spectrometry, *Rapid Commun. Mass Spectrom.* **11**, 1657–1660.

Wintzingerode, F., Bocker, S., Schlotelburg, C., Chiu, N.H.L., Storm, N., Jurinke, C., Cantor, C.R., Gobel, U.B., Boom, D.V. (2002) Base specific fragmentation of amplified 16S rRNA genes analyzed by mass spectrometry: a tool for rapid bacterial identification, *Proc. Natl. Acad. Sci.* **99**, 7039–7044.

Zhu, Y.F., Lee, K.L., Tang, K., Allman, S.L., Taranenko, N.I., Chen, C.H. (1995) Revisit of MALDI for small proteins, *Rapid Commun. Mass Spectrom.* **9**, 1315–1320.

Zhu, Y.F., Taranenko, N.I., Allman, S.L., Martin, S.A., Haff, L., Chen, C.H. (1996) The study of 2,3,4-trihydroxyacetophenone and 2,4,5-trihydroxyacetophenone as matrices for DNA detection in matrix-assisted laser desorption/ionization time-of-flight mass spectrometry, *Rapid Commun. Mass Spectrom.* **10**, 383–388.

Zhu, Y.F., Taranenko, N.I., Allman, S.L., Martin, S.A., Haff, L., Chen, C.H. (1996) The effect of ammonium salt and matrix in the detection of DNA by matrix-assisted laser desorption/ionization time-of-flight mass spectrometry, *Rapid Commun. Mass Spectrom.* **10**, 1591–1596.

Glossary of Basic Terms

The most basic terms in molecular cell biology are defined below. These, in combination with the key words listed at the head of each article, provide definitions of all essential terms in this Encyclopedia.

Alleles
Alternative forms of a given gene, inherited separately from each parent, differing in nucleotide base sequence and located in a specific position on each homologous chromosome, affecting the functioning of a single product (RNA and/or protein).

Amino Acid
An organic compound containing at least one amino group and one carboxyl group. In the 20 different amino acids that compose proteins, an amino group and carboxyl group are linked to a central carbon atom, the α-carbon, to which a variable side chain is bound (see pages at the back of each volume).

Amplification
The process of replication of specific DNA sequences in disproportionately greater amounts than are present in the parent genetic material, for example, PCR is an *in vitro* amplification technique.

Apoptosis
Regulated process leading to nonpathological animal cell death via a series of well-defined morphological changes; also called *programmed cell death*.

Bacteriophage (phage)
Any virus (containing DNA or RNA) that infects bacterial cells. Some bacteriophages are widely used as cloning vectors.

Base Pair
Association of two complementary nucleotides in a DNA or RNA molecule stabilized by hydrogen bonding between their base components. Adenine pairs with thymine or uracil (A–T; A–U) and guanine pairs with cytosine (G–C) (see pages at the back of each volume).

Bioinformatics
Computational approaches to answer biological questions and enhance the ability of researchers to manipulate, collect, and analyze data more quickly and in new ways. Experts predict that more biologists will do their work *in silico*, using the computer to synthesize, analyze, and interpret the many terabytes of data now being generated.

Encyclopedia of Molecular Cell Biology and Molecular Medicine, 2nd Edition. Volume 7
Edited by Robert A. Meyers.

ISBN: 3-527-30549-1

cDNA (complementary DNA)
A DNA copy of an RNA molecule synthesized from an mRNA template *in vitro* using an enzyme called *reverse transcriptase*; often used as a probe.

Cell Cycle
Ordered sequence of events in which a cell duplicates its chromosomes and divides itself into two. Most eukaryotic cell cycles can be commonly divided into four phases: G_1 (G1) period after mitosis but before DNA synthesis occurs; S-phase when most DNA replication occurs; G_2 (G2) phase period of cell cycle when cells contain twice the G1 complement of DNA; and M-phase when cell division occurs, yielding two daughter cells (mitosis) each with one complete genome.

Cell Differentiation
Progressive restriction of the developmental potential and increasing specialization of function that takes place during the development of the embryo and leads to the formation of specialized cells, tissues, and organs.

Cell Division
Separation of a cell into two daughter cells. In higher eukaryotes, it involves division of the nucleus (mitosis) and of the cytoplasm (cytokinesis); mitosis often is used to refer to both nuclear and cytoplasmic division.

Cell Line
A defined unique population of cells obtained by culture from a primary implant through numerous generations.

Chromatin
The complex of nucleic acids (DNA and RNA) and proteins (histones) comprising eukaryotic chromosomes.

Chromosome
In prokaryotes, the usually circular duplex DNA molecule constituting the genome; in eukaryotes, a threadlike structure consisting of chromatin and carrying genomic information on a DNA double helix molecule. A viral chromosome may be composed of DNA or RNA.

Cloning
Asexual reproduction of cells, organisms, genes, or segments of DNA identical to the original.

Cloning Vector *see* Vector

Codon
Sequence of three nucleotides in DNA or mRNA that specifies a particular amino acid during protein synthesis; also called *triplet*. Of the 64 possible codons, three are stop codons, which do not specify amino acids (see pages at the back of each volume).

Complementary Base Pairing
Nucleic acid sequences on paired polymers with opposing hydrogen-bonded bases adenine (designated A) bonded to thymine (T), guanine (G) to cytosine (C) in DNA and adenine to uracil (U) replacing adenine to thymine in RNA (see pages at the back of each volume).

Complementary DNA see cDNA

Dalton
Unit of molecular mass approximately equal to the mass of a hydrogen atom (1.66×10^{-24} g).

Deoxyribonucleic Acid *see* DNA

Diploid
The number of chromosomes in most cells except the gametes. In humans, the diploid number is 46.

DNA (Deoxyribonucleic Acid)
The molecular basis of the genetic code consisting of a poly-sugar phosphate backbone from which thymine, adenine, guanine, and cytosine bases project. Usually found as two complementary chains (duplex) forming a double helix associated by hydrogen bonds between complementary bases.

DNA Cloning (Gene Cloning)
Recombinant DNA technique in which specific cDNAs or fragments of genomic DNA are inserted into a cloning vector, which then is incorporated into cultured host cells (e.g., *E. coli* cells) and maintained during growth of the host cells.

DNA Library
Collection of cloned DNA molecules consisting of fragments of the entire genome (genomic library) or of DNA copies of all the mRNAs produced by a cell type (cDNA library) inserted into a suitable cloning vector.

DNA Polymerase
Enzymes that catalyze the replication of DNA from the deoxyribonucleotide triphosphates using single- or double-stranded DNA as a template.

DNA Transcription *see* Transcription

E. coli (Escherichia coli)
A colon bacillus, which is the most studied of all forms of life.

Embryonic Stem Cells (ES)
Cultured cells derived from the pluripotent inner cell mass of blastocyst-stage embryos.

Epigenetics
Mechanisms of storing and transmitting cellular information additional to those based on DNA sequences.

Escherichia coli see E. coli

Eukaryotes
Organisms whose cells have their genetic material packed in a membrane-surrounded, structurally discrete nucleus and with well-developed cell organelles. Eukaryotes include all organisms except *archaebacteria* and *eubacteria*.

Expression
The process of making the product of a gene, which is either a specific protein giving rise to a specific trait or RNA forms not translated into proteins (e.g. transfer ribosomal RNAs).

Functional Genomics
A discipline that aims to understand how genes are regulated and what they do, largely through massive parallel studies of gene expression over time and in a variety of tissues.

Gamete
Specialized haploid cell (in animals either a sperm or an egg) produced by meiosis of germ cells; in sexual reproduction, the union of a sperm and an egg initiates the development of a new diploid individual.

Gene Cloning *see* DNA Cloning

Gene
A DNA sequence, located in a particular position on a particular chromosome,

which encodes a specific protein or RNA molecule.

Genomics
Comparative analysis of the complete genomic sequences from different organisms; used to assess evolutionary relations between species and to predict the number and general types of proteins produced by an organism.

Genotype
Entire genetic constitution of an individual cell or organism; also, the alleles at one or more specific loci.

Haploid
The number of chromosomes in a sperm or egg cell, half the diploid number.

Heterozygous
Having two different alleles for a given trait in the homologous chromosomes.

Homologies
Similarities in DNA or protein sequences between individuals of the same species or among different species.

Homologous Chromosomes
Chromosome pairs, each derived from one parent, containing the same linear sequence of genes, and as a consequence, each gene is present in duplicate (e.g., humans have 23 homologous chromosome pairs, but the toad has 11 pairs, the mosquito has three pairs, and so on).

Homozygous
Having two identical alleles for a given trait in the homologous chromosomes.

Hybridization
The formation of a double-stranded polynucleotide molecule when two complementary strands are brought together at moderate temperature. The strands can be DNA or RNA or one of each; a technique for assessing the extent of sequence homology between single strands of nucleic acids.

Ligation
The formation of a phosphodiester bond to join adjacent terminal nucleotides (nicks) to form a longer nucleic acid chain (DNA of RNA); catalyzed by ligase.

Marker
A gene or a restriction enzyme cutting site with a known location on a chromosome and a clear-cut phenotype (expression), or pattern of inheritance, used as a point of reference when mapping a new mutant.

Meiosis
In eukaryotes, a special type of cell division that occurs during maturation of germ cells; comprises two successive nuclear and cellular divisions, with only one round of DNA replication resulting in production of four genetically nonequivalent haploid cells (gametes) from an initial diploid cell.

Messenger RNA *see* mRNA

Mitosis
In eukaryotic cells, the process whereby the nucleus is divided, involving condensation of the DNA into visible chromosomes, to produce two genetically equivalent daughter nuclei with the diploid number of chromosomes.

mRNA (messenger RNA)
RNA used to translate information from DNA to ribosome where the information is used to make one or several proteins.

Mutation
The heritable change in the nucleotide sequence of a chromosome.

Nucleotide
The monomer which, when polymerized, forms DNA or RNA. It is composed of a nitrogenous base bonded to a sugar (ribose or deoxyribose), bonded to a phosphate.

Oligonucleotide
A polynucleotide 2 to 20 nucleotide units in length.

Operon
A series of prokaryote genes encoding enzymes of a specific biosynthesis pathway and transcribed into a single RNA molecule.

Organelle
Any membrane-limited structure found in the cytoplasm of eukaryotic cells.

Phage *see* Bacteriophage

Phenotype
The observable characteristics of a cell or organism as distinct from it's genotype.

Plasmid
An extrachromosomal circular DNA molecule found in a variety of bacteria encoding "dispensable functions," such as resistance to antibiotics. Often found in multiple copies per cell and reproduces every time the bacterial cell reproduces. May be used as a cloning vector.

Polymorphism
Difference in DNA sequence among individuals expressed as different forms of a protein in individuals of the same interbreeding population.

Polynucleotide
The polymer formed by condensation of nucleotides.

Probe
A radioactively fluorescent or immunologically labeled oligonucleotide (RNA or DNA) used to detect complementary sequences in a hybridization experiment, for example, identify bacterial colonies that contain cloned genes or detect specific nucleic acids following separation by gel electrophoresis.

Procaryotes (Prokaryotes)
Typically unicellular or filamentous with DNA not located within a nuclear envelope. Prokaryotes include archaebacteria, eubacteria, cyanobacteria, prochlorophytes and mycoplasmas.

Programmed Cell Death *see* Apoptosis

Prokaryotes *see* Procaryotes

Protein
A linear polymer of amino acids linked together in a specific sequence and usually containing more than 50 residues. Proteins form the key structure elements in cells and participate in nearly all cellular activities.

Proteomics
A discipline that promises to determine the identity, function, and structure of each protein in an organelle or cell and to elucidate protein–protein interactions.

Replication
The copying of a DNA molecule duplex yielding two new DNA duplex molecules, each with one strand from the original DNA duplex. Single-stranded DNA

replication results in a single-stranded DNA molecule.

Repressor
A protein that binds to a specific location (operator) on DNA and prevents RNA transcription from a specific gene or operon.

Restriction Fragment Length Polymorphism *see* RFLP

Restriction Mapping
Uses restriction endonuclease enzymes to produce specific cuts (cleavage) in DNA, allowing preparation of a genome map describing the order and distance between cleavage sites.

Reverse Transcription
The synthesis of cDNA from an RNA template as catalyzed by reverse transcriptase.

RFLP (Restriction Fragment Length Polymorphism)
DNA fragment cut by enzymes specific to a base sequence (restriction endonuclease) generating a DNA fragment whose size varies from one individual to another. Used as markers on genome maps and for screening for mutations and genetic diseases.

Ribonucleic Acid *see* RNA

Ribosomes
Small cellular components composed of proteins plus ribosomal RNA that translate the genetic code into synthesis of specific proteins.

RNA (Ribonucleic Acid)
A single-stranded polynucleotide with a phosphate oxyribose backbone and four bases that are identical to those in DNA, with the exception that the base uracil is substituted for thymine.

RNA Interference (RNAi)
Intracellular degradation of RNA that removes foreign RNAs such as those from viruses. These fragments (small, micro, or mini RNA) cleaved from free double-stranded RNA (dsRNA) direct the degradative mechanism to other similar RNA sequences. Used as a technique to silence the expression of targeted genes in a sequence-dependent mode.

RNA Polymerase
The enzyme (peptide) that binds at specific nucleotide sequences, called promoters, in front of genes in DNA, which catalyze transcription of DNA to RNA.

RNA Translation *see* Translation

Stem Cell
A self-renewing cell that divides to give rise to a cell with an identical developmental potential and/or one with a more restricted developmental potential.

Structural Biology
The discovery, analysis and dissemination of three-dimensional structures of protein, DNA, RNA, and other biological macromolecules representing the entire range of structural diversity found in nature.

Transcription (DNA transcription)
Synthesis of an RNA molecule from a DNA template (gene) catalyzed by RNA polymerase.

Transfer RNA *see* tRNA

Translation (RNA translation)
The process on a ribosome by which the sequence of nucleotides in a mRNA

molecule directs the incorporation of amino acids into protein.

tRNA (transfer RNA)
RNA molecules that transport specific amino acids to ribosomes into position in the correct order during protein synthesis.

Vector
A DNA molecule originating from a virus, a plasmid, or a cell of a higher organism into which another DNA fragment can be integrated without loss of the vector's capacity for self-replication. Vectors introduce foreign DNA into host cells where it can be reproduced in large quantities.

Virus
A small parasite consisting of nucleic acid (RNA or DNA) enclosed in a protein coat that can replicate only in a susceptible host cell; widely used in cell biology research.

Wild type
Normal, nonmutant form of a macromolecule, cell or organism.

Zygote
A fertilized egg; a diploid cell resulting from fusion of a male and female gamete.

The Twenty Amino Acids that are Combined to Form Proteins in Living Things

Amino acids with nonpolar side chains

Glycine
Gly
G

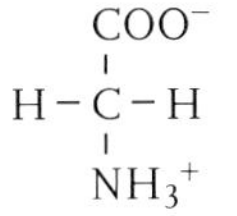

COO^-
H–C–H
NH_3^+

Alanine
Ala
A

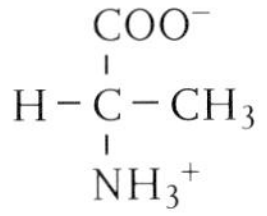

COO^-
H–C–CH_3
NH_3^+

Valine
Val
V

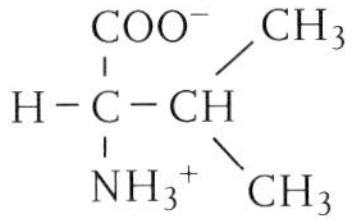

COO^- CH_3
H–C–CH
NH_3^+ CH_3

Leucine
Leu
L

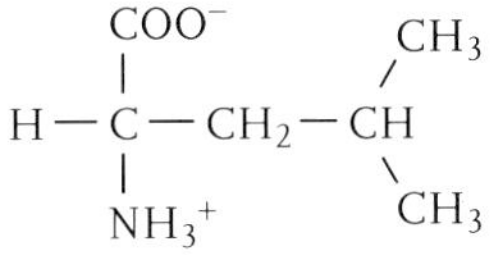

COO^- CH_3
H–C–CH_2–CH
NH_3^+ CH_3

Isoleucine
Ile
I

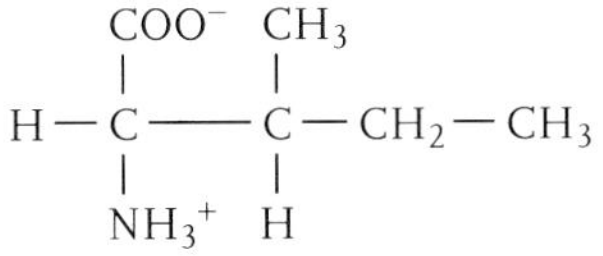

COO^- CH_3
H–C–C–CH_2–CH_3
NH_3^+ H

Methionine
Met
M

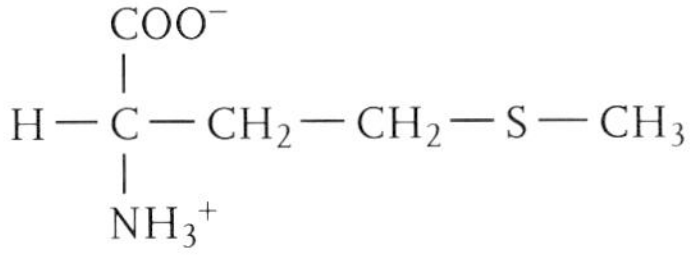

COO^-
H–C–CH_2–CH_2–S–CH_3
NH_3^+

Proline
Pro
P

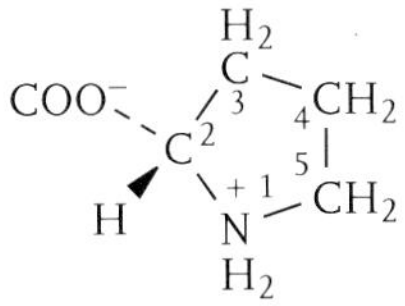

H_2
C 3 4 CH_2
COO^- C^2 5
H + 1 N CH_2
H_2

Encyclopedia of Molecular Cell Biology and Molecular Medicine, 2nd Edition. Volume 7
Edited by Robert A. Meyers.

ISBN: 3-527-30549-1

Amino acids with nonpolar side chains (continued)

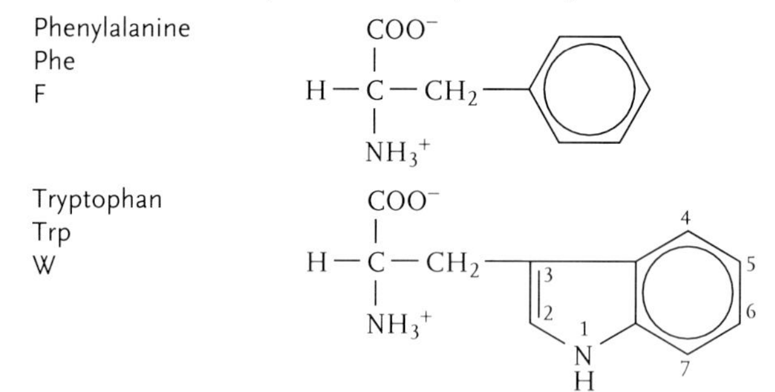

Amino acids with uncharged polar side chains

Serine
Ser
S

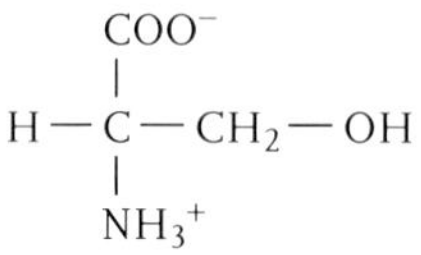

$$H-C(COO^-)(NH_3^+)-CH_2-OH$$

Threonine
Thr
T

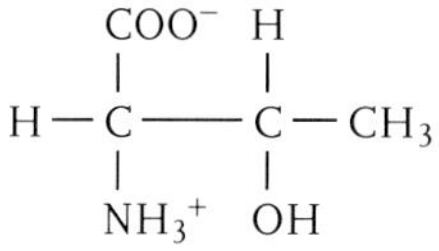

$$H-C(COO^-)(NH_3^+)-C(H)(OH)-CH_3$$

Asparagine
Asn
N

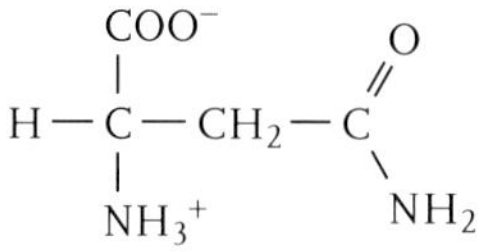

$$H-C(COO^-)(NH_3^+)-CH_2-C(=O)-NH_2$$

Glutamine
Gln
Q

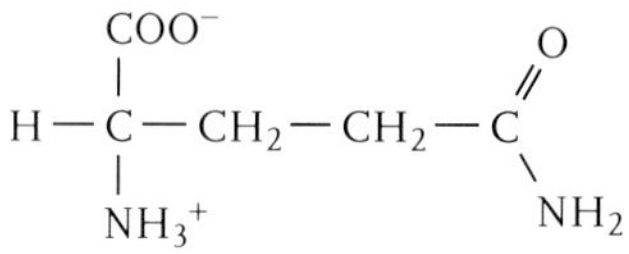

$$H-C(COO^-)(NH_3^+)-CH_2-CH_2-C(=O)-NH_2$$

Tyrosine
Tyr
Y

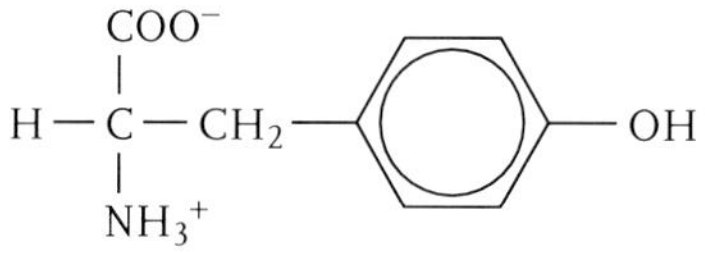

$$H-C(COO^-)(NH_3^+)-CH_2-C_6H_4-OH$$

Cysteine
Cys
C

$$H-C(COO^-)(NH_3^+)-CH_2-SH$$

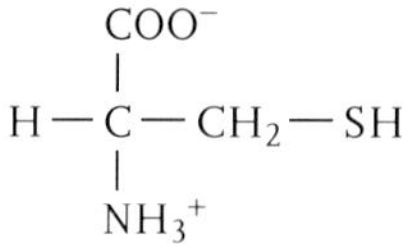

Amino acids with charged polar side chains

Lysine
Lys
K

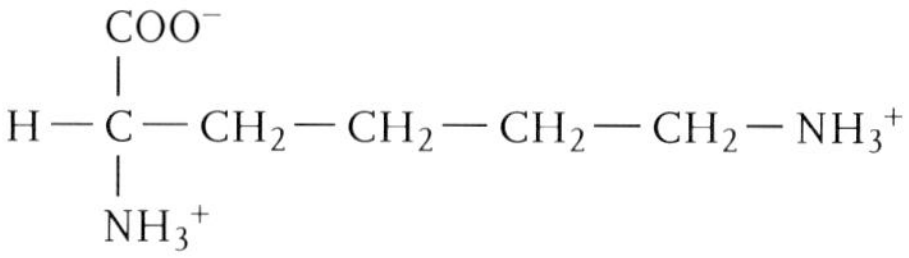

$$H-C(COO^-)(NH_3^+)-CH_2-CH_2-CH_2-CH_2-NH_3^+$$

Amino acids with charged polar side chains (continued)

Arginine
Arg
R

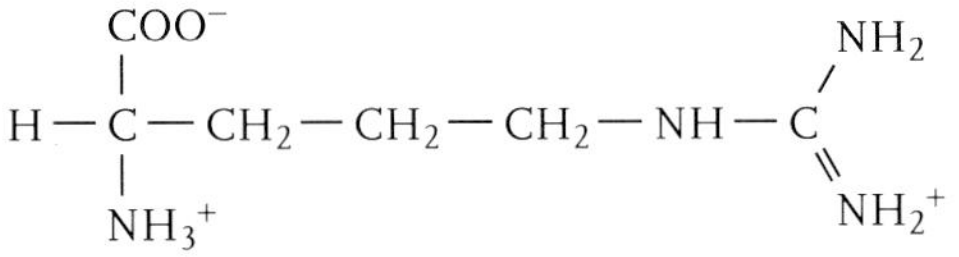

COO^- ; NH_2

$H-C-CH_2-CH_2-CH_2-NH-C$

NH_3^+ ; NH_2^+

Histidine
His
H

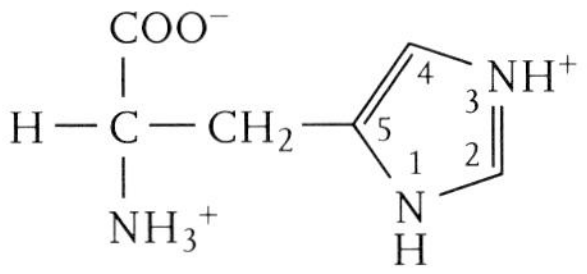

COO^-

$H-C-CH_2-$ (imidazole ring, positions 1–5: NH^+ at 3, $N-H$ at 1)

NH_3^+

Aspartic acid
Asp
D

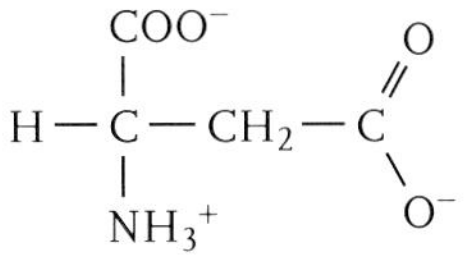

COO^- ; O

$H-C-CH_2-C$

NH_3^+ ; O^-

Glutamic acid
Glu
E

COO^- ; O

$H-C-CH_2-CH_2-C$

NH_3^+ ; O^-

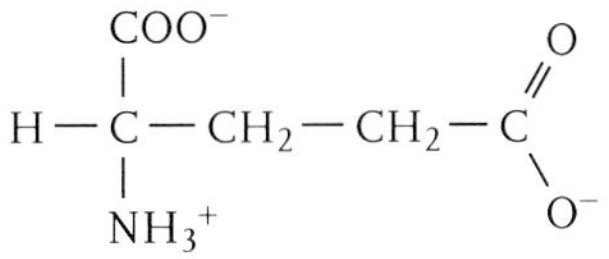

(Figures with kind permission from Voet, D., Voet, J.G., Pratt, C.W. (2001) *Fundamentals of Biochemistry*, Wiley, New York)

The Twenty Amino Acids with Abbreviations and Messenger RNA Code Designations

Amino acid	*One letter symbol*	*Three letter symbol*	*mRNA code designation*
alanine	A	ala	GCU, GCC, GCA, GCG
arginine	R	arg	CGU, CGC, CGA, CGG, AGA, AGG
asparagine	P	asn	AAU, AAC
aspartic acid	D	asp	GAU, GAC
cysteine	C	cys	UGU, UGC
glutamic acid	E	glu	GAA, GAG
glutamine	Q	gln	CAA, CAG
glycine	G	gly	GGU, GGC, GGA, GGG
histidine	H	his	CAU, CAC
isoleucine	I	ile	AUU, AUC, AUA
leucine	L	leu	UUA, UUG, CUU, CUC, CUA, CUG
lysine	K	lys	AAA, AAG
methionine	M	met	AUG
phenylalanine	F	phe	UUU, UUC
proline	P	pro	CCU, CCC, CCA, CCG
serine	S	ser	UCU, UCC, UCA, UCG, AGU, AGC
threonine	T	thr	ACU, ACC, ACA, ACG
tryptophan	W	trp	UGG
tyrosine	Y	tyr	UAU, UAC
valine	V	val	GUU, GUC, GUA, GUG

Complementary Strands of DNA with Base Pairing

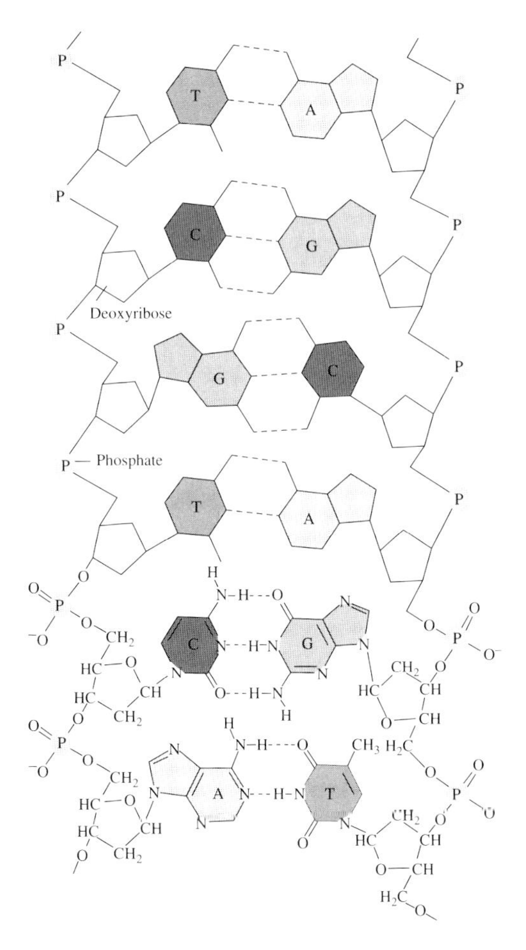

Two nucleotide chains associate by base pairing to form double-stranded DNA. A (Adenine) pairs with T (Thymine), and G (Guanine) pairs with C (Cytosine) by forming specific hydrogen bonds. (Figure with kind permission from Voet, D., Voet, J.G., Pratt, C.W. [2001]: Fundamentals of Biochemistry. Wiley: New York.)